Crystallisation – A Biological Perspective

University of Leeds, UK
23-25 July 2012

FARADAY DISCUSSIONS
Volume 159, 2012

RSC Publishing

The Faraday Division of the Royal Society of Chemistry, previously the Faraday Society, founded in 1903 to promote the study of sciences lying between Chemistry, Physics and Biology.

EDITORIAL STAFF

Editor
Philip Earis

Deputy editor
Jane Hordern

Development editor
Heather Montgomery

Senior publishing editor
Susan Weatherby

Publishing editors
Lucy Gilbert, Tanya Smekal

Publishing assistants
Aliya Anwar, Ella Mitchell, Claire Sissen

Publisher
Niamh O' Connor

Faraday Discussions (Print ISSN 1359-6640, Electronic ISSN 1364-5498) is published 6 times a year by the Royal Society of Chemistry, Thomas Graham House, Science Park, Milton Road, Cambridge, UK CB4 0WF. Volume 159 ISBN-13: 978-1-84973-449-3

2012 annual subscription price: print+electronic £709, US $1,322; electronic only £673, US $1,256. Customers in Canada will be subject to a surcharge to cover GST. Customers in the EU subscribing to the electronic version only will be charged VAT. All orders, with cheques made payable to the Royal Society of Chemistry, should be sent to RSC Distribution Services, c/o Portland Customer Services, Commerce Way, Colchester, Essex, UK CO2 8HP.
Tel +44 (0) 1206 226050;
E-mail sales@rscdistribution.org

If you take an institutional subscription to any RSC journal you are entitled to free, site-wide web access to that journal. You can arrange access *via* Internet Protocol (IP) address at www.rsc.org/ip. Customers should make payments by cheque in sterling payable on a UK clearing bank or in US dollars payable on a US clearing bank.

US Postmaster: send address changes to *Faraday Discussions*, c/o Mercury Airfreight International Ltd., 365 Blair Road, Avenel, NJ 07001. All despatches outside the UK by Consolidated Airfreight.

PRINTED IN THE UK

Faraday Discussions documents a long-established series of *Faraday Discussion* meetings which provide a unique international forum for the exchange of views and newly acquired results in developing areas of physical chemistry, biophysical chemistry and chemical physics.

SCIENTIFIC COMMITTEE, Volume 159

Chairs
Professor Fiona Meldrum (University of Leeds, UK)
Dr Yi-Yeoun Kim (University of Leeds, UK)

Dr Nico Sommerdijk (Eindhoven University of Technology, The Netherlands)
Professor Joanna Aizenberg (Harvard University, USA)
Professor Roger Davey (University of Manchester, UK)
Professor van der Schoot (Eindhoven University of Technology, The Netherlands)

FARADAY STANDING COMMITTEE ON CONFERENCES

Chair
A Mount (Edinburgh, UK)

W A Brown (UCL, UK)
I Hamley (Reading, UK)
J Hirst (Nottingham, UK)
Graham Hutchings (Cardiff University)
Carl Percival (University of Manchester)

© The Royal Society of Chemistry 2012. Apart from fair dealing for the purposes of research or private study, or criticism or review, as permitted under the Copyright, Designs and Patents Act 1988 and Related Rights Regulations 2003, this publication may only be reproduced, stored or transmitted, in any form or by any means, with the prior permission in writing of the Publishers or in the case of reprographic reproduction in accordance with the terms of licences issued by the Copyright Licensing Agency in the UK. US copyright law applicable to users in the USA. The Royal Society of Chemistry takes reasonable care in the preparation of this publication but does not accept liability for the consequences of any errors or omissions.

Royal Society of Chemistry:
Registered Charity No. 207890.

⊚The paper used in this publication meets the requirements of ANSI/NISO Z39.48-1992 (Permanence of Paper).

Crystallisation – A Biological Perspective

Faraday Discussions

www.rsc.org/faraday_d

A General Discussion on Crystallisation – A Biological Perspective was held in Leeds, UK on 23rd, 24th and 25th July 2012.

RSC Publishing is a not-for-profit publisher and a division of the Royal Society of Chemistry. Any surplus made is used to support charitable activities aimed at advancing the chemical sciences. Full details are available from www.rsc.org

CONTENTS

ISSN 1359-6640; ISBN 978-1-84973-449-3

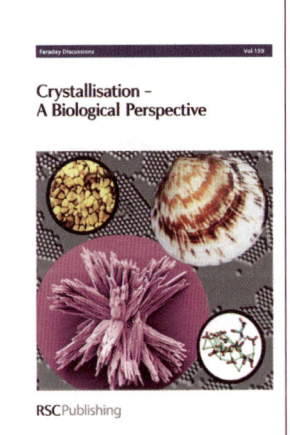

Cover
AFM image of the calcite prisms of *Pinna nobilis* and a shell of *Glycymeris glycymeris*, both from Dr Stephan Wolf, *Faraday Discuss.*, 2012, a cluster of aragonite needles, from Patricia Dove, *Faraday Discuss.*, 2012 and a cluster of 15 $CaCO_3$ units and one aspartate molecule, from Aaron Finney, *Faraday Discuss.*, 2012. The background image is of assembling colloidal particles from Dr Vinothan Manoharan, *Faraday Discuss.*, 2012.

INTRODUCTORY LECTURE

9   **Spiers Memorial Lecture: Effect of interaction specificity on the phase behaviour of patchy particles**
Nicolas Dorsaz, Laura Filion, Frank Smallenburg and Daan Frenkel

PAPERS AND DISCUSSIONS

23   **Amino acids form prenucleation clusters: ESI-MS as a fast detection method in comparison to analytical ultracentrifugation**
Matthias Kellermeier, Rose Rosenberg, Adrian Moise, Ulrike Anders, Michael Przybylski and Helmut Cölfen

47   **Probing the structure and stability of calcium carbonate pre-nucleation clusters**
Aaron R. Finney and P. Mark Rodger

Co-Sponsor

IOP | Institute of Physics
Molecular Physics Group

61	**Exploring the influence of organic species on pre- and post-nucleation calcium carbonate** Paolo Raiteri, Raffaella Demichelis, Julian D. Gale, Matthias Kellermeier, Denis Gebauer, David Quigley, Louise B. Wright and Tiffany R. Walsh
87	**Control of the nucleation of sickle cell hemoglobin polymers by free hematin** Veselina Uzunova, Weichun Pan, Vassiliy Lubchenko and Peter G. Vekilov
105	**Structural evolution, formation pathways and energetic controls during template-directed nucleation of $CaCO_3$** Michael H. Nielsen, Jonathan R. I. Lee, Qiaona Hu, Thomas Yong-Jin Han and James J. De Yoreo
123	**A two-step mechanism for crystal nucleation without supersaturation** Tamás Kovács and Hugo K. Christenson
139	**General discussion**
181	**Phase behavior of colloidal silica rods** Anke Kuijk, Dmytro V. Byelov, Andrei V. Petukhov, Alfons van Blaaderen and Arnout Imhof
201	**Inorganic salts direct the assembly of charged nanoparticles into composite nanoscopic spheres, plates, or needles** Bartosz A. Grzybowski, Bartlomiej Kowalczyk, István Lagzi, Dawei Wang, Konstantin V. Tretiakov and David A. Walker
211	**Real-space studies of the structure and dynamics of self-assembled colloidal clusters** Rebecca W. Perry, Guangnan Meng, Thomas G. Dimiduk, Jerome Fung and Vinothan N. Manoharan
235	**Aggregation of ferrihydrite nanoparticles in aqueous systems** Virany M. Yuwono, Nathan D. Burrows, Jennifer A. Soltis, Tram Anh Do and R. Lee Penn
247	**Biomimetic type morphologies of calcium carbonate grown in absence of additives** Jens-Petter Andreassen, Ralf Beck and Margrethe Nergaard
263	**Computer simulation of soft matter at the growth front of a hard-matter phase: incorporation of polymers, formation of transient pits and growth arrest** Richard P. Sear
277	**General discussion**
291	**A metastable liquid precursor phase of calcium carbonate and its interactions with polyaspartate** Mark A. Bewernitz, Denis Gebauer, Joanna Long, Helmut Cölfen and Laurie B. Gower
313	**The role of cluster formation and metastable liquid—liquid phase separation in protein crystallization** Fajun Zhang, Felix Roosen-Runge, Andrea Sauter, Roland Roth, Maximilian W. A. Skoda, Robert M. J. Jacobs, Michael Sztucki and Frank Schreiber
327	**Polymer-induced liquid precursor (PILP) phases of calcium carbonate formed in the presence of synthetic acidic polypeptides—relevance to biomineralization** Anna S. Schenk, Harshal Zope, Yi-Yeoun Kim, Alexander Kros, Nico A. J. M. Sommerdijk and Fiona C. Meldrum
345	**Precipitation of ACC in liposomes—a model for biomineralization in confined volumes** Chantel C. Tester, Ching-Hsuan Wu, Steven Weigand and Derk Joester

357 The role of the amorphous phase on the biomimetic mineralization of collagen
Fabio Nudelman, Paul H. H. Bomans, Anne George, Gijsbertus de With and Nico A. J. M. Sommerdijk

371 Revisiting geochemical controls on patterns of carbonate deposition through the lens of multiple pathways to mineralization
D. Wang, L. M. Hamm, A. J. Giuffre, T. Echigo, J. Donald Rimstidt, J. J. De Yoreo, J. Grotzinger and P. M. Dove

387 General discussion

421 Aragonite crystal orientation in mollusk shell nacre may depend on temperature. The angle spread of crystalline aragonite tablets records the water temperature at which nacre was deposited by *Pinctada margaritifera*
Ian C. Olson and Pupa U. P. A. Gilbert

433 Merging models of biomineralisation with concepts of nonclassical crystallisation: is a liquid amorphous precursor involved in the formation of the prismatic layer of the Mediterranean Fan Mussel *Pinna nobilis*?
Stephan E. Wolf, Ingo Lieberwirth, Filipe Natalio, Jean-Francois Bardeau, Nicolas Delorme, Franziska Emmerling, Raul Barrea, Michael Kappl and Frédéric Marin

449 Oligomer formation, metalation, and the existence of aggregation-prone and mobile sequences within the intracrystalline protein family, Asprich
Moise Ndao, Christopher B. Ponce and John Spencer Evans

463 GSP-37, a novel goldfish scale matrix protein: identification, localization and functional analysis
Kousei Miyabe, Hiroki Tokunaga, Hirotoshi Endo, Hirotaka Inoue, Michio Suzuki, Naoaki Tsutsui, Naoki Yokoo, Toshihiro Kogure and Hiromichi Nagasawa

483 $CaCO_3$/Chitin hybrids: recombinant acidic peptides based on a peptide extracted from the exoskeleton of a crayfish controls the structures of the hybrids
Hiromu Kumagai, Ryou Matsunaga, Tatsuya Nishimura, Yuya Yamamoto, Satoshi Kajiyama, Yuya Oaki, Kei Akaiwa, Hirotaka Inoue, Hiromichi Nagasawa, Kohei Tsumoto and Takashi Kato

495 General discussion

CONCLUDING REMARKS

509 The thermodynamics of calcite nucleation at organic interfaces: Classical *vs.* non-classical pathways
Q. Hu, M. H. Nielsen, C. L. Freeman, L. M. Hamm, J. Tao, J. R. I. Lee, T. Y. J. Han, U. Becker, J. H. Harding, P. M. Dove and J. J. De Yoreo

ADDITIONAL INFORMATION

525 Poster titles
529 List of participants
531 Index of contributors

Spiers Memorial Lecture: Effect of interaction specificity on the phase behaviour of patchy particles

Nicolas Dorsaz,[a] Laura Filion,[a] Frank Smallenburg[b] and Daan Frenkel[a]

Received 10th April 2012, Accepted 6th June 2012
DOI: 10.1039/c2fd20070h

We report a numerical study on the phase behaviour of a 'patch–anti-patch' model for particles with tetrahedrally arranged attractive spots. In particular, we compute the phase equilibria between the fluid and a low density diamond cubic (DC) crystal for different realizations of the patch–anti-patch interaction. By increasing the 'specificity' of the patches, i.e. lowering the number of corresponding attractive 'anti-patches' to a given patch, we find that the metastability gap between the DC freezing boundary and the liquid–gas critical point widens considerably. We argue that this effect of interaction specificity is relevant for the description of protein phase diagrams, as patch–anti-patch interactions can stabilise relatively open, ordered structures.

1 Introduction

Growing sufficiently high-quality protein crystals for use in X-ray crystallography is still a bottleneck to structure determination in protein science. Indeed, only a small fraction of all globular proteins crystallize readily and most that do, do so under conditions that are non-physiological: at high salt concentrations or in presence of a high depletant concentration.[1]

The poor crystallizability of proteins is usually attributed to two main causes. First of all, proteins might have directional interactions that are not compatible with simple crystal lattices.[2] In this context, it is interesting to note that the typical space groups of protein crystals are rather different form those of simple molecular crystals. There are 65 space groups compatible with crystal structures of chiral macromolecules, yet one-third of all protein crystals form in a single space group, $P2_1 2_1 2_1$. Conversely, many space groups that are common for simple molecular crystals have never been observed in protein crystals. The number of degrees of freedom for packing a low-symmetry molecule in different space groups seems to be the key: proteins crystallize preferentially in space groups where it is easier to achieve connectivity.[3] Secondly, kinetically trapped phases such as aggregates and gels that are frequently seen experimentally in protein suspensions, are enhanced by anisotropic, specific interactions and the formation of such structurally arrested states is likely to suppress the nucleation of protein crystals.[4-6]

Until recently, the search for suitable crystallization conditions was mainly based on trial and error.[7] In the last two decades there have been several attempts to tackle the problem of protein crystallization in a more systematic manner, trying to

[a] Department of Chemistry, University of Cambridge, Lensfield Road, Cambridge, CB2 1EW, UK
[b] Soft Condensed Matter, Deybe Institute for NanoMaterials Science, Utrecht University, Princetonplein 5, 3584CC, The Netherlands

understand the subtle interplay between protein interactions and their equilibrium phase diagram. In a seminal paper, George and Wilson[8] were able to rationalize the observed crystallization region for proteins: solvent conditions that were known to promote crystallization could be related to a particular range of the second osmotic virial coefficients B_2 of the protein solution. Hence, interactions should be attractive enough to promote crystallization, while not being so large that they result in disordered aggregation.[9] Colloidal models for globular proteins have also revealed that the fluid–fluid coexistence line might become metastable with respect to the solid–fluid boundary for a sufficiently small range of the attraction and that the presence of a metastable fluid–fluid critical point can lower substantially the free-energy barrier for nucleation of a crystalline germ and, thus, indirectly promote crystallization.[10] At the same time, the competition between glassy and crystal phases has been extensively investigated both in systems of purely hard spheres[11] and in systems of particles with short-ranged attraction.[12]

These investigations of the sphero-symmetric potential properties suggest that good knowledge of the equilibrium phase diagram is a prerequisite for materials design or protein crystal predictions. Such information is essential to predict not only if, but also how rapidly, a given crystal structure will assemble. A particular crystal structure may look promising on paper, but such information is irrelevant if it is not stable with respect to the liquid phase in the experimentally accessible temperature-density range. Similarly, even when an optimized interaction has been identified to produce a given crystal structure that is thermodynamically stable, the nucleation rate of the crystal might be so small that effective production of the crystal is impossible. To take a simple example, a suspension of monodisperse colloids brought to a sufficiently high density will form beautiful, iridescent crystals but the rate at which these crystals form has been shown to depend strongly on the steepness of the repulsive forces that act between the colloids: charged colloids with a soft, long-ranged repulsion tend for instance to crystallize much faster than hard-sphere colloids at the same supersaturation.[13] Kinetic effects are even more dramatic when it comes to proteins and their highly directional interactions.

Approaches based on orientationally averaged isotropic potentials have enriched our understanding of the general features of protein phase diagrams, and addressed, *via* ad-hoc coarse-grained models, the fluid phase behaviour of solutions of globular proteins more quantitatively.[14–16] However, they fail to predict quantitatively the fluid–solid equilibrium. In fact, isotropic interactions favor densely packed crystal phases of high symmetry such as face-centered cubic (FCC), body-centered cubic (BCC) or hexagonal close-packed (HCP), whereas protein crystals are typically much more open, with around 7 contacts on average per protein.[4] Clearly, where crystal structures are concerned, the patchiness of protein surfaces that is essential for their biological role, leads to strong directionality of the associated interactions that cannot simply be averaged out.

Patchy—or *aeolotopic*—models, which can be used to describe the directional nature of protein interactions have now been studied for a decade.[2,5,17] The first numerical studies of the Kern-Frenkel "patchy" potential have revealed that the fluid–fluid coexistence curve can be shifted to lower temperature by making the patches smaller or by decreasing the number of patches, and that anisotropy can stabilize multiple solid phases.[18–20] Investigations of a slightly different model in which each attractive spot is involved in not more than a single-bonded interaction have demonstrated that, by diminishing the number of bonded nearest neighbors, it is possible to generate liquid states (*i.e.*, states with temperature T lower than the liquid–gas critical temperature) with a very low, and eventually vanishing, packing fractions; a situation that cannot be realized with spherically interacting particles.[21] The resulting increase of the region of stability of the liquid phase is expected to favor considerably the formation of stable equilibrium gels at low densities.

Here, we study a slightly different class of patchy models in which each patch of a given particle interacts with a subset N_a of the N_p complementary anti-patches on the other particles. The introduction of such 'patch–anti-patch' (*p-ap*) models[2,20,22] was motivated by studies that compute the different contributions to the protein's osmotic second virial coefficients based on atomistic structural information[23,24] and has been used in other systems, such as supramolecular polymers.[25] The main contribution to the overall protein–protein interaction can be attributed to a small number of complementary, highly attractive configurations. The high-affinity configurations, denoted as 'patch–anti-patch' pairs,[22] result from complementary, opposing surface regions of the proteins in a particular arrangement.[26] Since the highest-energy bonds are formed by patch–anti-patch contacts, they are expected to play a fundamental role in protein aggregation and crystallization.

For a given protein structure, one can construct a coarse-grained patch–anti-patch model starting from atomistic structural information.[22] This approach would make it possible to arrive at a fair prediction of the full phase diagram of that specific protein. However, the objective of the present work is to derive some generic features of the *p-ap* model—in particular the similarities and differences with the better-known *p-p* models. We therefore focus on the *p-ap* version of the tetrahedral Kern-Frenkel patchy-sphere model. The phase diagrams of tetrahedral patchy particles have been determined by computer simulations.[27–29] These studies revealed the existence of an open, low density, diamond cubic (DC) and a diamond hexagonal (DH) crystals were found. The free-energy barriers to nucleate DC and/or DH crystals were calculated for different widths of the tetrahedral patches.

Knowledge of the phase behaviour of this reference system allows us to focus on the effects of higher patch specificity on the phase behaviour. From a geometric viewpoint, a fully bonded DC crystal can be satisfied by the patch-patch and the patch–anti-patch tetrahedral models, therefore both will have the same potential energy at zero temperature. However, from an entropic perspective, the two models are very different: N^2 different bonds can be formed between a pair of particles with N patches, whilst there are only N possible bonds between a pair of patch–anti-patch particles. The question is how highly specific, directional interactions modify the stability of the various crystal phases with respect to the liquid and the vapor.

The remainder of this paper is structured as follows. In Section 2 the tetrahedral patch–anti-patch model is defined. Section 3 gives a brief outline of the different Monte Carlo methods that we used to compute the free energy and phase equilibria of the liquid and the solid phases. In this section, we report the phase diagrams of the patch-patch *p-p* and the patch–anti-patch *p-ap* tetrahedral model and discuss the implications of our findings.

2 Model

To explore the effect of the patch specificity on the equilibrium phase behaviour, we consider the Kern-Frenkel model with four patches ($\alpha = 1$ to 4) placed in a tetrahedral arrangement on the surface of the particle as shown in Fig. 1. In addition, every patch p has a label α such that p^α only interacts with p^α. We examine three realizations of this tetrahedral patchy-sphere model:

(a) a 'patch-patch' model (*p-p*) where $p^\alpha = 1$ for all α and all patches on particle i interact with all patches of particle j (Fig. 1a).

(b) a 'patch–anti-patch' model (*p-ap$_1$*) where every particle has four identifiable patches labelled $p^\alpha = \alpha$; therefore a patch interacts only with a single ($N_a = 1$) complementary anti-patch on other particles (Fig. 1b).

(c) an intermediate model (*p-ap$_2$*) where every particle has two patches labelled 1 and two labelled 2 ($p^1 = p^2 = 1$ and $p^3 = p^4 = 2$); hence each patch has $N_a = 2$ complementary anti-patches (Fig. 1c).

In these models, two patchy particles i and j, located at r_i and r_j respectively, feel an attraction given by

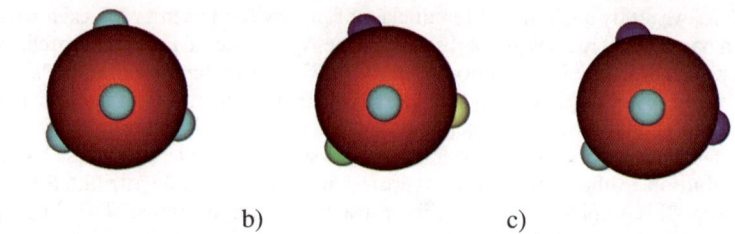

Fig. 1 Schematical view of the three different patchy models studied. Note that the shape of the patches is not precisely the shape of the Kern-Frenkel patches, and is only meant to identify the patch location. A patch on a particle interacts only with a complementary patch of the same color on another particle. (a) usual 'patch-patch' model (p-p) where all 4 patches on particle i interact with all patches on particle j. (b) 'patch–anti-patch' model (p-ap_1) where every particle has four identifiable patches ($p^\alpha = \{1, 2, 3, 4\}$) (c) intermediate 'patch–anti-patch' model (p-ap_2) where every particle has two distinct patch (anti-patch) types ($p^\alpha = \{1, 1, 2, 2\}$).

$$u_{\text{patch}}(i,j) = u_{sw}(r_{ij}) \sum_{\alpha,\beta=1}^{4} \delta_{p^\alpha,p^\beta} \Phi\left(r_{ij}, \hat{p}_i^\alpha\right) \Phi\left(r_{ji}, \hat{p}_j^\beta\right) \quad (1)$$

where $r_{ij} = r_j - r_i$, $\hat{p}_i^\alpha$ is the normalized vector pointing from the center of particle i towards patch α on the same particle, $\delta_{p^\alpha,p^\beta}$ is the Kronecker delta function which is equal to one if patch α and β are complementary patches ($p^\alpha = p^\beta$), and u_{sw} is a square-well potential of hard diameter σ, range δ and depth ε. The reduced temperature T is expressed in units ε/k_B, where k_B is Boltzmann's constant. The reduced pressure P and the number density ρ will be expressed in units ε/σ^3 and $1/\sigma^3$ respectively. The function $\Phi(r_{kl}, p^\alpha)$ is defined as

$$\Phi(r, p) = \begin{cases} -1 & \text{if } \hat{r} \cdot \hat{p} < \cos(\theta_m) \\ 0 & \text{otherwise} \end{cases} \quad (2)$$

with $\hat{k}$ a normalized vector in the direction of k.

We study the patchy models for $\cos(\theta_m) = 0.94$ and $\delta = 0.24\sigma$, parameters that ensure that no two patches on one particle can bind to the same patch on another particle. The equilibrium phase diagram and nucleation barriers of the usual 'patch-patch' realization with this set of parameters have been reported by Saika-Voivod et al.[30] It is worth noting that the p-ap_1 model can qualified as 'chiral' since there are two different ways to choose the label of the last two patches once the label of the first two have been fixed. Here, we study one realization of the p-ap_1 model, but it would be interesting to investigate if solutions with different mixing ratio of the two p-ap_1 'enantiomeric' forms have different phase behaviors.

3 Methods and results

In this section we examine the effect of the patch specificity on the equation of state of the liquid, the liquid–gas coexistence, and the equilibrium phase diagram.

3.1 Equation of state of the liquid phase

The equations of state for the liquid were calculated using isobaric NPT Monte Carlo simulations[31] of $N = 512$ particles, and using the second order virial expansion at very low density. The reduced second virial coefficient of the KF model reads:[18]

$$\frac{B_2}{B_2^{\text{HS}}} = 1 - \chi^2 \left[(1+\delta/\sigma)^3 - 1\right] \left(e^{\varepsilon/k_B T} - 1\right) \quad (3)$$

with $B_2^{HS} = (2/3)\pi\sigma^3$ the second virial coefficient of hard spheres and χ the fraction of the particle surface covered by the N_p attractive patches and is given by

$$\chi^{p-p} = \sqrt{N_p^2(1-\cos(\theta_m)/2)}. \tag{4}$$

The B_2 of the two *p-ap* models have the same functional form but with covering ratios

$$\chi^{p-ap_{N_a}} = \sqrt{N_p N_a (1-\cos(\theta_m)/2)}, \tag{5}$$

that account for the N_p*N_a ($N_a = 1,2$) possible patch–anti-patch pairs that can form between two particles, instead of N_p^2s for the fully patchy case. Equations of state for the *p-p* and the *p-ap₁* models are shown in Fig. 2 together with their compressibility factor $Z = \beta P/\rho$ and B_2 expansion at low density.

3.2 Liquid–gas coexistence

The liquid–gas phase coexistences were calculated using a Grand Canonical Wang-Landau based scheme[32] proposed by Ganzenmuller and Camp[33] and are presented in Fig. 3. Given the reduction in the effective coverage of the particles when considering patch–anti-patch interactions, the critical point of the liquid–gas coexistence curve gets shifted towards lower temperatures as expected, while the critical density is almost unaffected. Interestingly, the liquid–gas phase boundaries of the *p-p* and the *p-ap* models collapse on top of each other when plotted as a function of the reduced second virial coefficient defined by eqn (3) and (5). This observation suggests that there may be a generalised law of corresponding states[34,35] for patchy particles that share the same distribution of patches (here tetrahedral) but with a different number of corresponding anti-patchs to a given patch. This observation is potentially useful because it suggests that knowledge of the liquid–gas phase coexistence of the fully patchy model will be sufficient to predict the coexistence for other patch–anti-patch realizations. One could envisage inverting this argument as a method to classify "similarity" in protein–protein interactions.

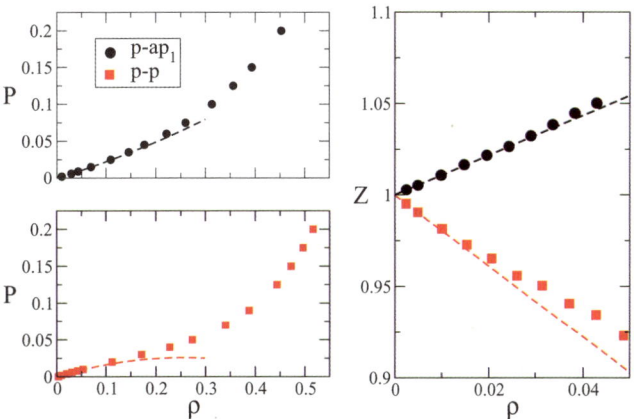

Fig. 2 (left) Fluid equations of state for the patch-patch *p-p* (squares) and the patch–anti-patch *p-ap₁* (dots) models at $T = 0.2$. The Monte Carlo results (symbols) are shown together with the virial expansion up to B_2 (dashed curves). (right) Corresponding compressibility factor $Z = \frac{\beta P}{\rho}$ at low density as obtained from MC simulations and B_2 expansions for the two models.

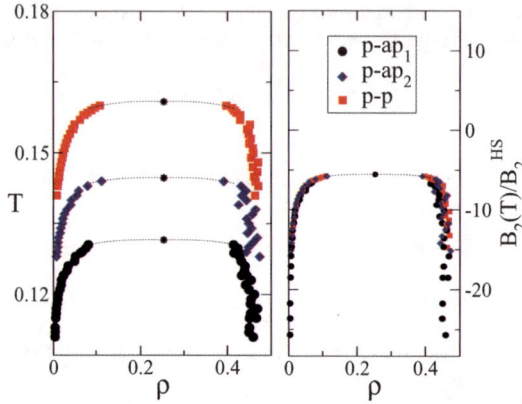

Fig. 3 (left) Liquid–gas coexistence curves for the patch-patch (squares) and the patch–anti-patch models $p\text{-}ap_1$ (dots) and $p\text{-}ap_2$ (diamonds) in the ρ–T plane. Small stars indicate the location of the critical points. The critical temperatures is reduced due to a lower effective coverage of the patchy particles when increasing the specificity of the patches from $p\text{-}p$ to $p\text{-}ap_1$; the critical densities are almost unaffected. (right) Liquid–gas coexistence boundary of the three models collapse on top of each other when mapped onto their reduced second virial coefficients $B_2(T)/B_s^{\text{HS}}$.

3.3 Phase diagrams

To calculate the phase diagrams for the three models, we first calculate the Helmholtz free energies of the potential phases, and then use common tangent constructions to determine the coexistence regions.

3.3.1 Gas free energy.
The Helmholtz free energy of the gas is given by[31]

$$\frac{\beta F(\rho)}{N} = \frac{\beta F_{\text{id}}(\rho)}{N} + \beta \int_0^\rho d\rho' \frac{P(\rho') - \rho'/\beta}{\rho'^2} \qquad (6)$$

where F_{id} is the ideal gas free energy. To determine the equation of state, $P(\rho)$, we used NPT Monte Carlo simulations with 512 particles.

3.3.2 Fluid free energy.
The free energy of the fluid at density ρ is calculated by thermodynamic integration using the hard sphere fluid as a reference system; hence the potential used for the thermodynamic integration is

$$U_{\text{liq}}(\lambda) = U_{\text{HS}} + \lambda U_{\text{patch}}. \qquad (7)$$

where λ is the coupling parameter, U_{HS} is the hard sphere potential and U_{patch} is the potential associated with the patches given by

$$U_{\text{patch}} = \frac{1}{2} \sum_{i,j=1}^{N} u_{\text{patch}}(i,j). \qquad (8)$$

Thus, for $\lambda = 0$ the system reduces to a hard sphere fluid and for $\lambda = 1$ we have the full potential energy. The free energy of the fluid is then given by[31]

$$F_{\text{liq}} = F_{\text{HS}} + \int_0^1 d\lambda \left\langle \frac{\partial U_{\text{liq}}(\lambda)}{\partial \lambda} \right\rangle_\lambda \qquad (9)$$

where F_{HS} is the free energy of the hard sphere fluid, which has been determined previously by Speedy,[36] and z_λ denotes an ensemble average for a system with potential energy $U(\lambda)$. The integrand was measured using NVT Monte Carlo simulations with $N = 512$ particles, and the integral was evaluated using a 20 point Gauss–Legendre quadrature.

Using the free energy determined at a reference density $F(\rho_0)$, the free energy as a function of density $F(\rho)$ can be determined using the equation of state. In this case, the free energy is given by[31]

$$\frac{\beta F(\rho)}{N} = \frac{\beta F(\rho_0)}{N} + \beta \int_{\rho_0}^{\rho} d\rho' \frac{P(\rho')}{(\rho')^2} \qquad (10)$$

where $\frac{\beta F(\rho_0)}{N}$ is the free energy at a reference density ρ_0. The equation of state $P(\rho)$ was determined using Monte Carlo NPT simulations with $N = 512$ particles.

3.3.3 Solid free energies.
Previous work on this model had identified three possible solid phases: face-centered-cubic, body-centered-cubic, and diamond. All three crystal structures are fully bonded, although the bonding in the FCC system does not have patches pointing directly towards each other. Snapshots of the three crystal phases, and their bonding is shown in Fig. 4.

We are specifically interested in the stability of the low density crystal phase, *i.e.* the diamond phase, and we will only examine the stability of the diamond and the associated bcc phases in the remainder of this paper. It should be noted that the FCC crystal will be stable for high densities (and hence, pressures). and the stable region will be strongly limited by the fact that the particles cannot rotate significantly while remaining bonded.

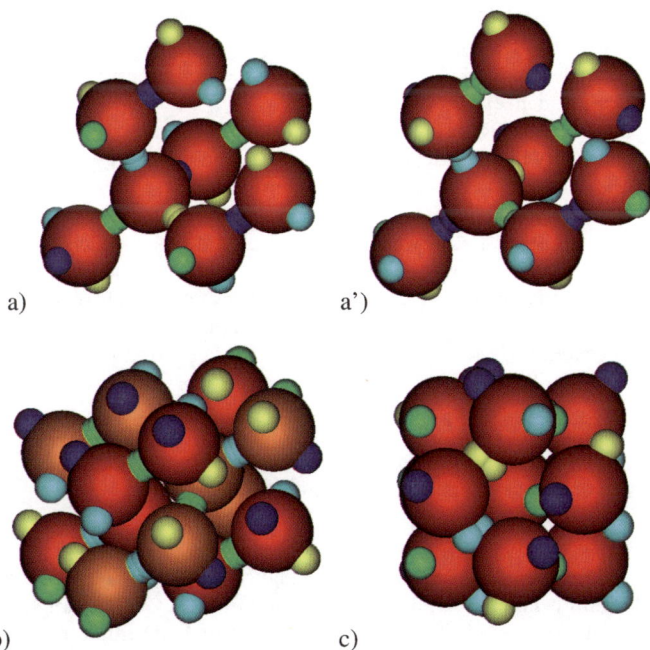

Fig. 4 Three possible fully bonded solid phases for the thetrahedral patch–anti-patch model p-ap_1: (a) cubic diamond (DC), (b) body-centered-cubic (BCC) and face-centered-cubic (FCC). Different patch realizations of DC exist (*e.g. a* and *a'*) while BCC is the result of two inter-penetrating DC lattices (darker and lighter red respectively in b). Note also in c) that patches cannot exactly aligned in the fully bonded FCC and therefore the angular stability range of this crystal is reduced. The particles and patches are not drawn to size.

In the case of the patch-patch model, there is trivially a single realization of the diamond, BCC and FCC phases. However, in the patch–anti-patch case, it is possible for the particles to be slocated on lattice sites, but with a different bonding structure. As shown in Fig. 4, in the case of diamond, there are at least two ways to bond this system. Further examinations of the system show that there are indeed a large number of different ways to bond the system, however, as soon as the orientations of the particles in a single plane are determined, there are no remaining degrees of freedom for the other particle in the system. Hence, this contribution to the free energy per particle will be of the form $\beta F_{bond}/N = 1/N \log(CN^{2/3})$, where C is a constant; this "surface" term vanishes for large N and hence will be set to zero for the remainder of this paper. This is also the case for the BCC and FCC phases.

In the p-ap_2 system, however, the situation is significantly different: the possibility of locally rearranging the bond identities adds extensively to the total entropy of the crystal phases. As each bond arrangement has the same energy, all possible bonding configurations are equally likely, and the entropy associated with rearranging the bonds can be calculated by estimating the number of possible bond configurations in the system:

$$\beta \delta F_{bond} = -\log N_{conf}. \tag{11}$$

The set of configurations in which the system is fully bonded can be sampled by starting from a 'blank' diamond crystal lattice, where the patches are not assigned an identity yet, and generating a bonded configuration according to the following scheme:

1. Select the first blank patch on the first particle in the system that still has blank patches.
2. Randomly assign one of the two possible identities to the selected patch.
3. If any blank patch is bonded with a non-blank patch, assign it the same identity as the patch it is bonded to.
4. If any particle already has two patches with the same identity, give all remaining patches the other identity.
5. Repeat steps 3 and 4 until the system no longer changes.
6. If there are any blank patches remaining, and no particles with more than two patches of the same kind, start again from step 1.

This strategy results in either a fully-bonded configuration, or a configuration with invalid particles (*i.e.* the wrong number of patches of each type). Since at each iteration of the algorithm there are only two choices, and each set of choices will result in a different configuration, this algorithm can be seen as a random walk in a binary tree, ending up at a specific configuration c after taking $d(c)$ decisions. The probability $P(c)$ of finding a specific configuration c is then given by:

$$P(c) = 2^{-d(c)}. \tag{12}$$

As a result, the total number of configurations can be calculated by:

$$N_{conf} = \sum_c f(c) = \sum_c P(c) 2^{d(c)} f(c) = \langle 2^{d(c)} f(c) \rangle, \tag{13}$$

where $f(c)$ is a function that equals 1 if a configuration is fully bonded, and 0 if it is invalid. The sum is taken over all possible configurations, and the angular brackets denote averaging over a set of configurations resulting from random walks through the binary tree, as described above.

Thus, the number of fully bonded configuration for a fixed system size can be calculated by sampling a large number of randomly generated configurations. Of course, the number of valid configurations is highly dependent on the number of particles. To estimate the contribution of the bonding entropy in the thermodynamic limit, we calculated the bonding free energy for a range of system sizes and extrapolated the result to an infinitely large system, starting from either a completely blank crystal, or one where only particles in a cube-shaped region were blanked. For each system size, we sampled at least 10^5 configurations. The resulting extrapolation is shown in Fig. 5, and yields a bonding entropy of $\beta F_{\text{bond}}/N = -0.34 k_B T$ per particle. While accurately sampling the number of configurations requires a large number of runs, we find that changes of around $0.05 k_B T$ in the bonding entropy do not significantly affect the phase diagram for the p-ap_2 system. Apart from local rearrangements, we expect the strongest contribution to the entropy for large systems to be either plane defects (for the fully free crystal), or surface effects (for the crystal with constrained boundaries). For both of these, the entropic contribution to the total free energy should be proportional to $N^{2/3}$. Thus, the leading finite-size correction term for the free energy per particle will be on the order of $N^{-1/3}$.

To calculate the complete free energies of the diamond and bcc solid phases we again used thermodynamic integration. In this case, we choose an Einstein crystal with a fixed center of mass and constrained orientation of the particles as the reference state:

$$U(\lambda) = U_{\text{HS}} + (1-\lambda)U_{\text{patch}} + \lambda U_{\text{Ein}} + \lambda U_{\text{rot}} \quad (14)$$

where U_{HS} is the hard sphere potential, U_{Ein} is the typical Einstein term which attaches particles to their "ideal" lattice sites denoted r_i^0 and is given by

$$U_{\text{Ein}} = \sum_{i=0}^{N}(r_i - r_i^0)^2,$$

and the fourth term ties the orientation of the particles to their "ideal" orientation; U_{rot} is given by

$$\beta U_{\text{rot}}^{p-p} = \sum_{i=1}^{N} \min_{k \neq l}\left\{2 - \text{sign}(c_{a,k})c_{a,k}^2 \delta_{p^a,p^k} - \text{sign}(c_{b,l})c_{b,l}^2 \delta_{p^b,p^l}\right\} \quad (15)$$

where $c_{a,k} = \hat{p}_i^k \cdot \hat{p}_i^{a,0}$, $\hat{p}_i^{a,0}$ is the "ideal" direction for patch a on particle i, and a and b correspond to two, nonidentical patches. In the tetrahedral case, any two patches are sufficient since all patches are not colinear. In a more general case, such as for

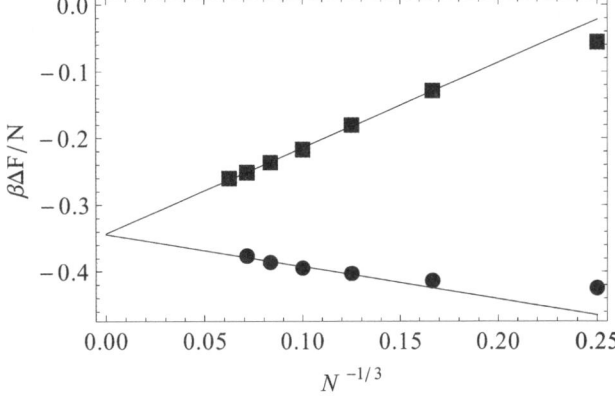

Fig. 5 Bonding entropy per particle as a function of the system size, both for a fully free system (circles), and a system where the bond identity for the particles at the edge is fixed (squares). The lines are linear fits, leading to an estimate of $\beta \delta F_{\text{bond}}/N = -0.34$ in the thermodynamic limit.

patches arranged in an octahedral symmetry, two non-colinear patches would need to be chosen. In the patch-antipatch case (p-ap_1), U_{rot} simplifies to

$$\beta U_{rot}^{p-ap} = \sum_{i=1}^{N}\left(2 - \text{sign}(c_{a,a})c_{a,a}^2 - \text{sign}(c_{b,b})c_{b,b}^2\right). \quad (16)$$

The free energy of the non-interacting system consists of the free energy of the Einstein crystal (F_{ein}) for symmetric particles with a fixed center of mass plus the non-interacting rotational free energy (F_{rot}). The expression for an Einstein crystal with a fixed center of mass is given by[37]

$$\frac{\beta F_{ein}^{CM}}{N} = -\frac{3}{2}\ln\frac{2\pi}{\lambda} - \frac{3}{2N}\ln\frac{\lambda}{2\pi} + \frac{\ln\rho}{N} - 2\frac{\ln N}{N} \quad (17)$$

while the rotational free energy is[19]

$$\frac{\beta F_{rot}}{N} = -\ln\left[\frac{\int d\Omega \exp(-\beta\lambda U_{rot})}{\int d\Omega}\right]. \quad (18)$$

Note that F_{Ein}^{CM} does not depend on the choice of model, i.e. patch-patch or patch–anti-patch. However, F_{rot} does depend on the model.

The full free energy of the system is given by

$$\frac{\beta F(\rho)}{N} = \frac{\beta F_{ein}^{CM}}{N} + \frac{\beta F_{rot}}{N} + \frac{\beta F_{int}}{N} + \frac{\beta F_{bond}}{N} \quad (19)$$

where

$$\frac{\beta F_{int}}{N} = -\frac{\beta}{N}\int\langle U_{Ein} + U_{rot} - U_{patch}\rangle_\lambda d\lambda. \quad (20)$$

Recall that $\beta F_{bond}/N$ is non-zero only in the p-ap_2 case. As in ref. 37, we calculate the integral in eqn (20) using 20 point Gauss–Legendre quadrature in combination with NVT MC simulations.

3.3.4 Phase coexistence. The phase diagrams of the three models presented in Fig. 6 summarize the effect of the interaction specificity on the equilibrium phase diagram of a colloidal model that include both a liquid–gas (L-G) and a fluid-diamond coexistence. The main result of our investigation is that, while the L-G boundaries shift towards lower temperatures when increasing the specificity of the patches from p-p to p-ap_1, the fluid-DC coexistence is less affected and does not follow the L-G critical point. The result is a large metastability gap that opens up between the liquid–gas critical temperature and the fluid-DC melting temperature at the critical density when increasing the specificity ot the interaction. The metastability gap M that measures the degree of metastability is usually the driving force for crystal nucleation at the critical point:

$$M = \frac{T_x - T_c}{T_c} \quad (21)$$

with T_x the freezing temperature of the solid (here DC) at the critical density and T_c the metastable L-G critical temperature.

In the case of isotropic potentials, T_c decreases faster than T_x upon reducing the interaction range and hence M increases as the range λ gets narrower, making crystallization just above or just below the critical point possible. However, Romano et al.[27] observed that for the the tetrahedral patch-patch model, the thermodynamic driving force for crystallisation at the critical point increases only slightly with decreasing λ. As a consequence, low-valence p-p particles have to be cooled well

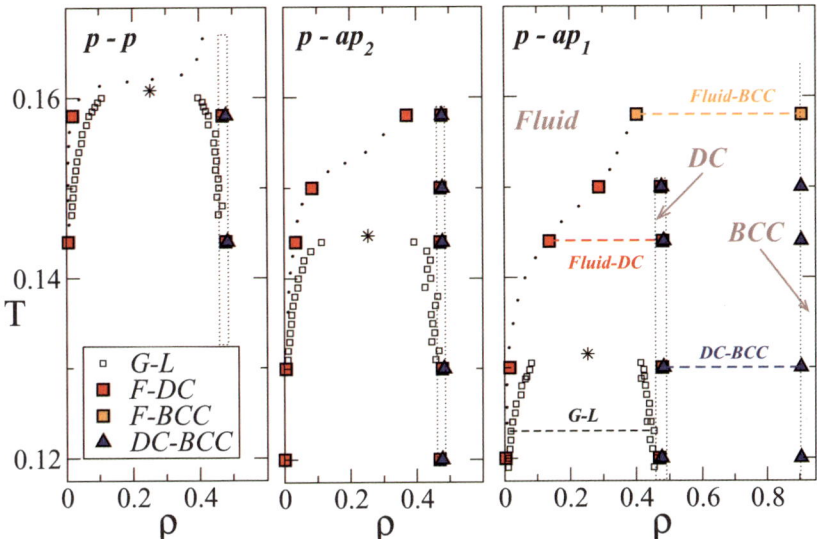

Fig. 6 Phase diagrams of the patch-patch and patch–anti-patch models. The fluid, diamond-cubic (DC) and body-centered-cubic BCC phases were considered. Increasing the specificity of the interaction from p-p to p-ap_1 enlarges the metastability gap M between the liquid–gas (G-L) critical temperature and the Fluid-DC melting temperature at the critical density. Dotted lines are guide to the eye for the boundary of the different phases and dashed lines account for phase coexistence.

below the critical temperature before there is a pronounced thermodynamic driving force for crystallisation. Yet, the viscosity of the (highly networked) liquid phase of patchy particles will increase very strongly with decreasing temperature and hence, when crystal nucleation becomes thermodynamically allowed it will be kinetically suppressed.

Surprisingly – and potentially relevant for real proteins – crystal nucleation in a system with the more specific p-ap interactions can occur in a regime close to the L-G critical point and is therefore more likely to be kinetically accessible: making the patchy interactions more specific opens up a gap between the freezing curve and the L-G coexistence region, in the same way that decreasing the range of attraction does for isotropic potentials.

There is, however, an important difference between the phases that crystallise in the isotropic and the p-ap case: the patch–anti-patch models that we consider freeze into low-density crystals, whilst purely isotropic interactions crystallize into densely packed phases. This observation seems highly relevant for protein crystallisation. We note that our results are also consistent with the phase diagrams obtained in a recent study on the effect of specific and non-specific interaction for the crystallization of a 2D model model of the SbpA surface-layer protein.[38]

The existence of a large metastability gap between the L-G and the Fluid-DC coexistence curves for p-ap models has important implications for the conditions under which crystal nucleation can take place: in the "symmetric" (patch-patch) case, the DC crystal can only nucleate from a fluid with a density that is close to that of the crystal phase. However, for the p-ap system it should be possible to nucleate the DC crystal from a low density fluid, which is the normal condition for protein crystallisation. Of course, thermodynamic driving force is not the whole story: as a liquid gets more supercooled, it will become more viscous and hence the kinetic pre factor for the nucleation rate would go down with supercooling. However, at the critical point, all patchy systems that we studied have the same reduced B_2 and this would suggest that the ease with which a bond can be broken in the liquid is the same for the *p-p* and *p-ap* models. If we assume that the time scale for bond breaking determines viscosity, then we expect the different patchy liquids to have very similar viscosities near T_c—however, we did not test this.

Of course, the structure of real proteins is such that they do not crystallize into DC structures.[3] They do, however, crystallise into open crystal structures. We believe that the tetrahedral patch–anti-patch model studied here gives valuable insights into their phase diagram and provides a better insight in the origin of the metastability gap that is typically found in real protein solutions: importantly, the short range of the attraction is not the whole story—anisotropy *and* specificity are important.

Many simulation papers on colloidal self assembly idly invoke the relevance of their results for the design of photonic band-gap materials. Although the authors are reluctant to join this bandwagon, the present study may actually be of real relevance for the strategies to make colloidal crystals with a diamond structure. Our message is: it is not enough to make patchy colloids with tetrahedral symmetry—rather, one should make tetrahedral colloids where some or all 4 patches have a different functionality (something that could, for instance, be achieved with DNA functionalization).

Acknowledgements

The authors would like to thank Seth Fraden and Bramie Lenhoff for inspiring discussions during their visits in Cambridge. This work was supported by ERC Advanced Grant 227758, Wolfson Merit Award 2007/R3 of the Royal Society of London and EPSRC Programme Grant EP/I001352/1. N.D. acknowledges financial support from the Swiss National Science Foundation (Project no. PBELP2-130895).

References

1 A. McPherson, *Crystallization of Biological Macromolecules*, Cold Spring Harbor Laboratory Press, 1999.
2 R. P. Sear, *J. Chem. Phys.*, 1999, **111**, 4800–4806.
3 S. W. Wukovitz and T. O. Yeates, *Nat. Struct. Biol.*, 1995, **2**, 1062–1067.
4 M. Muschol and F. Rosenberger, *J. Chem. Phys.*, 1997, **107**, 1953.
5 A. Lomakin, N. Asherie and G. B. Benedek, *Proc. Natl. Acad. Sci. U. S. A.*, 1999, **96**, 9465–9468.
6 F. Romano, E. Sanz and F. Sciortino, *J. Chem. Phys.*, 2011, **134**, 174502.
7 N. E. Chayen, *Curr. Opin. Struct. Biol.*, 2004, **14**, 577–583.
8 A. George and W. Wilson, *Acta Crystallogr., Sect. D: Biol. Crystallogr.*, 1994, **50**, 361, year.
9 Rosenbaum, Zamora and Zukoski, *Phys. Rev. Lett.*, 1996, **76**, 150–153.
10 P. R. ten Wolde and D. Frenkel, *Science*, 1997, **277**, 1975–1978.
11 P. N. Pusey and W. van Megen, *Nature*, 1986, **320**, 340–342.
12 E. Zaccarelli, G. Foffi, K. A. Dawson, S. V. Buldyrev, F. Sciortino and P. Tartaglia, *Phys. Rev. E: Stat. Phys., Plasmas, Fluids, Relat. Interdiscip. Top.*, 2002, **66**, 41402.
13 S. Auer and D. Frenkel, *Nature*, 2001, **409**, 1020–1023.
14 C. Liu, N. Asherie, A. Lomakin, J. Pande, O. Ogun and G. B. Denedek, *Proc. Natl. Acad. Sci. U. S. A.*, 1996, **93**, 377–382.
15 N. Dorsaz, G. M. Thurston, A. Stradner, P. Schurtenberger and G. Foffi, *Soft Matter*, 2011, **7**, 1763–1776.
16 P. R. Banerjee, A. Pande, J. Patrosz, G. M. Thurston and J. Pande, *Proc. Natl. Acad. Sci. U. S. A.*, 2011, **108**, 574–579.
17 C. D. Michele, S. Gabrielli, P. Tartaglia and F. Sciortino, *J. Phys. Chem. B*, 2006, **110**, 8064–8079.
18 N. Kern and D. Frenkel, *J. Chem. Phys.*, 2003, **118**, 9882.
19 E. G. Noya, C. Vega, J. P. K. Doye and A. A. Louis, *J. Chem. Phys.*, 2007, **127**, 054501.
20 J. Chang, A. M. Lenhoff and S. I. Sandler, *J. Chem. Phys.*, 2004, **120**, 3003–3014.
21 E. Bianchi, J. Largo, P. Tartaglia, E. Zaccarelli and F. Sciortino, *Phys. Rev. Lett.*, 2006, **97**, 168301.
22 A. L. M. Hloucha, J. F. M. Lodge and S. Sandler, *J. Cryst. Growth*, 2001, **232**, 195.

23 B. L. Neal, D. Asthagiri and A. M. Lenhoff, *Biophys. J.*, 1998, **75**, 2469–2477.
24 D. Asthagiri, B. L. Neal and A. M. Lenhoff, *Biophys. Chem.*, 1999, **78**, 219–231.
25 Y. Yan, A. de Keizer, M. Stuart and N. Besseling, in *Self organized nanostructures of amphiphilic block copolymers II*, Springer Verlag berlin, 2011, ch. From Coordination Polymers to Hierarchical Self-Assembled Structures, pp. 91–115.
26 S. Jones and J. M. Thornton, *Proc. Natl. Acad. Sci. U. S. A.*, 1996, **93**, 13–20.
27 F. Romano, E. Sanz and F. Sciortino, *J. Phys. Chem. B*, 2009, **113**, 15133–15136.
28 F. Romano, E. Sanz and F. Sciortino, *J. Chem. Phys.*, 2010, **132**, 184501.
29 E. G. Noya, C. Vega, J. P. K. Doye and A. A. Louis, *J. Chem. Phys.*, 2010, **132**, 234511.
30 I. Saika-Voivod, F. Romano and F. Sciortino, *J. Chem. Phys.*, 2011, **135**, 124506–124510.
31 D. Frenkel and B. Smit, *Understanding Molecular Simulations: From Algorithms to Applications*, Academic Press, London, UK, 2002.
32 F. Wang and D. P. Landau, *Phys. Rev. E: Stat. Phys., Plasmas, Fluids, Relat. Interdiscip. Top.*, 2001, **64**, 056101.
33 G. Ganzenmller and P. J. Camp, *J. Chem. Phys.*, 2007, **127**, 154504.
34 M. Noro and D. Frenkel, *J. Chem. Phys.*, 2000, **113**, 2941.
35 G. Foffi and F. Sciortino, *J. Phys. Chem. B*, 2007, **111**, 9702–9705.
36 R. J. Speedy, *J. Phys.: Condens. Matter*, 1997, **9**, 8591–8599.
37 J. M. Polson, E. Trizac, S. Pronk and D. Frenkel, *J. Chem. Phys.*, 2000, **112**, 5339.
38 T. K. Haxton and S. Whitelam, *Soft Matter*, 2012, **8**, 3558–3562.

PAPER   www.rsc.org/faraday_d | Faraday Discussions

Amino acids form prenucleation clusters: ESI-MS as a fast detection method in comparison to analytical ultracentrifugation†

Matthias Kellermeier,[a] Rose Rosenberg,[a] Adrian Moise,[b] Ulrike Anders,[ab] Michael Przybylski[b] and Helmut Cölfen[*a]

Received 30th March 2012, Accepted 11th May 2012
DOI: 10.1039/c2fd20060k

Electrospray ionisation mass spectrometry (ESI-MS) is a fast method which is able to provide molecular mass information with high precision. In this contribution, we show that prenucleation clusters—species recently found to play a pivotal role in crystallisation processes—are detected in addition to monomers by analytical ultracentrifugation (AUC) for the whole range of DL-amino acids, while higher oligomers are simultaneously observed in ESI-MS spectra. This suggests ESI-MS is a fast method to identify systems, which form prenucleation clusters. The occurrence of these clusters as relevant precursors in non-classical nucleation scenarios thus appears to be a more common phenomenon than so far assumed.

1 Introduction

Prenucleation clusters (PNCs) are a species currently receiving considerable interest because they are part of an alternative crystallisation pathway that opposes classical nucleation theory (CNT).[1] In contrast to the metastable clusters envisaged in CNT, PNCs exist in a reversible equilibrium with their dissolved components, which is characterised by an association constant and lies in a minimum of Gibbs energy.[2] Thus, they are stable and occur already in undersaturated solutions, where no thermodynamic driving force for crystallisation is expected. Increasing the concentration can lead to cluster aggregation and, eventually, precipitation of an amorphous phase, as shown for calcium carbonate[2,3] and phosphate.[4] PNCs are discussed to be relevant in biomineralisation processes like bone formation,[5,6] and also seem to be of importance for polymorph control[2,7] as well as the interaction of a precipitating system with crystallisation modifiers and scale inhibitors.[8] Recent theoretical work indicates that calcium carbonate PNCs are in fact highly dynamic and strongly hydrated polymeric species.[9] Experimental analyses of PNCs prove to be challenging, especially because the clusters are small and often present only in very low concentrations. Methods that were successfully applied to date for the detection of PNCs of inorganic minerals are analytical ultracentrifugation (AUC),[2] cryo-transmission electron microscopy (cryo-TEM),[3–5] and electrospray ionisation mass spectrometry (ESI-MS).[10] While AUC and cryo-TEM are expensive and time-consuming techniques, ESI-MS was so far only reported for CaCO$_3$ mineral clusters occurring during a crystallisation reaction.[10] Fast, robust and commonly available analytical methods for the characterisation of PNCs have thus not been established yet.

[a] University of Konstanz, Physical Chemistry, Universitätsstr. 10, D-78457 Konstanz. E-mail: helmut.coelfen@uni-konstanz.de; Fax: +49-7531-883091; Tel: +49-7531-884063
[b] University of Konstanz, Analytical Chemistry, Universitätsstr. 10, D-78457 Konstanz. E-mail: michael.przybylski@uni-konstanz.de; Fax: +49-7531-883097; Tel: +49-7531-882249

† Electronic supplementary information (ESI) available. See DOI: 10.1039/c2fd20060k

Interestingly, nanosized clusters have not only been detected in solutions of inorganic salts, but were also described for various organic molecules. For example, certain proteins tend to form dense, liquid-like clusters as precursors to final crystals;[11] however, as opposed to PNCs, these clusters appear to be metastable species and represent a nucleated phase. On the other hand, evidence for the existence of non-nucleated, solute clusters has been reported for supersaturated solutions of amino acids: for instance, Myerson and co-workers inferred clustering of glycine from a decrease of the mean diffusion coefficient with increasing supersaturation[12] and ageing time.[13] In undersaturated systems, such effects were in turn not observed. In addition to diffusivity studies, cluster formation in supersaturated binary and ternary amino acid solutions (glycine and/or valine) was further supported by activity measurements in levitated droplets,[14] as well as by the development of a concentration gradient in long columns,[15] which is conceptually related to AUC experiments. However, the vast majority of the traced species was rather small, comprising mainly dimers and trimers,[16] whereas larger aggregates appeared to be rare and formed in significant amounts only when the point of nucleation was approached. Similar observations were made also for other small molecules such as sucrose,[17] urea,[18] or citric acid,[19] which in part showed drastically higher average numbers of monomers (up to 100) in the detected clusters.[16]

While these findings were at that time interpreted as evidence for the presence of metastable pre-critical clusters in the framework of CNT, the observed species could indeed also have been stable PNCs. To address this question, it is necessary to study clustering in the undersaturated regime and at high dilution. Corresponding work has already been performed for amino acids by ESI-MS.[20–25] Thereby, oligomers consisting of up to more than thirty molecules were detected for arginine,[20,21] and different amino acids were ranked with respect to their tendency to form higher clusters.[23] However, these studies were generally focussed on the characterisation of aggregates held together by non-covalent interactions by means of ESI-MS, that is amongst others, preventing that such weak interactions are ruptured during ionisation and hence successfully transferring respective clusters from solution into the gas phase, as realised previously for a number of biopolymer complexes in our laboratory.[26] Any relevance of the observed species as solution precursors for crystallisation was in turn hardly noted. However, when considering the size of the oligomers with molar masses of up to $\sim$4200 g mol^{-1} (as reported for arginine), a relation of these clusters to PNCs in solution seems likely, given that the molar mass expected for calcium carbonate PNCs is in a similar range.[2] If such a relation would exist, ESI-MS could serve as fast and accurate method to identify systems in which PNC formation occurs. This would enable rapid screening of a large amount of crystallising systems, including organic as well as inorganic compounds, both of ionic or non-ionic nature and in different solvents. Corresponding results would certainly contribute to improve our understanding of the role of prenucleation clusters in crystallisation processes. In this work, we have investigated all 20 DL-amino acids by ESI-MS concerning the occurrence of higher oligomers in the gas phase and, in parallel, characterised the species present in solution by AUC, in order to be able to assess possible correlations between amino acid oligomer detection in ESI-MS and PNC formation in solution.

2 Materials and methods

2.1 Materials

The following amino acids were used as received: DL-alanine (Acros, 99%), D-arginine (Fluka, $\geq$99%), L-arginine (Acros, $\geq$98%), DL-asparagine monohydrate (Aldrich, 98%), DL-aspartic acid (Acros, $\geq$99%), D-cysteine (Aldrich, $\geq$99%), L-cysteine (Fluka, $\geq$99%), D-glutamine (Sigma, $\geq$98%), L-glutamine (Sigma, $\geq$99%), DL-glutamic acid monohydrate (Sigma, $\geq$98%), glycine (Aldrich, $\geq$99%),

DL-histidine (Sigma, ≥99%), DL-isoleucine (Alfa Aesar, 99%), DL-leucine (Aldrich, ≥99%), DL-lysine (Sigma, ≥98%), DL-methionine (Sigma, 99.5%), DL-phenylalanine (Aldrich, 99%), DL-proline (Aldrich, 99%), DL-serine (Aldrich, 99%), DL-threonine (Sigma, ≥98.5%), DL-tryptophan (Aldrich, ≥99%), DL-tyrosine (Acros, 99%), and DL-valine (Merck, 99%).

2.2 Sample preparation

First, 0.01 M stock solutions were prepared by dissolving appropriate amounts of amino acid in water of Milli-Q quality. Due to its poor solubility, the stock concentration had to be decreased to 0.001 M in the case of tyrosine. DL-Arg, DL-Cys and DL-Gln were obtained by mixing equimolar quantities of the respective L- and D-enantiomers. In a second step, the pH of the as-prepared solutions was adjusted to 3.1 (being the native pH of the most acidic amino acid in our study, aspartic acid) by adding aliquots of acetic acid (HOAc, VWR Prolabo, AnalaR, 100%). These samples were then directly used for the AUC measurements. For the ESI-MS screening experiments, solutions were in turn diluted 1 : 100 with water, giving typical final concentrations of around 10^{-4} M. Variations of the pH were done by doping neat 10^{-4} M solutions of DL-Arg (native pH: 9.3) with small volumes of ammonium hydroxide (25% ammonia solution, Merck, p.a., and dilutions thereof) or 0.01 M HOAc to cover a pH range of 10–12 and 5–9, respectively, while concentrated acetic acid had to be used for pH 2–4. The sample at pH 1 was obtained by dissolving DL-Arg directly in ~60% HOAc. To study the concentration dependence of arginine oligomers detected by ESI-MS, a stock solution at 0.1 M was prepared, set to pH 3.1 by addition of HOAc, and finally diluted with water to the desired concentration. Due to the high solubility of arginine in acidic media, the effect of concentration on the species observed by AUC was investigated by measuring different DL-Arg solutions at their respective native pH, so as to gain data also from supersaturated systems. The solubility of the arginine used for these studies (DL-arginine, Alfa Aesar, 98%) was determined to be about 1.3 M, and samples at concentrations in both the under- (0.01, 0.1, 0.2, 0.5, 0.8 and 1 M) and supersaturated regime (1.56 and 1.82 M, corresponding to supersaturation levels of *ca.* 20 and 40%) were investigated. Supersaturated solutions were prepared by gently heating (~30 °C) the samples for several hours until dissolution was completed.

2.3 Electrospray ionisation mass spectrometry (ESI-MS)

ESI-MS measurements were carried out by direct infusion of the samples at a flow rate of 5 μL s^{-1} into an Esquire 3000+ ion trap mass spectrometer (Bruker Daltonik, Bremen, Germany) operated in positive ion mode. Mass spectra were recorded by scanning from 50 to 2500 *m/z*. The ion source parameters were as follows: 15 psi nebulising gas (nitrogen), 6 L min^{-1} of drying gas (nitrogen) at a temperature of 200 °C, capillary voltage 3000 V and capillary exit 80 V. For some amino acids (*e.g.* glutamic acid), higher sample concentrations were found to result in ion signal saturation due to insufficient ionisation and limited ion sampling, as well as low transmission efficiencies as a consequence of space charge effects. Furthermore, deposition of solid material was observed on the needle tip and end cap under these conditions, leading to needle clogging and contamination. Comparative studies showed that none of the amino acids caused such problems at a concentration of 10^{-4} M, which therefore was chosen for the screening experiments.

2.4 Analytical ultracentrifugation (AUC)

AUC is a fractionating, absolute first-principle method for the detection of dissolved and dispersed species in solution. Its statistical basis is excellent since every molecule/particle is detected. It is therefore one of the few potential techniques which are able to fractionate and trace prenucleation clusters in solution.

Measurements were performed on an Optima XL-I (Beckman-Coulter, Palo Alto, CA, United States) using Rayleigh interference optics and 12 mm double-sector titanium centrepieces (Nanolytics, Potsdam, Germany). Samples were investigated at 20 °C and 60 000 rpm, corresponding to a centrifugal force as high as 280 000 g. Nevertheless, the effective force exerted on sedimenting clusters is very low and mainly results from the friction caused by Brownian motion, essentially because the clusters exhibit quite small sedimentation but very high diffusion coefficients. The true challenge is the evaluation of the AUC data, since the moving boundaries are extremely broad (due to diffusion) and certain scans need to be excluded from the analysis (*i.e.* those which do not describe a transport process anymore, but rather reflect sedimentation–diffusion equilibrium). Another challenge is the correct determination of the amount of clusters, which is very low as compared to the excess of monomeric units in solution. This has been tested and verified by numerous evaluations with simulated data for various noise levels.

Sedimentation coefficient distributions with diffusion corrections were determined using the program SEDFIT by Schuck.[27,28] In addition, sedimentation and diffusion coefficients as well as concentrations of the species present in solution were derived by Lamm equation modelling,[29] using the model of non-interacting species in SEDFIT as an independent evaluation approach. All experiments were carried out in triplicate to ensure statistical relevance. Sizes for the distinct species were calculated on the basis of the sedimentation coefficients *via* the following equation:[30]

$$d_H = \sqrt{\frac{18\eta s}{\rho_P - \rho_S}} \qquad (1)$$

where η and ρ_S are the viscosity and density of the medium, and ρ_P is the density of the sedimenting species (which are assumed to be solid hard spheres in this approach). Values for the latter were obtained by inverting the partial specific volumes of the amino acids, which were calculated using the SEDNTERP software.[31] Alternatively, hydrodynamic diameters were estimated with the simultaneously determined diffusion coefficient D *via* the Stokes–Einstein relation:

$$d_H = \frac{kT}{3\pi\eta D} \qquad (2)$$

2.5 Molecular modelling

Calculations of the size and structure of the investigated amino acids were performed using Materials Studio 5.5 (Accelrys Software Inc.). Thereby, the amino acid structure was simulated in vacuum with the aid of the discover module, leaving the minimiser setup at the default settings (*i.e.* smart minimiser, medium resolution, and a maximum of 5000 iterations using the compass force field). The longest measured molecule dimension was then taken as its predicted diameter. Even though this approach neglects the presence of a hydration layer around the charged amino acid molecules and will therefore underestimate their hydrodynamic radius (as determined by AUC), values calculated in this manner should at least be a rough approximate for the true size in solution and hence allow for a comparison of measured and predicted diameters (in addition to theoretical values derived from partial specific volumes (*cf.* eqn (4) and 5), which enable a direct estimation of the expected hydrodynamic size).

3 Results and discussion

3.1 Cluster detection by ESI-MS

In a first series of experiments, the occurrence of amino acid oligomers in the gas phase was studied by ESI-MS at a given set of conditions (10^{-4} M aqueous solutions of the DL-form, pH 3.1, positive ion mode). The collected data confirm the presence of higher oligomers for all amino acids (note that, hereinafter, the term "oligomer" refers to aggregates of molecules that are not covalently linked, *i.e.* intermolecular complexes and not peptides). Exemplarily, the mass spectra recorded for arginine, phenylalanine, and proline are shown in Fig. 1, while the full dataset is reproduced in Section S3.1 of the Supplementary Information (SI).†

In the case of DL-Arg, only singly charged species were observed, and the detected oligomers ranged from dimers to pentamers ($[(Arg)_n + H]^+$, $n = 1-5$). Thereby, isolation and collision-induced dissociation of an "n"-mer resulted in the "n − 1"-mer by neutral loss of one Arg molecule. Similar observations were made also for the other amino acids. In addition to protonated species, sodium and potassium adducts ($[(AA)_n + M]^+$, $M = $ Na or K) were formed in some cases (*cf.* Phe and Pro in Fig. 1; likely as a consequence of leaching of Na^+ (and K^+) from glass vessels and elution of K^+ from electrodes during pH adjustment). In general, the intensity of related signals differs between distinct amino acids, but was often found to exceed that of the corresponding protonated oligomer (*e.g.* $[(Phe)_n + M]^+$ and $[(Pro)_n + M]^+$, $n = 2-3$, $M = $ Na or K).

Comparing the spectra of the different amino acids suggests that, although oligomer formation appears to be a general phenomenon, the degree of aggregation varies from system to system. At 10^{-4} M and pH 3.1, the highest oligomer observed was a hexamer for DL-Ala, whereas merely dimers were detected for DL-Tyr. In an attempt to quantify the tendency of the different amino acids to undergo oligomerisation, average cluster sizes (ACS) were calculated from the ESI-MS data using the following equation, which was originally proposed by Nemes *et al.*:[23]

$$\mathrm{ACS} = \frac{\sum_n n \cdot I_n}{\sum_n I_n} \qquad (3)$$

wherein I_n represents the measured intensity of a cluster consisting of n monomer units. Values resulting for the average cluster size are listed in Table 1, together with the highest oligomer traced for a given amino acid and the respective most intense signal in the spectra.

The above findings evidence the presence of distinct amino acid oligomers in the gas phase. Essentially, this is well in line with previous ESI-MS studies, in which higher oligomers were observed for all amino acids.[20–25] It has further been noted that glycine is less prone to form (or maintain) oligomers in the gas phase due to destabilisation of its zwitterion under these conditions[21]—which is in good agreement with our results, given that we could hardly resolve any corresponding signals in this case and therefore did not consider resulting ACS values as meaningful (*cf.* Table 1).

However, there are also certain discrepancies between the data obtained in the present work and those reported previously. For instance, Nemes *et al.* provided a ranking of oligomer formation ability (in terms of ACS) for the 20 natural L-amino acids,[23] which differs in part from the order distinguished in our experiments (*cf.* Table 1 and Fig. S1 in the SI†). Although principal trends within distinct groups of amino acids are widely similar (*e.g.* Asp *vs.* Glu, Ser *vs.* Thr, or the rather weak clustering ability of basic amino acids), the position of individual amino acids in the sequence is subject to deviations (leucine, for example, showed the strongest propensity to form oligomers in the work of Nemes *et al.*,[23] whereas it exhibited only medium-high ACS values in our experiments). In addition, average cluster sizes calculated from the present data are generally rather small (ACS ≤ 2.7) as compared to previous results (ACS of up to 7.5).[23] These differences may be caused by the fact that DL-amino acids were used in this study; however, as described below, our analyses indicate that stereochemical configuration does not have a strong influence on the observed oligomerisation behaviour, at least for arginine. On the other hand, the present measurements were performed at relatively low concentrations and different

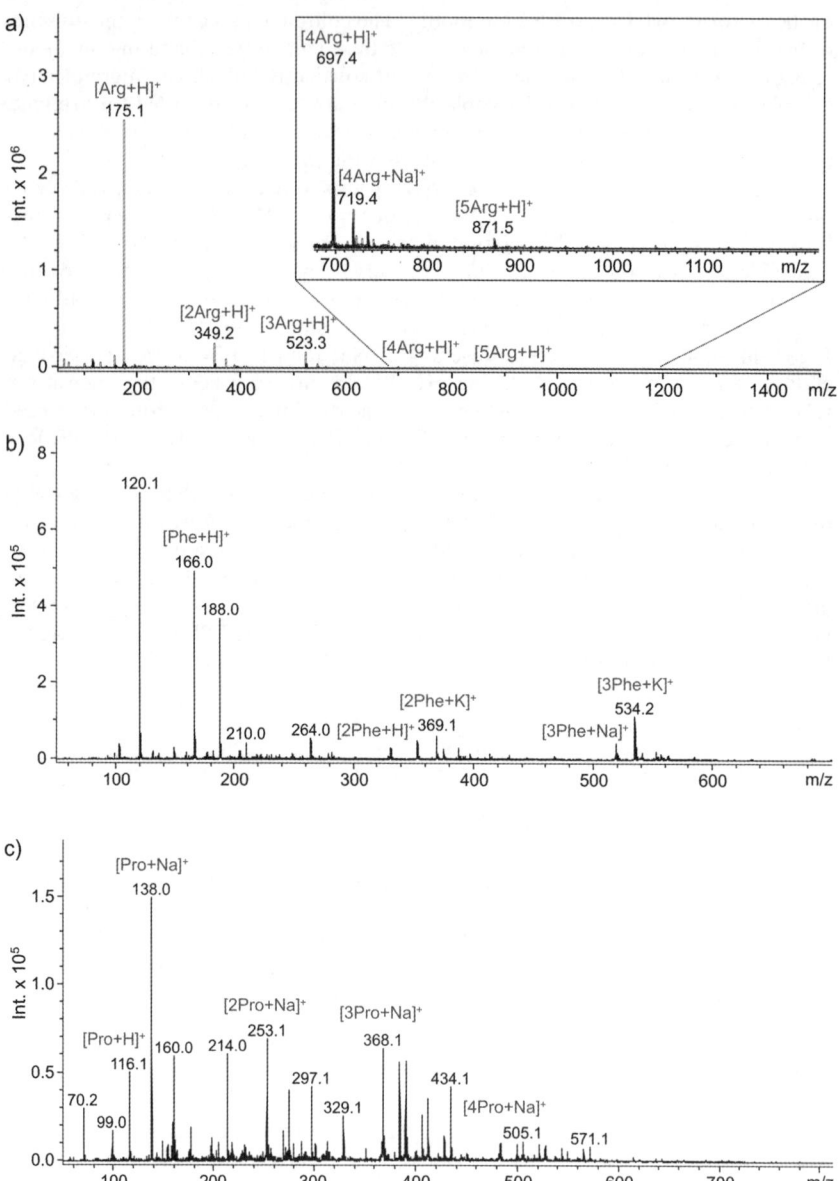

Fig. 1 ESI-MS spectra recorded for 10^{-4} M aqueous solutions of a) DL-arginine, b) DL-phenylalanine, and c) DL-proline. The pH of the samples was adjusted to 3.1. Note the pronounced occurrence of sodium and potassium adducts for both Phe and Pro. The various unlabelled peaks in the spectrum of proline (and other amino acids, cf. Section S3.1†) originate, amongst others, from loss of ammonia (99, 214, and 329 m/z in case of the Pro mono-, di-, and trimer, respectively, or generally −17 m/z relative to the regular protonated signal) or addition of two sodium ions with concurrent loss of a proton (e.g. accounting for the Pro peak at 160 m/z). For the sake of clarity, these signals are not assigned in the spectra.

pH (10^{-4} M and pH 3.1 vs. 0.01 M and native pH in the work of Nemes et al.), which may likewise explain the distinct results. It is moreover interesting to note that most previous MS analyses of amino acid oligomers were carried out using water–methanol mixtures (usually 1 : 1 per volume) as solvent,[20,21,23–25] and Meng et al. as well as

Table 1 Overview of amino acid oligomers detected by ESI-MS in the gas phase. Experiments were performed with 10^{-4} M solutions at pH 3.1. Values for the average cluster size (ACS) were calculated using eqn (3). Note that the amino acids are grouped according to their side chain functionality as non-polar/hydrophobic, polar/neutral, acidic, or basic (from top to bottom). Corresponding mass spectra are shown in the Supplementary Information (Section S3.1)[†]

Amino acid	Largest oligomer detected	Oligomer with highest abundance	ACS
DL-Ala	Hexamer	Trimer	2.67
DL-Ile	Trimer	Monomer	1.41
DL-Leu	Tetramer	Trimer	1.94
DL-Met	Trimer	Monomer	1.72
DL-Phe	Trimer	Monomer	2.49
DL-Pro	Tetramer	Dimer	1.89
DL-Trp	Trimer	Monomer	1.28
DL-Val	Trimer	Trimer	2.21
DL-Asn	Tetramer	Trimer	1.98
DL-Cys	Tetramer	Trimer	2.47
DL-Gln	Trimer	Monomer	1.57
Gly	Dimer	Monomer	—[a]
DL-Ser	Trimer	Trimer	2.66
DL-Thr	Trimer	Trimer	1.98
DL-Tyr	Dimer	Monomer	1.09
DL-Asp	Tetramer	Monomer	1.81
DL-Glu	Trimer	Monomer	1.64
DL-Arg	Pentamer	Monomer	1.25
DL-His	Trimer	Monomer	1.05
DL-Lys	Trimer	Monomer	1.42

[a] The intensities of both the monomer and dimer signal were very low in the case of 10^{-4} M glycine and corresponding peaks could hardly be distinguished next to impurities and unidentified species. Therefore, reliable ACS values cannot be given for this amino acid.

Takats et al. reported that oligomerisation was depressed when purely aqueous solutions (as in this work) were employed.[20,22] This would also rationalise the relatively low ACS values obtained in our study, and we therefore investigated the effect of methanol on the resulting mass spectra of DL-Arg. However, there were no significant changes discernible when water was replaced by 25 or 50% methanol, with respect to both the protonation level and the degree of aggregation (as reflected by ACS values). In this regard, the observed discrepancies are likely to originate either from distinct solution conditions (pH and concentration) or differences in the instrument settings, as it has previously been emphasised that the chosen ESI parameters and the type of spectrometer used can strongly affect the distribution of oligomers detected in the gas phase.[23] It is nevertheless clear that the size of the clusters does not correlate with the side chain functionality of the amino acid (cf. Fig. S1 in the SI†).

Moreover, our data do not provide unambiguous evidence for the existence of so-called "magic-number clusters", which show enhanced stability in the gas phase and would hence represent the most abundant species under various experimental conditions. This has been reported, amongst others, for tetramers of arginine[21] and especially the famous serine octamer, which aroused a great deal of interest in the context of homochirogenesis and chiral transmission during biomolecular evolution.[25,32-34] We cannot certainly explain why there was no preferential formation of Arg tetramers in our measurements and why the serine octamer could not be detected at all, but again speculate that distinct settings and possibly the rather low concentrations used in this study account for these circumstances.

In further experiments, different sample parameters were varied, including the pH as well as the concentration and configuration of the amino acid, using arginine as model system. The aim of these studies was to assess possible effects of solution conditions on the distribution of oligomeric species observed by ESI-MS, and also to be able to directly correlate clustering in the gas phase with the ultracentrifugation results for the state in solution, given that the AUC measurements were performed at higher concentrations (usually 0.01 M) and in part at different pH (in case of the arginine concentration series, which was carried out at the respective native pH). Stepwise increases of the amount of DL-Arg dissolved in the sampled solutions from 10^{-4} to 0.015 M did not induce the presence of oligomers larger than pentamers in the gas phase, with the monomer always being the most abundant species (see Section S3.2 for corresponding spectra†). However, the degree of aggregation as expressed by ACS values grew in a more or less linear fashion from initially 1.21 to about 1.8 in this concentration range (see Table S1 and Fig. S2 in the SI†). Starting from 0.025 M DL-Arg, hexa- and heptamers were detected in minor amounts, while the octamer was the largest observed oligomer between 0.045 and 0.075 M. At the highest concentration investigated (0.086 M), clusters consisting of up to 11 monomer units could be traced when fresh solutions were used, whereas only octamers remained upon ageing (this finding is difficult to explain at this stage and further experiments to clarify possible reasons are currently being carried out). Calculated average cluster sizes indicate a somewhat abrupt increase in the oligomerisation propensity between 0.025 and 0.045 M, where ACS values around 3.8 are reached (*cf.* Fig. S2 in the SI†); this corresponds to an estimated hydrodynamic diameter of 1.14 nm for the oligomers in average (see Section 3.2 for more details). Subsequently, ACS decreases gradually to about 3.0 at 0.086 M, suggesting that there is a maximum in the degree of aggregation at *ca.* 0.05 M. These observations agree well with previous results on the concentration dependence of oligomers detected by ESI-MS.[20,21,23] Using 0.1 M arginine solutions, Zhang *et al.* found undecamers in the gas phase and obtained an oligomer distribution roughly similar to our data at 0.086 M.[21] They reported a maximum in the degree of oligomerisation at about 0.01 M, which is in line with the study by Nemes *et al.*[23] and fairly close to the value determined in the present work (0.045 M). Likewise, it was noted that smaller aggregates become favoured at higher concentrations and that sodiated oligomers are markedly suppressed under these conditions,[21,23] both being confirmed by our results (signals for Na and K adducts were completely absent in spectra recorded from solutions in the millimolar range). As already mentioned above, there are certain, though rather minor differences between the present data and those reported by Zhang *et al.*[21] in terms of the dominant oligomer in the spectra at various concentrations. Up to 0.025 M, the Arg monomer signal remained the most intense peak in our measurements, accompanied by roughly equal amounts of the dimer and trimer (with corresponding signals exhibiting intensities between 10 and 40% of the monomer peak, which is in good agreement with literature[21]). At concentrations of 0.035 M or higher, the trimer was generally found to be the most abundant species, followed by the tetramer and pentamer (see Fig. S3 in the SI†). Although the relative frequency of the tetramer (which has previously been described as magic-number cluster with enhanced stability, thus giving the most intense signal at 0.1 M Arg)[21] is over a wide range close to that of the trimer and even slightly higher at 0.045 M, it does not represent a preferred state of aggregation in our experiments (which would rather apply for the trimer in the present case). In line with the calculated ACS values, peak intensities observed for higher oligomers decrease with respect to the monomer when increasing the concentration from 0.045 to 0.086 M, where the following order of abundance can eventually be discerned: trimer > tetramer > dimer > monomer > pentamer (*cf.* Fig. S3 in the SI†).

In turn, variations of the pH of 10^{-4} M arginine solutions (isoelectric point (IEP) at pH 9.675)[31] showed no systematic dependencies (see Table S2 and Fig. S4 in the SI,† and Section S3.3 for the raw spectra). Earlier work reported that oligomerisation of Arg was depressed when samples were acidified,[20,22] which is to some extent supported by the fact that, in this study, the tetramer was the highest detected oligomer at pH 1, whereas pentamers could be observed at pH 2 and 3.1. However, average cluster sizes determined from the data indicate slightly stronger degrees of aggregation at lower pH values, and thus contradict the above notion. Above pH 3, the pentamer was also no longer visible, with tetramers and trimers being the largest species traced in the near-neutral (pH 6–9) and alkaline (pH 10–12) range, respectively. Thereby, we could not distinguish any significant increase in the clustering propensity with pH, as observed by Zhang *et al.* for arginine solutions at pH 10.[21] Indeed, the monomer was always the most abundant species and the intensities of the dimer, trimer and tetramer, as well as corresponding ACS values, did not exhibit any consistent trend with pH (*cf.* Fig. S4 in the SI†), thus supporting the conclusions of Nemes *et al.* in that there is no clear correlation between the pH and the oligomers detected in the gas phase.[23]

Comparative studies using optically pure arginine samples (10^{-4} M D- and L-Arg) demonstrate that distinct chirality does not affect the resulting oligomer distribution to a noticeable extent. In fact, virtually identical mass spectra were obtained for both enantiomers and their racemic mixture (see Section S3.4 in the SI†), and the calculated average cluster size was about the same in all three systems (see Table S3 in the SI†). It is well known that certain amino acids can distinguish between enantiomers during oligomer formation, potentially displaying strong preference for homochiral aggregates as evidenced by isotope labelling for serine.[32–34] Similar, though less pronounced stereochemical selectivity was observed also for a series of other amino acids, including both homo- (*e.g.* alanine) and heterochiral preferences (*e.g.* valine), partially depending on the particular cluster size.[23] Arginine, however, was not among the amino acids showing this behaviour, as verified by the present experiments.

In essence, the results of the ESI-MS measurements performed in this work agree reasonably well with previously reported data and confirm that higher oligomers are observed in the gas phase for all of the investigated amino acid samples. This provides a sound basis for characterising and comparing corresponding species in solution by means of analytical ultracentrifugation.

3.2 AUC analyses of cluster formation in solution

A general problem associated with mass-spectrometric studies of molecular clustering refers to the critical question of whether the detected oligomers truly reflect the original state of species in solution, or if they possibly formed only upon solvent removal and ionisation during transfer into the gas phase, as outlined explicitly in many of the previous MS studies.[20,22,23] Therefore, if the physical chemistry of presumed clusters in solution is to be elucidated, a second independent technique is required to ascertain that oligomers traced by ESI-MS are relevant species also in the native solution state. Here, we have used analytical ultracentrifugation for this purpose.

Diffusion-corrected sedimentation coefficient distributions c(s) calculated for samples of DL-Arg demonstrate the existence of two populations in the solutions: individual amino acid monomers (s ≈ 0.05–0.25 S) as well as clusters with a sedimentation coefficient of about 1–2 S (Fig. 2a).

Extensive AUC analyses of the whole range of DL-amino acids prove that, indeed, species with s-values typical for clusters can be detected in all cases (see Section S4 in the SI†). Monomers generally exhibit sedimentation coefficients in the range of 0.05–0.25 S (see Table 2), which is similar to values known for single ions in solution.[2]

Estimations of relative concentrations show that the monomers always represent the vast majority of species occurring in the system, usually in fractions higher than 99 wt%. In turn, the amount of clusters in the samples is lower than 0.5 wt% for most amino acids. Corresponding sedimentation coefficients vary between 0.8 and 2 S,

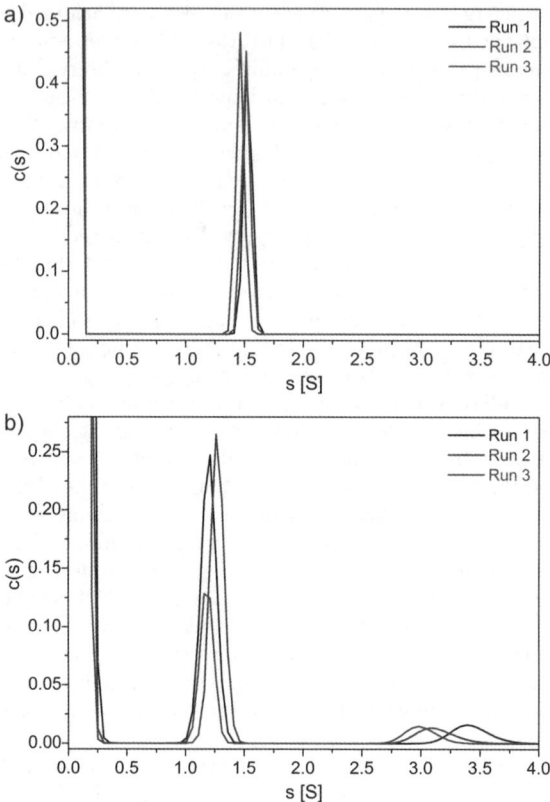

Fig. 2 Sedimentation coefficient distributions derived from AUC measurements using SED-FIT for 0.01 M solutions of a) DL-Arg and b) DL-Asn, both adjusted to pH 3.1. Note that the dominant signal for amino acid monomers at s < 0.5 S has been cut off so as clearly illustrate peaks belonging to clusters (1 ≤ s ≤ 1.5) and their aggregates (s > 2.5 in case of Asn).

which is considerably higher than those of the monomers and moreover agrees well with s-values reported for PNCs of calcium carbonate (~1.5 S).[2,3] This suggests that amino acids indeed form prenucleation clusters in undersaturated solutions. The low concentration of these clusters with respect to monomeric units indicates that their association constant is significantly smaller than in the $CaCO_3$ case (where typically 10–30% clusters occur in equilibrium with free ions at pH values between 9 and 10).[2] Nonetheless, they exhibit features that are characteristic of stable PNCs and distinct from what would be expected for metastable clusters in the framework of CNT. These are, on the one hand, the fact that measurable amounts of the clusters are present in solution far below the solubility limit and, on the other, the existence of a preferential cluster size that is clearly separated from monomers.[9] According to CNT, the probability of cluster formation decreases continuously with the number of monomer units in the cluster. Thus, one would expect a single peak in c(s) that is shifted slightly towards higher s-values if a minor fraction of the amino acids was involved in the assembly (and subsequent disintegration) of metastable clusters (given that AUC is not capable of resolving lower oligomers next to monomers, see below).

For some amino acids (*i.e.* Ile, Leu, Met, Phe, Pro, Trp, Asn, Gln, Glu and His), even larger or respectively more dense species are detected as a third component, with sedimentation coefficients ranging from about 2 to 3 S (*cf.* Table 2 and Fig. 2b). Again, this complies well with previous results for calcium carbonate,

Table 2 Summary of the AUC results for 0.01 M solutions of the investigated amino acids at pH 3.1, whereby $s_{Monomer}$ is the sedimentation coefficient of single amino acid molecules, $s_{Cluster\ I}$ that of presumed single PNCs, and $s_{Cluster\ II}$ that of PNC aggregates, all with corresponding relative concentrations c_i. Data were obtained by Lamm equation modelling of non-interacting species using the SEDFIT software

Amino acid	$s_{Monomer}$ [S]	$c_{Monomer}$ [wt%]	$s_{Cluster\ I}$ [S]	$c_{Cluster\ I}$ [wt%]	$s_{Cluster\ II}$ [S]	$c_{Cluster\ II}$ [wt%]
DL-Ala	0.18 ± 0.04	99.6 ± 0.3	2.0 ± 0.3	0.4 ± 0.3	—	—
DL-Ile	0.08 ± 0.03	99.91 ± 0.04	0.89 ± 0.07	0.05 ± 0.03	2.6 ± 0.2	0.03 ± 0.02
DL-Leu	0.10 ± 0.08	99.8 ± 0.2	0.78 ± 0.01	0.1 ± 0.2	1.4 ± 0.1	0.04 ± 0.05
DL-Met	0.12 ± 0.02	99.94 ± 0.03	1.2 ± 0.2	0.05 ± 0.02	2.3 ± 0.4	0.02 ± 0.02
DL-Phe	0.10 ± 0.01	99.98 ± 0.01	1.07 ± 0.03	0.01 ± 0.01	1.98 ± 0.03	0.01 ± 0.01
DL-Pro	0.05 ± 0.01	99.99 ± 0.01	0.94 ± 0.03	0.006 ± 0.002	2.46 ± 0.04	0.004 ± 0.004
DL-Trp	0.13 ± 0.01	99.91 ± 0.01	1.3 ± 0.2	0.009 ± 0.001	2.4 ± 0.2	0.002 ± 0.002
DL-Val	0.17 ± 0.09	99.91 ± 0.06	1.3 ± 0.2	0.09 ± 0.06	—	—
DL-Asn	0.13 ± 0.03	99.93 ± 0.04	0.90 ± 0.07	0.05 ± 0.03	3.1 ± 0.4	0.02 ± 0.01
DL-Cys	0.15 ± 0.05	99.89 ± 0.08	0.55 ± 0.06	0.08 ± 0.03	1.43 ± 0.07	0.02 ± 0.02
DL-Gln	0.15 ± 0.03	99.8 ± 0.2	0.57 ± 0.08	0.2 ± 0.1	1.37 ± 0.05	0.07 ± 0.07
Gly	0.19 ± 0.09	99.86 ± 0.06	1.5 ± 0.2	0.14 ± 0.06	—	—
DL-Ser	0.11 ± 0.02	99.99 ± 0.01	1.4 ± 0.2	0.004 ± 0.002	—	—
DL-Thr	0.09 ± 0.03	99.98 ± 0.01	1.4 ± 0.1	0.02 ± 0.01	—	—
DL-Tyr	0.24 ± 0.09	98 ± 2	1.9 ± 0.1	2 ± 2	—	—
DL-Asp	0.22 ± 0.03	99.8 ± 0.1	2.1 ± 0.1	0.2 ± 0.1	—	—
DL-Glu	0.17 ± 0.03	99.8 ± 0.1	1.49 ± 0.03	0.07 ± 0.09	2.91 ± 0.07	0.17 ± 0.05
DL-Arg	0.05 ± 0.01	99.99 ± 0.01	1.05 ± 0.04	0.002 ± 0.001	—	—
DL-His	0.11 ± 0.03	99.99 ± 0.01	0.8 ± 0.2	0.005 ± 0.001	1.8 ± 0.3	0.002 ± 0.001
DL-Lys	0.08 ± 0.02	99.99 ± 0.01	1.6 ± 0.2	0.005 ± 0.004	—	—

where species with similar s-values were identified as aggregates of PNCs.[2] The concentration of such cluster assemblies in the present samples is even lower (<0.2 wt%) than that of individual PNCs (*cf.* Table 2).

If the amino acids are sorted according to their functionality (Fig. 3), it is obvious that a species with a sedimentation coefficient typical for PNCs is found regardless of the structure of the amino acid. This is fully in line with the ESI-MS data, which show that at least dimers (Gly and Tyr) and in most cases oligomers of three or more molecules are detected in the gas phase, thus consistently giving ACS values higher than 1 (which implies significant oligomerisation, *cf.* Table 1). The larger aggregates observed for some systems by AUC (labelled Cluster II in Table 2) do also not evidently depend on the type of amino acid, although such species are particularly common in the group of non-polar/hydrophobic amino acids (*cf.* Fig. 3). Furthermore, we could not trace any coherent trend of cluster size (and abundance) with the saturation of the solutions, which can thus likely be ruled out as a possible driving force for clustering (note that, due to the distinct solubility of the amino acids, the degree of (under)saturation at 0.01 M varies over two orders of magnitude; see Table S4 in the SI†).

Interestingly, the signals of both the amino acid monomers and the PNCs are very sharp in the sedimentation coefficient distributions, while those of the presumed

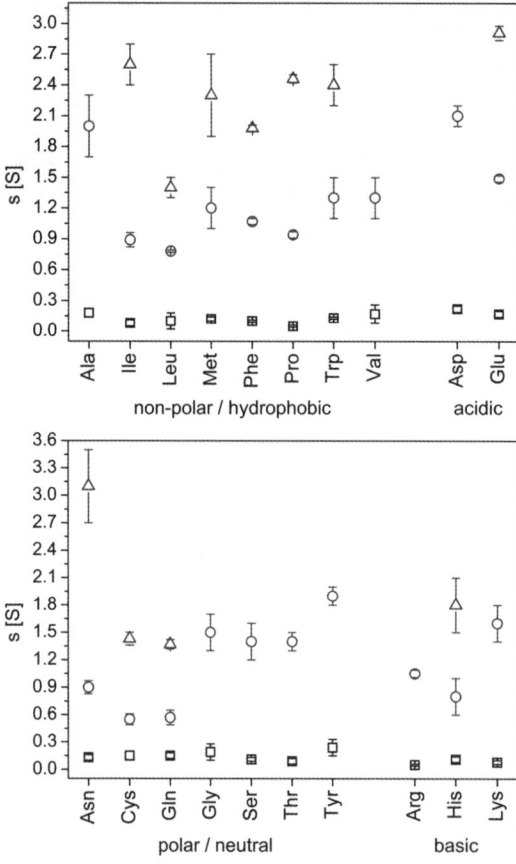

Fig. 3 Comparison of the sedimentation coefficients determined for the different amino acids. Fitting the experimental data required prescription of 2–3 distinct components (depending on the system), as indicated by different symbols (squares: Species 1/monomer, circles: Species 2/Cluster I, triangles: Species 3/Cluster II) and according to Table 2.

cluster aggregates are markedly broader (cf. Fig. 2b). This supports the notion that the larger species are assemblies of prenucleation clusters with varying aggregation numbers, whereas the amino acid molecules are by nature monodisperse. In light of recent modelling results, prenucleation clusters appear to be very dynamic, liquid-like polymeric structures (called "dynamically ordered liquid-like oxyanion polymers" (DOLLOP) in the case of $CaCO_3$),[9] which steadily form and disintegrate on timescales much faster than that of an AUC experiment (which is hours). If such a rapid exchange takes place, only the time-averaged signal will be detected by AUC, as is well known from the analysis of fast reversible interactions between biopolymers.[35] This could explain the sharp signal in the sedimentation coefficient distribution for the species denoted Cluster I (for a dynamic structure like DOLLOPs, fairly broad size distributions would in principle be expected).[9] Assuming that amino clusters form on the basis of a multiple-binding equilibrium as observed for $CaCO_3$,[2] the detected signal would then correspond to the weighted-average species between the monomer and the largest occurring PNC (i.e. polymer), as resulting from an evaluation of the datasets with a non-interacting model.

Hydrodynamic diameters can be derived for the different species either from the sedimentation or diffusion coefficient using eqn (1) and 2, respectively. Further, it is possible to calculate d_H via the volume V of the amino acids and their oligomers, which can be estimated with the partial specific volume $\bar{v}$ of hydrated amino acids and the molar mass M according to:[36]

$$V = \frac{\bar{v} \cdot M}{N_A} \qquad (4)$$

The hydrodynamic diameter d_H is then readily calculated as follows:

$$d_H = 2 \cdot \left(\frac{3V}{4\pi}\right)^{1/3} \qquad (5)$$

Values obtained in this manner for the smallest detected species are listed in Table 3 and compared to those calculated from s and D as well as diameters estimated on the basis of molecular modelling of amino acids in vacuum (as described in Section 2.5).

It is evident that the sizes predicted via the partial specific volumes are generally similar to maximum extensions measured in vacuum (which, thus, appear to be reasonable estimates for molecular dimensions in solution even though hydration layers were neglected in the modelling approach). While corresponding values nearly coincide for certain amino acids (Pro, Val and Cys), $\bar{v}$-based diameters are typically somewhat smaller than the simulated molecule length, although an opposite trend is observed in some cases (Ala, Ile, Gly, Ser and Thr). This indicates that the structure in solution is different from that in vacuum (likely due to hydration, cf. above). Nevertheless, the fairly good agreement between the results from calculations according to eqn (4) and 5 and those obtained from modelling show that sizes derived from the partial specific volume can be considered realistic.

The experimental hydrodynamic diameters resulting from the sedimentation coefficient and those determined via the diffusion coefficient differ slightly, with s-based values generally exhibiting larger errors than those originating from D. This is reasonable since the sedimentation coefficients are close to the resolution limit and sedimentation is so slow under the given conditions that diffusion is the dominating transport process. We therefore regard the hydrodynamic diameters calculated from the diffusion coefficients to be more reliable in this case. The fact that D-based sizes comply well with values predicted with the aid of eqn (4) and 5 for single amino acid molecules suggest that all amino acids exist predominantly as

Table 3 Estimated hydrodynamic diameters of the smallest species observed by AUC. Experimental values were derived from both the sedimentation (s) and the diffusion coefficient (D) via eqn (1) and 2, respectively, using densities resulting from partial specific volumes ($\bar{v}$) calculated with SEDNTERP.[31] For comparison, predicted diameters obtained via the molar mass and the partial specific volumes (eqn (4) and 5) as well as maximum diameters determined by measuring the longest dimension in simulated amino acids molecules are also listed

Amino acid	$\bar{v}$ [cm³ g⁻¹]	$d_{H,Monomer}$ [nm]			
		Experimental (s)	Experimental (D)	Predicted (specific volume)	Predicted (modelling)
DL-Ala	0.7379	1.0 ± 0.2	0.55 ± 0.05	0.59	0.48
DL-Ile	0.8979	1.1 ± 0.3	0.55 ± 0.01	0.72	0.64
DL-Leu	0.8979	1 ± 1	0.60 ± 0.04	0.72	0.79
DL-Met	0.7479	0.8 ± 0.1	0.67 ± 0.01	0.71	0.80
DL-Phe	0.7679	0.76 ± 0.07	0.66 ± 0.01	0.74	0.92
DL-Pro	0.7579	0.5 ± 0.1	0.49 ± 0.01	0.65	0.65
DL-Trp	0.7379	0.81 ± 0.02	0.74 ± 0.01	0.78	0.95
DL-Val	0.8579	1.4 ± 0.8	0.49 ± 0.07	0.68	0.67
DL-Asn	0.6179	0.6 ± 0.1	0.58 ± 0.02	0.64	0.74
DL-Cys	0.6279	0.7 ± 0.2	0.61 ± 0.02	0.62	0.60
DL-Gln	0.6679	0.7 ± 0.2	0.65 ± 0.03	0.68	0.86
Gly	0.6379	0.8 ± 0.4	0.41 ± 0.01	0.53	0.47
DL-Ser	0.6279	0.6 ± 0.1	0.55 ± 0.01	0.56	0.46
DL-Thr	0.6979	0.6 ± 0.2	0.58 ± 0.01	0.64	0.55
DL-Tyr	0.7079	1.0 ± 0.4	0.53 ± 0.04	0.74	1.06
DL-Asp	0.5979	0.8 ± 0.1	0.61 ± 0.07	0.63	0.67
DL-Glu	0.6579	0.8 ± 0.1	0.69 ± 0.01	0.67	0.88
DL-Arg	0.6979	0.45 ± 0.08	0.52 ± 0.01	0.73	1.07
DL-His	0.6679	0.6 ± 0.2	0.52 ± 0.02	0.69	0.90
DL-Lys	0.8179	0.8 ± 0.2	0.49 ± 0.01	0.72	1.00

monomers in solution (especially when considering that experimental d_H values derived from D are by trend even lower than predicted ones). Diameters obtained from the sedimentation coefficient for the smallest species are often larger than those expected for the respective monomer (in case of Ala, Ile, Leu, Val, Gly, Tyr, Asp, Glu). However, for many of the concerned amino acids, the determined size of the monomer is within the limits of accuracy (Leu, Val, Gly, Tyr), and the difference in case of the remaining amino acids (Ala, Ile, Asp, Glu) is not too significant with respect to the relatively large experimental error of such low sedimentation coefficients.

Thus, the diameters obtained from the diffusion coefficient likely represent a realistic estimate of the true hydrodynamic size, and resulting values strongly suggest that the major fraction of the amino acids are monomeric in solution, with a certain amount of larger oligomers co-existing. Using glycine as a model system, we have tested the capability of AUC to distinguish the presence of dimers next to a ca. 90% excess of monomers and traces of larger clusters. Datasets simulated for a corresponding mixture (89.87 wt% monomer, 9.99 wt% dimer, and 0.14 wt% PNCs) were found to be almost equally well described by fits according to a monomer–cluster and monomer–dimer–cluster model, respectively. Therefore, we cannot unequivocally prove that all amino acids are present as monomers in our samples. However, since experimentally determined d_H values are usually lower than diameters expected for monomers, the fraction of dimers (if any) cannot be high.

This is an interesting finding, given that there has been much debate in the past years about whether glycine occurs preferentially as monomer or dimer in aqueous solution. While early studies indicated considerable dimerisation (and in part also higher-order clustering)[12–16,37] and related observations were later taken as evidence to explain polymorph selection during Gly crystallisation,[38,39] more recent work argued that glycine indeed exists mainly in its monomeric form in water. The latter conclusion was supported by cryoscopy measurements,[40] diffusion coefficients determined *via* solution NMR spectroscopy,[40,41] dielectric relaxation studies,[42] as well as molecular dynamics (MD) simulations,[41,43] yielding approximate dimer fractions between 5 and 30%, depending on concentration and pH. Our results confirm this notion and, in addition, demonstrate that monomers are the dominant solution species for all of the investigated amino acids.

Apart from that, the MD simulations performed by Hamad *et al.* also suggested the presence of a small fraction of glycine clusters (including species up to pentamers) next to a large excess of monomers in aqueous solutions.[43] Indeed, size distributions calculated in their work bear obvious resemblance to the sedimentation coefficient distributions obtained in our study. Typical radii of gyration (R_G) resulting for glycine oligomers from the simulations ranged from 0.2 (dimer) to more than 0.75 nm (pentamer). Assuming these clusters to be random-coil polymeric structures in analogy to $CaCO_3$ DOLLOPs,[9] the reported R_G values would correspond to hydrodynamic diameters of about 1 nm in case of the pentamer ($R_H/R_G \approx 0.66$ for a random coil),[44] which is in fair agreement with cluster sizes estimated on the basis of the present AUC and ESI-MS data (as described below, *cf.* Table 4), and nearly coincides with predictions made for Gly pentamers by eqn (4) and 5 (giving $d_H = 0.91$ nm). Moreover, Hamad *et al.* found that clustering of glycine relies on hydrogen bonds between monomers, which continuously break and re-form on timescales of picoseconds.[43] This implies that the structure of the clusters is highly dynamic and hence, by nature, to some degree similar to the polymer-like DOLLOPs described for calcium carbonate.[9] While it remains to be proven that glycine clusters are indeed thermodynamically stable species (as their $CaCO_3$ counterparts), it seems clear that, if so, the equilibrium constant underlying their formation is much lower than in the $CaCO_3$ case (as judged from relative cluster fraction in equilibrium with monomers, *cf.* above). Differences in the clustering behaviour appear plausible when considering that interactions between monomer units are not (only) based on electrostatics, but (additionally) involve hydrogen bonding in the case of glycine. In this context, it is finally worth noting that very recent modelling results corroborate clustering of amino acids in solution, as aspartate was shown to assemble into supramolecular polymers under conditions close to those used in the present AUC analyses.[45]

The sedimentation data also allow determination of the hydrodynamic diameters of the detected larger species. As discussed above and in light of the DOLLOP concept,[9] resulting values should be considered as an average of the supposedly polymeric PNCs. Under this assumption, the employed partial specific volumes should represent a good estimate for the cluster density and, thus, the sizes calculated from the sedimentation coefficient (Table 4) are expected to be more realistic in case of the amino acid clusters. Hydrodynamic diameters derived from the diffusion coefficient (see Table S5 in the SI†) are in turn judged to be less reliable and, accordingly, exhibit relatively high errors and deviate significantly from the values obtained *via* the sedimentation coefficient.

The data given in Table 4 show that the s-based diameters of individual PNCs (Cluster I) vary between 1.4 and 3.7 nm, with Ile and Lys apparently forming the largest clusters. Mean sizes determined for the PNC aggregates (Cluster II) range from 2.1 nm for Cys over 4.7 nm for Leu up to more than 6 nm in the case of Ile, and thus fall into the same order of magnitude as reported for aggregates of $CaCO_3$ prenucleation clusters.[2] To directly correlate these results with the observations made by ESI-MS, we have evaluated the dimensions of oligomers identified in the mass spectra by converting corresponding molar masses into approximate diameters *via* eqn (4) and 5, utilising the partial specific volumes listed in Table 3. This was done for both the highest detected oligomer and the average number of monomer units in the gas-phase clusters (as given by the ACS value), yielding sizes respectively denoted d_{Max} and d_{Av} in Table 4. It is evident that, in all of the investigated systems, (maximum) cluster diameters indicated by ESI-MS are considerably smaller (around 1 nm) than what is obtained by AUC for single PNCs on average. Probably, this implies that larger clusters occurring in solution are less effectively transported into the gas phase than species of low molecular weight, or that they dissociate (*i.e.* break apart) upon ionisation. In either of these two scenarios, the oligomer distribution detected by ESI-MS would not correctly represent the chemical speciation in solution, but rather underestimate the mean size of the clusters (as observed).

Interestingly, much larger oligomers were traced in some of the previous MS studies on amino acids.[20,21,23] For example, Zhang *et al.* detected a 36-mer in their work on arginine, presumably as a consequence of more suitable experimental conditions regarding the ionisation of higher oligomers.[21] According to calculations based on the partial specific volume of hydrated amino acids, such a cluster should have a diameter of about 2.4 nm, which is indeed close to the average value determined by AUC in this case (2.1 nm). These considerations lead to the conclusion that although ESI-MS can be used to rapidly screen distinct systems concerning the formation of clusters, it is not capable of delivering quantitative information on the size and relative frequency of such species, unless instrument settings (and probably also sample parameters like concentration) are fine-tuned and optimised

Table 4 Comparison of sizes estimated for the different cluster species observed by AUC and the amino acid oligomers traced by ESI-MS. Hydrodynamic diameters d_H were derived from the sedimentation coefficient *via* eqn (1) in the case of the AUC data, while the values quoted for the ESI-MS experiments were obtained by eqn (4) and 5 using the partial specific volumes listed in Table 3 and the molar mass corresponding to the highest detected oligomer (d_{Max}, number of monomers indicated in brackets) and the average number of monomer units in the clusters (d_{Av}, as deduced from ACS values), respectively. Also included are the apparent percentages of clusters in the solutions (X_C, calculated according to eqn (6)) as resulting from the AUC experiments as well as, for direct comparison, the average cluster size obtained from ESI-MS analyses (*cf.* Table 1)

Amino acid	AUC			ESI-MS		
	$d_{H,Cluster\ I}$ [nm]	$d_{H,Cluster\ II}$ [nm]	X_C [%]	d_{Max} [nm]	d_{Av} [nm]	ACS
DL-Ala	3.2 ± 0.4	—	2.3 ± 0.1	1.08 (6)	0.82	2.67
DL-Ile	3.7 ± 0.3	6.4 ± 0.5	1.6 ± 0.3	1.04 (3)	0.81	1.41
DL-Leu	3.49 ± 0.04	4.7 ± 0.4	1.3 ± 0.8	1.14 (4)	0.90	1.94
DL-Met	2.6 ± 0.4	3.5 ± 0.7	0.3 ± 0.1	1.02 (3)	0.85	1.72
DL-Phe	2.52 ± 0.08	3.44 ± 0.06	0.5 ± 0.1	1.06 (3)	1.00	2.49
DL-Pro	2.30 ± 0.07	3.73 ± 0.05	0.5 ± 0.1	1.03 (4)	0.81	1.89
DL-Trp	2.5 ± 0.5	3.5 ± 0.3	0.2 ± 0.1	1.13 (3)	0.85	1.28
DL-Val	3.8 ± 0.7	—	0.9 ± 0.4	0.99 (3)	0.89	2.21
DL-Asn	1.6 ± 0.1	3.0 ± 0.4	0.4 ± 0.2	1.01 (4)	0.80	1.98
DL-Cys	1.3 ± 0.1	2.1 ± 0.1	1.0 ± 0.2	0.99 (4)	0.84	2.47
DL-Gln	1.4 ± 0.2	2.18 ± 0.08	1.0 ± 0.1	0.98 (3)	0.79	1.57
Gly	2.2 ± 0.3	—	1.7 ± 0.1	0.77 (3)	—	—
DL-Ser	2.0 ± 0.2	—	0.3 ± 0.2	0.80 (3)	0.77	2.66
DL-Thr	2.4 ± 0.2	—	0.3 ± 0.1	0.92 (3)	0.81	1.98
DL-Tyr	2.9 ± 0.2	—	0.9 ± 0.5	0.93 (2)	0.76	1.09
DL-Asp	2.4 ± 0.1	—	0.7 ± 0.2	1.00 (4)	0.77	1.81
DL-Glu	2.28 ± 0.04	3.18 ± 0.07	1.2 ± 0.1	0.97 (3)	0.80	1.64
DL-Arg	2.09 ± 0.09	—	0.1 ± 0.1	1.24 (5)	0.78	1.25
DL-His	1.7 ± 0.5	2.5 ± 0.5	0.1 ± 0.1	1.00 (3)	0.70	1.05
DL-Lys	3.6 ± 0.4	—	2.4 ± 0.2	1.04 (3)	0.81	1.42

for a given substance. In other words, while large clusters appear to generally exhibit low intensities in mass spectra due to insufficient transfer into the gas phase, they may well represent (one of) the most abundant species in solution (as it is possibly the case for the 36-mer of arginine).

In analogy to the ACS values used to quantify the degree of aggregation in the ESI-MS experiments, apparent percentages of clusters (X_C) out of the total amount of amino acid molecules present were determined from the sedimentation coefficient distributions. This was achieved by relating the integrated area of cluster peaks in the c(s) diagrams to the sum of the integrals of both monomer and cluster signals, according to:

$$X_C = 100 \cdot \frac{A_{\text{Cluster peaks}}}{A_{\text{Cluster peaks}} + A_{\text{Monomer peak}}} \quad (6)$$

Corresponding results are listed in Table 4 and confronted with the average cluster size derived from ESI-MS. Obviously, there is no coherent correlation between the two techniques in terms of the indicated clustering propensities. While fairly good agreement is observed for some amino acids (*e.g.* Ala or Tyr), ACS and X_C differ substantially for others (*e.g.* Ser, Phe, or Lys) and trends found with one method do not coincide with those obtained from the other (*cf.* Fig. S1 in the SI†). This suggests that the extent of clustering seen by ESI-MS in the gas phase does not necessarily reflect the state in solution as probed by AUC, again likely due to distinct abilities (or probabilities) for the different amino acid oligomers to be ionised and transferred successfully into the gas phase. Nevertheless, these data confirm that ESI-MS is able to trace systems which form clusters, while a detailed and quantitative analysis of cluster sizes and distributions requires a more elaborate technique such as AUC.

In this regard, and given that clusters were observed for all studied amino acids, it would be of particular interest to find and investigate a compound that does not form PNCs but prefers to stay monomeric. Such a counter-example would serve as a further proof of concept concerning the potential of the two methods as outlined above. Therefore, we performed AUC measurements with several other water-soluble organic molecules and traced clustering in most cases—with the exception of guanidinium hydrochloride, a strong water-structure breaker and denaturant of protein folding,[46] for which only monomeric species with $s = 0.2 \pm 0.2$ S was detected (at a concentration of 0.02 M). Indeed, ESI-MS spectra recorded from a 10^{-4} M solution at pH 3.1 confirm the absence of higher-order clusters (see Fig. S5 in the SI†), but a distinct peak of the dimer can nonetheless clearly be distinguished. Overall, the signals of both the monomer and the dimer are quite weak in this case (intensities being comparable to those of impurities, decomposition products and certain adducts), such that the clustering propensity appears to be roughly similar to what has been found for glycine (which, by contrast, forms clusters according to AUC results). This implies that somewhat more appropriate criteria (or maybe simply other ESI parameters) must be defined in future work to render mass spectrometry a reliable screening method for the presence of PNCs, especially when it comes to systems in which no clustering takes place. On the other hand, it is important to stress that the absence of higher oligomers in the sedimentation data does not necessarily mean that PNCs do not exist. Under certain conditions (*i.e.* at very low species concentrations), this could also be an issue of the detection limit, which is in the picomolar range in case of ESI-MS, while the Rayleigh interference optics used in AUC can resolve refractive index differences (Δn) on the order of *ca.* 10^{-6}.[47] For an amino acid with a molar mass of 100 g mol^{-1} and a refractive index increment (dn/dc) equal to that of a typical protein (*ca.* 0.19×10^{-3} mL mg^{-1}), this would correspond to a detectable concentration threshold of about 19 nM. Thus, AUC is capable of tracing PNCs down to quite low concentrations; however,

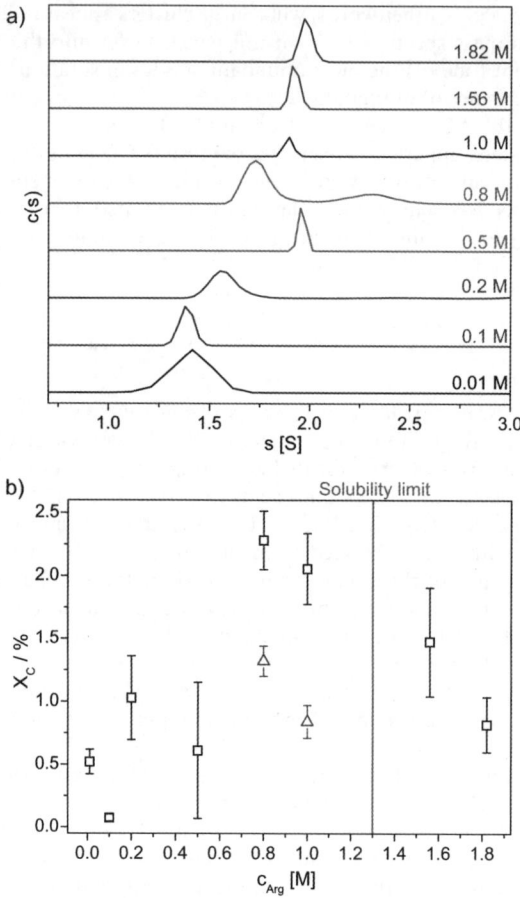

Fig. 4 AUC results for solutions of DL-arginine at various concentrations (all at their respective native pH, ranging from about 10 to 12). a) Selected sedimentation coefficient distributions c(s) for both under- (0.01–1 M) and supersaturated systems (1.56 and 1.82 M, the solubility being around 1.3 M). Note the shift of the cluster signal towards higher values between 0.2 and 0.5 M, as well as the occurrence of a broad peak at s > 2 S for 0.8 and 1.0 M, which is in turn absent in supersaturated solutions. b) Plot of the estimated total percentage of clusters at different concentrations (squares, calculated according to eqn (6), *i.e.* comprising both single clusters and cluster aggregates) and the fractional percentage of single clusters (triangles) at those concentrations where also cluster aggregates were detected (0.8 and 1.0 M).

if the equilibrium constant of cluster formation is very small, these species might indeed not be detectable anymore. ESI-MS is a more sensitive and hence promising alternative in this respect, but obtained results may heavily depend on the chosen settings and state of the used samples, as evidenced by the present and previous findings. Therefore, the collected data do not allow us to ascertain unequivocally whether guanidinium hydrochloride forms clusters (which would be rare, though) or not.

Finally, we performed another set of AUC experiments in which solutions with various arginine contents (each left at its respective native pH) were investigated concerning the size and relative amount of occurring clusters, complementary to the concentration series carried out by ESI-MS. Resulting sedimentation coefficient distributions (Fig. 4a) and cluster percentages calculated according to eqn (6) (Fig. 4b) both show clear trends with increasing concentration. At Arg contents of up to 0.1 M, the c(s) diagrams exhibit a single cluster species with s ≈ 1.4 S next to a large excess of monomers (peak not shown in Fig. 4a). This is generally in line with the data obtained for a 0.01 M Arg solution at pH 3.1 (although s-values of the cluster component differ to some degree, *cf.* Table 2 and Fig. 4), hence confirming the ESI-MS measurements in that the pH does not strongly influence the clustering behaviour of arginine. As opposed to that, the clusters traced by AUC are again significantly larger (*ca.* 2.4 nm) than the highest oligomers detected in ESI-MS for solutions at similar concentrations (11-mer with d ≈ 1.6 nm at 0.086 M), due to the reasons discussed above.

Raising the amount of dissolved Arg leads to a gradual shift of the maximum in c(s) to ~1.6 S at 0.2 M and ~2.0 S at 0.5 M, indicating that the mean (equilibrium) size of the clusters increases with the concentration. This is in line with considerations made for DOLLOPs of $CaCO_3$, where more frequent collisions between clusters and monomeric units (ions or ion pairs) at higher concentrations were proposed to account for a change in the size distribution of the clusters.[9] The estimated percentage of clusters varies inconsistently up to 0.5 M and appears to be constant within the limits of error (0–1%, *cf.* Fig. 4b). Starting from 0.8 M (*i.e.* when the solubility limit of ~1.3 M is slowly approached), X_C adopts considerably higher values (2–2.5% at 0.8 and 1 M). Simultaneously, a second population can be discerned in the s-distributions, centred at about 2.3 and 2.7 S for 0.8 and 1 M, respectively. Corresponding peaks are markedly broadened as compared to the signal at lower s, suggesting that these species are cluster aggregates of variable size. Their enhanced occurrence at these concentrations rationalises the parallel increase in the cluster percentage, as new PNCs will be generated from Arg monomers upon aggregation of others—provided that cluster formation is governed by equilibrium thermodynamics as in the case of calcium carbonate.[2]

This may also explain why the average size of the PNCs is smaller at 0.8 M than at 0.5 M (where pronounced aggregation does not yet take place). Further increasing the Arg concentration to 1 M leads to higher s-values for both individual PNCs (~1.9 S) and their aggregates (~2.7 S), while the total percentage of clusters is more or less the same (note that X_C is not weighted by the number of monomers in the clusters). In the supersaturated regime, the relative amount of clusters drops significantly as aggregates are no longer detected and only the signal of single PNCs at ~1.9 S is left at both 1.56 and 1.82 M. This suggests that cluster aggregates were removed from the system by nucleation of a solid phase during (or even prior to) the AUC measurement, yielding particles that are too large to be observed in c(s) at the given rotor speed.

These concentration-dependent data indicate that the traced clusters are relevant species for crystallisation: we found that aggregation of clusters occurs progressively when the saturation limit is approached and, therefore, speculate that nucleation proceeds *via* (or at the expense) of such aggregates—both being directly analogous to $CaCO_3$.[1–3] Our results further evidence that at least the arginine clusters discussed in detail herein have characteristics similar to those of calcium carbonate,[2] since they appear to form on the basis of an equilibrium between monomeric units and PNCs, too. This is supported by the determined percentages of clusters as a function of concentration: as soon as significant amounts of cluster aggregates are formed in solution (*i.e.* at 0.8 and 1.0 M), X_C rises noticeably. Thereby, it is crucial to realise that cluster percentages calculated according to eqn (6) include contributions of both single clusters and aggregates, that is, $A_{Cluster\ peaks}$ is the sum of peak areas at s > 1 S. Thus, the observed increase in X_C is essentially due to the incorporation of clusters into aggregates which, in line with what is expected for an equilibrium reaction, causes new clusters to be formed from monomeric units (which then increase X_C). In fact, calculations of the actual percentage of single clusters (*i.e.* using only the area of the peak at 1–2 S in the numerator of eqn (6)) show that their fraction in equilibrium with monomers remains constant within the limits of error over the entire range of concentrations investigated (corresponding values are represented by triangles for 0.8 and 1.0 M in Fig. 4b). This behaviour agrees well with results reported for calcium carbonate,[2] and suggests that amino acid clustering in solution relies on equilibrium thermodynamics.

We note that nucleation of glycine has previously been examined by means of small-angle X-ray scattering (SAXS) experiments using concentrated solutions that were gradually cooled in order to achieve increasing levels of supersaturation.[48,49] Analyses of the data indicated that a significant fraction of dimers was present at neutral pH from the beginning on (as judged from calculated radii of gyration) and that further oligomerisation took place with decreasing temperature (in line with early diffusivity and sedimentation studies),[12–16,48] although this interpretation has later been debated.[43] Based on these findings and evaluations of the fractal behaviour of the samples, it was proposed that glycine nucleation is triggered by the formation of a liquid-like cluster consisting of dimer building units and their subsequent reorganisation,[48] according to a typical two-step nucleation model which assumes the intermediate clusters to be metastable.[50] The present results confront this notion in that amino acid PNCs were shown to exist already in undersaturated systems and therefore likely do not represent metastable crystallisation precursors. Our observations rather agree with preliminary neutron scattering (SANS) data on crystallising glycine solutions, which suggested the occurrence of larger aggregates (several nm and thus similar to the species denoted Cluster II in our analyses) before nucleation.[41] A general disadvantage of scattering techniques in this context is the fact that they probe the whole range of co-existing species at once and often only yield average (size) information for the entire ensemble,[43] such that small fractions of possibly relevant crystallisation precursors may well be overlooked. This is by nature different in AUC, as it separates populations with sufficiently distinct sedimentation coefficient prior to detection. Consequently, even species with very low concentrations can be traced and a full distribution of sizes is obtained. On that basis, the measurements performed in this work have revealed the presence of nanosized clusters next to a huge excess of monomers for all amino acids in undersaturated solutions. Beyond that, we propose that the crystallisation of arginine relies on these clusters and essentially involves their aggregation as a crucial step towards nucleation; this, however, remains to be definitely proven, and related studies are required to shed light on the crystallisation pathways of the other amino acids as well.

4 Conclusions

Our findings demonstrate the existence of a qualitative correlation between amino acid oligomers observed in ESI-MS and PNCs traced by AUC: when clusters are detected by AUC, higher oligomers occur in the mass spectra. Thus, it is likely that the species observed in the gas phase are in fact present in solution and not the result of artefacts during evaporation and ionisation. This implies that ESI-MS may very well serve as rapid technique for the identification of systems in which prenucleation clusters occur. On the other hand, comparison of the oligomer distributions and clustering propensities indicated by the two methods reveals marked discrepancies, which we ascribe to distinct probabilities for the different amino acid oligomers to be transferred successfully into the gas phase. In particular, our data suggest that larger clusters are much less effectively transported and/or ionised, such that their contribution to size distributions is greatly underestimated by ESI-MS. Moreover, experimental parameters and the type of instrument used seem to be further delicate factors influencing the final output. Therefore, even though mass spectrometry appears to be a promising alternative to search for prenucleation clusters, quantitative evaluation of cluster formation in solution by this technique remains challenging.

The AUC experiments carried out in this work have shown that measurable amounts of amino acid clusters occur even in dilute solutions. This—together with size distributions estimated for these species (discrete peak separated from monomers), the traced concentration dependence of cluster formation (constant percentage of single clusters), and the aggregation behaviour observed at higher

Table 5 Sedimentation coefficients of ions and PNCs found for different salts. Measurements were carried out at a concentration of 0.02 M (and hence below the saturation limit) for ammonium acetate (NH$_4$OAc), MgSO$_4$, LiI, CsI, CsF, and Na$_2$CO$_3$, whereas saturated solutions were used in all other cases

Substance	s_{Ions} [S]	$s_{Cluster\ I}$ [S]	$s_{Cluster\ II}$ [S]	$s_{Cluster\ III}$ [S]
NH$_4$OAc	0.11 ± 0.07	0.4 ± 0.2	3 ± 1	—
MgSO$_4$	0.23 ± 0.02	1.0 ± 0.2	—	—
NaCl	0.04 ± 0.01	1.5 ± 0.2	2.75 ± 0.02	—
LiI	0.14 ± 0.01	0.93 ± 0.04	3.27 ± 0.04	—
CsI	0.81 ± 0.02	2.4 ± 0.3	5.1 ± 0.4	—
CsF	0.4 ± 0.1	1.01 ± 0.03	—	—
Na$_2$CO$_3$	0.19 ± 0.06	1.00 ± 0.34	3.1 ± 0.1	—
MgCO$_3$	0.09 ± 0.01	1.0 ± 0.6	3.0 ± 0.1	—
CaCO$_3$	0.11 ± 0.05	1.4 ± 0.1	5 ± 1	9 ± 2
SrCO$_3$	0.14 ± 0.04	0.9 ± 0.2	—	—

concentrations—serves as evidence that the detected clusters form on the basis of equilibrium thermodynamics and, hence, that they truly are PNCs in the sense of what has been described previously for calcium carbonate.[2] An obvious difference to the CaCO$_3$ case is the relatively low fraction of amino acid clusters existing in equilibrium with monomers (<1%, as opposed to several tens of percents for CaCO$_3$). However, due to the fairly high solubility of amino acids, the factual concentration of clusters (and aggregates) in solution amounts to the millimolar range when the point of nucleation is approached, as demonstrated for the example of arginine in the present study (for which the cluster concentration is on the order of 0.01–0.02 M close to the solubility limit). In this regard, it seems possible that these clusters (or respectively their aggregates) are precursors of initially nucleated particles and, thus, that they are relevant species for nucleation. Although the data obtained in this work support such a scenario, further insight to the nucleation process is required to draw definite conclusions.

Essentially, cluster formation was confirmed for all amino acids regardless of their functionality (anionic, cationic, hydrophilic or hydrophobic). This raises questions about the driving force of amino acid clustering, given that their solubility (and thus degree of saturation in this study) is quite distinct and, moreover, the hydration of the different amino acids should vary depending on their charge and hydrophilicity. Although we cannot answer this question on the basis of the present data, it seems evident that PNC formation is a more common phenomenon than hitherto believed. This notion is corroborated by additional AUC measurements of various inorganic salts other than the previously studied biominerals calcium carbonate, oxalate and phosphate.[2] The results shown in Table 5 evidence that PNCs were detected for all investigated salts which, remarkably, include both well and hardly soluble compounds and even salts of simple monovalent ions such as alkali halogenides.

We note that ion clustering has been reported for soluble salts in earlier work as well, but only concerning solutions at fairly high supersaturation.[19,51] First attempts to obtain good ESI-MS spectra for ion pairs of inorganic salts and higher oligomers thereof were not yet successful in our lab, whereas meaningful results could be achieved in a previous study on calcium carbonate.[10] Most likely, our experimental parameters will have to be fine-tuned to enable mass-spectrometric determination of clusters also for inorganic systems. If a correlation similar to that presented here for the amino acid case can be established, ESI-MS would be a fast method of choice to screen a large amount of samples for the presence of PNCs, which could then be precisely characterised for example by AUC. This combination of techniques would help to clarify how general prenucleation cluster formation is. The data collected in this work indicate that it is more widespread than previously assumed.

Acknowledgements

The authors thank Dr Denis Gebauer (University of Konstanz) for valuable discussions. MK is grateful to BASF SE for funding a postdoc position.

References

1. D. Gebauer and H. Cölfen, *Nano Today*, 2011, **6**, 564.
2. D. Gebauer, A. Völkel and H. Cölfen, *Science*, 2008, **322**, 1819.
3. E. M. Pouget, P. H. H. Bomans, J. A. C. M. Goos, P. M. Frederik, G. de With and N. A. J. M. Sommerdijk, *Science*, 2009, **323**, 1455.
4. A. Dey, P. H. H. Bomans, F. A. Müller, J. Will, P. M. Frederik, G. de With and N. A. J. M. Sommerdijk, *Nat. Mater.*, 2010, **9**, 1010.
5. F. Nudelman, K. Pieterse, A. George, P. H. H. Bomans, H. Friedrich, L. J. Brylka, P. A. J. Hilbers, G. de With and N. A. J. M. Sommerdijk, *Nat. Mater.*, 2010, **9**, 1004.
6. H. Cölfen, *Nat. Mater.*, 2010, **9**, 960.
7. D. Gebauer, P. N. Gunawidjaja, J. Y. P. Ko, Z. Bacsik, B. Aziz, L. Liu, Y. Hu, L. Bergström, C. W. Tai, T. K. Sham, M. Eden and N. Hedin, *Angew. Chem., Int. Ed.*, 2010, **49**, 8889.
8. D. Gebauer, H. Cölfen, A. Verch and M. Antonietti, *Adv. Mater.*, 2009, **21**, 435; A. Verch, D. Gebauer, M. Antonietti and H. Cölfen, *Phys. Chem. Chem. Phys.*, 2011, **13**, 16811.
9. R. Demichelis, P. Raiteri, J. D. Gale, D. Quigley and D. Gebauer, *Nat. Commun.*, 2011, **2**, 590.
10. S. E. Wolf, L. Müller, R. Barrea, C. J. Kampf, J. Leiterer, U. Panne, T. Hoffmann, F. Emmerling and W. Tremel, *Nanoscale*, 2011, **3**, 1158.
11. O. Gliko, W. Pan, P. Katsonis, N. Neumaier, O. Galkin, S. Weinkauf and P. G. Vekilov, *J. Phys. Chem. B*, 2007, **111**, 3106.
12. Y. C. Chang and A. S. Myerson, *AIChE J.*, 1986, **9**, 1567.
13. Y. C. Chang and A. S. Myerson, *AIChE J.*, 1987, **33**, 697; A. S. Myerson and P. Y. Lo, *J. Cryst. Growth*, 1990, **99**, 1048.
14. H. S. Na, S. Arnold and A. S. Myerson, *J. Cryst. Growth*, 1994, **139**, 104.
15. A. S. Myerson and P. Y. Lo, *J. Cryst. Growth*, 1991, **110**, 26.
16. R. M. Ginde and A. S. Myerson, *J. Cryst. Growth*, 1992, **116**, 41.
17. A. T. Allen, R. M. Wood and M. P. McDonald, *Sugar Tech. Rev.*, 1974, **2**, 165.
18. L. S. Sorell and A. S. Myerson, *AIChE J.*, 1982, **28**, 772.
19. M. A. Larson and J. Garside, *Chem. Eng. Sci.*, 1986, **41**, 1285.
20. C. K. Meng and J. B. Fenn, *Org. Mass Spectrom.*, 1991, **26**, 542.
21. D. Zhang, L. Wu, K. J. Koch and R. G. Cooks, *Eur. J. Mass Spectrom.*, 1999, **5**, 353.
22. Z. Takats, S. C. Nanita, R. G. Cooks, G. Schlosser and K. Vekey, *Anal. Chem.*, 2003, **75**, 1514.
23. P. Nemes, G. Schlosser and K. Vekey, *J. Mass Spectrom.*, 2005, **40**, 43.
24. N. Toyama, J. Kohno, F. Mafune and T. Kondow, *Chem. Phys. Lett.*, 2006, **419**, 369.
25. P. Yang, R. Xu, S. C. Nanita and R. G. Cooks, *J. Am. Chem. Soc.*, 2006, **128**, 17074.
26. M. Przybylski and M. O. Glocker, *Angew. Chem., Int. Ed. Engl.*, 1996, **35**, 806; A. Marquardt, B. Bernevic and M. Przybylski, *J. Pept. Sci.*, 2007, **13**, 803; L. J. Deterding, J. Kast, M. Przybylski and K. B. Tomer, *Bioconjugate Chem.*, 2000, **11**, 335; H. Wendt, E. Durr, R. M. Thomas, M. Przybylski and H. R. Bosshar, *Protein Sci.*, 1995, **4**, 1563; M. Przybylski, M. O. Glocker, C. Maier, C. Borchers, E. Dürr, W. Fiedler, J. Kast, H. Wendt and H. R. Bosshard, in *Peptides*, ed. H. L. S. Maia, Escom Science Publications, Leiden, 1994, pp. 42.
27. P. Schuck, *Biophys. J.*, 2000, **78**, 1606.
28. http://www.analyticalultracentrifugation.com/default.htm.
29. B. Demeler and H. Saber, *Biophys. J.*, 1998, **74**, 444.
30. H. Cölfen and A. Völkel, *Prog. Colloid Polym. Sci.*, 2004, **127**, 31.
31. http://jphilo.mailway.com/download.htm.
32. K. J. Koch, F. C. Gozzo, D. Zhang, M. N. Eberlin and R. G. Cooks, *Chem. Commun.*, 2001, 1854; R. G. Cooks, D. Zhang, K. J. Koch, F. C. Gozzo and M. N. Eberlin, *Anal. Chem.*, 2001, **73**, 3646; K. J. Koch, F. C. Gozzo, S. C. Nanita, Z. Takats, M. N. Eberlin and R. G. Cooks, *Angew. Chem., Int. Ed.*, 2002, **41**, 1721.
33. A. E. Counterman and D. E. Clemmer, *J. Phys. Chem. B*, 2001, **105**, 8092; R. R. Julian, R. Hodyss, B. Kinnear, M. F. Jarrold and J. L. Beauchamp, *J. Phys. Chem. B*, 2002, **106**, 1219.
34. C. A. Schalley and P. Weis, *Int. J. Mass Spectrom.*, 2002, **221**, 9.
35. J. L. Cole, J. W. Lary, T. Moody and T. M. Laue, *Methods Cell Biol.*, 2008, **84**, 143.

36 H. P. Erickson, *Biol. Proced. Online*, 2009, **11**, 32.
37 W. C. M. Lewis, *Chem. Rev.*, 1931, **8**, 81; G. A. Anslow, M. L. Foster and C. Klingler, *J. Biol. Chem.*, 1933, **103**, 81.
38 D. Gidalevitz, R. Freidenhans'l, S. Matlis, D. F. Similgies, M. J. Christensen and L. Leiserowitz, *Angew. Chem., Int. Ed. Engl.*, 1997, **36**, 955; I. Weissbuch, V. Y. Torbeev, L. Leiserowitz and M. Lahav, *Angew. Chem., Int. Ed.*, 2005, **44**, 3226; V. Y. Torbeev, E. Shavit, I. Weissbuch, L. Leiserowitz and M. Lahav, *Cryst. Growth Des.*, 2005, **5**, 2190.
39 C. S. Towler, R. J. Davey, R. W. Lancaster and C. J. Price, *J. Am. Chem. Soc.*, 2004, **126**, 13347.
40 J. Huang, T. C. Stringfellow and L. Yu, *J. Am. Chem. Soc.*, 2008, **130**, 13973.
41 C. E. Hughes, S. Hamad, K. D. M. Harris, C. R. A. Catlow and P. C. Griffiths, *Faraday Discuss.*, 2007, **136**, 71.
42 T. Sato, R. Buchner, S. Fernandez, A. Chiba and W. Kunz, *J. Mol. Liq.*, 2005, **117**, 93.
43 S. Hamad, C. E. Hughes, C. R. A. Catlow and K. D. M. Harris, *J. Phys. Chem. B*, 2008, **112**, 7280.
44 J. G. Kirkwood and J. Riseman, in *Rheology: theory and applications*, ed. F. Eirich, Academic Press, New York, 1956, pp. 495.
45 P. Raiteri, R. Demichelis, J. D. Gale, M. Kellermeier, D. Gebauer, D. Quigley, L. B. Wright and T. R. Walsh, *Faraday Discuss.*, 2012, DOI: 10.1039/C2FD20052J.
46 S. Lapanje, *Physicochemical aspects of protein denaturation*, Wiley, New York, 1978.
47 T. M. Laue, *Choosing which optical system of the OptimaTM XL-I analytical ultracentrifuge to use*, Application Information A-1821A, Beckman Instruments, Fullerton (CA), 1996.
48 S. Chattopadhyay, D. Erdemir, J. M. B. Evans, J. Ilavsky, H. Amenitsch, C. U. Segre and A. S. Myerson, *Cryst. Growth Des.*, 2005, **5**, 523.
49 D. Erdemir, S. Chattopadhyay, L. Guo, J. Ilavsky, H. Amenitsch, C. U. Segre and A. S. Myerson, *Phys. Rev. Lett.*, 2007, **99**, 115702.
50 P. G. Vekilov, *Cryst. Growth Des.*, 2004, **4**, 671.
51 Y. C. Chang and A. S. Myerson, *AIChE J.*, 1985, **31**, 890; R. Mohan, O. Kaytancioglu and A. S. Myerson, *J. Cryst. Growth*, 2000, **217**, 393.

Probing the structure and stability of calcium carbonate pre-nucleation clusters†

Aaron R. Finney and P. Mark Rodger*

Received 22nd March 2012, Accepted 6th June 2012
DOI: 10.1039/c2fd20054f

Recent advances in our understanding of the emergence of biomineral phases from solution has provoked new and challenging questions. A consensus is beginning to form, attesting to the existence of pre-nucleation clusters of calcium carbonate in solution, which subsequently aggregate to initiate solid growth. The structure and stability of these clusters has not yet been fully determined; this needs to be addressed if biomineralisation mechanisms are to be exploited. Here, we present the results of an exhaustive computational study in the search for possible candidate pre-nucleation clusters that might arise in clusters up to 80 ions in size. Both anhydrous and hydrated clusters have been studied. A significant sample of the clusters were simulated, using molecular dynamics, to elucidate the metastability of these nanoclusters in water. An analogous study was conducted for hydrated clusters containing aspartate, to observe the effects of this amino acid on the structure and stability of candidate pre-nucleation clusters. Our results suggest that pre-nucleation cluster stability is a balance between ionic coordination and ion hydration. We find that clusters are generally dynamic in the lower limit of stability, forming chains to which ions frequently aggregate or dissolve. Larger calcium carbonate clusters retain a higher level of coordination in solution but swell to maximise hydration. The effect of additives on the structure and stability of clusters as a function of cluster size is intriguing, with trends in our data suggesting that aspartate can limit ion dissolution, but still allow for dynamic ordering and increasing ion hydration. Finally we find a bias in the ionic charge distributions for relatively dense clusters, indicating that these clusters sustain a negatively charged surface.

1 Introduction

The formation of crystalline calcium carbonate is a necessity for many living organisms, principally to impart structural or defensive support, as found in the mineralised exoskeletons of crustaceans and insects. The ability of natural organisms to direct the crystal growth of a particular polymorph of calcium carbonate with high selectivity, whilst lending these materials specialised mechanical properties and functionality, is a consequence of both specific environmental conditions and the presence of directing organic frameworks to crystallisation.[1,2] The level of selectivity and intricacy displayed in biomineral formations is currently inaccessible to scientists in the laboratory. In contrast to nature, the formation of calcium based biominerals in many industrial settings, is often undesirable. Scale build up can impede production efficiency, and also detrimentally affect product quality in many process industries (*e.g. calcium oxalate precipitation in the paper and brewing industries*).[3,4] In

Centre for Scientific Computing and Department of Chemistry, University of Warwick, Coventry, United Kingdom. E-mail: p.m.rodger@warwick.ac.uk

† Electronic supplementary information (ESI) available. See DOI: 10.1039/c2fd20054f

both applications—biominerals and scale—it is crucial to understand the role of organic additives in affecting nucleation and early growth. The focus of this paper is on the biomineralisation context, but the generality of the concepts developed is worth stressing.

Proteins and polymers have been shown to direct the growth of particular crystalline polymorphs from amorphous clusters of calcium carbonate, as well as stabilising amorphous phases in organisms.[5–7] Of particular interest is the role of aspartate (ASP), which can be found in calcite promoting proteins *in vivo*.[8,9] However, the role of peptides and proteins in the very early stages of ionic coordination in solution remains unclear.

In order for scientists and engineers to mimic nature and produce detailed mineral microstructures with tailored mechanical properties, or to inhibit the crystallisation of calcium based materials from solution, a greater understanding of the nucleation and growth mechanisms for calcium carbonate is required. It is now widely accepted that crystalline polymorphs of calcium carbonate are produced from a non-classical nucleation mechanism, which proceeds *via* an intermediate, amorphous calcium carbonate (ACC), phase.[10–12] A full understanding of the very early stages of nucleation from solution remains elusive.

A growing body of evidence has recently emerged, supporting the concept of stable pre-nucleation clusters of calcium carbonate, which are thought to aggregate in solution before spontaneous growth of solids may occur.[13,14] The composition of these clusters is not yet established; however it is expected that clusters are both amorphous and hydrated (*cf.* hydrated ACC which is known to exist, post-nucleation).[15,16] Estimates for the size of pre-nucleation clusters place them in the limit of 1.1–2 nm hydrodynamic diameter. The upper limit of this estimate is thought to correspond to a 70 calcium and carbonate ion spherical cluster with a density comparable to ACC.[13] It is expected that pre-nucleation clusters have a lower density than ACC; therefore one could expect a cluster containing $35CaCO_3$ to be in the limit of stability for pre-nucleation clusters. It must be stated that these estimates are approximate and that cluster size distributions have not yet been determined.[17]

Probing the internal structure of pre-nucleation clusters and developing an understanding of their stability has proven to be difficult using analytical techniques, partly due to the length scales upon which these clusters are thought to exist. Theoretical methods have been able to elucidate some of these features, and in particular, molecular dynamics (MD) calculations have provided candidate structures to represent pre-nucleation clusters. Significant insight for the nature of clusters has been provided by Gale and co-workers. Early investigations indicated that $CaCO_3$ addition to a hydrated amorphous cluster involved no free energy penalty, and suggested that amorphous clusters were the pre-nucleation species before crystallisation.[18] Later work, involving simulation over the pH range 8.5–11.5 (defined by the ratio, $[CO_3^{2-}]:[HCO_3^-]$) showed the dynamic coordination of ions in aqueous solution, producing a polymeric species labelled, dynamically ordered liquid-like oxyanion polymer (DOLLOP).[19]

While Gale *et al.* have established a possible state for small clusters, DOLLOP has been formed from the spontaneous aggregation of dispersed ions in solution, and its observation does not prove that there are no, more compact, clusters of greater stability. It is therefore of interest to undertake a comprehensive search of possible pre-nuclear cluster structures to determine the range of (*meta*)stability that can be exhibited. Herein, we present the results of an extensive study into the structure of anhydrous and hydrated $CaCO_3$ clusters up to ~2 nm. Candidate pre-nucleation clusters have been generated for clusters containing up to $40CaCO_3$ units, from which a sample has been studied dynamically in water to investigate structural transformations and stability. Potential structures for hydrated clusters, as well as hydrated clusters containing the organic additive ASP, have also been generated.

2 Methods

To produce candidate structures for possible pre-nucleation clusters up to 40CaCO$_3$ units in size, random structure searches were conducted. Ten thousand initial structures were condensed from random configurations of nCa^{2+} and nCO$_3^{2-}$ in the gas phase using a conjugate gradient optimisation (CGO) algorithm, for each value of n formula units in the range 1–40. Ions were positioned randomly in a sphere, with a density of 0.04 atoms Å^{-3}. This density was chosen pragmatically to ensure reasonable optimisation time, but did not limit condensation to spherical clusters: a wide range of aspect ratios was observed amongst the clusters generated for any given range of n. Although the net charge of clusters here is neutral, we cannot be sure that this is the case in nature. However, owing to a limitation in the current understanding of cluster composition, and the necessity to dynamically simulate systems with a net neutral charge, significant insight into the stability of cluster species can still be gained by studying cluster systems as a function of ion pair composition.

CGO of ten thousand random configurations was also conducted for hydrated clusters with the composition n(CaCO$_3 \cdot$H$_2$O) for n = 4, 6, 10, and 15. A 1 : 1 Ca-CO$_3$: H$_2$O stoichiometry was chosen, as previous studies have indicated this to be the composition of stable hydrated ACC.[20,21] These particular cluster sizes were chosen for our initial hydrated study, because the corresponding anhydrous clusters showed interesting features when immersed in water. Finally, CGO was repeated for ten thousand hydrated clusters of the same H$_2$O : CaCO$_3$ composition but with one ASP molecule also included (where the conformation of the amino acid was its minimum energy conformation in water). The protonation state of ASP was chosen to correspond to that of the amino acid at high pH; the acidity level where carbonate species dominate that of equivalent bicarbonate in aqueous solution. The known protonation state of the amino acid when bound in CaCO$_3 \cdot$H$_2$O clusters is unconfirmed, but the one adopted here is a reasonable assumption to make for the case of pre-nucleation species emerging from solution.

The force field employed in these studies has been shown to predict accurately the structure and energetics of stable CaCO$_3$ polymorphs in solution,[22] and has been adapted to include distortion of the CO$_3^{2-}$ plane.[18] In this model, water was described by a modified SPC/Fw potential.[23] ASP was modelled using the AMBER forcefield,[24] re-parameterised for compatibility with this model force-field for CaCO$_3$.[25,26] Although this re-parametrisation was conducted with a TIP3P water model,[27] studies have indicated little structural and energetic changes in peptides when modelled with TIP3P and SPC/Fw water.[28]

A cluster analysis was performed on the set of structures for each cluster size n, using a range of order parameters to quantify the distance between clusters. These order parameters included: cluster potential energy (U); cluster volume; CO$_3^{2-}$ orientation; cluster moment of inertia (I); radial distribution functions for atom pairs Ca–C and Ca–O ($g(r)_{Ca-C}$ and $g(r)_{Ca-O}$); Ca^{2+} coordination to C in the first coordination shell (n_{Ca}); cluster asphericity ($asph$); cluster acylindricity ($acyl$); cluster radius of gyration (R_g); and finally cluster relative shape anisotropy (κ^2). As indicated in Fig. 1, of these order parameters, U, n_{Ca} and one shape descriptor (of $asph$, R_g or κ^2) provided a successful set of classes to guide the selection of candidate pre-nucleation clusters. These shape order parameters have been successfully applied to analyse dendrimers and polymers in previous computational studies, and to characterise calcium carbonate clusters in solution.[19,29] From this discrete set, five structures were taken as candidate pre-nucleation clusters for (1–40) CaCO$_3$ unit clusters in the context of biomineralisation. The five candidates were minimum energy structures taken for particular values of n_{Ca} at regular intervals of the total distribution of coordination. For many structure sets, the cluster with the highest value of n_{Ca} was also the minimum energy cluster, but for instances where this was not the case, both the minimum energy structure and structure with highest coordination were taken as

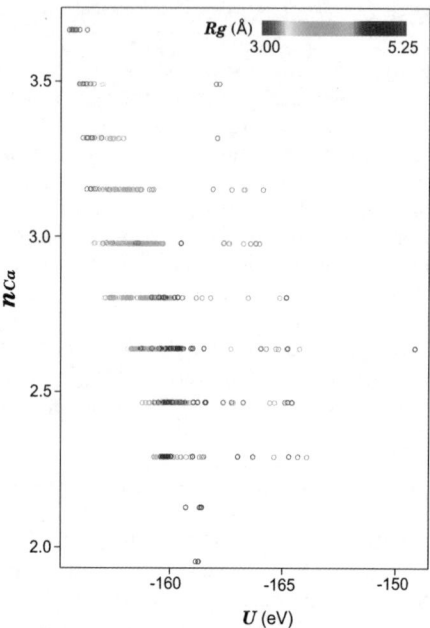

Fig. 1 A plot of average calcium coordination number (n_{Ca}) as a function of potential energy (U) for ten thousand minimised random structures of 6CaCO$_3$ from ions in the gas phase. The points are coloured according to a gradient representing cluster radius of gyration (R_g). The minimum energy data points for each discrete level of n_{Ca}, span the n_{Ca}, U and R_g distributions.

candidate clusters, with the remainder candidates taken by sampling the distribution of n_{Ca}.

The candidate structures extracted from cluster analyses were subsequently simulated in SPC/Fw water using MD. Throughout this paper we shall refer to simulations in which a cluster—whether hydrous or anhydrous as generated in the random structure search—is immersed in bulk water as a *solvated* system. The amount of water required to ensure bulk solvent behaviour sufficiently far from the cluster was calculated by ensuring that the change in total energy of a solvated cluster system, on addition of extra solvent water, converged to the energy of the equivalent amount of bulk water. The number of water molecules was kept constant for simulations of the same nCaCO$_3$. All systems were simulated in the *NPT* ensemble at 298 K and 1 atm, employing a Nosé–Hoover thermostat and barostat with 0.1 ps and 1.0 ps relaxation times, and cubic periodic boundaries were used throughout. A short, 20 ps simulation was initially carried out to enable relaxation of water whilst keeping solute atoms fixed in space. A 1 fs timestep was used throughout, and trajectories of 10 ns for minimum energy structures and 5 ns for other candidate clusters were generated, with all atoms mobile. The cut off for short range interactions was set to 9.0 Å, with a tapering function applied to the Ca, C and O calcium carbonate atom pairwise interactions from 6.0–9.0 Å, as recommended in the chosen force field. Electrostatics were treated using the smooth particle mesh Ewald (SPME) method.[30] The simulation package used to run the MD simulations was *DL_POLY Classic*.[31]

3 Results

3.1 Anhydrous CaCO$_3$ clusters

3.1.1 Random structure optimisation. To facilitate the further discussion of the dynamics of candidate CaCO$_3$ clusters in water, it is first useful to discuss the structures that have been found from the aforementioned random structure generation minimisations. Fig. 2 shows a selection of minimised structures for $7CaCO_3$, from the ten thousand structures generated, and provides an example of configurations which span: U, n_{Ca} and shape parameter (R_g or κ^2) distributions. As might be expected, configurations with lowest n_{Ca} are those of highest energy in the gas phase, while the opposite is true for structures of highest coordination, and in this case density. The lowest energy structure found for this particular composition of cluster was -194.223 eV; the energies are low as they are in comparison to constituent ions *in vacuo*, for which energies will be approaching zero. Between the bounds of lowest to highest energy, the coordination of ions decreases, giving rise to particular coordination motifs that can be found in all minimised structures of $nCaCO_3$. For small n, many minima for low coordination are visited during the random structure optimisation, giving rise to ring and offset dual-chain type motifs. However, as the value of n increases, so does the level of coordination in clusters, and therefore although low coordination motifs found for small numbers of n are replicated in these larger systems, the average value of n_{Ca} for each particular cluster is generally higher than for smaller ones.

Clusters minimised from the gas phase, in their native state, have an interesting property in that the distribution of charge within the clusters is non-uniform in many cases (see Fig. 3). An excess concentration of anions was usually observed at the surface of clusters, and this was seen in both minimum energy clusters (which usually showed the highest density) and for clusters with lower coordination (albeit to a much lesser extent for low coordination clusters of less than ten $CaCO_3$ units). This can be explained by considering the packing of ions in anhydrous clusters. To achieve a higher density for the cluster and therefore increase ionic coordination, smaller cationic species, Ca^{2+}, will favourably occupy sites below the cluster surface, with larger anionic species surrounding cations, reducing the potential energy of the system *via* electrostatic attraction. This will result in carbonate ions more often residing at the periphery of the cluster, and hence introduces the negative surface charge. The data in Fig. 3 represents structures minimised in vacuum; this will accentuate the surface charge effect as carbonate ions tend to orient with the carbonate and surface planes aligned, in order to maximise coordination.

3.1.2 Cluster dynamics. The minimum energy structures of $nCaCO_3$ clusters for each n in the range 1–40 were simulated for 10 ns in water, and the mean potential energy taken from the latter regions of the trajectories are shown in Fig. 4. The underlying trends in the data show a rapid monotonic decrease in potential energy with increasing cluster size for clusters up to around nine formula units. For larger clusters the figure shows some scatter superimposed on a general increase of U with cluster size, up to a plateau around $n = 30$. The difference in U between the minimum around $n = 9$ and the largest cluster sizes is on the order of ~ 0.1 eV per $CaCO_3$ ion pair, which is comparable in strength to a moderately strong

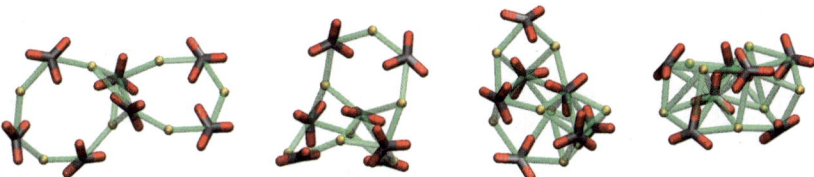

Fig. 2 Representative structures from random structure minimisations of $7CaCO_3$ from ions in the gas phase. Red, black and yellow atoms represent oxygen, carbon and calcium atoms respectively, while green lines represent distances between carbon and calcium atoms below 3.825 Å, indicating atoms connected in the first coordination shell. Potential energies decrease from left to right between ~ -185 to -194 eV, while n_{Ca} increases from 2.3 to 3.6.

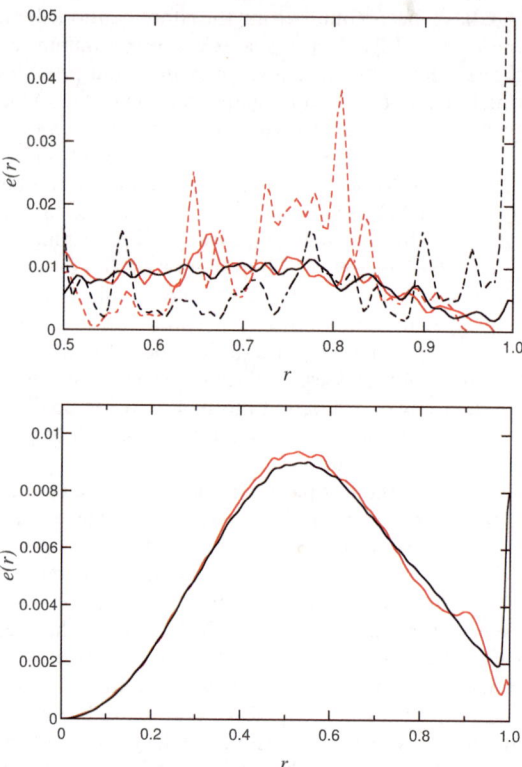

Fig. 3 Plots showing ionic charge distributions for minimised anhydrous nCaCO$_3$ clusters; red lines indicate cationic charge distributions from cluster centres of mass to the surface, while black lines are the equivalent anionic distributions. Top: mean ionic charge distributions for minimum energy clusters of $n = 1$–10 (dotted lines) and 31–40 (solid lines) for nCaCO$_3$. Bottom: mean charge distributions for 20CaCO$_3$, taken from ten thousand minimised structures.

hydrogen bond. The general trends observed here contrast with those seen from cluster optimisations in vacuum (see Fig. 4), where extra stability is found from increasing ionic coordination in dense amorphous clusters up to the largest cluster size considered in this study (~2 nm in diameter).

The difference can be ascribed to ion hydration in solution, and can be explained by considering the time evolution of n_{Ca} for the various system sizes. In small clusters (nCaCO$_3$ for $1 \leq n \leq 10$) nearly every ion is in contact with the solvent—even for particles with maximum n_{Ca}. The presence of the cluster–solvent boundary introduces an energy penalty (*i.e.* the interfacial energy), and the clusters are too small to be stabilised by a favourable bulk crystal energy contribution. Energetically it is more favourable for the clusters to partially dissolve, allowing retention of limited ionic coordination, but also greatly increasing ion stabilisation through solvation. In the case of 9CaCO$_3$, by $t = 10$ ns, the cluster dissociated and small clusters of ions dynamically dissolved and aggregated during the simulation.

As cluster size increases, the stabilisation from high ionic coordination within the cluster is sufficient to avoid complete cluster dissolution over a 10 ns trajectory. The internal core sub-lattice structure of ions in the cluster over 10 ns is retained for clusters of initial high density with n larger than 20. Surface ions in larger particles coordinate strongly with solvent molecules, and although this binding is, on occasion, sufficient to disassemble surface ions, creating dendritic arms and rings which

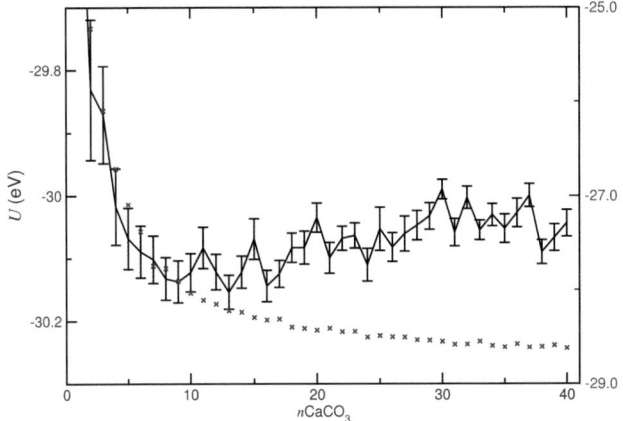

Fig. 4 Average potential energy, U, per formula unit of calcium carbonate calculated for clusters of n formula units. Average values were calculated from the final 2 ns of a 10 ns MD trajectory of clusters simulated in water. The energy values are exclusive of bulk water energies in respective systems. Error bars indicate uncertainties relating to one standard deviation of the mean. The potential energy of the clusters in the gas phase are also given by the blue data points, with the shifted scale (in blue) on the right of the graph applicable to these data points only.

protrude into solution and, on occasion, dissociate small clusters from the main cluster surface. However, this is insufficient to "break" the cluster on the time scale studied in these simulations. This result appears to contrast from the findings of earlier simulation studies reporting the structural stability of highly ordered nanoparticles of $CaCO_3$ in water, where calcite nanoparticles (of comparable size to the highly ordered clusters in this study) were found to be stabilised by the effect of strongly coordinating surface water.[32,33] Although we have not considered calcite stability at this scale, the general differing result can most probably be ascribed to the choice of force field.

As small clusters partially dissolved in solution, n_{Ca} was observed to decrease over time, resulting in structures with coordination motifs that were found in higher energy clusters from gas phase optimisation. Oligomers of ions formed with solvent stabilisation of ions at the chain ends of dendritic arms. The coordination between monomeric ions in the chain was observed to be dynamic, with ion–ion separation reaching considerable distances, and ion pair loss and recombination prevalent over long time scales. This behaviour is very much analogous to the that of Ca^{2+}, CO_3^{2-} and HCO_3^- ions in aqueous solution, and is further evidence for the stability of DOLLOP over more ordered structures in solution.[19]

To investigate whether larger cluster sizes are likely to become dynamically ordered structures when immersed in water, clusters of $nCaCO_3$, but with lower initial n_{Ca} than those of maximum density clusters, have also been simulated in water. Fig. 5 shows the potential energy of clusters containing 20–29 formula units sampled from clusters across the whole range of initial n_{Ca}. It is clear that lower initial configuration energy does lead to a solvated system with lower potential energy with the difference being ~0.1 eV per $CaCO_3$ unit, *i.e.* comparable with the potential energy barrier evident in Fig. 4 for compact clusters.

At this point we should state that in this study we have considered the potential energy of the system as a measure of cluster stability, and hence entropic effects are unaccounted for. However, monotonically decreasing free energy has been found for amorphous particles with increasing numbers ion pairs of calcium carbonate, indicating that larger amorphous clusters are thermodynamically more stable than smaller ones.[18]

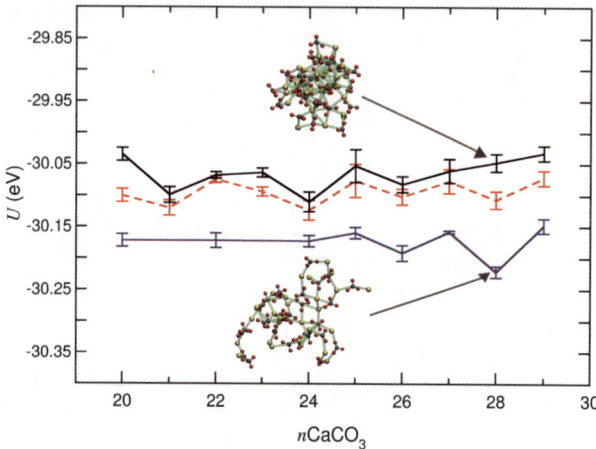

Fig. 5 Potential energy, U, per formula unit, n, of clusters in aqueous solution for (20–29) CaCO$_3$ with maximum (black) and minimum (blue) initial n_{Ca}. The red-dashed line gives the average U for the five samples simulated, where these samples span the n_{Ca} distribution, as described in section 2. Data points were calculated from the final 1 ns of respective 10 or 2.5 ns trajectories, with uncertainty of one standard deviation of the mean indicated. Final configurations for clusters of 28CaCO$_3$ are provided. Atom and bond colours are as for Fig. 2.

The behaviour of the larger low-coordination clusters in water are more comparable with those of smaller sizes, in that they display dynamic coordination of ionic species throughout the duration of the trajectories. Again, coordination is broken and reformed continually, with clusters displaying polymer chain-like motion, but the ions in the cluster do not disassemble as readily as for the cluster sizes in the potential minimum of Fig. 4. This can be explained by considering the dynamics of breaking and reforming ionic connections within the clusters. For very small clusters, when a connection is broken, there is a limited, small number of possible connections that can be made to reproduce a cluster of equal size to the initial state. However, for larger clusters, there are many possible ionic connections apparent upon cluster dissociation. Furthermore, the lifetime of an insufficiently coordinated ion in larger clusters is reduced compared with smaller clusters of comparable density, as ionic concentration in the local proximity of cluster constituents increases.

The mobility of ions around the centre of mass of clusters with low coordination, is higher than for equivalently sized clusters with more dense packing. The final configurations for a 28CaCO$_3$ cluster have been included in Fig. 5; these show a much less ordered structure after 5 ns of simulation for a cluster with a low level of initial coordination, than for the corresponding high coordination structure (simulated from the minimum energy structure in vacuum) after 10 ns. The average U of all five samples considered appears to converge over the range of cluster sizes considered.

The strength of the negative shell observed for clusters of ions in the gas phase (particularly those of high density) is reduced when clusters are immersed and relaxed in water. This can be attributed to ion hydration and partial dissolution of clusters. However, the nature of a charge distribution bias does persist in these systems, as shown in Fig. 6, which provides the final configurations for an open and a compact cluster of 27CaCO$_3$ when simulated in water. The cluster with maximum initial coordination remained relatively dense by the end of a 5 ns simulation, and the majority of positively charged species resided below the surface of the cluster. In contrast, ions in the sampled cluster with lowest coordination became further exposed to solvent molecules over the course of the simulation, and density continued to reduce as the structure disassembled, resulting in charge distributions

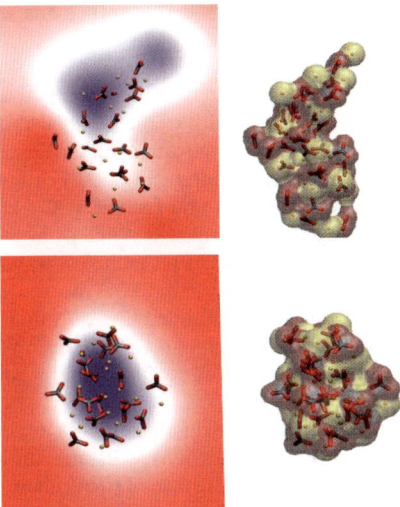

Fig. 6 Final configurations for clusters of 27CaCO$_3$ after 2.5 ns of simulation in water. Top and bottom show clusters with n_{Ca} = 2.5 (t_0, n_{Ca} = 3.1) and n_{Ca} = 3.8 (t_0, n_{Ca} = 4.7); the explicit structure is shown on the right (see the caption in Fig. 2 for atom identities), while the image on the left is an electrostatic potential map indicating the electrostatic charge distribution half way through the simulation cell z axis for Ca, C and O atoms only.[34] Here shading represents regions of negative (red) and positive (blue) charge, ranging from -100–$280 \frac{k_B T}{e}$.

approaching uniformity. Although data is presented here for one size of cluster, similar behaviour was commonly found for all of the larger cluster sizes studied.

3.2 Hydrated clusters

3.2.1 Optimisation of vacuum structures for CaCO$_3$ with included water and additives. In the biomineralisation process, ions of Ca^{2+} and CO$_3^{2-}$ associate in solution. During this process, water could become kinetically trapped during transformation to a more dense CaCO$_3$ phase.[35] Indeed, the presence of internal water may stabilise the emerging phase from solution.[18] Furthermore, the effect of organic additives on the stability of clusters is of interest. With this in mind, random structure searches have been conducted on n(CaCO$_3$·H$_2$O) with and without one inclusive ASP, for n = 4, 6, 10 and 15. Comparison of hydrous and anhydrous clusters after immersion in water also provides a useful check on the possibility of metastable states when water is included within the cluster.

For these cluster sizes, the structures found for hydrated clusters were similar to those found for anhydrous clusters. There was little variation in the average coordination numbers for any of the cluster sizes, as shown in Table 1. In fact, the minimum energy structures found for both four- and six-unit hydrated clusters

Table 1 Average n_{Ca} values for nCaCO$_3$ (anhydrous), n(CaCO$_3$·H$_2$O) (hydrated) and n(CaCO$_3$·H$_2$O)·ASP(hydrated + ASP) optimised clusters

n	Anhydrous	Hydrated	Hydrated + ASP
4	2.38	2.39	1.80
6	2.55	2.62	2.18
10	2.90	2.93	2.61
15	3.21	3.17	2.91

are essentially identical to those found in the anhydrous optimisations. In these systems, the role of water appears to be to solvate the anhydrous particles found in the anhydrous optimisations.

As the cluster size increases, the amount of structures with inclusive water also increases. This is expected, as water will become trapped during the minimisation. However, the structures containing large quantities of internal water molecules were generally found to have higher potential energies. For instance, the water molecules in the minimum energy structure found for $10CaCO_3 \cdot H_2O$ were located at the surface of the cluster, forming the first solvation shell as in the case for four and six unit clusters. It is only for $15CaCO_3 \cdot H_2O$ that inclusive water was found within the minimum energy structure. In this system, two water molecules were found embedded in the cluster, completely surrounded by Ca^{2+} and CO_3^{2-} ions; the remaining water molecules again formed the first solvation shell. We further suspect that the global minimum for this composition, would contain no inclusive water molecules.

For these hydrated systems, the energy of clusters was again observed to correlate with n_{Ca}; hence water favourably solvates dense clusters, allowing for the maximum coordination of ions below the cluster surface. The result is a structure sample set which is very similar to those found for anhydrous clusters (for the sizes investigated).

The structures minimised with ASP differed from those of anhydrous or hydrated clusters. The carboxylate oxygens and amine hydrogens of ASP bound to the Ca^{2+} ions and oxygens of CO_3^{2-} in the cluster, respectively. This resulted in a more distorted network amongst the remaining calcium carbonate than was observed in the anhydrous and hydrous clusters. Fig. S1† (supplementary information) shows the minimum energy structures found for four and ten calcium carbonate unit clusters; these clusters have greater asphericity than equivalent anhydrous or hydrated structures. The distortion induced in $6CaCO_3$ clusters by ASP can be clearly seen in the minimum energy structures (see Fig. S2†). While water of inclusion had little effect on the cluster configuration with respect to Ca^{2+} and CO_3^{2-}, ASP displaced ions within the cluster as functional groups of the organic species competed to coordinate with ions. While n_{Ca} values shown in Table 1 suggest a less dense structure for clusters conataining ASP, these data do not include the coordination between Ca^{2+} and CO_3^{2-} ions and the functional groups of ASP.

It is interesting to note that both hydrated and hydrated + ASP clusters showed the same non-uniform charge distributions observed in the anhydrous clusters, with a bias in the charge distribution and a significant concentration of negative charge at the surface of clusters. For small clusters, the strength of the surface charge was smaller in the low density (low coordination) structures.

3.2.2 Effects of inclusive water and ASP on cluster stability.
The hydrated clusters, with and without ASP, were immersed in water and MD simulations conducted. The behaviour of the smaller solvated hydrous clusters was very similar to that observed with anhydrous clusters. Small hydrated clusters were found to dissociate into smaller clusters and ion pairs over the course of the simulation; these subsequently aggregated and re-dissolved, displaying dynamic ordering. Larger (dense) clusters tended not to disassemble in solution and remained particulate throughout the trajectory, albeit with some reconfiguration apparent inside, and at the surface of, the cluster. The case of the $15CaCO_3$ cluster with maximum initial density, is particularly interesting as this contained two embedded water molecules at $t = 0$. After approximately 80 ps of simulation, a third water molecule was absorbed from the surface and became integrated internally within the cluster (see Fig. S3†). All three water molecules remained embedded within the cluster for the remainder of the trajectory. The rate of disassembly in this cluster, as indicated by the fraction of ions with maximum n_{Ca} in Fig. S4,† is comparable with that of the dense anhydrous cluster.

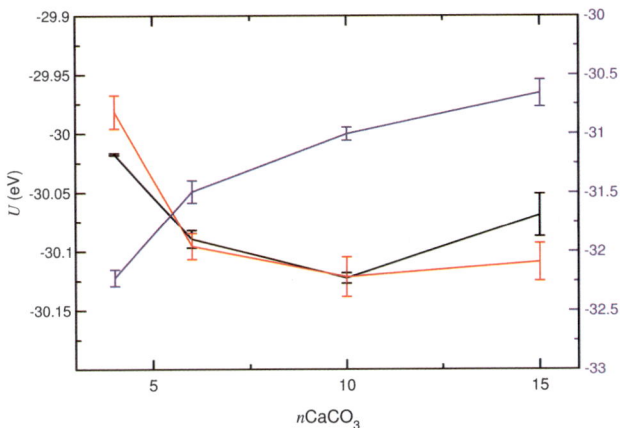

Fig. 7 Average U is plotted as a function of cluster size for the four hydrated cluster sizes considered. The black data is taken from Fig. 4 for solvated, anhydrous clusters and is plotted along with similar data for hydrated (red) clusters. The blue curve shows the average U for the clusters containing one ASP molecule during optimisation, the shifted scale (due to the energy associated with ASP inclusion) for which is shown on the right in blue. Mean values were taken from the final 1 ns of 10 (anhydrous) and 5 ns (hydrated) trajectories, with uncertainty in the data representing one standard deviation of the mean.

Fig. 7 shows the mean potential energy, U, for hydrated systems over the final portions of a 10 ns trajectory. The data is similar to that for anhydrous clusters, and there is little difference between the structures and energetics found between anhydrous and hydrous clusters in this size regime. From this data, we can expect the number of ions within a cluster to define the limit of cluster stability, rather than the level of hydration for dense cluster systems, especially when one considers the nature of the dynamics seen for clusters in the size range studied here.

The potential energy data for clusters containing one ASP molecule does not follow the trends found for other solvated cluster systems. Alternatively, U increases as the number of ions in the cluster increases. This can be explained by considering the interaction of ASP functional groups to ions of Ca^{2+} and CO_3^{2-} in solvated clusters. The binding between ASP and ions in the cluster is sufficiently strong to limit the amount to which the clusters may dissociate. Although disassembly may occur for ASP-bound clusters (of initial maximum density) at a faster rate than in similar anhydrous and hydrous clusters (see Fig. S4†), with n_{Ca} decreasing and R_g increasing at early times during the trajectory, cluster separation into smaller species is less likely in the presence of ASP.

The cluster size distributions, shown in Fig. 8, highlight the possible cluster sizes accessible during the final half of a 10 ns trajectory for clusters containing ASP, and those for the equivalent control experiments, for which ASP is absent. For small cluster sizes, with all ions initially in contact with solution, distributions indicate many possible cluster sizes, and the influence of the organic species is unclear. However, as cluster size increases, the effect of ASP is to reduce the amount of cluster dissociation to smaller species. For instance, when simulating clusters containing $15CaCO_3$ ASP, distributions indicate dissolution into approximately two smaller clusters when ASP is bound in the initial state; on the other hand, a wide range of accessible cluster sizes is displayed in the control system, over the timescales studied here.

Initial and relaxed structures for clusters containing 15 $CaCO_3$ units, with and without ASP, are depicted in Fig. 9. Note that although the cluster with ASP was optimised with a 1 : 1 ratio of $CaCO_3 : H_2O$, water was not found below the cluster surface, and instead this water formed a solvation shell. The images show that after 10 ns of simulation, the cluster without ASP has disassembled, and ions have

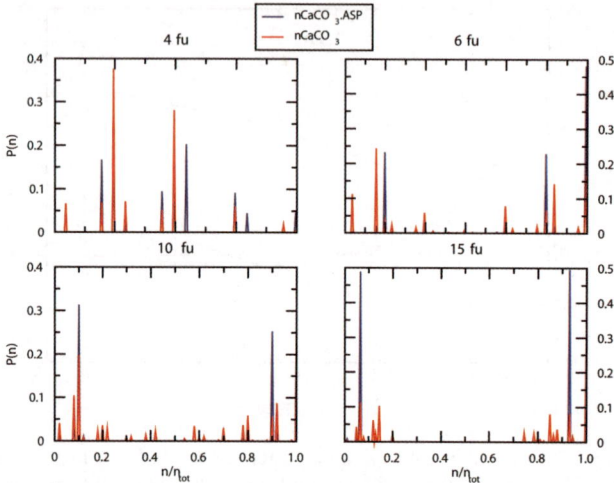

Fig. 8 Cluster size distributions for simulations of clusters containing CaCO$_3$ (red) and CaCO$_3 \cdot$ASP (blue) for n = 4, 6, 10 and 15. Distributions have been sampled from the final 5 ns of a 10 ns trajectory, and have been normalised against maximum cluster size for ease of comparison.

partially dissociated to increase hydration, with dynamic ordering also observed. This results in the polymer-like species described in section 3.1.2. However, for the cluster which includes ASP, the change in coordination of ions in this cluster has not approached the same level as for the control system. During the course of a 10 ns simulation, the strength of the binding between ASP and ions in the cluster is such that the organic molecule does not readily desorb from the cluster. Furthermore, because at $t = 0$ the cluster configuration has maximum n_{Ca}, and that ASP is strongly coordinated to ions across the surface of a cluster, the organic molecule reduces the possibility for the cluster to dissociate into smaller species.

Our studies do not fully explain the role of ASP on prenuclear species; this is due to the fact that we have investigated clusters with bound ASP in the initial state. In

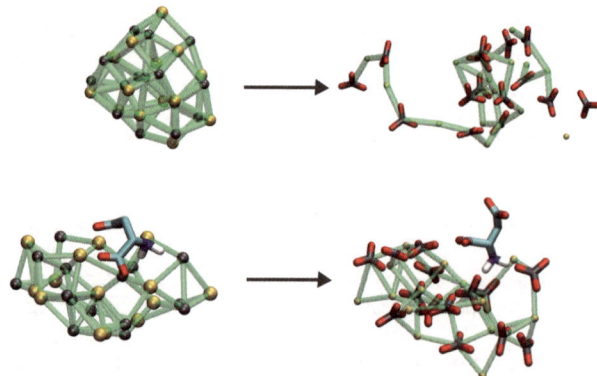

Fig. 9 Configurations are provided for clusters containing 15 formula units of CaCO$_3$ (top), and CaCO$_3$ with one molecule of ASP (bottom). Calcium carbonate atoms are coloured as described for previous figures (see Fig. 2 caption), while nitrogen, oxygen, carbon and hydrogen of ASP are coloured: blue, red, cyan and white respectively. The structures on the left indicate the configuration of clusters at $t = 0$ and do not show carbonate oxygen atoms for clarity (top: $n_{Ca} = 4.5$, bottom: $n_{Ca} = 3.9$). Structures on the right show the systems at $t = 10$ ns (top: $n_{Ca} = 2.4$, bottom: $n_{Ca} = 3.0$).

reality, ASP is more likely to associate with an emerging cluster from a dilute ionic solution. In this environment, the influence of ASP on cluster stability may be different from that found here. However, we can infer some information about the role of ASP on pre-nucleation species. As ASP was shown to limit the dissolution of clusters, this could be interpreted as ASP limiting the DOLLOP nature of pre-nucleation species, and hence stabilising more dense clusters in solution. One could further hypothesise that this effect could lead to a reduction in the frequency of dissolution/re-aggregation of cluster species, and an overall inhibition of the growth of pre-nucleation species. Conversely, the multiple binding sites associated with ASP could lead to the organic species bridging clusters in solution and leading to more rapid cluster–cluster aggregation, and therefore promoting cluster growth. Further studies are necessary for a full explanation of the influence of ASP on clusters in this size regime.

4 Summary and conclusions

A rigorous search has been conducted for possible candidate pre-nucleation clusters which are thought to be the emergent species in the precipitation of $CaCO_3$ from aqueous solution.[13] Structures were generated from random starting configurations of Ca^{2+}, CO_3^{2-}, H_2O and ASP in the gas phase, which were then condensed to produce configurations for anhydrous and hydrous clusters of calcium carbonate, with and without ASP. From the thousands of possible structures generated for each target cluster size, we performed cluster analyses and have identified order parameters which can successfully group the structures into similarity classes and hence identify a representative subset of candidate clusters.

The resulting samples were simulated in aqueous solution along with the minimum energy configurations found for structures containing up to 40 formula units of calcium carbonate. The results indicate that calcium–carbon coordination can aid in understanding the stability of structures, both in vacuum and in solution. Structures with maximum coordination were found to be stable in vacuum, however in solution, stabilisation of ions was greatest for small dynamically ordered clusters, which is in agreement to the studies of Demichelis *et al.* and the existence of DOLLOP.[19]

Although solvent-induced deformation is seen at the surface of the largest, most dense, clusters in solution, these clusters remained highly coordinated by the end of a 10 ns trajectory, with the internal structure of the clusters appearing to be quite stable. Samples with lower levels of coordination, and therefore lower density, tended to disassemble more readily in solution, with regular breaking of the cation–anion network at chain ends of dendritic clusters. These lower coordinated clusters were found to have a lower potential energy than structures with maximum coordination, and this has been ascribed to ion solvation, which is maximised in clusters with a large surface area.

The simulations for hydrous clusters, containing a 1 : 1 ratio of $CaCO_3$: H_2O, were analogous to the simulations for anhydrous clusters of the same size. Solvated clusters which also included ASP did not show the same trends in solution as other clusters of equivalent composition. The potential energy of these clusters (per formula unit) increased as the number of ions in the cluster increased. It was shown that ASP reduced the amount of ion dissolution and aggregation in solution, and limited the internal mobility of clusters of more than $\sim$6 formula units. Hence, ions in clusters could not dissociate into smaller species, as those in clusters without ASP.

The density and structure of clusters in solution was found to affect their stability according to both ionic coordination and ion hydration. In aqueous solution, we expect clusters of low coordination to preferentially form over ones with high coordination. This ensures a high level of hydration, and allows for a limited level of ionic coordination.

For all clusters studied, the majority of the more dense clusters were found to have a bias in the charge distribution from cluster centres of mass, due to a relatively high concentration of anions at the surface of the clusters. This was seen, both in minimised structures in the gas phase, and for systems that had been relaxed in water. This effect could be a contributing factor to the (*meta*)stability of small clusters of calcium carbonate in solution, if relatively dense structures (of the size range considered here) form in solution.

Acknowledgements

We thank D. J. Sparkes, C. L. Freeman and J. H. Harding (Deptartment of Materials Science and Engineering, University of Sheffield, UK) for supplying the force field details for ASP, and A. M. Bano and D. Quigley for helpful discussions. The research was supported under the EPSRC grant EP-I001514. Computational resources were provided by the Centre for Scientific Computing, University of Warwick.

References

1 H. A. Lowenstam and S. Weiner, *On biomineralization*, Oxford University Press, New York, 1989.
2 F. C. Meldrum, *Int. Mater. Rev.*, 2003, **48**, 187–224.
3 P. Sayan, S. T. Sargut and B. Kran, *Cryst. Res. Technol.*, 2009, **44**, 807.
4 M. Masár, M. Žúborová, D. Kaniansky and B. Stanislawski, *J. Sep. Sci.*, 2003, **26**, 647–652.
5 C. L. Freeman, J. H. Harding, D. Quigley and P. M. Rodger, *Angew. Chem., Int. Ed.*, 2010, **49**, 5135.
6 F. C. Meldrum and H. Cölfen, *Chem. Rev.*, 2008, **108**, 4332.
7 S. E. Wolf, J. Leiterer, V. Pipich, R. Barrea, F. Emmerling and W. Tremel, *Journal of the American Chemical Society*, 2011, **133**, 1264212649.
8 J. Aizenberg, G. Lambert, S. Weiner and L. Addadi, *J. Am. Chem. Soc.*, 2002, **124**, 32.
9 S. Weiner, *J. Struct. Biol.*, 2008, **163**, 229–234.
10 L. Brečević and A. E. Nielsen, *J. Cryst. Growth*, 1989, **98**, 504.
11 L. Addadi, S. Raz and S. Weiner, *Adv. Mater.*, 2003, **15**, 959.
12 A. Xu, Y. Ma and H. Cölfen, *J. Mater. Chem.*, 2007, **17**, 415.
13 D. Gebauer, A. Völkel and H. Cölfen, *Science*, 2008, **322**, 1819–1822.
14 E. M. Pouget, P. H. H. Bomans, J. A. C. M. Goos, P. M. Frederik, G. de With and N. A. J. M. Sommerdijk, *Science*, 2009, **323**, 1455–1458.
15 S. Raz, P. Hamilton, F. Wilt, S. Weiner and L. Addadi, *Adv. Funct. Mater.*, 2003, **13**, 480–486.
16 A. V. Radha, T. Z. Forbes, C. E. Killian, P. U. P. A. Gilbert and A. Navrotsky, *Proc. Natl. Acad. Sci. U. S. A.*, 2010, **107**, 16438–16443.
17 D. Gebauer and H. Cölfen, *Nano Today*, 2011, **6**, 564.
18 P. Raiteri and J. D. Gale, *J. Am. Chem. Soc.*, 2010, **132**, 17623.
19 R. Demichelis, P. Raiteri, J. D. Gale, D. Quigley and D. Gebauer, *Nat. Commun.*, 2011, **2**, 590.
20 Y. Levi-Kalisman, S. Raz, S. Weiner, L. Addadi and I. Sagi, *Adv. Funct. Mater.*, 2002, **12**, 43.
21 M. Neumann and M. Epple, *Eur. J. Inorg. Chem.*, 2007, **2007**, 1953–1957.
22 P. Raiteri, J. D. Gale, D. Quigley and P. Rodger, *J. Phys. Chem. C*, 2010, **114**, 5997.
23 Y. Wu, H. L. Tepper and G. A. Voth, *J. Chem. Phys.*, 2006, **124**, 024503.
24 D. A. Case, T. E. Cheatham, T. Darden, H. Gohlke, R. Luo, K. M. Merz, A. Onufriev, C. Simmerling, B. Wang and R. J. Woods, *J. Comput. Chem.*, 2005, **26**, 1668.
25 C. Freeman, J. Harding, D. Cooke, J. Elliott, J. Lardge and D. Duffy, *J. Phys. Chem. C*, 2007, **111**, 11943–11951.
26 D. J. Sparkes, C. L. Freeman and J. H. Harding, To be published.
27 W. L. Jorgensen, J. Chandrasekhar, J. D. Madura, R. W. Impey and M. L. Klein, *J. Chem. Phys.*, 1983, **79**, 926.
28 J. L. Desmond, P. M. Rodger and T. R. Walsh, To be published.
29 J. T. Bosko, B. D. Todd and R. J. Sadus, *J. Chem. Phys.*, 2006, **124**, 044910.
30 U. Essmann, L. Perera, M. L. Berkowitz, T. Darden, H. Lee and L. G. Pedersen, *J. Chem. Phys.*, 1995, **103**, 8577.
31 W. Smith, C. Yong and P. Rodger, *Mol. Simul.*, 2002, **28**, 385–471.
32 S. Kerisit and S. C. Parker, *J. Am. Chem. Soc.*, 2004, **126**, 10152–10161.
33 D. J. Cooke and J. A. Elliott, *J. Chem. Phys.*, 2007, **127**, 104706.
34 A. Aksimentiev and K. Schulten, *Biophys. J.*, 2005, **88**, 3745–3761.
35 G. A. Tribello, F. Bruneval, C. Liew and M. Parrinello, *J. Phys. Chem. B*, 2009, **113**, 11680–11687.

PAPER | www.rsc.org/faraday_d | Faraday Discussions

Exploring the influence of organic species on pre- and post-nucleation calcium carbonate†

Paolo Raiteri,[a] Raffaella Demichelis,[a] Julian D. Gale,[*a] Matthias Kellermeier,[b] Denis Gebauer,[b] David Quigley,[c] Louise B. Wright[d] and Tiffany R. Walsh[e]

Received 16th March 2012, Accepted 1st June 2012
DOI: 10.1039/c2fd20052j

Organic additives are well known to influence the nucleation and growth of minerals. A combination of experimental and theoretical methods has been used to probe how three simple additives, containing varying numbers of carboxylate groups, influence the early stages of the growth of calcium carbonate. Computationally, the free energy landscape has been examined for each additive binding to Ca^{2+}, the calcium carbonate ion pair, the surface of an amorphous calcium carbonate nanoparticle, and the basal plane of calcite. The different influence of the three organic ligands on the early stages of growth of calcium carbonate observed experimentally can be rationalised in terms of the degree of association of each anion with the species present prior to, and immediately after nucleation.

1 Introduction

Calcium carbonate occurs naturally in the structures of many organisms as a result of biomineralisation.[1] Aside from playing either a protective role in shells[2] or a support function in skeletons,[3] it has also been found in primitive vision systems as an optical material.[4] Unsurprisingly, this diversity of purpose and wide occurrence in nature has led to calcium carbonate being one of the most extensively studied substances in the field of biomineralisation.[5,6] Despite this, there remains much to learn regarding the pathways by which organisms can selectively choose

---

[a] Nanochemistry Research Institute, Department of Chemistry, Curtin University, PO Box U1987, Perth, WA 6845, Australia. E-mail: J.Gale@curtin.edu.au; Fax: +61 9266 4699; Tel: +61 9266 7800
[b] Department of Physical Chemistry, University of Konstanz, Universitätsstrasse 10, Box 714, D-78457 Konstanz, Germany. E-mail: Denis.Gebauer@uni-konstanz.de; Fax: +49 7531 883898; Tel: +49 7531 882169
[c] Department of Physics, Centre for Scientific Computing, University of Warwick, Gibbet Hill Road, Coventry, CV4 7AL, UK. E-mail: d.quigley@warwick.ac.uk; Fax: +44 2476 573133; Tel: +44 24765 74580
[d] Department of Chemistry, Centre for Scientific Computing, University of Warwick, Gibbet Hill Road, Coventry, CV4 7AL, UK. E-mail: louise.wright@warwick.ac.uk; Fax: +44 2476 524112; Tel: +44 2476 523653
[e] Institute for Frontier Materials, Deakin University, Geelong, Vic, 3217, Australia. E-mail: tiffany.walsh@deakin.edu.au; Fax: +61 3522 71103; Tel: +61 3522 73116

† Electronic Supplementary Information (ESI) available: parameters for conventional unreactive force field (Tables S1, S2, S3); schematics of anions showing atom types and charges for the conventional force field (Fig. S1–S3); reproducibility of the titration experiments (Fig. S4); evolution of the amount of free calcium upon dosing calcium chloride into aqueous additive solutions at different concentrations (Fig. S5, Table S4); effect of the different additives on the slope of the increase during the pre-nucleation stage in carbonate buffer (Fig. S6). See DOI: 10.1039/c2fd20052j

between polymorphs and organise crystallites into complex morphologies.[7] Factors including the control of ion concentration,[5] use of confined environments,[8] and the interaction of organic species during crystal growth[9] are all possible influences, though a clear picture of the atomic level detail remains elusive.

Even without the full complexity of an *in vivo* environment, understanding the crystallisation of calcium carbonate presents many challenges. Three crystalline polymorphs of $CaCO_3$, namely calcite, aragonite and vaterite, in order of stability at ambient conditions, compete with each other, in addition to two hydrates (monohydrocalcite and ikaite) under certain conditions.[10,11] While the structures of calcite and aragonite are well known, the nature of the disorder within vaterite continues to be debated.[12,13] Under conditions appropriate to biomineralisation there is increasing evidence that formation of crystalline polymorphs occurs *via* the initial nucleation of amorphous calcium carbonate (ACC), which subsequently transforms upon agglomeration of nanoparticles.[6] ACC itself is a complex material that exhibits variable characteristics depending on whether it is biogenic or not.[14] This may reflect varying degrees of water content from anhydrous to 1–2 waters per formula unit,[15] the presence of impurities, most notably magnesium,[16] but potentially even incorporation of organics,[17] as well as perhaps intrinsic polyamorphism.[18]

In the last few years it has become apparent that the fascinating growth mechanisms of calcium carbonate can challenge our understanding even prior to the appearance of ACC. In contravention of classical nucleation theory, the presence of stable pre-nucleation clusters beyond simple ion pairs can be detected experimentally.[19] Through a combination of computer simulation and experiment, these precursor species have been shown to be a supramolecular polymer of calcium and carbonate ions that is constantly changing structure and remains in equilibrium with the solution, thereby avoiding a phase boundary.[20] Furthermore, such clusters can change from rings to linear to branched chains all with a free energy difference that is comparable to ambient thermal energy. Because of these characteristics, this new species has been named as a dynamically ordered liquid like oxyanion polymer (DOLLOP). The dominant coordination numbers for calcium and carbonate ions by other anions or cations, respectively, in DOLLOP are 1, 2 and 3, thereby making it quite distinct from ACC.[20,21]

While the influence of the biological environment on ACC remains the focus of many studies,[22] the identification of pre-nucleation species raises the possibility that the control of biomineralisation might begin even earlier in the crystal growth processes.[23,24] Indications that this may be possible come from the prior discovery of polymer-induced liquid-precursor (PILP).[6,25] Here, the addition of polyaspartate to a solution of calcium carbonate is found to generate a phase separated liquid mineral precursor whose shape can be manipulated. Inspired by this possibility, the aim of the present study is to perform a first examination of how the presence of organic species may alter the nature of pre-nucleation species of calcium carbonate. Rather than attempt to probe an *in vivo* system with all its inherent complexity, we begin by exploring the effects of several simple organic additives under controlled conditions. Specifically, we consider the three anions of acetate ($CH_3CO_2^-$), aspartate ($^-O_2CCH(NH_3^+)CH_2CO_2^-$) and citrate ($^-O_2CC(OH)(CH_2CO_2^-)_2$) that contain a varying number of carboxylate groups from one to three, as well as different net charges. It is hoped that this information may offer preliminary insights as to how larger and more complex organics with similar functional groups may direct the nucleation and growth of calcium carbonate.

2 Methodology

2.A Experimental

All chemicals were used without further purification. Aqueous solutions were prepared with water taken from a Milli-Q system. Details regarding our pH-constant experimentation have already been described extensively in the literature.[19,23,26] However, we now utilize vessels that are surrounded by a jacket fed with oil from a thermostat, keeping the solution temperature constant at 25 ± 0.2 °C. Titrations were carried out by dosing 10 mM calcium chloride solution (Ca ion standard, Metrohm) into 50 mL of a 10 mM carbonate buffer (pH = 9.00, prepared using sodium hydrogen carbonate from Riedel-de Haen (ACS reagent, ≥99.7%) and anhydrous sodium carbonate from Sigma-Aldrich (ACS reagent, ≥99.9%) at a constant rate of 10 μL min^{-1}, while the calcium potential was recorded with an ion-selective electrode and the pre-set pH was kept constant by means of counter-titration with 10 mM NaOH (Alfa Aesar, standard solution). This procedure allows us to determine the concentration of free calcium and carbonate ions throughout the different early stages of precipitation (*i.e.* pre-nucleation, nucleation, post-nucleation).[19,23] The reference experiment without additives was compared to assays in which 1 mM, 5 mM, and 10 mM of L-aspartic acid (Acros, ≥98%) or sodium acetate (NaOAc, Merck, anhydrous, p.a.) were added to the carbonate buffer of the same pH prior to titration. Additive-containing buffers were prepared by mixing 20 mM carbonate buffer with an equal volume of a 2, 10, or 20 mM solution of the respective additive in water, which had previously been adjusted to pH 9.00 with NaOH. For further comparison and to elucidate the role of ionic strength, additional reference runs were carried out in which carbonate buffers containing different amounts of sodium chloride (VWR Prolabo, AnalaR, ≥99.9%) were used. Finally, in order to assess binding of Ca^{2+} ions by the additives, independent titration experiments were conducted where $CaCl_2$ was dosed into aqueous solutions of the additives (as well as NaCl) at corresponding concentrations (likewise set to pH 9.00), but in the absence of carbonate.

2.B Simulation methods

In order to probe the atomistic detail of what is occurring when organic additives are present during the pre- and early post-nucleation period of calcium carbonate formation from aqueous solution we have performed molecular dynamics simulations based on force field methods.

The key to reliable simulations is the quality of the force field parameterisation. We have extensively studied this aspect in previous works, to ensure that important aspects of the thermochemistry of calcium carbonate are correctly described.[27] In particular, a model has been developed that accurately reproduces the free energy difference between calcite and aragonite at ambient conditions, while also capturing the metastability of the disordered vaterite phase. Furthermore, the interactions of the component ions, Ca^{2+} and CO_3^{2-}, with water, as represented by the SPC/Fw force field,[28] have been fitted in order to reproduce experimental free energies of solvation of the species. Here we adopt the latest version of the force field that includes anharmonic bond–bond and bond–angle coupling within the carbonate anion. A full description of the force field, including the parameters can be found elsewhere.[20] By combining the quasiharmonic lattice free energy with the solvation free energies of the ions, as well as correcting for the gas phase translational and rotational entropy (for carbonate) of the ions, the free energy of dissolution can be estimated. Here our computed value for calcite of +46.3 kJ mol^{-1} compares favourably with the experimental value of +48.4 kJ mol^{-1}.

In the present work it was necessary to extend the parameterisation of the force field to include organic species and their interaction with Ca^{2+} and the other species present in solution. There are many different organic force fields already available from the biological community and there have already been numerous works that have used several of these models to consider the interaction of organic molecules with calcium carbonate. One of the earliest was that of Griffiths and Heyes[29] who used the CVFF force field[30] to study the binding of sulfonates and phenates to small amorphous $CaCO_3$ clusters. Later de Leeuw and Cooper[31] also employed the same organic force field to examine the binding of small C1 & C2 organics on calcite, though in combination with a shell model for water that was subsequently found to freeze at longer timescales. Recently Aschauer et al.[32] have extended this combined shell model water approach to the study of polyacrylic and polyaspartic acids on calcite, but now using Dreiding[33] instead of CVFF. Duffy, Harding and co-workers have examined the influence of self-assembled monolayers on the nucleation and growth of calcium carbonate,[34,35] in particular for stearic acid, using the CHARMM[36] force field for the organic part. A recent study of the binding of alkaline metal cations to the aspartate anion by Hamm et al.[37] also favours the use of CHARMM for the organic interactions. In contrast, Metzler et al.[38] have used the AMBER[39] force field to examine peptide-induced aggregation of calcium and carbonate ions in solution. An attempt to produce a general force field for the simulation of biomineralisation was made by Freeman et al.[40] based on the $CaCO_3$ model of Pavese and co-workers[41] combined again with the use of AMBER for the organic part. Surveying the trends in the above literature, it is clear that the trend is towards the use of more extensive and specific organic force fields, such as AMBER and CHARMM, as opposed to more generic parameterisations like CVFF and Dreiding. Here we choose to adopt the CHARMM force field for the intramolecular organic interactions.

When considering the organic–water interactions it is necessary to refit the interaction parameters, rather than taking them unmodified from the existing CHARMM force field, for two reasons. Firstly, we are using the SPC/Fw force field for water, whereas CHARMM parameters were derived for the TIP3P water model. Secondly, an important part of our philosophy is to try to ensure that the experimental thermodynamics are reproduced where possible, and in particular, the free energy of solvation of all ions. Failure to consider this may lead to incorrect binding energies in the simulations. Hence the organic–water parameters have been varied to ensure the solvation properties are correct (final parameters are given in the Electronic Supplementary Information as Tables S1–3†).

One of the issues that must be addressed in the above approach is to determine the reference value of the free energy of solvation. While there are numerous experimental and theoretical studies available for the solvation of the acetate anion, values for L-aspartate and citrate in the protonation state appropriate to the pH of interest are harder to find. In order to supplement the experimental information available, which will be discussed during the results section, we have also performed quantum mechanical calculations to estimate the free energy of solvation. Here the relevant anions were optimised using density functional theory with the M06 meta hybrid exchange–correlation functional of Zhao and Truhlar,[42] in conjunction with a 6-31+G** Gaussian basis set. These calculations were performed within the framework of the SM8[43] implicit solvation model with the parameters, including dielectric constant, set appropriate to water. All calculations have been performed using the QChem software.[44] While inclusion of an explicit solvation shell of water is often required in order to obtain an accurate description of species in solution, the SM8 model has been calibrated to yield accurate solvation free energies without this.

Interactions between the organics and water were initially taken from the bicarbonate–water interactions taken from our previous study with rescaling of the Buckingham A parameter by the ratio of the charges. The parameters for the water–carboxylate interactions were then refined against the target free energy of solvation. All initial free energies of solvation were computed using a modified version of DL_POLY_2.19[45,46] by use of a two-stage perturbation approach, as described in previous work. Subsequent refinement was performed by using the perturbation between two sets of interatomic parameters to compute the change in the free energy of solvation.

The final set of force field parameters required for the present work involves the interaction between Ca^{2+} and the organic species. Given that there is limited crystallographic or other experimental evidence to fit against, we have taken a similar approach to previous works by using scaling relationships to determine the coefficients for a Buckingham potential in the case of the Ca–O interaction:

$$U_{ij} = A_{ij} \exp\left(-\frac{r_{ij}}{\rho_{ij}}\right) - \frac{C_{ij}}{r_{ij}^6}$$

Here the leading coefficient is scaled by the charge of the oxygen with which the calcium is interacting (given that the calcium charge is fixed at +2) relative to the parameter fitted for Ca–O in bulk calcium carbonate:

$$A_{ij} = A_{ij}^{Ca-CO_3} \left(\frac{q_O}{q_{O(CO_3)}}\right)$$

While this scaling relationship was found to work well in transferring parameters from carbonate to bicarbonate, evaluation of the ion pair free energies based on the resulting short-range potentials showed a systematic over-binding. To correct for this, a uniform scaling was applied of 1.34 to the Buckingham A parameters between calcium and carboxylate oxygen. The final set of intermolecular force field parameters used in the present work, including those previously published for aqueous calcium carbonate systems, are given as Electronic Supporting Information† (Tables S1–S3 with Fig. S1–S3 showing the charges used). Intramolecular interactions are generated from the CHARMM force field[47] using the CGenFF program.[48] Note that long-range interactions, including Coulomb terms, are excluded for 1–2 and 1–3 connected atoms, with the exception of bicarbonate and acetate, where the 1–4 interactions were also excluded.

One of the issues that must be addressed when simulating carbonate speciation in water, as well as that of polybasic organics, is that of pH. The pK_a of acetic acid is 4.76 and so we can consider it as being essentially fully dissociated under the basic conditions of interest here. Aspartic acid has three pK_a values of approximately 2.1, 3.9 and 9.8 and so will have lost both carboxylate protons under mildly basic conditions. Hence we consider aspartate as being in the zwitterionic form in which the amine group remains protonated to give a net charge of -1. Citric acid is also tribasic, but with pK_a values of 3.2, 4.8 and 5.2 the citrate anion can be regarded as completely deprotonated under relevant conditions.

In the conventional molecular mechanics force field derived above the protonation state of all functional groups must be specified *a priori* and thus this represents an approximation. In order to test the validity of the assumptions made we have also chosen to perform simulations using a reactive force field model. Here we use the ReaxFF formalism of van Duin and co-workers[49] as implemented in the program GULP.[50] Parameters for aqueous calcium carbonate systems in ReaxFF have been determined by Gale *et al.*[51] using a fixed charged model for the calcium cation, which essentially remains fully ionic in aqueous conditions. A fully reactive model for calcium-containing systems has subsequently been derived,[52] though this appears not to offer substantial benefits for the present system and yields inferior results for calcite. Thus the unreactive calcium model is retained in the present study. To allow for the incorporation of organics the parameters developed for the simulation of peptides[53] in water have been used.

All molecular dynamics simulations using ReaxFF have been performed using a domain decomposition algorithm with iterative charge solution. A time step of 0.5 fs was employed in combination with a stochastic integrator.[54,55] For each combination of ions studied, the species were placed in a system containing 308 water molecules within a cubic cell of just over 20 Å along each side. An *NPT* ensemble was used with the cubic cell being allowed to vary isotropically at 298.15 K. Where the system was charged, a neutralising background charge was applied, as well as a Madelung correction, to remove the spurious interaction between images. All simulations were run for 100 ps of equilibration followed by 2.5 ns for production.

All MD simulations using the standard unreactive force field approach have been performed with LAMMPS[56] (version as of 27th October 2011) using a 1 fs time step and the PPPM algorithm for the calculation of the reciprocal space contribution to the electrostatics. A chain of 5 Nosé–Hoover thermostats and barostats with relaxation times of 0.1 and 1.0 fs, respectively, has been used in all simulations to control temperature and pressure ensuring that the correct ensemble was sampled. The free energy calculations have been performed using the PLUMED 1.3 plug-in[57] in the NVT ensemble. The distance, or minimum distance, was used as a collective variable, depending on the number of carboxylate groups that were present in the binding species. Well-tempered metadynamics[58] with a bias factor equal to 11 and 50 multiple-walkers[59] have been used in all calculations. The initial Gaussian height and width were k_BT and 0.1 Å, respectively, with a new Gaussian being added every 0.5 ps during a total simulation length of 130 ns.

Car–Parrinello molecular dynamics (CPMD) simulations of the citrate–water system were also carried out to provide additional points of comparison for the force fields developed in this work. Due to the computational expense inherent to first-principles simulations, we considered only modest length-scales and time-scales here, with a system comprising one citrate molecule (featuring three carboxylate anionic groups), 128 water molecules, and three Na^+ counter ions (405 atoms in total). To perform these simulations we used the CPMD software package[60] version 3.13.2, with the BLYP[61,62] functional and ultrasoft[63] pseudopotentials, along with a plane wave cut-off of 340.1 eV. The system was contained in a cubic simulation cell of side-length 16.1 Å, with 3-D periodic boundary conditions used throughout. Prior to the CPMD run, the system was geometry optimised to a threshold of 2.57 eVÅ^{-1}. All atoms were free to move in the simulation, which was conducted in the NVT ensemble at 300 K, using a Nosé–Hoover thermostat with a coupling-constant of 2915 cm^{-1}. A time-step of 0.0968 fs and a fictitious electron mass of 400 atomic units were used, and k-point sampling included the Γ-point only. The duration of the CPMD simulation was 5 ps in total, of which the first 2 ps was treated as equilibration and only the last 3 ps was used for subsequent analysis.

For the purpose of our analysis of the first principles molecular dynamics, we define two types of carboxylate present in citrate, denoted as OM and OE, signifying the central and end carboxylates respectively, with the label OH indicating the oxygen of the central hydroxyl group. We have calculated radial distribution functions (RDFs) around the OM, OE and OH sites using a bin width of 0.02 Å.

3 Results

3.A Pre-nucleation clusters in the absence of organics

The nature of pre-nucleation clusters in the absence of additives other than a background electrolyte has already been previously described in earlier work.[20] Here we describe the salient features that are relevant to any discussion of the influence of additives. In our previous studies we have demonstrated that calcium and carbonate ions come together in solution to create both ion pairs and higher aggregates known as DOLLOP. These DOLLOP structures involve dynamically evolving chains of alternating calcium and carbonate ions that can branch, form rings, lose and gain ions. Because of the rapid exchange of ions with solution there is no phase boundary between these species. The combination of strong hydration, leading to enthalpic stabilisation, and favourable entropic contribution, resulting from the large number of microstates available within an almost degenerate energy, means that these species are thermodynamically stable with respect to ions and ion pairs in solution at mM concentrations and alkaline pH. The DOLLOP species remain structurally distinct from ACC, including hydrous ACC; whether there is an activated process that connects DOLLOP to ACC or whether conversion occurs by dissolution and regrowth remains unknown at present.

3.B Solvation of isolated organic anions

Following the approach taken for the derivation of a force field model to describe both Ca^{2+} and CO_3^{2-} in aqueous solution, the interaction of the three organic anions with water has been parameterised in order to reproduce the free energy of hydration, while also trying to ensure a reasonable solvation environment about the key functional groups. Any error in the thermodynamics of solvation is likely to impact on the competition between the anions binding to cationic species *versus* remaining isolated in aqueous solution.

In the case of acetate there are numerous values for the free energy of hydration available in the literature, as summarised in Table 1. Because the experimental values for this quantity can only be inferred indirectly, in the case of an anion *via* a thermodynamic cycle, this leads to a degree of variability. Theoretically, there is also some scatter in the values with Gao *et al.*[64] demonstrating that the value obtained from *ab initio* calculations is sensitive to the combination of the level of quantum mechanical theory, basis set and solvation model chosen. Similarly within force field studies the results will depend on the parameterisation of the water model and whether polarisation effects are included.[65] Here the value for the free energy of hydration from the model adopted spans the range of several of the less extreme values, to within the statistical uncertainty in the final result, while being slightly less exothermic than the value obtained from the quantum mechanical calculations performed in the present work.

There are also many studies of the aspartate anion in the literature because of its relationship to biological molecules. This includes examinations of conformational and tautomeric equilibria,[66] and calculation of the pK_a values.[67] Despite this, there appears to be no explicit estimate of the free energy of hydration available. Furthermore, because the zwitterionic configuration that is adopted by the anion in solution is unstable in the gas phase, it is not possible to construct a purely experimental thermodynamic cycle to obtain the free energy of hydration. Consequently we have adopted the SM8/M06/6-31+G** value for this quantity as an appropriate target. Although it would be possible to precisely reproduce the quantum mechanical free

Table 1 Comparison of literature values (experiment and theory) for the free energies of hydration of the acetate, aspartate and citrate anions with the computed values at the M06/6-31+G** level of theory within the SM8 solvation model (QM) and force field calculations, as determined in the present work. All values are in kJ mol^{-1}

	Acetate	Aspartate	Citrate
Experiment	-337.4^a		
	-322.2^b		
	-313.8^c		
Theory	-343.8^a		
	-341^d		
	-278 to -338^e		
	-320 to -325^b		
	-324.3^f		
QM (this work)	-334.4	-418.9	-1774
Force field (this work)	-326 ± 3	-414 ± 3	-1771 ± 3

^{a} from ref. 68. ^{b} from ref. 69. ^{c} from ref. 70. ^{d} from ref. 71. ^{e} from ref. 64. ^{f} from ref. 65.

energy of solvation by fine-tuning of the water interactions, it was considered that a slight overestimate was acceptable given the uncertainties in the target value and that the acetate value also deviates from the quantum mechanical estimate in the same direction.

For both citric acid and citrate there is also a dearth of information in the literature regarding the thermodynamics of hydration, again leading us to take the SM8/M06/6-31+G** value as the best available estimate for evaluation of the force field. As per acetate and aspartate, the absolute magnitude of the solvation free energy with the force field model is slightly under-estimated, but only by a few kJ mol^{-1}, which is well within the uncertainty of the target quantity.

Aside from considering the thermodynamics of solvation, it is also important to examine whether the structure of water around the organics is reasonable. While both acetate and aspartate have been studied previously in the literature, the least well-calibrated system is that of citrate. Given that it is the most highly charged of the three anions, it is also represents the most severe test of the degree of water ordering in the first hydration shell. As there is no experimental data to compare against, that we are aware of, first principles simulations of citrate in water have been performed to provide comparative information. Sodium counter-ions were employed for charge neutrality as they have a low tendency to associate with the citrate anion. It is important to note that there are limitations on the interpretation of the data obtained from the first principles molecular dynamics too, including the systematic errors of current GGA functionals for the structure and dynamics of water, and also the short time scale that is accessible. We note that inclusion of van der Waals interactions appears to improve some aspects of the description of water,[72] and artificially increasing the temperature may be a pragmatic approach to improve the results. However, neither of these approaches was used in the present work.

In Fig. 1 the radial distribution functions for selected interactions between citrate and water are given, as determined according to the conventional force field, ReaxFF and first principles molecular dynamics (CPMD). Starting with the carboxylate groups, all three methods show good agreement as to the position of the first peaks for OE to both the hydrogen and oxygen of water. Even the subsequent solvation shells seen in the OE–OW distribution are in good agreement, allowing for the inherent noise present in the CPMD data due to the restricted sampling. The main discrepancy is that the ReaxFF model indicates some splitting of the first peak for both OE–OW and OE–HW, while this is absent for the other two methods. This is due to the fact that the ReaxFF citrate RDFs are computed from a simulation that also contains a Ca^{2+} counter ion. Although this cation is initially separated by half of a box length, during the course of the 2.5 ns run the ions diffuse together to form a solvent separated ion pair, thereby perturbing the distribution of hydrogen bonds around citrate.

Turning to consider the radial distribution functions for the hydroxyl group of citrate, as shown in Fig. 2, here the different techniques are less in accord. In comparison to the first principles data, the hydroxyl group of citrate interacts too strongly with water as an acceptor of hydrogen bonds. However, the aforementioned caveats must be borne in mind before deciding where the correct result lies. In particular, it was observed in the ReaxFF simulations that an intramolecular hydrogen bond between the OH and a carboxylate group of citrate could persist for considerably longer than the time scale of the first principles run. Consequently the initial configuration will influence the outcome, as there is insufficient time to equilibrate the intra- and inter-molecular configurations with the proper statistical weight. Given that the first principles dynamics starts from a conformation that has an intramolecular hydrogen bond, which lasts for 85% of the production phase, while the conventional force field shows little or no evidence of the same interaction in the long time limit, the deviations in the radial distribution function can be apportioned to the different conformations being sampled.

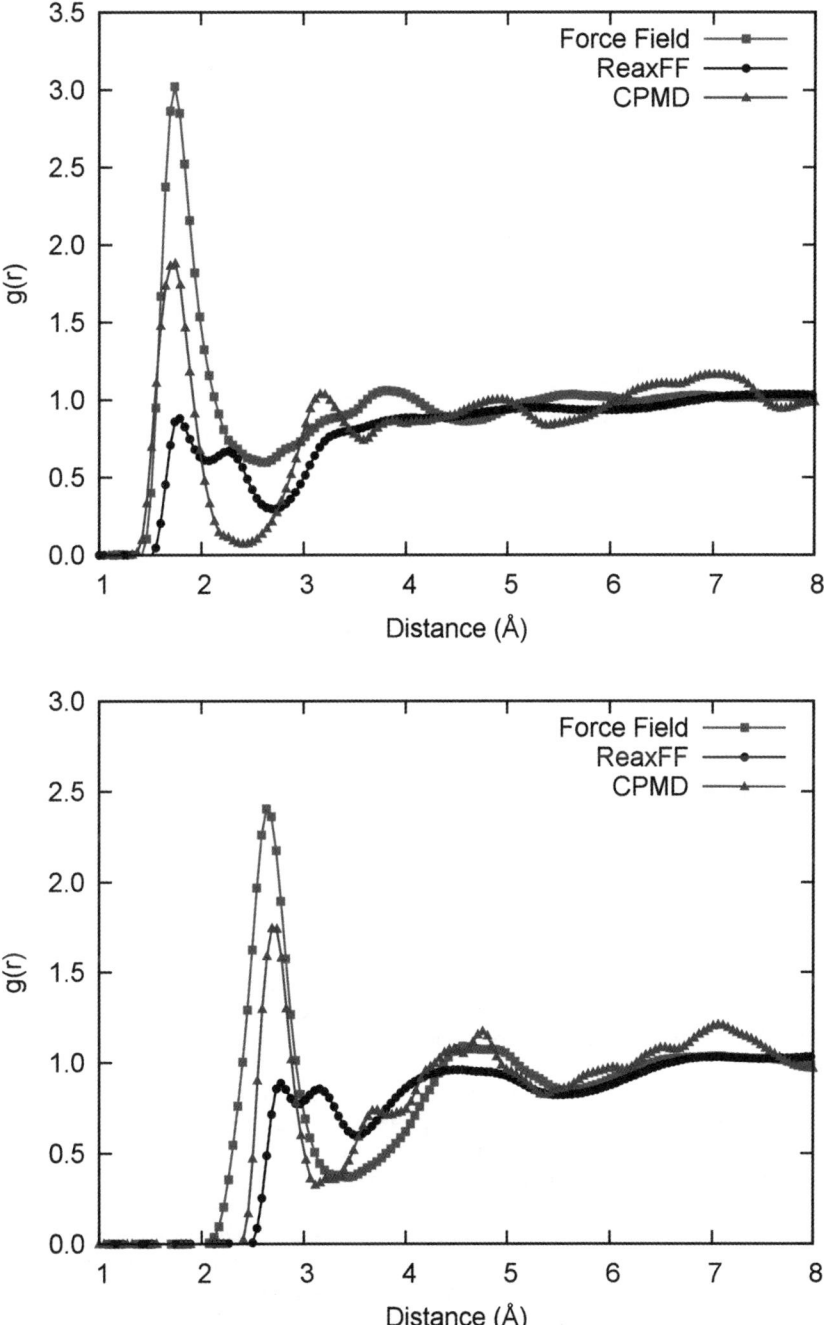

Fig. 1 Radial distribution function (g(r)) for the oxygens (OE) of the two equivalent carboxylate groups of citrate to H (top) and O (bottom) of water.

3.C Influence of organics on pre-nucleation clusters

Having calibrated the free energies of hydration for acetate, aspartate and citrate, we now turn to consider the simulation of the association between these anions and species that would be present in solution prior to nucleation. In particular, the ion

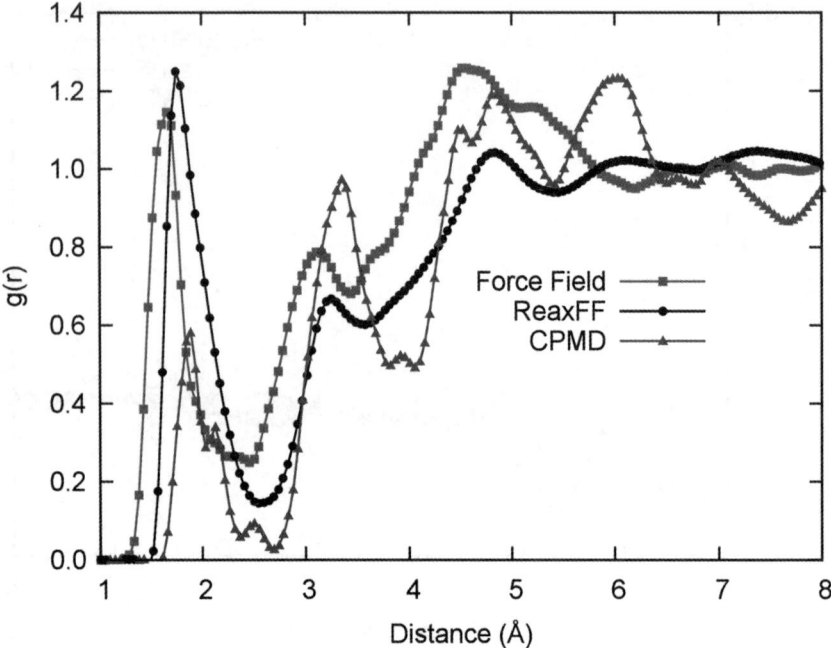

Fig. 2 Radial distribution function for oxygen of the hydroxyl group of the citrate anion to the hydrogens of water molecules as computed with three different methods.

pairing between Ca^{2+} and the above organics is likely to be the initial mode of association:

$$Ca^{2+}_{(aq)} + X^{n-}_{(aq)} \Leftrightarrow CaX^{(2-n)+}_{(aq)}$$

Using metadynamics, we have computed the free energy profiles for each of the above ion pairing reactions, as shown in Fig. 3. The overall free energies of binding, after correction for the radial dependence of the entropy, are given in Table 2 along with literature values.

It should be noted that computed values are in effect for the system at high dilution since the difference is taken between the free energy minimum and the asymptotic limit for long-range separation. Under standard conditions of 1 M the average separation between ions is reduced and so taking the commensurate free energy at this distance reduces the binding strength by up to 5 kJ mol^{-1}, depending on the species. Hence the discrepancy with the experimental values is actually less than that apparent from Table 2 when this is kept in mind. In the case of citrate we note that the interaction with Ca^{2+} has a longer range than for the other organics and so the free energy curve has not quite reached the plateau by the maximum distance sampled. However, the error is comparable with the statistical uncertainty.

As initially computed with the straight scaling relationship for the repulsive Ca–O interaction, all of the ion pairs were found to be over bound by a significant amount. Consequently, a single scaling factor of 1.34 was introduced for the Buckingham A parameter of this interaction. Following this modification the general trend in the ion pairing free energies was found to be in good agreement with experiment, with citrate being the most strongly bound by a significant margin. Whether or not acetate and aspartate have the correct relative order depends on the choice of

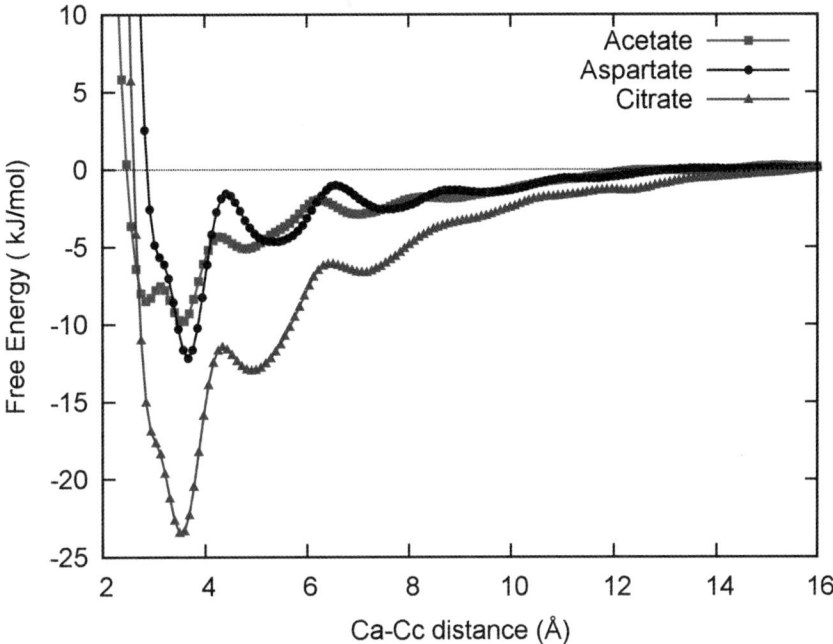

Fig. 3 Free energy as a function of the minimum distance between calcium (Ca) and the nearest carbon of any carboxylate group (Cc) of the organic anions, acetate, aspartate and citrate, in aqueous solution. The zero of free energy is specified with reference to the fully separated ions.

Table 2 Comparison of literature values (experiment and simulation) for the free energies of ion pairing of calcium with the acetate, aspartate and citrate anions with the computed values from force field simulation in the present work. Experimental values for acetate and aspartate–citrate are taken from ref. 73 and ref. 74, respectively, while previous simulation is from ref. 37. All values are in kJ mol^{-1}

	Acetate	Aspartate	Citrate
Experiment	−4.4 to −7.1	−6.7	−19.0
Previous simulation		−16.7	
Simulation	−9.5	−12.2	−20.7

experimental value, but the upper bound to the ion pairing free energy would be most consistent with the simulation data. In terms of the absolute free energy values, once the correction for the concentration back to standard conditions is allowed for, all of the simulation values are within ambient thermal energy from experiment. It should be noted that although the well-tempered metadynamics is converged to ∼0.001 kJ mol^{-1}, the remaining discrepancy in the absolute values is hard to remove since it is within the residual statistical uncertainty due to fluctuations.

Although the above results justify the introduction of a modified scaling procedure, there is a risk that fitting of free energy differences for solution species might lead to a model that yields unreasonable results for other properties. This would especially be the case if the original error in the binding data was as a consequence of an incorrect free energy of solvation for one or more of the component species. Finding data to validate against is difficult though. One option might be to use gas phase quantum mechanical data. However, the charge distribution would be radically different from the condensed phase unless the ion pair is surrounded by at least the first solvation shell of water. An alternative choice is to use crystallographic data for solid forms as a reference point. Taking the calcium–aspartate interaction as our test case, there is a known crystal structure for Ca(L-Asp)·2H$_2$O,[75] which adopts an orthorhombic cell containing four formula units. Two complications exist in using this as a reference structure. Firstly, only the heavy atom positions are reported as X-ray diffraction was used. This is easily resolved as the majority of the approximate hydrogen positions are unambiguous given the heavy atom geometry. Secondly, the structure contains the aspartate dianion, rather than the zwitterionic singly charged anion being considered in the present study. To handle this, the charge of the –NH$_3^+$ group was redistributed in the –NH$_2$ moiety to achieve the correct overall charge and the Ca^{2+}–N interaction was set equal to that of the calcium to oxygen of the carboxylate groups given that the resulting charges of N and O are similar.

In the experimental crystal structure for Ca(L-Asp)·2H$_2$O, the calcium–oxygen bond lengths fall in the range of 2.429–2.608 Å. Based on the original scaled force field parameters there is a systematic underestimation of these distances (2.324–2.540 Å), while using the parameters corrected against the ion pairing free energy the bond lengths (2.439–2.601 Å) are now in excellent agreement. Hence, while this is a somewhat indirect validation, it suggests that increasing the Ca–O repulsive interaction for the ion pairs is indeed justified and likely to improve structural properties.

Beyond just the overall free energy of ion pair formation, we can analyse the nature of the complete free energy landscape, as shown in Fig. 3. Both aspartate and citrate show qualitatively similar free energy profiles, though they are quantitatively different. There is a minimum in which the anion binds in a monodentate configuration to calcium, at 3.5–3.6 Å, with a point of inflection at shorter distances corresponding to the bidentate arrangement. In effect, the bidentate form is a transition state for exchanging the oxygen atom of the organic that is coordinated to the calcium cation. Beyond this inner minimum there are discernable minima corresponding to solvent separated states, though the binding in these configurations is weaker than for the contact ion pair. Furthermore, the barrier to go from the solvent separated minima to the directly coordinated state is generally less than ambient thermal energy and so there is little in the way of an activation barrier to contact ion pair formation. In line with experiment, we find that citrate exhibits much stronger binding to calcium than for aspartate, as would be expected from the charge of the anion being three times as large.

Acetate is found to have a quite different free energy profile of binding to that for aspartate and citrate with the curve exhibiting two distinct minima for the bi- and mono-dentate configurations. Although the monodentate form remains marginally more stable than the bidentate configuration, the free energy difference is small. In terms of rationalising this different behaviour for acetate relative to the other two anions, the most probable factor is the different charge distribution for the carboxylate groups; in acetate the carboxylate is less polar than in aspartate or citrate, where the charges are similar. The rationale for this is that in the case of acetate it proved difficult to adopt the same degree of polarity while still obtaining both a reasonable free energy of solvation and a good description of water structure around the carboxylate group.

For the case of aspartate, free energy profiles for ion pairing have been computed by Hamm et al.[37] for the binding with Mg^{2+}, Ca^{2+} and Sr^{2+}. In this prior work the binding free energy for aspartate to Ca^{2+} is found to be exothermic to the extent of ~16 kJ mol^{-1}, making it more strongly bound than in the present study and further from the experimental values.

As an additional qualitative validation of the results obtained with the conventional force field model, we have also performed simulations of the association of the organics with calcium ions using the reactive ReaxFF approach. Although obtaining quantitative free energies is more complex within this model, it does allow some of the assumptions regarding protonation state to be tested, while providing an indication of whether a totally different parameterisation and functional form of interaction would alter any of the behaviour observed for this system.

Considering first the case of acetate binding with Ca^{2+}, the reactive model indicates that monodentate coordination by the carboxylate group is strongly preferred with a Ca–O distance of 2.29 Å. Bidentate coordination appears to be predominantly a transition state during swapping of the oxygen that is bound to calcium, which occurs on 8–9 occasions during 2.5 ns. Very similar behaviour is observed for aspartate binding to Ca^{2+}. The fact that the reactive model shows consistent modes of coordination in the ion pair for acetate and aspartate may suggest that the results from the conventional force field overestimate the presence of the bidentate configuration for acetate as a result of the charge distribution.

The ReaxFF simulations for citrate were run from several different initial configurations to probe the binding with the distinct functional groups. Some starting geometries led to the ion pair dissociating, though the ions tended to remain in a solvent separated state rather than diffusing apart, consistent with the free energy profile computed with the conventional model. In other cases a carboxylate group that was not initially bound to calcium would enter its coordination environment such that the cation was chelated by either two carboxylate groups or one carboxylate and the hydroxyl oxygen. Even when only one carboxylate group is directly attached to Ca^{2+}, a second one is clearly hydrogen bonded to a water molecule from the first solvation shell of calcium.

Having considered the ion pairing between Ca^{2+} and the three organics alone, it is now possible to examine whether there may be some association between the additives and stable pre-nucleation clusters. In our previous work we have shown that it is possible to simulate the formation of DOLLOP species through the use of a "brute force" approach, provided that the concentration is elevated to overcome the time scale for diffusion and collision of ions. DOLLOP formed in this way can then be examined at realistic concentrations in the presence of a background electrolyte, as per the experiments of Gebauer et al.,[19] and be shown to be stable.[20] Here we performed a limited number of simulations following this same strategy but now in the presence of organic anions. Unfortunately, the timescale accessible at present was insufficient to be able to identify any influence within the statistical uncertainty given that the binding of the organics is generally weaker than that of carbonate with calcium. While some association between aspartate and citrate was observed with DOLLOP species, a more systematic approach is required to quantify any possible binding. That said, some interesting behaviour was seen for the case of aspartate, where the organic anions began to associate with each other to form a supramolecular polymer. While this can be ascribed to the particular concentration of aspartate present in the simulations, we believe that the binding is not an artefact of the force field parameters for two reasons. First, the interaction distances found between two aspartate anions held together by two N–H–O hydrogen bonds *in vacuo* agree well between the force field and a quantum mechanical calculation at the M06/6-31+G** level (we note that any discrepancy in the distances would be reduced by improving the basis set quality further). Second, there is evidence from experiments that such association can and does occur under certain conditions.[76]

As a first attempt to shed some light on the possible binding of the three organics to DOLLOP-like species, we have quantified the thermodynamics of association with a single $CaCO_3^{(0)}$ ion pair. Unfortunately this is not as straight forward as for the previous ion pairing calculation because of the non-negligible probability that the ion pair will dissociate during the metadynamics simulation. To circumvent this problem we applied the following two-step procedure. First of all we performed a metadynamics simulation as above to construct a free energy profile for the collective variable s subject to an artificial harmonic restraint ($K = 2$ eV Å^{-2}, $r_0 = 3.4$ Å) between the Ca and carbonate carbon to keep the ion pair together. The resulting free energy profile was then used as a fixed bias for a set of 50 independent 5 ns unrestrained simulations, in which any deviations from a uniform sampling in s result from differences between the restrained and unrestrained free energy landscapes. During these simulations we calculated the probability distribution for the system along the biased collective variable and then extracted the free energy contribution due to the removed harmonic bond using;

$$dA = -k_B T \ln(P(s))$$

where $P(s)$ is the calculated probability distribution along s. This free energy contribution was then added to the one previously calculated with metadynamics to obtain the total pairing free energy profile for the $CaCO_3^{(0)}$ ion pair to the organic molecules. This procedure did not completely eliminate the probability that the ion-pair was broken during the calculations but it has the advantage that the parts of the trajectories not relevant to the calculations (when the Ca–Cc distance was longer than 4 Å) could be easily removed during the post processing. At large distances the effect of the restraint on the Ca–Cc distance on the free energy profile is limited and this indeed was reflected by the fact that the probability distributions obtained during the second set of simulations was flat within thermal noise, as it should be once the free energy profile is exactly compensated by the metadynamics bias.

Results for the free energy of interaction between the three organic anions and the calcium carbonate ion pair in solution are given in Fig. 4. In comparison to the binding of the organics to the Ca^{2+} ion alone, the free energy is less exothermic, as would be expected, due to the repulsion between the two anions that are within the first coordination sphere. What is less expected is that the effect is non-uniform and that citrate is the least affected, despite being the most highly charged anion. Clearly the presence of the aqueous environment screens the Coulomb repulsion between carbonate and more remote charged functional groups of the other ligand coordinated to calcium. In contrast, the electrostatic binding between Ca^{2+} and the more highly charged citrate is less screened since the interactions are through the ligand, which has a much lower dielectric constant than water.

While citrate and aspartate maintain their similarity in having a minimum for direct coordination to the calcium of $CaCO_3^{(0)}$, the minimum free energy for acetate occurs at a longer distance intermediate between the contact and first solvent separated state for the Ca–Ace$^+$ ion pair. Even though aspartate has a minimum that is at a shorter distance than acetate, the strength of binding surprisingly goes as citrate $\gg$ acetate > aspartate. Based on this, it would be likely that acetate has more influence on ion pairing and perhaps DOLLOP formation than aspartate. However, the minimum in which aspartate coordinates to the calcium carbonate ion pair has a larger activation energy between it and the solvent separated state. Therefore aspartate may exert a similar influence because of this kinetic factor. One thing is definitely clear from the simulation results—citrate is the only ligand that is likely to show significant association with pre-nucleation species in calcium carbonate solution.

Reactive force field simulations have also been performed for the ion pair + organic systems described above. The most important finding was that the organic anions remained in the same protonation state as per the conventional model throughout the molecular dynamics in all cases. In order to verify that this was not just an issue of timescale or an overly high barrier to protonation, a simulation was performed in which an excess proton was added to the simulation of the $(CaCO_3–citrate)^{3-}$ complex. Although the proton was initially part of a hydronium ion that was as far as possible from the ion pair within the simulation box, it rapidly diffuses *via* a Grotthus mechanism and leads to protonation of a carboxylate group of citrate. Hence the system appears to have sufficient opportunity to sample the protonation states during 2.5 ns.

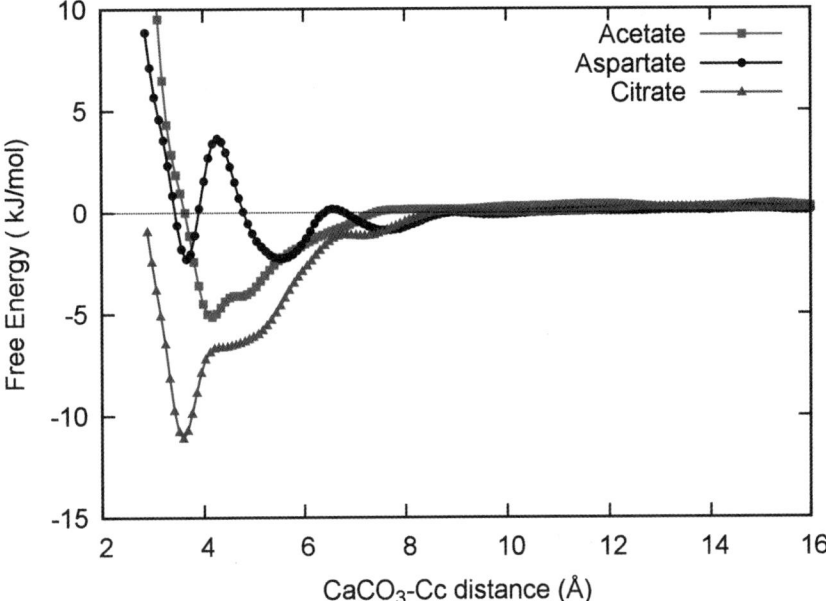

Fig. 4 Free energy surface for association of the acetate, aspartate and citrate anions with the $CaCO_3^{(0)}$ ion pair in aqueous solution as a function of the minimum distance between Ca or C of carbonate ($CaCO_3$) and the carbon (Cc) of the nearest carboxyl group of the organic species.

3.D Binding of organics to calcium carbonate

Aside from the influence of organic anions on pre-nucleation clusters of calcium carbonate, there is also the possibility that additives may alter the growth process *via* binding to post-nucleation species. In order to probe this possibility we have used metadynamics to determine the free energy profiles for binding of acetate, aspartate and citrate to two forms of post-nucleation calcium carbonate. First, given that under conditions relevant to biomineralisation the initial product is amorphous calcium carbonate, we have considered binding to a 36 formula unit ACC nanoparticle taken from our previous work.[21] Since we predict that water content increases with increasing particle size, we have taken a nominally dry ACC nanoparticle for this smaller nanoparticle. Second, as the ultimate macroscopic product will be a crystalline phase, we have also examined the binding of the organics to the basal ($10\bar{1}4$) surface of calcite as the dominant facet of the most stable phase.

The resulting free energy curves for acetate, aspartate and citrate approaching the surface of the above two calcium carbonate materials are shown in Fig. 5 and 6. Considering first the basal surface of calcite, all three ligands show slightly distinct behaviour. Acetate has a negligible degree of association with the basal surface of calcite as all minima are barely above the level of statistical noise. Experiments and simulations have both shown that there are two relatively strongly-ordered water layers over the basal face of calcite,[77] and it appears that acetate is unable to penetrate this interfacial region. Aspartate on the other hand is initially slightly repelled by the surface, but can attain a solvent separated state that may be marginally stable at 5–6 Å above the surface. This parallels the behaviour previously seen for the carbonate anion.[27] While the absolute depth of the minimum is no greater than for acetate, the barrier to exit the state is more pronounced. Finally, citrate exhibits no binding at the flat calcite surface as a result of the strong hydration experienced in aqueous solution.

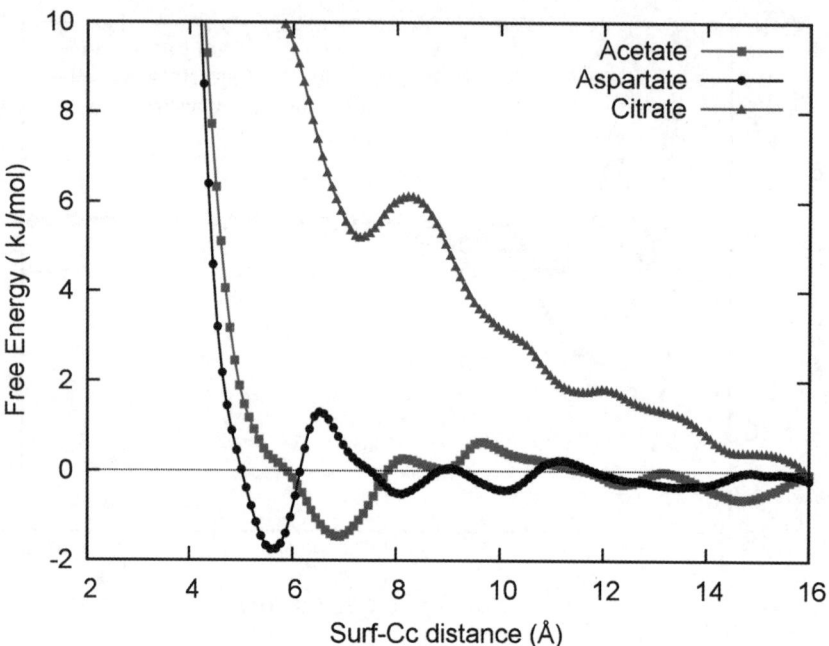

Fig. 5 Free energy profile as a function of the normal distance above the calcite (10$\bar{1}$4) surface to the nearest carbon of any carboxyl group of the three organics, acetate, aspartate and citrate. On the right hand side of the figure the free energy becomes asymptotic to zero, which is set as the free energy of the anion in bulk solution.

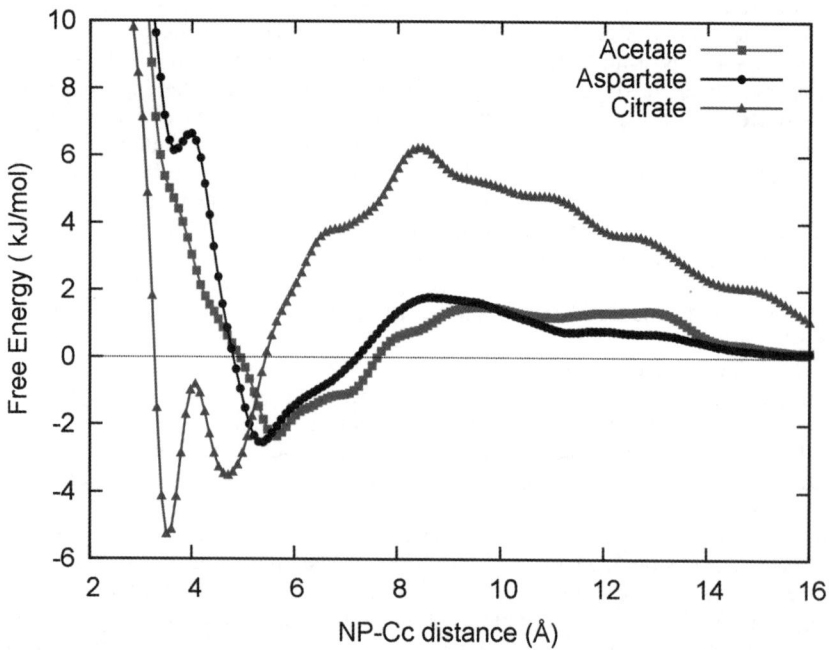

Fig. 6 Free energy profile for bringing an organic anion from solution (right-hand side) towards a 36 formula unit nanoparticle (NP) (left-hand side) of dry amorphous calcium carbonate.

Overall, it is found that none of the anions show any real association with the basal surface of calcite. This does not preclude any possible influence of additives on calcite growth, but we predict this would require either binding to steps or kink sites, or expression of other surfaces where calcium is more exposed.[78] Indeed, Elhadj et al.[79] have computed average binding enthalpies of over 400 kJ mol^{-1} for the aspartate anion at steps on the calcite basal surface using semi-empirical cluster calculations in combination with a mix of explicit and continuum water models. Although we have yet to examine the influence of steps on binding with the present model, it appears unlikely that the free energy of binding of aspartate to calcite steps will be as exothermic as the above value based on the strength of interaction with the flat surface alone. Aside from the quantitative details, based on the indications from the flat surface, it can be speculated that the order of influence for the additives considered would be aspartate > acetate ≫ citrate for processes involving calcite surfaces or interfacial environments where water is strongly ordered.

Turning now to consider the interaction of the three organics with ACC, a more complex situation is found to occur. Both acetate and aspartate exhibit a similar minimum that corresponds to a solvent separated complex with the nanoparticle. Although the bound minimum only has a slightly lower free energy than the anions in bulk solution, there is a small activation barrier to separation from the nanoparticle, and so there is the possibility for acetate and aspartate to spend some time associated with ACC. In the case of citrate, the free energy profile of binding is radically different from that for the calcite surface, and to the profile for the other two organics. At first sight there is now a significantly exothermic state corresponding to citrate coordinating directly to the ACC nanoparticle. However, the unusual variation with distance indicates that something unexpected is occurring. Further analysis of the trajectories for the multiple walkers indicates that in some cases the citrate ligand pulls a calcium ion from the surface of ACC and so as it detaches the asymptotic limit is for formation of the Ca–(Cit)$^-$ ion pair. Associated with this fragmentation, some times a carbonate anion is also ejected from the ACC nanoparticle in order to remove the excess negative charge, as shown in Fig. 7. Because of these observations we can deduce that the curve for citrate in Fig. 6 is actually the convolution of the calcium–citrate ion pair and the underlying curve actually being sought. This illustrates the difficulty of trying to map the free energy surface for complex systems in which there are alternative competing reactions to that of the collective variable under consideration. Part of the problem here may be because one of the smallest possible nanoparticles of ACC was employed as the model and so the more open structure makes it easier for calcium ions to be removed by water in combination with the ligand. Although it has not been possible to quantify the interaction of citrate with ACC in the present work, it is also clear that citrate is not repelled by ACC in the same manner as the basal surface of calcite. The absence of a strongly ordered water layer makes it possible for citrate to occasionally reach the surface of ACC and bind to calcium, though this sometimes leads to removal of the ion from the surface. Preliminary unbiased simulations for larger nanoparticles of ACC suggest that the tendency to remove calcium is a consequence of the small particle size for 36 formula units.

3.E Influence of acetate, L-aspartate and citrate on *in vitro* precipitation

In order to assess the effect of the different additives on *in vitro* crystallisation, quantitative assays with and without the different molecular additives have been conducted. The experiments were carried out at constant pH = 9.00, starting with pure carbonate buffer. Upon slow addition of dilute calcium solution into the buffer, supersaturation was generated until nucleation occurred, and solid forms of calcium carbonate were precipitated. This process was traced utilising a calcium ion selective electrode and, as discussed in detail in previous work,[23] the calcium curves can be used to quantitatively characterise the influence of additives on the early stages of

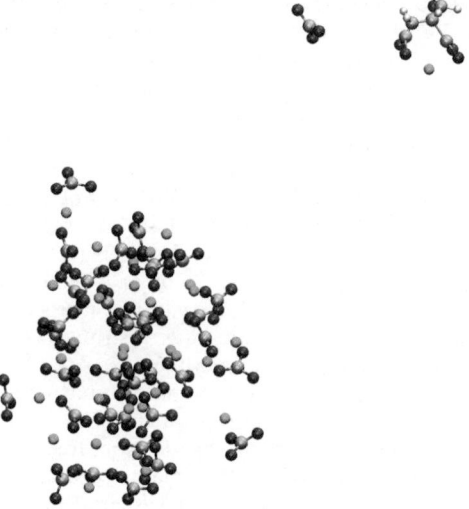

Fig. 7 Configuration taken from the simulation of citrate in contact with an amorphous calcium carbonate nanoparticle initially containing 36 formula units. The citrate anion (top right) has detached taking a calcium cation with it. A carbonate ion (top centre right) then also dissociates from the nanoparticle. Surrounding water molecules are hidden for clarity.

calcium carbonate precipitation through comparison with the reference experiment carried out in the absence of additives. In principle, there are five different effects that can be identified and quantified based on the calcium curves; (i) complexation of calcium ions, which is apparent from a delayed increase in calcium; (ii) influence of the additives on the pre-nucleation equilibria, which is associated with an altered slope for the pre-nucleation calcium development; (iii) inhibition of nucleation can be identified by a shift of the maximum in the curves to later times (more addition of calcium required) and to higher amounts of detected calcium; (iv) adsorption of the additives on nucleated particles can become apparent from a new increase of detected calcium after nucleation, *i.e.* a constant solubility is not maintained; (v) the solubility can be directly determined and shows whether the additives influence the type of amorphous phase or crystalline polymorph formed.

In the reference experiment (Fig. 8 A), the amount of free calcium in the carbonate buffer at first increases linearly upon constant addition of calcium chloride solution, while less calcium is detected than actually added. This is the pre-nucleation stage, and the calcium binding in combination with the pH-titration can be used to quantitatively characterise the pre-nucleation cluster equilibrium.[19] When a critical point is reached, nucleation occurs and the amount of free calcium drops to a value that corresponds to the solubility of the precipitated phase (which is proto-calcite ACC at this pH-level, and relates to calcite in terms of its short-range order).[18,19]

Titration curves obtained in the presence of 1, 5 and 10 mM acetate show no distinct deviation from the reference experiment within the common limits of accuracy and reproducibility (Fig. 8 A). The slight tendency towards inhibition of nucleation seen at 10 mM (*i.e.* when acetate is equimolar to the carbonate buffer and thus is at a large excess over calcium carbonate species) cannot be regarded as significant (see ESI Fig. S4†). Indeed, similar observations were made when 10 mM sodium chloride was added to the buffer and therefore, if anything, corresponding effects are likely to originate from the higher salinity of the system (ESI Fig. S4†). At 1 mM L-aspartate, the delay of nucleation is somewhat more pronounced than at 10 mM acetate. With increasing concentration of aspartate, inhibition of nucleation

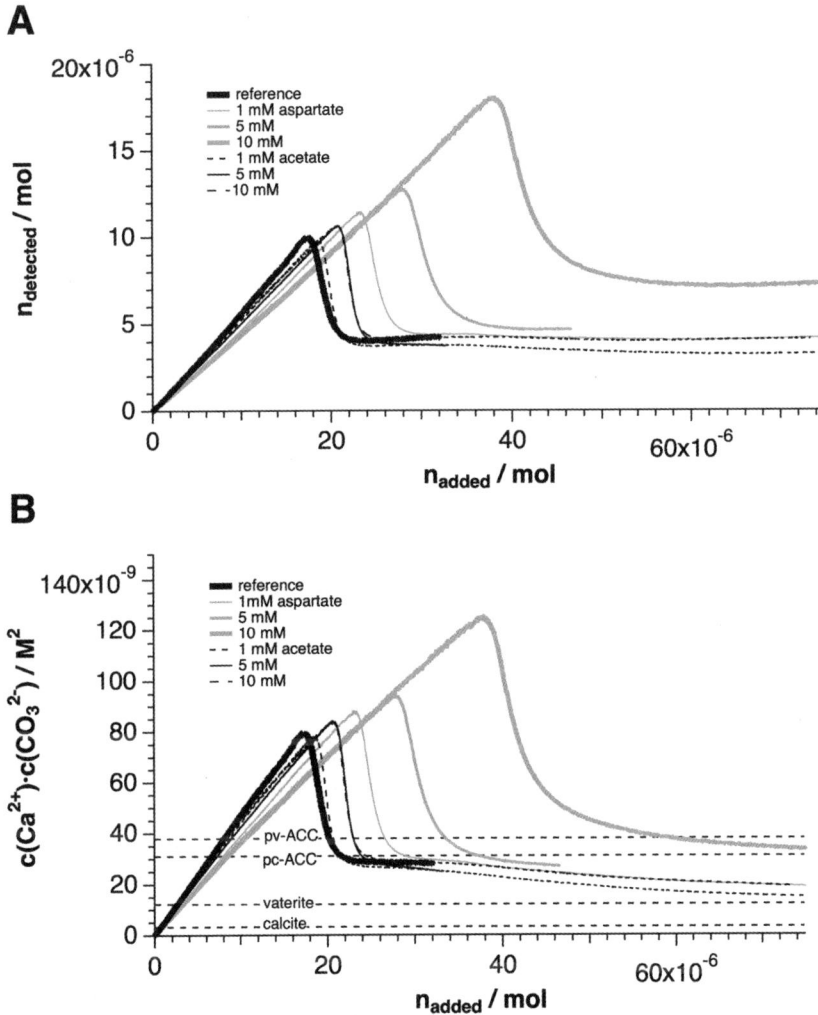

Fig. 8 The influence of acetate and aspartate on the early stages of calcium carbonate precipitation. A: development of the amount of free calcium in 50 mL carbonate buffer (10 mM, pH = 9.00) upon addition of dilute calcium chloride solution, in the absence (reference) and presence of different amounts of additives as indicated. The steep drop in the detected free calcium identifies the nucleation event. B: corresponding free ion product as a function of the amount of calcium added. Dashed lines mark the solubilities of proto-vaterite ACC (pv-ACC), proto-calcite-ACC (pc-ACC),[18] vaterite, and calcite.[18,80] Note that the curves of 5 and 10 mM acetate largely overlap.

becomes much more distinct and approximately twice the amount of calcium has to be added at 10 mM aspartate for nucleation to occur. Furthermore, the level of free calcium after nucleation—which reflects the solubility of the precipitated calcium carbonate phase—appears to be higher than in the reference experiment (Fig. 8 A). This indicates that L-aspartate stabilises calcium carbonate phases that are less stable than those precipitated without additives or in the presence of acetate. This is consistent with the findings of the simulations that aspartate has a stronger association with more disordered and solvated species (i.e. the stability of binding is Ca^{2+} > $CaCO_3^{(0)}$ > ACC ≫ calcite).

In addition to the above, we find that the pre-nucleation development in terms of the amount of free calcium ions (Fig. 8 A) is slightly flatter in both aspartate- and acetate-containing samples than in the reference case. This effect can be attributed to a decrease in ionic activity caused by the charged additives, as evidenced by the fact that these differences become larger when the concentration of the additives is increased. Moreover, flattening of the pre-nucleation regime was also observed when sodium chloride was added to the buffer (cf. ESI Fig. S4†), and titrations of aqueous additive solutions without carbonate confirm their non-ideality by revealing a consistent influence of ionic strength on the measured calcium potential (see ESI Fig. S5 & Table S4†). However, it is worth noting that these changes in the pre-nucleation slope may not exclusively be due to activity effects when it comes to the experiments in carbonate buffer. Comparing the data recorded for aspartate and acetate with those obtained in the presence of sodium chloride suggests that average slopes at a given concentration of aspartate and acetate are by trend smaller than those found for NaCl at the same ionic strength (see ESI Fig. S6 & Table S4†). We note that the differences do only slightly exceed the common limits of experimental error, while we regard the rather high number of experiments that would be required to achieve good statistical quantification of this minor effect unjustified. Therefore, we can only speculate that there may be a stabilising influence of these two additives on pre-nucleation clusters (apparent from the fact that more calcium is bound in clusters in their presence), which appears to be somewhat more distinct in the case of aspartate. This indicates that the degree of cluster stabilisation increases with the number of carboxylate functional groups in the molecular additives, given that citrate shows a pronounced effect (see below). However, such a stabilising effect cannot be found for the polyacids.[23,26] These findings agree with the results of the simulations for the association of the three organics with the $CaCO_3^{(0)}$ ion pair; acetate and aspartate have at best a weak association with this species, while citrate definitely is capable of binding. Given that larger pre-nucleation clusters adopt the flexible DOLLOP structure that includes ion pair chains, we would expect the influence of the organic species to be similar to that for the neutral ion pair, which is in accordance with the experimental findings.

When dilution effects are corrected for by calculation of actual free ion products (Fig. 8 B), it becomes evident that in the reference experiment, as well as in the presence of acetate, proto-calcite ACC (pc-ACC) is initially nucleated, in agreement with earlier findings.[19,23] During the early post-nucleation stage, this metastable intermediate appears to transform into more stable species, as the calculated ion product gradually approaches the solubility of vaterite (note that the concurrent formation of more stable species such as calcite is not necessarily reflected by the measured solubility, which is governed by the most soluble phase, because the system has not reached equilibrium according to Gibbs' phase rule). X-ray diffraction analyses of precipitates isolated after completed titration show that calcite and vaterite co-exist in all samples, at varying relative amounts (data not shown). We could not discern any clear trend that would indicate polymorph selection. Only in the case of aspartate, a distinct influence on the stability of the initially precipitated phase can be detected (Fig. 8 B). At 5 and 10 mM aspartate, the free ion product decreases very slowly to the solubility value corresponding to pc-ACC after nucleation: again, this suggests that more soluble (i.e. less stable) amorphous phases are temporarily stabilised by the additive. It remains unclear, though, whether (energetically downhill) transitions between different forms of ACC occur, or interactions of ACC with the additive impede the establishment of a constant solubility product via inhibition of growth (e.g. by adsorption onto nucleated particles).

The influence of citrate ions on the early stages of calcium carbonate crystallisation has been discussed already in previous studies,[26,81] and we summarise the main observations here. First, it was found that citrate and calcium ions associate in solution (complexation), which is the reason for the common application of citrate as an antiscalant. This effect manifested through a distinct *upward* bend of the pre-nucleation curve prior to becoming asymptotically linear, as a result of complexation of calcium by citrate retarding the onset of pre-nucleation cluster formation by reducing the concentration of free calcium.[23,26] In the present work, such association equilibrium between calcium and acetate, as well as aspartate, could neither be detected in carbonate buffers (Fig. 8 A) nor in pure water (ESI Fig. S5 and Table S4†). Experiments with citrate further revealed a distinct flattening of the pre-nucleation slope already at very low concentrations (50 mg L^{-1}, corresponding to ca. 0.26 mM), which was much more pronounced than what has been observed for 10 mM acetate and aspartate in this work. Although citrate carries three carboxylate groups that all are completely deprotonated at the investigated pH level, simple activity effects can be neglected due to the small concentration of additive used. Therefore, the flatter slope in the pre-nucleation stage indicated that more calcium was bound in clusters than in the absence of citrate or, in other words, that the clusters were stabilised by the additive. Quantitative evaluation suggested that this stabilisation was ca. 1–2 kJ mol^{-1} (relative to the additive-free case and in terms of the free enthalpy change for the formation of a CaCO$_3$ unit within a pre-nucleation cluster).[26] In contrast to aspartate, citrate was shown to only slightly inhibit nucleation, mainly as a consequence of calcium ion complexation, and moreover did not stabilise nucleated particles to any noticeable extent. In turn, it appeared to control the polymorphism of the final solid product, since pure-phase calcite was found in presence of citrate.[26]

4 Discussion

Surprisingly, the investigated molecular additives show quite marked differences with respect to their influence on the early stages of *in vitro* calcium carbonate crystallisation, although they essentially differ merely by the number of charged carboxylate functions (at 25 °C: acetate, pK_a = 4.76; L-aspartate, pK_a (α-COOH) = 1.95; pK_a (β-COOH) = 3.71; citrate, $pK_a(1)$ = 3.13, $pK_a(2)$ = 4.76, $pK_a(3)$ = 6.40).[82]

In terms of our recently introduced categorisation of additives,[23] L-aspartate can be classified as a type III (inhibition of nucleation), type IV (stabilisation of amorphous intermediates), and possibly also weak type II additive (stabilisation of pre-nucleation clusters). The latter could also apply for acetate, which otherwise did not show any significant effect. Citrate belongs to types I/II/III/V (complexation of calcium, stabilisation of pre-nucleation clusters, inhibition of nucleation, polymorph selection toward calcite). Interestingly, while both aspartate and citrate are capable of inhibiting nucleation, this effect is not necessarily based on ion complexation, as classically assumed. The titration data recorded in the present work demonstrates that aspartate does not form complexes with calcium under the given experimental conditions, whereas citrate–calcium association is evident already at the lowest concentrations investigated.[26] Finally, our results suggest that of the three additives considered here only aspartate can stabilise intermediate phases. In this regard, already monomeric L-aspartate appears to exhibit additive properties that are related to its macromolecular analogue, poly(aspartic acid), which has proven to be capable of stabilising an unstable intermediate[26] that presumably corresponds to PILP.[25] The atomistic basis underlying the substantially different effects exerted by these rather simple molecules is at first puzzling.

The experimental finding that acetate does not interact with the species present during the early stages of precipitation is largely in accordance with the computer simulations. When it comes to interactions with the surface of ACC, the simulations do not show a distinct difference between acetate and aspartate since both only show a weak solvent separated association. Given that the strongest difference in behaviour between acetate and aspartate occurs when they bind to calcium, either alone or in an ion pair (or potentially DOLLOP through association with other ion pairs), this suggests that the stabilisation of nucleated amorphous nanoparticles may be based upon additional interaction with other species. In previous work[21] it was shown that ion pairs of calcium carbonate readily bind to the surface of ACC nanoparticles, and so it may be possible that calcium–organic ion pairs may also do so.

Because the association of aspartate with $Ca^{2+}_{(aq)}$ is stronger than for acetate, while it is the opposite for $CaCO_3^{(0)}$, this suggests that the altered stabilisation of amorphous nanoparticles is more likely to occur through attachment of calcium–organic ion pairs to the nanoparticle surface.

The simulations show that citrate forms complexes with calcium ions, and moreover indicate that it interacts more strongly than the other additives with pre-nucleation clusters, but not with the ordered surfaces of crystalline forms of calcium carbonate. The strong interaction of citrate with pre-nucleation clusters may rationalize why the clusters are experimentally found to be stabilised, while the weaker interaction with aspartate might be reflected in the experimental findings as well. Experimentally, both aspartate and citrate do interact with pre-nucleation clusters and amorphous nanoparticles, however the effects are distinct (aspartate: inhibition of nucleation and stabilisation of amorphous forms; citrate: minor inhibition of nucleation, mostly based on a lowered level of supersaturation due to complexation of ions). That the inhibition of nucleation and the stabilisation of amorphous intermediates would be more efficient at intermediate strength of interactions is counter-intuitive. Since the mechanism of calcium carbonate nucleation has yet to be determined within the simulations, it is difficult to unambiguously explain this effect based on atomistic considerations. However, there are at least two possible hypotheses. Firstly, citrate binds more strongly to pre-nucleation species than post-nucleation forms, while aspartate binds with a similar exothermicity to all species. Hence the concentration of aspartate on ACC nanoparticles may be higher due to less competition. Secondly, since nucleation of amorphous calcium carbonate nanoparticles appears to proceed *via* aggregation of pre-nucleation clusters,[83] aspartate may be effective through colloidal stabilisation. Binding of the additives to DOLLOPs may render their aggregation to yield larger ACC nanoparticles difficult. In this context it is important to note that aspartate appears to be distinct from citrate as it readily assembles into supramolecular polymers too. These species may be much more efficient in colloidal stabilisation than monomers, perhaps in coarse analogy to the steric stabilisation of nanoparticle dispersions by polymers. The tendency of aspartate to form clusters may moreover explain the fact that it can stabilise amorphous nanoparticles efficiently. However, it should be noted that the mere occurrence of clusters of an organic additive is not necessarily connected to a clear and distinct influence on the early stages of calcium carbonate crystallisation. The geometry and energetics of all possible interactions (that is, with ions, ion pairs, DOLLOP, ACC and crystalline forms) are expected to be different and dependent on the particular additive. This is comparable to the situation for "classical" polymers that are joined by covalent bonds; polymeric additives with the same chemical function but different backbones can have distinct "fingerprints" during the early stages of calcium carbonate crystallisation, for instance, poly(aspartic acid) and poly(acrylic acid).[23,26] Thus it is also possible that certain supramolecular polymers of organic additives do not exhibit any influence on the early stages of precipitation of calcium carbonate at all. Of the two hypotheses mentioned above, the simulation data is consistent with the first proposal in showing that competition between species may result in the selectivity for post-nucleation species, while the second hypothesis has yet to be tested theoretically.

Last, but not least, it may be speculated that the ability of citrate to interact with all species that occur during the early stages of calcium carbonate crystallisation is the basis of polymorph selection toward calcite by this additive; the interactions may lead to the stabilisation of proto-calcite[18] structural forms throughout the different stages of precipitation. Indeed the scenario here may have parallels with that proposed by Freeman *et al.*[22] for the egg shell protein ovocledin-17 and ACC. Citrate initially interacts favourably with the pre-nucleation clusters and perhaps the early amorphous phases, and may preferentially alter their stability. However, when nucleation to the crystalline phase occurs then citrate would be expelled since it fails to bind to ordered flat surfaces. Given the lack of binding to the majority of the basal surface of calcite (*i.e.* away from steps and kinks) substantial energy would be released through the dissolution of citrate from this polymorph that might drive its formation.

PR, RD and JDG thank the Australian Research Council and Curtin University for funding, as well as iVEC and NCI for the provision of computer time. MK appreciates funding of a postdoc position by BASF SE. DG thanks Markus Antonietti for a loan titration system, and Helmut Cölfen for his support. TRW and LBW acknowledge the EPSRC Programme Grant 'Hard–Soft Interfaces: From Understanding to Engineering' (EP/I001514/1), and HECToR, the national high-performance computing facility of the U.K. LBW thanks the EPSRC for PhD support *via* a DTA grant.

References

1 H. Lowenstam and S. Weiner, *On biomineralization*, Oxford University Press, New York, 1989.
2 L. Addadi, D. Joester, F. Nudelman and S. Weiner, *Chem.–Eur. J.*, 2006, **12**, 980–987.
3 J. Aizenberg, G. Lambert, S. Weiner and L. Addadi, *J. Am. Chem. Soc.*, 2002, **124**, 32–39.
4 J. Aizenberg, A. Tkachenko, S. Weiner, L. Addadi and G. Hendler, *Nature*, 2001, **412**, 819–822.
5 F. C. Meldrum and H. Cölfen, *Chem. Rev.*, 2008, **108**, 4332–4432.
6 L. B. Gower, *Chem. Rev.*, 2008, **108**, 4551–4627.
7 N. A. J. M. Sommerdijk and H. Cölfen, *MRS Bull.*, 2010, **35**, 116–121.
8 C. J. Stephens, S. F. Ladden, F. C. Meldrum and H. K. Christenson, *Adv. Funct. Mater.*, 2010, **20**, 2108–2115.
9 R.-Q. Song and H. Cölfen, *CrystEngComm*, 2011, **13**, 1249–1276.
10 C. C. Tang, S. P. Thompson, J. E. Parker, A. R. Lennie, F. Azough and K. Kato, *J. Appl. Crystallogr.*, 2009, **42**, 225–233.
11 M. Neumann and M. Epple, *Eur. J. Inorg. Chem.*, 2007, **2007**, 1953–1957.
12 J. Wang and U. Becker, *Am. Mineral.*, 2009, **94**, 380–386.
13 R. Demichelis, P. Raiteri, J. D. Gale and R. Dovesi, *CrystEngComm*, 2012, **14**, 44–47.
14 L. Addadi, S. Raz and S. Weiner, *Adv. Mater.*, 2003, **15**, 959–970.
15 A. V. Radha, T. Z. Forbes, C. E. Killian, P. U. P. A. Gilbert and A. Navrotsky, *Proc. Natl. Acad. Sci. U. S. A.*, 2010, **107**, 16438–16443.
16 E. Loste, R. M. Wilson, R. Seshadri and F. C. Meldrum, *J. Cryst. Growth*, 2003, **254**, 206–218.
17 R. S. K. Lam, J. M. Charnock, A. Lennie and F. C. Meldrum, *CrystEngComm*, 2007, **9**, 1226–1236.
18 D. Gebauer, P. N. Gunawidjaja, J. Y. P. Ko, Z. Bacsik, B. Aziz, L. J. Liu, Y. F. Hu, L. Bergström, C. W. Tai, T. K. Sham, M. én and N. Hedin, *Angew. Chem., Int. Ed.*, 2010, **49**, 8889–8891.
19 D. Gebauer, A. Völkel and H. Cölfen, *Science*, 2008, **322**, 1819–1822.
20 R. Demichelis, P. Raiteri, J. D. Gale, D. Quigley and D. Gebauer, *Nat. Commun.*, 2011, **2**, 590.
21 P. Raiteri and J. D. Gale, *J. Am. Chem. Soc.*, 2010, **132**, 17623–17634.
22 C. L. Freeman, J. H. Harding, D. Quigley and P. M. Rodger, *Angew. Chem., Int. Ed.*, 2010, **49**, 5135–5137.
23 D. Gebauer, H. Cölfen, A. Verch and M. Antonietti, *Adv. Mater.*, 2009, **21**, 435–439.
24 D. Gebauer, A. Verch, H. G. Börner and H. Cölfen, *Cryst. Growth Des.*, 2009, **9**, 2398–2403.
25 L. B. Gower and D. J. Odom, *J. Cryst. Growth*, 2000, **210**, 719–734.
26 A. Verch, D. Gebauer, M. Antonietti and H. Cölfen, *Phys. Chem. Chem. Phys.*, 2011, **13**, 16811–16820.
27 P. Raiteri, J. D. Gale, D. Quigley and P. M. Rodger, *J. Phys. Chem. C*, 2010, **114**, 5997–6010.
28 Y. Wu, H. L. Tepper and G. A. Voth, *J. Chem. Phys.*, 2006, **124**, 024503.
29 J. A. Griffiths and D. M. Heyes, *Langmuir*, 1996, **12**, 2418–2424.
30 P. Dauber-Osguthorpe, V. A. Roberts, D. J. Osguthorpe, J. Wolff, M. Genest and A. T. Hagler, *Proteins: Struct., Funct., Genet.*, 1988, **4**, 31–47.
31 N. H. de Leeuw and T. G. Cooper, *Cryst. Growth Des.*, 2004, **4**, 123–133.
32 U. Aschauer, D. Spagnoli, P. Bowen and S. C. Parker, *J. Colloid Interface Sci.*, 2010, **346**, 226–231.
33 S. L. Mayo, B. D. Olafson and W. A. Goddard, *J. Phys. Chem.*, 1990, **94**, 8897–8909.

34 D. M. Duffy and J. H. Harding, *J. Mater. Chem.*, 2002, **12**, 3419–3425.
35 D. Quigley, P. M. Rodger, C. L. Freeman, J. H. Harding and D. M. Duffy, *J. Chem. Phys.*, 2009, **131**, 094703.
36 B. R. Brooks, R. E. Bruccoleri, B. D. Olafson, D. J. States, S. Swaminathan and M. Karplus, *J. Comput. Chem.*, 1983, **4**, 187–217.
37 L. M. Hamm, A. F. Wallace and P. M. Dove, *J. Phys. Chem. B*, 2010, **114**, 10488–10495.
38 R. A. Metzler, G. A. Tribello, M. Parrinello and P. U. P. A. Gilbert, *J. Am. Chem. Soc.*, 2010, **132**, 11585–11591.
39 W. D. Cornell, P. Cieplak, C. I. Bayly, I. R. Gould, K. M. Merz, D. M. Ferguson, D. C. Spellmeyer, T. Fox, J. W. Caldwell and P. A. Kollman, *J. Am. Chem. Soc.*, 1995, **117**, 5179–5197.
40 C. L. Freeman, J. H. Harding, D. J. Cooke, J. A. Elliott, J. S. Lardge and D. M. Duffy, *J. Phys. Chem. B*, 2007, **111**, 11943–11951.
41 A. Pavese, M. Catti, S. C. Parker and A. Wall, *Phys. Chem. Miner.*, 1996, **23**, 89–93.
42 Y. Zhao and D. G. Truhlar, *Theor. Chem. Acc.*, 2007, **120**, 215–241.
43 A. V. Marenich, R. M. Olson, C. P. Kelly, C. J. Cramer and D. G. Truhlar, *J. Chem. Theory Comput.*, 2007, **3**, 2011–2033.
44 Y. Shao, L. F. Molnar, Y. Jung, J. Kussmann, C. Ochsenfeld, S. T. Brown, A. T. B. Gilbert, L. V. Slipchenko, S. V. Levchenko, D. P. O'Neill, R. A. DiStasio Jr, R. C. Lochan, T. Wang, G. J. O. Beran, N. A. Besley, J. M. Herbert, C. Yeh Lin, T. Van Voorhis, S. Hung Chien, A. Sodt, R. P. Steele, V. A. Rassolov, P. E. Maslen, P. P. Korambath, R. D. Adamson, B. Austin, J. Baker, E. F. C. Byrd, H. Dachsel, R. J. Doerksen, A. Dreuw, B. D. Dunietz, A. D. Dutoi, T. R. Furlani, S. R. Gwaltney, A. Heyden, S. Hirata, C.-P. Hsu, G. Kedziora, R. Z. Khalliulin, P. Klunzinger, A. M. Lee, M. S. Lee, W. Liang, I. Lotan, N. Nair, B. Peters, E. I. Proynov, P. A. Pieniazek, Y. Min Rhee, J. Ritchie, E. Rosta, C. David Sherrill, A. C. Simmonett, J. E. Subotnik, H. Lee Woodcock III, W. Zhang, A. T. Bell, A. K. Chakraborty, D. M. Chipman, F. J. Keil, A. Warshel, W. J. Hehre, H. F. Schaefer III, J. Kong, A. I. Krylov, P. M. W. Gill and M. Head-Gordon, *Phys. Chem. Chem. Phys.*, 2006, **8**, 3172.
45 W. Smith, *Mol. Simul.*, 2006, **32**, 933–933.
46 W. Smith and I. T. Todorov, *Mol. Simul.*, 2006, **32**, 935–943.
47 K. Vanommeslaeghe, E. Hatcher, C. Acharya, S. Kundu, S. Zhong, J. Shim, E. Darian, O. Guvench, P. Lopes, I. Vorobyov and A. D. Mackerell, *J. Comput. Chem.*, 2010, **31**, 671–690.
48 http://dogmans.umaryland.edu/~kenno/cgenff/index.html, 2012.
49 A. C. T. van Duin, S. Dasgupta, F. Lorant and W. A. Goddard, *J. Phys. Chem. A*, 2001, **105**, 9396–9409.
50 J. D. Gale and A. L. Rohl, *Mol. Simul.*, 2003, **29**, 291–341.
51 J. D. Gale, P. Raiteri and A. C. T. van Duin, *Phys. Chem. Chem. Phys.*, 2011, **13**, 16666.
52 H. Manzano, R. J. M. Pellenq, F.-J. Ulm, M. J. Buehler and A. C. T. van Duin, *Langmuir*, 2012, **28**, 4187–4197.
53 O. Rahaman, A. C. T. van Duin, W. A. Goddard and D. J. Doren, *J. Phys. Chem. B*, 2011, **115**, 249–261.
54 G. Bussi, D. Donadio and M. Parrinello, *J. Chem. Phys.*, 2007, **126**, 014101.
55 P. Raiteri, J. D. Gale and G. Bussi, *J. Phys.: Condens. Matter*, 2011, **23**, 334213.
56 S. Plimpton, *J. Comput. Phys.*, 1995, **117**, 1–19.
57 M. Bonomi, D. Branduardi, G. Bussi, C. Camilloni, D. Provasi, P. Raiteri, D. Donadio, F. Marinelli, F. Pietrucci, R. A. Broglia and M. Parrinello, *Comput. Phys. Commun.*, 2009, **180**, 1961–1972.
58 A. Barducci, G. Bussi and M. Parrinello, *Phys. Rev. Lett.*, 2008, **100**, 020603.
59 P. Raiteri, A. Laio, F. L. Gervasio, C. Micheletti and M. Parrinello, *J. Phys. Chem. B*, 2006, **110**, 3533–3539.
60 CPMD, IBM Corp. and Max-Planck-Institut für Festkörperforschung, Stuttgart, 2000.
61 A. D. Becke, *Phys. Rev. A: At., Mol., Opt. Phys.*, 1988, **38**, 3098–3100.
62 C. T. Lee, W. T. Yang and R. G. Parr, *Phys. Rev. B*, 1988, **37**, 785–789.
63 D. Vanderbilt, *Phys. Rev. B: Condens. Matter*, 1990, **41**, 7892–7895.
64 D. Gao, P. Svoronos, P. K. Wong, D. Maddalena, J. Hwang and H. Walker, *J. Phys. Chem. A*, 2005, **109**, 10776–10785.
65 E. C. Meng, P. Cieplak, J. W. Caldwell and P. A. Kollman, *J. Am. Chem. Soc.*, 1994, **116**, 12061–12062.
66 P. I. Nagy and B. Noszál, *J. Phys. Chem. A*, 2000, **104**, 6834–6843.
67 W. Sang-Aroon and V. Ruangpornvisuti, *Int. J. Quantum Chem.*, 2008, **108**, 1181–1188.
68 J. M. Wang, W. Wang, S. H. Huo, M. Lee and P. A. Kollman, *J. Phys. Chem. B*, 2001, **105**, 5055–5067.
69 A. M. Toth, M. D. Liptak, D. L. Phillips and G. C. Shields, *J. Chem. Phys.*, 2001, **114**, 4595.

70 R. G. Pearson, *J. Am. Chem. Soc.*, 1986, **108**, 6109–6114.
71 M. V. Fedotova and S. E. Kruchinin, *J. Mol. Liq.*, 2011, **164**, 201–206.
72 J. Wang, G. Román-Pérez, J. M. Soler, E. Artacho and M.-V. Fernández-Serra, *J. Chem. Phys.*, 2011, **134**, 024516.
73 E. Shock and C. M. Koretsky, *Geochim. Cosmochim. Acta*, 1995, **59**, 1497–1532.
74 A. K. Covington and E. Y. Danish, *J. Solution Chem.*, 2009, **38**, 1449–1462.
75 H. Schmidbaur, I. Bach, D. L. Wilkinson and G. Müller, *Chem. Ber.*, 1989, **122**, 1439–1444.
76 M. Kellermeier, R. Rosenberg, A. Moise, U. Anders, M. Przybylski and H. Cölfen, *Faraday Discuss.*, 2012, DOI: 10.1039/c2fd20060k.
77 P. Geissbühler, P. Fenter, E. DiMasi, G. Srajer, L. B. Sorensen and N. C. Sturchio, *Surf. Sci.*, 2004, **573**, 191–203.
78 H. H. Teng and P. M. Dove, *Am. Mineral.*, 1997, **82**, 878–887.
79 S. Elhadj, E. A. Salter, A. Wierzbicki, J. J. De Yoreo, N. Han and P. M. Dove, *Cryst. Growth Des.*, 2006, **6**, 197–201.
80 L. Brečević and A. E. Nielsen, *J. Cryst. Growth*, 1989, **98**, 504–510.
81 K.-J. Westin and Å. C. Rasmuson, *J. Colloid Inter. Sci.*, 2005, **282**, 370–379.
82 D. R. Lide, *CRC handbook of chemistry and physics: a ready-reference book of chemical and physical data 2002–2003*, CRC Press, Boca Raton, 2002.
83 D. Gebauer and H. Cölfen, *Nano Today*, 2011, **6**, 564–584.

PAPER

Control of the nucleation of sickle cell hemoglobin polymers by free hematin

Veselina Uzunova,[a] Weichun Pan,[a] Vassiliy Lubchenko[b] and Peter G. Vekilov*[ab]

Received 27th March 2012, Accepted 24th April 2012
DOI: 10.1039/c2fd20058a

The polymerization of sickle cell hemoglobin (HbS) in the erythrocytes of sickle cell anemia patients is the primary event in the pathophysiology of this debilitating and deadly disease. Correspondingly, the majority of the current clinical treatments rely on delaying HbS polymerization. In search of pathways towards novel, more efficient treatment strategies, we explore the mechanism of nucleation of the HbS polymers. Previous work has shown that this nucleation follows a two-step mechanism, whereby the polymers nucleate inside dense liquid clusters suspended in the solution and occupying about 10^{-5} of the solution volume. We show that free hematin, which is spontaneously released by HbS due to its intrinsic instability to autoxidation, accelerates by $\sim 100\times$ the rates of both nucleation and growth of the polymers and that its removal leads to complete arrest of HbS polymerization. Exploring the mechanism underlying these hematin effects, we show that hematin enhances the attraction between the hemoglobin molecules in the solution and this yields $\sim 100\times$ higher volume of the dense liquid clusters in which nucleation occurs. These findings suggest that the nucleation of sickle cell hemoglobin polymers and the ensuing pathology can be suppressed by controlling the release and concentration of hematin in the red blood cells.

Introduction

Sickle cell anemia is a hemolytic anemia with acute and chronic manifestations.[1] The current treatment options are administration of hydroxyurea and symptomatic drugs, blood transfusions, and bone marrow transplantation. All these therapeutic approaches have downsides (*e.g.*, less than half of the treated patients benefit from hydroxyurea[2]) or significant adverse effects—especially the bone marrow transplantation—and the search for alternative therapeutic options continues.

Many cellular and molecular factors are implicated in the pathogenesis of sickle cell anemia, but the polymerization of a mutant hemoglobin, called sickle cell hemoglobin (HbS, in which a glutamic acid residue at the sixth position of the β-chain is replaced with non-polar valine) expressed in the erythrocytes of sickle cell patients, is considered the primary event.[3] The polymers stretch and rigidify the erythrocytes and alter the normal composition of the erythrocyte membrane.[4] Results with transgenic mice expressing human HbS[5] have demonstrated that delaying polymerization prevents sickling crises. The mouse HbS was genetically modified: valine at the β6 position was left intact, but the residues, with which it forms hydrophobic contacts in the fiber, were modified. It was found that the additional mutations inhibit

[a]*Department of Chemical and Biomolecular Engineering, University of Houston, Houston, Texas, 77204-4004, USA*
[b]*Department of Chemistry, University of Houston, Houston, Texas, 77204-5003, USA*

incorporation of the modified HbS into the polymer. The modification was found to delay polymerization, strongly reduce the fraction of sickled red blood cells, and reduce the severity of sickle crises. Significantly, two features of the disease, red cell dehydration and the count of irreversibly sickled cells, sometimes considered as independent factors for the disease, were also reduced.[5] These results provide a cure for sickle cell anemia, at least in mice, other than bone marrow transplantation. The remarkable fact is that the cure works through delay of HbS polymerization. This delay was achieved through a genetic modification of hemoglobin, and it is likely that such gene therapy in humans will be delayed by many years. Still, this finding highlights the validity of suppression of HbS polymerization as a potential sickle cell cure.

In this paper, we discuss the mechanism of by which the presence of hematin in HbS solutions delays the nucleation of HbS polymers. Hematin may be released in sickle erythrocytes after autoxidation of hemoglobin to methemoglobin.[6] Hence, instead of Fe^{2+}, as in the heme hosted in hemoglobin, hematin contains Fe^{3+}. After the release of the oxidized heme from the globin chain, the ligand site occupied by histidine in native hemoglobin is taken by a hydroxyl ion. In the commercial preparation, that ligand site is occupied by a chloride anion and this form is referred to as hemin. In what follows, we add hemin to HbS solutions to model the effects of the presence of hematin produced after the release of heme from the hemoglobin. The differences between the effects of the two species are expected to be small.

The presence of hematin in the erythrocytes has been implicated in damage of the red cell membrane, leading to higher red cell adhesion to the endothelium.[7] Existing evidence suggests that higher frequency and severity of sickle cell crises in patients with equal expression of HbS, such as monozygotic twins, may be related to enhanced release of heme, exhibited as membrane-bound iron.[8] Thus, the instability of sickle cell hemoglobin, leading to enhanced hematin concentrations in the red cell cytosol has been identified as the second consequence of the sickle cell gene related to sickle cell pathology. In this way, the results presented here demonstrate the interactions between these two disease factors, which have hereto been considered independently.

Experimental methods

Determination of the rate of nucleation of sickle cell hemoglobin polymers

The kinetics of nucleation of sickle cell hemoglobin polymers was characterized by direct monitoring of a supersaturated solution of HbS. If the solution is held in a thin slide, individual fibers as short as 0.5–1 μm can be detected by differential interference contrast microscopy (DIC).[9,10] A typical sequence of images of the evolution of HbS polymerization is shown in Fig. 1, where the determinations of the fiber length, fiber orientation, and the time of appearance of individual fibers is also illustrated.

To characterize the kinetics of nucleation of HbS polymers, a sequence of images, such as those in Fig. 1, is collected at defined times after the solution is supersaturated. If polymerization is driven by low-to-moderate supersaturations, the fibers do not branch and the solid phase consists of separate HbS fibers.[11] Slightly higher supersaturations ensure spherulitic morphology of the HbS polymer phase, which are easier to detect. About 20–50 images, which span over ~10–20 s, are collected at each spot. The number of HbS spherulitic polymers in the illuminated area is counted in each image. The illuminated spot is moved to another, randomly selected location on the slide and the time evolution of the number of spherulites is recorded again.

Test showed that the individual nucleation events in this set-up are independent of one another, and the error introduced by the assumption that each spherulite is generated by a single homogeneous nucleation event is insignificant.[11] The statistics of nucleation of the sickle-cell hemoglobin polymers indicate that under the probed

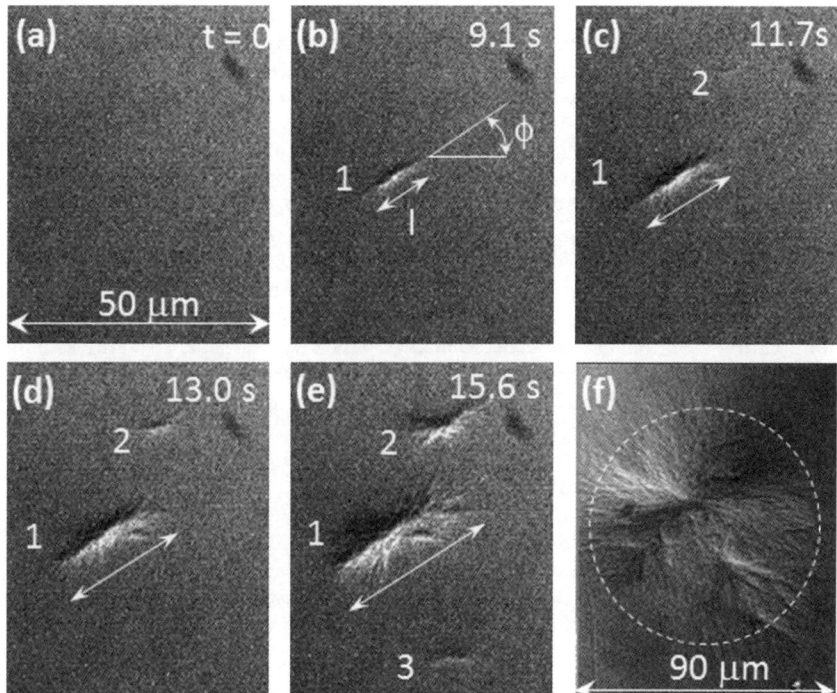

Fig. 1 Evolution of HbS polymerization monitored by differential interference contrast microscopy in a 25 μm thick slide of supersaturated HbS solution. Times for (a)–(e) indicated on panels, image in (f) corresponds to equilibrium between polymers and solution reached after ~1 min of polymerization. Elongated spherulites in (b)–(e) evolve into isometric spherulites in (f). Width of panels (a)–(e) is shown in (a). Determinations of fiber length L and orientation angle ϕ are illustrated in (b). Individual spherulites traced through (b)–(e) are labeled with numbers.

conditions it follows the general laws of nucleation of first-order phase transitions. The numbers of spherulites appearing at certain conditions for a certain time t after supersaturation was imposed were averaged. Since nucleation is intrinsically stochastic, for experimental statistics, 81 to 400 image sequences, similar to the one illustrated in Fig. 1 were collected, for details, see ref. 11,12. From each image, the number of HbS polymer spherulites, and the length and orientation of each spherulite were determined.

From the dependence of the mean number $\bar{N}$ of spherulites for each t, we determined the nucleation rate J and the nucleation delay time θ, as discussed in detail in ref. 11, 12. The delay time θ, defined from the dependencies of the number of nuclei on elapsed time after supersaturation is imposed, is a characteristic of the homogeneous nucleation only and is unaffected by the kinetics of growth and branching.[13] This delay time is modeled by nucleation theories and data on it allow critical test of these theories and the models that they represent. The orientation of a spherule was persistent between images belonging to the same sequence. The increase in length of the individual spherulites ΔL was used to determine the growth rate R: if both edges were at identical transport conditions, $R = \Delta L/2\Delta t$, where Δt is the time between the capture of the two images.

Static and dynamic light scattering

The formation of dense liquid clusters in deoxy-HbS solutions, held in sealed cuvettes, was characterized by dynamic light scattering;[14] interactions between deoxy-HbS molecules in solution and their modifications by free heme were characterized by static light scattering.[15,16] Both methods employed a ALV-5000/EPP static and dynamic light scattering device (ALV-Gmbh, Langen, Germany) with a 35 mW He–Ne laser operating at wavelength $\lambda = 632.8$ nm (Uniphase), for further experimental details, see ref. 14, 15.

The two-step mechanism of nucleation of HbS polymers

Nucleation of ordered solid phases in solution

Historically, it has been implicitly or explicitly assumed that the nucleation of sickle cell hemoglobin polymers is a one-step process: the disordered HbS molecules from the solution assemble into an ordered nucleus which has the same structure as long HbS fibers. An alternative mechanism was suggested by recent results on another first-order phase transition with proteins: formation of crystals of the protein lysozyme. Both experiment and theory revealed that, in this case, the formation of dense liquid droplets may precede and facilitate the formation of ordered nuclei as illustrated in Fig. 2.[17–22] The dense liquid may in some cases be stable with respect to the dilute solution,[23,24] or, in other cases, it may be metastable, Fig. 2b.[20,25,26]

The evidence for the applicability of this mechanism to the nucleation of crystals of other proteins is less direct. In ref. 27, crystals of several intact immunoglobins were found to coexist for extended lengths of time with dense liquid droplets without the droplets generating additional crystal nuclei. The crystals that were nucleated on

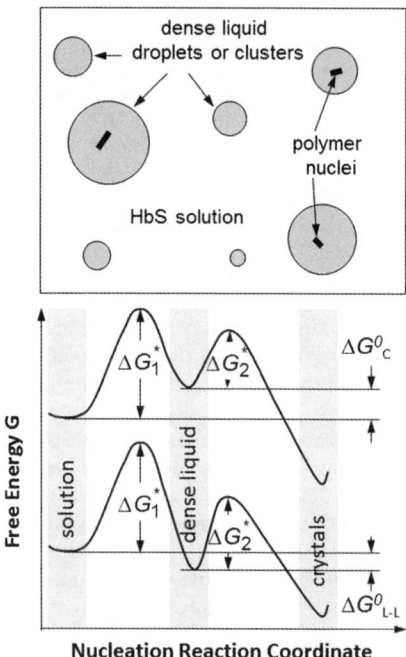

Fig. 2 (a) Schematic illustration of the two-step mechanism of nucleation of HbS polymer fibers. Step 1 is the formation of dense liquid droplets. Step 2 is the formation of fiber nuclei within these droplets. (b) The free-energy ΔG along two possible versions of the two step nucleation mechanism. If dense liquid is unstable and $\Delta G^0_{L-L} > 0$ (ΔG^0_{L-L}—standard free energy of formation of dense liquid phase), dense liquid exists as mesoscopic clusters, ΔG^0_{L-L} transforms to ΔG^0_C, and upper curve applies; if dense liquid is stable, $\Delta G^0_{L-L} < 0$, reflected by lower curve. ΔG^*_1 is the barrier for formation of a cluster of dense liquid, ΔG^*_2—for the formation of an ordered nucleus inside the dense liquid.

the droplet boundaries grew into the dilute solution, rather than into the dense liquid. This was interpreted in favor of nucleation of the crystals within dense liquid clusters suspended in the solution.

Besides the nucleation of protein crystals, studies have shown that the nucleation of amyloid fibrils of several proteins and peptide fragments, such as Alzheimer-causing A–β peptide or the yeast prion protein follows a variant of the two-step mechanism in which the role of the intermediate liquid state is played by a molten globule of consisting of unfolded protein chains.[28,29]

The applicability of the two-step mechanism to the nucleation of crystals of urea and glycine was deduced in a series of experiments, in which high power laser pulses were shone on supersaturated solutions.[30,31] It was fond that the nucleation rate increases as a result of the illumination by eight–nine orders of magnitude and that by using elliptically or linearly polarized light, α- or γ- glycine crystals could be preferentially nucleated. Since glycine does not absorb the illumination wavelength, and the electric field intensity was insufficient to orient single glycine molecules, it was concluded that the elliptically or linearly polarized pulses stabilize the structure fluctuations within the dense liquid, which lead to the respective solid phases.[30,32]

Colloid systems are the ones for which the evidence in favor of the applicability of the two-step mechanism is the strongest. By tracking the motions of individual particles of size a few microns by scanning confocal microscopy, the nucleation of crystals in colloidal solutions was directly observed.[33–35] These experiments revealed that the formation of crystalline nuclei occurs within dense disordered and fluid regions of the solution.[36]

The role of an amorphous precursor in the nucleation of crystal of biominerals has been speculated for a long time, for a historic overview, see ref. 37. However, it was envisioned that the precursor does not facilitate the formation of the crystalline nuclei, but only serves as a source of material for re-precipitation into a crystalline phase. Only recently it was shown that amorphous or liquid clusters of calcium and carbonate ions are present in calcium carbonate solutions and facilitate the nucleation of calcite crystals, in a manner similar to the role of the mesoscopic clusters in lysozyme crystallization discussed above.[37–41] The free energy landscape along the nucleation reaction pathway in Fig. 2b was used to characterize kinetics of the process of calcite crystallization.[39]

A two-step nucleation mechanism going through metastable clusters (in this case, swollen micelles) has also been theoretically predicted for a ternary system of two homopolymers and their block-copolymer.[42]

Stable dense liquid was found to exist in solutions of organic materials and serve as location where crystals nucleate and grow.[43] The existence of the dense liquid in these solutions has been attributed to the same fundamental physical mechanism as the one acting in protein solutions: the size of the solute molecules is larger than the characteristic lengthscale of the intermolecular interactions in the solution.[44] On the other hand, unpublished evidence from the pharmaceutical industry suggests that in many other cases the stable dense liquid, referred to as "oil" by the practitioners in the field, is so viscous that no crystals can form in it. While this has not been tested, it is possible that the two-step mechanism operates in these organic systems by utilizing dense liquid clusters, similar to those seen in protein, colloid, and calcium carbonate solutions.

The broad variety of systems in which the two-step mechanism operates suggests that its selection by the crystallizing systems in preference to the nucleation of ordered phases directly from the low-concentration solution may be based on general physical principles. This idea is supported by two examples of physical theory: by Sear[45] and by Lutsko and Nicolis.[21] Of particular interest is the latter work. It treated a range of points in the phase diagram of two different model systems which likely encompass a broad variety of real solutions and demonstrated that the two-step formation of crystalline nuclei, *via* a dense liquid intermediate, encounters a significantly lower barrier than the direct formation of an ordered nucleus and should be faster. Interestingly, the intermediate state resulting from the theory was not stabilized and represents a just a well-developed density fluctuation.

In view of the broad applicability of the two-step mechanism of nucleation of ordered solid phases form solution, it is not surprising that the this mechanism also applies to the nucleation of sickle cell hemoglobin fibers.[12,46] Below, we discuss the most important aspect of the applicability of this mechanism to HbS polymer nucleation: the nature of the metastable dense liquid clusters, in which the HbS polymers nucleate.

Dense HbS liquid and metastable dense liquid clusters

The first question on the applicability of the two-step nucleation mechanism to HbS polymers is whether dense liquid droplets exist in HbS solutions. Initially, stable dense liquid phases were sought. A stable phase would form macroscopic, observable domains, which would reversibly appear and disappear upon, *e.g.*, raising or lowering of temperature. Dense liquid phases were directly observed in solutions of oxy, carbomonoxy, and deoxy hemoglobin S, and of oxy- and deoxy-HbA containing 0.1, 1, or higher % PEG. In such solutions, in a few seconds after the temperature was raised to a value around 41–42 °C, droplets of a few tenths of a micrometer were detected and quickly grew to sizes of about 1–5 μm, Fig. 3a. The formation of a dense liquid phase is reversible and represents a true thermodynamic phase transition. It was found that it is very sensitive to the solution composition. Tests revealed that in solutions of any of the tested Hb variants liquid–liquid separation cannot be observed unless at least 0.1% (w/v) of polyethylene glycol (PEG) is present. The need for PEG was attributed to the net repulsion between the Hb molecules in pure Hb solution.[16,47] Addition of low concentrations of PEG introduces attraction into the intermolecular interaction potentials.[17,48]

When samples with deoxy-HbS in concentrations above ~180 mg ml^{-1} are rapidly cooled to ~30 °C after a dense liquid phase has formed at $T > 40$ °C, the droplets that have not disappeared serve as an initiator for the formation of deoxy-HbS polymers, Fig. 3b and c. Cooling to temperatures around 5 °C dissolves the deoxy-HbS polymers. Rapid re-heating to $T > 40$ °C leads to dissolution of the HbS polymers and new liquid–liquid phase separation. The observations illustrated in Fig. 3 suggest that if dense liquid droplets form in a supersaturated deoxy-HbS solution, they facilitate the formation of the HbS polymers.

The above observations show that a *stable* dense liquid phase does not exist in solutions of deoxy-HbS without PEG, in which the nucleation rate measurements are typically carried out and which are similar to the red cell cytosol. This conclusion may have put an end to the discussion of the two-step mechanism for HbS polymers,

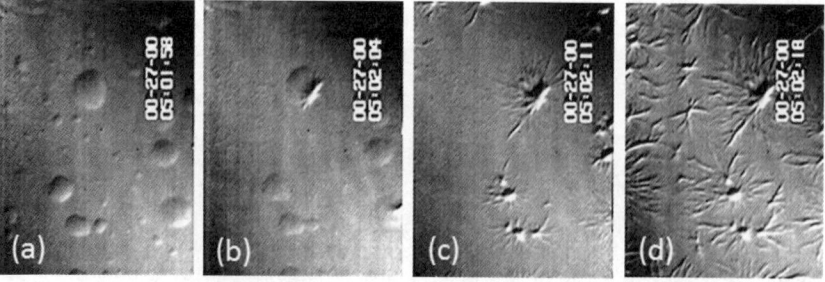

Fig. 3 Droplets of dense deoxy-HbS liquid disappear or convert to polymer spherules upon cooling from 40 °C to 35 °C in (b)–(d). HbS concentration in the starting solution is 220 mg ml^{-1} + 1% (w/v) PEG 8000. Vertical dimension is 105 μm.

if it were not for recent findings of *metastable* dense liquid phases in protein solutions: of the proteins lumazine synthase[49,50] and lysozyme.[51]

To test for metastable dense liquid clusters, within which the HbS polymer nucleation may occur, deoxy-HbS solutions were monitored by dynamic light scattering.[14] Fig. 4a shows a typical intensity correlation function of such solution (correlation functions with similar features were also seen in oxy-HbS and oxy-HbA solutions). The correlation function reveals two processes: the one with characteristic time of ~0.04 ms is the Brownian motion of single HbS molecules and it is present at all solution concentrations. A second process has a longer characteristic time and its amplitude increases with higher hemoglobin concentrations. It was shown that this slower time corresponds to HbS clusters suspended in the HbS solution, and not to single HbS molecules embedded in a loose network structure constraining their free diffusion.

The mean hydrodynamic radius of the cluster population R_2 and the volume fraction ϕ occupied by the clusters are determined from the correlation functions; the time-dependence of ϕ is shown in Fig. 4b. It was found that with all three Hb variants, the clusters exist in broad temperature and Hb concentration ranges. The lower bond of the cluster lifetime was found to be 15 milliseconds. The HbS clusters occupy $\phi = 10^{-6}$–10^{-5} of the solution volume, Fig. 4.

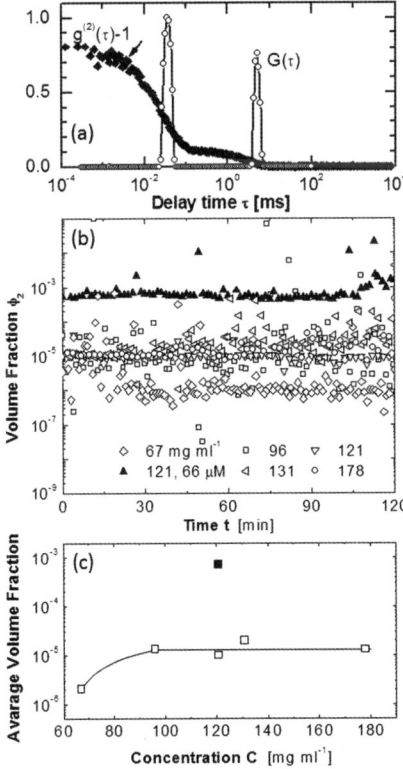

Fig. 4 Light scattering characterization of dense liquid clusters. (a) A correlation function (diamonds) and the respective delay time distribution function (open circles) of a deoxy-HbS solution with $C_{HbS} = 67$ mg ml^{-1}. (b) Time dependence of volume fraction ϕ_2 occupied by clusters, determined as in ref. 14 at five HbS concentrations, shown in legend, and in the presence of shown concentration of heme at one of these HbS concentrations. (c) The dependence of the average volume fraction from (b) on HbS concentration. Solid symbol at 121 mg ml^{-1}: in the presence of 66 μM of free heme; open symbols: in the absence of heme. Line is just a guide for the eye.

The existence of the clusters in illustrated in Fig. 4 resolves the controversy between the kinetic evidence, discussed above and in ref. 12,46, in favor of a nucleation mechanism of HbS polymers involving dense liquid droplets, and a previous observation of a lack of dense liquid phases in HbS solutions.[47]

Similar clusters of protein-rich liquids have been detected in solutions of several proteins: lumazine synthase,[49,50] lysozyme,[52,53] and others, at electrolyte concentrations ranging from 20 mM[53] to 1.3 M,[49,50] by atomic force microscopy,[49] dynamic light scattering,[14,50,52,53] and UV resonance Raman spectroscopy.[54] The clusters are mesoscopic in size, from under one hundred to several hundred nanometers,[50,53,55] and are liquid in nature.[50,56] The clusters occupy a low fraction of the solution volume: from $\sim 10^{-7}$ (below which they are not reliably detectable) to $\sim 10^{-3}$. The cluster volume fraction remains at these low levels even as the protein concentration in the bulk solution approaches that of the liquid within the clusters.[53,55] The mesoscopic clusters exist with similar characteristics both in the homogeneous region of the phase diagram of the protein solution (where no condensed phases, liquid or solid, are stable or present as long-lived metastable domains) and under conditions supersaturated with respect to ordered solid phases, such as crystals[50] or, as here, sickle cell hemoglobin polymers.[14,46] According to the two-step nucleation mechanism, in supersaturated solutions the clusters are crucial sites for the nucleation of ordered solid phases.[46,57,58] The protein clusters are similar to, but larger than clusters found in solutions of biominerals,[37–39] organic molecules,[30,59] polymers,[42] and colloids,[33–35] where they also play a crucial role in crystal nucleation.

The anomalous mesoscopic clusters are of interest because their existence challenges our understanding of phases and phase equilibria. Furthermore, insight into cluster behavior may provide a means to control nucleation, one of the most secretive processes in physics, chemistry, and materials science. Recent studies reveal that the small fraction of protein contained in the clusters reflect the significant free energy cost of increasing the protein concentration.[53] The mesoscopic size of the clusters is significantly greater than the predictions of several mechanisms of cluster formation in protein and colloid solutions.[53,60] If these mechanisms are inactive, due to the free energy excess of high concentration protein liquid, the clusters should consist of just a few protein molecules, as indicated by a straightforward thermodynamic evaluation.[53] To solve the puzzle of the cluster size, we argued in ref. 53 that the clusters largely consisted of transient protein complexes that formed at high protein concentrations. Several types of interactions: hydration, electrostatic, and the formation of domain-swapped dimers after partial protein misfolding, were considered and tested as possible bonds between the protein molecules in the transient complexes.[53]

Several of the predictions of the theory can be tested against the experimental observations of clusters in HbS solutions. The theory predicts that the cluster size is of order several hundred nanometers, and the clusters are composed of dense protein liquid: both of these predictions agree with the observations with HbS.[14] A important conclusion of the theory is that if the solution, in which the clusters form, has a concentration <100 mg ml^{-1}, the cluster volume fraction should be of the order 10^{-6} and it should increase as HbS concentration increases. However, if the HbS concentration is above 100 mg ml^{-1}, the cluster volume fraction ϕ_2 should be constant. Fig. 4b and c demonstrate that as HbS concentration increases from 67 to 96 mg ml^{-1}, ϕ_2 increases from $\sim 10^{-6}$ to $\sim 10^{-5}$. Further increase in HbS concentration to 121, 131, and 178 mg ml^{-1} (HbS solubility at 22 °C is 185 mg ml^{-1}) does not lead to increasing φ_2, *i.e.*, the cluster volume fraction follows the non-trivial prediction of the theory.

The lack of increase of the cluster volume fraction at solution concentrations above 100 mg ml^{-1} is similar to the observation with the protein lysozyme and was attributed to the similarity of the solution concentration to that in the clusters, which leads to a spinodal-like regime of cluster generation.[53]

Another prediction of the cluster theory in ref. 53, which can be verified using the data in Fig. 4 is that cluster formation is reversible and the clusters are in metastable equilibrium with the solution. Correspondingly, the fraction of the hemoglobin in the clusters ν_2 complies with the Boltzmann distribution and can be evaluated as $\nu_2 \approx \exp(-\Delta G(C_L,C_H)/k_B T)$, where $\Delta G(C_L,C_H)$ is the free energy excess of the dense liquid with concentration C_H over the HbS solution, with concentration C_L; $\nu_2 = C_H\varphi_2/(C_L\varphi_1 + C_H\varphi_2) \cong C_H\varphi_2/C_L\varphi_1$ is of the same order of magnitude, but somewhat higher than the ratio $\varphi_2 : \varphi_1$.

To evaluate ΔG, we use the dependence of the KC/R_θ ratio on protein mass concentration C (K is a device constant, R_θ is the Raleigh ratio of the scattered to incident light[61]), determined by static light scattering[15,16] and displayed in Fig. 5; for details about static light scattering in solutions which absorb the wavelength used in the determinations, see ref. 53. This ratio is directly related to the inverse osmotic compressibility of the solution: $KC/R_\theta = (\partial\Pi/\partial C)/RT$ (Π is the contribution of the protein to the osmotic pressure, and R is the universal gas constant). After obtaining the compressibility $\partial\Pi/\partial C$, we can integrate it to compute the free energy $\Delta G = -\int_{C_L}^{C_H} \Pi dV + \Delta(\Pi V)$ needed to increase the concentration of N protein molecules from that in the dilute solution C_L to that in the dense liquid C_H. This allows us to determine ΔG as the free energy difference between states with densities C_L and C_H, per particle

$$\frac{\Delta G(C_L, C_H)}{N k_B T} = \int_{C_L}^{C_H} \frac{d\rho}{C^2}\int_0^C M_W\left(\frac{KC'}{R_\theta}\right)dC' + \Delta\left[\frac{1}{C}\int_0^C M_W\left(\frac{KC'}{R_\theta}\right)dC'\right]_{C_L}^{C_H}, \quad (1)$$

where M_W is the molar mass of the protein.

The experimentally determined dependence $KC/R_\theta(C)$ is fitted by a cubic polynomial and the integrated using eqn (1). The resulting $\Delta G/k_B T$, shown in Fig. 5, varies between $-3.7\ k_B T$ at $C = 67$ mg ml^{-1} and $-1.5\ k_B T$ at $C = 96$ mg ml^{-1}. Thus, the free energy of the solution would increase by $2.2\ k_B T$ upon this concentration increase. Since the individual clusters are not affected by the concentration of the solution which generates them,[53] this increase would lead to a decrease of the excess free energy of the clusters above that of the solution. According to the Boltzmann distribution, the corresponding decrease in cluster volume fraction would be about $9\times$, in agreement with the data in Fig. 4b and c. The absorbance of hemoglobin of the wavelength used in the determinations of $KC/R_\theta(C)$ does not allow

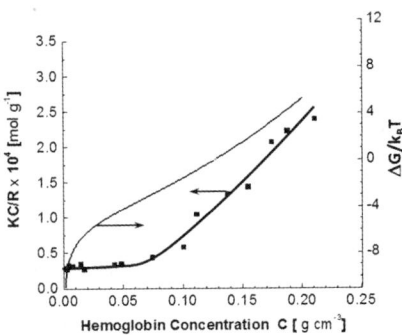

Fig. 5 (a)The dependence of the ratio $KC : R_\theta$ on hemoglobin concentration C. K—instrument constant, $R_\theta = I_\theta/I_0$—Raleigh ratio of intensities of scattered at angle $\theta = 90°$ I_θ and incident light I_0. (b) Integration of data in (a) according to eqn (1) to determine the solution free energy ΔG.

determinations at concentrations above 200 mg ml^{-1}. Hence, extrapolation to the suspected concentration in the clusters, ~400 mg ml^{-1}, would be unjustified and comparisons of the cluster volume fraction in Fig. 4b and c to the predictions of the Boltzmann law are not feasible.

The role of the hematin in polymerization

Enhanced polymerization of HbS upon ageing

If supersaturated solutions of freshly isolated and purified deoxy-HbS are held in slides and monitored, polymerization does not occur. Solutions prepared from oxy-HbS stock stored under liquid nitrogen at −198.5 °C for longer than a week exhibit slow nucleation and growth. Stocks older than a month yield nucleation and growth rates which are higher, compare Fig. 6a and c, reproducible, and equal to those in ref. 11, 12, 62 and very close to those in ref. 62, 63. In solutions prepared from oxy-HbS stock stored at room temperature for one day, more polymer fibers form and they grow larger for shorter time, i.e., the nucleation and growth rates are faster, compare Fig. 6a and b. The enhanced rates of nucleation and growth is solutions prepared from oxy-HbS stock kept at room temperature, is similar to observation in ref. 64. If CO-HbS or deoxy-HbS stock solutions are stored at room temperature overnight, both acceleration and slow-down of polymerization was recorded, but the differences with the freshly prepared sample were small.

Free hematin and ageing

Characterization of the aged HbS solution by mass spectroscopy, Fig. 7, reveals that no unexpected species, e.g., pieces of protein chains, appear in the solution upon ageing, and that hematin is the only solution component, which appears to increase its concentration (mass spectroscopy cannot be used for quantification of the detected species). Fe^{3+}, while not seen in the mass spectra, has solubility in the attomolar range at the solution pH.

To remove the hematin released in HbS solutions during storage, we dialyzed the solutions. Spectroscopic characterization of the dialyzed HbS solution showed spectra identical to those prior to dialysis, except at $\lambda > 600$ nm, discussed in ref. 55. In the dialyzed solution, no polymers form, even at temperatures as high as 36.5 °C (higher concentrations and temperatures lead to faster polymerization); an illustrative example is shown in Fig. 6d. Eight combinations of HbS concentration and temperature were tested at which polymers appear within a few seconds in normally prepared HbS solutions, but did not appear after dialysis. The addition of 0.26 mM of hemin to the dialyzed solution led to nucleation and growth rates significantly faster than before dialysis, Fig. 6e.

Kinetics of nucleation and growth of HbS polymer fibers in the presence of hemin

Quantification of the kinetics of nucleation and growth in Fig. 8 shows that (i) the rates of nucleation and growth in solutions prepared from one-week old stock are lower and the nucleation delay times are longer than in solutions prepared from one-month old stock in ref. 11, 12, 46. As in Fig. 6, we conclude that the faster rates after longer storage are due to release of hematin, likely during thawing and re-freezing. (ii) After dialysis, no HbS polymers nucleate, i.e., $J = 0$, $\theta = \infty$, and the fiber growth rate R is undefined. (iii) The addition of 160 or 260 μM of hemin to dialyzed solutions enhances J and R by more than two orders of magnitude and shortens θ by approximately the same factor in comparison with the rates and times prior to hematin removal. Since according to the above determinations, in the undialyzed solutions hematin concentration is ~80 μM,

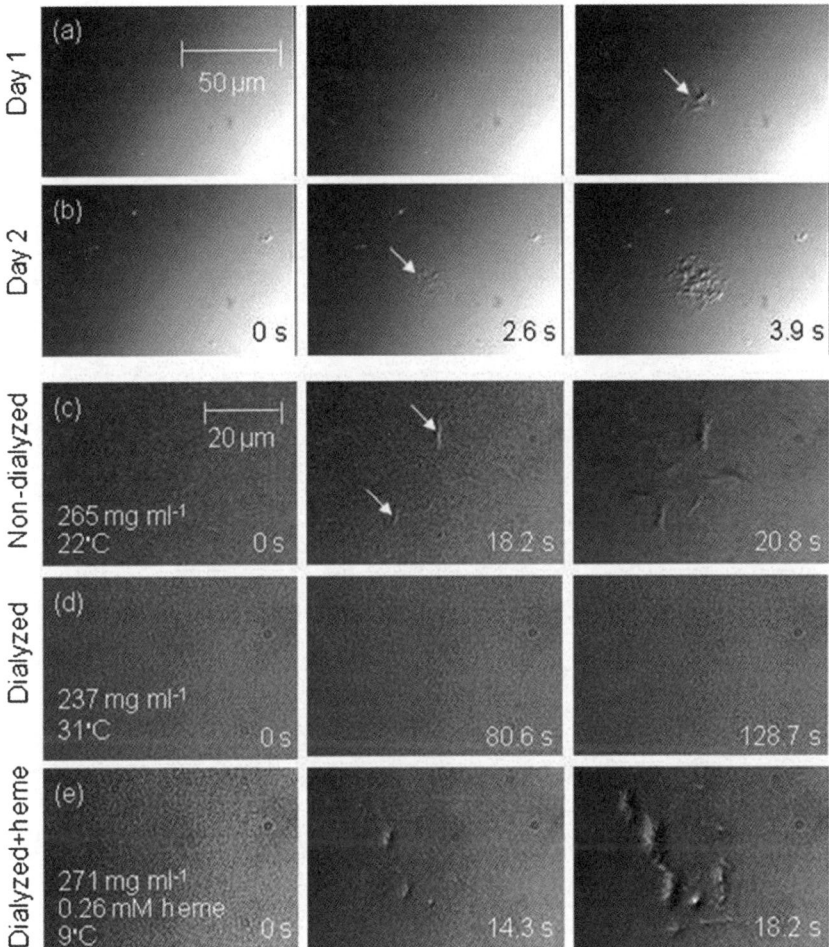

Fig. 6 Effects of aging and free heme on the polymerization of sickle cell hemoglobin. Solution of CO-HbS is held in slides of uniform 5 mm thickness. The panels in each row represent the evolution of polymerization under a set of conditions. Time after start of photolysis is indicated in each panel. HbS concentration and temperature for each row are indicated in left panel. Experiments in (a) and (b) are at 22 °C and with HbS concentration of 197 mg ml^{-1}. Scales for (a)–(b) and (c)–(e) are different. White arrows point at HbS polymer domains. (a)–(b) Evolution of polymerization in solution prepared from one-month old stock stored under liquid nitrogen: (a) freshly prepared solution; (b) solutions prepared from stock stored at room temperature for one day. (c)–(e) The effect of hematin: (c) HbS polymerization in a solution prepared from one-week old stock; (d) removal of small molecules by dialysis with molecular weight cut-off 2000 g mol^{-1} prevents polymerization even at higher temperature; (e) hemin at shown concentration was added, temperature was lowered to suppress polymerization.

this acceleration is consistent with higher polymerization rates at higher hematin concentrations.

These observations show that hematin accelerates HbS polymerization. Significantly, from the point of view of potential clinical applications, in the absence of hematin, polymerization is prevented under the conditions tested here. The low concentration of free heme, 160 or 260 μM, required for significant effect on nucleation and growth, would be released if, *e.g.*, <2% of HbS molecules, whose concentration is ~4 mM, lose their four hemes.

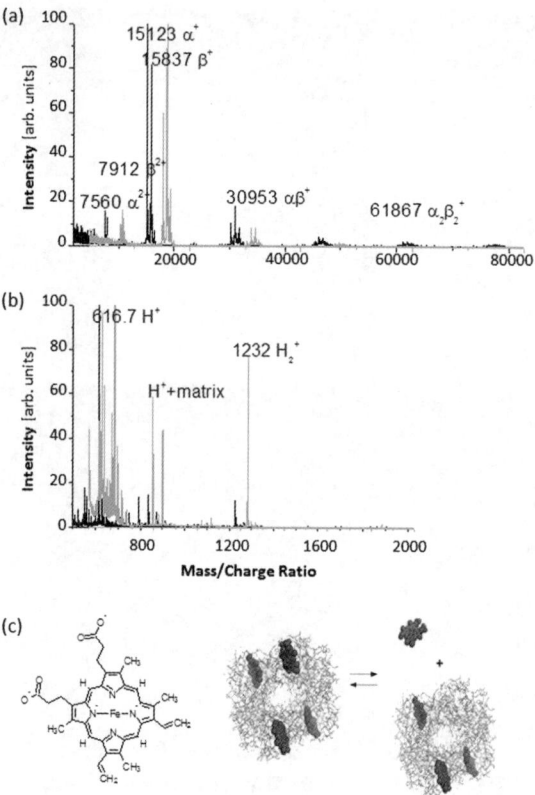

Fig. 7 Species released upon ageing of HbS at 22 °C. (a) Cation mass spectra of oxy-HbS, mass-to-charge ratio region 2000–80 000 g mol^{-1}. (b) Cation mass spectra of HbS in the mass-to-charge ratio region 450–2000 g mol^{-1}. The peaks corresponding to α and β-chains and their complexes and to hemin and its dimer are labeled. Black and grey denote, respectively, solutions prepared immediately after thawing of the stock and solutions kept at 22 °C for one day. The second-day spectra are shifted to the right for clarity. (c) The structure of heme and schematic of its release from hemoglobin.

The mechanism of free heme action

Free heme and the precursors for HbS polymer nuclei

Fig. 4b and c above show that the addition of hemin leads to an increase in the cluster volume fraction ϕ_2 by two orders of magnitude. The cluster radius is the same as in solutions without hematin, *i.e.*, the cluster number density also increases by two orders of magnitude. Since the nuclei of HbS fibers form inside the metastable dense liquid clusters,[46,47] the faster J and shorter θ recorded in the presence of hematin directly correlate to the increase in ϕ_2.

The factor of increase in the volume of the dense liquid clusters is equal to the factors of acceleration of HbS polymer nucleation and extension of the nucleation delay time. This correspondence renders powerful support for the action of the two-step nucleation mechanism in HbS polymer nucleation: since the clusters are the locations in which the polymers nucleate, this mechanism predicts that increasing their volume would lead to similar increase in the nucleation rate.

Thus, we have elucidated the first step in the mechanism of action of free heme. The next question, addressed below, is how does the free heme induce such increase in the cluster volume?

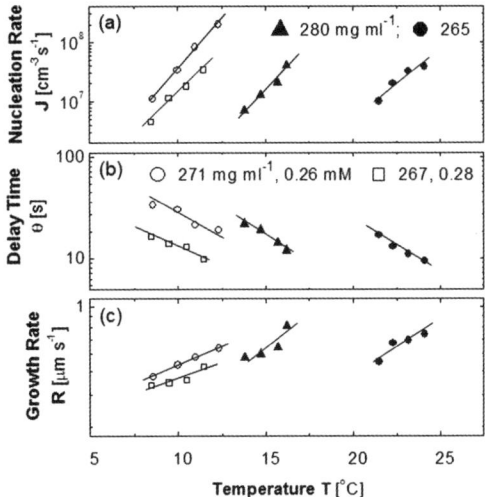

Fig. 8 The effect of free hemin on the nucleation rate J in (a) and delay time θ in (b), and growth rate R in (c) of HbS polymer fibers. HbS concentrations in mg ml^{-1} are shown in (a) and (b). Solid symbols: HbS solution in 0.15 M phosphate buffer, no hemin added. Open symbols: hemin at concentrations in mM shown in (b) was added to dialyzed solutions. Lines are just guides for the eye.

The role of hemin in the interactions between Hb molecules

Static light scattering is a powerful method to study interactions between protein molecules in solutions. Light scattering from multicomponent systems containing non-electrolytes[65,66] was described by Kirkwood and Goldberg by an expression similar to the virial expansion, Eq. (20) in ref. 65. In this expression, we neglect the quadratic terms C^2 and C_{hematin}^2, but leave the cross product $C\,C_{\text{hemin}}$. To justify these assumptions, we note that the molar concentrations $m_i = c_i/M_i$, i = hemin, HbS, of both hemin and hemoglobin are low: $m_{\text{hemin}} = 50$ μM, 80 μM $\leq m_{\text{HbS}} \leq$ 300 μM. However, due to its two carboxylic groups, at the chosen pH hemin has a greater charge, $-2e$ molecule^{-1}, than hemoglobin, which has about $-0.5e$ molecule^{-1} (e—elementary charge). As a result, the electrostatic repulsion from the hemin is stronger than from hemoglobin, but since $m_{\text{HbS}} > m_{\text{heme}}$, this repulsion is mostly felt by the Hb molecules. With these assumptions and with $\alpha = \dfrac{(\mathrm{d}n/\mathrm{d}C_{\text{hemin}})}{(\mathrm{d}n/\mathrm{d}C)} = const(C_{\text{hemin}}, C) = 5.6$ from determination in our laboratory (plots not shown), the expression of the dependence of the scattered light intensity on the mass concentrations C_{heme} and C becomes

$$\frac{KC}{\Delta R_\theta} = \frac{1}{M_{\text{HbS}}} + 2\alpha A_{12} C_{\text{heme}} + 2\left\{ A_{22} + \left[\left(1 + 2\alpha \frac{M_{\text{HbS}}}{M_{\text{heme}}}\right) A_{212} - \frac{M_{\text{HbS}}}{2 M_{\text{heme}}} A_{12}^2 \right.\right.$$
$$\left.\left. + \alpha M_{\text{HbS}} A_{12} A_{22} \right] C_{\text{heme}} \right\} C \qquad (2)$$

where $\Delta R_\theta = R_{\theta(1+2)} - R_{\theta 1} = (I_{\theta(1+2)} - I_{\theta 1})/I_0$: $R_{\theta(1+2)}$ and $I_{\theta(1+2)}$ are, respectively, the Rayleigh ratio and scattered intensity at angle θ for a solution containing both species and $R_{\theta 1}$ and $I_{\theta 1}$ are these variables in a the solution containing only hemin. A pair of subscripts of the virial coefficients A denotes that this coefficient relates to interactions between two molecules of the respective species. Thus, A_{22} characterizes interactions between two hemoglobin molecules; since the hemin concentration is low, its effects on these interactions are negligible and we assume $A_{22} \cong A_2$. The coefficient A_{212} characterizes three-body interactions between two hemoglobin and one hemin molecules.

If the dependence $KC/\Delta R_\theta(C)$ is linear, according to eqn (2), its intercept defines an apparent molecular mass $M_{HbS,app}$, and the slope—an effective second virial coefficient $A_{2,eff}$.

Both hemoglobin and free heme strongly absorb the illuminating wavelength and this absorption attenuates the intensity of the scattered light. It was shown that the errors in static and dynamic light scattering results due to absorption of light by the solution are minor.[67,68] The absorption modifies the Debye equation in a solution containing both hemoglobin and heme, but the absorption by these two molecules can be directly accounted for using the published values of their respective extinction coefficients at the wavelength of employed for light scattering: for oxy-HbA $\varepsilon_2 = 561$ M^{-1} cm^{-1}.[69] The extinction coefficient of heme at 632 nm, determined in our laboratory as $\varepsilon_1 = 5.89$ mM^{-1} cm^{-1}.

To probe the effects of the hemin on the intermolecular interactions, we first remove by dialysis any free hemin that may have been released by the hemoglobin molecules. Hemin at molar concentration $m_1 = 50$ μM ($c_1 = 0.032$ mg ml^{-1}) was added to HbS solutions, whose concentration c_2 range corresponds to 80 μM < m_2 < 300 μM. The Debye plots for deoxy-HbS are shown in the main plot of

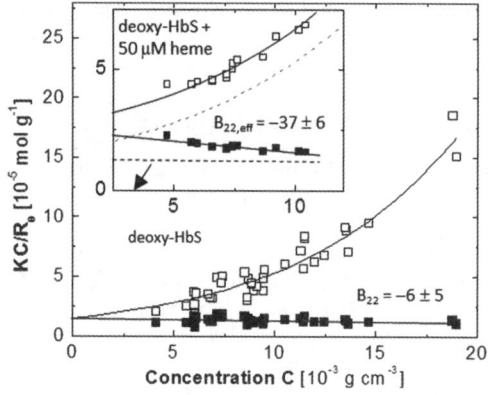

Fig. 9 Characterization of the interactions between HbS molecules in deoxy-HbS solutions in terms of second virial coefficients B_{22}, extracted from Debye plots $KC/R_\theta(C)$. K: Instrument constant, $R_\theta = I_\theta/I_0$: ratio of scattered at angle θ to incident intensity, C: HbS concentration. Open symbols: measured I_θ is lowered because of absorption of light by HbS, line through points is a fit to a product of Debye and Beer equations: KC/R_θ exp($-2.3\varepsilon Cl$) = $1/M_w$ + $2A_{22}C$, where $\varepsilon = 4930.8$ cm^{-1} mol^{-1} l is the extinction coefficient of Hb at 628 nm, the wavelength used, $l = 0.8$ cm is the optical pathway in the cuvette, and M_w is the hemoglobin molecular weight. Solid symbols: I_θ is corrected with exp($-2.3\varepsilon Cl$) and the data are fit with $KC/R_\theta = 1/M_w + 2A_{22}C$. The slope of this dependence is the second virial coefficient $A_{22} = -7.8 \times 10^{-5}$ mol cm^{-3} g^{-2}, nearly equal to a determination in ref. 83. The dimensionless B_2, shown in plots, is determined from $A_{22} = B_{22}V_mM_w^{-2}$, where $V_m = 500\,560$ cm^3 mol^{-1} is the HbS molar volume. Values of $B_{22} < 4$, the value for non-interacting hard spheres, indicate intermolecular attraction. Main plot: deoxy-HbS solution without the addition of hemin. At $C = 0$ g cm^{-3}, the data extrapolate to $1/M_w$, from which $M_w = 68\,000 \pm 7000$ g mol^{-1}, close to the actual 64 000 g mol^{-1}. Insert: 50 mM of hemin added after dialysis. The additional absorption of light by hemin is accounted by calibrating against solution with same hemin concentration. The concentration of HbS at which the molar concentration of the HbS tetramer is equal to that of the added hemin is marked with an arrow. Dashed lines indicate locations of data in the main plot. According to eqn (2), the intercept of this straight line depends on the hemoglobin molecular mass M_w and on the virial coefficient A_{12} accounting for the interaction between hemin and hemoglobin. The dimensionless slope $B_{22,eff}$ is more negative than B_{22} in solutions without hemin in the main plot. This indicates stronger effective attraction between the hemoglobin molecules in the presence of hemin than in the solutions without hemin.

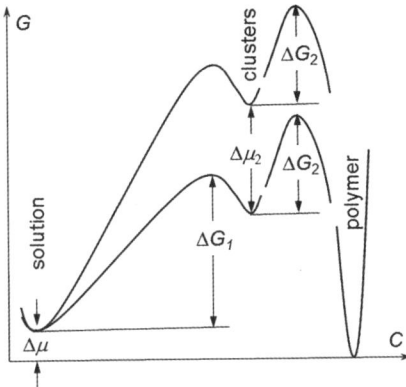

Fig. 10 Towards understanding of the effects of hemin on the volume fraction of the nucleation precursors. $\Delta\mu$ is the supersaturation of the solution with respect to the polymers; $\Delta\mu_2$ is the change of the chemical potential of HbS in the clusters in the presence of hemin. ΔG_1 and ΔG_2 denote same the variables as in Fig. 2.

Fig. 9, and those in the presence of hemin are shown in its inserts. Form these Debye plots, we determine A_2 and M_2, or, in the case of heme–hemoglobin solutions, $A_{2,\text{eff}}$ and $M_{2,\text{app}}$. M_2 Is compared to the known molecular mass of hemoglobin for a partial verification of the determinations, while $M_{2,\text{app}}$ is used to determine A_{12}, using eqn (2) above. From $A_{2,\text{eff}}$, using A_{22}, A_{12}, and α, we determine A_{212}.

These data show that $M_{\text{HbS,app}}$, and $A_{2,\text{eff}}$ decrease upon the addition of hemin. Straightforward thermodynamic analyses reveal that the decrease in $A_{2,\text{eff}}$ indicates that hemin enhances the attraction between the HbS molecules.

The enhanced attraction leads to slower increase of the chemical potential μ_2 of HbS as its concentration is increased. Quantification of this effect in ref. 15 using an expression similar to eqn (1) yields μ_2 lowered by $\Delta\mu_2 \cong 4k_\text{B}T$ in the presence of 50 μM hemin from its value in a solution without hemin. According to the theory of metastable clusters,[53] the excess chemical potential of high concentration liquid determines the volume fraction occupied by the clusters. As illustrated in Fig. 10, a lower chemical potential of HbS in the clusters directly leads to a cluster volume fraction increased by a factor of $\exp(\Delta\mu_2/k_\text{B}T) \cong \exp(4) \cong 50$. This is consistent with the 80 fold increase of cluster volume fraction induced by 67 mM of hemin in Fig. 4.

The known biophysical mechanisms require, in the case of bridging, $m_{\text{hemin}}/m_{\text{HbS}} \approx 1$, or, in the cases of depletion or screening, orders of magnitude higher m_{hemin} than m_{HbS}.[70,71] A qualitative phenomenological model based on the osmotic virial coefficients in a three component solutions, A_{12} and A_{212}, suggests that repulsion by the hemin prevents the hemoglobin molecules from accessing parts of the volume, which acts as effective attraction. Thus, the intuitively expected stronger attraction between HbS than between HbA molecules, which in reality only occurs in the presence of hemin, is not a simple consequence of the sickle cell mutation: it is a result of a combination of electrostatic and hydrophobic behaviors, accounted in A_{12} and A_{212}, in which the hemin plays crucial role. The high efficacy of the free hemin is a consequence of two factors, which allows a single hemin molecule to interact with many hemoglobin molecules: the long-range nature of the electrostatic forces, and the high mobility of the relatively small hemin.

Conclusions and implications

We have demonstrated that the nucleation of the HbS polymers follows a two-step mechanism, consisting of the formation of a metastable dense liquid cluster and the nucleation of an ordered polymer nucleus within this cluster. The applicability of the two-step nucleation mechanism for the nucleation of HbS polymers has fundamental and clinical consequences. It provides an example of the applicability of this mechanism for a new class of systems and suggests that the presence of a disordered precursor may be a general feature of the self-assembly of ordered structures. The huge clinical variability of sickle cell anemia has been attributed to processes occurring independently of the HbS polymerization: erythrocyte and endothelial wall adhesion, erythrocyte deformability, erythrocyte membrane damage, and others.[72–74] If polymerization follows the accepted scenario of one-step homogeneous nucleation, followed by growth, branching, and gelation, the rate of HbS polymerization is entirely determined by the HbS activity. The HbS activity is mostly determined by the HbS concentration since the HbS molecules only interact through their excluded volume[75,76] and can vary only mildly in response to variations in the concentration of solution components other than HbS. This is hard to reconcile with the clinical variability among patients with identical expression of HbS in the erythrocytes.[8,77] On the other hand, the two-step mechanism of homogenous nucleation of the HbS polymers allows reconciliation between the primary role of HbS polymerization and the clinical variability: the sizes, properties, and volume fractions of the dense liquid precursors likely are strongly modified by solution components at micromolar concentrations.[17,47]

We have shown that free heme is a powerful accelerant of sickle cell hemoglobin polymerization and that in its absence nucleation of sickle cell hemoglobin polymers is completely suppressed.

Exploring the mechanism of action of free heme, we have shown that in solutions of sickle cell hemoglobin, it enhances the attraction between the hemoglobin molecules. The enhanced attraction leads to greater volume fraction of the metastable dense liquid clusters, which, in turn, leads to faster nucleation of the sickle cell hemoglobin polymers. We have demonstrated a quantitative correlation between the hemoglobin chemical potential change due to the enhanced attraction, the increase in the volume occupied by the nucleation precursors, and the acceleration of polymer nucleation in the presence of free heme. Micromolar amounts of hematin, which could be released by a minor fraction of the hemoglobin molecules, lead to orders-of-magnitude acceleration of the nucleation and growth of the sickle cell polymers.

Since the experiment conditions mimic those inside the red cell cytosol it is likely that hematin released from hemoglobin in the red cells would have similar effects on the nucleation of sickle cell polymers *in vivo*. The concentration of free hematin in the red cells is lowered by three processes: diffusion thorough the cell membrane,[78] association to the membrane or the cytoskeleton,[79] or intracellular degradation.[80] Their combined action could lead to significant variability of the hematin concentration in the red cells. Because of the high sensitivity of the rates of nucleation and growth of sickle cell polymers to hematin concentration, this variability could lead to significant irreproducibility of polymerization between the red cells of identical HbS activities of a patient. The variability of hematin release and removal between patients could be the factor underlying the variability in the prognosis of patients with identical HbS expression.

A more significant conclusion from our results is that if the free hematin concentration in the erythrocytes is lowered, this could result in lower sickling and less severe sickle cell symptoms.

Last, it is possible that the hematin released in infected red blood cells by the malaria plasmodia as a byproduct of their metabolism of hemoglobin[81] causes the known sickling of sickle-trait erythrocytes (which would not sickle in the absence of hematin because of lower HbS activity) and enables their selective removal from the circulation.[82] This may be a novel explanation of the link between malaria and sickle cell disease.

References

1 E. Beutler, in *Williams Hematology*, ed. E. Beutler, M. A. Lichtman, B. S. Coller, T. J. Kipps and U. Seligsohn, McGraw Hill, New York, 6th edn, 2001, pp. 581–605.
2 K. R. Bridges, G. D. Barabino, C. Brugnara, M. R. Cho, G. W. Christoph, G. Dover, B. M. Ewenstein, D. E. Golan, C. R. Guttmann, J. Hofrichter, R. V. Mulkern, B. Zhang and W. A. Eaton, *Blood*, 1996, **88**, 4701–4710.
3 P. Vekilov, *Br. J. Haematol.*, 2007, **139**, 173–184.
4 W. A. Eaton, *Biophys. Chem.*, 2003, **100**, 109–116.
5 R. Pawliuk, K. A. Westerman, M. E. Fabry, E. Payen, R. Tighe, E. E. Bouhassira, S. A. Acharya, J. Ellis, I. M. London, C. J. Eaves, R. K. Humphries, Y. Beuzard, R. L. Nagel and P. Leboulch, *Science*, 2001, **294**, 2368–2371.
6 R. P. Hebbel, W. T. Morgan, J. W. Eaton and B. E. Hedlund, *Proc. Natl. Acad. Sci. U. S. A.*, 1988, **85**, 237–241.
7 R. P. Hebbel, *J. Clin. Invest.*, 1997, **99**, 2561–2564.
8 B. R. Amin, R. M. Bauersachs, H. J. Meiselman, N. Mohandas, R. P. Hebbel, P. E. Bowen, R. A. Schlegel, P. Williamson and M. P. Westerman, *Hemoglobin*, 1991, **15**, 247–256.
9 R. E. Samuel, E. D. Salmon and R. W. Briehl, *Nature*, 1990, **345**, 833–835.
10 R. W. Briehl, *J. Mol. Biol.*, 1995, **245**, 710–723.
11 O. Galkin and P. G. Vekilov, *J. Mol. Biol.*, 2004, **336**, 43–59.
12 O. Galkin, R. L. Nagel and P. G. Vekilov, *J. Mol. Biol.*, 2007, **365**, 425–439.
13 D. Kashchiev, *Nucleation. Basic theory with applications*, Butterworth, Heinemann, Oxford, 2000.
14 W. Pan, O. Galkin, L. Filobelo, R. L. Nagel and P. G. Vekilov, *Biophys. J.*, 2007, **92**, 267–277.
15 W. Pan, V. V. Uzunova and P. G. Vekilov, *Biopolymers*, 2009, **91**, 1108–1116.
16 P. G. Vekilov, A. R. Feeling-Taylor, D. N. Petsev, O. Galkin, R. L. Nagel and R. E. Hirsch, *Biophys. J.*, 2002, **83**, 1147–1156.
17 O. Galkin and P. G. Vekilov, *Proc. Natl. Acad. Sci. U. S. A.*, 2000, **97**, 6277–6281.
18 V. J. Anderson and H. N. W. Lekkerkerker, *Nature*, 2002, **416**, 811–815.
19 P. R. ten Wolde and D. Frenkel, *Science*, 1997, **277**, 1975–1978.
20 P. G. Vekilov, *Cryst. Growth Des.*, 2004, **4**, 671–685.
21 J. F. Lutsko and G. Nicolis, *Phys. Rev. Lett.*, 2006, **96**, 046102.
22 A. Shiryayev and J. D. Gunton, *J. Chem. Phys.*, 2004, **120**, 8318–8326.
23 A. Lomakin, N. Asherie and G. B. Benedek, *Proc. Natl. Acad. Sci. U. S. A.*, 2003, **100**, 10254–10257.
24 D. Vivares, E. Kaler and A. Lenhoff, *Acta Crystallogr., Sect. D: Biol. Crystallogr.*, 2005, **61**, 819–825.
25 W. Pan, A. B. Kolomeisky and P. G. Vekilov, *J. Chem. Phys.*, 2005, **122**, 174905.
26 L. F. Filobelo, O. Galkin and P. G. Vekilov, *J. Chem. Phys.*, 2005, **123**, 014904.
27 Y. G. Kuznetsov, A. J. Malkin and A. McPherson, *J. Cryst. Growth*, 2001, **232**, 30–39.
28 A. Lomakin, D. S. Chung, G. B. Benedek, D. A. Kirschner and D. B. Teplow, *Proc. Natl. Acad. Sci. U. S. A.*, 1996, **93**, 1125–1129.
29 R. Krishnan and S. L. Lindquist, *Nature*, 2005, **435**, 765–772.
30 J. E. Aber, S. Arnold and B. A. Garetz, *Phys. Rev. Lett.*, 2005, **94**, 145503.
31 B. Garetz, J. Matic and A. Myerson, *Phys. Rev. Lett.*, 2002, **89**, 175501.
32 D. W. Oxtoby, *Nature*, 2002, **420**, 277–278.
33 M. E. Leunissen, C. G. Christova, A.-P. Hynninen, C. P. Royall, A. I. Campbell, A. Imhof, M. Dijkstra, R. van Roij and A. van Blaaderen, *Nature*, 2005, **437**, 235–240.
34 J. R. Savage and A. D. Dinsmore, *Phys. Rev. Lett.*, 2009, **102**, 198302.
35 T. H. Zhang and X. Y. Liu, *J. Phys. Chem. B*, 2007, **111**, 14001–14005.
36 T. Kawasaki and H. Tanaka, *Proc. Natl. Acad. Sci. U. S. A.*, 2010, **107**, 14036–14041.
37 L. B. Gower, *Chem. Rev.*, 2008, **108**, 4551–4627.
38 E. M. Pouget, P. H. H. Bomans, J. A. C. M. Goos, P. M. Frederik, G. de With and N. A. J. M. Sommerdijk, *Science*, 2009, **323**, 1455–1458.
39 D. Gebauer, A. Volkel and H. Colfen, *Science*, 2008, **322**, 1819–1822.
40 F. Nudelman, K. Pieterse, A. George, P. H. H. Bomans, H. Friedrich, L. J. Brylka, P. A. J. Hilbers, G. de With and N. A. J. M. Sommerdijk, *Nat. Mater.*, 2010, **9**, 1004–1009.
41 A. Dey, P. H. H. Bomans, F. A. Müller, J. Will, P. M. Frederik, G. de With and N. A. J. M. Sommerdijk, *Nat. Mater.*, 2010, **9**, 1010–1014.
42 J. F. Wang, M. Muller and Z. G. Wang, *Journal of Chemical Physics*, 2009, 130.
43 P. E. Bonnett, K. J. Carpenter, S. Dawson and R. J. Davey, *Chem. Commun.*, 2003, 698–699.
44 N. Asherie, A. Lomakin and G. B. Benedek, *Phys. Rev. Lett.*, 1996, **77**, 4832–4835.
45 R. P. Sear, *J. Chem. Phys.*, 2009, **131**, 074702.

46 O. Galkin, W. Pan, L. Filobelo, R. E. Hirsch, R. L. Nagel and P. G. Vekilov, *Biophys. J.*, 2007, **93**, 902–913.
47 O. Galkin, K. Chen, R. L. Nagel, R. E. Hirsch and P. G. Vekilov, *Proc. Natl. Acad. Sci. U. S. A.*, 2002, **99**, 8479–8483.
48 A. M. Kulkarni, A. P. Chatterjee, K. S. Schweitzer and C. F. Zukoski, *Phys. Rev. Lett.*, 1999, **83**, 4554–4557.
49 O. Gliko, N. Neumaier, W. Pan, I. Haase, M. Fischer, A. Bacher, S. Weinkauf and P. G. Vekilov, *J. Am. Chem. Soc.*, 2005, **127**, 3433–3438.
50 O. Gliko, W. Pan, P. Katsonis, N. Neumaier, O. Galkin, S. Weinkauf and P. G. Vekilov, *J. Phys. Chem. B*, 2007, **111**, 3106–3114.
51 A. Stradner, H. Sedgwick, F. Cardinaux, W. C. K. Poon, S. U. Egelhaaf and P. Schurtenberger, *Nature*, 2004, **432**, 492–495.
52 Y. Georgalis, P. Umbach, W. Saenger, B. Ihmels and D. M. Soumpasis, *J. Am. Chem. Soc.*, 1999, **121**, 1627–1635.
53 W. Pan, P. G. Vekilov and V. Lubchenko, *J. Phys. Chem. B*, 2010, **114**, 7620–7630.
54 K. M. Knee and I. Mukerji, *Biochemistry*, 2009, **48**, 9903–9911.
55 V. V. Uzunova, W. Pan, O. Galkin and P. G. Vekilov, *Biophys. J.*, 2010, **99**, 1976–1985.
56 P. G. Vekilov, W. Pan, O. Gliko, P. Katsonis and O. Galkin, in *Lecture notes in physics: Aspects of physical biology: Biological water, protein solutions, transport and replication*, ed. G. Franzese and M. Rubi, Springer, Heidelberg, 2008, vol. 752, pp. 65–95.
57 P. G. Vekilov, *Cryst. Growth Des.*, 2010, **10**, 5007–5019.
58 W. D. Brubaker, J. A. Freites, K. J. Golchert, R. A. Shapiro, V. Morikis, Douglas J. Tobias and R. W. Martin, *Biophys. J.*, 2011, **100**, 498–506.
59 D. Erdemir, A. Y. Lee and A. S. Myerson, *Acc. Chem. Res.*, 2009, **42**, 621–629.
60 S. B. Hutchens and Z.-G. Wang, *J. Chem. Phys.*, 2007, **127**, 084912.
61 K. S. Schmitz, *Dynamic light scattering by macromolecules*, Academic Press, New York, 1990.
62 Z. Cao and F. A. Ferrone, *Biophys. J.*, 1997, **72**, 343–352.
63 F. A. Ferrone, H. Hofrichter and W. A. Eaton, *J. Mol. Biol.*, 1985, **183**, 611–631.
64 F. A. Ferrone, H. Hofrichter and W. A. Eaton, *J. Mol. Biol.*, 1985, **183**, 591–610.
65 J. G. Kirkwood and R. J. Goldberg, *J. Chem. Phys.*, 1950, **18**, 54–57.
66 J. Bloustine, T. Virmani, G. M. Thurston and S. Fraden, *Phys. Rev. Lett.*, 2006, **96**, 087803.
67 A. Sehgal and T. A. P. Seery, *Macromolecules*, 1999, **32**, 7807–7814.
68 R. S. Hall, Y. S. Oh and C. S. Johnson Jr., *J. Phys. Chem.*, 1980, **84**, 756–767.
69 S. Prahl, *Optical absorption of hemoglobin*, http://omlc.ogi.edu/spectra/hemoglobin/, 1999.
70 D. Leckband and J. Israelachvili, *Q. Rev. Biophys.*, 2001, **34**, 105–267.
71 J. N. Israelachvili, *Intermolecular and surface forces*, Academic Press, New York, 1995.
72 R. P. Hebbel, *Blood*, 1991, **77**, 214–237.
73 S. H. Embury, *Microcirculation*, 2004, **11**, 101–113.
74 M. H. Steinberg, *Br. J. Haematol.*, 2005, **129**, 465–481.
75 P. D. Ross and A. P. Minton, *J. Mol. Biol.*, 1977, **112**, 437–452.
76 W. A. Eaton and J. Hofrichter, in *Advances in protein chemistry*, ed. C. B. Anfinsen, J. T. Edsal, F. M. Richards and D. S. Eisenberg, Academic Press, San Diego, 1990, vol. 40, pp. 63–279.
77 M. A. el-Hazmi, *J Trop Pediatr*, 1992, **38**, 106–112.
78 J. B. Cannon, F. Kuo, R. F. Pasternack, N. M. Wong and U. Muller-Eberhard, *Biochemistry*, 1984, **23**, 3715–3721.
79 B. H. Rank, J. Carlsson and R. P. Hebbel, *J. Clin. Invest.*, 1985, **75**, 1531–1537.
80 H. Atamna and H. Ginsburg, *J. Biol. Chem.*, 1995, **270**, 24876–24883.
81 D. Jani, R. Nagarkatti, W. Beatty, R. Angel, C. Slebodnick, J. Andersen, S. Kumar and D. Rathore, *PLoS Pathog.*, 2008, **4**, e1000053.
82 R. L. Nagel, in *Disorders of hemoglobin: genetics, pathophysiology and clinical management*, ed. M. H. Steinberg, B. G. Forget, D. R. Higgs and R. L. Nagel, Cambridge University, Cambridge, 2001, pp. 832–860.
83 D. Elbaum, R. L. Nagel, R. M. Brookchin and T. T. Herskovits, *Proc. Natl. Acad. Sci. U. S. A.*, 1974, **71**, 4718–4722.

Structural evolution, formation pathways and energetic controls during template-directed nucleation of CaCO$_3$†

Michael H. Nielsen,[ab] Jonathan R. I. Lee,[c] Qiaona Hu,[bd] Thomas Yong-Jin Han[c] and James J. De Yoreo[*b]

Received 15th March 2012, Accepted 7th June 2012
DOI: 10.1039/c2fd20050c

Through the process of biomineralization living organisms use macromolecules to direct nucleation and growth of nanophases of a variety of inorganic materials. Evidence shows this is a widespread strategy for controlling the timing, polymorphism, morphology, and crystallographic orientation of CaCO$_3$ nuclei. In the past decade, self-assembled monolayers (SAMs) of alkanethiols have been used as a simple model to reproduce the controls of organic substrates. However, despite the importance of nucleation phenomena in the crystallization of inorganic materials our understanding of the reaction dynamics is extremely limited because, until recently, there was no experimental tool that possessed the spatial and temporal resolution needed to capture the formative events in the process. Issues such as the formation of amorphous precursors, and polymorph selection during the initial stages of nucleation, as well as the structural relationships and energetic controls of the inorganic matrix on the emerging nucleus have not been fully explored. To address these gaps in our understanding we have developed a suite of *in situ* methods and applied them, along with synchrotron-based X-ray spectroscopies, to CaCO$_3$ nucleation. We used these methods to observe CaCO$_3$ nucleation rates on alkanethiol SAMs. We found that for two carboxyl-terminated alkanethiol SAMs with odd (mercaptoundecanoic acid) and even (mercaptohexadecanoic acid) carbon chains, the effective interfacial energy is reduced from about 109 mJ m^{-2} in solution to 81 mJ m^{-2} and 72 mJ m^{-2}, respectively, showing that templating is driven by a reduction in the thermodynamic barrier to nucleation. We also report *in situ* transmission electron microscopy (TEM) observations of crystal nucleation and growth in solution at the nanometre scale and video rates. This capability is enabled by the combination of a custom designed TEM stage and fluid cell. Significantly, the design of the cell and holder ensures temperature and electrochemical control over the reaction environment, allowing for direct investigation of nucleation. Our first results show that CaCO$_3$ nucleates *via* nanoparticles of an apparent metastable precursor— most likely ACC—followed by consolidation and faceting. Finally, we report insights from the use of synchrotron-based near-edge X-ray absorption fine structure spectroscopy (NEXAFS) into the evolution of

[a]*Department of Materials Sciences and Engineering, University of California, Berkeley, Berkeley, CA, 94720*
[b]*Molecular Foundry, Lawrence Berkeley National Laboratory, Berkeley, CA, 94720. E-mail: jjdeyoreo@lbl.gov*
[c]*Lawrence Livermore National Laboratory, Livermore, CA 94550*
[d]*Department of Geological Sciences, University of Michigan, Ann Arbor, MI, 48105*

† Electronic supplementary information (ESI) available. See DOI: 10.1039/c2fd20050c

the SAM during the nucleation process. Based on measurements of SAM monomer orientation, we argue that the ability of the SAM to reorganize during the nucleation process is a key feature of an organic matrix that successfully directs mineralization.

1. Introduction

Biological control over crystallization of hard materials commonly relies on a scaffold of macromolecules that is often referred to as the organic matrix. The matrix is believed to play an active role in directing nucleation of the mineral phase by presenting a "template" that controls the location and orientation of the individual crystallites.[1-3] The hierarchical structure and mechanical properties made possible through matrix-mediated mineralization are beyond the reach of laboratory synthesis.[4] Consequently, achieving a mechanistic understanding of this phenomenon is of interest for development of new approaches to materials synthesis, thus a great deal of research has been directed towards creating model matrices for that purpose.

Early studies of crystal templating[5-8] used Langmuir monolayers with specific head-group chemistries as simple biomimetic systems to look at the affect of organic films on nucleation of inorganic crystals. These studies found that compressed monolayers induced oriented nucleation of $CaCO_3$ and $BaSO_4$ on specific faces dependent on chemical functionality at the organic–inorganic interface. In the case of vaterite on stearate there was no epitaxial match between the nucleating face and the organic template and the authors concluded that stereochemical and electrostatic matching override the lattice mismatch. In addition they observed that partially compressed monolayers produced the best control over vaterite nucleation, and suggested that scaffolds that can undergo conformational changes in the presence of the mineral may provide the highest control over mineralization. For calcite on stearate and sulfonated films, and $BaSO_4$ on sulfate and phosphonate monolayers, although they observed lattice matching at the interface of the organic and mineral phases, selection of a single crystal nucleation plane was attributed to stereochemical matching between the template head-group chemistries and the nucleating face. Adding to these findings Berman et al.[9] demonstrated that calcite nuclei were co-aligned with respect to the conjugated backbone of a PDA Langmuir monolayer on a solid support, and that the alkyl side chains of the supported PDA film reorganized during mineralization to optimize the stereochemical fit to the calcite structure.

Aizenberg et al.[10,11] showed that organothiol SAMs exhibiting surface functional groups commonly expressed on biomolecules generated well-defined calcite crystal orientations, narrowly distributed crystal sizes, and a pattern of nucleation dependent on the location of these functional groups. They demonstrated that by varying either the terminal groups on SAMs of alkane thiols or the underlying noble metal substrate one could precisely control the plane of nucleation of calcite, and found that the geometry of the SAM functional groups was the primary factor in controlling the orientation of the calcite nuclei. Travaille et al.[12] showed a 1 : 1 relationship along one direction between the underlying Au structure and in-plane orientation of the templated crystals. Han and Aizenberg[13] extended Aizenberg's earlier studies to systematically explore the effects of SAM alkyl chain parity and material of the underlying substrate on calcite nucleation. Through these studies they showed the nucleation plane depended strongly on a combination of three aspects of the template: packing geometry of the templating surface, head-group chemistry and orientation of the functional group.

These observations led to the idea that directed growth is determined at the initial stage of nucleation where the nucleus contacts specifically with the functional groups of the organic interface to develop into a specific polymorph and orientation. Fig. 1

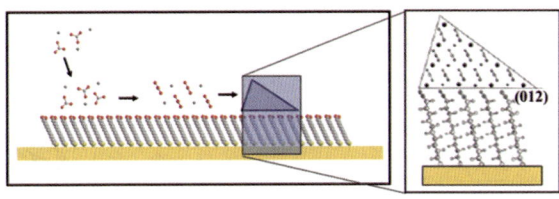

Fig. 1 Schematic of classical pathway, wherein the initial state consists of a well-ordered SAM assembled on an underlying substrate submerged in a supersaturated calcium carbonate solution. A suitable number of ions to constitute a critical nucleus organize at the organic–solution interface into an energetically favorable arrangement and crystallizes, with the final phase oriented by the underlying SAM. Inset modified from ref. 14.

shows this schematically, with the SAM a well-ordered template with which the solvated mineral comes into close proximity, aligns into some energetically favorable configuration with respect to the organic template, and crystallizes into its energetically favorable phase. However, direct evidence for the structural relationship between film and crystal and the pathway of calcite formation, as well as a quantitative understanding of the energetic drivers underlying the process were lacking. Thus the above studies set the context for three key questions regarding oriented nucleation on templates: 1) What is the structural relationship between organic template and nucleus that leads to directed crystallization, and how does that relationship evolve during the process of mineralization? 2) What is the pathway taken by the mineralizing constituents from the solvated state to the final, energetically stable phase? 3) How does the presence of the organic surface affect the energetics of mineralization; that is, is templated nucleation driven by thermodynamics through reduction of interfacial energy or kinetics through reduction of activation barriers, and by how much are they reduced?

Partial answers to some of these questions came from a combination of theory and experiment. Simulations[15–17] of calcite nucleating on defect-free, rigid, COOH-terminated SAMs predicted that calcite should preferentially nucleate on the (001) face rather than the (012) face as had been seen in experiments. The (001) face provided a strong structural match for the surface of the monolayer whereas the (012) face matched only in one direction along the surface. To get the correct nucleating face the researchers had to introduce packing defects in the SAM or crystal, allow the SAM to reorganize through the natural thermal vibrations of the monomers common at experimental temperatures, or add bicarbonate ions into the simulation to balance the charge mismatch between the crystal and SAM. In addition, more recent work[18] indicated that the orientation of the headgroup has little effect on face selectivity. Instead a close match between both the charge density and the degree of epitaxy of the film and that of the crystal was required for crystallization on a particular face to occur. Such a close match could initially extend over only a small region as long as the SAM's structure was relaxed and allowed to reorganize to improve the match over a larger area as crystallization progressed.

Using synchrotron-based X-ray absorption spectroscopy—primarily near edge X-ray absorption fine structure (NEXAFS) spectroscopy—Lee et al.[19] investigated the evolution of SAM monomer orientation during templating. The experimental procedure is extensively detailed in the reference, and we direct the interested reader there for a comprehensive account of the experiments. The salient point for the current discussion is that, through the acquisition of carbon K-edge absorption spectra for a range of incident angles of the X-rays relative to the sample ranging from grazing to normal (Fig. 2a), one can determine the degree to which the SAM head-groups exhibit a well-defined orientation and quantify how far off surface normal the SAM head-group is tilted (Fig. 2b).

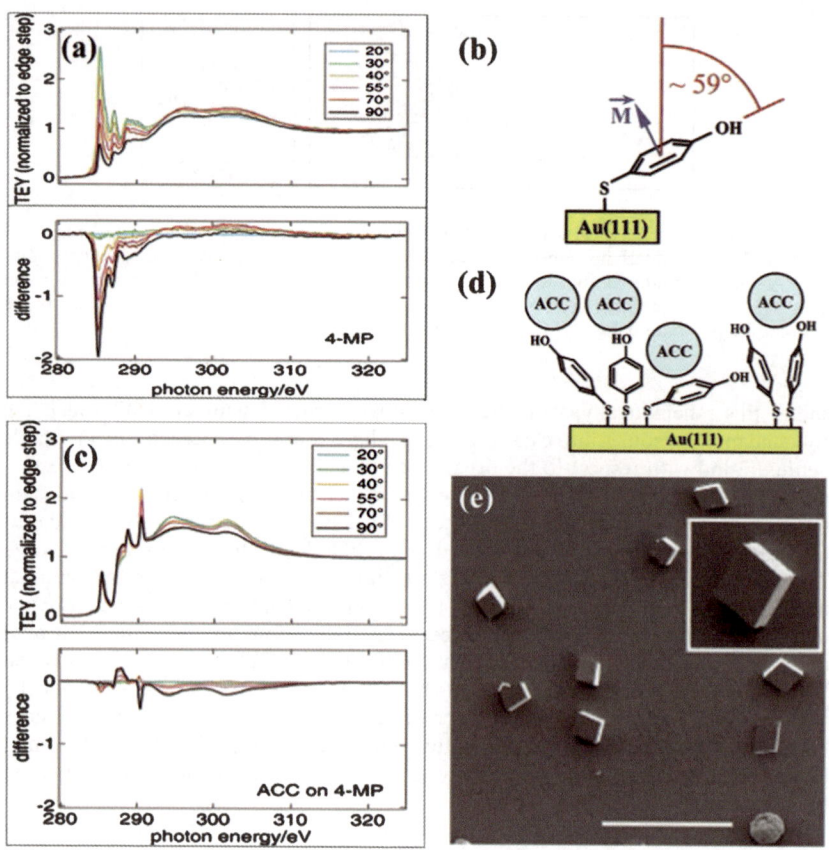

Fig. 2 Normalized carbon K-edge NEXAFS spectra for 4-MP in its (a) as-prepared state and (c) in the presence of ACC, wherein an angular-dependence of the resonance peak at 285 eV signifies a well-defined aryl ring orientation. Schematics showing the 4-MP adoping (b) a well-defined orientation with the OH functionality tilted off surface normal by 59° when in its native state, and (d) a disordered state when covered by a film of ACC. (e) SEM image of final (104)-oriented crystals on 4-MP film. Adapted from ref. 19.

Applying this technique to SAMs comprising isomers of OH-terminated mercaptophenol (MP), Lee et al.[19] demonstrated that the as-grown SAM was ordered with a well-defined tilt angle regardless of which isomer was used (Fig. 2a and 2b). However, highly oriented calcite crystals were only obtained with the 3-MP and 4-MP SAMs, which expose the OH groups, while low coverage of randomly oriented crystals was observed on the 2-MP film, presumably because the hydroxyl functionality was buried beneath the surface of the SAM. This result showed that the headgroup chemistry was critical for successful templating. Moreover, in these experiments the authors found that amorphous calcium carbonate (ACC) was the first phase to form, either in the surrounding solution or directly on the films, before transforming into calcite via a dissolution–reprecipitation mechanism. During the ACC formation process, the SAM monomers became randomly oriented (Fig. 2c and 2d). Nonetheless, they produced highly oriented calcite (Fig. 2e). This result connects to more recent findings by Gebauer et al.[20] who showed that bulk solutions of $CaCO_3$ contain an equilibrium population of neutral $CaCO_3$ clusters prior to homogeneous nucleation of ACC and by Pouget et al.[21] who found that sub-nm $CaCO_3$ clusters aggregated to form ACC which, upon contact with a Langmuir monolayer, converted to oriented calcite.

Further evidence for the connection between templated nucleation and the evolution of the organic template's structure throughout the process was found in studies of calcite nucleation on Langmuir monolayers.[22,23] Using amino acid functionalized surfactants with exposed carboxyl groups they showed that all of their compounds were active in calcium carbonate nucleation. However, only the surfactants flexible enough to allow for restructuring of the organic layer during calcite mineralization had an observable measure of control over the plane of nucleation. Rigid monolayers nucleated calcite at higher densities than the control, but with little effect on the crystal habit or orientation.

Taken together, this collection of studies suggest that during nucleation of calcite on organothiol SAMs: 1) templating is a dynamic process requiring an evolving interface in order for the crystal to be adequately directed by the organic film into a specific orientation, and 2) the pathway to the mineral phase passes through ACC, which nucleates in bulk solution before formation of the crystalline phase. However, in all of these experiments nucleation occurred either at supersaturations known to be in excess of the solubility limit for ACC or during diffusion of gas from an ammonium carbonate source into a $CaCl_2$ solution to create an unknown, but extremely high supersaturation relative to calcite. Moreover, the careful measurements of SAM orientation were carried out using very short monomers, which are generally expected to be less rigid and less ordered than long-chain SAMs. Also, the OH head-group of the MP molecule leads to templating of calcite on what is already its natural (104) face, perhaps diminishing the degree to which the film can be said to control orientation. Moreover, these studies leave open the questions concerning the energetic controls on the templating process.

2. SAM structural evolution

In order to investigate the interaction between nucleating $CaCO_3$ crystals and SAMs that are similar to the conventional ω-substituted alkanethiols commonly used in studies of calcite templating, we applied the NEXAFS spectroscopic methods described above[19] to SAMs of *para*-(p), *meta*-(m), and bimeta-(bm-)mercaptodecylbenzoic acid (MDBA) (Fig. 3a). The full details and results of the experiment will be discussed in an upcoming paper.[25] In contrast to the MP monomers discussed above, MDBA molecules have a long hydrocarbon chain that yields SAMs with low packing defects. Furthermore, MDBA has carboxyl functionality at the surface for specific interactions, rather than the hydroxyl chemistry of MP that interacts non-specifically with calcite. Mg-stabilized ACC[26] was prepared to produce amorphous material with a lifetime on the carboxylated surface of the MDBA SAMs sufficient for NEXAFS measurements. As with the MP SAMs, during mineralization through exposure of $CaCl_2$ solution to ammonium carbonate, the MDBA SAMs became covered with a film of ACC nanoparticles (Fig. 3b), which then transformed into calcite.

We found that of the three (p-, m-, and bm-MDBA) chemistries examined, only p-MDBA produced highly oriented calcite crystals, which were nucleated on the (018) plane (Fig. 3c). Both the m- and bm-MDBA films produced a random distribution of crystal orientations when the crystallization ran to completion (Fig. 3d). Examination of the NEXAFS spectra for the three templates under as-prepared conditions and in the presence of the ACC film highlighted a key difference. All three SAMs had well-defined aryl ring orientations in both the as-prepared state and in the presence of ACC. However, while the p-MDBA SAM underwent a significant change in its aryl ring orientation between the as-prepared state and in contact with ACC (Fig. 3e), both the m- and bm-MDBA SAMs exhibited no significant change in their orientations between the two states (Fig. 3f). The hydrophobicity of the m-MDBA SAM indicated that, as with the 2-MP SAM, its functional group was buried beneath the SAM surface and therefore unable to interact with the mineral. In contrast, bm-MDBA was found to be roughly as hydrophilic as the p-MDBA,

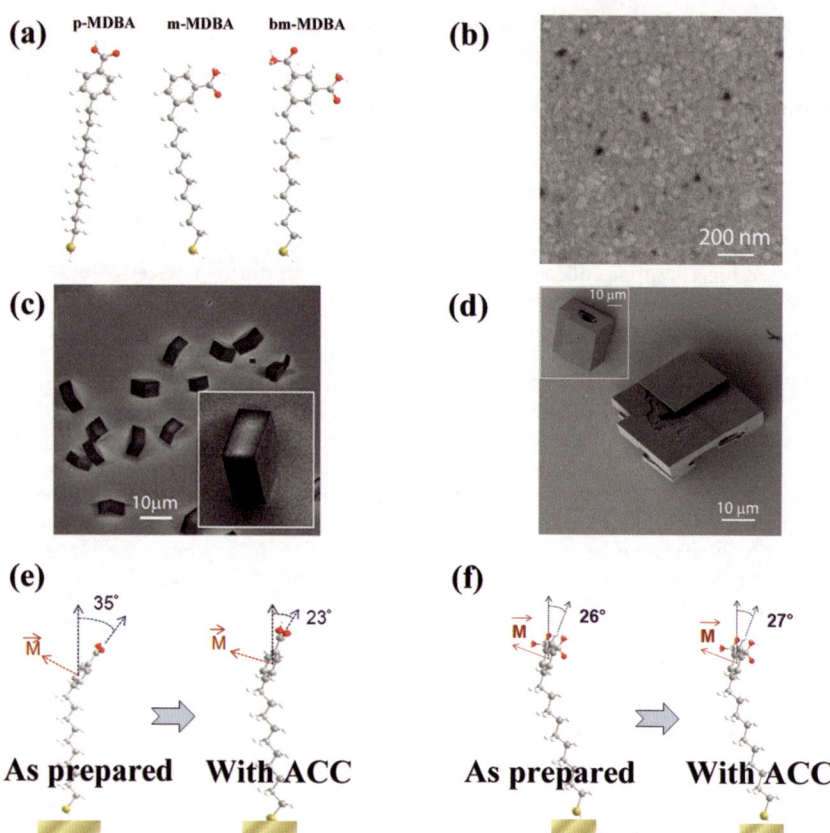

Fig. 3 (a) Structures for the three MDBA monomers. SEM micrographs of (b) Mg-stabilized ACC on a p-MDBA SAM on Au(111), (c) calcite prepared on a p-MDBA SAM on Au(111), and (d) calcite prepared on an m-MDBDA SAM on Au(111). Magnified image (c inset) of calcite on p-MBDA SAM, and additional example (d inset) of calcite on an m-MDBA SAM. Schematics of structural evolution for (e) p-MDBA and (f) bm-MDBA SAMs, from as-prepared state to the presence of Mg-stabilized ACC. Panels b–d adapted from ref. 25.

suggesting that at least one of the carboxyl groups on each monomer was exposed at the surface and available for interaction with the mineral. Nonetheless, this SAM neither reoriented during mineralization nor directed the final crystal orientation.

Taken together, the findings of these NEXAFS studies imply the following about the structural and chemical requirements for a template to act as a directing agent in calcite nucleation. While exposure of an appropriate head-group chemical functionality is a necessary condition, it is insufficient to ensure templating. In order to produce nucleation and growth on specific crystallographic planes, structural flexibility is an essential feature. Thus the SAM structure evolves as mineralization proceeds and it may be appropriate to consider templated crystallization as a co-templating process whereby the mineral and the organic phases each direct the structural evolution of the other phase, as shown schematically in Fig. 4.

3. Energetic controls on templated nucleation

From the classical thermodynamic expression for the change in free energy during nucleation, one can show[27] that there is a nucleation barrier Δg_n associated with a critical nucleus size that is a function of the supersaturation σ, and the effective

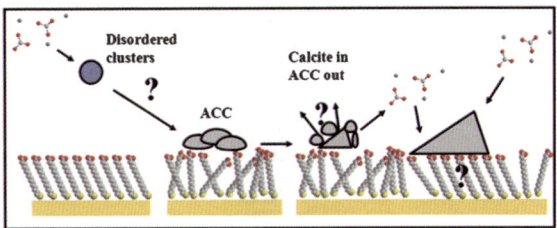

Fig. 4 Schematic of non-classical process of templated mineralization. Disordered clusters form in a supersaturated solution covering an ordered SAM. The exact steps in the phase pathway from clusters to ACC on the SAM remain unclear. At a later stage in the process, ACC has settled on the organic template, causing a restructuring of the SAM, either to another ordered configuration or to a disordered state. The ACC then transforms to the oriented crystals of the energetically favorable calcite phase, either by nucleation of the crystalline phase within the ACC or *via* dissolution of the ACC and reprecipitation of calcite. It is unclear whether this causes a further evolution in the structure of the underlying SAM.

interfacial energy, α between the nucleus and the surrounding medium (which includes the substrate). This barrier is given by:

$$\Delta g_n = f\alpha^3(\Omega/k_B T\sigma)^2 \propto \alpha^3/\sigma^2 \qquad (1)$$

where f is a numerical factor that depends on the geometry of the nucleus and Ω is the volume per molecule of the solid phase. Nucleation is also a kinetically controlled process and occurs at a rate J_n (number of nuclei per unit area per unit time) proportional to the exponential of the nucleation barrier divided by $k_B T$ according to:

$$J_n = A\exp(-E_A/k_B T)\exp(-\Delta g_n/k_B T) = A\exp(-E_A/k_B T)\exp(-B\alpha^3/\sigma^2) \qquad (2)$$

where A is a pre-factor that is independent of σ and E_A is an effective activation barrier that captures the kinetic barriers to reactions such as desolvation of solute ions, attachment to the forming nucleus and structural rearrangements. Rewriting this expression provides us with:

$$\ln(J_n) = -B\alpha^3(1/\sigma^2) + \ln(A\exp(-E_A/k_B T)) \qquad (3)$$

This shows that the absolute value of α can be determined from the slope of $\ln(J_n)$ *vs.* σ^{-2} and the relative value of activation barrier for two different SAMs can be obtained from the intercept.

To investigate the relative role of thermodynamic and kinetic controls and quantify the interfacial energies and activation barriers created by SAM templates, we developed an *in situ* continuous-flow fluid cell (Fig. 5a) to optically measure calcite nucleation rates on carboxyl-functionalized alkanethiol SAMs of 16-mercaptohexadecanoic acid (MHA) and 11-mercaptoundecanoic acid (MUA) as a function of supersaturation. To ensure that the solution had a well-defined supersaturation relative to calcite and was under-saturated with respect to ACC, both the Ca^{2+} and CO_3^{3-} sources were in solution form and were mixed in equal and known concentrations prior to entering the fluid cell. The substrates were secured face-down in the flow through cell to ensure that only heterogeneous nucleation on the SAM would give rise to the crystals observed with the inverted microscope. Care was taken to pump the salt solution into the cell at a high enough rate to ensure nucleation was limited by the surface nucleation kinetics rather than by bulk diffusion. Because the measurements were made optically, the moment of nucleus formation could not be observed. However, for the relatively low nucleation densities of our experiments

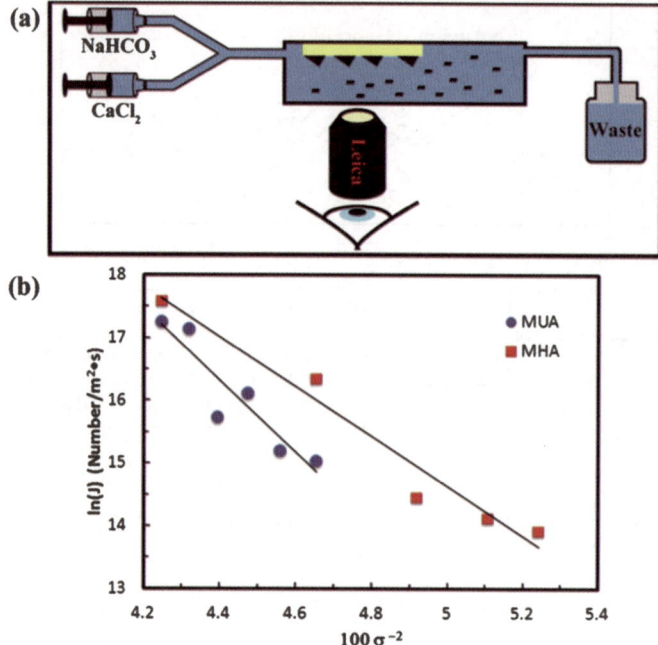

Fig. 5 (a) Experimental set-up for measurements of nucleation rates. The SAM is suspended upside-down in a flow-through cell and imaged using an inverted optical microscope to measure nucleation rates as supersaturated calcium carbonate solutions are continuously pumped through the cell to maintain constant supersaturation. (b) Plot showing the natural log of the nucleation rates on MHA and MUA *versus* the inverse square of supersaturation at room temperature. In the surface-assisted nucleation regime, the slope of the trend line yields effective interfacial energies (α), which is directly proportional to the free energy of nucleus formation Δg. Adapted from ref. 24.

we assumed that each optically-resolved crystal was the product of a single nucleus, as the separation between crystals was on the order of tens of microns. The full details and results of the experiment are discussed elsewhere.[24] These experiments constitute the first direct measurements, to our knowledge, of SAM-templated calcite nucleation rates and subsequent determination of interfacial energies and activation barriers.

In accord with previous reports, nucleation occurred on distinct crystallographic planes for the two different SAMs. The 16-carbon chain MHA SAM induced calcite nucleation almost exclusively on the non-natural (012) face, while on the 11-carbon chain MUA SAM, nucleation occurred primarily on the (013) face, demonstrating the so-called odd/even effect of SAM templating. However, on the MUA SAM two other orientations were also common. Despite these differences in nucleation plane the same value of B in eqn (3) can be used to analyze the data. For example, for nucleation on the (012) plane, $f = 19.71$. However, for a large range of nucleation planes, its value differs by no more than about 10%. Because the interfacial energy is raised to the third power, this variation has a negligible effect on the value of α extracted from the data.

Measured rates of nucleation on the two films are plotted *vs.* σ^{-2} in Fig. 5b, which shows that the measured rates exhibit the expected dependence on supersaturation. From the slope we obtain values for the interfacial energy of 72 mJ m² for MHA and 81 mJ m² for MUA. The value for MHA is in good agreement with Duffy *et al.*'s calculated 74 mJ m².[17] Our relative values for the odd/even SAMs also match well with the calculated 95 mJ m² (odd) and 74 mJ m² (even), although we used an odd-chain SAM of different carbon length. Both of these measured values are substantially smaller than the value expected for the interfacial energy of calcite in bulk solution. The expected value is surprisingly an ill-defined quantity. Sohnel and Mullin[28–30] made measurements of induction time and homogeneous nuclei to produce values of 83, 85, and 98 mJ m². However, based on their theoretical model they expected a value of 124 mJ m². They attribute the difference to a mixture of vaterite and calcite crystals forming in their experiment, and conclude that the estimation of 120 mJ m² is valid. Using the approach of Sohnel and Mullin, we expect a value of 109 mJ m² for the interfacial energy of calcite in bulk solution for our conditions. Even accounting for a substantial uncertainty in this estimation, there is still a marked decrease in the interfacial energy for calcite nucleating on the SAM. These differences in interfacial energy have a dramatic impact on nucleation rates. For example, in the middle of the investigated supersaturation range, the corresponding free energy barriers for MHA, MUA and bulk solution are 19 kT, 27 kT, and 105 kT, which would alone correspond to relative nucleation rates $J_{MHA} : J_{MUA} : J_{sol}$ of $1 : 3.4 \times 10^{-4} : 4.5 \times 10^{-38}$. Although the advantage of the MHA film over that of the MUA film is somewhat reduced because it also produces a large value of E_A (as can be seen from the smaller value of the y-intercept for MHA when extrapolated to $\sigma^{-2} = 0$) by about 7 kT, these results clearly show that both the enhancement of nucleation on the SAMs relative to bulk solution and the advantage of the SAM with the even number of carbons over that with the odd number can be explained by classical nucleation theory through differences in the interfacial energy.

4. Pathways of nucleation on SAMs

4.1 *Ex situ* studies

In attempt to determine whether the first phase to form in these nucleation rate experiments was still ACC despite the fact that the solution was undersaturated with respect to that phase, we carried out a series of micro-Raman, TEM and *in situ* atomic force microscopy (AFM) experiments under identical solution conditions. Samples were prepared in two ways. First, using the same experimental set-up as for the nucleation rate measurements, we arrested nucleation by changing the flow stream from a $CaCO_3$ solution to pure ethanol. Here we used OH-terminated SAMs to maximize the likelihood that any ACC would be preserved. This approach had proven successful in preserving the amorphous phase in the case of calcium phosphate nucleation on collagen.[31] Optical images and micro-Raman measurements revealed only calcite up to concentrations well in excess of the ACC solubility limit (supersaturation with respect to ACC of 0.4). However, close examination of such samples by SEM revealed the presence of nearly unresolvable particles in addition to the calcite rhombohedrons that were not suitable for Raman analysis. Second, we collected samples using a filtration and drying method that is commonly used to preserve ACC formed from much more concentrated solutions. The collected samples were dispersed on TEM grids and subsequent TEM analysis revealed sporadic occurrence of spherical ACC nanoparticles in addition to calcite rhombohedrons. However, we could find no conclusive evidence that these particles served as precursors to the calcite crystals and their rare occurrence raised the possibility that they were artifacts of the collection process.

4.2 *In situ* TEM: cell design

The above experiments highlight a gap in our understanding of templated nucleation due to the lack of a platform with which to observe crystallization *in situ* at high temporal and spatial resolution. Many studies have been carried out on templated calcite mineralization aimed at gaining information at high spatial resolution through the use of electron microscopes and synchrotron-based X-ray techniques. While these methods provide many useful pieces of information, as *ex situ* approaches they only provide time slices of the dynamic nucleation process. Thus it is very possible that critical steps in the process are not observed and any attempt at reconstructing, for instance, the complete phase pathway for calcite mineralizing on SAMs will yield an incomplete picture. In addition, other experiments, like those described above to optically measure nucleation densities, allow for observations at high temporal resolution at the expense of resolving details at sub-micron length scales. In order to have a system for both high spatial and high temporal resolution, *in situ* studies of biomineral nucleation and growth, we developed a fluid cell TEM system that allows for nanometer-scale measurements of dynamic processes with image collection at video rates. The system consists of a hermetically sealed cell on a custom TEM stage containing electrical feed-throughs for connection to external electronics, and is compatible with commercial electron microscopes.

TEM is an attractive system for studying template-directed biomineral formation as it allows for the concurrent observation of many different aspects of the process. In addition to very high spatial resolution, TEM can provide information on crystal phase and orientation through electron diffraction, and can further supply chemical mapping *via* energy dispersive X-ray spectroscopy (EDS) and electron energy loss spectroscopy (EELS). With this suite of tools, TEM can potentially allow for studying, among other things, phase pathways of nucleating biominerals and nucleation densities at the point of nucleation in contrast to the optical approach discussed above.

Among the challenges of using TEM to study biomineral formation, however, is timing the reaction of interest to occur when the setup is ready for measurement in the microscope. One must have a stable solution, or one that reacts at a slow enough rate, such that the system can be prepared for data acquisition prior to the start of the reaction. Clearly a trigger mechanism is necessary to efficiently run experiments of this nature in the TEM. To this end, our system has two design elements (Fig. 6a) for manually providing a driving force to initiate the mineralization process when the fluid cell is in the microscope and the microscope is aligned and ready for imaging.

The first method of control is electrochemical, using an approach reported by Gabrielli *et al.*[32] A sufficiently high electrical bias applied to the working electrode of an electrochemical cell reduces the dissolved molecular oxygen in the solution, producing hydroxide ions at the metal/liquid interface according to the following reactions.

$$O_2 + 2H_2O + 2e^- \rightarrow H_2O_2 + 2OH^- \tag{4}$$

$$H_2O_2 + 2e^- \rightarrow 2OH^- \tag{5}$$

$$O_2 + 2H_2O + 4e^- \rightarrow 4OH^- \tag{6}$$

This increases the pH at the surface of the electrode. In the case of calcium carbonate, this increases the supersaturation and drives mineralization. The results of a bench-top electrodeposition test on an unassembled cell component are shown in Fig. 6b. The optical image shows that the driving force is highly localized and that mineralization occurs primarily on the electrode, so with proper fluid cell design we can restrict the active area to regions observable in the TEM.

The second trigger mechanism that we have built into our system is temperature control. Our custom stage built by Hummingbird Scientific has a built-in Peltier device and our cell design has an attachment site for a thermocouple to provide a measurement of the fluid cell's temperature. The on-stage Peltier allows temperature

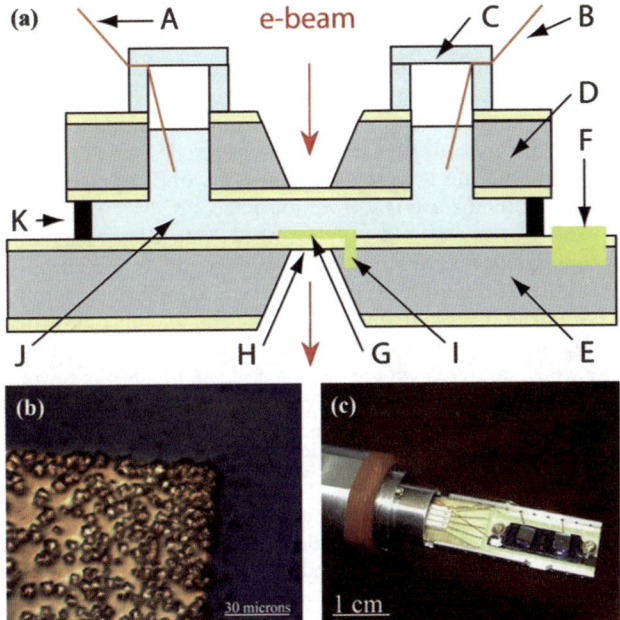

Fig. 6 (a) Schematic cross-section of fluid cell. A, B: reference and counter electrodes, C: glass cap, D, E: 100 nm Si_xN_y/300 μm Si(100)/100 nm Si_xN_y wafers, F, G: 5 nm/20 nm Ti/Au working electrode, H: Si_xN_y window, I: electrical contact between Si(100) and Au, J: solution reservoir, K: 200–500 nm Si_xN_y spacer. (b) Optical micrograph of electrodeposited calcite crystals on bench-top test of fluid cell components. (c) Photograph of fluid cell wired for electrodeposition and secured onto TEM stage.

modulation in the range of 5 °C < T < 70 °C and has been tested to be stable within ± 0.05 °C over the course of at least one hour. Based on the temperature dependence of a material's solubility, we can therefore modulate the temperature of the system to suppress the reaction until ready for imaging and then drive the process forward while synchronizing image acquisition.

Our cell design is based on the pioneering work of researchers at IBM[33] who used fluid cell TEM to quantitatively study the nucleation and growth of electrodeposited Cu islands on Au electrodes.[33–36] A side-view schematic of the initial design can be seen in Fig. 6a. The cell is composed of two 300 micron thick silicon wafers with silicon nitride membranes coating both sides of each wafer. The silicon and nitride is etched away to produce nitride membranes that constitute the electron-transparent windows for imaging in the TEM, as well as solution reservoirs to either side of the imaging area. On the lower wafer, vias are etched through the nitride to the underlying highly doped silicon wafer, and Ti/Au pads constituting the working electrode are evaporated on the wafer, one external to the fluid chamber and the other internal, extending over the imaging window. Additionally, on the lower wafer a spacer layer of silicon nitride is deposited to create the gap to support a fluid layer of an appropriate thickness. This spacer layer can be tailored to the experiment; to date our cells have been fabricated with spacer thicknesses ranging between 200–500 nm.

The two wafers are glued together by drawing a small bead of epoxy (M-Bond 610) around the outer edges of the aligned wafers and curing at high temperature. Two silicon towers that act as solution reservoirs are glued to the upper wafer using the same epoxy. The vast bulk of the fluid cell's reaction volume (total volume is 1–4 μL, depending on specific geometry) resides in these solution reservoirs, with less than 1% of the total volume expected in the thin layer between the silicon wafers.

When the cell is assembled to this point, solution is injected into one of the reservoirs, where it is drawn through the cell *via* capillary action. The second reservoir is then filled with solution. Wires constituting reference and counter electrodes are each inserted into one of the reservoirs, glass caps are placed over the openings and sealed to the towers using a UV-curable epoxy (Norland Opticure 63). This procedure results in a hermetically sealed, three-electrode fluid cell vacuum compatible with the TEM. A fully assembled *in situ* cell wired for electrochemistry and secured on the TEM fluid stage is shown in Fig. 6c.

Over the course of extensive testing of the system detailed above, a number of issues came to light that significant hindered the ability to conduct experiments in the TEM. One of the major issues was with the design of the electrochemical set-up, with problems arising both from the use of wires as counter and reference electrodes as well as the design of the working electrode patterned onto the lower cell component. We found that the wires interfered with achieving a secure seal at the towers, often with the result that pin-hole leaks developed in the glue line and resulted in the solution leaking out of the cell over time. In addition, the reaction solution tended to wick out of the reservoir along the wire, lowering the liquid level to the point that the wire was no longer submerged, thus disconnecting the electrode from the system.

The working electrode was found to be problematic due to electrical connectivity issues and an excess of active area outside of the imaging window. To be able to observe nucleation and growth of materials on the electrode at the initial stages of the process the electrode must be suitably thin. As such we fabricated the electrodes by depositing 5 nm Ti as an adhesion layer and 20 nm Au for the electrode. We found, however, that this thin layer of metal often developed a break between the regions covering the nitride and the *via* into the underlying silicon, causing the electrode section extending over the imaging window to be disconnected from external control.

Even when the electrode had continuity out to the imaging area, the portion of the electrode visible in the microscope constituted only 15–20% of the total active area. This resulted in a sizable probability of nucleation events occurring on regions of the electrode not visible in the microscope, decreasing the likelihood of making successful observations in a given experiment. In addition to the aforementioned design issues, investigating SAM-templated calcite nucleation would be problematic, because the thiol-gold bond strength is similar in magnitude to the applied electrical bias needed to drive mineralization.[37]

There was also a significant design issue with the size of the nitride windows. The window on the upper part was greatly elongated transverse to the body of the fluid stage to facilitate EDS analysis with minimal tilting of the stage. The large size of the window, however, made the window both more fragile and subject to a large amount of flex. The window can bow out from the cell during fluid injection and sealing of the cell, and can also flex inward following the curing of the sealant. In the event that it bows out, the resultant fluid layer thickness can be microns thick rather than the desired, nominal sub-micron thickness. Further complicating matters, when filling the cell with solution it is not uncommon for a pocket of air to get trapped under the window as it bows out. Such air bubbles are difficult to remove from the cell and seem to promote the trapping of more bubbles if an attempt is made to dry the cell and refill it. When a bubble covers the imaging area it renders the cell useless for *in situ* imaging. When the window flexes inward, the two windows can press together and force out any liquid that was previously in the imaging area.

The design issues raised above led us to make a number of modifications to the cell. All three of the electrodes are now patterned on the surface of the nitride membrane on the lower wafer, eliminating the need to feed wires into the cell and eliminating continuity problems from breaks in the electrode conduction path. In addition to depositing the spacer layer for separation of the two major cell components, we now deposit a passivation layer over much of the electrode surface area internal to the cell. The result is that the section of the working electrode visible in the TEM is around 80% of the total active area of the electrode, greatly decreasing the chance that reactions occur outside the visible area. For studies that involve SAMs, the working electrode is now patterned across the window with e-beam lithography to produce patches of gold electrically isolated from the electrode but separated by no more than a few microns, so that they should still be in the region of increased pH as a bias is applied to the nearby electrode.[38,39] To mitigate window flex, the area of the upper window has been greatly reduced, which also decreases the likelihood of bubbles becoming trapped between the windows. Although the window is now a fraction of its original size, EDS is still possible as the stage has a tilt range of ±25°.

4.3 In situ TEM: initial results for CaCO$_3$ nucleation on Au

TEM imaging in the liquid cell can be conducted in conditions such that the electron beam is not the primary driving force for the reaction,[33–36] but imaging can also be carried out with the intent of having the electron beam initiate the process, as demonstrated in Zheng et al.'s study of the formation of platinum nanoparticles.[40] There are a variety of beam effects that might occur from interaction with the liquid layer, both thermal and non-thermal. Beam heating has been calculated[41] to produce a small temperature rise of at most a few degrees, due to the high thermal conductance of the liquid and silicon wafers. Non-thermal effects include secondary electron generation, charging, and production of reactive species as a result of interactions between the electron beam and the aqueous liquid.[42] Although our goal is to control nucleation and growth of CaCO$_3$ by the application of a measured thermal or electrochemical driving force, we found that beam effects under conditions of high electron beam intensity could also initiate CaCO$_3$ nucleation and growth in the fluid cell.

Imaging was conducted with a JEOL 2100F TEM at an accelerating voltage of 200 kV using the free VirtualDub software to capture video from the live view images in Digital Micrograph (Gatan). For file size considerations images were acquired every 0.2 s, although image acquisition can be as rapid as 30 frames per second. The liquid cell was filled with aqueous solution containing 5 mM calcium chloride hydrate (Sigma-Aldrich 99.999%) and 5 mM sodium bicarbonate (Alfa Aesar 99.998%), and sealed as described. The cell was not connected to any external electronics for electrochemical control or measurement. During TEM alignment there was no observable reaction occurring in the cell. Magnification and beam current density were increased until CaCO$_3$ mineralization became visible.

Fig. 7 shows the initial stages of CaCO$_3$ nucleation on the Au electrode. In Fig. 7a, nucleation has not yet occurred; the Au electrode can be seen as the dark structure extending across the bottom and right side of the image, and the silicon nitride window as the light gray area. In the first few seconds numerous particles with no discernable structure nucleate on the surface and along the edge of the electrode (Fig. 7b–d). These particles are presumably ACC, given their lack of structure, and can be seen as the small round protrusions around 15 nm in size in the first panel. After about 60 s of growth the ACC particles merge together and begin conversion into faceted crystals (Fig. 7e–g). Following the conversion of ACC to the crystalline phase the initially formed crystals merged and continued to grow into larger faceted crystals (Fig. 7h–i). The result of this coarsening process is seen by comparing Fig. 7g and 7i. (Note that due to instrument drift the imaged area is slightly shifted from one image to the next.)

Later on in the growth process, we also observed CaCO$_3$ nanophase stability and ripening phenomena as shown in Fig. 8. In Fig. 8a, there are three relatively large crystals hundreds of nanometers in size, as well as a small aggregate of particles drifting in solution near the top of the image. The lack of faceting and absence of growth suggests that this aggregate is a less stable phase of CaCO$_3$, such as ACC, vaterite or aragonite. While the large crystals are growing at a steady rate, the size of the isolated aggregate remains unchanged. With time, it drifts towards one of the large

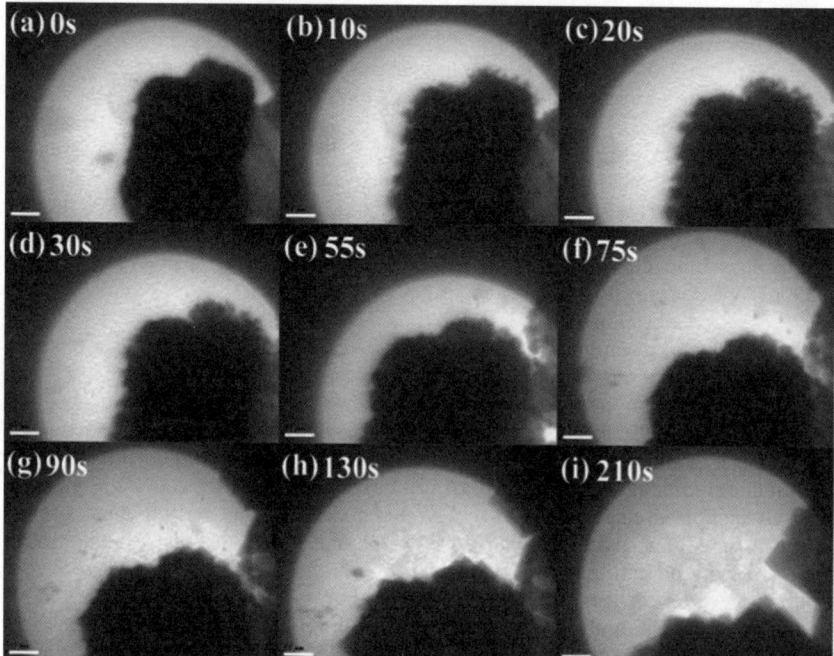

Fig. 7 Selected images from video of CaCO$_3$ nucleation, with 50nm scale bars. (a) The edge of the Au electrode before CaCO$_3$ deposition. (b–d) Formation of small particles, presumably ACC, with no discernible structure. (e–g) Merging of ACC and conversion into faceted crystals. (h–i) Continued growth into large, faceted crystals. Change in imaging area between panels is a combined result of sample drift and repositioning.

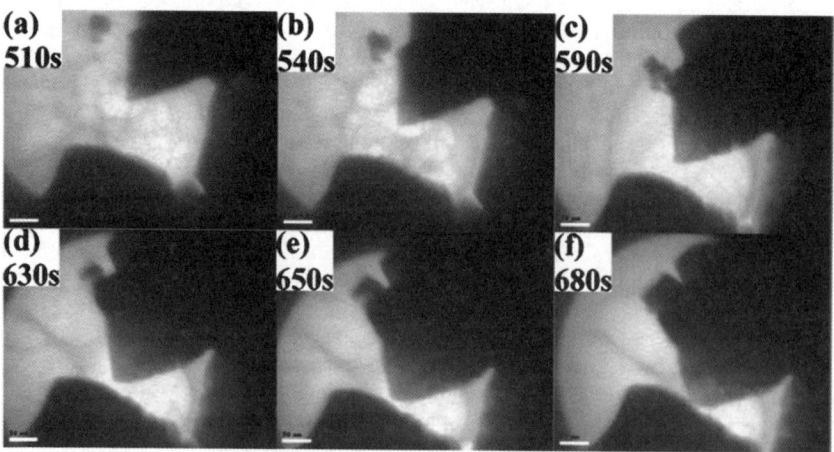

Fig. 8 Selected images from CaCO$_3$ video, with 50nm scale bars. (a) Three large crystals dominate imaging area, with detached aggregate in solution. (b–c) While crystals grow, aggregate maintains size and drifts towards nearby crystal. (d–f) Attachment of aggregate to crystal, followed by rapid growth and development of facets as aggregate transforms into crystal.

growing crystals and attaches to it (Fig. 8b–c). Immediately upon attachment, it begins to rapidly grow into a single faceted extension of the main crystal (Fig. 8d–f), presumably due to its transformation into the more stable crystalline phase.

5. Conclusion

The findings reported here provide insights into the structural and energetic aspects of SAM-directed nucleation of $CaCO_3$. The combined NEXAFS results from the previous work on MP SAMs and the findings contained herein on MDBA SAMs constitute experimental evidence of, in these systems at least, organic templates that direct the orientation of calcite nucleation undergoing a reorganization of the interfacial structure of the monolayer during the mineralization process. Furthermore, the results suggest that the capacity to relax the interfacial structure and reorganize during nucleation is a necessary feature for an organic scaffold to exert a templating control over nucleation. As such, it may be appropriate to think of calcite nucleating on SAMs as a co-evolution of structure between the two phases wherein both the organic and mineral phases direct the structure of the other phase throughout the process.

Measurements of nucleation rates have allowed for quantification of thermodynamic parameters for calcite nucleating on two carboxyl-terminated SAMs, showing that the organic templates exert a thermodynamic control over calcite nucleation. Differences in interfacial energy and corresponding free energy barriers greatly impact nucleation rates. These differences can be used to explain, in terms of classical nucleation theory, the enhancement of calcite nucleation on SAMs over bulk solution and the differences between SAMs with even and odd numbers of carbons in the alkane chain.

The question of nucleation pathways for different solution conditions is still a little unclear. While there exist conditions such that pre-nucleation clusters form initially and transform into ACC and on to crystalline phases, it is unknown whether or not this holds true for conditions undersaturated with respect to ACC. Our *ex situ* measurements did not yield definitive answers as to the presence or absence of ACC as a precursor to calcite for conditions ranging from unsaturated to moderately saturated with respect to ACC. The direct, *in situ* TEM observation of calcium carbonate nucleating on gold suggests that ACC is a precursor to the final crystalline phase. However, the use of the electron beam as driving force for mineralization in this observation precludes knowledge of the conditions in the reaction volume. As such, we cannot, as of yet, make concrete conclusions about direct *vs.* multi-step phase pathways for calcite nucleation varying supersaturations.

Acknowledgements

This research was supported by the U.S. Department of Energy, Office of Basic Energy Sciences through Lawrence Berkeley National Laboratory and as part of the Center for Nanoscale Control of Geologic CO_2, an Energy Frontier Research Center, both under contract No. DE-AC02-05CH11231, and through Lawrence Livermore National Laboratory under Contract DE-AC52-07NA27344. Additional government support under and awarded by DoD, Air Force Office of Scientific Research, National Defense Science and Engineering Graduate (NDSEG) Fellowship, 32 CFR 168a. Measurements were performed at the Molecular Foundry, and at the Stanford Synchrotron Radiation Laboratory, national user facilities operated by Lawrence Berkeley National Laboratory and Stanford University, respectively, on behalf of the U.S. Department of Energy, Office of Basic Energy Sciences. The authors thank the SSRL staff, particularly Dan Brehmer and Curtis Troxel, for their assistance during the course of these experiments.

References

1 J. R. Young, J. M. Didymus, P. R. Bown, B. Prins and S. Mann, *Nature*, 1992, **356**, 516–518.
2 S. Mann, *Nature*, 1993, **365**, 499–505.

3 J. R. Young and K. Henriksen, in *Biomineralization*, ed. P. M. Dove, J. J. De Yoreo, and S. Weiner, Mineralogical Society of America, Washington D. C., 2003, pp. 189–215.
4 J. Aizenberg, *Adv. Mater.*, 2004, **16**, 1295–1302.
5 S. Mann, B. R. Heywood, S. Rajam and J. D. Birchall, *Nature*, 1988, **334**, 692–695.
6 S. Mann, D. D. Archibald, J. M. Didymus, T. Douglas, B. R. Heywood, F. C. Meldrum and N. J. Reeves, *Science*, 1993, **261**, 1286–1292.
7 B. R. Heywood, S. Rajam and S. Mann, *J. Chem. Soc., Faraday Trans.*, 1991, **87**, 735–743.
8 S. Rajam, B. R. Heywood, J. B. A. Walker, S. Mann, R. J. Davey and J. D. Birchall, *J. Chem. Soc., Faraday Trans.*, 1991, **87**, 727–734.
9 A. Berman, D. J. Ahn, A. Lio, M. Salmeron, A. Reichert and D. Charych, *Science*, 1995, **269**, 515–518.
10 J. Aizenberg, A. J. Black and G. H. Whitesides, *J. Am. Chem. Soc.*, 1999, **121**, 4500–4509.
11 J. Aizenberg, A. J. Black and G. M. Whitesides, *Nature*, 1999, **398**, 495–498.
12 A. M. Travaille, J. Donners, J. W. Gerritsen, N. Sommerdijk, R. J. M. Nolte and H. van Kempen, *Adv. Mater.*, 2002, **14**, 492.
13 Y. J. Han and J. Aizenberg, *Angew. Chem., Int. Ed.*, 2003, **42**, 3668–3670.
14 Y. J. Han and J. Aizenberg, *J. Am. Chem. Soc.*, 2003, **125**, 4032–4033.
15 D. M. Duffy and J. H. Harding, *Langmuir*, 2004, **20**, 7637–7642.
16 D. M. Duffy and J. H. Harding, *Surf. Sci.*, 2005, **595**, 151–156.
17 D. M. Duffy, A. M. Travaille, H. van Kempen and J. H. Harding, *J. Phys. Chem. B*, 2005, **109**, 5713–5718.
18 C. L. Freeman, J. H. Harding and D. M. Duffy, *Langmuir*, 2008, **24**, 9607–9615.
19 J. R. I. Lee, T. Y. J. Han, T. M. Willey, D. Wang, R. W. Meulenberg, J. Nilsson, P. M. Dove, L. J. Terminello, T. van Buuren and J. J. De Yoreo, *J. Am. Chem. Soc.*, 2007, **129**, 10370–10381.
20 D. Gebauer, A. Volkel and H. Colfen, *Science*, 2008, **322**, 1819–1822.
21 E. M. Pouget, P. H. H. Bomans, J. A. C. M. Goos, P. M. Frederik, G. de With and N. A. J. M. Sommerdijk, *Science*, 2009, **323**, 1555–1458.
22 D. C. Popescu, M. M. J. Smulders, B. P. Pichon, N. Chebotareva, S. Y. Kwak, O. L. J. van Asselen, R. P. Sijbesma, E. DiMasi and N. Sommerdijk, *J. Am. Chem. Soc.*, 2007, **129**, 14058–14067.
23 E. DiMasi, S. Y. Kwak, B. P. Pichon and N. Sommerdijk, *CrystEngComm*, 2007, **9**, 1192–1204.
24 Q. Hu, M. H. Nielsen, C. L. Freeman, C. M. Hamm, J. Tao, J. R. I. Lee, T. Y. J. Han, U. Becker, J. H. Harding, P. M. Dove and J. J. De Yoreo, *Faraday Discuss.*, 2012, **159**, in press.
25 J. R. I. Lee, T. Y. J. Han, T. M. Willey, M. H. Nielsen, L. M. Klivansky, Y. Liu, S. W. Chung, L. J. Terminello, T. van Buuren, and J. J. De Yoreo, in prep.
26 E. Loste, R. M. Wilson, R. Seshadri and F. C. Meldrum, *J. Cryst. Growth*, 2003, **254**, 206–218.
27 J. J. De Yoreo, and P. G. Vekilov, in *Biomineralization*, ed. P. M. Dove, J. J. De Yoreo, and S. Weiner, Mineralogical Society of America, Washington D. C., 2003, pp. 57–93.
28 O. Sohnel and J. W. Mullin, *J. Cryst. Growth*, 1978, **44**, 377–382.
29 O. Sohnel, *J. Cryst. Growth*, 1982, **57**, 101–108.
30 O. Sohnel and J. W. Mullin, *J. Cryst. Growth*, 1982, **60**, 239–250.
31 J. Tao, *personal communicaiton*.
32 C. Gabrielli, G. Maurin, G. Poindessous and R. Rosset, *J. Cryst. Growth*, 1999, **200**, 236–250.
33 M. J. Williamson, R. M. Tromp, P. M. Vereecken, R. Hull and F. M. Ross, *Nat. Mater.*, 2003, **2**, 532–536.
34 A. Radisic, F. M. Ross and P. C. Searson, *J. Phys. Chem. B*, 2006, **110**, 7862–7868.
35 A. Radisic, P. M. Vereecken, J. B. Hannon, P. C. Searson and F. M. Ross, *Nano Lett.*, 2006, **6**, 238–242.
36 A. Radisic, P. M. Vereecken, P. C. Searson and F. M. Ross, *Surf. Sci.*, 2006, **600**, 1817–1826.
37 H. Gronbeck, A. Curioni and W. Andreoni, *J. Am. Chem. Soc.*, 2000, **122**, 3839–3842.
38 C. Deslouis, I. Frateur, G. Maurin and B. Tribollet, *J. Appl. Electrochem.*, 1997, **27**, 482–492.
39 M. M. Tlili, M. Benamor, C. Gabrielli, H. Perrot and B. Tribollet, *J. Electrochem. Soc.*, 2003, **150**, C765–C771.
40 H. M. Zheng, R. K. Smith, Y. W. Jun, C. Kisielowski, U. Dahmen and A. P. Alivisatos, *Science*, 2009, **324**, 1309–1312.
41 H. M. Zheng, S. A. Claridge, A. M. Minor, A. P. Alivisatos and U. Dahmen, *Nano Lett.*, 2009, **9**, 2460–2465.

42 B. C. Garrett, D. A. Dixon, D. M. Camaioni, D. M. Chipman, M. A. Johnson, C. D. Jonah, G. A. Kimmel, J. H. Miller, T. N. Rescigno, P. J. Rossky, S. S. Xantheas, S. D. Colson, A. H. Laufer, D. Ray, P. F. Barbara, D. M. Bartels, K. H. Becker, H. Bowen, S. E. Bradforth, I. Carmichael, J. V. Coe, L. R. Corrales, J. P. Cowin, M. Dupuis, K. B. Eisenthal, J. A. Franz, M. S. Gutowski, K. D. Jordan, B. D. Kay, J. A. LaVerne, S. V. Lymar, T. E. Madey, C. W. McCurdy, D. Meisel, S. Mukamel, A. R. Nilsson, T. M. Orlando, N. G. Petrik, S. M. Pimblott, J. R. Rustad, G. K. Schenter, S. J. Singer, A. Tokmakoff, L. S. Wang, C. Wittig and T. S. Zwier, *Chem. Rev.*, 2005, **105**, 355–389.

PAPER

A two-step mechanism for crystal nucleation without supersaturation

Tamás Kovács and Hugo K. Christenson*

Received 19th March 2012, Accepted 4th April 2012
DOI: 10.1039/c2fd20053h

There is currently considerable interest in two-step models of crystal nucleation, which have been implicated in a number of systems including proteins, colloids and small organic molecules. Classical nucleation theory (CNT) postulates the formation of an ordered crystalline nucleus directly from dilute vapour or solution. By contrast, the new models explain how crystallisation *via* a more concentrated but still fluid (disordered) phase can lead to a significant enhancement of nucleation rates. In this article, we extend recent work showing that crystal deposition from vapour can also be greatly accelerated by the operation of a two-step mechanism. The process relies on a very acute, annular wedge, in which restricted amounts of liquid condense below the bulk melting point T_m. Crystals then nucleate in the liquid condensates at sufficient temperature depressions ΔT (typically $\geq$30 K) below T_m, followed by rapid growth of these crystals from the saturated vapour. By using a range of model substances (neopentanol, norbornane, hexamethylcyclotrisiloxane, hexachloroethane, menthol, cyclooctane and pinacol) we show that this is a viable mechanism for substances with reasonably high absolute vapour pressures (>*ca.* 1 mm Hg). The lack of appreciable crystal deposition with substances of significantly lower vapour pressures (<*ca.* 0.01 mm Hg) is most likely due to geometric restrictions impeding diffusion in our experimental set-up. The results confirm the feasibility of a mechanism for atmospheric ice nucleation that has been suggested in the literature. Furthermore, there are thermodynamic analogies with the crystallisation of biominerals *via* amorphous or fluid-like precursor phases and protein nucleation in surface topographical features.

1 Introduction

The nucleation of solid from vapour is of great importance in cloud formation and atmospheric precipitation, both on earth and in other planetary atmospheres. It also plays an essential role in industrial technologies that use chemical or physical vapour deposition, and an understanding of this process is essential for its control.

The formation of crystals from vapour typically starts with heterogeneous nucleation on a surface such as an impurity particle or the container wall. Although the free energy barrier is much reduced compared to the case of homogeneous nucleation, supersaturations of 25–50% are often required, *e.g.* for the nucleation of ice on solid aerosol particles.[1] Classical nucleation theory (CNT) relates the free energy barrier towards nucleation to the surface free energy cost of forming the nucleus of a new phase. Although originally devised for the case of liquid condensing from vapour, it has routinely been applied to the nucleation of crystals both from solution and from vapour. Despite the added complications due to the anisotropy of the

University of Leeds, School of Physics and Astronomy, Leeds, United Kingdom. E-mail: h.k.christenson@leeds.ac.uk; Fax: +44 113 3433879; Tel: +44 113 3433900

crystalline state, CNT has had some success in qualitatively accounting for experimental observations.

However, it has become clear that there are many instances where the simple picture provided by CNT does not agree with experiment or the results of computer simulations.[2] The new phase does not form simply as a result of a random fluctuation that brings together a sufficient number of molecules to create a nucleus. Rather, the nucleation occurs in two stages; first a denser aggregate of molecules forms and then within this aggregate the actual nucleation takes place. Further growth of this crystal nucleus then reduces its free energy. In many systems the formation of a denser, metastable fluid state from a dilute solution is kinetically favoured. The crystalline phase may then nucleate from this metastable state, and the phase transition to a thermodynamically stable phase thus proceeds *via* a two-step mechanism.

The idea of two-step nucleation has been successfully applied to solutions like proteins,[2-4] small organic molecules[2,4,5] and colloids.[6] According to simulations of protein crystallization,[7,8] the first step is the formation of the solute cluster which is then followed by its reorganisation into a more ordered structure. As the rearrangement time increases significantly with molecular complexity it was suggested that the rate-determining step is the reorganisation.

Crystallisation from vapour requires the surmounting of a particularly high free energy barrier due to the large free energy of a crystalline surface in vapour. However, in the one-step mechanism of CNT, changes in the surface topography that increase the surface-nucleus area can facilitate solid nucleation.[9,10] On the other hand, as liquids usually have lower surface energies than solids undercooled liquid often deposits from vapour below the bulk melting point T_m. In wedge-shaped (grooves or conical pits) condensation of liquid, both above and below T_m, may take place without any free energy barrier whatsoever, provided that the contact angle of the liquid on the solid is less than half the wedge or cone angle.[10]

Condensation of liquid may occur even from undersaturated vapour in fine pores and is termed capillary condensation. It takes place because the liquid phase is always stabilised relative to the vapour phase in a sufficiently narrow pore as long as the liquid has a contact angle θ below 90° on the pore walls. This is in practice not a very restrictive condition and most substances that are liquid under ambient conditions have low contact angles. The important exceptions are water on hydrophobic surfaces and metals like mercury. A θ < 90° implies a liquid–vapour interface that is concave towards the vapour phase, and the vapour pressure over this is lower than over a flat surface. Consequently, the pore-held liquid is in equilibrium with undersaturated vapour, which is quantitatively described by the Kelvin equation,

$$r = \frac{V_M \gamma_{lv}}{RT \ln(p/p_s)} \quad (1)$$

where T is the temperature, p_s is the saturation vapour pressure, V_M is the molar volume of the condensing substance, γ_{lv} the surface tension of the liquid and r the total radius of curvature of the interface—negative for a concave interface. When $p = p_s$ the liquid–vapour interface has zero curvature ($r = \infty$) and we have the case of bulk liquid. Capillary condensation leads to the absorption of vapours by porous media and hence plays an important role in the processes of drying and moisture retention in soils, construction materials and other porous bodies.[11] It leads to clogging of finely divided grains and powders at high humidities.

Capillary condensation is a first-order phase-transition and is in general subject to hysteresis like condensation of bulk liquid and crystallisation. In a wedge-shaped, or conical pit, however, a capillary condensate can grow continuously and reversibly from the inside of the vertex—as discussed above in the context of the free energy barrier towards nucleation in a surface cavity.

Below T_m the stable bulk phase is usually a crystalline solid. Just like in the case of liquid and vapour the proximity of two surfaces favours the liquid and is, for small enough pores, sufficient to stabilise the liquid relative to the crystal. The liquid condensate grows until gains in interfacial energy by keeping the condensate liquid is balanced by the unfavourable entropy of melting. One of us has shown that the equilibrium radius of curvature r of the liquid–vapour interface of the supercooled condensate is given by[12]

$$r = \frac{V_M \gamma_{lv}(T) T_m}{\Delta T \left(\Delta H_{fus} - T_m \Delta C_p \left(1 + \frac{T}{\Delta T} \ln\left(\frac{T}{T_m}\right) \right) \right) + RTT_m \ln\left(\frac{p}{p_s}\right)} \quad (2)$$

Here $\Delta T = T_m - T$ (the undercooling), ΔH_{fus} is the heat of fusion, ΔC_p is the difference in the molar heat capacity of the liquid and solid, γ_{lv} is the (temperature-dependent) surface tension of the liquid. At saturation ($p = p_s$) this expression reduces to

$$r = \frac{V_M \gamma_{lv}(T) T_m}{\Delta T \left(\Delta H_{fus} - T_m \Delta C_p \left(1 + \frac{T}{\Delta T} \ln\left(\frac{T}{T_m}\right) \right) \right)} \quad (3)$$

If the temperature dependence of γ_{lv} and ΔH_{fus} is neglected the expression becomes very simple;

$$r = \frac{V_M \gamma_{lv} T_m}{\Delta T \Delta H_{fus}} \quad (4)$$

So below T_m the radius of curvature of the condensate is inversely proportional to the undercooling, instead of being inversely proportional to the $\ln[p/p_s]$ as above T_m (the Kelvin equation). In both cases r is equal to $V_M \gamma_{lv}$ divided by the difference in free energy between the condensed liquid and the bulk liquid, *i.e.* $RT \ln[p/p_s]$ above T_m and $\Delta T \Delta H_{fus}/T = \Delta T \Delta S$ below T_m.

The above relationships for the curvature of the liquid–vapour interface both above and below T_m have been experimentally verified in a series of experiments with the surface force apparatus (SFA).[12–14] This instrument was originally designed to measure forces between two molecularly smooth mica surfaces in air[15] and liquids.[16–18] Multiple-beam interferometry, which is used to measure the surface separation in the SFA also permits the refractive index of the medium between the mica surfaces to be determined.[19,20] This makes the instrument ideal for the study of phase changes in pores, such as capillary condensation or freezing-point depression. The two mica surfaces are in a crossed-cylinder configuration, which is equivalent to a sphere-on-a-flat, so that when the two surfaces are in contact an annular, wedge-like pore is created around the contact point (Fig. 1). In this pore capillary condensation may be studied under hysteresis free conditions, but if the surfaces are separated the local environment approximates a slit-pore, which allows hysteresis effects to be investigated. Both types of pore coexist with a bulk reservoir in a sealed and temperature-controlled chamber.

One of us has considered previously the conditions under which the size-limited liquid capillary condensates formed below T_m might freeze.[12,13] A simple argument based on a comparison of surface and interfacial free energies and reasonable assumptions about the wetting behaviour of the liquid and the crystalline solid has suggested that a crystalline condensate would be thermodynamically stable in the outer part of the wedge as long as $\gamma_{lv} > 2\gamma_{sl}$, the interfacial tension between the liquid and the crystalline solid. A kinetic argument based on homogeneous nucleation according to CNT would change this to $\gamma_{lv} > 3\gamma_{sl}$. These conditions should be easily met by most liquids, except possibly the second one for water.[13] Experimentally, however, no freezing could be observed for cyclohexane down to 13 K below T_m,[12] or for water down to 9 K below T_m.[13] This is hardly surprising as undercoolings of 30–40 K are frequently required for homogeneous nucleation in pure liquids, and the mica surface is unlikely to promote significantly the nucleation of water or cyclohexane.

Very recently[21] we have shown that capillary condensates will indeed freeze for large enough ΔT, and that this is followed by the deposition of large amounts of solid, since there is no limit on the growth of the stable bulk phase, unlike the case with the liquid condensates. In our preliminary study[21] deposition of solid from vapour was observed at large enough undercoolings ($\Delta T > 18$ K for neopentanol, $\Delta T > 33$ K for HMCTS and $\Delta T > 37$ K for norbornane) and it was concluded that deposition of crystals from vapour *via* liquid condensates is possible when the absolute vapour pressure is high enough (>7 mm Hg)—one of the reasons why the above substances were chosen for study.

In an earlier study one of us has shown that capillary condensates of long-chain *n*-alkanes confined between mica surfaces will freeze quite easily if the mica surfaces are separated and the condensate forms a liquid bridge.[22,23] This can be related to the surface freezing of these alkanes at the liquid–vapour interface, which means that this interface very readily nucleates the bulk crystal. For the same reason these long-chain alkanes will not supercool to any significant extent and their nucleation behaviour is anomalous.

We here present further investigations of crystal nucleation from vapour *via* capillary condensates using four additional substances (cyclooctane, menthol, pinacol, and hexachloroethane), some with significantly lower absolute vapour pressures (down to *ca*. 10^{-3} mm Hg at the lowest experimental temperatures). These experiments confirm the generality of the phenomenon but suggest that diffusion-related effects play a key role in limiting this type of nucleation if the geometry is too restricted.

2 Experimental

The experiments were carried out with a simplified surface force apparatus (SFA). A detailed description of this instrument has been given[24–26] and we here restrict ourselves to a brief summary.

For each experiment 2–5 μm thick mica sheets are cleaved from a 0.1–0.5 mm thick mica block (Paramount Corp., New York) and cut into approximately 1 cm^2 squares with a hot platinum wire.[27] The pieces are then coated with 50 nm Ag (99.99%, Advent) by thermal evaporation at $p = 10^{-6}$ mbar. The silvered mica sheets are glued onto cylindrically polished silica discs (radius of curvature $R = 2$ cm) using an epoxy resin (Epikote 1004), with the lower disk attached to a rigid support. White light from a 150 W 21 V halogen bulb is directed onto the opposing, back-silvered mica surfaces using an optical fibre and the transmitted light is resolved into discrete wavelengths (FECO—fringes of equal chromatic order[19]) with a monochromator (McPherson, model No. 2035). The interference fringes are recorded with a CCD camera (Perkin-Elmer Pixcellent) and stored digitally. The apparatus is housed in a thermostated enclosure that allows the temperature to be controlled and measured to within ±0.1 °C over the range of −10 to +50 °C, with a platinum thermometer placed 2 cm from the surfaces. The separation of the surfaces is controlled with a piezoelectric device to within ±0.2 nm and measured with multiple-beam interferometry.

At the beginning of each experiment the surfaces are brought into contact in a N$_2$ atmosphere at 22 °C in order to establish the zero of separation, or mica–mica contact. Empirical corrections are made to account for the thermal expansion of mica (more properly changes in the optical path length—typically corresponding to 10^{-5} K^{-1}), by determining mica–mica contact as a function of temperature. The substance to be studied is then introduced, with at least five times the amount required to saturate the chamber added. In order to achieve faster equilibration the temperature is usually set to 30 °C and 2–4 days are allowed for equilibration.

Analytical purity substances were used (cyclooctane: 99%, neopentanol: 99%, Sigma-Aldrich, norbornane: 98%, Sigma-Aldrich, hexamethylcyclotrisiloxane, henceforth abbreviated HMCTS: 98%, Arcos Organics, menthol: 99%, Sigma-Aldrich, hexachloroethane: 98%, Alfa Aesar, pinacol: 99%, Arcos Organics). All experiments were carried out in the presence of drying agent (molecular sieves, 3A, Sigma-Aldrich).

A typical experiment involves bringing the surfaces together slowly until they are pulled together by the formation of a capillary condensate at separations of about 15 nm, although this was not accurately measured. The growth of any annular condensate around the flattened contact zone is then followed, while the contact separation and the refractive index of the film between the two surfaces is measured. The condensate size h is determined from the discontinuities in the interference fringes due to refractive index changes at the condensate vapour interface (Fig. 2). The radius of curvature of the condensate–vapour interface r is given by[12,28]

$$2r + 3t = h \qquad (5)$$

where t is the adsorbed film thickness of the substance on the surfaces at large separations. It has been shown that an average adsorbed film thickness for many substances below T_m is about 1 nm,[12,29] and we used this value, except where otherwise noted.

After varying times (from a few s to 3 h) in contact the surfaces are separated and the outwards jump measured. The surfaces are then left well apart to let deposited material evaporate, and a number of repeat cycles carried out. When measurements are concluded at one temperature overnight equilibration is allowed at each new temperature.

The phase state of the condensate is established by monitoring the behaviour of the condensate on attempting to separate the surfaces. With a liquid condensate the contact diameter decreases as the load on the surfaces is decreased and as the surfaces finally jump apart the annular condensate becomes a bridge joining the surfaces. The adhesion or pull-off force F, when normalised by R, is related to the surface tension of the liquid γ_{lv} by $F/R = 4\pi\gamma_{lv}$,[24] i.e. which translates to typical outward jumps of 40–60 nm and normalised forces of 0.25–0.4 Nm^{-1}. On further separation the bridge snaps and droplets spread on each of the two surfaces. By contrast, a solid condensate does not flow and the contact diameter cannot decrease as the surfaces are effectively bonded together by the solid. Instead, noticeable deformations are evident on the fringes, at separations beyond the location of the condensate. The force that has to be used to pull the surfaces apart is an order of magnitude larger than with the liquid condensates.

In order to study the effect of supersaturating the vapour phase a specially designed side plate to be bolted to the main chamber was constructed. The crystals

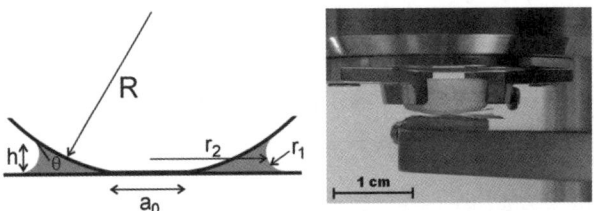

Fig. 1 Left: schematic cross-section of mica surfaces shown in the equivalent sphere-on-a-flat configuration, with a liquid capillary condensate around the flattened contact region. Typically, $R = 2$ cm, $a_0 \sim 25$ μm, $r_2 \sim +13$–50 μm, $h/2 \approx r_1 \sim -(5$ nm $- 1$ μm). Right: close-up of the surfaces in contact, with the lower surface on a rigid support and the upper on a piezoelectric actuator.

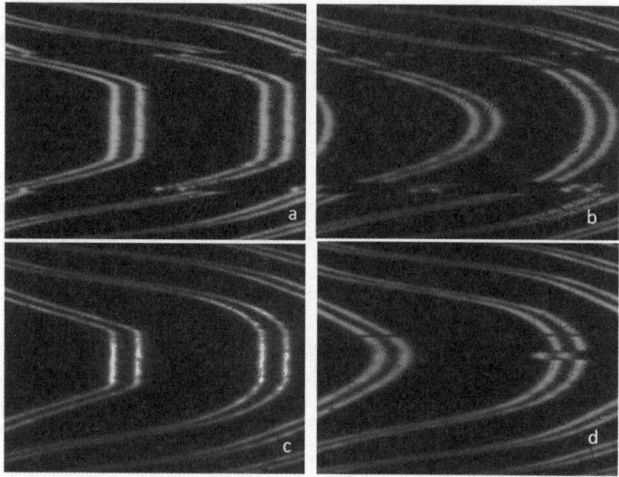

Fig. 2 Interference fringes of mica surfaces in neopentanol vapour. The fringes are doublets due to the birefringence of mica, and the wavelength increases towards the left. a) After 2 h in contact in neopentanol vapour at $\Delta T = 33$ K with solid annulus of $h \approx 200$ nm showing discontinuous shifts in the wavelength at the condensate–vapour interface b) after large jump apart, showing annular, solid residue on the surfaces, c) after 2 h in contact in neopentanol vapour at at $\Delta T = 28$ K with discontinuities barely visible (only on the right-hand fringe) due to the small size of the liquid annulus, d) after separation from contact with a very small liquid bridge in the centre.

of the substance under study were placed in a stainless steel vessel on the inside of this plate, about 6 cm from the surfaces. The vessel was heated by external thermistors that allowed temperature control to ±0.1 °C. The experiments were carried out in the same way as in the case of saturation but with the source crystals at a positive temperature difference with respect to the surfaces. The vapour pressure at the surfaces p was calculated using the Clausius–Clapeyron equation,

$$\ln\left(\frac{p}{p_0}\right) = \frac{\Delta H_{\text{sub}}}{R}\left(\frac{1}{T_0} - \frac{1}{T}\right) \qquad (6)$$

where (p/p_0) is the ratio of the vapour pressure at the surface to that at the source crystal, T is the temperature at the surface and T_0 the temperature of the vapour source. The supersaturation (S) is then $S = 1 - (p/p_s)$. ΔH_{sub} is the sublimation enthalpy, which is 81 kJ mol^{-1} for pinacol and 96 kJ mol^{-1} for menthol (see Table 1). A temperature differential of 2–6 K between the bulk phase and the surfaces were used and this corresponds to supersaturations of 0.3–1.2.

3 Results

In what follows observations with the different substances will be described sequentially. Table 1 lists some relevant physical properties of the substances used.

3.1 Cyclooctane

The cyclooctane condensates remained liquid down to −7.5 °C, or $\Delta T = 22$ K, and in view of results with other substances it is likely that the temperature depression was insufficient to give nucleation of solid. However, the vapour pressure at the lowest T used is an order of magnitude less than which gave solid deposition with HMCTS.[21] The condensate size h (290 nm) determined at 18 °C, above the T_m of 14.3 °C, was used to calculate a minimum relative vapour pressure $p/p_0 = 0.993$ using

Table 1

Substance	T_m (°C)	p_{sat} (mm Hg)	ΔH_{fus} (kJ kg^{-1})	ΔH_{subl} (kJ kg^{-1})	γ (mNm^{-1})	Crystal structure
Cyclooctane	14.530	3.0–5.4,31 15–25 °C	21.531	523.132	31.014	Simple cubic33 (plastic)
Neopentanol	52.530	9.9–63.0,34 25–56 °C	47.522	635.135	14.836 (53 °C)	fcc37 (plastic)
Hexamethyl-cyclotrisiloxane	64.530	4.2–8.9,38 24–34 °C	69.638	248.139	16.440	Trigonal41
Norbornane	87.530	27.7–47.0,42 25–35 °C	45.242	415.943	31.344	hcp45 (plastic)
(-)-Menthol	43.030	2 × 10^{-3}–0.07,46 0–26 °C	76.122	613.047	28.348	Trigonal49
Racemic menthol	34.050	3 × 10^{-3}–0.06,46 0–26 °C	6646	50446		
Pinacol	43.330	0.37 at 25 °C51	124.552	682.052,53	28.154	Monoclinic55
Hexachloro-ethane	186.830	0.32–1.4956 at 25–45 °C	41.257	248.758	37.5 at 20 °C59	Orthorhombic (<45 °C)60

the Kelvin equation. Given the slow rate of growth of these condensates it is difficult to ascertain whether or not the equilibrium condensate size has been reached, as found in previous studies of cyclooctane below T_m.[14] The condensate sizes at six different temperatures below T are plotted in Fig. 3. Eqn (3) accounts accurately for the r of the condensate for $\Delta T < 15$ K, but the reason for the deviation for larger ΔT is uncertain.

3.2 Racemic menthol

The condensate sizes measured at five different temperatures below T_m with racemic menthol are also shown in Fig. 3. The condensates remained liquid down to 10 °C ($\Delta T = 25$ K) as shown by their flow and deformation properties, and the adhesion forces $F/R \approx 0.4$ Nm^{-1}. However, the fringes showed that the surfaces became more deformed during separation than at higher temperatures. This might indicate differences in the state of the adsorbed films on the surfaces—perhaps they became more solid-like, therefore altering somewhat the adhesion–deformation properties of the surfaces. At 1 °C ($\Delta T = 33$ K) no liquid condensate or bridge could be identified at any time, although F/R increased to $ca.$ 0.8 Nm^{-1}. A separation–approach cycle indicated only the presence of traces of material trapped between the surfaces at contact. The radius of curvature of the condensate–vapour interface decreased from 13 to 3.5 nm as ΔT increased from 8 to 27 (Fig. 3). The r values at small ΔT are close to the theoretical predictions but at larger ΔT values they are smaller than expected (larger $1/r$), possibly due to a smaller adsorbed film thickness at the lower temperatures, which would influence r according to eqn (5). Indeed, decreasing the estimated adsorbed film thickness per surface gives better agreement for the lower temperatures (Fig. 2).

3.3 (1R, 2S, 5R)–menthol

Optically pure (1R, 2S, 5R)–(-)-menthol (L-menthol), which has a higher T_m (42 °C) and a lower p_{sat} than the racemic compound (Table 1) was investigated over the range $17 < \Delta T < 46$ K. No condensates could be observed from the interference fringes, but the vapour pressure at $\Delta T = 46$ K is only about 1×10^{-3} mm Hg. When the apparatus was opened at the end of the experiment $ca.$ 1 mm long,

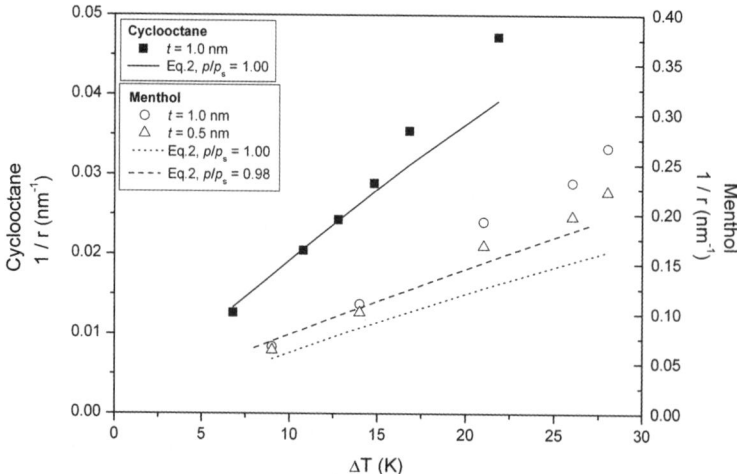

Fig. 3 Inverse radii of curvature $1/r$ ($=2/[h - 3t]$, see text) of the condensate–vapour interface for liquid menthol and cyclooctane condensates as a function of temperature ΔT below the bulk melting point. Note the different scales for menthol (right) and cyclooctane (left).

needle-like menthol crystals were found to have deposited at random places on stainless steel surfaces, suggesting that the atmosphere in the chamber was indeed saturated. No crystals of menthol (or any other substance) were ever observed to deposit on the flat mica surfaces, away from the annular wedge. The fact that crystal nucleation from saturated vapour occurs on the inside of the chamber walls is not surprising given that the vapour will there be locally supersaturated during cooling cycles.

3.4 Neopentanol

Experiments were carried out for 8 K< ΔT < 36 K, and have been described previously.[21] For 8 K < ΔT < 18 K the condensates were always liquid (Fig. 4), and for ΔT > 34 K the nucleation and subsequent deposition of crystalline material was so rapid that any liquid condensate cannot be detected. Over an intermediate temperature range (18 < ΔT < 34 K) the liquid condensates could be induced to solidify by mechanical perturbation of the surfaces. The most effective way was found to be the initial application of an additional load on the surfaces, thereby increasing the contact diameter, followed by a reduction in load. The consequent reduction in contact diameter was then immediately followed by a dramatic increase in the rate of deposition from vapour, and the subsequent properties of the condensate showed clearly that it was solid. The normalised adhesion with liquid condensates was typically 0.3 Nm^{-1}, and 2.5 Nm^{-1} with the solid condensates. After the experiment small crystals of neopentanol were found to have deposited on the inside of the chamber top, showing that the neopentanol vapour was saturated.

3.5 Norbornane

As reported[21] rapid deposition of solid occurred with norbornane (bicyclo[2.2.1]heptane) as soon as contact was achieved at all ΔT > 33 K (the maximum attainable temperature). The adhesion between the surfaces was large, with outward jumps of 750 ± 70 nm, which corresponds to $F/R \approx 4\text{–}5$ Nm^{-1}.

These solid condensates grew at an increasing rate with temperature (Fig. 5), although clearly not just in proportion to the absolute vapour pressure. Even after

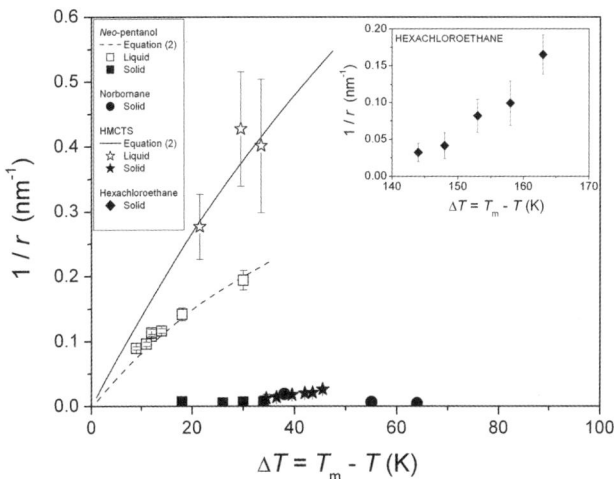

Fig. 4 The inverse of the radius of curvature, $1/r$ ($=[2/h - 3t]$) of the condensate–vapour interface for condensates of neopentanol HMCTS, norbornane and hexachloroethane (inset) as a function of the temperature ΔT below the bulk melting point. Note the dramatic decrease in $1/r$ (increase in r) for the solid deposits compared to the liquid condensates.

40 min there was no sign of a levelling out of the curves. Note that the absolute vapour pressure of norbornane is higher than that of any of the other substances studied. The condensate volumes, V were calculated from h using[22]

$$V \approx \pi R h^2 \qquad (7)$$

Observation of the surfaces from above through a microscope (Fig. 6) showed that the norbornane condensates were often overall hexagonal in shape, albeit with a locally rough outline. This could be due to epitaxy directed by the pseudohexagonal symmetry of the mica basal plane, and/or the close-packed structure of the norbornane, which is hcp below a transition temperature of 33 °C, and fcc above.[45] On attempting to separate the surfaces the solid norbornane condensate broke into fragments which then evaporated, confirming that the vapour phase was not supersaturated. The plastic crystal phase that norbornane forms is quite deformable, hence the rather rounded shapes of the fragments. Note that due to surface tension effects a liquid bridge of this size always snaps into two distinct droplets and never fragments.

3.6 Hexamethylcyclotrisiloxane

As reported previously,[21] both liquid and solid condensates were observed with hexamethylcylotrisiloxane. At undercoolings $\Delta T < 33$ K very small liquid condensates formed around the contact zone, while for $\Delta T \geq 34$ K solid nucleated in the condensates. The size h and interfacial radius of curvature r of the liquid condensates could only determined by estimating the size of liquid bridge directly after separation and then applying eqn (7) on the assumption of unchanged volume (hence the large error bars in Fig. 4). The solid HMCTS deposits gave a very large adhesion between the surfaces with outward jumps of 1.1–1.9 μm measured, corresponding to normalised adhesion forces of 7–10 Nm^{-1}.

3.7 Pinacol

Pinacol (2,3-dimethyl-2,3-butanediol, $T_m = 43$ °C) was investigated over the ΔT range of 10–44 K at four different temperatures. As with menthol no condensate could be observed from the fringes, even after 3 h in contact, although the shift in

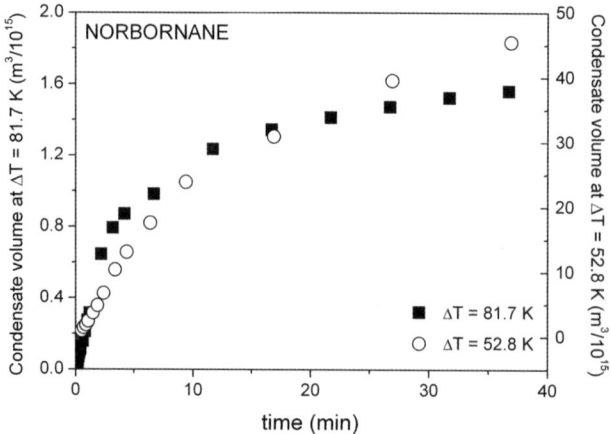

Fig. 5 Condensate growth rates in norbornane vapour. The left hand scale refers the condensate at 5.8 °C ($\Delta T = 81.7$ K), while the right-hand scale to the condensate at at 34.7 °C ($\Delta T = 52.8$ K).

contact from the value in nitrogen and the shapes of the fringes during separation pointed to the presence of some adsorbed material on the nanometre scale. The vapour pressure of pinacol at $-1\ °C$ ($\Delta T = 44$ K) is estimated from eqn (6) and the value at 25 °C (Table 1) to be only 10^{-2} mm Hg, but as with menthol crystals were found to have deposited on the inside of the chamber at the end of the experiment.

3.8 Hexachloroethane

Hexachloroethane has the highest melting point (186 °C) of the substances studied so these experiments could only be carried out for $\Delta T > 143$ K (23–43 °C). Over this entire range reasonably large condensates formed immediately after contact was achieved and their size (inverse interfacial radius of curvature) after 3 h in contact is shown in Fig. 4. Their solid nature was demonstrated by the lack of flow of the condensates on decreasing the load and the large jump out distances of 220–340 nm ($F/R \approx 1.3$–2.0 Nm^{-1}). The condensate size increased with temperature, as found with norbornane.

3.9 Condensation from supersaturated vapour

Slight supersaturations of about 0.3 were induced by applying a 2 °C temperature difference between the bulk crystal phase (the vapour source) and the surfaces. At these supersaturations deposition of solid was observed with pinacol at $\Delta T = 38$ K, but only after perturbing the surfaces by first increasing the load on them and then unloading them, as described earlier with neopentanol. The condensate grew to $h \approx 130$ nm after 30 min and then remained unchanged for up to 3 h. No solid deposition was observed with menthol, even at supersaturations of up to 1.2 at $\Delta T = 42$ K (6 K temperature differential).

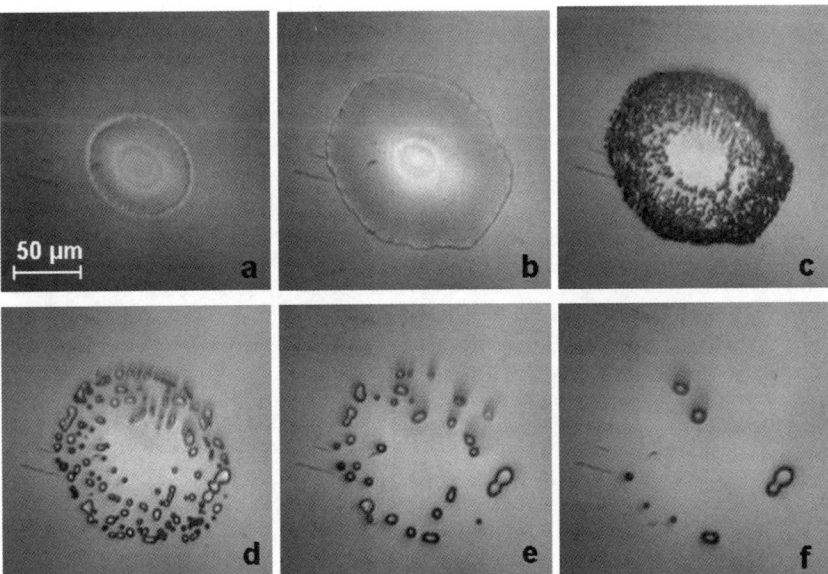

Fig. 6 Microscope image from above of a solid norbornane deposit in the annular wedge between crossed mica cylinders formed in saturated vapour at $\Delta T = 62$ K, showing growth after initial contact (a,b), rupture of crystal on separation (c) and subsequent evaporation of the norbornane fragments (d–f). The brighter centre region is mica–mica contact, and the norbornane vapour interface is an irregular hexagon.

4 Discussion

The results show beyond doubt that nucleation of crystals in capillary condensates provides a means of facilitating crystal deposition from vapour. No crystals were ever observed on the flat mica surfaces outside the annular condensate. There is a very definite correlation between the absolute vapour pressure p and the amount of crystalline solid that deposits. By far the fastest deposition occurs with norbornane, which has the highest p, followed by neopentanol and then HMCTS, in order of decreasing p. The precise value of ΔT necessary for nucleation is lower at 18 K for neopentanol than for the other substances, where it is 33 K or larger, although no lower bound could be determined for norbornane and hexachloroethane as solid deposits at all experimentally accessible temperatures. The substances with the lowest vapour pressure, +(-)-menthol and pinacol did not give rise to deposition of anything more than nanometre-thick films, even at substantial undercoolings ΔT. At around 0 °C, where nucleation might reasonably be expected to occur, their vapour pressure is less than 10^{-2} mm Hg, or three orders of magnitude less than that of HMCTS when it nucleates in liquid condensates at 30 °C ($\Delta T = 34$). The lack of solid deposition with cyclooctane and racemic menthol is most likely due to a large enough ΔT not being attainable, as liquid condensates were in evidence except possibly at the lowest temperature with racemic menthol. It is noteworthy that we have observed that both menthol and cyclooctane supercool very easily in the bulk.

Detailed investigation of the nucleation process in neopentanol, norbornane and HMCTS has indicated the existence of three temperature regimes.[21] For undercoolings of $\Delta T \leq 18$ °C (neopentanol) or $\Delta T \leq 33$ °C (HMCTS) a liquid condensate with limited size initially forms around the contact zone. No liquid condensates were found with norbornane as the minimum undercooling achievable with our system was still 37 K. At the largest undercoolings the behaviour was very different, and with all of these substances the condensate grew rapidly immediately after contact, and definitive evidence for the presence of a liquid condensate could not be obtained. This is the behaviour that we have now found for C_2Cl_6 as well—the large ΔT prevents the detection of any liquid condensate. In the intermediate regime nucleation of solid appeared to be promoted by mechanical perturbation of the condensates, by loading and unloading the surfaces or separating these from contact.

The temperatures at which nucleation occurs in the liquid capillary condensates is, with the exception of neopentanol, similar to typical homogeneous nucleation temperatures in bulk liquids. Naturally, most experiments have been carried out with water, and they point to typical nucleation rates of 10^{11} to 10^{13} m^{-3}s^{-1} at $\Delta T = 35$ K and 10^{13} to 10^{15} m^{-3} s^{-1} at $\Delta T = 38$ K.[61] The volume V of the annular capillary condensates given by eqn (7) is approximately 10^{-17} m^3 for $\Delta T = 35$–38 K, which translates to nucleation times of the order of μs–ms. It is hence possible that classical homogeneous nucleation of crystals in the liquid capillary condensates explains our results. Heterogeneous nucleation at the mica surface appears much less likely, although it could perhaps account for the higher nucleation temperatures (smaller undercooling) observed with neopentanol.

According to CNT the radius $r*$ of a critical nucleus for nucleation from the melt is given by

$$r* = -\frac{2V_M \gamma_{sl}}{\Delta G_v} = -\frac{2V_M \gamma_{sl} T_m}{\Delta T \Delta H_{fus}} \qquad (8)$$

This value of $r*$ differs from the r given by eqn (4) by a factor of 2, and the replacement of γ_{lv} with γ_{sl}. This is related to the discussion of metastable liquid condensates in the introduction, and it appears that our experiments vindicate the thermodynamic argument that solid capillary condensates are stable in most cases, and may under favourable conditions lead to deposition of large amounts of crystalline material from vapour.

The temperature of the experimental chamber is controlled and measured to an accuracy of ±0.1 K at best. We cannot guarantee that temperature gradients of this magnitude are not present, and this leads to possible uncertainties in p/p_0. Using the Clausius–Clapeyron (eqn (6)) with typical values of the enthalpy of sublimation of 40–80 kJ mol^{-1} an error in T of 0.1 K translates into an error in p/p_s of 0.006 to 0.013. A liquid capillary condensate above T_m would grow infinitely large at saturation, but with typical values of V_M (150 cm^3) and γ_{lv} (30 mJ m^{-2}) would be limited to a radius of curvature r ($\approx h/2$) of 140–300 nm at 300 K. Since thermodynamically, undersaturation should prevent the deposition of bulk solid as well, this may well be part of the explanation for why the size and growth rate of the solid deposits varies.

Such an uncertainty cannot, of course explain why nucleation and growth of crystals was not observed with some substances. Strictly speaking our technique does not allow us to detect a nucleation event as such—it is rather the accelerated deposition that follows the nucleation that we observe. We can also note the increased adhesion that the solid gives rise to, but if only a very small amount of solid is deposited this will not be readily evident. Clearly, if there is insufficient material in the vapour phase growth will be severely curtailed, whatever the mechanism behind the nucleation. More importantly, perhaps, the capillary condensates in the annular wedge are in a very inaccessible location, with vapour molecules having to diffuse long distances between two walls that are closer together than the mean free path of the vapour molecules. In such a case the diffusing molecules collide with the wall more frequently than with one another, leading to a substantial slowing down of diffusion. The Knudsen number Kn is a relevant measure of this type of restricted diffusion. It is defined as the ratio of the molecular mean free path λ to the representative physical length scale L,[62]

$$Kn = \frac{\lambda}{L} = \frac{k_B T}{\sqrt{2}\pi\sigma^2 p_{tot} L} \qquad (10)$$

where σ is the molecular diameter and p_{tot} is the total pressure (here $ca.$ 101 kPa). If Kn is much greater than one Knudsen diffusion is important. In our case it is reasonable to take L as twice the condensate radius, since this represents the final width of the space through which the molecules have to diffuse before the condensate proper is reached. L is then of the order of 5 nm, and with $\sigma = 0.5$ nm we obtain $Kn \approx 10$, meaning that Knudsen diffusion is important and material transport into the wedge pore is limited. We can estimate from the annular wedge geometry that the last 10–20 μm of diffusion before the condensate is reached occurs through a slit-pore of less than twice the width of L assumed above. This approximates a slit-pore with an aspect ratio of about 10^3, and the diffusion must be impeded in the extreme. It is very likely that this accounts for why no capillary condensates appeared to form with (1R, 2S, 5R)-menthol, pinacol, and the racemic menthol at the lowest temperatures.

With a supersaturation of the vapour phase by 30% it was necessary to perturb the surfaces to initiate the deposition of solid pinacol ($h = 140$ nm after 30 min) but with (1R, 2S, 5R)-menthol even 120% supersaturation gave no discernible condensation or deposition, even after attempts were made to perturb the surfaces. The pronounced difference caused by a relatively small increase in the vapour-phase concentration of pinacol is perhaps surprising, as is the complete lack of an effect with the menthol.

Over eighty years ago Volmer pointed out in a discussion of heterogeneous nucleation[63] that "embryos" (*i.e.* nuclei) could be retained in conical cavities on surfaces at temperatures above a phase transition and subsequently seed renewed growth as the phase boundary was recrossed. Similar ideas were later discussed in the context of atmospheric nucleation and it was proposed that capillary condensation in conical surface pits could lead to ice deposition on solid aerosol particles,[64] although no direct experimental evidence appears to have been published. We believe that we have now shown that this may be a viable mechanism of nucleation on insoluble aerosol particles, although further studies will have to be carried out with better model systems. The vapour pressure over ice at temperatures relevant to atmospheric nucleation at high altitudes varies from about 0.1 mm Hg at $-40\,°C$ to 0.01 mm Hg at $-60\,°C$,[65] so the restricted diffusion in our present experimental set-up would be problematic.

The obvious shortcoming of our experiments is that the annular wedge is too acute and too narrow. In practice such very narrow pits or grooves in a surface would be rare, so our model is not a completely realistic one. However, it is likely that the possibility of nucleation *via* capillary condensation could only be greater with a more open cavity. A larger wedge angle would not reduce significantly the possibility of capillary condensation without any energy barrier, but make transport of material for growth of the solid phase much easier. Future experiments should concentrate on providing a better model system with more obtuse but more accessible surface cavities. It should then be easier to establish definitively whether nucleation in capillary condensates is a general phenomenon that explains the enhanced nucleation often seen with rough surfaces.[66]

Crystal deposition from vapour *via* liquid condensates has many similarities to recently discussed models of two-step nucleation from solution. In contrast to CNT these models propose the initial formation of a denser, liquid-like phase (or clusters) which then crystallises. This was originally found in simulations that showed that a metastable critical point in colloidal systems with short-range attractive forces could greatly enhance nucleation rates.[7] This metastable critical point and the associated denser fluid phase are remnants from systems with longer-range attractive forces where a thermodynamically stable liquid-like phase does exist. The analogy to this denser, liquid-like phase is obvious in our case. Liquid, which is unstable in bulk below T_m is stabilised by the wedge and leads to nucleation of solid from vapour without the requirement for any supersaturation. The nucleation rate is thus significantly enhanced. Recent simulations have suggested that the rate of crystal nucleation from vapour can increase by proceeding *via* liquid droplets[67,68]—an analogous mechanism in homogeneous nucleation.

The formation of biominerals like calcite,[69] aragonite, hydroxyapatite and even calcium sulphate[70] is known to proceed *via* an amorphous phase, *i.e.* a much denser but in many ways still liquid-like phase. It would be interesting to investigate whether or not suitable surface cavities with affinity for these precursor phases might act to enhance nucleation rates of such crystalline biominerals from solution. Simulations have suggested that protein nucleation is facilitated by "capillary condensation" from solution in surface pits.[71]

5 Conclusions

Nucleation of crystals occurs readily in liquid capillary condensates formed from saturated vapour below the bulk melting point of the substance, provided that the undercooling is large enough and that the condensates are of a reasonable size. The crystal nucleation is followed by rapid growth of these crystals by deposition from vapour if the absolute vapour pressure is high enough (above approximately 1–4 mm Hg). In our specific experimental set-up it is likely that restricted diffusion is hindering crystal nucleation and growth when the absolute vapour pressure of the substance is too low (of the order of 1 mm Hg or less).

Acknowledgements

The Leverhulme Trust is thanked for supporting this project.

References

1. B. J. Murray, T. W. Wilson, S. Dobbie, Z. Q. Cui, S. Al-Jumur, O. Mohler, M. Schnaiter, R. Wagner, S. Benz, M. Niemand, H. Saatfoff, V. Ebert, S. Wagner and B. Karcher, *Nat. Geosci.*, 2010, **3**, 233–237.
2. D. Erdemir, A. Y. Lee and A. S. Myerson, *Acc. Chem. Res.*, 2009, **42**, 621.
3. R. P. Sear, *J. Phys.: Condens. Matter*, 2007, **19**, 033101.
4. P. G. Vekilov, *Cryst. Growth Des.*, 2010, **10**, 5007–5019.
5. B. A. Garetz, J. Matic and A. S. Myerson, *Phys. Rev. Lett.*, 2002, **89**, 175501.
6. J. R. Savage and A. D. Dinsmore, *Phys. Rev. Lett.*, 2009, **102**, 198302.
7. P. R. ten Volde and D. Frenkel, *Science*, 1997, 277.
8. D. Frenkel, *Nature*, 2006, **443**, 641.
9. D. Turnbull, *J. Chem. Phys.*, 1950, **18**, 198–203.
10. C. A. Scholl and N. H. Fletcher, *Acta Metall.*, 1970, **18**, 1083–1086.
11. M. Tuller, D. Or and L. M. Dudley, *Water Resour. Res.*, 1999, **35**, 1949–1964.
12. D. Nowak, M. Heuberger, M. Zach and H. K. Christenson, *J. Chem. Phys.*, 2008, **129**, 154509.
13. D. Nowak and H. K. Christenson, *Langmuir*, 2009, **25**, 9908.
14. P. Barber, T. Asakawa and H. K. Christenson, *J. Phys. Chem. C*, 2007, **111**, 2141–2148.
15. D. Tabor and R. H. S. Winterton, *Proc. R. Soc. London, Ser. A*, 1969, **312**, 435.
16. J. N. Israelachvili and G. E. Adams, *J. Chem. Soc., Faraday Trans. 1*, 1978, **74**, 975–1001.
17. H. K. Christenson and R. G. Horn, *Chem. Scr.*, 1985, **25**, 37–41.
18. H. K. Christenson, P. M. Claesson and R. M. Pashley, *Proc. Indian Acad. Sci. (Chem. Sci.)*, 1987, **98**, 379–389.
19. J. Israelachvili, *J. Colloid Interface Sci.*, 1973, **44**, 259.
20. S. Tolansky, *Multiple-beam interferometry of surfaces and films*, Oxford University Press (Clarendon), London, 1949.
21. T. Kovács, F. C. Meldrum and H. K. Christenson, *J. Phys. Chem. Lett.*, 2012, **3**, 1602–1606.
22. N. Maeda and H. K. Christenson, *Colloids Surf., A*, 1999, **159**, 135–148.
23. N. Maeda, M. M. Kohonen and H. K. Christenson, *J. Phys. Chem. B*, 2001, **105**, 5906–5913.
24. H. K. Christenson and V. V. Yaminsky, *Langmuir*, 1993, **9**, 2448–2454.
25. E. J. Wanless and H. K. Christenson, *J. Chem. Phys.*, 1994, **101**, 4260–4267.
26. J. E. Curry and H. K. Christenson, *Langmuir*, 1996, **12**, 5729–5735.
27. S. Ohnishi, M. Hato, K. Tamada and H. K. Christenson, *Langmuir*, 1999, **15**, 3312–3316.
28. R. Evans and U. Martini Bettolo Marconi, *Chem. Phys. Lett.*, 1985, **114**, 415.
29. Y. Qiao and H. K. Christenson, *Phys. Rev. Lett.*, 1999, **83**, 1371–1374.
30. *CRC Handbook of chemistry and physics*, CRC Press, 2009.
31. H. L. Finke, D. W. Scott, M. E. Gross, J. F. Messerly and G. Waddington, *J. Am. Chem. Soc.*, 1956, **78**, 5469.
32. A. Bondi, *J. Chem. Eng. Data*, 1963, **8**, 371.
33. D. E. Sands and V. W. Day, *Acta Crystallogr.*, 1965, **19**, 278–279.
34. *Chemical properties handbook*, McGraw-Hill, 1999.
35. Chemical Dictionary Online, http://www.chemicaldictionary.org/dic/N/Neopentyl-alcohol_3573.html, date accessed: July 26, 2012.
36. *Kirk-Othmer encyclopedia of chemical technology*, John Wiley and Sons, New York, 1991.
37. G. B. Carpenter, *Acta Cryst.*, 1969, **163**, B25.
38. R. C. Osthoff, W. T. Grubb and C. A. Burkhard, *J. Am. Chem. Soc.*, 1953, **75**, 2227–2229.
39. N. W. Luft, *Ind. Chem.*, 1955, **31**, 502–504.
40. C. Wohlfarth, in *The Landolt-Börnstein database*, Springer-Verlag, Berlin Heidelberg, 2008.
41. G. Peyronel, Crystal structure of hexamethylcyclotrisiloxane, Brooklyn Polytechnic Institute, 1954.
42. S. P. Verevkin and V. L. Emel'yanenko, *J. Phys. Chem. A*, 2004, **108**, 6575–6580.
43. X. An, I. Zhu and R. Hu, *Thermochim. Acta*, 1987, **121**, 473.
44. ChemSpider, The free chemical database, http://www.chemspider.com/Chemical-Structure.8878.html, date accessed: July 26, 2012.
45. A. N. Fitch and H. Jobic, *J. Chem. Soc., Chem. Commun.*, 1993, 1516–1517.
46. J. S. Chickos, D. L. Garin, M. Hitt and G. Schilling, *Tetrahedron*, 1981, **37**, 2255–2259.
47. J. S. Chickos and W. E. Acree, *J. Phys. Chem. Ref. Data*, 2002, **31**, 537–698.
48. C. Becker, H. Reiss and R. H. Heist, *J. Chem. Phys.*, 1978, **68**, 3585–3594.
49. P. Bombicz, J. Buschmann, P. Luger, D. Nguyen Xuan and C. Ba Nam, *Z. Kristallogr.*, 1999, **214**, 420–422.
50. X.-Y. Su, A. Li Wan Po and J. S. Millership, *Chirality*, 1993, **5**, 58–60.
51. ChemSpider, The free chemical database, http://www.chemspider.com/Chemical-Structure.21109330.html, date accessed: July 26, 2012.

52 J. G. Priest, E. M. Woolley, J. B. Ott and J. R. Goates, *J. Chem. Thermodyn.*, 1983, **15**, 357–366.
53 J. P. Guthrie, *Can. J. Chem.*, 1977, **55**, 3562–3574.
54 S. N. Omenyi, A. W. Neumann and C. J. van Oss, *J. Appl. Phys.*, 1981, **52**, 789–795.
55 M. Dahlqvist and R. Sillanpaa, *J. Mol. Struct.*, 2000, **524**, 141–149.
56 O. A. Nelson, *Ind. Eng. Chem.*, 1930, **22**, 971–972.
57 *VDI Heat atlas*, Springer, Dusseldorf, Germany, 1993.
58 R. M. Stephenson and S. Malanowski, *Handbook of the thermodynamics of organic compounds*, Elsevier, New York, 1987.
59 R. P. Pohanish, *HazMat data: for first response, transportation, storage, and security*, John Wiley and Sons, 2005.
60 J. Sasada and M. Atoji, *J. Chem. Phys.*, 1953, **21**, 145–152.
61 B. J. Murray, S. L. Broadley, T. W. Wilson, S. Bull, R. H. Wills, H. K. Christenson and E. J. Murray, *Phys. Chem. Chem. Phys.*, 2010, **12**, 10380.
62 R. W. Barber and D. R. Emerson, *Advanced fluidmechanics IV*, WIT Press, Southampton, UK, 2002.
63 M. Volmer and A. Weber, *Z. Phys. Chem.*, 1926, **119**, 227–301.
64 N. Fukuta, *J. Atmos. Sci.*, 1966, **23**, 741–750.
65 A. Wexler, *J. Res. Nat. Bur. Stand. (US)*, 1977, **81A**, 5–20.
66 J. L. Holbrough, J. M. Campbell, F. C. Meldrum and H. K. Christenson, *Cryst. Growth Des.*, 2012, **12**, 750–755.
67 B. Chen, H. Kim, S. J. Keasler and R. B. Nellas, *J. Phys. Chem. B*, 2008, **112**, 4067–4078.
68 J. A. van Meel, A. J. Page, P. R. Sear and D. Frenkel, *J. Chem. Phys.*, 2008, **129**, 204505.
69 F. C. Meldrum, *Int. Mater. Rev.*, 2003, **48**, 187–224.
70 Y. Wang, Y.-Y. Kim, H. K. Christenson and F. C. Meldrum, *Chem. Commun.*, 2012, **48**, 504–506.
71 J. A. van Meel, R. P. Sear and D. Frenkel, *Phys. Rev. Lett.*, 2010, **105**, 205501.

General discussion

Professor Vekilov opened the discussion of the paper by Professor Daan Frenkel: In the dependence of the nucleation rate on the solute volume fraction, the nucleation rate becomes a weak function of the volume fractions at high values of the latter. Is this an indication of vanishing nucleation barriers, *i.e.*, of a spinodal regime? How can we understand spinodal decomposition for the formation of crystals, which seems to go beyond the classical van der Waals definition?

Professor Frenkel responded: Indeed, in high volume fractions crystallisation is no longer an activated process. Although crystallisation in this regime is sometimes called 'spinodal crystallisation', I would be reluctant to use this term because the phenomenon is qualitatively different from, say, spinodal demixing.

Professor Roberts remarked: What effect, if any, do you feel there might be due to molecular polarisation and changes to it at the early stages of cluster and nanocrystal formation? Also, do you think that changes in molecular conformation during the same process might also play a role? Intuitively, I suppose that one might expect changes to both polar and dispersive intermolecular interactions as the cluster grows in size during both the pre-nucleation and post-nucleation stages associated with the crystallisation process due to the increasing role played by inter- with respect to the intramolecular forces in defining the cluster structure. Beyond the 3-D (nucleation) to 2-D (growth) transition which, for stable growth, leads to a facetted crystal, molecular modelling studies[1,2] have shown that the molecular polarisation can be distinctly different between to bulk and surface co-ordinated molecules. This is presumably due to their lower coordination, and hence reduced intermolecular interactions for the latter species. Parallel studies of conformational variations molecule to molecule within clusters have tended to reveal more conformational variability in the smaller *versus* larger clusters and within the surface *versus* bulk regions, respectively[3-5] thus hinting at an impact upon a given material's "crystalisability".

Hence, whilst the hard sphere models are very attractive in terms of the fundamental insight they provide, do you think that perhaps the inherent granularity of the models may miss the more subtle effects related to molecular polarisation and any associated structural changes associated with it during growth?

Such effects may be more pronounced at small cluster sizes where the mechanical strength, with respect to the bulk crystal structure, may be much lower and hence the deformability of the clusters/nano-crystals may be much more pronounced. Particularly for the lower symmetry piezoelectric crystals such polarisation change effects could potentially play a role not only in mediating the nucleation process but also in effecting inter-particle interactions and "agglomeratability". In this regard, for completeness, it is probably helpful to note that symmetry reduction at crystal surfaces tends to make the surfaces of most materials piezoelectric. Do you think that including a consideration of these factors has a role to play in providing a better insight into the nucleation and growth processes?

1. R. Docherty, K. J. Roberts, V. Saunders, S. Black and R. J. Davey, Theoretical analysis of the polar morphology and absolute polarity of crystalline urea, *Faraday Discussions*, 1993, **95**, 11–25.
2. G. Clydesdale, K. J. Roberts, G. Telfer, V. Saunders, D. Pugh, R. A. Jackson and P. Meenan, An analysis of the crystal morphology of sodium chlorate using a polar attachment energy model, *J. Phys. Chem. B*, 1998, **102**, 7044–7049.
3. G. P. Hastie, J. Johnstone, E. M. Walker and K. J. Roberts, Direct evidence for surface reconstruction on organic solid surfaces: Benzil 0001, *Perkin Trans. 2*, 1996, 2049–2050.

4. R. B. Hammond, K. Pencheva and K. J. Roberts, An examination of the polymorphic stability and molecular conformational flexibility as a function of crystal size associated with the nucleation and growth of benzophenone, *Faraday Discuss.*, 2007, **136**, 87–102.
5. R. B. Hammond, K. Pencheva and K. J. Roberts, On the relative polymorphic structural stability, as a function of particle size and shape, for nano-sized molecular clusters: studies of L-glutamic acid and D-mannitol, *CrystEngComm*, 2012, **14**, 1069–1082.

Professor Frenkel answered: There are many effects that have barely been investigated. Molecular polarisation is one example. Another one is the change of the charge state of the molecule to be incorporated in a crystal (*e.g.*, from zwitterionic to uncharged). At this stage, I cannot speculate what effects will be observed, but I agree with Professor Roberts that the question is both interesting and important.

Professor De Yoreo enquired: What are the prospects for incorporating protein conformational changes into theoretical models and simulations of protein assembly and crystallization and how do you think it will change our current picture of these processes?

Professor Frenkel replied: At present, it would still be very challenging to use fully atomistic models. However, intermediate levels of coarse graining can be considered. Including conformational changes in the study would be extremely interesting because the incorporation of a protein in a crystal clearly puts strong constraints on the allowed conformations. At present, we have no clear idea how this affects the pathway and rate of crystal nucleation.

Mr Rosbottom asked: What effect, if any, have you considered for variation in shape of the nuclei, *i.e.* that all nuclei are not spherical as assumed by the classical nucleation theory?

Professor Frenkel responded: The simulations make no assumptions about the shape of the nuclei. However, they do yield information about shapes and, in this way, allow us to test the assumption of the classical nucleation theory that nuclei are spherical.

In the case of simple colloidal systems, we find that crystal nuclei—although fluctuating in shape due to their small size—are, on average, spherical. However, for NaCl, we find that the critical nucleus has the same morphology as a macroscopic NaCl crystal, namely a cube.[1] An interesting morphology was observed in the case of pre-critical nuclei of polar liquids nucleating from the vapour: the nuclei that are not at all droplet like—rather, they had the appearance of 'living' chains of polar molecules.[2]

1. C. Valeriani, E. Sanz and D. Frenkel, Rate of homogeneous crystal nucleation in molten NaCl, *J. Chem. Phys.*, 2005, **122**, 194501.
2. P. R. ten Wolde, D. W. Oxtoby and D. Frenkel, Chain formation in homogeneous gas–liquid nucleation of polar fluids, *J. Chem. Phys.*, 1999, **111**, 4762–4773.

Mrs Virone remarked: In case of seeding, is nucleation at the surface of a crystal a secondary nucleation mechanism (such as initial breeding, shear forces, attrition, *etc.*) or an activated (primary) heterogeneous mechanism?

Professor Frenkel replied: In my lecture (though not in the paper) I discussed the case where crystal nucleation on a rough surface is preceded by capillary condensation (*Phys. Rev. Lett.*, 2010, **105**, 205501). If capillary condensation is the primary nucleation event, then the subsequent crystal nucleation may be viewed as secondary nucleation. However, in the case of crystal nucleation on the surface of 'dry' seeds, activated crystal formation would be a primary nucleation event.

Professor Vekilov commented: Not always is nucleation the rate-determining step in the formation of a new phase. In many cases, a supersaturated solution is seeded, *i.e.*, a large number of small crystals are added to it, which then grow and in this way supersaturation is lowered.

Dr Sear commented: You discussed George and Wilson's observation[1] that proteins frequently crystallise in a 'window' or 'slot' of values of the second virial coefficient, B_2, where B_2 is negative but not too large in magnitude. This 'window' is typically around its value at a dilute-solution/concentrated-solution critical point. So, perhaps this is evidence that a metastable dilute-solution/concentrated-solution transition plays a role in the crystallisation of many proteins.

However, perhaps there is an alternative rationalisation of George and Wilson's observation. Proteins are crystallised from dilute solution, and so some attractions are needed to drive crystallisation. However, strong attractions lead to gelation that would interfere with crystallisation. These two considerations also suggest that there should be a relatively narrow 'slot' of attraction strengths over which crystallisation is possible. Could this be behind the observation the crystallisation frequently occurs in a 'slot' of values of B_2?

Could you comment on these two alternatives? Could we distinguish between these two possibilities?

1. A. George and W. W. Wilson, *Acta Crystallogr., Sect. D*, 1994, **50**, 361.

Professor Frenkel answered: I agree that this is a possibility. A way to distinguish is to study the crystallisation kinetics of patchy particles. For these particles, the value of B_2 at T_c is different from that of spherically symmetric particles. The question is then: is the crystallisation slot for these particles at the same value of B_2 as spherical particles (hich would rule out any effect of T_c), or is it at the B_2 of the critical temperature, or somewhere else.

Professor Vekilov commented: The rule that moderately negative values of the second osmotic virial coefficient B_2 are required for crystallization is just what it says: an empirical rule, formulated on the basis of observations with a limited number of proteins. Its theoretical justification is under extremely constraining assumptions. There are by now numerous published examples of excellent protein crystals obtained at positive values of B_2.

Professor Frenkel answered: Crystal nucleation at positive B_2 is very common in the colloidal regime (just think of hard, colloidal spheres). The observation that a moderately negative B_2 that correlates with the crystallisation window of some (but not all) globular proteins has greatly helped to shape our (partial) understanding of the phenomenon, but it is obviously not the last word on this matter.

Professor De Yoreo opened the discussion of the paper by Professor Helmut Cölfen: How does one reconcile the dynamic nature of DOLLOP and the lack of a clear minimum in the free energy *vs.* number of ions with monodispersity of pre-nucleation clusters based on the sharp peak in the AUC data and the inferred stability of the pre-nucleation clusters based on their concentration relative to the free ions?

Dr Gebauer replied: Indeed the dynamic nature of DOLLOP[1] (*i.e.*, the structural form of pre-nucleation clusters), and the proposed formation mechanism imply a rather broad size distribution of pre-nucleation clusters that should relate to those typical for the outcome of polycondensation reactions.[2] At first sight, the virtually monodisperse cluster size determined by means of analytical ultracentrifugation (AUC) appears to be a contradiction. But actually, it is a strong proof-of-concept:

AUC is an absolute method that does not depend on assumptions to begin with. It is crucial to note that AUC experiments take several hours, because the small species (ions and clusters) sediment slowly even in very strong sedimentational fields. While AUC can separate ions and clusters, it can only determine actual size distributions for distinct species, *e.g.* for conventional polymers held together by covalent bonds. However, when the species can interconvert and dynamically change their size, AUC can only resolve an actual size distribution if the dynamics of the species were slower than the AUC experiment. Computer simulations suggest that DOLLOPs (the structural form of pre-nucleation clusters) can change conformation within hundreds of picoseconds,[1] *i.e.* orders of magnitude faster than the timescale of AUC experimentation (which, again, takes several hours). AUC does not give a snapshot like microscopic techniques, here especially cryo-TEM, but the sedimentation coefficient reflects an average value for all clusters in the system that are interconverting during the time of experiment. This is well known for reversibly interacting biopolymers.[3] It is, in principle, very similar to NMR signals, by the way. Again, the very narrow, virtually monodisperse size determined for pre-nucleation clusters by means of AUC thus reflects the time average of all clusters in the system during the duration of the AUC experiment, and does not represent the actual size distribution of pre-nucleation clusters. Thus, the AUC results are a proof-of-concept supporting the notion of highly dynamic, interconverting equilibrium species, where the dynamics are much quicker than the duration of AUC experiments.

We are not sure what is meant by "their concentration relative to free ions" in the question. This part of the question appears to imply that the concentration of calcium carbonate pre-nucleation clusters was very low compared to relevant free ions. It is important to note, though, that the systems investigated in this context always contain a significant amount of ions, which are entirely irrelevant for pre-nucleation cluster formation. These are the counter ions of the buffer, essentially sodium cations, and there is always a rather large excess of bi/carbonate species in the buffer with respect to calcium ions. Unfortunately, there is no way around this, owing to the rather low solubility of calcium carbonate and concurrent technical restrictions of the experiments, especially of potential measurements. Anyway, when the relative amount of all ions in a typical system is 100, the relative amount of calcium is only approximately one.[4,5] Depending on pH, 30–70% of this calcium is bound in pre-nucleation species and the thermodynamic model of speciation shows that the clusters are stable. Compared to all free ions, the concentration of clusters is certainly low indeed; but most ions are completely irrelevant and are true "spectators". That said, this misunderstanding is intimately connected with the point above. If there were only ions relevant for pre-nucleation cluster formation in the system, we could not separate ions and clusters in AUC, as they dynamically interconvert. The average value of the size of pre-nucleation clusters includes the size of the relevant ions, because they are a part of the dynamics. The signal of the single ions in AUC comes from the irrelevant ions, which are the excess species in these systems anyway.

1. R. Demichelis, P. Raiteri, J. D. Gale, D. Quigley and D. Gebauer, *Nat. Commun.*, 2011, **2**, 590.
2. P. J. Flory, *J. Am. Chem. Soc.*, 1936, **58**, 1877–1885.
3. J. L. Cole, J. W. Lary, T. Moody and T. M. Laue, *Meth. Cell Biol.*, 2008, **84**, 143.
4. D. Gebauer, A. Völkel and H. Cölfen, *Science*, 2008, **322**, 1819–1822.
5. M. A. Bewernitz, D. Gebauer, J. Long, H. Cölfen and L. B. Gower, *Faraday Dicsuss.*, 2012, **159**, DOI: 10.1039/C2FD20080E.

Professor Gale commented: To add to the previous reply, the simulations from ref. 1 show that there is a stable equilibrium distribution of species, below a certain concentration threshold, despite the underlying dynamic exchange of ions. The AUC reflects the well defined nature of this distribution and the association with a specific size is only a guide based on approximations. As for the "inferred stability" it should be made clear that the simulations directly show that the free energy of DOLLOP species are lower than the separate ions at mM concentrations. This comes from calculations of the free energy profile for ion pair association as shown elsewhere[2] as well as the speciation models of ref. 1. There is often a misconception that the observation that the concentration of DOLLOP species is lower than that of ion pairs implies that they are less stable. Because the concentrations of ions are small and they are raised to the power of the number of formula units in the cluster, the equilibrium constant can easily be ~10 000 (*i.e.* signifying a clearly negative free energy) without the concentration of DOLLOP exceeding that of ion pairs.

1. R. Demichelis, P. Raiteri, J. D. Gale, D. Quigley and D. Gebauer, *Nat. Commun.*, 2011, **2**, 590.
2. P. Raiteri and J. D. Gale, *J. Am. Chem. Soc.*, 2010, **132**, 17623.

Professor Frenkel remarked: There need not be a contradiction between the comment of Professor De Yoreo that the low concentration of pre-nucleation clusters indicates that the free energy of formation of these clusters is likely to be positive and the statement of Professor Gale that the enthalpy of formation is negative, as long as the entropy of formation is sufficiently negative.

Professor De Yoreo enquired: If the free energy of pre-nucleation clusters as calculated based on multi-ion reactions and the assumption that all bound Ca is in the clusters, then what other assumptions do we need to make to have them play a significant role in nucleation?

Professor Cölfen replied: The evaluation of calcium and carbonate binding based on equilibrium thermodynamics does not require the assumption indicated by Professor De Yoreo.

Free ions cannot disappear according to the law of mass conservation, thus they must be bound within chemical species. Classically, the observed binding has been assigned to ion pair formation. Since analytical ultracentrifugation shows that the species formed in solution prior to nucleation are distinctly larger than ion pairs on average, it is justified to add another parameter, in accord with the notion of Occam's razor. This parameter allows for evaluating multi-ion reactions that are necessary to form species larger than simple ion pairs. In this context, it is crucial to note that the treatment of the ion binding on the basis of equilibrium thermodynamics is based upon the observation that the binding of ions prior to nucleation is reversible;[1,2] there are no assumptions required at this point.

Also, the multiple-binding equilibrium is a speciation model, which does not rely on assumptions to begin with. These equilibria must occur when clusters form. The only assumption that we have made subsequently is that binding sites for calcium ions on carbonate ions in the clusters are considered to be equal and independent,[2] or in other words, that equilibrium constants for successive additions of ions are all equal. While this might be a drastic approximation at first sight, computer simulations have shown that this particular assumption is indeed sustainable[3,4]

Evaluation of the multiple-binding equilibrium on the basis of measured concentrations of free ions, and respective amounts of bound carbonate and calcium, shows that the formed clusters are thermodynamically stable, as they are associated with a negative free enthalpy. Larger clusters will be more stable than smaller ones (and ion pairs), while the increase in stability for each addition of an ion pair to a cluster lies within few kT.[4] This is the reason why the clusters do not grow without limit, and their size distribution is similar to the outcome typical of polycondensation reactions.[5] Note that any such cluster species is more stable than free ions.

While the treatment outlined above does not include any major assumptions, the classical nucleation theory indeed does, as it literally assumes that pre-critical clusters have to be unstable owing to an excess surface energy. Our analyses show that these assumptions are obviously not sustainable in case of calcium carbonate. There is not a single critical cluster size above which nuclei become able to grow, but rather a broad population of clusters that are more stable than the free ions. Moreover, it has been directly observed that the clusters are important species during nucleation,[6–8] as nucleation can proceed *via* cluster aggregation under certain conditions. However, this does not always have to be the preferred pathway of nucleation, as for example the presence of suitable interfaces like SAMs can obviously change the situation.[9] Still, assuming that the stable clusters would sit in a "trap" of free energy as suggested by Professor De Yoreo is based on the notions of CNT and relies on multiple assumptions. We cannot answer with certainty what the physicochemical basis of the barrier separating prenucleation clusters (which are solutes) and particles actually is (*i.e.* the barrier of nucleation), but simulations indicate that cluster dehydration may be the key[4] (also consider that binding of ions in clusters is already accompanied by some loss of hydration water, which may reduce the later barrier as compared to direct nucleation from ions). Clearly, this would be a completely different barrier than the one envisaged by CNT, such that the clusters would not be trapped and can be direct precursors of amorphous calcium carbonate.

1. D. Gebauer and H. Cölfen, *Nano Today*, 2011, **6**, 564–584.
2. D. Gebauer, A. Völkel and H. Cölfen, *Science*, 2008, **322**, 1819–1822.
3. P. Raiteri and J. D. Gale, *J. Am. Chem. Soc.*, 2010, **49**, 17623–17634.
4. R. Demichelis, P. Raiteri, J. D. Gale, D. Quigley and D. Gebauer, *Nat. Commun.*, 2011, **2**, 590.
5. P. J. Flory, *J. Am. Chem. Soc.*, 1936, **58**, 1877–1885.
6. E. M. Pouget *et al.*, *Science*, 2009, **323**, 1455–1458.
7. A. Dey *et al.*, *Nat. Mater.*, 2010, **9**, 1010–1014.
8. M. Kellermeier *et al.*, *Adv. Funct. Mater.*, 2012, DOI: 10.1002/adfm.201200953
9. M. H. Nielsen, J. R. I. Lee, Q. N. Hu, T. Y. J. Han and J. J. De Yoreo, *Faraday Discuss.*, 2012, DOI: 10.1039/C2FD20050C.

Professor Gale replied: Before addressing the question of nucleation, it is important to note that there is no assumption made that all Ca ions are in the clusters. As in any speciation situation, there is a set of equilibrium constants and so some proportion of the Ca ions will not be in clusters. That said, under the particular experimental conditions used in the work of Gebauer *et al.*[1] calcium is being added to an excess of carbonate/bicarbonate solution and so the equilibria will favour the situation in which most of the Ca is in ion pairs and DOLLOP. However, this is not an assumption, but a result of the free energies for the reactions involved. Turning to the question of what role these species may have during nucleation, at present we cannot say anything regarding this from the simulations since this event has yet to be characterised. What we can be certain of is that association of the ions in the solution will explore this region of configuration space either prior to a liquid precursor[2] or formation of a critical nucleus by some other pathway, such as concentration fluctuations, as yet unknown. While we can only currently speculate regarding the nucleation event itself, the observation that calcium and carbonate ions prefer to come together in a stable disordered state at small cluster sizes, rather than wait for a fluctuation to reach a metastable crystalline nanoparticle, is at least suggestive of a rationale for why amorphous calcium carbonate is often the first phase to appear under appropriate conditions, even though it is the least stable in terms of bulk thermodynamics.

1. D. Gebauer, A. Voelkel and H. Coelfen, *Science*, 2008, **322**, 1819–1822.
2. M. A. Bewernitz, D. Gebauer, J. Long, H. Coelfen and L.B. Gower, *Faraday Discuss.*, 2012, **159**, DOI: 10.1039/C2FD20080E.

Dr Christenson asked: It would seem that according to the authors one would be able to define any association of ions or molecules in solution as a pre-nucleation cluster, however weak or transient the association might be. Surely one should require some evidence that these clusters are actually involved in nucleation events.

Professor Cölfen replied: It is not true that any association of ions or molecules in solution can be defined as a prenucleation cluster. The answer to the question is already included in the term "prenucleation cluster". That is, it must be a cluster which is observed in solution prior to nucleation and, as correctly stated by Dr Christenson, it must be directly involved in the nucleation event. Hence, nucleation of another phase from such species is indeed a necessary prerequisite for an associate of solutes to meet the definition of prenucleation clusters. This was for example shown by means of electron microscopy for calcium carbonate[1,2] or calcium phosphate.[3] In turn, if associates occur in solution but are not relevant for nucleation—which is for example true for many reversible protein interactions—they can of course not be referred to as prenucleation clusters. Generally speaking, if a nucleation event is observed and associates of ions or molecules can be reliably detected in solution before, it is likely that these associates are prenucleation clusters. However, I agree that it has to be demonstrated for each system that these clusters are truly involved in nucleation. One possible way to do this (apart from cryo-TEM) is to perform direct solution analysis by means of analytical ultracentrifugation. Here, concentration-dependent data may reflect the formation of larger aggregates from prenucleation clusters as shown for calcium carbonate[4] and for arginine in our *Faraday Discussions* paper.[5] In the case of calcium carbonate, this process leads to nucleation of ACC.[1,2] In our paper, we propose that a similar mechanism also underlies nucleation of arginine, but this matter is of course still to be demonstrated.

1. E. M. Pouget *et al.*, *Science*, 2009, **323**, 1455–1458.
2. M. Kellermeier *et al.*, *Adv. Funct. Mater.*, DOI: 10.1002/adfm.201200953.
3. A. Dey *et al.*, *Nat. Mater.*, 2010, **9**, 1010–1014.
4. D. Gebauer, A. Völkel and H. Cölfen, 2008, *Science*, 322, 1819.
5. M. Kellermeier *et al.*, *Faraday Discuss.*, 2012, **159**, DOI: 10.1039/C2FD20060K, p. 20, line 20 to p. 22 line 11, including Fig. 4.

Professor Gilbert asked: Is it possible that the free Ca^{2+} ions missing from solution in Fig. 1 of ref. 1 are in fact not aggregated into prenucleation clusters but in $CaCO_3$ ion pairs? If analytical ultra-centrifugation (AUC) cannot detect clusters as small as a single ion pair, what is the smallest cluster it can detect?

Since the AUC experiment takes 10 h, it is possible that clusters with fast dynamics won't be detected. Assume, for instance, that there exists a distribution of clusters, from zero to infinity, that is, from a single $CaCO_3$ ion pair, 2 ion pairs, 3,4... up to all remaining ions aggregated into the largest of clusters. Further assume that small cluster sizes occur more frequently than large ones, hence a mass spectrometry experiment would see a cluster distribution that decays as cluster size increases. Last, assume that all these polydispersed clusters convert into one another quickly at all times, that is, they are not stable. AUC would detect their existence, and assign to them a narrowly distributed size, perhaps the same 70 ion pairs, 2 nm size detected in ref. 1 but this is an artifact, as that size is only an average, which does not at all represent a mono-dispersed population of 2-nm clusters. Can you exclude this hypothetical scenario as an alternative interpretation of the data in Fig. 1 in ref. 1?

1. D. Gebauer, A. Volkel and H. Cölfen, *Science*, 2008, 322.

Professor Cölfen replied: First of all, it is important to note that the dynamics of a species does not reflect its thermodynamic stability. Kinetics and thermodynamics cannot be related directly here. This is explained in somewhat more detail in the reply to the question asked by Michael Nielsen in the discussion of session 3.

That said, it is not possible that the free Ca^{2+} ions missing in solution (in Fig. 1 of the Gebauer *et al.* 2008 *Science* paper) are bound only in ion pairs. Although the used ion-selective electrode can only distinguish free from bound ions, analytical ultracentrifugation (AUC) data clearly show that the average size of the bound species is much larger than that of ions or ion pairs, in case of a fast reversible interaction. What we see in the AUC experiment is the following: one sharp peak at low sedimentation coefficients representing those ions (or molecules), which do not participate in the reversible interaction; a second sharp peak at distinctly higher sedimentation coefficients characterizing the species involved in the prenucleation cluster binding equilibrium; a third peak at still higher s-values, which is much broader and corresponds to cluster aggregates. Since these aggregates do apparently not distinctly take part in the fast dynamics of the primary pre-nucleation cluster equilibrium, they may relate to the larger DOLLOPs referred to by Prof. Gale in his reply to Dr Sear and Professor Gilbert's questions earlier in this discussion.

If a reversible association/dissociation reaction is faster than the timescale of an experiment, in principle, the experiment can only detect an average response. In case of AUC, this means that the shape of the sedimenting boundary (which is what is detected) is a function of the sedimentation coefficient of the involved species, their concentrations as well as the equilibrium constants underlying exchange between these species. This is well known for the analysis of biopolymers by AUC,[1] and can in turn be used to work out stoichiometries and interaction constants between the interacting species. We have already mentioned this in our *Faraday Discussions* paper and also noted that the signals for the less dynamic cluster aggregates are not sharp anymore.[2] This means: the sharp signal observed for the prenucleation clusters is an average over all species participating in the reversible interaction, ranging from single ions over ion pairs to larger clusters. The fact that the average size was estimated to *ca.* 35 ion pairs for $CaCO_3$ shows that the amount of larger clusters must be very significant in this case. This definitely excludes that the missing Ca^{2+} is only bound in ion pairs, because the average sedimentation coefficient obtained for the clusters does not at all agree with the size expected for these species.

Since the AUC can detect single ions, it can certainly also detect anything bigger, including ion pairs and clusters. For a hypothetical scenario of a cluster size distribution ranging from "0 to infinity" as suggested by Prof. Gilbert, with smaller clusters being more frequent than larger ones and a fast interconversion of all species, the AUC analysis will give a sharp peak at the average sedimentation coefficient, as discussed above. The occurrence of a sharp signal for the mean size results, of course, from the underlying fast reversible equilibria and is definitely not an artifact. In addition, the estimated average size of *ca.* 35 ion pairs definitely shows that the above hypothetical model suggested by Prof. Gilbert is not consistent with the results in the Gebauer *et al.* 2008 *Science* paper, or in other words, the missing Ca^{2+} is bound predominantly in clusters with average sizes clearly exceeding those of ion pairs.

The below calculation for a typical $CaCO_3$ experiment shows that the concentration of clusters detected in AUC actually agrees well with the amount of bound Ca^{2+} found in the titration experiments. It is very important to note that the ions bound in clusters make up only a very small fraction out of the total amount of ions present. However, they contain most of the ions that are relevant for pre-nucleation cluster formation (*i.e.* 30–70% of the available Ca^{2+} depending on pH). Therefore, most of the calcium carbonate species present in the prenucleation stage are bound in the clusters, which are thus stable species.

AUC: Clusters are 0.1–1% of all ions. From the composition of a 10 mM carbonate/bicarbonate buffer, pH 9, made from sodium carbonate and bicarbonate, it follows:

Total carbonate/bicarbonate 10 mM; total Na^+ 10.45 mM; carbonate at pH 9 450 μM. Solution taken for AUC experiment after addition of 250 μM Ca^{2+} to

20.45 mM total ions. Detected amounts of bound ions by potential measurements and titration: 100 μM of both Ca^{2+} and carbonate. Thus, free Ca^{2+} = 150 μM, and free carbonate ~420 μM (restoration of the carbonate/bicarbonate equilibrium) 100 μM bound ions at a total ion concentration of *ca*. 21 mM (considering all added and bound ions) corresponds to a fraction of about 0.48%, which agrees well with the 0.1–1% of clusters as detected by AUC. This supports that AUC and titration experiments detect the same species.

1. J. L. Cole, J. W. Lary, T. Moody and T. M. Laue, *Meth. Cell Biol.*, 2008, **84**, 143.
2. M. Kellermeier *et al.*, *Faraday Discuss.*, 2012, **159**, 13, line 32 to 43

Miss Barber asked: Universal definition of prenucleation stage.
What's the difference between prenucleates and other active ions within the solution?
Emphasise the relevance of these prenucleating cluster to define the final crystal structure.
How long can these prenucleation stage last? Would that be part of induction time?

Professor Cölfen responded: The distinction of prenucleation clusters from other associated solution species is detailed in the reply to the question asked earlier by Dr Christenson regarding our paper. As to the difference between prenucleates and other 'active ions' in solution, it is clear that the species we discuss are clusters of the relevant ions, and not simple ions or ion pairs (see also the reply to Professor Gilbert's question). The crucial feature rendering solute associates prenucleation clusters is the fact that they play a direct role in nucleation. The prenucleation stage, on the other hand, means any point in time before nucleation, *i.e.* the state where all species are still in solution and there is no phase boundary present. There is strong evidence that prenucleation clusters can have distinct structural characteristics, which depend on intensive solution parameters, and may be carried over to amorphous calcium carbonate (ACC) upon nucleation.[1] Regarding the relevance of these proto-structures of ACC for the final crystals, please see Dr Gebauer's reply to the question asked by Johannes Ihli regarding Professor Gale's paper.

The prenucleation stage ends with nucleation, that is, when the solution becomes critically metastable and the rate of nucleation is measureable. From our point of view, this would include induction times, which likely represent the final part of the prenucleation stage. However, the prenucleation stage also includes the undersaturated regime as well as supersaturated solution states corresponding to the Ostwald-Miers region, whereas induction times can only be measured for critically supersaturated solutions.

1. D. Gebauer and H. Cölfen, *Nano Today*, 2011, **6**, 564–584.

Mr Davis opened the discussion of the paper by Professor Mark P. Rodger: What lifetime do you think these clusters exist for, what causes these clusters to go on to form ACC; and assuming they exist, are these clusters really pre-nucleation clusters or just a separate artifact of ions in solution, what is the link to ACC? Is there any scope to be able to model the formed clusters-ACC transition?

Professor Rodger responded: The lifetime of our dense clusters is an open question. At present we can only say they are stable for the lifetime of our simulations ($\sim 10^{-8}$ s), and we are working on placing a more useful upper bound to that value. The low density clusters do appear to establish an equilibrium on the simulation timescale, and this would argue for there being a stable population of such clusters, although the simulations probably underestimate the cluster size distribution due to finite size effects in the simulations.

Methods for modelling an pre-nucleation cluster? ACC transition is a subject of current work. The major hurdle appears to be the dehydration of the cluster, with consequent loss of lability in the cluster dynamics.

The clusters we see in the simulations are clearly "pre-nucleation" clusters that already exist in solution before any nucleation of a recognisable solid phase for $CaCO_3$ has occurred. This does not, however, imply that they are direct precursors to nucleation; the existing evidence is not sufficiently unequivocal to distinguish between a direct transition from cluster's critical nucleus on the one hand, and a coupled dissolution/reprecipitation mechanism on the other. The simulation methods alluded to the above would be very useful in helping to clarify this.

Dr Sear enquired: In your computer simulations, do you find evidence of a characteristic size of the clusters, as Professor Cölfen finds in his analytic ultracentrifugation experiments?

Professor Rodger answered: We do not yet have a free energy landscape connecting both the different cluster sizes and geometries seen in our random structure searches. Our multi-nanosecond MD simulations do show dynamical interconversion of cluster sizes, particularly with the low density clusters. This would be consistent with there being a distribution of pre-nucleation cluster sizes rather than one predominant cluster size.

Dr Sear addressed Professors Rodger, Cölfen and Gale: All three of you have presented results on clusters of calcium carbonate in solution. Professor Rodger in his paper, Professors Cölfen[1] and Gale[2] in earlier work. Are your clusters the same? In other words if you compare the results for your clusters with those of the others, do you think the results are consistent? If not, where are the results possibly incompatible?

1. D. Gebauer, A. Völkel and Helmut Cölfen, *Science*, 2008, **322**, 1819.
2. R. Demichelis, P. Raiteri, J. D. Gale, D. Quigley and D. Gebauer, *Nat. Commun.*, 2011, **2**, 590.

Professor Gale responded: While direct structural comparison with the experimental study of Gebauer *et al.* is clearly difficult, since the structure of dynamic clusters is hard to obtain, there is strong quantitative agreement between the speciation and thermodynamics obtained. Furthermore, as noted in Demichelis *et al.*, the effective coordination numbers obtained from the experimental speciation match those observed in the simulation. Hence everything that can be tested agrees strongly between experiment and simulation for the first formed stable species in solution. Both also agree that there is association beyond ion pairing occurring, yet without forming a phase boundary. The simulations explain how this is possible. As to whether the the clusters observed in the paper presented by Professor Rodger are the same as we have observed, given that both studies use our force field parameters, the results for small DOLLOP-like clusters will be identical to within statistical variation. At large cluster sizes there may be differences in the structures characterised since our work focuses on free energy minima, whereas the paper under discussion here targets the internal energy of clusters, which may not always represent a reliable guide to probable states of the system.

Professor Cölfen replied: Our experimental results[1] on prenucleation clusters compare very well (and almost quantitatively in terms of thermodynamics) to those of Professor Gale, as also noted in the corresponding publication of Demichelis *et al.*[2] Moreover, the narrow peaks detected for the prenucleation clusters in AUC point towards a fast reversible equilibrium, which is perfectly in line with the dynamic nature of prenucleation clusters found in the simulations of both Professors

Gale and Rodger. We are not sure if our results are entirely compatible with those of Professor Rodger, but would like to note that Professor Rodger finds the same DOLLOP-like structures for the smaller clusters as Professor Gale.

1. D. Gebauer, A. Völkel and Helmut Cölfen, *Science*, 2008, **322**, 1819.
2. R. Demichelis, P. Raiteri, J. D. Gale, D. Quigley and D. Gebauer, *Nat. Commun.*, 2011, **2**, 590.

Professor Rodger answered: The low density clusters identified in our study appear to be consistent with those reported by Gale and co-workers at high pH. In general terms, they also appear to be consistent with those reported by Cölfen and co-workers, though a definitive comparison is difficult given (1) the different time and length-scales addressed by the experiments compared with the simulations and (2) the fact that the stability of the simulated clusters is still being assessed kinetically. The high density clusters we report are different from those reported by Gale and co-workers, but we stress that we have not yet determined neither the thermodynamic stability nor the lifetime (beyond about 10 ns) of these dense clusters.

Professor Gilbert opened the discussion of the paper by Professor Julian D. Gale: You said that you don't know if the DOLLOP chains resulting from your simulations could be the same objects observed experimentally by Dr Sommerdijk and his group in ref. 1. Is this because the sizes differ?

1. E. M. Pouget, P. H. H. Bomans, J. Goos, P. M. Frederik, G. de With and N. Sommerdijk, *Science*, 2009, 323.

Professor Gale answered: Because of the dynamic nature of the structure of DOLLOP and the fact that there is no specific uniform stable size it would seem unlikely that the species observed in the cryo-TEM are of this nature. However, in the simulations we see that beyond a certain critical concentration larger species emerge in the size distribution that give rise to discrete peaks (see Fig. 3b in ref. 1). In my opinion it is more probable that these post-DOLLOP species that are no longer in fast equilibrium with the solution are most likely to be what is observed in the work of Dr Nico Sommerdijk and co-workers.

1. Demichelis, P. Raiteri, J. D. Gale, D. Quigley and D. Gebauer, *Nat. Commun.*, 2011, **2**, 590.

Dr Gebauer remarked: During the discussions, the characteritics of pre-nucleation clusters for the case of calcium carbonate and amino acids were mixed up. It is important to realize that calcium carbonate pre-nucleation clusters bind between 30–70% of the available calcium (depending on pH),[1] and the model of speciation shows that they are thermodynamically stable with respect to ions in the single-phase solution system before nucleation (equilibrium constant $K > 1$, $\Delta G = -RT \ln K$).[1,2]

In case of the amino acids, it was found that the clusters make up approximately 0.1–0.5 wt% of the species present in under-saturated solutions.[3] Based on these numbers, it was argued during the discussions that the amino acid pre-nucleation clusters have to be metastable, if not unstable, since $K < 1$. However, this is a gut-feeling that is not based on quantitative considerations. If we assume oligomerization equilibria to be valid speciation models, we can evaluate them based on actual data. The initial monomer concentration was 10 mM, and analytical ultracentrifugation showed that around 0.1 wt% clusters formed on average.[3] Taking a molar mass of approximately 100 g mol^{-1} for monomeric amino acids, this corresponds to a concentration of roughly 1 g L^{-1} monomers and 1×10^{-3} g L^{-1} oligomers in the solutions on average. For dimerization, we can write (monomer Mo; dimer D, equilibrium constant K_D), $K_D = [D][Mo]^{-2}$; $[Mo] = 1 \times 10^{-2}$ M; $[D] = 10^{-3}$ g L^{-1}/(200 g mol^{-1}) $= 5 \times 10^{-6}$ M, $K_D = 0.05$ M^{-1} (unstable). For trimerization, we can write (trimer T, equilibrium constant K_T); $K_T = $ [T][Mo]$^{-3}$, [T] = 10^{-3} g L^{-1}/(300 g mol^{-1}) = 3.3 × 10−6 M; K_T = 3.3 M^{-2} (stable). For tetramerization, we can write (tetramer Te, equilibrium constant K_{Te}); $K_{Te} = $ [Te][Mo]$^{-4}$, [Te] = 10^{-3} g L^{-1}/(400 g mol^{-1}) = 2.5 × 10^{-6} M; K_{Te} = 250 M^{-3} (stable).

These considerations show that, if the speciation models are valid, the measured concentrations correspond to the formation of oligomers, which are thermodynamically stable if they are larger than trimers on average. Indeed, the experimental data indicate that the amino acid clusters are distinctly larger than trimers on average.[3] It is important to realize that higher oligomers become more and more stable, based on this speciation model, because the degree of oligomerization enters the exponent in the law of mass action. Metastable, or unstable, species of higher clusters would be present at much lower concentrations, rendering them experimentally non-detectable in most cases.

1. D. Gebauer, A. Völkel and H. Cölfen, *Science*, 2008, **322**, 1819–1822.
2. R. Demichelis, P. Raiteri, J. D. Gale, D. Quigley and D. Gebauer, *Nat. Commun.*, 2011, **2**, 590.
3. M. Kellermeier, R. Rosenberg, A. Moise, U. Anders, M. Przybylski and H. Cölfen, *Faraday Discuss.*, 2012, **159**, DOI: 10.1039/C2FD20060K.

Dr Christenson answered: Dr Gebauer is technically quite correct, of course. The question is whether or not this criterion of thermodynamic stability is actually meaningful. It is always possible to get an equilibrium constant larger than one as long as one postulates a large enough cluster in a small concentration. In Dr Gebauer's notation $K_{Poly} = $ [Poly][Mo]$^{-n}$, where Poly is a "polymeric" cluster of n monomers Mo, which is larger than one provided that n is large enough. I do not believe that this criterion alone means anything.

Professor De Yoreo answered: I agree with this analysis. If multi-ion equilibria apply, then the consequence is that one finds the oligomers are stable. One then has to address the other consequence: they are much less likely to play a role in nucleation without invoking an as yet unidentified low barrier pathway and unless one can account for the small percentage of oligomer–oligomer interaction events relative to those of monomers.

Mr Cantaert addressed Dr Kellermeier and Professor Cölfen: I was wondering if you considered using some other ionisation technique? Is there any particular reason you used ESI, and why no other technique such as ICP?

Professor Cölfen replied: 1) We used electrospray ionisation, because ESI is the only ionization method that is depicting "true", *i.e.* authentic 3-dimensional structures, in this case structures and composition of clusters from solution phase; ESI only reveals structural details qua the emitted multi-charged ions from solution droplets (a process called Taylor-cone emission); 2) Other "soft- ionization" methods such as MALDI (matrix-assisted-laser desorption-ionization) can also, in principle , depict intact noncovalent assemblies such as clusters, but it is highly uncertain if MALDI will reveal authentic cluster compositions.

ICP (inductively-coupled-plasma ionization) is an element (atom)-ionization method, which for molecular clusters would be entirely unfeasible, since the high plasma energy of this method is, in principle, structure-destroying.

Dr Kellermeier answered: ICP is a technique that is commonly used for elemental analyses, in particular of metals. It uses a very hot flame that would in fact burn up any organic clusters.

ESI, in turn, is a soft ionization method that is much less likely to fragment or destroy species which are present in solution (although the distribution of clusters in the gas phase will depend on the particular sample and instrument settings). Also note that ESI has been applied successfully to study amino acid clustering in previous work, as is documented, among others, in papers by Zhang et al. (*Eur. Mass Spectrom.*, 1999, **5**, 353) and Nemes et al. (*J. Mass Spectrom.*, 2005, **40**, 43). Thus, it appeared to be a quite natural choice for our work, which was aimed at correlating species observed by MS in the gas phase with those seen by AUC in the solution state.

Professor Davey asked: Given that you find clusters at all concentrations and that there appears no proven link between them and the ultimate process of nucleation, do you regret calling them 'prenucleation' clusters?

Professor Cölfen replied: The observation that clusters can be found at all concentrations is in accord with the notion that they form based on thermodynamic equilibrium. The analytical ultracentrifugation data as discussed in our paper[1] strongly suggests that amino acid prenucleation clusters do play a role during nucleation. Also in case of calcium carbonate, it has been shown that calcium carbonate prenucleation clusters do play a role during nucleation of ACC too.[2,3] Here, the pathway from prenucleation clusters *via* their aggregates, ACC and final crystalline calcium carbonate was experimentally demonstrated.

That said, we do not regret calling the clusters "prenucleation", however, we agree that the term might imply that they are not present after nucleation. This is one of the reasons why the term DOLLOP[4] has been introduced, while it is important to note that this name refers to a specific structural form. This form may not necessarily represent that of prenucleation clusters of other compounds than calcium carbonate.

1. M. Kellermeier, R. Rosenberg, A. Moise, U. Anders, M. Przybylski and H. Cölfen, *Faraday Discuss.*, 2012, **159**, DOI: 10.1039/C2FD20060K.
2. E. M. Pouget et al., *Science*, 2009, **323**, 1455–1458.
3. M. Kellermeier et al., *Adv. Funct. Mater.*, 2012, DOI: 10.1002/adfm.201200953.
4. R. Demichelis, P. Raiteri, J. D. Gale, D. Quigley and D. Gebauer, *Nat. Commun.*, 2011, **2**, 590.

Professor De Yoreo said: Would the fall-off in instrument sensitivity lead to a peak in the measured distribution of cluster sizes even if that distribution was, in fact, growing exponentially down to zero size and would cryoEM be able to see the difference between an ACC particle and a dense liquid droplet? Also, could one run a cryoEM experiment under identical conditions to a titration experiment to reconcile the seeming discrepancy between the size distribution inferred from each?

Dr Sommerdijk answered: In a cryoTEM experiment the detection limit of small objects is determined by the pixel size of the images recorded, as well as by the contrast of the object against the background. In the cryoTEM experiments performed in our studies[1] a pixel size of 0.15 nm was used and a threshold of 0.45 nm (3*pixel size) was taken as the detection limit. This experimental limit is above the ~0.12 nm information limit of the microscope used (the TU/e cryoTitan). Using a detection threshold of 0.45 nm cuts all the data below this value.

Further, the low electron dose protocols used in our cryoTEM experiments provide images in which indeed the detection of objects with decreasing sizes becomes more and more difficult below 1 nm due to their vanishing contrast with respect to the vitrified solution. It is therefore correct to state that the observed peak in Fig. 1C of ref. 1 could be possibly caused by a decrease in contrast going to smaller particle sizes rather than by a decrease in number. Due to the high degree of hydration dense liquid droplets show far greater electron beam sensitivity than amorphous particles. This was demonstrated in a recent paper of Cantaert et al.[2] in which we used CryoTEM to distinguish between a polymer induced liquid precursor phase and the amorphous calcium carbonate that formed from it. Recently we have run cryoTEM experiments taking samples directly from a titration experiment to study in parallel the early stage development of morphology and structure of calcium carbonate.[3]

1. E. M. Pouget, P. H. H. Bomans, J. A. C. M. Goos, P. M. Frederik, G. de With and N. A. J. M. Sommerdijk. The initial stages of template controlled $CaCO_3$ formation revealed by CryoTEM, Science, 2009, **323**, 1455.
2. B. Cantaert, Y.-Y. Kim, H. Ludwig, F. Nudelman, N. A. J. M. Sommerdijk, F. C. Meldrum, *Adv. Funct. Mater.*, 2012, **22**, 907–915.
3. P. J. Smeets, F. Nudelman, W. J. E. M. Habraken and N. A. J. M. Sommerdijk, in preparation.

Dr Andreassen said: Prenucleation clustering has been detected in both undersaturated and supersaturated solutions of vanillin previously, with small angle neutron scattering and photon correlation spectroscopy.[1] Maybe the experimental techniques and analyses from that investigation can be of help in the search for stable and unstable clusters in other systems.

1. T. J. Sorensen, P. C. Sontum, J. Samseth, G. Thorsen and D. Malthe-Sorenssen, *Chem. Eng. Technol.*, 2003, **26**, 307–312.

Professor Cölfen replied: Thank you for this hint and reference.

Professor Vekilov remarked: The mechanism that you describe seems to consist of the following steps: homogeneous solution–liquid clusters suspended in the solution–amorphous calcium carbonate–calcite. I have several issues with this mechanism: (1) Why would the nucleation of amorphous calcium carbonate require a precursor, in the form of the clusters? Such precursor are needed for the nucleation of crystals, which, due to their high order, is slow. However, the nucleation of an amorphous phase should be sufficiently fast to proceed without a precursor. (2) How can a crystal nucleate in a solid amorphous substance? In such substances, the motion of the molecules and ions is fully arrested. How could the calcium and carbonate ions in the amorphous state move towards their crystalline positions?

Professor Cölfen responded: Prenucleation clusters are considered to be solutes and thus represent species of the homogeneous solution. They do not constitute a second phase (as the—much larger—liquid clusters found as precursors in protein crystallization), and hence cannot be regarded to be 'suspended' in the solution. They form in any solution state, both under- and supersaturated.[1] We cannot follow the argumentation that an amorphous particle would not need such a precursor (point 1), but this statement may be related to the misunderstanding that the prenucleation clusters would already represent a nucleated phase. It has been shown experimentally that aggregation of solute prenucleation clusters leads to the formation of nanoparticles of amorphous calcium carbonate (ACC),[2,3] and that finally a crystal can nucleate within the ACC phase.[3] This does not mean that ACC cannot also nucleate directly from ions, but aggregation of pre-nucleation clusters is certainly an experimentally observed formation pathway of ACC.

For the second point, we do not see any reason to suppose that the motions of ions within ACC would be fully arrested. Besides the fact that the direct transformation of ACC has been observed experimentally,[3,4] we note that the density of ACC is approximately 1.5 g cm^{-3}, whereas the density of crystalline calcium carbonate is around 2.5–3 g cm^{-3} depending on the polymorph.[5] Why shouldn't there be a possibility for density fluctuations in the short- to medium-range structure of ACC resulting in the development of small parts with higher density, which subsequently develop crystalline long-range order propagating through the material? The differences in density of the amorphous and crystalline states would clearly allow for the room that ions would need to arrange into crystalline lattices. Moreover, the actual structure of ACC may contain nanopores and significant amounts of water,[6] and likely many defects.

Besides the particular case of ACC, it is well known that ions can diffuse also in crystalline lattices.[7] This is actually just one example among many others; modern lithium batteries without memory effects but with fast charge/recharge cycles would be impossible if the motion of ions was fully arrested in solids. While the movement of protein molecules may or may not be arrested completely in amorphous states, this is actually one example stressing that observations made in protein systems are not necessarily transferable to ionic systems in a straightforward manner.

1. D. Gebauer and H.Cölfen, *Nano Today*, 2011, **6**, 564–584.
2. M. Kellermeier *et al.*, *Adv. Funct. Mater.*, DOI: 10.1002/adfm.201200953.
3. E. M. Pouget *et al.*, *Science*, 2009, **323**, 1455–1458.
4. B. P. Pichon, P. H. H. Bomans, P. M. Frederik and N. A. J. M. Sommerdijk, *J. Am. Chem. Soc.*, 2008, **130**, 4034–4040.
5. H. Cölfen and A. Völkel, *Progr. Colloid Polym. Sci.*, 2006, **131**, 126–128.
6. A. L. Goodwin *et al.*, *Chem. Mater.*, 2010, **22**, 3197–3205.
7. M. Wilkening, D. Gebauer and P. Heitjans, *J. Phys. Condens. Matter*, 2008, **20**, 022201.

Dr Kellermeier asked: There has been some debate during the meeting as to whether values reported for the solubility of ACC in the early literature are reliable. This is an important issue with respect to the notion that non-classical pathways of $CaCO_3$ crystallization involving an initially amorphous phase should not be accessible in solutions that are undersaturated with respect to ACC (but supersaturated relative to calcite or other crystalline polymorphs).

The most frequently cited values for the solubility product of ACC are probably those determined by Brecevic *et al.*[1] and Clarkson *et al.*,[2] which are about 4×10^{-7} and 9×10^{-7} M^2 at 25 °C, respectively. By contrast, recent titration-based studies gave solubilities that were more than one magnitude lower than those reported previously (3-4 $\times$ 10^{-8} M^2, depending on the type of ACC produced).[3] Given this significant discrepancy, it is worthwhile considering the experimental methodologies by which these values were determined. Brecevic *et al.* used turbidity measurements to monitor dissolution of ACC in water, whereby the solubility was derived from the point at which a "clear solution" was obtained.[1] Here, one may wonder if indeed all of the ACC had actually dissolved at this point or if, possibly, the turbidity sensor might have missed small nanoparticles (any size threshold being related to the wavelength of the used light), thus leading to a higher apparent solubility. On the other hand, the results of Clarkson *et al.* were obtained by recording the pH and Ca^{2+} activity (using ion-selective electrodes) during and after precipitation of $CaCO_3$ from solution[2]—which is conceptually related to the more recent work by Gebauer *et al.*[3] We note, however, that Clarkson *et al.* conducted their experiments in the presence of a few ppm of triphosphate species, under the assumption that these additives would not influence nucleation (and solubility) of ACC, but only extend its lifetime in solution and hence facilitate measurements.[2] Interestingly, titration assays performed by Verch *et al.* showed that triphosphate ions can very well affect the nucleation process and, even more importantly, that the solubility of the initially precipitated amorphous phase was increased by a factor of around 10 (relative to the reference) at additive concentrations in the ppm range.[4] These values agree very well with the data of Clarkson *et al.*[2] and demonstrate that solubilities determined in the presence of triphosphate do not reflect that of pure ACC.

In this regard, we propose that the true solubility of ACC is significantly lower than what has been reported in the early literature—at least for ACC materials that were formed at modest levels of supersaturation (as per ref. 3 and 4) and in the absence of additives (note that there appears to be a family of various types of ACC, which differ with respect to their short-range structure and may exhibit distinct stability and, with it, solubility;[3] particular crystallization conditions might very well influence the type of ACC occurring, and thus higher supersaturation and/ or the presence of additives can lead to materials with different solubility, possibly reaching values found by Brecevic et al. and Clarkson et al.).

This should be taken into account when designing experiments meant to shed light on crystallization pathways in solutions at supersaturations below ACC solubility. We further believe that more effort should be spent to verify literature data in this respect and measure ACC solubilities under various experimental conditions.

1. L. Brecevic and A. E. Nielsen, *J. Cryst. Growth*, 1989, **98**, 504–510.
2. J. R. Clarkson, T. J. Price and C. J. Adams, *J. Chem. Soc., Faraday Trans.*, 1992, **88**, 243–249.
3. D. Gebauer, A. Völkel, H. Cölfen, *Science*, 2008, **322**, 1819–1822.
4. A. Verch, D. Gebauer, M. Antonietti and H. Cölfen, *Phys. Chem. Chem. Phys.*, 2011, **13**, 16811–16820.

Dr Beck answered: It may be of importance that the solubility product for amorphous calcium carbonate is measured by different methods, and by different research groups. It is, however, important to keep in mind that the solubility product should be reported in terms of activities, since the activity-based (unlike the concentration-based solubility product) is only dependent on temperature. Also (activity-based) solution speciation is required. Thus carbon, for example, can be found in the form of dissolved CO_2, CO_3^{2-}, HCO_3^-, $CaHCO_3^+$ and as $CaCO_3$-complex $CaCO_3^0$.

Dr Kellermeier replied: It is of course true that solubility products should be quoted in terms of activities rather than concentrations. However, doing so in the titration experiments introduced by Gebauer *et al.* results in a still lower solubility product (*ca.* 2.5×10^{-8}) as compared to what is obtained under the assumption of ideal conditions, and hence the discrepancy with respect to other literature values is even bigger.

If we take into account the presence of solute clusters (*i.e.* $CaCO_3^0$ and its polymeric forms) after nucleation, values calculated for the solubility product would indeed slightly increase, but usually by no more than around 10% under typical conditions. Obviously, this cannot explain the observed difference of about one order of magnitude between the present solubility products and those reported previously.

Further note that the existence of carbonate species other than CO_3^{2-} (*i.e.* CO_2 and HCO_3^-) in solution has been considered in the evaluation of solubility values based on titration data, and that association of Ca^{2+} with HCO_3^- can be neglected under the given conditions (as shown by equimolar binding of Ca^{2+} and CO_3^{2-} throughout the experiments for pH values equal to or higher than 9; *cf.* Gebauer *et al.*, *Science*, 2008, **322**, 1819).

Dr Beck asked: Our current research from August 2012 measuring the solubility product of the amorphous phase of calcium carbonate in water indicates values that are close to the values measured by Brecevic and Nielsen[1] and thus higher than the values reported in ref. 2. Whether there exist more amorphous phases with even lower solubility products will have to be studied before the results can be published.

We have precipitated calcium carbonate until equilibrium has been achieved in order to determine the solubility product. We followed the crystallization process and determined the solubility product of the amorphous phase by logging the solution pH on-line, together with thermodynamic modelling. Input parameters in the thermodynamic model are temperature, pressure, total alkalinity, total carbon concentration, chloride-, sodium-, and total calcium concentration (all in mmol kg^{-1} solvent), and the fact that solid calcium carbonate consumes stochiometric amounts of calcium and carbonate ions. The thermodynamic (activity-based) equilibria involving the species Ca^{2+}, $CaHCO_3^+$, $CaCO_3^0$; HCO_3^-, CO_3^{2-}, H^+, OH^-, H_2O are considered and the activity coefficients are calculated.

We found that it is crucial, when using pH, to determine the amount of CO_2-ingress from the atmosphere to specify the amount of total carbon concentration. When the ingress of CO_2 into the carbonate source is not considered, the calculated solubility product is significantly lower.

1. L. Brecevic, A. E. Nielsen, *J. Cryst. Growth*, 1989, **98**, 504–510.
2. D. Gebauer, A. Völkel and H. Cölfen, *Science*, 2008, **322**, 1819–1822.

Dr Gebauer remarked: The solubilities that we have measured for the different forms of amorphous calcium carbonate, ACC (ACC I/pc-ACC, $\sim$3.1 $\times$ 10^{-8} M^2; ACC II/pv-ACC, $\sim$3.8 $\times$ 10^{-8} M^2),[1,2] are approximately one order of magnitude lower than the commonly used literature value determined *via* dissolution of ACC that was precipitated from high levels of supersaturation.[3] Owing to the low ionic strength in the solutions of our experiments,[1] we neglected activity effects as they lie within experimental error of the potential measurements anyway (*i.e.* the determined solubilities quoted above could be too high by *ca.* 5–10% if we take activities into account). This alone would clearly affect classical assignments as to whether a given system was supersaturated with respect to ACC or not.[4]

Moreover, the solubilities of calcite, vaterite and aragonite that we typically show in the developments of ion products (*e.g.* Fig. 3 in ref. 1), represent the stability of macroscopic phases. However, if we consider nanoparticles, Gibbs-Thomson effects will play a role, and the crystalline nanoparticles will have a higher solubility than macroscopic crystals that may reach, or even exceed, the solubilities that we have determined for the ACC phases initially precipitated (quoted above). When we refer to macroscopic solubilities calculating the level of supersaturation, we literally assume that macroscopic crystals dropped out of solution, do we not? From my point of view, this is essentially based on the capillary assumption underlying the classical nucleation theory, that nanoparticles behaved as if they were macroscopic, not only with respect to surface properties, but also with respect to the bulk. The particles that are formed first upon nucleation must be nanoparticles, and as we know, essentially everything can and does change at the nanoscale. What happens if there is a crossover in thermodynamic stability at the nanoscale, as supported by recent theoretical work?[5] When the crystalline polymorphs of calcium carbonate are metastable with respect to ACC at small particle sizes, ACC would literally always form first based upon *thermodynamic* reasons rather than kinetic ones.

1. D. Gebauer, A. Völkel and H. Cölfen, *Science*, 2008, **322**, 1819–1822.
2. D. Gebauer, P. N. Gunawidjaja, J. Y. P. Ko, Z. Bacsik, B. Aziz, L. Liu, Y. Hu, L. Bergström, C.-W. Tai, T.-K. Sham, M. Edén and N. Hedin, *Angew. Chem. Int. Ed.*, 2010, **49**, 8889–8891.
3. L. Brecevic and A. E. Nielsen, *J. Cryst. Growth*, 1989, **98**, 504–510.
4. J.-P. Andreassen, R. Beck and M. Nergaard, *Faraday Discuss.*, 2012, **159**, DOI: 10.1039/C2FD20056B.
5. P. Raiteri and J. D. Gale, *J. Am. Chem. Soc.*, 2010, **132**, 17623–17434.

Professor De Yoreo answered: There are three points being made here, each of which is considered to impact nucleation and lead to a multistage pathway. The first is that the solubility limit of ACC is much lower than previously published. The second is that the Gibbs-Thomson effect changes the solubility of the crystalline phases. The third is that the usual capillary assumption, which is embodied in the interfacial energy term is not applicable to nanoparticles.

Knowing the true solubility of ACC is important, as is understanding why the values reported by you are substantially less than those reported previously. However, the only consequence of that would be to better constrain the supersaturation with respect to the crystalline phase when that limit is surpassed. Given that, even in classical nucleation theory (CNT), the thermodynamic barrier to calcite nucleation is enormous when taking the previous literature value for ACC solubility and that the barrier can only get bigger if the solubility limit is lowered to the value you report, the conclusion that homogeneous nucleation of calcite is extremely unlikely does not change.[1] This certainly makes it more likely that ACC will form at a given supersaturation, but only if the solution is above the solubility limit of ACC. It has no bearing on whether a direct pathway is followed below that limit or whether the process of calcite nucleation can be described by CNT. Simply put, I do not see how there can be homogeneous nucleation of calcite on any reasonable timescale below the solubility limit of ACC, wherever it lies. Having said that, the situation changes completely when surfaces are introduced, as is shown in ref. 1.

The Gibbs Thomson effect is already taken into account in CNT, so invoking it as a reason that the nucleation pathway will change is not appropriate. (To see the connection between the Gibbs-Thomson effect and the critical size, see ref. 2.) There is no question that the dependence of excess free energy on size is a critical factor in determining the pathway of nucleation and whether or not a classical description applies. (When divided by the surface area, this excess free energy becomes the interfacial free energy at macroscopic size.) There is also no question that phases can invert their relative stabilities at small size. The many studies of A. Navrotsky clearly demonstrate this. Ironically, this effect is completely associated with surface energy, so to use this fact as an argument that surface energy does not apply is, if nothing else, ironic. One can not have it both ways. Having said that, there is simply no data that tells us how the excess free energy depends on size or at what size it begins to fall off. This is more the case for calcite than for many other materials. So to assume that nanoscale particles of calcite have a lower value of excess free energy than bulk calcite is simply unjustified at this time. In fact, theoretical treatments suggest that, for ionic solids, even a single formula unit already possesses much of the energetic features of the bulk.[3] However, whether or not this is true does not immediately lead to a conclusion about pathways or the validity of CNT. It depends on the length scale over which excess free energy rises to the bulk value and the supersaturation applied to the system. If it turns out to be true that there is an inversion in the relative stability of amorphous and crystalline phases of calcium carbonate, then this will have a strong effect on pathways and, while it would not necessarily invalidate CNT, it makes its application much more complicated. Having said that, one can not make a bulk mass of ACC below its solubility limit, so even if at small enough size ACC became the more stable form, it would be a transient unstable state on the way to calcite in the manner that bassanite was recently shown to be a transient phase that forms below its bulk solubility limit on the way to forming gypsum.[4] However, while this may well be the case for calcium carbonate, I do not believe that we are justified in coming to that conclusion based solely on an MD simulation, given the limitations that such simulations face in accurately modeling large systems of ions in water. Experimental evidence is needed.

1. Q. Hu et al., *Faraday Discuss.*, 2012, **159**
2. H. H. Teng et al., *Science*, 1998, **282**, 724
3. G. V. Gibbs et al., *J. Phys. Chem. A*, 2011, **115**, 12933.
4. A. E. S. Van Driessche et al., *Science*, 2012,**366**, 69.

Dr Andreassen replied: We performed our experiments[1] at low activity based supersaturation and found it instructive to show their position with regard to activity based solubility products of ACC as well as the crystalline polymorphs that are available in the literature. We were aware that you have determined ACC solubility products,[2] but since these were based on concentrations and hence relative to the ionic strength applied, we did not use them in our analysis (Fig. 1). I agree that when converted to an activity basis they become even lower. The large discrepancy between the literature values seems to illustrate that the composition and nature of

ACC is varying, and your comments regarding the difference in stability between different amorphous and crystalline phases at small sizes and the deficiency of the capillary assumption propose that ACC may be present initially also in our investigation. That may well be. We did not probe our system to look for such structures. Our concern was to show that morphologies explained by the presence of additive could also be made without these additives. Whether initially precipitated ACC is also present at these low supersaturation values, and which role it plays in the particle formation of the crystalline polymorphs, will have to be a matter of future investigations.

1. J.-P. Andreassen, R. Beck and M. Nergaard, *Faraday Discuss.*, 2012, **159**, DOI: 10.1039/C2FD20056B.
2. D. Gebauer, A. Völkel and H. Cölfen, *Science*, 2008, 322, 1819–1822.

Professor Gale replied: With regard to the stability of nanoparticles, our previous calculations have shown indeed that ACC has a lower free energy and therefore is more stable than calcite when the size of the nanoparticles is less than approximately 4 nm.[1] Furthermore, the thermodynamically favoured composition of ACC with respect to the water content is also found to evolve with the particles changing from being drier at small sizes to a water content in excess of one molecule per formula unit of calcium carbonate by 4 nm. Of course, kinetic considerations are likely to influence whether the thermodynamic minimum with respect to water content is achieved based on the relative rate of growth to water diffusion within the ACC nanoparticles. This may lead to an inhomogeneous distribution of water, as proposed by experimental PDF studies.[2]

1. P. Raiteri and J. D. Gale, *J. Am. Chem. Soc.*, 2010, **132**, 17623–17434.
2. A. L. Goodwin, F. M. Michel, B. L. Phillips, D. A. Keen, M. T. Dove and R. J. Reeder, *Chem. Mater.*, 2010, **22**, 3197–3205.

Mr Ihli asked: The question concerns the importance of pre-structures detected in amorphous calcium carbonate (proto-vaterite/calcite), resembling the later crystalline polymorphs found in solution.[1] Considering that most crystalline species in solution seemingly form by dissolution and reprecipitation mechanisms from an amorphous intermediate as elaborated at the meeting, is it really possible to distinguish between ACC dictating the resulting polymorph or a coincidental resemblance in structure based on environmental conditions, as all pre-structures in the ACC should have disappeared upon dissolution?

1. D. Gebauer, P. N. Gunawidjaja, J. Y. P. Ko, Z. Bacsik, B. Aziz, L. Liu, Y. Hu, L. Bergström, C.-W. Tai, T.-K. Sham, M. Edén and N. Hedin, *Angew. Chem., Int. Ed.*, 2010, **49**, 8889.

Dr Gebauer replied: The direct transformation of biogenic amorphous calcium carbonate (ACC) into a crystalline phase *in vivo*, that is, *via* a solid/solid state transformation, was suggested for the first time by Beniash *et al.*[1] Many authors took up on this idea, but the first direct evidence for this pathway to occur in biomineralization was obtained by Politi *et al.*[2] Also in *in vitro* experiments, the solid-state transformation of ACC into crystalline states has been directly observed.[3,4]

The statement that the crystallization of ACC followed a dissolution/re-crystallization pathway in most cases is based upon an assessment without any statistical justification. Obviously, the crystalliztion of ACC can proceed *via* dissolution/re-crystallization or solid/solid state transformation pathways, depending on the conditions. These will likely include intensive parameters like temperature, pH, or salinity, but also extensive parameters like convection, confinement, the presence of suitable interfaces/additives, or ACC particle size. We agree that in cases where crystallization of ACC proceeds *via* dissolution/re-crystallization, distinct ployamorphic structures will arguably play no major role.

It is important to note that the different pre-structures in ACC do not dictate the resulting crystalline polymorph. In the original publication, we pointed out the the different structurings were only one out of many factors (see above) that influence crystallization.[5] The notion of proto-calcite and proto-vaterite ACC was introduced based upon spectroscopic evidence (MAS NMR, IR, EXAFS).[5] Moreover, the distinct pre-structures are not coincidental, but their development depends on the set of intensive parameters during precipitation.[5,6]

1. E. Beniash, J. Aizenberg, L. Addadi and S. Weiner, *Proc. R. Soc. London B*, 1997, **264**, 461–465.
2. Y. Politi, T. Arad, E. Klein, S. Weiner and L. Addadi, *Science*, 2004, **306**, 1161–1164.
3. B. P. Pichon, P. H. H. Bomans, P. M. Frederik and N. A. J. M. Sommerdijk, *J. Am. Chem. Soc.*, 2008, **130**, 4034–4040.
4. E. M. Pouget, P. H. H. Bomans, J. A. C. M. Goos, P. M. Frederik, G. de With and N. A. J. M. Sommerdijk, *Science*, 2009, **323**, 1455–1458.
5. D. Gebauer, P. N. Gunawidjaja, J. Y. P. Ko, Z. Bacsik, B. Aziz, L. Liu, Y. Hu, L. Bergström, C.-W. Tai, T.-K. Sham, M. Edén and N. Hedin, *Angew. Chem. Int. Ed.*, 2010, **49**, 8889–8891.
6. D. Gebauer, H. Cölfen, 6. D. Gebauer and H. Cölfen, *Nano Today*, 2011, **6**, 564-584.

Dr Beck asked: In response to Dr Gebauer's reply:

It has been reported that the transformation from ACC to crystalline phases can proceed *via* solid-state transformation.[1,2] When I perform experiments in aqueous solutions which yield amorphous calcium carbonate first (and would transform to a crystalline phase within minutes when left in contact with water), then take out a sample, filter the particles, wash them with ethanol and then dry them at 60 °C, I cannot observe any transformation to a crystalline phase. From that observation it could be concluded that contact with the solvent (also possible *via* humidity from the surrounding air) is needed to induce phase transformation, and that transformation only proceeds *via* a solvent-mediated mechanism. Which parameters do you need to induce a solid-state transformation from ACC to a crystalline phase?

1. B. P. Pichon, P. H. H. Bomans, P. M. Frederik and N. A. J. M. Sommerdijk, *J. Am. Chem. Soc.*, 2008, **130**, 4034–4040.
2. E. M. Pouget, P. H. H. Bomans, J. A. C. M. Goos, P. M. Frederik, G. de With and N. A. J. M. Sommerdijk, *Science*, 2009, **323**, 1455–1458.

Dr Gebauer responded: The questioner's argumentation implies that the particular ACC would not have changed during the drying procedure. We do not think that this point is sustainable. Another assumption is that it was exactly the same ACC with respect to water content, particle size, structure, *etc.*, as the one that, for example, Sommerdijk *et al.* have investigated.

Having said that, perhaps one has to simply wait longer for solid/solid state transformations if the ACC particles are isolated in the dry state; certain molecular conditions, which could be required for the activation of solid/solid-state transformations in solution, could be absent. For example, the crystallization of the bulk could be activated on the surface and rely on hydration (please consider the nanoscale of most relevant ACC particles).

The question for the parameters determining the type of transformation is highly relevant, however, to the best of our knowledge, it can not be answered at present. There are no data. We would like to point out, however, that detailed investigations utilizing techniques with high resolution appear to be a fundamental requirement. Indirect observations are not very useful in this context.

Dr Beck asked Dr Gebauer: How would the measured solubilities of ACC change when activities of both calcium ions and carbonate ions would be considered instead of concentrations?

Activity coefficients start to deviate from ideality at very low ionic strengths. The Debye–Hückel equation is valid until solutions of 0.001 mol L^{-1}. For higher concentrated solutions up to approximately 0.1 mol L^{-1} the Davies modification to the Debye–Hückel equation can be used. At even higher concentrations different models like the Pitzer equations have to be used. In ref. 1 the Davies modification is used to calculate activity coefficients. In ref. 2 the commercial software Multiscale to calculate activity coefficients based on the Pitzer equations is used. In the calculations, association or reactions of ions with solution constituents are considered. Carbonate, for example, can associate to $CaHCO_3$, to the calcium carbonate complex $CaCO_3^0$ and to HCO_3^+, and can react to CO_2, reducing the free concentration of carbonate. Concentration-based solubility products depend on the ionic strength pH, the total amount of dissolved carbon species (total amount of dissolved carbon species: CO_2 that comes in and out from the atmosphere, already dissolved CO_2, HCO_3^-, $CaCO_3^0$, CO_3^{2-}, $CaHCO_3^+$), temperature and other parameters. The big advantage of using activity-based solubility producs is that they do not depend on the ionic strength, pH, the total amount of dissolved carbon species, *etc.*, they only depend on temperature.

1. R. Beck, M. Seiersten and J.-P. Andreassen, The Constant Composition Method for Crystallization of Calcium Carbonate at Constant Supersaturation, *manuscript in preparation*, 2012.
2. J.-P. Andreassen, R. Beck and M. Nergaard, *Faraday Discuss.*, 2012, **159**, DOI: 10.1039/C2FD20056B.

Dr Gebauer replied: Ion selective electrodes (ISEs) measure activities by principle, that is, they detect only free ions, not ions bound in complexes, or pre-nucleation clusters. All different complexes and species indicated by the questioner are thereby accounted for in ISE measurements *a priori*.

That said, the ionic strength of the solutions, which we typically investigate,[1–3] can be in turn accounted for during calibration by means of additions of appropriate amounts of NaCl. This is an approach to experimentally deal with activity upon increased ionic strength that does not rely on any assumptions, except that NaCl would only influence ionic strength alone (please note that the calculated values as oulined by the questioner in fact do rely on various assumptions). When this is done, we find an activity effect (due to increased ionic strength) that can lie within 5–10% in terms of the amount of bound calcium during the prenucleation stage. This is within experimental error of typical potential measurements in terms of reproducibility. Therefore, we have neglected these effects in studies performed at comparably low salinity.[2,3]

Note that after nucleation, owing to the precipitation of solid $CaCO_3$, the ionic strength of the system is further reduced distinctly, as compared to the prenucleation stage. Thus, the effect of ionic strength on activity-based solubilies, which are naturally determined in the post-nucleation stage, is even lower than stated above, and actually becomes negligible.

In summary, the activity effects outlined by the questioner are either accounted for, or negligible. The values that we have reported for the solubilities of ACCI and ACCII agree with the respective activity products to within a deviation of maximal 5–10%.

1. D. Gebauer, A. Völkel and H. Cölfen, *Science*, 2008, **322**, 1819–1822.
2. D. Gebauer *et al.*, *Adv. Mater.*, 2009, **21**, 435–439.
3. A. Verch *et al.*, *Phys. Chem. Chem. Phys.*, 2011, **13**, 16811–16820.

Dr Beck remarked: Yes, I agree. The ion selective electrode, if calibrated (correlating the mV signals from the electrode with calcium activities) delivers calcium activities.

I want to add that activity-based solubilities do not depend on ionic strength or pH. Concentration-based solubilities, however, do. Let us consider two examples at 25 °C and 1 bar, one solution of 1 mmol kg^{-1} solvent $CaCl_2$ and 1 mmol kg^{-1} solvent Na_2CO_3, and one other solution consisting of 1 mmol kg^{-1} solvent $CaCl_2$ and 1 mmol kg^{-1} solvent Na_2CO_3 and 200 mmol kg^{-1} solvent NaCl.

Considering all activity-based equilibrium constants for the autoprotolysis of water, carbon dioxide, hydrogen carbonate, calcium hydrogen carbonate, the calcium carbonate complex and activity coefficients calculated by the Pitzer equations (commercial software MultiScale used), the free concentrations of calcium and carbonate, and the mean activity coefficients at Sc=1 (equilibrium with the calcite phase = all crystalline mass of calcite has precipitated) are:

Solution 1 at Sc = 1:
Ca^{2+}: 1.28×10^{-4} mol kg^{-1} solvent
CO_3^{2-}: 3.90×10^{-5} mol kg^{-1} solvent
γ: 0.80
$K_{sp} = \gamma^{2}*c, Ca^{2+}*c, CO_3^{2-}$: 3.2×10^{-9}
$[K_{sp}] = c, Ca^{2+}*c, CO_3^{2-}$: 4.99×10^{-9}

Solution 2 (with 200 mmol kg^{-1} solvent NaCl) at Sc = 1:
Ca^{2+}: 2.81×10^{-4} mol kg^{-1} solvent
CO_3^{2-}: 1.51×10^{-4} mol kg^{-1} solvent
γ: 0.275
$K_{sp} = \gamma^{2}*c, Ca^{2+}*c, CO_3^{2-}$: 3.2×10^{-9}
$[K_{sp}] = c, Ca^{2+}*c, CO_3^{2-}$: 42.8×10^{-9}

K_{sp} denotes the activity-based solubility product and $[K_{sp}]$ denotes the concentration-based solubility product. It can be seen that the concentration-based solubility product is approximately 10 times higher in the solution with 200 (mmol kg^{-1} solvent) NaCl. However, the activity-based solutility product is constant in both cases (since γ is 0.275 at equilibrium in solution 2 as compared to 0.8 in solution 1. This is the big advantage of activity-based solubility values.

The free concentration of calcium was determined by an ion selective electrode. I did not understand completely how the free concentration of carbonate ions in equilibrium was determined in order to obtain the reported solubility products?[1,2]

1. D. Gebauer, A. Völkel and H. Cölfen, *Science*, 2008, **322**, 1819–1822.
2. M. A. Bewernitz, D. Gebauer, J. Long, H. Cölfen and L. B. Gower, *Faraday Dicsuss.*, 2012, **159**, DOI: 10.1039/c2fd20080e.

Dr Sear asked Professor Cölfen: Is the amorphous calcium carbonate (ACC) you study an equilibrium thermodynamic phase that coexists at metastable thermodynamic equilibrium with the surrounding solution? Here a thermodynamic phase is something like the, say, decane-rich phase of a phase-separated mixture of decane and water, where the composition of the decane-rich (and also of the water-rich) phases is exactly determined and fixed at a fixed temperature and pressure (and pH *etc.*). The decane-rich and water-rich phases can come into thermodynamic equilibrium because the decane and water molecules can diffuse inside both phases, and can rapidly move from one phase to another, allowing the system to relax to the unique (here metastable) equilibrium free-energy minimum at the particular values of temperature, pressure, *etc.*

If ACC is not a thermodynamic phase, what is it?

Professor Cölfen answered: Yes, ACC is a thermodynamic phase, which is metastable with respect to crystalline $CaCO_3$. It has a solubility product which can be measured and the type of ACC precipitated depends on experimental variables like pH.[1] The measured solubility product remains constant upon further calcium addition into the carbonate buffer, which means that these added ions get incorporated into ACC, and that the level of free ions is dictated by the solubility of the

ACC phase. This is exactly what would be expected for a solid phase in equilibrium with its surrounding solution. As already noted above, however, the solid phase does not necessarily have to be the stable solid form, *i.e.* the whole system can still be on its way to thermodynamic equilibrium and thereby seemingly violate Gibbs' phase rule; the initial formation of ACC is in no way different from the initial nucleation of the unstable polymorph vaterite in this respect.

1. D. Gebauer, A. Völkel and Helmut Cölfen, 2008, *Science*, **322**, 1819.

Professor Van Blaaderen addressed Professors Rodger, Cölfen and Gale: How does one know/prove/make likely in an experimental crystallization study (and also for simulation studies that try to exactly mimic experimental studies) that intermediate species observed (such as prenucleation clusters), are 'thermodynamically stable' before a final crystal phase has established itself? Of course I am well aware of the fact that almost all processes in especially biology, and also chemistry and even physics are in some way or other 'out-of-equilibrium'. For instance, to take the extreme version of this: if one allows for fusion reactions to occur, ALL atoms are metastable with respect to only one isotope of iron (and if the proton is not infinitely stable, possibly not even that isotope of iron)! Clearly, it still makes sense to define and study 'equilibrium' systems as one can of course exclude certain pathways (*e.g.* nuclear reactions) and still define some kind of 'local' equilibrium state. Also, if one goes through a set of transition states in a reversible way from one state to another state using, *e.g.*, an external field that controls the path and makes it reversible, local equilibrium can be defined. However, it is not clear to me what is meant by a local equilibrium of for instance a glassy or amorphous cluster state when a crystalline state is lower in free energy and thus when the end-result of the transformations taking place in the system results in crystals. Is it not true that the 'in-between states' can only be called being in (local) equilibrium if the constraints are mentioned and defined as well; or said differently: if the reverse set of states can be travelled and be observed as well? If one cannot make the system travel (in time) through the reverse order of states, there is not a reversible path to the structures and their free energy cannot be defined. For instance, in the umbrella sampling technique used in computer simulations a local potential field makes sure the intermediate structures have a well-defined free energy. However, I would not know how to assure that this is the case in experiments without the use of also an external field (except for trying to grow structures as slowly as possible, which of course cannot be done for all super saturations). This is why in several of the studies and presentations I question whether the structures observed and reported are indeed in some kind of (local) equilibrium. For instance, in case of a size distribution, following a Boltzmann type of distribution is certainly not enough as the amount of disorder could be different depending on the path taken or the time scale of the process.

Professor Gale responded: It is true that simulations can obviously suffer from issues of timescale and so may be out of equilibrium for the duration of the run or only establish a local equilibrium that omits a slow reaction. In the specific case of calcium carbonate speciation being discussed here it is clear that under many of the conditions simulated the speciation is at least in local equilibrium as demonstrated by the Boltzmann distribution of probabilities. This is a dynamic equilibrium as ions are rapidly attaching and detaching, and so both forward and backward reactions are observed many times. The best evidence that this agrees with experiment is that the detectable free calcium concentration observed by Gebauer *et al.*[1] is quantitatively reproduced by the simulation derived speciation model over a wide range of concentrations and pH values, demonstrating that the underlying free energies for competing reactions are sound.[2] These values also agree with those derived by either umbrella sampling or metadynamics in which timescale is not an issue providing further validation. Given that the experiments are on a timescale many orders of magnitude longer than the simulation, but still match the simulation results, this suggests that timescale is not an issue when small DOLLOPs predominate. Obviously at the point of nucleation this is no longer true and direct unbiased simulation cannot hope to describe this process for calcium carbonate.

As for the need to apply the same potential in the experiments as used in the umbrella sampling, this is absolutely not the case. Central to all of the free energy determination methods is that any bias applied to sample regions of the probability distribution is ultimately removed to obtain the unbiased distribution and thereby the free energy landscape. Consequently the free energy difference between the end states should match the experiment without any modification of the experiment. The caveat that should be applied is that such simulations require careful choice of an appropriate set of collective variables.

1. D. Gebauer, A. Voelkel and H. H. Coelfen, *Science*, 2008, **322**, 1819–1822.
2. R. Demichelis, P. Raiteri, J. D. Gale, D. Quigley and D. Gebauer, *Nature Commun.*, 2011, **2**, 590.

Professor Cölfen replied: The notion of the thermodynamic stability is based on the observation of a negative value of ΔG associated with cluster formation. In the single-phase solution system, the solute clusters are more stable than free ions as well as ion pairs (for more details, also see my replies to questions referring to our paper, and Dr Gebauer's comments concerning our papers in sessions 1 and 3).

Here, one misunderstanding appears to be based on an issue of boundary conditions. It is important to note that prenucleation clusters are certainly not stable with respect to amorphous calcium carbonate or crystalline polymorphs once the solution has become supersaturated. But in this case, we consider at least two co-existing phases, where all relevant chemical species in the supersaturated phase are metastable with respect to the second phase. Once thermodynamic equilibrium has been reached in such multi-phase systems, it must conform to Gibbs' phase rule. In other words, in equilibrium, the chemical potential of all species in the different phases is equal.

However, in presence of a second phase (regardless whether the second phase is more stable than the solution phase if supersaturated, or whether both phases are in equilibrium as indicated above), the clusters always remain thermodynamically stable in the single-phase solution system, that is, within the phase boundary. This is in accord with all experimental observations detailed in the replies and comments referred to above.

Professor Gale remarked: The question was asked as to how we define equilibrium in the context of our simulations of pre-nucleation speciation. For our simulations we observe a Boltzmann distribution of cluster sizes that does not evolve with time despite the fact that there is dynamic exchange of ions with solution and other clusters.

This can be seen, for example, in Fig. 3a of Demichelis *et al.*, *Nat. Commun.*, 2012, **2**, 590. This we take to demonstrate that we have an equilibrium distribution within the local region of the free energy configuration space that is accessible on the timescale of the simulation. It is important to stress the fact that the probability of finding larger clusters decays with increasing size does not imply that they are not stable. It is important to remember that the equilibrium constant for oligomer formation contains the concentrations of the monomers to the power of the oligomer size. Hence if the concentrations are in the millimolar range then the equilibrium constant can be very much greater than 1 (*i.e.* ΔG is negative) and yet the concentration of the oligomer need not necessarily exceed that of the ions or ion pairs. Finally, Fig. 3b of the above article shows that beyond a certain critical size and concentration there is a point where DOLLOP is no longer in equilibrium with solution and this is then clearly observed in the speciation.

Professor Roberts said: Citric acid is an interesting species in its own right in that for crystallisation from aqueous solution, it has a rather high metastable zone width and thus does not easily nucleate.[1] This probably reflects its propensity for strong solute/solvent binding and perhaps conformational variation within the incipient citric acid clusters formed close to nucleation. In your Fig. 7 you appear to show a cluster of citrate, Ca and CO_3 ions and I was wondering if you can provide any information concerning the citrate ion's molecular structure and salvation properties with respect to those present in "mother" compounds such as calcium citrate, citric acid *etc.* as this may provide some further molecular-scale insight into not just this particular study but also related crystallisation process involving these species.[1] An examination of the batch crystallisation behaviour of citric acid as monitored *in situ* by ATR-FTIR spectroscopy, is ref. 1.

1. H. C. Groen and K. J. Roberts, *J. Phys. Chem.*, 2001, **105**, 10723–10730.

Professor Gale replied: Because the protonation state of our citrate model is fixed we haven't been able to compare to citric acid in this work. However, we can compare the conformation of the citrate anion between the solution and solid state. Unfortunately the crystal structure of anhydrous calcium citrate appears to be unknown, though there are hydrates that have been solved.[1] When Ca cations coordinate with the citrate anion in solution we observe a range of configurations that are explored involving both mono- and bidentate coordination that involves combinations of all three carboxylate groups.

In the crystal structure of the tetrahydrate[1] the calcium is simultaneously coordinated to three distinct carboxylate groups and the oxygen of a hydroxyl group from three neighbouring citrate anions. The coordination here represents a superposition of binding modes observed separately for the ion pair in solution. This crystal structure also contains an intramolecular hydrogen bond between the hydroxyl group and one of the carboxylate oxygens of the two equivalent side chains; something also observed for the anion in solution.

One difference between the solid state structure of the hydrate and the citrate anion in solution is the conformation of the organic. In the solid there is a single configuration in which the CH_2 groups are in close proximity, whereas in solution other torsional angles are possible and are explored during the free energy calculations.

1. E. Herdtweck, T. Kornprobst, R. Sieber, L. Straver and J. Plank, *Z. Anorg. Allg. Chem.*, 2011, **637**, 655–659.

Dr Zhang opened the discussion of the paper by Professor Peter G. Vekilov: In the previous session, we have seen that the clusters are also formed in the calcite and the amino acid systems, similar to the protein clusters showed in your paper, they all have a very low volume fraction and limited lifetime which make them difficult to characterize experimentally. From these common features, could we say that they share similar mechanisms of clustering?

From the view of interactions between molecules, the clusters can be formed by the balance of repulsive and attractive interactions. While the protein–protein interactions are more complex, interactions in calcite/amino acid systems are relatively simple. Can we get some clue of clustering in a protein system? What kind of interaction plays the most important role?

Professor Vekilov answered: I think too few of the properties and behaviors of the clusters in calcite and other organic and inorganic solution have been determined and monitored experimentally to be able to make justified conclusions about their mechanism. As far as trying to guess their mechanism, the insights about the protein clusters suggest that the cluster existence may be due to processes much more complex than the simple balance between attraction and repulsion. While such balance considerations adequately describe the clusters in colloid solutions,[1–3] they were shown to fail to predict the characteristics of the dense liquid protein clusters.[4,5] It is possible that a complex mechanism, similar to the one suggested for dense liquid protein clusters,[5] may be needed to explain clustering in solutions of calcium carbonate and other organic and inorganic compounds.

1. J. Groenewold and W. K. Kegel, *J. Phys. Chem. B*, 2001, **105**, 11702–11709.
2. F. Sciortino, S. Mossa, E. Zaccarelli and P. Tartaglia, *Phys. Rev. Lett.*, 2004, **93**, 055701.
3. A. Stradner, H. Sedgwick, F. Cardinaux, W. C. K. Poon, S. U. Egelhaaf and P. Schurtenberger, *Nature*, 2004, **432**, 492–495.
4. S. B. Hutchens and Z.-G. Wang, *J. Chem. Phys.*, 2007, **127**, 084912.
5. W. Pan, P. G. Vekilov and V. Lubchenko, *J. Phys. Chem. B*, 2010, **114** 7620–7630.

Dr Andreassen addressed Professor Vekilov: In the opening lecture by Professor Frenkel it was suggested that heterogeneous nucleation provides a much better fit to experimental data than homogeneous nucleation and in your paper nucleation was also well described by heterogeneous nucleation in the classical sense. In the two-step mechanism, is nucleation in the concentrated droplets or condensed phase heterogeneous or homogeneous and is the formation of the concentrated (condensed) phase always a requirement for nucleation?

Professor Vekilov answered: Heterogeneous nucleation, provided that proper substrates are available, is always faster than homogeneous nucleation. J. W Gibbs was the first to estimate the barrier for heterogeneous nucleation and to conclude that it could be significantly lower than the homogeneous nucleation barrier.[1–3] The degree of attenuation of the barrier is determined by the contact angle between the substrate and the nucleating cluster (Gibbs was considering the nucleation of liquids). In Gibbs's model, this contact angle depends on the ratios of the surface free energies of the three interfaces existing between the substrate, the nucleating substance, and the environment. In subsequent work, this formalism was modified to account for the interaction of a crystal with a substrate, but the general conclusion, that the attenuation of the barrier by a foreign substrate depends on the interactions of this substrate with the crystalizing substance, remained.[4–6] Volmer's expression linking the nucleation rate to the barrier *via* an Arrhenius-type equation predicts that heterogeneous nucleation will be faster by orders of magnitude.[7]

In our experiments we encounter both homogeneous and heterogeneous nucleation: despite our efforts to minimize the latter, it is always present with some proteins (heterogeneous nucleation appears not to occur with crystals of hemoglobin C and of polymers of hemoglobin S for reasons that we do not fully understand[8,9]). If the nucleation rate is determined from the evolution of the average number of nucleated crystals in time, the homogeneous and heterogeneous nucleation events can be separated: the latter are much faster and are thus constrained to early times. Since homogeneous nucleation could be overwhelmed by heterogeneous nucleation events, such separation is only possible after minimization of heterogeneous nucleation with careful experiment preparation.

Since the homogeneous nucleation mechanism is simpler—it depends on fewer parameters—we use data on the homogenous nucleation rate to deduce the nucleation mechanisms, including its two most striking features: its two steps[10–14] and the disappearance of the nucleation barrier at certain moderate supersaturations; we call the latter "solution-to-crystal spinodal".[14,15] In our investigations of the two-step mechanism, we found that the quantitative parameters of the generation of the clusters, its first step, are reproducible in numerous experiments, with protein of different sources, with slight variations of the solution composition, and with the use of two different techniques.[16,17] Thus, we conclude that the clusters are homogenously nucleated. If a cluster lands on a surface or encapsulates a foreign particle, the nucleation of a crystal inside it may be heterogeneous. We think that these are the sources of the heterogeneous nucleation events occurring in the system. In the other cases, crystal nucleation inside the cluster is homogenous.

As far as the generality of the two-step mechanism, we have studied in detail the nucleation of two protein systems: crystals of lysozyme[10,11] and polymers of sickle cell hemoglobin.[18,21] Both of them follow this mechanism. There is significant evidence that other protein, small molecule inorganic and organic substances, colloids and polymers also follow it, for a brief review see ref. 14. To access the applicability of this mechanism to the overwhelming majority of untested systems, we note that its action relies of the availability of disordered liquid or amorphous metastable clusters in the homogeneous solutions prior to nucleation. While such clusters have been demonstrated in solutions of numerous compounds of widely variable chemical and physical properties, it is feasible that not all solutions would support the existence of such clusters with properties allowing the nucleation of crystals in them. In such systems the action of the direct nucleation mechanism might be the only option.

1. J. W. Gibbs, *Trans. Connect. Acad. Sci.*, 1876, **3**, 108–248.
2. J. W. Gibbs, *Trans. Connect. Acad. Sci.*, 1878, **16**, 343–524.
3. J. W. Gibbs, *The scientific papers of J. W. Gibbs. Volume One: Thermodynamics*, Oxbow Press, Woodbridge, Connecticutt, 1993.
4. I. N. Stranski and R. Kaischew, *Z. Phys. Chem., Sect. B*, 1934, **26**, 100–113.
5. I. N. Stranski and R. Kaischew, *Z. Phys. Chem., Sect. B*, 1934, **26**, 114–116.
6. A. A. Chernov, *Modern Crystallography III, Crystal Growth*, Springer, Berlin, 1984.
7. M. Volmer, *Kinetik der Phasenbildung*, Steinkopff, Dresden, 1939.
8. P. G. Vekilov, A. R. Feeling-Taylor, D. N. Petsev, O. Galkin, R. L. Nagel and R. E. Hirsch, *Biophys. J.*, 2002, **83**, 1147–1156.
9. O. Galkin and P. G. Vekilov, *J. Mol. Biol.*, 2004, **336**, 43–59.
10. O. Galkin and P. G. Vekilov, *Proc. Natl. Acad. Sci. USA*, 2000, **97**, 6277–6281.
11. P. G. Vekilov, *Crystal Growth and Design*, 2004, **4**, 671–685.
12. D. Kashchiev, P. G. Vekilov and A. B. Kolomeisky, *J. Chem. Phys.*, 2005, **122**, 244706.
13. W. Pan, A. B. Kolomeisky and P. G. Vekilov, *J. Chem. Phys.*, 2005, **122**, 174905.
14. P. G. Vekilov, *Cryst. Growth Des.*, 2010, **10**, 5007–5019.
15. L. F. Filobelo, O. Galkin and P. G. Vekilov, *J. Chem. Phys.*, 2005, **123**, 014904.
16. Y. Li, V. Lubchenko and P. G. Vekilov, *Rev. Sci. Instrum.*, 2011, **82**, 053106.
17. Y. Li, V. Lubchenko and P. G. Vekilov, *J. Phys. Chem.*, 2012, submitted.
18. O. Galkin, R. L. Nagel and P. G. Vekilov, *J. Mol. Biol.*, 2007, **365**, 425–439.
19. O. Galkin, W. Pan, L. Filobelo, R. E. Hirsch, R. L. Nagel and P. G. Vekilov, *Biophys. J.*, 2007, **93**, 902–913.
20. W. Pan, O. Galkin, L. Filobelo, R. L. Nagel and P. G. Vekilov, *Biophys. J.*, 2007, **92**, 267–277.
21. V. V. Uzunova, W. Pan, O. Galkin and P. G. Vekilov, *Biophys. J.*, 2010, **99**, 1976–1985.

Dr Christenson replied: In the case of nucleation in the capillary condensates (shown in our paper) we cannot be absolutely sure whether the nucleation is homogeneous or heterogeneous. We do not observe the actual nucleation event itself, only the fact that it is followed by the deposition of effectively macroscopic amounts of crystalline material from the vapour. That the undercooling in most cases is comparable to that found for homogeneous nucleation (30-40 K), and that the crystalline mica surface is very unlikely to promote the nucleation of another crystalline substance suggest that we are dealing with heterogeneous nucleation. Furthermore, homogeneous nucleation is also almost invariably the assumption when considering freezing of liquids confined to porous materials,[1,2] where the pore diameters are comparable to those of our experiments. The experimental results are very often consistent with the presence of a liquid film between the frozen solid and the pore walls, a situation that would not be consistent with heterogeneous nucleation. We never observe nucleation on flat mica surfaces at saturation, in the case of hexachloroethane not even when the undercooling is over 160 K!

1. H. K. Christenson, *J. Phys.: Condensed Matter*, 2001, **13**, R95–R133.
2. C. Alba-Simionesco, B. Coasne, G. Dosseh, G. Dudziak, K. E. Gubbins, R. Radhakrishnan and M. Sliwinska-Bartkowiak, *J. Phys.: Condens. Matter*, 2006, **18**, R15–R68.

Dr Andreassen said: The two-step nucleation mechanism requires the formation of a dense liquid phase from which the crystals nucleate. In the case of calcium carbonate the precursor phases are either small prenucleation clusters, PILP, DOLLOP or ACC. In the concept of the two-step mechanism, can any of these structures qualify as the dense liquid phase and as such facilitate the nucleation of calcium carbonate?

Professor Vekilov answered: The dense liquid within which the crystals nucleate can exist in two firms: as a macroscopic phase stable with respect to the dilute solution, and as clusters of limited size which are metastable with respect to the solution. Some of the clusters detected in solutions of calcium carbonate are so small that they could just be oligomers of several ions stabilized by respective chemical bonds. It seems that the available experimental data are insufficient to decide which of these three forms is chosen by the clusters detected in solutions of calcium carbonate. It also appears that the role of these clusters in nucleation is also far from proven: are they just a depot for extra calcium and carbonate ions, or do they serve as sites of facilitated nucleation of calcite crystals?

Dr Zhang commented: The two-step nucleation and crystallization mechanism so far discussed are all related to two order parameters, *i.e.* the density and 3D-crystal structure, which can possibly be developed separately during a nucleation process. In your system, the polymerization of the sickle cell hemoglobin from the metastable dense liquid phase leads to 1D structure. My question is whether the structure order parameter (3D *vs.* 1D) affects the two-step nucleation?

Professor Vekilov replied: The applicability of the two-step mechanism to the nucleation of the sickle cell hemoglobin (HbS) polymers is supported by five sets of experimental results: (1) direct observation in two-phase (HbS) solutions;[1] (2) the existence of dense liquid clusters of size several hundred nanometers in under- and supersaturated HbS solutions;[2] (3) the correlation between the nucleation delay time and the molecular attachment frequency, which contradicts the predictions of the classical, direct nucleation, and agrees with the predictions of a two-step model;[3–5] (4) the orientation of the HbS polymers, which is always normal to the plane of polarization of the illuminating light;[5] and (5) the equivalent $\sim 100\times$ response of both the volume fraction of the HbS dense liquid clusters and the HbS polymer nucleation rate to the addition of free heme to the solution.[5] Thus, even though the HbS polymers represent ordered structures with only one direction of translational symmetry, their nucleation follows a mechanism identical to the one established for crystals with 3-D translational symmetry. The only experimentally observed specificity of HbS nucleation is the significantly faster, by seven–eight orders of magnitude, nucleation rate. Whether this is due to the lower dimensionality of the nucleating structures or to other thermodynamic or kinetic factors is a big open question.

1. O. Galkin, K. Chen, R. L. Nagel, R. E. Hirsch and P. G. Vekilov, *Proc. Natl. Acad. Sci. USA*, 2002, **99**, 8479–8483.
2. W. Pan, O. Galkin, L. Filobelo, R. L. Nagel and P. G. Vekilov, *Biophys. J.*, 2007, **92**, 267–277.
3. D. Kashchiev, P. G. Vekilov and A. B. Kolomeisky, *J. Chem. Phys.*, 2005, **122**, 244706.
4. O. Galkin, R. L. Nagel and P. G. Vekilov, *J. Mol. Biol.*, 2007, **365**, 425–439.
5. O. Galkin, W. Pan, L. Filobelo, R. E. Hirsch, R. L. Nagel and P. G. Vekilov, *Biophys. J.*, 2007, **93**, 902–913.

Dr Zhang commented: Dynamic protein clusters, mainly in lysozyme solutions, with short relaxation time have been recently explored using the neutron spin echo technique.[1,2] The question is: are those dynamic clusters the same thing as presented in your work by DLS?

1. Cardinaux, F.d.r., *et al.*, Cluster-Driven Dynamical Arrest in Concentrated Lysozyme Solutions, *J. Phys. Chem. B*, 2011. **115**(22), 7227–7237.
2. L. Porcar *et al.*, Formation of the Dynamic Clusters in Concentrated Lysozyme Protein Solutions. *J. Phys. Chem. Lett.*, 2009. **1**(1): 126–129.

Professor Vekilov responded: The clusters observed in these and some other papers differ in several important aspects from the mesoscopic dense liquid clusters studied by ours and other groups. (1) The former clusters are indeed dynamic, *i.e.*, they are detected because they exist for times longer than the characteristic diffusion time of the monomers, but still macroscopically short. This is highlighted as the explanation of their absence in some studies.[1] The dense liquid clusters have lifetimes of order seconds with some proteins[2,3] or with a lower bound of 15 ms with others.[4–6] (2) The dynamic clusters are small, of order of 10 nm or smaller,[7] while the dense liquid clusters are about 100 nm in size.[2–6] (3) The dynamic clusters engage a significant fraction of the protein in solution, while the dense liquid clusters occupy only a very small volume fraction, from 10^{-7} to 10^{-4} and hold a comparably small fraction of the solution volume.[2–6] (4) Because of this high volume fraction, the dynamic clusters can affect the macroscopic solution properties, such as solution viscosity[8] and even induce viscoelasticity.[9] The low volume occupied by the dense liquid clusters precludes any effects on the solution macroscopic properties. (5) Because of their purported low density and loose structure, the dynamic clusters do not support the nucleation of other phases in the solution and, because of the associated viscosity increase,[9,10] likely slow down the growth of new phases. The dense liquid clusters are crucial sites for the nucleation of crystals or other ordered solid phases.[10–17]

1. A. Shukla, E. Mylonas, E. Di Cola, S. Finet, P. Timmins, T. Narayanan and D. I. Svergun, *Proc. Natl. Acad. Sci.*, 2008, **105**, 5075–5080.
2. O. Gliko, N. Neumaier, W. Pan, I. Haase, M. Fischer, A. Bacher, S. Weinkauf and P. G. Vekilov, *J. Am. Chem. Soc.*, 2005, **127**, 3433–3438.
3. O. Gliko, W. Pan, P. Katsonis, N. Neumaier, O. Galkin, S. Weinkauf and P. G. Vekilov, *J. Phys. Chem. B*, 2007, **111**, 3106–3114.
4. W. Pan, O. Galkin, L. Filobelo, R. L. Nagel and P. G. Vekilov, *Biophys. J.*, 2007, **92**, 267–277.
5. Y. Li, V. Lubchenko and P. G. Vekilov, *Rev. Sci. Instrum.*, 2011, **82**, 053106
6. W. Pan, P. G. Vekilov and V. Lubchenko, *J. Phys. Chem. B*, 2010, **114** 7620–7630.
7. A. Stradner, H. Sedgwick, F. Cardinaux, W. C. K. Poon, S. U. Egelhaaf and P. Schurtenberger, *Nature*, 2004, **432**, 492–495.
8. Cardinaux, F.d.r., *et al.*, Cluster-Driven Dynamical Arrest in Concentrated Lysozyme Solutions. *J. Phys. Chem. B*, 2011. **115**(22): 7227–7237.
9. W. Pan, L. Filobelo, N. D. Q. Pham, O. Galkin, V. V. Uzunova and P. G. Vekilov, *Phys. Rev. Lett.*, 2009, **102**, 058101.
10. P. G. Vekilov, *Cryst. Growth Des.*, 2004, **4**, 671–685.
11. W. Pan, A. B. Kolomeisky and P. G. Vekilov, *J. Chem. Phys.*, 2005, **122**, 174905.
12. P. G. Vekilov, *J. Cryst. Growth*, 2005, **275**, 65–76.
13. O. Galkin, W. Pan, L. Filobelo, R. E. Hirsch, R. L. Nagel and P. G. Vekilov, *Biophys. J.*, 2007, **93**, 902–913.
14. P. Vekilov, *Br. J. Haematol.*, 2007, **139**, 173–184.
15. V. V. Uzunova, W. Pan, O. Galkin and P. G. Vekilov, *Biophys. J.*, 2010, **99**, 1976–1985.
16. P. G. Vekilov, *Soft Matter*, 2010, **6**, 5254–5272.
17. P. G. Vekilov, *Cryst. Growth Des.*, 2010, **10**, 5007–5019.

Mr Rosbottom asked: How long do these large nucleation clusters exist in the conformation detected before they undergo a conformational change to the conformation seen in the crystal structure?

Professor Vekilov responded: Even though the cluster concentration is very low, only very few of the clusters serve as locations where crystal nuclei appear. While the clusters contain from 10^5–10^6 molecules, the crystal nuclei consist of just 10–100 molecules. The remaining molecules in a cluster are gradually incorporated into the crystal as it grows. If these molecules are in a different conformation that the one found in the crystal, they need to convert to the conformation before getting incorporated. The clusters in which nucleation does not occur have limited lifetime: for some proteins it is of the order of 10 s,[1–3] for others we only know that their lifetime has a lower bound of 15 ms.[4,5] At the end of their lifetime such clusters decay either by the dissociation of single molecules, or by breaking into smaller clusters.[5]

1. O. Gliko, N. Neumaier, W. Pan, I. Haase, M. Fischer, A. Bacher, S. Weinkauf and P. G. Vekilov, *J. Am. Chem. Soc.*, 2005, **127**, 3433–3438.
2. O. Gliko, W. Pan, P. Katsonis, N. Neumaier, O. Galkin, S. Weinkauf and P. G. Vekilov, *J. Phys. Chem. B*, 2007, **111**, 3106–3114.
3. Y. Li, V. Lubchenko and P. G. Vekilov, *Rev. Sci. Instrum.*, 2011, **82**, 053106
4. W. Pan, O. Galkin, L. Filobelo, R. L. Nagel and P. G. Vekilov, *Biophys. J.*, 2007, **92**, 267–277.
5. W. Pan, P. G. Vekilov and V. Lubchenko, *J. Phys. Chem. B*, 2010, **114** 7620–7630.

Dr Sear commented: I believe you estimate nucleation rates, for example those in Fig. 8, by counting nuclei. Theories such as classical nucleation theory predict that at early times when the supersaturation is constant, that the rate at which nuclei appear is constant. Have you checked that you observe nuclei appearing at an approximately constant rate over your observation time? The reason I ask is that deviations from a constant rate may provide information on the precise nature of the nucleation dynamics.

Professor Vekilov responded: We have determined the rates of nucleation of crystals of lysozyme,[1,2] hemoglobin C (a mutant which crystallizes in the red blood cells[3]),[4] and insulin,[5] and polymers of sickle cell hemoglobin.[6,7] In all of these cases, the nucleation rates are determined from the evolution curves of the average (over a large ensemble of independent identical experiments) number of nuclei. The nuclei are detected after their growth to macroscopic crystals and tests are carried out to ensure the needed equivalency of nuclei and crystals.[8] The evolution curves trace this average number of nuclei as a function of the time allowed for nucleation; the means to vary this time are system-specific. Steady supersaturation during the time allowed for nucleation is maintained by keeping the crystals small and few.

With this, we look for periods of the evolution curves during which the average number of nuclei increases linearly with time. Typically, these periods were long: they comprised a significant fraction of the total experimental time. The slope of this straight line was taken as the nucleation rate. The behaviors of the evolution curve before and after this steady regime are indeed highly informative. In the cases of crystal nucleation, the intercept between the linear evolution and the number of nuclei axis allows the quantification of the heterogeneously nucleated crystals,[1,2,9] in the case of sickle cell hemoglobin nucleation, the evolution curves start with a period in which the number of nuclei is zero. The length of this period is the nucleation delay time. The nucleation delay time is evaluated in various models[10–12] and its determination allows discrimination between different possible mechanisms of nucleation.[13–15]

1. O. Galkin and P. G. Vekilov, *J. Phys. Chem.*, 1999, **103**, 10965–10971.
2. O. Galkin and P. G. Vekilov, *J. Am. Chem. Soc.*, 2000, **122**, 156–163.
3. L. S. Lessin, W. N. Jensen and E. Ponder, *J. Exp. Med.*, 1969, **130**, 443–466.
4. P. G. Vekilov, A. R. Feeling-Taylor, D. N. Petsev, O. Galkin, R. L. Nagel and R. E. Hirsch, *Biophys. J.*, 2002, **83**, 1147–1156.
5. L. Bergeron, L. Filobelo, O. Galkin and P. G. Vekilov, *Biophys. J.*, 2003, **85**, 3935–3942.
6. O. Galkin and P. G. Vekilov, *J. Mol. Biol.*, 2004, **336**, 43–59.

7. O. Galkin, R. L. Nagel and P. G. Vekilov, *J. Mol. Biol.*, 2007, **365**, 425–439.
8. P. G. Vekilov and O. Galkin, *Colloids Surf. A*, 2003, **215**, 125–130.
9. L. F. Filobelo, O. Galkin and P. G. Vekilov, *J. Chem. Phys.*, 2005, **123**, 014904.
10. D. Kashchiev, *Surf. Sci.*, 1969, **14**, 209–220.
11. D. Kashchiev, *Nucleation. Basic theory with applications*, Butterworth, Heinemann, Oxford, 2000.
12. D. Kashchiev, P. G. Vekilov and A. B. Kolomeisky, *J. Chem. Phys.*, 2005, **122**, 244706.
13. O. Galkin, R. L. Nagel and P. G. Vekilov, *J. Mol. Biol.*, 2007, **365**, 425–439.
14. O. Galkin, W. Pan, L. Filobelo, R. E. Hirsch, R. L. Nagel and P. G. Vekilov, *Biophys. J.*, 2007, **93**, 902–913.
15. P. Vekilov, *Br. J. Haematol.*, 2007, **139**, 173–184.

Professor Roberts asked: I was wondering whether the effects of secondary nucleation and/or growth rate dispersion might be playing some kind of a role in the analysis of the nucleation data? In terms of secondary nucleation effects and trying to assess its effect on the nucleation mechanism, it is sometimes hard to separate out the effects of where either attrition fragments or contact breeding contribute to the formation of crystals and hence to the measured nucleation rate. In terms of attrition, where this results in large(ish) crystal fragments in the micron size range, these can be easy to separate from the effects of primary nucleation in experiments. However, when the attrition fragment size is in the nm size range, the effect is hard to characterise due to difficulties in *in situ* crystal detection size detection in this size range. In the case of contact breeding, hydrodynamic forces shear off "swarms" of oriented molecular clusters from the boundary layer region thus obviating the need for cluster formation. As with the case of nano-sized crystals it can be challenging to separate this effect of the appearance of primary nuclei. Do you feel that there are situations in macromolecular crystallisation processes where consideration of these factors in assessing nucleation behaviour might be important? Growth rate dispersion involves different crystals growing at different rates under the same crystallisation supersaturation. There was quite a lot of studies into the origins of this in the late 1980s to early 1990s period[1,2] where, for example, Sherwood and co-workers were *e.g.* able to demonstrate[3] a direct relationship between the crystal lattice strain, as assessed through mosaic spread measurements, and the growth rate of the crystals. Given the molecular complexity of biological macromolecules such as proteins, do you think growth rate dispersion effects are playing a role in terms of not only the effect of assessing the nucleation kinetics of such systems but also in terms of establishing appropriate conditions for the growth of high quality single crystals?

1. D. A. H. Cunningham, A. R. Gerson, K. J. Roberts, J. N. Sherwood and K. Wojciechowski, *Quantifying some of the structural aspects of crystallisation processes: Experiments using synchrotron radiation*, in "Advances in Industrial Crystallisation", ed. J. Garside, R. J. Davey and A. G. Jones, Butterworths, London, 1991, p. 105.
2. J N Sherwood, Fifty years as a crystal gazer? Life as an imperfectionist, *Cryst. Growth Des.*, 2004, **4**, 863.
3. R. I. Ristic, J. N. Sherwood and K. Wojciechowski, Assessment of the Strain in Small Sodium Chlorate Crystals and its Relation to Growth Rate Dispersion, *J. Cryst. Growth*, 1988, **91**, 163.

Professor Vekilov answered: The first half of this question is whether secondary nucleation could be happening in protein solutions. Secondary nucleation typically happens under the influence of solution flow, both externally induced and due to the sedimentation of a crystal in the earth's gravity field. In research laboratories, protein crystals are typically grown in very small containers, with a characteristic dimension of the order of 10 to at most 1000 μm. Under these conditions, solution flows should be slow (for simulations and characterizations of the rates of convections in such systems, see ref. 1–5) and unlikely to induce secondary nucleation. On the other hand, proteins are sometimes crystallized in large amounts in the pharmaceutical and biochemical industries.[6–12] Since the size of the crystallizers in these cases is from one to tens of liters, and crystallization often occurs under constant stirring, secondary nucleation in these cases is likely a common mode of crystal generation. Another potential problem related to secondary nucleation is if its occurrence could have biased the conclusions about the nucleation mechanism drawn from dependencies of the nucleation rate on supersaturation and temperature. Although these determinations were carried out in very small volumes, which minimizes secondary nucleation, tests for its action were carried out. It was found that the distribution of the probability of finding a certain number of crystals in the experiments follows Poison's law at all times.[13–16] This observation demonstrates that all nucleation events are independent of one another, *i.e.*, secondary nucleation does not occur. The second part of this question is about growth rate dispersion. Growth rate dispersion has been observed in protein crystals as a result of several factors: lattice strain caused by the accumulation of defects and impurities,[17–19] direct action of the impurities on the kinetics,[20,24] variations in solute supply,[25,27] and step bunching. In many cases the variations in growth rate are accompanied by varying impurity and defect density incorporation,[32] leading to strained lattices, poor X-ray diffraction resolution, and hence low utility of these crystals for X-ray structure determinations.[33] Thus, growth rate variation should be avoided for higher crystal quality. It has even been speculated that the enhancement of the quality of crystals of some proteins observed in microgravity grown crystals might be due to the achievement of steady growth.[34,35] On the other hand, the effects of the growth rate dispersion on the determinations of the nucleation rate should be insignificant: while this dispersion could lead to the coexistence of smaller and larger crystals in the experimental cells, the nucleation rate is determined from the number of crystals and not from their sizes.[13–16]

1. L. Carotenuto, J. H. Cartwright, D. Castagnolo, J. M. Garcia Ruiz and F. Otalora, *Microgravity Sci. Technol.*, 2002, **13**, 14–21.
2. H. Lin, D. N. Petsev, S.-T. Yau, B. R. Thomas and P. G. Vekilov, *Cryst. Growth Des.*, 2001, **1**, 73–79.
3. A. Penkova, O. Gliko, I. L. Dimitrov, F. V. Hodjaoglu, C. Nanev and P. G. Vekilov, *J. Cryst. Growth*, 2005, **275**, e1527–e1532.
4. H. Lin, F. Rosenberger, J. I. D. Alexander and A. Nadarajah, *J. Cryst. Growth*, 1995, **151**, 153–162.
5. M. Pusey, W. Witherow and R. Naumann, *J. Cryst. Growth*, 1988, **90**, 105–111.
6. M. Y. Rose, R. A. Thompson, W. R. Light and J. S. Olson, *J. Biol. Chem.*, 1985, **260**, 6632–6640.
7. S.-H. Kim, D. H. Shin, J. Liu, V. Oganesyan, S. Chen, Q. S. Xu, J.-S. Kim, D. Das, U. Schulze-Gahmen, S. R. Holbrook, E. L. Holbrook, B. A. Martinez, N. Oganesyan, A. DeGiovanni, Y. Lou, M. Henriquez, C. Huang, J. Jancarik, R. Pufan, n.-G. Choi, J.-M. Chandonia, J. Hou, B. Gold, H. Yokota, S. E. Brenner, P. D. Adams and R. Kim, *J. Struct. Funct. Genom.*, 2005, **6**, 63–70.
8. J. Brange, *Galenics of Insulin*, Springer, Berlin, 1987.
9. M. L. Long, J. B. Bishop, T. L. Nagabhushan, P. Reichert, G. D. Smith and L. J. DeLucas, *J. Cryst. Growth*, 1996, **168**, 233–243.
10. S. Matsuda, T. Senda, S. Itoh, G. Kawano, H. Mizuno and Y. Mitsui, *J. Biol. Chem.*, 1989, **264**, 13381–13382.
11. S. Peseta, J. A. Langer, K. C. Zoon and C. E. Samuel, in *Annual Review of Biochemistry*, ed. C. C. Richardson, P. D. Boyer, I. B. Dawid and A. Meister, Annual Reviews, Palo Alto, 1989, vol. 56, pp. 727–778.
12. P. Reichert, C. McNemar, N. Nagabhushan, T. L. Nagabhushan, S. Tindal and A. Hruza, US Patent Pat., 5,441,734., 1995.
13. O. Galkin and P. G. Vekilov, *J. Phys. Chem.*, 1999, **103**, 10965–10971.
14. O. Galkin and P. G. Vekilov, *J. Am. Chem. Soc.*, 2000, **122**, 156–163.
15. O. Galkin and P. G. Vekilov, *J. Mol. Biol.*, 2004, **336**, 43–59.
16. O. Galkin, R. L. Nagel and P. G. Vekilov, *J. Mol. Biol.*, 2007, **365**, 425–439.
17. T. A. Nyce and F. Rosenberger, *J. Cryst. Growth*, 1991, **110**, 52–59.
18. S.-T. Yau, B. R. Thomas, O. Galkin, O. Gliko and P. G. Vekilov, *Prot. Struct. Funct. Genet.*, 2001, **43**, 343–352.

19. P. G. Vekilov, L. A. Monaco, B. R. Thomas, V. Stojanoff and F. Rosenberger, *Acta Crystallogr., Sect. D*, 1996, **52**, 785–798.
20. P. G. Vekilov and F. Rosenberger, *J. Cryst. Growth*, 1996, **158**, 540–551.
21. P. G. Vekilov, B. R. Thomas and F. Rosenberger, *J. Phys. Chem. B*, 1998, **102**, 5208–5216.
22. A. J. Malkin, Y. G. Kuznetsov and A. McPherson, *J. Struct. Biol.*, 1996, **117**, 124–137.
23. A. McPherson, A. J. Malkin, Y. G. Kuznetsov and S. Koszelak, *J. Cryst. Growth*, 1996, **168**, 74–92.
24. A. J. Malkin and R. E. Thorne, *Methods*, 2004, **34**, 273–299.
25. P. G. Vekilov, L. A. Monaco and F. Rosenberger, *J. Cryst. Growth*, 1995, **156**, 267–278.
26. H. Lin, P. G. Vekilov and F. Rosenberger, *J. Cryst. Growth*, 1996, **158**, 552–559.
27. H. Lin, D. N. Petsev, S.-T. Yau, B. R. Thomas and P. G. Vekilov, *Cryst. Growth Des.*, 2001, **1**, 73–79.
28. P. G. Vekilov, J. I. D. Alexander and F. Rosenberger, *Phys. Rev. E*, 1996, **54**, 6650–6660.
29. P. G. Vekilov, H. Lin and F. Rosenberger, *Phys. Rev. E*, 1997, **55**, 3202–3214.
30. P. G. Vekilov and F. Rosenberger, *Phys. Rev. Lett.*, 1998, **80**, 2654–2656.
31. P. G. Vekilov and J. I. D. Alexander, *Chem. Rev.*, 2000, **100**, 2061–2089.
32. P. G. Vekilov and F. Rosenberger, *Phys. Rev. E*, 1998, **57**, 6979–6981.
33. V. Stojanoff, D. P. Siddons, L. A. Monaco, P. G. Vekilov and F. Rosenberger, *Acta Crystallogr., Sect. D*, 1997, **53**, 588–595.
34. F. Rosenberger, P. G. Vekilov, H. Lin and J. I. D. Alexander, *Microgravity Sci. Tech.*, 1997, **10**, 29–35.
35. P. G. Vekilov, *Adv. Space Res.*, 1999, **24**, 1231–1240.

Professor Frenkel remarked: Professor Vekilov's comments suggest that crystal nucleation in a glassy phase is impossible. Whilst it is undoubtedly true that high viscosity tends to slow down crystal nucleation, there are examples of nucleation in a glass. A case in point is the study by Sanz *et al.* of crystal nucleation in a glass of hard, spherical colloids.[1]

1. E. Sanz, C. Valeriani, E. Zacarelli, W. C. K. Poon, P. N. Pusey and M. E. Cates, *Phys. Rev. Lett.*, 2011, **106**, 215701.

Professor Vekilov replied: As the paper by Sanz *et al.* demonstrates, the crystallization from glasses requires the particles or molecules to move to distances comparable to their size. This is possible in colloid glasses, which are the subject of that paper, where the interactions are often characterized with broad and shallow minima. On the other hand, in solid phases of small organic or inorganic molecules and proteins, the interaction minima are narrow and allow neither rotational nor translation motions. An example is the pair of proteins apoferritin and ferritin, in which the bonds leading to crystallization consist of pairs of salt bridges built of an aspartic acid residue, a Cd^{2+} ion, and a glutamic acid residue.[1–3] Clearly, these bonds are translationally and rotationally rigid. Their free energy has been determines as $3.2kBT = 7.8$ kJ mol^{-1}.[3–6] Even if the bonds are hydrophobic, as in the sickle cell hemoglobin polymers, they require a particular arrangement and separation of the molecules and leave little freedom of motion. In this case, the bonds consist of a valine residue at the β6 site (the location of the sickle cell mutation) and alanine, phenylalanine, and leucine from an adjacent HbS molecule.[7–10] The formation of this bond is accompanied by an entropy gain of 102 J mol^{-1} K^{-1}.[11,12] While little is known about protein gels, it is unlikely that they are built on bonds different than those employed in the ordered solids: most protein were optimized by evolution to resist precipitation, crystallization, or aggregation by making their surfaces repulsive.[13,14] A limited number of studies have demonstrated that protein gels suppress crystallization.[15,16] As far as the crystallization of small molecules organics is concerned, the general knowledge of the practitioners in the field is that if in a crystallizing solution a second liquid phase, often called "oil", emerges, crystals will not form. The suppression of crystallization is usually assigned to the high viscosity of the dense liquid. Exceptions to this rule exist, but are rare.[17]

1. E. C. Theil, *Ann. Rev. Biochem.*, 1987, **56**, 289–315.

2. D. M. Lawson, P. J. Artymiuk, S. J. Yewdall, J. M. A. Smith, J. C. Livingstone, A. Trefry, A. Luzzago, S. Levi, P. Arosio, G. Cesareni, C. D. Thomas, W. V. Shaw and P. M. Harrison, *Nature*, 1991, **349**, 541–544.
3. W. H. Massover, *Micron*, 1993, **24**, 389–437.
4. S.-T. Yau, D. N. Petsev, B. R. Thomas and P. G. Vekilov, *J. Mol. Biol.*, 2000, **303**, 667–678.
5. P. G. Vekilov, A. R. Feeling-Taylor, D. N. Petsev, O. Galkin, R. L. Nagel and R. E. Hirsch, *Biophys. J.*, 2002, **83**, 1147–1156.
6. P. G. Vekilov, A. R. Feeling-Taylor, S.-T. Yau and D. N. Petsev, *Acta Crystallogr. Sect. D*, 2002, **58**, 1611–1616.
7. B. Wishner, K. Ward, E. Lattman and W. Love, *J. Mol. Biol.*, 1975, **98**, 179–194.
8. C. Fronticelli and R. Gold, *J. Biol. Chem.*, 1976, **251**, 4968–4972.
9. G. W. Dykes, R. H. Crepeay and S. J. Edelstein, *J. Mol. Biol.*, 1979, **130**, 451–472.
10. B. Carrager, D. A. Bluemke, B. Gabriel, M. J. Potel and R. Josephs, *J. Mol. Biol.*, 1988, **199**, 315–331.
11. P. Vekilov, *Br. J. Haematol.*, 2007, **139**, 173–184.
12. P. G. Vekilov, O. Galkin, B. M. Pettitt, N. Choudhury and R. L. Nagel, *J. Mol. Biol.*, 2008, **377**, 882–888.
13. J. P. K. Doye, A. A. Louis and M. Vendruscolo, *Phys. Biol.*, 2004, **1**, 9–13.
14. J. P. K. Doye and W. C. K. Poon, *Curr. Opin. Colloid Interface Sci.*, 2006, **11**, 40–46.
15. O. Galkin and P. G. Vekilov, *Proc. Natl. Acad. Sci. USA*, 2000, **97**, 627–6281.
16. P. G. Vekilov, *Cryst. Growth Des.*, 2004, **4**, 671–685.
17. P. E. Bonnett, K. J. Carpenter, S. Dawson and R. J. Davey, *Chem. Commun.*, 2003, 698–699.

Professor Van Blaaderen opened the discussion of the paper by Professor James J. De Yoreo: It was mentioned frequently during the meeting that crystallization from an amorphous phase would not be possible or very slow. From the field of colloidal crystallization it is known that in gels or glassy/amorphous phases (using a very local definition of amorphous based on bond order parameters) crystals can nucleate and grow on time scales comparable to those for dense liquid systems. A recent example from an amorphous phase formed between particles that interacted by screened Coulomb interactions between oppositely charged particles is shown in the following reference: A qualitative confocal microscopy study on a range of colloidal processes by simulating microgravity conditions through slow rotations: D. El Masri, T. Vissers, S. Badaire, J. C. P. Stiefelhagen, H. R. Vutukuri, P. Helfferich, T. H. Zhang, W. K. Kegel, A. Imhof, and A. van Blaaderen, *Soft Matter*, 2012, **8**, 6979–6990.

Professor Vekilov commented: I want to remark that stable liquid phases do not always support nucleation. This is well known in protein crystallization, where it has been attributed to the high viscosity of the stable dense liquid. In pharmaceutical crystallization, the general knowledge of the practitioners in the field is that if in a crystallizing solution a second liquid phase, often called "oil", emerges, crystals will not form. The suppression of crystallization is usually assigned to the high viscosity of the dense liquid. Exceptions to this rule exist, but are rare.

Professor Gilbert remarked: If selected area electron diffraction (SAED) or monochromatic X-ray diffraction (XRD) do not show sharp rings, it does not mean that the material is amorphous, it only means that the material lacks long-range order, thus it could be either amorphous or nanocrystalline. Nanocrystalline ferrihydrite, precursor to magnetite in chiton teeth[1] is not amorphous as widely but incorrectly stated in the biomineralization literature: it is nanocrystalline.[2] Can we all agree that nanocrystalline is still crystalline, and not amorphous?

1. K. M. Towe and Lowenstam, *J. Ultrastruct. Res.*, 1967, **17**, 1.
2. F. M. Michel, L. Ehm, S. M. Antao, P. L. Lee, P. J. Chupas, G. Liu, D. R. Strongin, M. A. A. Schoonen, B. L. Phillips and J. B. Parise, *Science*, 2007, **316**, 5832.

Dr Gower answered: I thought the definition of crystalline was that it has a long range order. Is there a cutoff point when we describe something as nanocrystalline *versus* amorphous? This reminds me of the popular old terminology of paracrystalline. Given the techniques that are available to most of us (XRD and SAED), it is convenient to be able to describe something as amorphous if it does not show sharp rings, so how else do we go about describing it? Do we just say that diffraction shows that it lacks long range order?

Dr Rodríguez Blanco responded: I think this is just a problem of definition and characterization. Materials can be amorphous all the way to crystalline... with different degrees of ordering.

a) Micrometric particles of a mineral will produce X-ray diffraction peaks, but if the same particles are nanometric below a certain size (10 nm) they will not. That is the reason some amorphization methods consist of decreasing the particle size *via* mechanical or other methods.

b) However, a (truly) amorphous material will never produce X-ray diffraction peaks, regardless of the size of the particles. This is what happens with ACC. So in my opinion there are two questions which need to be addressed:

1. How can we name/refer to a sample which has nanometric particles, with sizes below 10 nm, has a highly ordered structure but does not produce diffraction peaks with XRD? My answer would be 'non-diffracting nanosized crystals'.
2. How can we know if the sample is truly amorphous or not? My answer would be: by using electron diffraction at the TEM.

Professor Roberts replied: I agree that the bio-mineralisation literature seem to have developed their own definition of an amorphous material as being one which is "X-ray amorphous", *i.e.* having no detectable X-ray diffraction peaks but this does not mean the sample IS truly amorphous. As an example, platinum black catalyst powders with a 10 nm crystal size will show an X-ray diffraction pattern indicating no long range order but examination in a TEM using electron diffraction will reveal a highly crystalline face centred cubic material. In this case it would be a nano-crystalline solid with high crystallinity. Thus, X-ray and electron diffraction have though different sensitivities to crystal size which reflects their different interactions with condensed matter: strong with electron and weak with X-rays. Thus, I would think that if you observed no electron diffraction, then you would be pretty safe that the material would be truly amorphous.

Dr Wolf replied: I think this discussion is very comparable to the discussion how to define a quasicrystal, thus an aperiodic crystal lacking three-dimensional translational periodicity. In 1991, the Ad Interim Commission on Aperiodic Crystals stated—for the case of quasicrystals—the following: "...by 'crystal' we mean any solid having an essentially discrete diffraction diagram...". I think we should rely on this definition of a crystal by its spectral properties and call a material a crystal when it shows "an essentially discrete diffraction diagram" during X-ray diffraction **and** electron diffraction.[1]

1. *Acta Cryst., Sect. A*, 1992, **48**, 922–946, relevant paragraph on p. 928.

Professor Penn answered: I would argue that one should state that no long range order is detected by XRD. It is certainly plausible that long range order could exist but at a mass concentration that is insufficient to detect or the crystal size might be so small that peak broadening leads to a diffraction pattern that appears consistent with an amorphous material.

Professor Roberts commented: It is hard to fully understand where ACC is a truly amorphous phase using X-ray diffraction as, due to size dependent line broadening considerations, crystals which are much smaller than about 10 nm do not give very recognisable diffraction paterns. Whilst challenging as a technique, would it not be better to use selected area electron diffraction to characterise the 3D order present (or not) in ACC due to the fact that the electron scattering is much stronger than with X-rays and thus the peak reflection widths in electron diffraction are much narower and potentially the analysis is much less ambiguous in terms of assessing the crystallinity of ACC?

Mr Nielsen replied: With the fluid cell TEM we can, when the cell spacing between the two windows is small enough, get electron diffraction data from the solution. Ideally electron diffraction will be used to make the determination you discuss in the question, and we are working towards utilizing that technique during mineralization experiments in the TEM.

Professor Van Blaaderen enquired: What is the definition of 'amorphous' in this community? I know of definitions that define amorphous from a background of scattering and that it usually involves the presence or absence of delta peaks. However, I am also familiar with a more local and real space definition which is usually made in terms of bond order parameters and a definition that involves neighboring particles. It is especially important to make the distinction clear in case of nucleation and crystal growth studies. For instance, I performed a spacial Fourier transformation of a collection of small crystals with roughly sizes of 5×5×5 particles (more or less randomly oriented). The scattering pattern of such a collection does not have any 'sharp' (Bragg or delta) peak-like features, nevertheless the local definition of crystallinity would still identify many particles (atoms) as being in a crystalline environment. It is also clear that crystal nuclei of 5×5×5 particles would significantly decrease crystal growth (or nucleation) with these as seeds, while if the same particles are amorphous, also by a local definition the crystal nucleation/growth would be quite different. In short, what is called amorphous or an amorphous cluster needs to be defined well in order to avoid confusion.

Professor Roberts replied: For a glassy structure we regard an amorphous structure as having no order beyond the first co-ordination sphere and overall this is the definition that I would use. If it has long range order, *i.e.* order that extends to the crystal size then it is crystalline.

If the crystalline domain size is small say in the 1–10 nm scale size then it would be generally referred to as nano-crystalline but crystalline it still is. X-Rays are weakly scattered by materials and hence XRD peaks start to broaden at crystallite sizes much less than 1 μ due to Fourier truncation effects. Thus, nanocrystalline phases are often characterised as being amorphous as they do not have any defined XRD pattern. In contrast, electrons are strongly scattered by materials and so electron diffraction of a nanocrystalline phase will show strong peaks. For example, decompose chloroplatinic acid and you will get fine powder of platinum black. From powder XRD it will look amorphous as there will be no well-defined scattering peaks. However, analysis by electron diffraction, using an electron microscope, will show sharply defined peaks typical of an FCC structure.

Crystalline phases, though, can exhibit less than perfect 3D order and this shows up by XRD peak broadening due to lattice strain and/or atomic/molecular groups occupying several orientations around their lattice site.[1] Such effects become more severe at higher wave vectors (Q), *i.e.* higher scattering angles and can be seen experimentally by XRD pattern intensities dropping away significantly as a function of Q. Due to the effects of thermal vibrations, all patterns do this naturally, *e.g.* organic XRD patterns rarely produce much scattering beyond a 2?o of about 40o. Nonetheless disordering can be seen as an accentuation of this thermal vibration damping effect and the effect can be assessed and quantified.[2]

1. See *e.g.* X-ray analysis of changes to the atomic structure around Ni associated with the interdifffusion and mechanical alloying of pure Ni and Mo powders, G. Cocco, S. Enzo, N. T. Barrett and K. J. Roberts, *Phys. Rev. B*, 1992, **45**, 7066–7076.
2. See *e.g.* Investigation into the structures of binary-, ternary-, and quinternary mixtures of n-alkanes and real diesel waxes using high resolution synchrotron X-ray powder diffraction, S. R. Craig, G. P. Hastie, A. R. Gerson, K. J. Roberts, J. N. Sherwood and R. D. Tack, *J. Mater. Chem.*, 1997, **8**, 859–869.

Dr Andreassen remarked: The *in situ* TEM study shows that calcite is nucleating and growing in the presence of what is presumably ACC aggregates. Is it possible to quantify the supersaturation in the TEM-cell solution environment? Is it likely that the level of supersaturation has risen to values above the solubility of ACC and that what you observe is a solution mediated transformation of ACC to calcite, where initially formed ACC is dissolving as calcite is consuming the supersaturation?

Mr Nielsen answered: Currently there is no *in situ* probe that allows us to quantify the supersaturation in the illuminated region of the TEM fluid cell. We fill the fluid cell with a solution with a well-defined supersaturation, but in the data shown the mineralization is purposefully induced with the electron beam. As such, there is a profound effect on the reaction environment by the imaging technique, and it is unclear if and how the supersaturation might be modified from its initially set value. When the electrochemical system is functional, we can provide a driving force for mineralization that will likely be orders of magnitude larger than any driving force that might be supplied by the electron beam, as suggested by the studies by Radisic and Ross cited in the paper. As long as we can characterize the resulting pH change induced near the electrode surface upon biasing the electrode, we should be able to quantify the supersaturation at the point of nucleation. However, this is a task for the future, because a method for measuring the pH in that space has not yet been developed by us or, to the best of our knowledge, by anyone else currently performing *in situ* TEM experiments with a fluid cell.

Professor De Yoreo remarked: *In situ* TEM has some serious limitations when applied to nucleation pathways. The beam has significant effects, including changes in pH, which we currently have no way to measure in the confines of the TEM fluid cell.

Miss Barber enquired: Explain how did you count the particles within the microscope cell. To what extent is this approach feasible?!

Mr Nielsen responded: The constant composition flow-through cell is used in conjunction with an inverted optical microscope with an attached digital camera that allows for image capture at well-defined intervals up to a rate of ten frames per second. We captured movies with this set-up and manually measured micronscale crystals from selected frames. This approach has utility when working in supersaturation ranges that result in nucleation events spaced far enough apart that a given nucleus doesn't affect neighboring nuclei, so that each large scale crystal can be said to have grown from individual nuclei. If this regime is achieved, the data will exhibit a period of time during which the number of nuclei increases linearly with time. The data are only valid during this period and, as a consequence, the method has limited applicability. At low supersaturations, the number of nuclei is too small to give decent statistics, while at high supersaturation, there is no sufficient period of time when the nucleation events are independent of one another to obtain decent statistics. Because the nucleation rate is exponential in the inverse supersaturation squared, the range of supersaturation that lies between these two limits is small.

Professor Penn addressed Professor De Yoreo: These high-resolution *in situ* series are simply fantastic. I realize that, at this point, the number of observations is still relatively small. However, do you have data tracking the frequency of oriented *versus* slightly misoriented attachment?

Professor De Yoreo answered: We do not have enough data to state that with confidence. The main problem is that not all attachment events occur in an orientation relative to the beam for us to state with certainty whether or not there was some misorientation. It will take some time before we work out the statistics.

Dr Andreassen remarked: During the presentation some TEM videos were shown where calcite aggregated, resulting in oriented attachment. A limited number of these events were observed within a time frame of tens of seconds. These observations are of very high relevance for the ongoing discussion of crystal growth by assembly of precursor crystals or by ion by ion attachment. In the former case, for a crystal of some ten μm to be composed of 10 nm precursor crystals would require efficient aggregation by some eight orders of magnitude number of crystals during the minutes it takes for this final crystal to form.[1] Based on the aggregation frequency of nano-particles you observe in TEM, does this seem likely?

1. J. P. Andreassen, *J. Cryst. Growth*, 2005, **274**, 256–264.

Mr Nielsen answered: The number and rate of attachment events observed is much lower than would be expected to account for a crystal of many microns to be an aggregate of such crystals. However the TEM observations are likely made in a highly constrained system and there is a very limited data set, so the initial data may not be fully representative of more widely used experimental systems.

Professor Penn said: These *in situ* experiments provide so much insight into how these crystals grow. In Fig. 8 and the movie shown during your lecture, does the aggregate reorient once it has made a close approach to the larger crystal? Is it known whether the aggregate is composed of crystalline or amorphous material?

Mr Nielsen replied: The resolution in the movie in question is such that we can't make any statements that would satisfactorily answer either question. In theory we could answer both of these questions with electron diffraction, but those data were not collected for this study.

Miss Barber asked: Any X-ray evidences of polymorphic transition from ACC to calcite?
Very nice work!

Mr Nielsen responded: There are a couple of different experiments that comprise the content of this paper. The X-ray studies were of SAMs in the presence of Ca^{2+} solution and Mg-stabilised ACC. In some of the spectra we see peaks corresponding to CO_3^{2-} in calcite. This does not suggest, however, direct transition from ACC to calcite, as the X-ray signal comes from a large area. All we can conclude from this is that by the time we collect spectra from the sample, some calcite has already formed, despite the stabilization effect of Mg. The signal gives us no information about potential pathways that led to the observable presence of calcite.

The templated nucleation experiments on carboxyl-terminated SAMs were carried out in pure $CaCO_3$ solutions with well-defined characteristics and measured optically. Without the stabilization effect of Mg used in the X-ray spectroscopic experiments, the formation of calcite on COOH SAMs occurs rapidly, rendering it difficult to ascertain the presence of ACC prior to the formation of calcite. We have done further *in situ* microscopic studies, both optical and in the atomic force microscope, as well as micro-Raman, and *ex situ* SEM and TEM, to try to detect the presence of ACC in this system and determine whether or not ACC that might exist serves as a precursor to calcite during templated nucleation. These further studies are presented in detail in ref. 1, and provide evidence that suggests, although does not provide irrefutable proof, that ACC does not precede calcite during nucleation on COOH SAMs from pure $CaCO_3$ solutions. The *in situ* TEM observation of $CaCO_3$ mineralization detailed in the paper had known initial solution conditions, but the reaction was induced *via* the TEM electron beam. It should also be noted that mineralization was observed on a gold surface, in contrast to the COOH SAMs in the above experiments. In the paper we write that the initially formed material is presumably ACC and, as the experiment continues, the development of well-defined crystalline edges suggests a transformation from ACC to a crystalline phase. In fact there are a few possibilities for the apparent phase progression observed in these studies: (i) the pathway discussed in the paper; (ii) the initially observed particles are in fact the same phase as the mature crystals but are small enough initially that it isn't clearly resolved; or (iii) the initial phase is one crystalline phase that then converts into a more energetically favorable phase.

To make a sufficiently convincing statement as to which of the possible pathways accurately describes such observations, one might switch between imaging and electron diffraction modes in the TEM as the process is underway. Such a procedure is possible with the fluid cell TEM system, and will be done in the future.

1. Q. Hu *et al.*, *Faraday Discuss.*, 2012, **159**.

Professor Vekilov asked: I would like to disagree with the statement that classical nucleation theory successfully predicts some of the features of nucleation in solution. Classical nucleation theory does not account for several essential features of the nucleation processes: nuclei consisting of a few molecules, discrete transitions between nucleus sizes, the presence of a solution to crystal spinodal, and, in particular, the presence of a dense liquid precursor for the formation of the crystalline nuclei. Hence, any correspondence between the predictions of CNT and experimental results is at best fortuitous. Often such correspondence is only qualitative, in the shape of the kinetic curves, with severe deviations in the predicted and measured nucleation rates.

Professor De Yoreo responded: The data on calcite nucleation on alkyl thiol SAMs has many features that make it consistent with CNT, all of which are described in a combination of papers, one of which will be that associated with my closing lecture in this volume. It is not possible to go into the data in detail in the context of this answer. I will just say that the failure of CNT to explain what happens on OH films and in solutions reinforces the degree to which it works for the carboxyl and sulfonated films. Having said that, clearly CNT is not predictive because, even if it can correctly account for the thermodynamic barrier, it does not predict the pre-factor. Moreover, clearly CNT is also irrelevant in many systems, because even if the CNT pathway is open, there are far more likely processes available and so it is not seen. The bottom line is that CNT works when it works and fails when it fails. There simply is no single answer.

Professor De Yoreo remarked: Classical nucleation theory may not work in pure, bulk solution for a given material. However, that doesn't mean it fails everywhere for all materials. It depends on many factors, including the interfacial energy of the surfaces exposed and the solubility of the crystal, and there is no single answer.

Dr Beck responded: Classical nucleation theory has been proven to work satifactory well in many cases. From a chemical engineering point of view it is very important that the classical nucleation theory is not abandoned unless a better theory has emerged that can explain the observations in a better way. Whether classical nucleation theory (concerning heterogeneous and homogeneous nucleation) works depends also on the present nucleation mechanism.

If classical nucleation theory is abandoned totally and no better theory is proposed and used it is not possible to predict nucleation rates that deviate from the nucleation rates determined at certain, specific experimental conditions. This makes the amount of work for a chemical engineer who, for example, has to design equipment or construction sites by far more time-consuming, and costs would increase tremendously.

Professor Kato asked: I have a question about the surface of SAM membranes for the crystallization. You introduced a phenyl benzoic acid moiety. The group is steric. When you compare the results of phenyl benzoic acid and simple aliphatic acid, steric effects are very different. This difference may cause different ordering of the functional group on the surface of the SAM and the distance of carboxylic acids. It seems for normal SAM experiments, simple aliphatic acid is introduced on the surface. But you introduced phenyl benzoic acid, which is more steric than carboxylic acid. This difference might result in different crystallization. Do you have reasons to use this acid or do you have some design for the surface? It would be interesting to compare the results of the benzoic acid and carboxylic acid.

Professor De Yoreo answered: The phenyl ring was introduced intentionally in order to induce a polarzation-dependent intensity in the NEXAFS spectrum. By then collecting NEXAFS data as a function of the angle between the beam polarization and the normal to the plane of the phenyl ring, we were able to determine the orientation of the SAM monomers and the degree of order in the SAM. From published data on similarly designed SAMs and our own AFM and NEXAFS data, we conclude that these SAMs have a typical packing structure. The only difference between the *para*, *meta* and bi-*meta* versions is either the accessibility of the carboxyl group (accessible to the solution in the case of *para* and bi-*meta* and inaccessible in the case of *meta*) and the ability of the SAMs to reorient (possible in the case of *para* and not possible in the case of bi-*meta*). Some of this is discussed in more detail in Nielsen *et al.*, within this Faraday Discussions volume. The details are in a manuscript submitted for publication.

Professor Vekilov opened the discussion of the paper by Dr Hugo Christenson: I want to propose an explanation of the observed vapor–liquid–solid transformation in a saturated vapor, in the absence of supersaturation. It seems to me that the role of the curved surfaces is to shift the equilibria between the crystal, liquid and vapor, according to the Laplace law. I would think that both crossed rods partake in this role. With the shifted equilibrium, the vapor is supersaturated with respect to the liquid, and the liquid is supersaturated with respect to the crystals. When the two crossed rods are separated, this lowers the effective curvature of the system. Since the crystalline material deposited over a curved rod is strained, it evaporates. This explanation would not violate the second law of thermodynamics (such violation is suggested by the title). Is this a reasonable explanation? I do not have the system thermodynamic and geometric parameters to judge that.

Dr Christenson replied: It is indeed the curved liquid–vapour interface that lowers the chemical potential of the pore-held liquid. Above the bulk melting temperature T_m of the substance this leads to equilibrium between pore-held liquid and undersaturated vapour—classical capillary condensation. Below T_m it leads to equilibrium between saturated vapour, solid and pore-held liquid with a liquid–vapour interface with finite curvature. A surface invariably leads to a much richer phase diagram.[1]

However, the vapour, which is saturated with respect to a flat solid surface (bulk), is not supersaturated with respect to the pore-held liquid, since the equilibrium vapour pressure of liquid is always higher than that of a solid. The fact that the condensate is liquid increases its vapour pressure as compared to a solid condensate, but the curved interface lowers it back to the original value of the solid. This can be illustrated schematically as in the attached figure, based on a 1944 paper,[2] and on a later publication dealing with atmospheric nucleation.[3] The full derivation, including heat capacity effects, a temperature-dependent surface tension and vapour pressures below solid coexistence has been given.[4–6]

The outer region of the liquid condensate does in general have a higher chemical potential (I prefer this term here as we are dealing with differences in temperature, not concentrations or pressures) than the solid; in other words it is metastable towards freezing, What we have not addressed thus far is whether or not the difference in chemical potential of the capillary-condensed liquid as compared to bulk supercooled liquid has a significant effect on the nucleation behaviour.

I do not understand why the title of our paper would suggest any violation of the second law of thermodynamics. If a crystal showed complete wetting on a flat surface it would also deposit without the need for supersaturation.

The role of the curved mica surfaces is merely to provide a wedge for the capillary condensation to take place in, and their curvature is unimportant. A linear groove or a number of conical pits would do equally well as long as the total volume of condensed liquid was of the same order.

The strain caused by the curved mica surfaces is negligible because of the macroscopic radius of curvature (2 cm). On such curved mica surfaces we can grow 10 μm epitaxial calcite crystals of the same type as on flat surfaces.[7] The crystalline fragments evaporate after separation because their curvature is convex and they are hence in equilibrium with slightly supersaturated vapour. They sublime for the

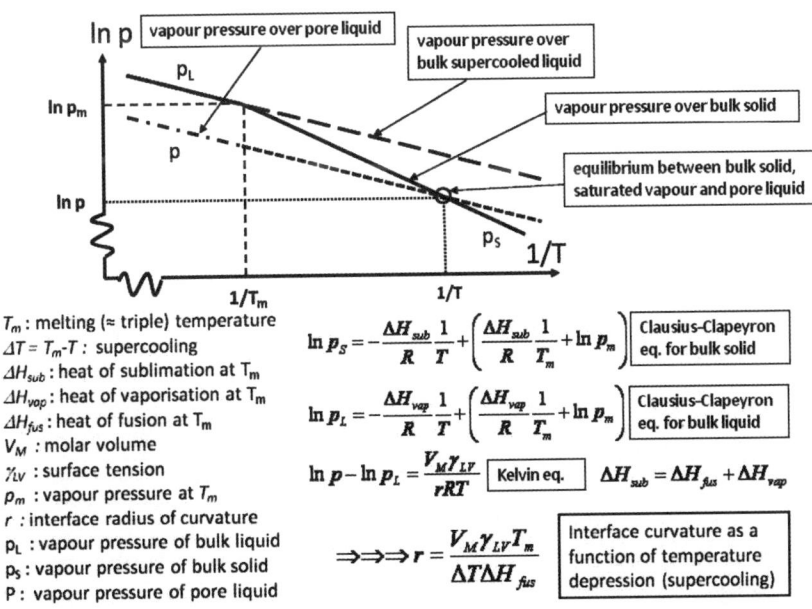

Fig. 1 Schematic plot of ln[p] vs $1/T$ showing the Clausius–Clapeyron equation for a liquid (p_L – dashed below the bulk melting point T_m) and for a solid (p_S). The Kelvin equation shifts ln[p_L] towards smaller values (p) due to the curved liquid–vapour interface. Where this line intersects the solid–vapour coexistence line p_S there is equilibrium between bulk solid, vapour and capillary-held liquid—a capillary triple-point. The plot is effectively a phase diagram in ln[p] and $1/T$ coordinates.

same reason that crystals cannot deposit on a flat surface from vapour that is not supersaturated!

In both the Clausius–Clapeyron equation and the Kelvin equation the logarithm of the vapour pressure is inversely proportional to the absolute temperature, which may be conveniently illustrated in the schematic diagram above. If the saturation vapour pressure is eliminated from the equations the result is a relation describing the radius of a pore-held liquid–vapour interface in equilibrium with bulk solid and its saturated vapour.

1. N. Maeda and H. K. Christenson, *Coll. Surf. A: Physicochem. Eng. Asp.*, 1999, **159**, 135–148.
2. R. W. Batchelor and A. G. Foster, *Trans. Faraday Soc.*, 1944, **40**, 300.
3. N. Fukuta, *J. Atmos. Sci.*, 1966, **23**, 741–750.
4. P. Barber, T. Asakawa and H. K. Christenson, *J. Phys. Chem. C*, 2007, **111**, 2141–2148.
5. D. Nowak, M. Heuberger, M. Zach and H. K. Christenson, *J. Chem. Phys.*, 2008, **129**, 154509.
6. D. Nowak and H. K. Christenson, *Langmuir*, 2009, **25**, 9908–9912.
7. C. J. Stephens, Y. Mouhamad, F. C. Meldrum and H. K. Christenson, *Cryst. Growth Des.*, 2010, **10**, 734–738.

Professor Frenkel asked: Dr Christenson expressed the opinion that not much useful information can be extracted for computed nucleation rates as these are 'all over the place' Whilst it is certainly true that in some cases there are gigantic discrepancies between computed nucleation rates and the corresponding experimental estimates, it should be stressed that a) this is not always the case (*e.g.* in the case of NaCl, there is no evidence for a serious discrepancy between experiment and simulation,[1] b) the computed nucleation rates are extremely sensitive to the functional form of the intermolecular interactions, and c) for a given, well-studied system (hard spheres) all nucleation rates computed using different techniques agree in the regime where they can be compared.[2] To illustrate point b), for a system of hard colloids with a diameter of 400 nm, the nucleation rate changes by more than an order of magnitude if one takes into account the thin (approximately 20 nm) layer of grafted molecules that provide steric stabilisation.[3]

1. C. Valeriani, E. Sanz and D. Frenkel, *J. Chem. Phys.*, 2005, **122**, 194501.
2. L. Filion, R. Ni, D. Frenkel and M. Dijkstra, *J. Chem. Phys.*, 2011, **134**, 134901.
3. S. Auer, W. C. K. Poon, and D. Frenkel, *Phys. Rev. E*, 2003, **67**, 020401.

Dr Christenson replied: I am sorry to say that this was a bit of a careless statement by me and I thank Professor Frenkel for pointing this out. The best I can say is perhaps that it was easier for me to use experimental data since I had a recently published paper of which I was a minor co-author.[1]

1. B. J. Murray, S. L. Broadley, T. W. Wilson, S. Bull, R. H. Wills, H. K. Christenson and E. J. Murray, *Phys. Chem. Chem. Phys.*, 2010, **12**, 10180–10187.

Phase behavior of colloidal silica rods

Anke Kuijk,[a] Dmytro V. Byelov,[b] Andrei V. Petukhov,[b] Alfons van Blaaderen[a] and Arnout Imhof[a]

Received 26th April 2012, Accepted 6th June 2012
DOI: 10.1039/c2fd20084h

Recently, a novel colloidal hard-rod-like model system was developed which consists of silica rods [Kuijk *et al.*, *JACS*, 2011, **133**, 2346]. Here, we present a study of the phase behavior of these rods, for aspect ratios ranging from 3.7 to 8.0. By combining real-space confocal laser scanning microscopy with small angle X-ray scattering, a phase diagram depending on concentration and aspect ratio was constructed, which shows good qualitative agreement with the simulation results for the hard spherocylinder system. Besides the expected nematic and smectic liquid crystalline phases for the higher aspect ratios, we found a smectic-*B* phase at high densities for all systems. Additionally, real-space measurements on the single-particle level provided preliminary information on (liquid) crystal nucleation, defects and dynamics in the smectic phase.

1 Introduction

The phase behavior of anisotropic particles is richer than that of spherical particles. In addition to the gas, liquid, crystal and glass phases that were observed for spheres, anisotropic particles such as rods can form liquid crystal phases: additional phases between the liquid and crystal phase that possess long-range orientational and positional order in less than the three dimensions of a 3D crystal. One of the first theoretical explanations for the formation of a nematic liquid crystal phase was provided by Onsager in 1949.[1] He explained why the transition from the isotropic to the nematic phase for long, hard rods could be purely entropy based. Later, computer simulations showed that not only nematic, but also smectic and crystalline phases can occur in systems of hard spherocylinders (HSC).[2,3] The formation of liquid crystal phases depends on the concentration and aspect ratio of the rodlike particles. Also, the shape of the particles is an important parameter. Ellipsoids, for example, only show a nematic liquid crystal phase, while for spherocylinders a nematic and a smectic phase were found.[4]

The experimental verification of the simulation results requires a system of hard rods. The majority of experimental systems that have been used to study liquid crystalline phases only allows for measurements on the many-particle level, either because of their small dimensions or due to their high refractive index. Therefore, research on rod-like particles has been limited to the many-particle level and reciprocal space for a long time. Even though it is possible to distinguish smectic phases of, for instance, *fd*-virus using light microscopy, the individual particles can only be imaged this way when a doped system is used (in which 1 on every 30 000 particles is labeled).[5,6] Imaging all individual particles is not possible in this system using light

---

[a] *Soft Condensed Matter, Debye Institute for Nanomaterials Science, Department of Physics, Utrecht University, Princetonplein 5, 3584 CC Utrecht, The Netherlands. E-mail: A.Imhof@uu.nl*
[b] *Van 't Hoff Laboratory for Physical and Colloid Chemistry, Debye Institute for Nanomaterials Science, Department of Chemistry, Utrecht University, Padualaan 8, 3584 CH Utrecht, The Netherlands*

microscopy. Studies involving mineral liquid crystals, like boehmite or goethite are also based on many-particle effects such as birefringence and scattering.[7-10] The first real-space data on 3D systems of rods on the single-particle level were reported by Maeda et al. in 2003.[11] They showed the process of self-ordering for several aspect ratios of inorganic rods. Although their observations already provided a lot of information that is hard or impossible to obtain from scattering experiments, the amount of information was limited by the systems that were used. The systems that were studied by Maeda consisted of rods with high refractive indices. The resulting strong scattering makes it hard to obtain 3D data for these systems.

Systems of anisotropic model colloids that do allow for 3D data acquisition, because their refractive index can be matched to that of the solvent, include PMMA ellipsoids,[12,13] silica dumbbells[14] and silica rods produced either by strong anisotropic etching of structured silicon wafers[15] or by our recently developed 'bulk' synthesis method.[16] Of these systems, the system of 'bulk' silica rods is the only one that has aspect ratios high enough to form liquid crystal phases, a shape that allows for both nematics and smectics to form and that can be produced in sufficient quantity to perform phase separation experiments. Furthermore, research has shown that silica colloids are a good model system for the hard sphere system if they are always present and attractive van der Waals interactions can be sufficiently suppressed.[17,18] Therefore, we expect the system of silica rods to be a good model system for the hard rod system as well. However, there are some differences between our experimental rods and a perfect HSC-system. Firstly, the rods are initially bullet-shaped. Nevertheless, they resemble the shape of a spherocylinder (a cylinder capped with a hemisphere on both ends) more and more after coating with several extra layers of silica, including fluorescent labeling, to make them suitable for confocal laser scanning microscopy (CLSM).[18,19] Secondly, experimental colloidal systems always have an inherent polydispersity which influences their phase behavior. For HSC-systems it was shown that a smectic phase can only form if the length polydispersity is below 18%.[20] Since the polydispersity of the silica rods is around 10%, the system is expected to be able to form smectic phases. Thirdly, our colloidal dispersion of silica rods is a charge-stabilized system. It is known that charges increase the effective diameter of needles,[21] and also that colloidal charges and weak-screening conditions lead to a decreased aspect ratio of charged rods.[22] For equivalent systems of charged spheres, it was shown that the effective increased diameter due to charge can be used to map results on phase behavior on the hard sphere model.[23,24] However, since theoretical studies showed that charge tends to stabilize the positionally ordered phases, especially the columnar phase, and decrease the nematic and smectic phase,[25,26] it is not obvious that the same strategy can be followed for rod-systems unless the Debye screening length is very small as compared to the smallest particle dimension.

In this paper, we describe the phases found in sediments of rods using small angle X-ray scattering (SAXS) and real space CLSM measurements. By combining the results of both methods a phase diagram was constructed depending on aspect ratio and volume fraction. Furthermore, we show real-space measurements on the single-particle level that provide detailed information on (liquid) crystal nucleation, defects and dynamics in concentrated phases.

2 Experimental

2.1 Dispersions

The properties of the systems used for the SAXS and CLSM measurements are summarized in Table 1. All silica rods, which are shown in Fig. 1, were prepared as described by Kuijk et al.[16,19] Systems B31 and B35 consist of a non-fluorescent core, a 30 nm fluorescein isothiocyanate (FITC) labeled inner shell and a 190 nm non-fluorescent outer shell. The rods of B36 were not coated after synthesis. Systems

Table 1 Properties of the systems of colloidal silica rods that were used in this work. Here, L is the length of the rods, D the diameter, σ the polydispersity, φ the volume fraction and l_g the gravitational height

	$L/\mu m$	$\sigma_L/\%$	D/nm	$\sigma_D/\%$	L/D	φ	$l_g/\mu m$
B31	2.37	10	640	7.5	3.7	0.10	0.7
B35	3.3	10	550	11	6.0	0.10	0.7
B36	1.9	15	235	17	8.0	0.128	6.5
B48	2.6	8.5	630	6.3	4.1	0.10	0.7
N51	2.66	10	530	6.3	5.0	0.105	0.9

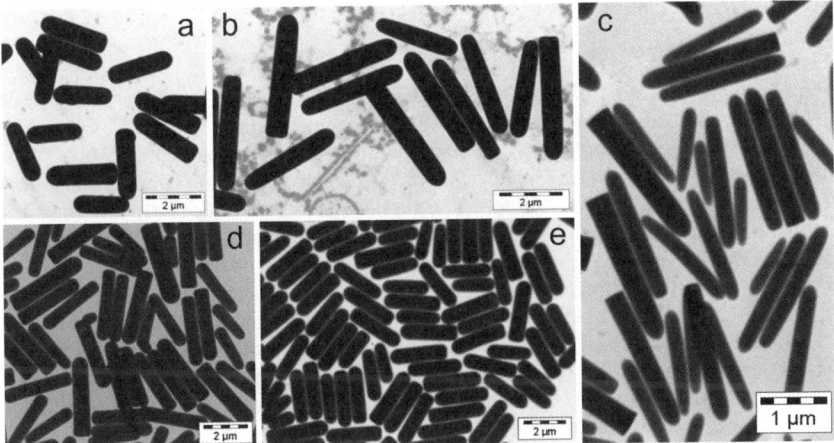

Fig. 1 TEM images of the systems of silica rods used. (a) B31. (b) B35. (c) B36. (d) N51. (e) B48. The dimensions of the rods are listed in Table 1.

B48 and N51 consist of a rhodamine isothiocyanate (RITC) labeled core and a 175 or 150 nm non-fluorescent shell, respectively.

Rods with $L = 1.9$ μm and $D = 420$ nm were used in the optical Bragg-reflection measurements. A 6 cm high capillary was filled with an initial volume fraction of 0.2 and afterwards left to sediment.

For the diffusion measurements in the smectic phase, a system of rods with length $L = 1.4$ μm, diameter $D = 280$ nm and polydispersities of $\sigma_L = 6\%$ and $\sigma_D = 10\%$ was used. This system was coated with a 20 nm FITC-labeled layer only. For single particle observation, a system of FITC labeled rods of $L = 2.2$ μm, $D = 340$ nm, $\sigma_L = 10\%$ and $\sigma_D = 15\%$ was mixed with a system of unlabeled rods of $L = 2.1$ μm, $D = 250$ nm, $\sigma_L = 9\%$ and $\sigma_D = 17\%$. The ratio of labeled to unlabeled particles was 1 : 100.

Particle size distributions were determined by transmission electron microscopy (TEM), using a Technai 10 or 12 electron microscope (FEI company). The average length $\langle L \rangle$ and diameter $\langle D \rangle$ of the rods, as well as their standard deviation δ, were measured using iTEM imaging software. The polydispersity is defined as $\sigma_L = \delta_L/\langle L \rangle$. For each sample 50 to 100 particles were measured.

The solvent mixture of all systems consisted of dimethylsulfoxide (DMSO, ≥99.9%, Sigma-Aldrich) and ultrapure water (Millipore system). The particles were dispersed in DMSO first, after which water was added until the refractive index was matched by eye. This resulted in a 10/0.85 volume ratio of DMSO/water ($n = 1.47$).

2.2 Confocal microscopy

Confocal microscopy measurements were performed with a Leica SP2 or a Nikon C1 confocal, of which the Nikon was used to study samples that were positioned in a vertical position (gravity along the length of the capillary) using a 90° tilted Leica TCS NT inverse microscope frame. A 63× oil immersion objective with a numerical apperture of 1.4 was used (Leica PLAN APO). The dispersions were studied in capillaries of 1 or 2 mm wide and 0.1 mm high with glass walls of about 100 μm thick (Vitrotubes). The capillaries were sealed with candle wax first and two-component epoxy glue (Bison Kombi rapide) on top of that.

The coordinates of the rods in confocal microscopy images in which the rods were oriented perpendicular to the plane of view (so that they look like spheres in the image, see for instance Fig. 8) were obtained by a method similar to that of Crocker and Grier.[27]

2.3 X-Ray experiments

Small angle X-ray scattering (SAXS) measurements were performed at the DUBBLE beamline of the European Synchrotron Radiation Facility (ESRF, Grenoble, France). Flat glass capillaries (Vitrocom) with internal dimensions of 0.1 × 1 × 100 mm^3 were filled with several dispersions of rods (see Table 1). The glass thickness of these capillaries is about 100 μm. The capillaries were closed by melting the ends and covering them with two-component epoxy glue (Bison Kombi rapide) to ensure full closure. After filling, the capillaries were kept in a vertical position to allow the establishment of a sedimentation equilibrium profile. The samples were prepared in a period of three weeks before measuring. For these measurements the microradian resolution setup was used.[28] Here, a CCD detector (Photonic Science Ltd) with pixel size 9 × 9 μm was placed at a distance of 7.4 m from the sample. The selected wavelength of the X-rays was 0.095 nm, the beam size at the sample position about 0.3 mm. The intensity profiles shown in Fig. 4 were further calculated by integrating over a circular sector containing the area of interest in the scattering patterns.

3 Results and discussion

The phase behavior of colloids depends strongly on the volume fraction of the dispersion. Since a system of silica rods in DMSO/water was used, the volume fraction in our experiments was not constant, but changed during sedimentation. When the sedimenting system had reached an equilibrium, the balance between the gravitational pressure and the osmotic pressure caused a gradient in volume fraction depending on the height in the sample. As a consequence, a range of volume fractions, and therefore multiple phases, could be studied in one sediment.[29]

The formation of a sediment is displayed in Fig. 2, which shows photographs of a typical sample after several sedimentation times. The dark area at the bottom was caused by a drop of glue, which covered the lowest 5 mm of the capillary and deflected the light from the white-light source that illuminated the sediment from behind. From Fig. 2 we calculated the sedimentation velocity v_{sed} by measuring the position of the interface (indicated by the upper line) in time. Based on a model for the sedimentation of cylinders,[30,31] a theoretical v_{sed} of 6.5 μm min^{-1} was calculated in the dilute limit for hard rods. Experimentally, we measured a much lower and decreasing v_{sed} of 0.20 μm min^{-1} for days 11 to 14, to 0.17 μm min^{-1} for the next two days and 0.06 μm min^{-1} for days 16 to 18. The lower experimentally measured v_{sed} as well as its decrease in time are caused by the increasing volume fraction in the sediment.[32]

The second line from above in Fig. 2 denotes the interface between a region that shows Bragg reflections (layered phase) and a region that does not (isotropic or nematic phase). Bragg reflections occur when there is periodicity in the sample, so they are an indication of order. The first Bragg reflections were observed after 5

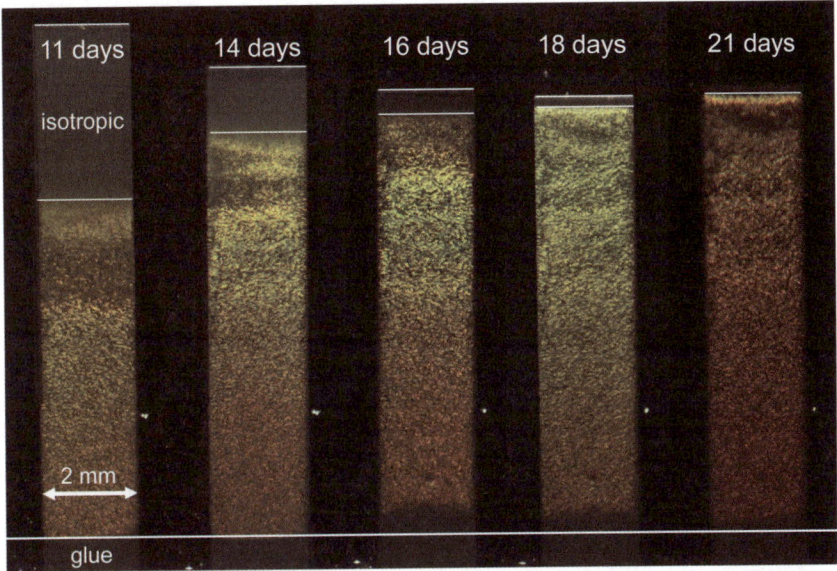

Fig. 2 A sedimenting sample of $L = 1.9$ μm, $D = 420$ nm rods followed in time. Photographs were taken while illuminating the sample from behind with white light. After several days, a Bragg-reflecting ordered region began to form in the bottom of the sample. In time, this region grew, while the interface between suspension and supernatant came down, indicating that rods are still sedimenting. After about 20 days the ordered region almost touched the interface, which stayed stable for the next weeks.

days of sedimentation. The formation of this ordered phase was initially faster than the sedimentation velocity (0.36 compared to 0.20 μm min^{-1} from day 11 to 14). The following days this region grew until it filled almost the whole sediment (after 21 days). Note that the density of Bragg reflecting areas in this phase changed in time; the top part of the Bragg reflecting area after 11 days showed less reflections than the same part of the sample after 14 days. The formation of ordered regions thus continued for several days.

The origin of the Bragg reflections was found using confocal microscopy (Fig. 3). In the area of little Bragg reflecting spots, small layered domains were found with dimensions of 20 to 50 μm in diameter and around 5 layers deep. Towards the bottom, where we can see more and stronger Bragg reflections, these areas increased to hundreds of microns and around 20 layers deep. The Bragg reflections thus result from the layered structures inside the sample. The fact that these do not span the whole capillary is due to the increasing concentration and multiple domains. Sedimentation is so fast that rotational and translational diffusion that facilitate ordering are hindered by the increasing density of the sample.

3.1 Sediments of rods studied by SAXS and CLSM

The phases in sediments of rod-like colloids with aspect ratios ranging from 3.7 to 8.0 were studied in more detail by SAXS and CLSM. Based on computer simulations for the HSC-system, a smectic liquid crystalline phase is expected for rods with an aspect ratio higher than 4.1, and a nematic phase for aspect ratios higher than 4.7.[33]

The sediment of the system with the largest aspect ratio ($L/D = 8.0$) shows, from top to bottom, an isotropic phase of about 0.5 mm high, followed by 3 mm of nematic phase and a smectic phase in the bottom 11 mm. SAXS patterns of these phases are shown in Fig. 4a–c. The scattering pattern of the isotropic phase shows

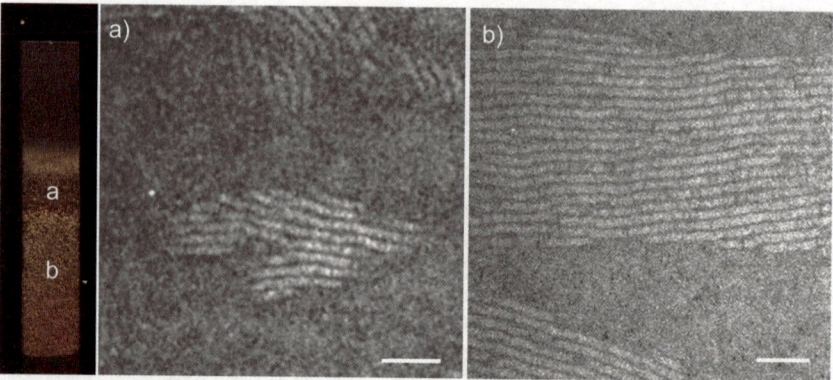

Fig. 3 Confocal microscopy images taken in the Bragg reflecting areas shown in Fig. 2. (a) Top part of the sediment with small layered domains. (b) Bottom part of the sediment with large layered domains. Scale bars indicate 10 μm.

weak signs of preferred orientation. This is caused by the size of the beam (around 0.3 mm), which is roughly equal to the height of the isotropic phase. Therefore, part of the nematic phase was also hit while imaging the isotropic phase. For the nematic phase, the intensity profile of the SAXS pattern shows a characteristic butterfly pattern with one broad peak that corresponds to the liquid-like ordering of neighboring rods (Fig. 4e), while no correlations were found in the length direction of the rods, corresponding to the literature.[34] Intensity profiles of the SAXS patterns of the smectic phase show very sharp peaks, up to the sixth order, due to diffraction from the layers of this phase that have a spacing of 2.4 μm. One broad peak at higher q-range originates from the liquid-like ordering of the rods within the layers

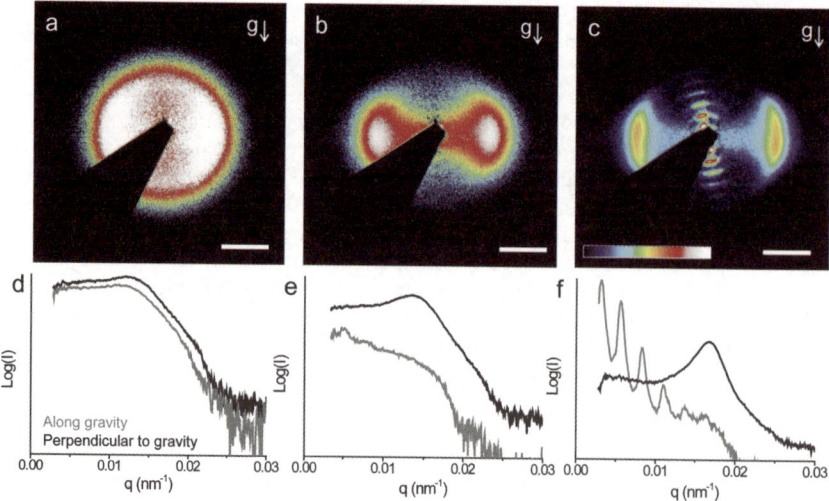

Fig. 4 SAXS measurements of a sediment of $L/D = 8.0$ rods. (a) Scattering pattern of the isotropic phase on a log-scale at a height of 14 mm from the bottom of the sample. (b) Scattering pattern of the nematic phase at a height of 12 mm. (c) Scattering pattern of the smectic-C phase at a height of 9 mm. (d–f) Intensity profiles of the patterns shown in (a–c) in the horizontal (black) and vertical (grey) direction, resp. perpendicular and along gravity g. Scale bars in (a), (b) and (c) indicate 0.01 nm^{-1}. The colored bar shows the false color - log-(intensity) scale with white color corresponding to the highest intensity.

(Fig. 4f). In this sample, the peaks originating from correlations in length are not oriented exactly perpendicularly to the peaks originating from correlations in diameter. This implies that the system formed a smectic-C phase, in which the rods within the layers are oriented at an angle of about 12° with respect to the layer's normal. This is probably caused by gravitational compression. Since the $L/D = 8.0$ rods were not fluorescently labeled, there are no real-space confocal microscopy images of this system available.

For slightly shorter rods ($L/D = 6.0$), all three liquid crystal phases were found as well (Fig. 5), but in this case also a phase that showed order within the layers was observed. In the bottom 11 mm of this sample a layered phase was found that shows stronger correlations within the smectic layers than the rods of $L/D = 8.0$. Three peaks were found at relative distances of 1, 1.8 and 2.8 ($q = 0.0098$, 0.0175 and 0.0274 nm^{-1}). These peaks suggest hexagonal ordering of the rods within the layers. For a perfect 2D hexagonal lattice, peaks are expected at relative distances of 1, $\sqrt{3}$, $\sqrt{4}$, $\sqrt{7}$ and $\sqrt{9}$. An explanation for the experimentally found distances is offered by the merging of the $\sqrt{3}$-peak and the $\sqrt{4}$-peak due to broadening, as well as the $\sqrt{7}$ and $\sqrt{9}$-peak. Hexagonal scattering patterns from such a structure were observed in a study on fd-virus as well.[35] The merging of the peaks is probably caused by the polydispersity of the sample, causing small variations in inter-particle distances. Because of its long-range hexagonal order within layers, but absence of crystalline correlations between layers, we identify this phase as a smectic-B phase.

On top of the smectic-B phase, a 3 mm high smectic-A phase was found (Fig. 5c,g,l). In the smectic-A phase there is no long-range positional order of rods within the layers. The peaks in the intensity profile of this phase are broader than those of the phase below, representing the more liquid-like ordering of this phase. The peaks that result from periodicity along the length of the rods, although present, are less well defined than those in the sample of longer rods. This corresponds well to the real-space measurements of this sample, depicted in Fig. 5l, which indicate that the layers show rather large fluctuations.

Higher up in the sample, a 1 mm high nematic phase was found (Fig. 5b,f,k). This phase also shows more correlations than the nematic phase of the longer rods: two peaks were found that correspond to correlations in the plane perpendicular to the nematic director. These peaks are related to presmectic ordering, which was also observed in computer simulations.[29] Compared to the layered phases of this sample, the peaks are broader, which indicates liquid-like order with more variations in inter-particle distances. The real-space confocal microscopy image of the nematic phase (Fig. 5k) shows the origin of the second peak: the rods showed a beginning of ordering into layers. Peaks that are caused by periodicities along the length of the rods are absent, which distinguishes a nematic from a smectic phase.

In the about 0.5 mm high isotropic phase that was found above the nematic all liquid crystalline order disappeared, represented by the isotropic scattering and the broad peaks of Fig. 5a and e. Very weak correlations are still present, as indicated by the small second order peak, but these cannot be found in the real-space image (Fig. 5j).

The more monodisperse system with $L/D = 5.0$ shows even stronger signs of hexagonal ordering (Fig. 6). The scattering pattern of this phase shows strong, sharp peaks for correlations within the layers that are spaced at relative distances of 1 (for the first peak), $\sqrt{3}$, $\sqrt{4}$ and $\sqrt{7}$ ($q = 0.0098$, 0.0169, 0.0195 and 0.0262 nm^{-1}), which indicates that the rods are hexagonally ordered within the layers. This is confirmed by confocal microscopy images of the phase. Fig. 7 shows the hexagonal layers from the top and two sides. Due to the lower resolution of the confocal microscope in the z-direction, the rods appear stretched in this direction. The hexagonal order within the layers is clearly visible in Fig. 7a. A cut through the xy-plane shows disorder in the stacking of the layers themselves, which is expressed in broad and small peaks in the SAXS pattern originating from correlations along the length. Although these observations hint towards a crystalline phase, no peaks due to correlations between

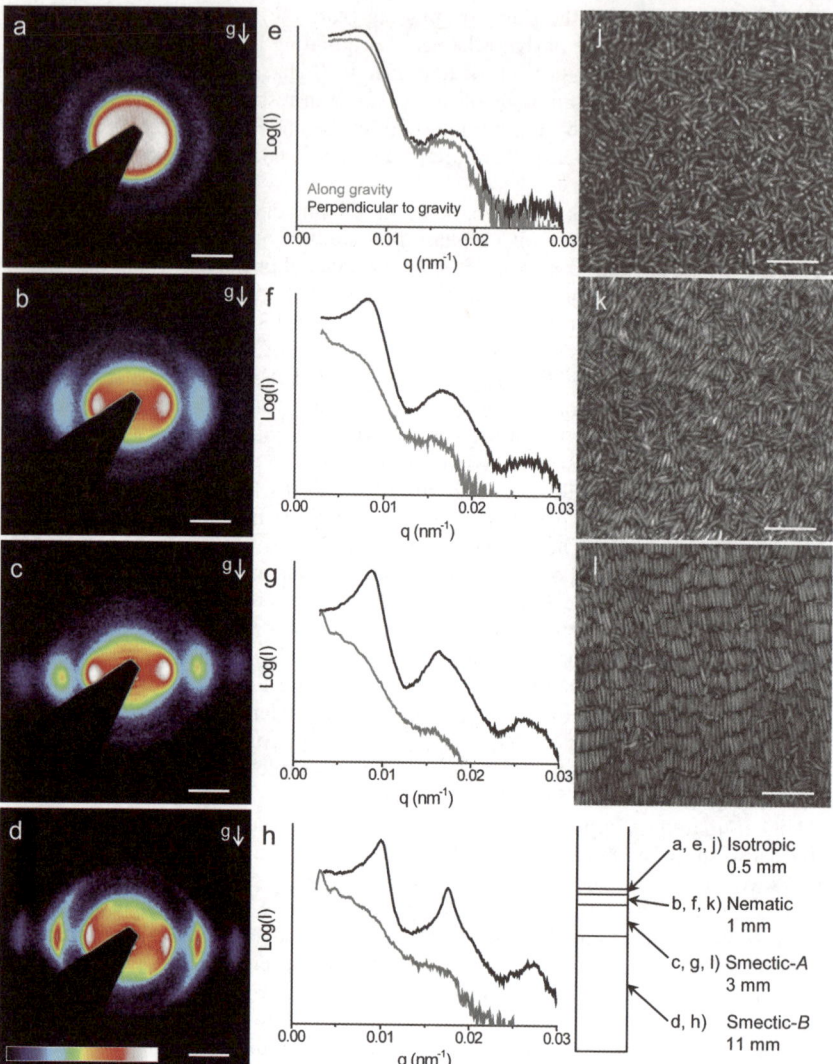

Fig. 5 SAXS measurements of a sediment of $L/D = 6.0$ rods. (a–d) SAXS patterns on a log-scale of the isotropic, nematic, smectic-A and smectic-B phase, respectively. Scale bars are 0.01 nm^{-1}. (e–h) Intensity profiles of the patterns shown in (a–d) in the horizontal (black) and vertical direction (grey). (j–l) Confocal microscopy images of the isotropic, nematic and smectic phase of the same sample. Scale bars indicate 10 μm. The colored bar shows the false color - log(intensity) scale with white color corresponding to the highest intensity.

the hexagonal patterns of subsequent layers were found, which excludes the structure being fully crystalline in three dimensions. Therefore, we identified the structure as a smectic-B phase. Higher up in this sample we found a smectic-A, a nematic and finally an isotropic phase.

For the system of rods with the shortest aspect ratio ($L/D = 3.7$), hexagonally ordered rods were found in domains that extended tens of microns, as shown in Fig. 8a. The degree of hexagonal order in these layers is shown in Fig. 8b, which shows the 2D radial distribution function of the rods as compared to that of a perfect 2D hexagonal lattice. The experimental peaks were scaled to fit the first peak of the theoretical $g(r)$ and extend over 10 particle diameters, comparing well

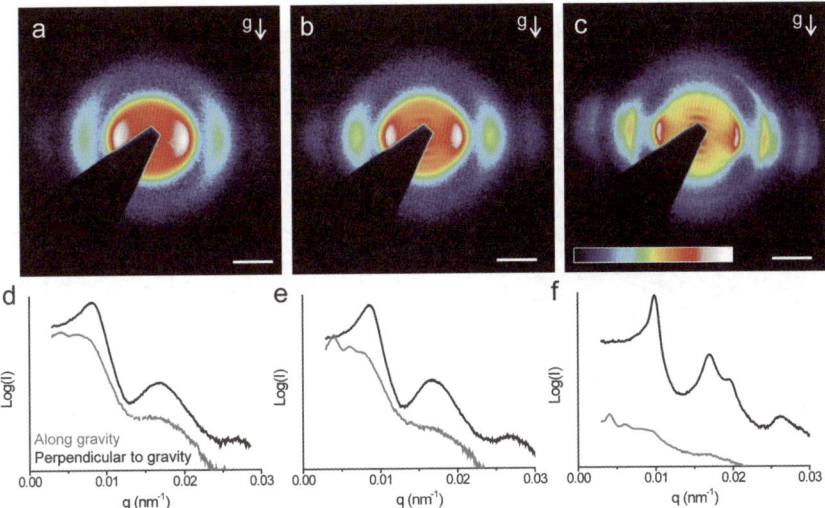

Fig. 6 SAXS measurements of $L/D = 5.0$ rods. (a–c) Scattering patterns of the nematic, smectic-A and smectic-B phase, respectively. (d–f) Intensity profiles of the patterns shown above in the horizontal (black) and vertical direction (grey). Scale bars indicate 0.01 nm^{-1}. The colored bar shows the false color - log(intensity) scale with white color corresponding to the highest intensity.

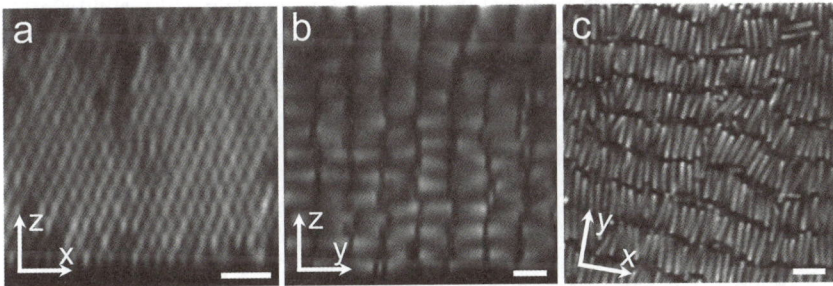

Fig. 7 Confocal microscopy images of hexagonally ordered layers in a sediment of $L/D = 5.0$ rods through different planes. Scale bars indicate 3 µm.

to the $g(r)$ of the perfect lattice. The average distance between the centers of the rods was 910 nm, while the diameter measured by TEM was 640 nm. The difference is caused mostly by the negative charge on the silica rods, causing repulsive interactions, and partly by the fact that these measurements were not performed at the densest packing. The degree of hexagonal order in the sample was quantified by the 2D local hexagonal bond orientational order parameter ψ_6:[36]

$$\psi_6(r_j) = \frac{1}{N} \sum_k \exp(i6\theta(r_{jk})) \tag{1}$$

where k runs over all N neighboring particles (determined by Delaunay triangulation) of particle j and $\theta(r_{jk})$ is the angle between the vector between particles j and k and a fixed reference axis. The value of $|\psi_6|$, which is a number between 0 (no order) and 1 (perfectly crystalline), gives the degree of hexagonal order around each particle, $\langle \psi_6 \rangle$ denotes this value as averaged over all particles. In the crystalline sheets found in sediments of short rods, $\langle \psi_6 \rangle$ was around 0.75. Confocal images of larger areas are needed to determine whether the ordering in these samples is

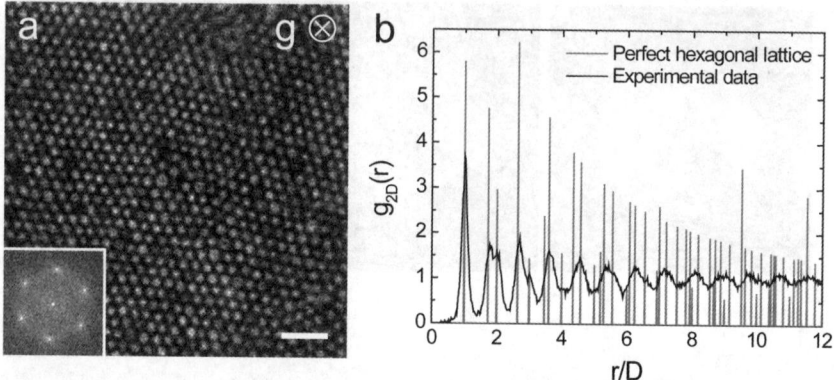

Fig. 8 (a) Confocal microscopy image of short rods (L D^{-1} = 3.7) sedimented in a flat capillary. Large areas of hexagonally ordered rods are visible. In the inset a Fourier transform of the image is shown. Scale bar indicates 5 μm. (b) 2D Radial distribution function of a hexagonally ordered sheet of rods in black. The equivalent for a perfect hexagonal lattice is shown in grey.

really hexagonal or hexatic. Fig. 8a further shows the Fourier transform of the confocal image, in which the first six peaks corresponding to hexagonal ordering are visible. This image corresponds well to the SAXS patterns of the sample, which shows six peaks spaced at approximately 60° up to two scattering orders (Fig. 9a). Although in most of the sample the layers were oriented with rods parallel to the wall of the capillary (Fig. 9c,d), we were able to find some domains with the rods perpendicular to the wall, as in Fig. 9a,b. The crystalline phase continued higher up in the sample until the scattering pattern of an isotropic phase was seen. Smectic-A or nematic phases were not observed in the L/D = 3.7 sample.

3.2 Phase diagram

The measurements described in the previous section can be summarized in a rough phase diagram (Fig. 10). To estimate the volume fractions of the phases observed, inter-particle distances L_s and D_s were determined at different heights in the capillary from the SAXS measurements. Subsequently, the volume fractions of the different phases were calculated as follows. In the smectic-B phase, the volume fraction was calculated by dividing the volume of the rods that was measured by TEM (assuming the shape of a cylinder with one hemispherical cap) by the available volume per particle measured by SAXS: $\frac{1}{2}\sqrt{3}D_s^2L_s$. For a smectic-$A$ phase there is no hexagonal order between the rods and the available volume per particle was estimated by $D_s^2L_s$. In case of a nematic, the extra distance along the length of the rods was assumed to be equal to the calculated difference for the diameter. In order to compare the experimental data with computer simulations, the volume fractions were normalized with respect to the volume fraction at close packing φ_{cp}. Since there were no correlations found between neighboring layers, we assume $\varphi_{cp} = \frac{\pi}{2\sqrt{3}}\left(1 - \frac{D}{6L}\right)$ (hexagonally close-packed rods in one layer). The resulting experimental phase diagram is shown in Fig. 10a.

Our experimental results show many similarities with the phase diagram computed by Bolhuis and Frenkel.[33] Isotropic phases were found at low volume fractions, and for higher volume fractions, nematic and smectic phases were found. At the highest volume fractions, however, Bolhuis and Frenkel found a crystalline phase, while we identified our phase as a smectic-B phase. These phases are actually very much alike. The difference between the two is the presence (for the crystal) or

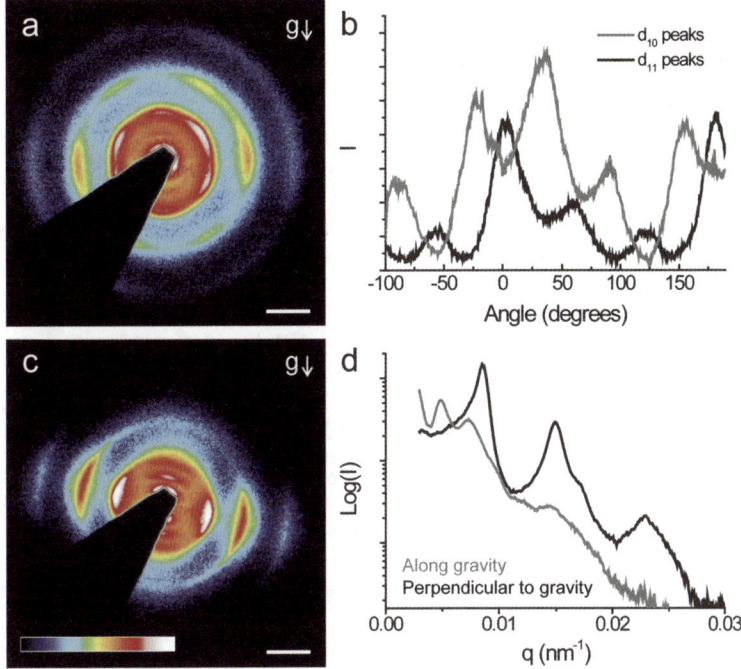

Fig. 9 SAXS pattern of hexagonally ordered rods ($L/D = 3.7$) in two orientations. (a) Rods perpendicular to the capillary wall. (b) Intensity profile of the hexagonal peaks integrated over the circles shown in (a). (c) Rods parallel to the capillary wall. (d) Intensity profiles of the pattern shown in (b). Scale bars indicate 0.01 nm^{-1}. The colored bar shows the false color - log(intensity) scale with white color corresponding to the highest intensity.

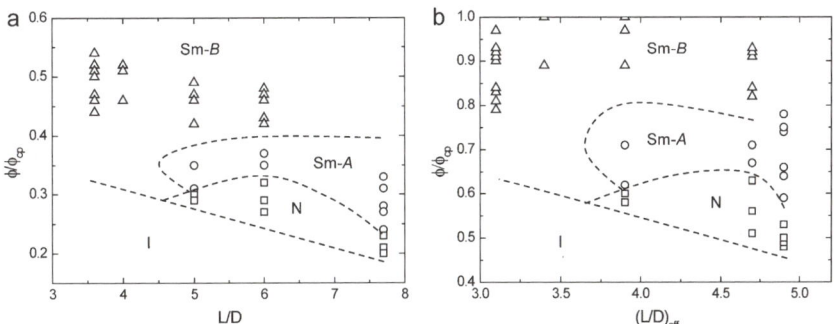

Fig. 10 (a) Experimental phase diagram for silica rods depending on aspect ratio and density. The volume fractions of the smectic-B phase are depicted as triangles, circles denote the smectic-A phase and squares the nematic phase. Lines were drawn as a guide to the eye. (b) Phase diagram corrected for electric double layer repulsions by taking an effective hard-rod diameter and length.

absence (for the smectic-B) of correlations between subsequent layers.[37] Since we did not find proof of correlations between the layers, we refer to the structure as smectic-B. The formation of a smectic-B instead of a full crystalline phase is probably caused by the polydispersity of the systems used (around 10%). Also, charge may play a role. According to simulations, the layer spacing is less than 1.048 times the length of the rods for aspect ratios from 3 to 5 in the smectic phase.[38] In our systems, the layer spacing was often larger (on the order of 1.1 to 1.2 times L_{eff}), which may be the reason that the layers were not correlated.

Also similar to the simulations, the isotropic-nematic phase boundary of the experimental phase diagram shows the same dependence on the aspect ratio: the volume fraction at which the transition occurs decreases with increasing aspect ratio. Furthermore, smectic and nematic phases were found only at aspect ratios of 5 and higher, which is in correspondence with the simulations that predicted nematics for $L/D > 4.7$ and smectics for $L/D > 4.1$.

A large discrepancy between experiments and simulations was found, as fully expected, in the volume fraction of the phases. The experimentally determined volume fractions are significantly lower than those calculated by computer simulations. This is caused by the fact that the experimental system is not a perfect hard rod system. The silica particles are negatively charged and therefore experience repulsive interactions. In literature, it was shown for virus suspensions that the isotropic-nematic volume fractions are described well by an increased effective diameter.[39] The nematic-smectic volume fractions rescaled in this way disagreed with the hard rod model, however, except at high ionic strength.[40] Since in our case the double layer thickness is much smaller than the rod diameter a rescaling should be acceptable. The effective diameter D_{eff} was determined by measuring the minimum inter-particle distances between the rods at the bottom of the sediment, where we assume the rods to be close packed. The effective length was assumed to be $L + D_{eff} - D$, where L and D are the hard-rod dimensions, measured by TEM. The hard-rod and effective dimensions are listed in Table 2 and the rescaled phase diagram is shown in Fig. 10b. For this "effective hard-rod" phase diagram, the volume fractions correspond well with the simulations, but the effective aspect ratio's do not. Nematic and smectic phases occur now for aspect ratios of 3.9 and higher. Apparently, this method of correcting for the double layer, that worked well for spheres, does not map our results onto the HSC phase diagram exactly.

In literature, several other examples of short rods forming smectic phases exist. Maeda et al., for example, observed isotropic-smectic transitions for selenium and β-FeOOH rods of $L/D = 3.5$–8.0.[11,41] Their observations for these low aspect ratios correspond well with our observations, including hexagonal ordering of the rods with $L/D = 3.5$ and side-by-side ordering in clusters before forming a smectic(-B) phase. Nematic phases, however, were only observed for $L/D = 10$–35 in this study. Maeda et al. did not correct for repulsions or polydispersity in their work, which they claim could be the cause of the difference with the simulations, which predicted nematics from $L/D = 4.7$. In our experience, the nematic phase was hard to find by real-space measurements, because it occurs only in a small range of densities. Guided by the SAXS-measurements and having the ability to image 2D slices in dense sediments by confocal microscopy, we were able to find a thin layer of rods in a nematic phase (Fig. 5k) where they might have been missed by Maeda et al., who used a standard optical microscope.

Table 2 Hard rod dimensions as measured by TEM and effective rod dimensions as measured by SAXS. Here, L is the length of the rods, D the diameter, and L/D the aspect ratio

	Hard rod dimensions			Effective rod dimensions		
	$L/\mu m$	D/nm	L/D	$L_{eff}/\mu m$	D_{eff}/nm	$(L/D)_{eff}$
B31	2.37	640	3.7	2.56	830	3.1
B35	3.3	550	6.0	3.49	740	4.7
B36	1.9	235	8.0	2.1	430	4.9
B48	2.6	630	4.1	2.8	830	3.4
N51	2.66	530	5.0	2.86	730	3.9

Other well known liquid crystal systems that show nematic phases, such as TMV or *fd*-virus all have larger aspect ratios than the rods studied in this chapter ($L/D <$ 8).[5,39,40] Also for boehmite rods, nematic phases were only observed for larger aspect ratios.[42] For goethite particles, nematic phases were observed for $L/D = 3.5$, but these particles are much thinner in the third dimension ($L/T \sim 10$, with T as their thickness).[43]

3.3 Influence of the flat wall

In experimental studies of colloidal dispersions, boundary effects (such as the flat wall of the cell that contains the dispersion) are always present. For spheres, it was found that the presence of a flat wall can lead to layering and eventually pre-freezing of a liquid phase.[44,45] Also, the effect of a flat wall on crystallization has been studied intensively.[23,46] The effect of a flat wall on the ordering of rods was examined using computer simulations.[47,48] Here, it was found that a thick nematic film forms on the wall in the isotropic phase.

For our system of hard rods, we experimentally confirmed that the rods tend to align with their long axis parallel to the flat surface. In capillaries positioned with their flat wall to the bottom, we found that the rods always lay flat on the bottom. This resulted in layered phases that were always oriented with their layers perpendicular to the bottom plane. In standing capillaries, for which the flat wall was positioned in the vertical direction, the rods were found to orient again parallel to the flat wall, pointing down this time. This orientation was found in both the confocal and the scattering data, *e.g.* in Fig. 5. Additionally, we found that the influence of the wall extended tens of microns into the sample, after which domains of random orientations of the formed phase were observed.

3.4 Nucleation and defects in smectic phases

Although we did not specifically investigate the phenomenon of the nucleation of crystal or liquid crystal phases, we can indirectly infer some conclusions with respect to the nucleation of the smectic phase as can be seen in Fig. 11a and b. In Fig. 11a a smectic-*B* phase consisting of just a few layers can be seen below an isotropic phase in a sedimenting system of rods with an aspect ratio of 3.7. Fig. 11b shows that adjacent planes tended to stack perpendicularly. Ni *et al.* described the crystal nucleation of a system of hard rods with an aspect ratio of 3 using computer simulations.[49] At lower supersaturation, the nucleation of multilayered crystalline structures was observed much like those visible in Fig. 11a/b (compare with Fig. 11c). At higher supersaturation, small randomly oriented small crystallites were observed that got kinetically trapped (Fig. 11d). The cubatic order of these crystallites corresponds well with the experimentally observed perpendicular orientation of domains like those in Fig. 11b.

The agreement with the computer simulations, which try to identify the equilibrium path, and our preliminary experimental results on the nucleation of the smectic and/or crystal phases of the rods is quite satisfactory. However, it is also a recent finding of our group, based on experiments on hard sphere-like particles, that even for crystal nucleation and growth that is done relatively slowly as to stay close to equilibrium, the system does not follow equilibrium growth with respect to *e.g.* the number of defects quite quickly after the nucleation phase.[50] Also, these experiments and findings on growing 3D crystals on preformed 2D nuclei arranged by optical tweezers were supported by computer simulations. According to us, a clear way on how these non-equilibrium aspects can be adequately incorporated in a theory, or, related to this, how can be judged how far away from equilibrium the crystal nucleation and growth takes place in experiments, are still important open questions. Perhaps colloidal model systems are a way in which these questions can

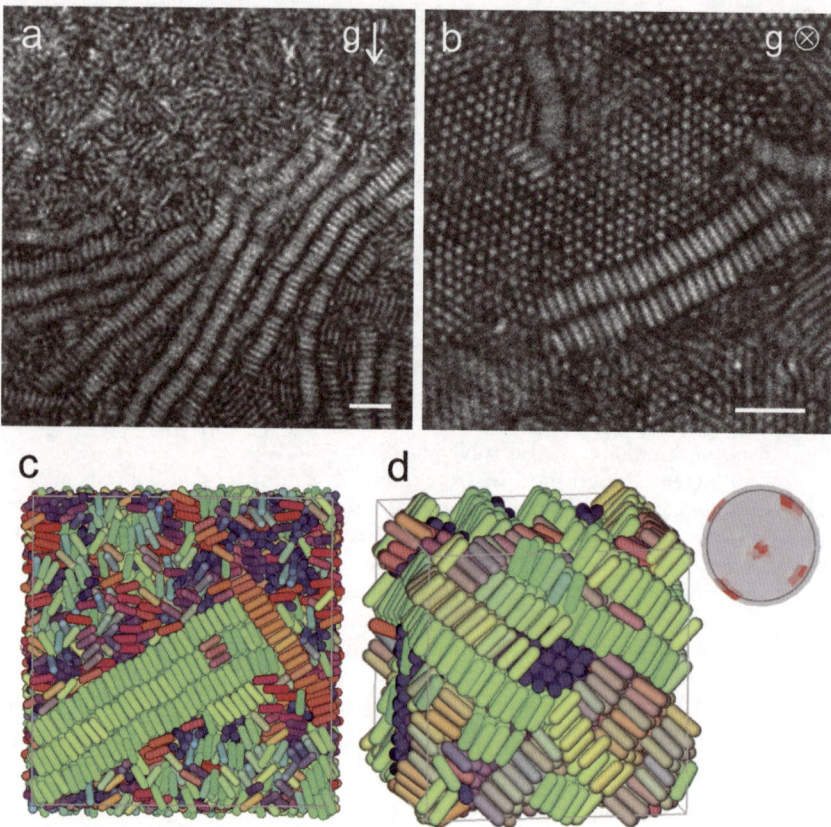

Fig. 11 Nucleation of short rods. (a) Confocal microscopy image of rods with an aspect ratio of 3.7. An isotropic phase is visible in the top of the image, with a smectic-B phase below. (b) Image inside a smectic-B plane, showing the perpendicular stacking of the domains. Scale bars indicate 5 μm. (c) Simulation result at a pressure $p^* = 7.6$. A smectic nucleus is formed in the isotropic phase. The different colors denote the different orientations of the rods. (d) Simulation result at a pressure $p^* = 8$. The system ends up in a jammed state with cubatic order. The angular distribution of the particles' orientation is shown in the unit sphere beside the configuration. The simulation images c and d were provided by the authors of ref. 49.

also be addressed for atomic and molecular systems, even though the dynamics in the two systems are different.

The appearance of the smectic phases found varied strongly as a function of the aspect ratio of the rods. For long aspect ratios (>6), layer formation occurred more readily and large domains formed that extended over the whole sample. Defects that were found in these samples include mainly edge dislocations and splay distortions (Fig. 12a). In samples of shorter rods, defects were observed more frequently. Also, the domains found in these samples were smaller than for long rods. Fig. 12b shows three types of defects. The first type, inside circle 1, is that neighboring domains or defect layers are positioned perpendicularly with respect to each other. Inside circle 2 the ending of a layer is shown, called an edge dislocation. Both types of defects have been described for other liquid crystal systems.[51] The third type of defect, however, has not been observed before experimentally as far as we know. This defect, which is called a transverse interlayer (TI) particle, is described as a rod positioned in between the layers of a smectic and oriented with its long axis parallel to the layers. TI particles have a higher free energy than particles inside the layers. Since the presence of TI particles minimizes the free energy of the

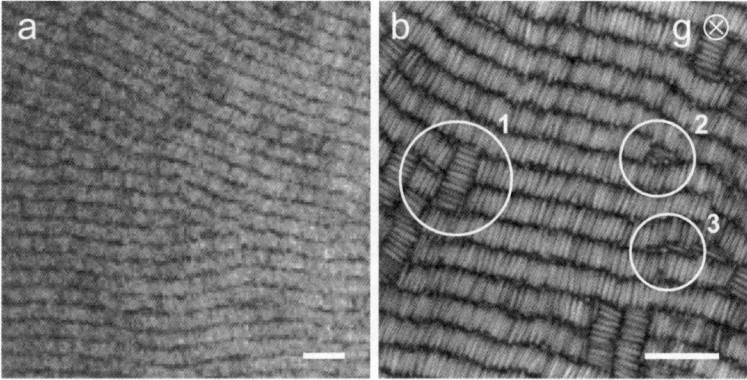

Fig. 12 Confocal microscopy images showing several types of defects. (a) Edge dislocations and splay distortions in a sample with rods of aspect ratio 8.7. (b) Sediment of $L/D = 6$ rods. 1) Small domain oriented perpendicular to its neighbor-domains. 2) Edge dislocation. 3) Transverse-interlayer (TI) particles. Scale bars indicate 10 μm.

configuration, this type of defects cannot be annealed out. The existence of TI particles was predicted by computer simulations already in 1995,[52,53] but no experimental confirmation of the phenomenon has been reported up to now. In the simulations it was found that the abundance of TI-particles depends on the smectic layer spacing λ. The simulations were done for $\lambda = 1.03$, while in our experiments $\lambda \sim 1.13$. This might explain why we found more TI particles than expected; we found on the order of a few percent of the rods to be TI particles, while the simulations resulted in a fraction on the order of 10^{-5}.

3.5 Dynamics in the smectic phase

Beside the previously described structural aspects of the different liquid crystal phases, there are also dynamical differences. The diffusion of rods in liquid crystalline phases was found to be greatly affected by the increased structural order. Computer simulations and experiments showed that in nematic phases the long-time self-diffusion coefficient along the director $D_\parallel$ is higher than the one perpendicular to the director $D_\perp$.[54] A ratio of $D_\parallel/D_\perp = 2$ to 4 was found near the isotropic-nematic transition, which increased with increasing concentration. In a smectic, however, $D_\parallel$ is expected to decrease dramatically, to even below $D_\perp$. Below, we describe some preliminary observations on dynamics in one of our smectic phases.

A dynamic difference between the smectic and the crystalline phase is that long-time self-diffusion only occurs in smectics and not in crystals. To determine whether the layered phase formed by rods with a length of 1.4 μm and a diameter of 280 nm is a smectic, we studied the diffusion of the rods mostly qualitatively. Due to their small diameter, single particle imaging of these rods was not possible, but the layers of the structure are clearly visible in Fig. 13. A 50×50 μm² x-z plane was bleached perpendicular to the orientation of the layers by scanning this plane 10 times in short succession (Fig. 13 top row). In time, the bleached particles diffused to the non-bleached area and *vice versa*, resulting in a spread of the dark area in the images. Since there is clearly diffusion of rods, the layered phase is indeed a smectic. When a plane parallel to the layers (in other words: one of the layers) was bleached, the spreading was slower and the boundary between the bleached and non-bleached area stayed much sharper. These observations confirm the simulation results on hard spherocylinders by Löwen, who found that $D_\perp > D_\parallel$ in smectics.[54] Intuitively, this is easy to understand since the rods are ordered liquid-like inside the layers and diffusion is therefore easy. In the direction perpendicular to the layers the structure is

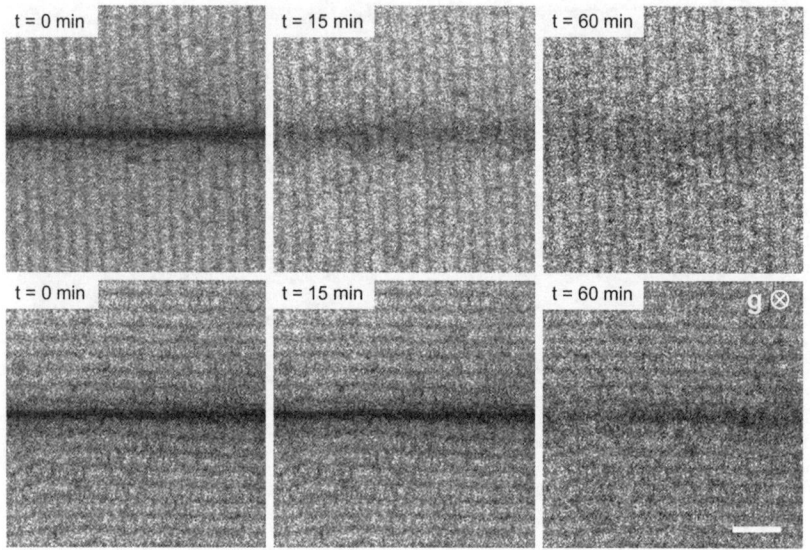

Fig. 13 Diffusion of bleached particles with aspect ratio 5.0 within smectic planes (upper row) and from plane to plane (lower row). Scale bar indicates 5 μm.

ordered, which makes diffusion slower because an energy barrier has to be overcome. However, exactly the opposite was found in diffusion studies on the *fd*-virus, where self-diffusion takes place preferentially in the direction perpendicular to the smectic layers and occurs by quasiquantized steps of one rod length.[55,56] This is possibly caused by the flexibility of the virus particles. In a rough estimation from Fig. 13 we found $D_\perp \sim 0.2$ μm^2 s^{-1} in our system.

Diffusion on the single particle level for thin rods was studied using a mixture of labeled and unlabeled rods, similarly as was done for *fd*-virus.[6] The labeled rods were thicker than the unlabeled rods in this case (340 *versus* 250 nm). Sedimentation of a sample with a 1 : 100 ratio of labeled : unlabeled rods and an initial volume fraction of 0.17 resulted in the formation of a smectic phase. In this tracer system, it is possible to observe single particles with small diameters in the smectic phase.

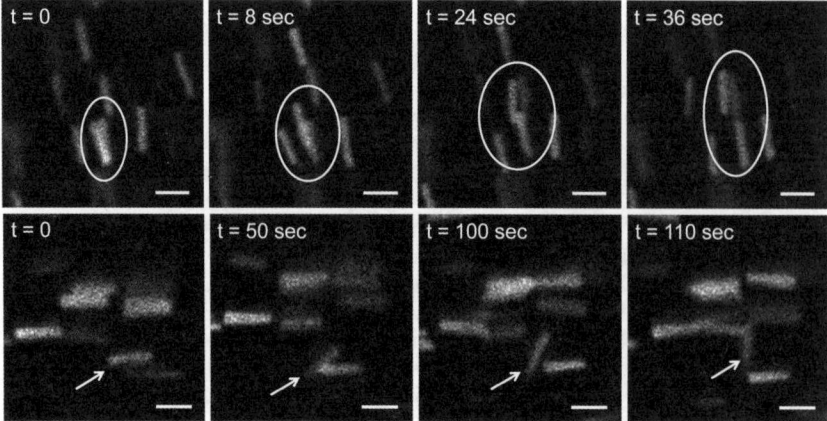

Fig. 14 On a single particle level: diffusion from one smectic plane to another (upper row) and formation of a transverse-interlayer particle (lower row). Scale bars indicate 2 μm.

From measurements just above the nematic/smectic transition we found $D_\perp \sim 0.5$ µm^2 s^{-1}. Besides diffusion within the layers, we also found more rare events like hopping from one layer to another and the transformation from normal rod to TI-particle (Fig. 14). The single particle hopping from one layer to another or from in-layer to transverse-interlayer particles shows a striking resemblance with the particle trajectories as observed in computer simulations by Patti et al.[38] The hopping phenomenon was studied in more detail by computer simulations,[38,57,58] as well as experiments on fd-virus.[55]

4 Conclusions

The phase behavior of rod-like silica particles with aspect ratios smaller than 8 was determined using small angle X-ray scattering and confocal laser scanning microscopy. The results were summarized in a phase diagram, which corresponds well with computer simulations on HSC-systems. In correspondence with the simulations, no liquid crystalline phases were found for silica rods with small aspect ratios. Nematic and smectic phases were found for aspect ratios of 5.0 and higher, while simulations predict nematics for $L/D > 4.7$ and smectics for $L/D < 4.1$. Instead of forming a fully crystalline phase, the rods ordered into a hexagonally ordered smectic-B phase at high volume fractions, probably due to their length polydispersity ($\sim 10\%$) and/or their charge which causes larger layer spacings. The experimentally found volume fractions of the different phases were significantly lower than those calculated in the simulations, which is caused by the negative charge of the rods. The resulting electric double layer increased the effective dimensions of the rods and decreased the volume fraction.

The system of colloidal silica rods that was used in this work is unique because it allows us to examine liquid crystalline phases at the single particle level in 3D. Thus, we have not only studied the structure of fully developed liquid crystal phases with both scattering and microscopy, but also identified a number of different defect structures that we encountered. Though admittedly more preliminary, we found, using real space techniques, strong indications that confirmed the nucleation process for short rods that was found by computer simulations. Also, several types of defects in the smectic phase were found. The amount of defects decreased with increasing aspect ratio. Because the rods could be imaged on the single particle level, the existence of transverse interlayer particles was observed experimentally for the first time. Also, dynamical processes such as diffusion within and between layers as well as the transition from in-layer to transverse-interlayer particle were observed. The order of magnitude of the diffusion coefficient, determined by these preliminary measurements corresponds to measurements on other rod-like colloidal systems. We believe that our rod-like silica colloids form an excellent system to study these properties of liquid crystalline suspensions in more detail.

Acknowledgements

We would like to thank Bart de Nijs for his help with the SAXS measurements. The staff of the BM26 beamline of the ESRF is acknowledged for their excellent support. NWO is acknowledged for the beam time. Furthermore, we thank Ran Ni for providing the simulation images and Marjolein Dijkstra for critical reading of the manuscript. This work was funded by the High Potential Program of Utrecht University and by the EU (NANODIRECT, grant number CP-FP 213948-2).

References

1 L. Onsager, *Ann. N. Y. Acad. Sci.*, 1949, **51**, 627–659.
2 D. Frenkel and B. M. Mulder, *Mol. Phys.*, 1985, **55**, 1171.
3 D. Frenkel, H. N. W. Lekkerkerker and A. Stroobants, *Nature*, 1988, **332**, 822–823.

4 G. J. Vroege and H. N. W. Lekkerkerker, *Rep. Prog. Phys.*, 1992, **55**, 1241–1309.
5 Z. Dogic and S. Fraden, *Curr. Opin. Colloid Interface Sci.*, 2006, **11**, 47–55.
6 M. P. Lettinga, E. Barry and Z. Dogic, *Europhys. Lett.*, 2005, **71**, 692–698.
7 B. J. Lemaire, P. Davidson, J. Ferré, J. P. Jamet, D. Peterman, P. Panine, I. Dozov and J. P. Jolivet, *Eur. Phys. J. E*, 2004, **13**, 291–308.
8 B. J. Lemaire, P. Davidson, J. Ferré, J. P. Jamet, D. Peterman, P. Panine, I. Dozov and J. P. Jolivet, *Eur. Phys. J. E*, 2004, **13**, 309–319.
9 E. van den Pol, A. V. Petukhov, D. M. E. Thies-Weessie, D. V. Byelov and G. J. Vroege, *J. Colloid Interface Sci.*, 2010, **352**, 354–358.
10 E. van den Pol, A. A. Verhoeff, A. Lupascu, M. A. Diaconeasa, P. Davidson, I. Dozov, B. W. M. Kuipers, D. M. E. Thies-Weesie and G. J. Vroege, *J. Phys.: Condens. Matter*, 2011, **23**, 194108.
11 H. Maeda and Y. Maeda, *Phys. Rev. Lett.*, 2003, **90**, 018303.
12 K. M. Keville, E. I. Franses and J. M. Caruthers, *J. Colloid Interface Sci.*, 1991, **144**, 103.
13 C. C. Ho, A. Keller, J. A. Odell and R. H. Ottewill, Colloid Polym, *Sci.*, 1993, **271**, 469.
14 A. F. Demirörs, P. M. Johnson, C. M. van Kats, A. van Blaaderen and A. Imhof, *Langmuir*, 2010, **26**, 14466–14471.
15 C. M. van Kats, P. M. Johnson, J. E. A. M. Meerakker and A. van Blaaderen, *Langmuir*, 2004, **20**, 11201–11207.
16 A. Kuijk, A. van Blaaderen and A. Imhof, *J. Am. Chem. Soc.*, 2011, **133**, 2346–2349.
17 A. Vrij, J. W. Jansen, J. K. G. Dhont, C. Pathmamanoharan, M. M. Kops-Werkhoven and H. M. Fijnaut, *Faraday Discuss. Chem. Soc.*, 1983, **76**, 19–35.
18 A. van Blaaderen and P. Wiltzius, *Science*, 1995, **270**, 1177–1179.
19 A. Kuijk, A. Imhof and A. van Blaaderen, *Langmuir*, 2012, submitted.
20 M. A. Bates and D. Frenkel, *J. Chem. Phys.*, 1998, **109**, 6193–6199.
21 A. Stroobants, H. N. W. Lekkerkerker and T. Odijk, *Macromolecules*, 1986, **19**, 2232.
22 E. Eggen, M. Dijkstra and R. van Roij, *Phys. Rev. E: Stat., Nonlinear, Soft Matter Phys.*, 2009, **79**, 041401.
23 J. P. Hoogenboom, P. Vergeer and A. van Blaaderen, *J. Chem. Phys.*, 2003, **119**, 3371–3383.
24 A. van Blaaderen, J. Peetermans, G. Maret and J. K. G. Dhont, *J. Chem. Phys.*, 1992, **96**, 4591–4603.
25 E. M. Kramer and J. Herzfeld, *Phys. Rev. E: Stat. Phys., Plasmas, Fluids, Relat. Interdiscip. Top.*, 2000, **61**, 6872–6878.
26 H. Graf and H. Löwen, *Phys. Rev. E: Stat. Phys., Plasmas, Fluids, Relat. Interdiscip. Top.*, 1999, **59**, 1932.
27 J. C. Crocker and D. G. Grier, *J. Colloid Interface Sci.*, 1996, **179**, 298.
28 A. V. Petukhov, J. H. H. Thijssen, D. C. 't Hart, A. Imhof, A. van Blaaderen, I. P. Dolbnya, A. Snigirev, A. Moussaid and I. Snigireva, *J. Appl. Crystallogr.*, 2006, **39**, 137–144.
29 S. V. Savenko and M. Dijkstra, *Phys. Rev. E: Stat., Nonlinear, Soft Matter Phys.*, 2004, **70**, 051401.
30 S. Boersma, *J. Chem. Phys.*, 1960, 32.
31 M. M. Tirado, C. L. Martinez and J. Garcia de la Torre, *J. Chem. Phys.*, 1984, **81**, 2047–2052.
32 Z. Dogic, A. P. Philipse, S. Fraden and J. K. G. Dhont, *J. Chem. Phys.*, 2000, **113**, 8368–8380.
33 P. G. Bolhuis and D. Frenkel, *J. Chem. Phys.*, 1997, **106**, 666.
34 K. R. Purdy, Z. Dogic, S. Fraden, A. Rühm, L. Lurio and S. G. J. Mochrie, *Phys. Rev. E: Stat. Phys., Plasmas, Fluids, Relat. Interdiscip. Top.*, 2003, **67**, 031708.
35 E. Grelet, *Phys. Rev. Lett.*, 2008, **100**, 168301.
36 C. Murray, Experimental studies of melting and hexatic order in two-dimensional colloidal suspensions, in: *Bond-orientational order in condensed matter systems*, Springer-Verlag, New York, 1992, pp. 137–215.
37 P. G. de Gennes and J. Prost, *The physics of Liquid Crystals*, Oxford University Press, 2nd edn, 1995.
38 A. Patti, D. El Masri, R. van Roij and M. Dijkstra, *J. Chem. Phys.*, 2010, **132**, 224907.
39 K. R. Purdy and S. Fraden, *Phys. Rev. E: Stat., Nonlinear, Soft Matter Phys.*, 2004, **70**, 061703.
40 K. R. Purdy and S. Fraden, *Phys. Rev. E: Stat., Nonlinear, Soft Matter Phys.*, 2007, **76**, 011705.
41 H. Maeda and Y. Maeda, *J. Chem. Phys.*, 2004, **121**, 12655.
42 P. A. Buining, A. P. Philipse and H. N. W. Lekkerkerker, *Langmuir*, 1994, **10**, 2106–2114.
43 E. van den Pol, D. M. E. Thies-Weesie, A. V. Petukhov, G. J. Vroege and K. Kvashnina, *J. Chem. Phys.*, 2008, **129**, 164715.
44 D. J. Courtemanche, T. A. Pasmore and F. van Swol, *Mol. Phys.*, 1993, **80**, 861–875.
45 M. Dijkstra, *Phys. Rev. Lett.*, 2004, **93**, 108303.

46 S. Dittrich, *Phase transitions and critical phenomena*, Academic Press, London, 1988.
47 M. Dijkstra, R. van Roij and R. Evans, *Phys. Rev. E*, 2001, **63**, 051703.
48 L. Harnau and S. Dietrich, *Phys. Rev. E*, 2002, **66**, 051702.
49 R. Ni, S. Belli, R. van Roij and M. Dijkstra, *Phys. Rev. Lett.*, 2010, **105**, 088302.
50 M. Hermes, E. C. M. Vermolen, M. E. Leunissen, D. L. Vossen, P. D. J. van Oostrum, M. Dijkstra and A. van Blaaderen, *Soft Matter*, 2011, **7**, 4623–4628.
51 Y. Iwashita and H. Tanaka, *Phys. Rev. E*, 2008, **77**, 041706.
52 R. van Roij, P. G. Bolhuis, B. Mulder and D. Frenkel, *Phys. Rev. E*, 1995, **52**, 1277–1281.
53 J. S. van Duijneveldt and M. P. Allen, *Mol. Phys.*, 1997, **90**, 243–250.
54 H. Löwen, *Phys. Rev. E*, 1999, **59**, 1989–1995.
55 E. Grelet, M. P. Lettinga, M. Bier, R. van Roij and P. van der Schoot, *J. Phys.: Condens. Matter*, 2008, **20**, 494213.
56 M. P. Lettinga and E. Grelet, *Phys. Rev. Lett.*, 2007, **99**, 197802.
57 R. Matena, M. Dijkstra and A. Patti, *Phys. Rev. E*, 2010, **81**, 021704.
58 A. Patti, D. El Masri, R. van Roij and M. Dijkstra, *Phys. Rev. Lett.*, 2009, **103**, 248304.

PAPER

Inorganic salts direct the assembly of charged nanoparticles into composite nanoscopic spheres, plates, or needles

Bartosz A. Grzybowski,[*ab] Bartlomiej Kowalczyk,[ab] István Lagzi,[ab] Dawei Wang,[bc] Konstantin V. Tretiakov[d] and David A. Walker[b]

Received 15th April 2012, Accepted 1st June 2012
DOI: 10.1039/c2fd20074k

Oppositely charged, nanoionic nanoparticles can act as "universal surfactants" regulating the growth of ionic microcrystals. This phenomenon derives from a subtle interplay between crystal growth and cooperative electrostatic adsorption of the nanoparticles onto crystal faces. In addition to the electrostatic interactions acting in the system, the nature of salts is also important in the sense that for the same Debye screening length, different salts can mediate formation of markedly different assemblies including supraspheres, nanoneedles, or nanoplates. The method can be further extended to coat non-ionic crystals with appropriately functionalized nanoparticles.

1 Introduction

Self-assembly, SA, of nanoscopic components into "superstructures" of various morphologies is of interest for their size- and shape dependent optical, electronic and magnetic properties[1–3] as well as for the potential applications in the fabrication of magnetic[4] and electronic devices,[5] water remediation,[6] biosensing[7] and energy storage.[8] Such structures have been prepared by a host of methods and include 3D nanoparticle crystals,[9] planar sheets[10] and ribbons,[11] hollow spheres[8] and molecule-like[12,13] and polymer-like[14] assemblies. Preparation of these assemblies entails skilful use and control of nanoscale interactions including van der Waals, steric/entropic, magnetic and electrostatic (for review, see ref. 15). Electrostatic interactions are quite versatile in SA applications since nanoobjects of various material properties can be made charged—to various degrees[16]—by functionalization with charged ligands. We have previously used this strategy to assemble oppositely-charged nanoparticles,[17,18] nanotriangles[19] and nanorods[20] into crystals,[9] heterodimers,[20] "nanomolecules,"[12,13] or dynamic aggregates exhibiting oscillations and/or propagation of chemical waves.[21] Most recently, we have combined the self-assembly of oppositely-charged, (+)/(−) NPs with the growth of microcrystals of inorganic salts ($CaCO_3$, K_2SO_4, Na_2SO_4, and more) as well as charged organic molecules (*e.g.*, (L)-Lysine or vitamin B_5).[24] These compounds were crystallized from

[a]*Department of Chemistry, Department of Chemical and Biological Engineering, Northwestern University, Evanston, Illinois, 60208, USA. E-mail: grzybor@northwestern.edu; Fax: +01 847 491 3024; Tel: +01 847 491 3024*
[b]*Department of Chemical and Biological Engineering, Northwestern University, Evanston, Illinois, 60208, USA. E-mail: grzybor@northwestern.edu; Fax: +01 847 491 3024; Tel: +01 847 491 3024*
[c]*School of Materials Science and Engineering, Northwestern Polytechnical University, Xi'an 710072, China*
[d]*Institute of Molecular Physics, Polish Academy of Sciences, Smoluchowskiego 17, 60-179 Poznan, Poland*

water–DSMO mixtures in the presence of the (+)/(–) nanoparticles. As the crystals grew, the oppositely charged NPs adsorbed onto them in a cooperative fashion to form thin coatings that retarded crystal growth. By increasing the proportion of the NPs to the salt precursors, we were able to favor coating formation *vs.* crystal growth and ultimately control crystal size from tens of microns to ~100 nanometers. In all these experiments, the (+)/(–) NPs acted as a polyvalent "surfactant" effective toward various types of crystallizing materials.

In describing these systems, we found that a continuum description of electrostatic interactions was sufficient—that is, the magnitudes and the ranges of electrostatic forces were regulated by the concentration of inorganic salts present in solution and did not depend on the salt's specific nature. Here, we describe two systems in which the mode of nanoscale electrostatic self-assembly, nESA,[22] is salt specific. We show that depending on the salt's nature and solubility, the same charged nanoparticles can give rise to either supraspheres (using Ni^{2+} salts) or into low symmetry nanoplates or nanoneedles (using Cu^{2+} salts). In the first case, nickel cations facilitate bridging between the assembling NPs; in the latter, particle assembly is templated by the formation of nanoscopic calcium hydroxide seeds. While these examples illustrate the experimental latitude of the nESA method, the general approach can be extended even further, beyond electrostatic and ion-bridging interactions. To outline the opportunities that lie ahead, we provide one illustrative example in which uncharged nanoparticles interact with and coat organic donor–acceptor crystals.

2 Methods

Synthesis and functionalization of AuNPs

(i) **AuDDA NPs.** Dodecylamine (DDA) coated gold NPs were prepared according to a modified literature procedure[16,17] and had average diameters, 5.0 ± 0.8 nm, as estimated from TEM images of at least 200 NPs from each batch used.

(ii) **Ligand exchange on gold NPs.** All ultrapure-grade thiols [11-mercaptoundecanoic acid, MUA, and *N,N,N*-trimethyl(11-mercaptoundecyl)ammonium chloride, TMA] were obtained from ProChimia Surfaces (www.prochimia.com) and used as received. A toluene solution of DDA-capped AuNPs (7 μmol mL^{-1}, 20 mL, 140 μmol) was quenched with 100 mL of methanol to give a black precipitate. The supernatant solution containing excess of capping agent and surfactant was decanted, and the precipitate was washed with methanol (50 mL) and then dissolved in toluene (100 mL) to which a thiol solution (140 μmol) in 10 mL of CH_2Cl_2 was added upon stirring. Thus prepared AuNPs functionalized with thiolate SAMs[23] were allowed to settle down, the mother-liqueur solution was decanted, and the solid was washed with CH_2Cl_2 (3 × 30 mL). The precipitate was then dissolved in 5 mL of methanol by sonication. To introduce negative charges on the AuMUAs, these particles were deprotonated with 25% methanolic solution of NMe_4OH (70 μL, 165 μmol), precipitated with acetone (30 mL), and washed with acetone (2 × 30 mL). The zeta potential of these particles was measured on a Malvern Nano Zeta-Sizer instrument and had an average value of -42.0 ± 4.6 mV. The positively charged AuTMAs were precipitated with ethyl acetate (100 mL) and washed with CH_2Cl_2 and acetone; the zeta potential of these particles was $+38.2 \pm 3.9$. Finally, all the precipitates of thiol-coated AuNPs were dried and dissolved in 13 mL of water to obtain ~10 mM (in terms of numbers of metal atoms) NP solutions. In the case of AuMUA, the pH of the solution was adjusted to ~11 with 0.2 M solution of NMe_4OH.

Preparation of a mixture of oppositely charged nanoparticles

Solution of TMA-coated NPs was titrated with ~100 μL aliquots of fully deprotonated (at pH = 11) MUA-coated NPs (note: titrating MUA particles with TMA particles gave identical results, see ref. 17). As described in our earlier publications, the oppositely charged NPs precipitate at the point of electroneutrality, whereby the sum of charges on these particles is zero, $\sum Q_{NP(+)} + \sum Q_{NP(-)} = 0$. The precipitated NPs were collected, centrifuged and washed with DI water to remove excess salts. Finally, NPs were redissolved in DI water upon gentle heating. This solution was then stable for at least several months. For further details, see ref. 9, 16 and 17.

General procedure for crystallization of salts in the presence of NPs

In a typical experiment, 1 μmol (in terms of Au atoms) of the oppositely charged NPs was dissolved in 3 mL of water and DMSO (2 : 1 v/v). To this solution, either copper sulfate (n_{CuSO4} = 0.3–1.0 μmol) or nickel sulfate (n_{NiSO4} = 0.15–0.18 μmol) was added, the pH was adjusted to ~8 by addition of tetramethylammonium hydroxide, and then water (the "good" solvent) was evaporated slowly, over 16–24 h at T = 65 °C. In the last stages of evaporation, a dark brown solid precipitated from DMSO. This solid was washed several times with acetonitrile, deposited onto a TEM grid, dried under high vacuum, and then characterized by electron microscopy (TEM and SEM); composition scans were taken on a TEM equipped with an electron diffraction spectroscopy, EDS, detector. All nanoparticle assemblies characterized in this way were stable under organic solvent (DMSO) or in the air for at least several months.

3 Results and discussion

In our previous work on microcrystallization controlled by oppositely charged NPs, we delineated two main regimes: (1) one in which the salts are readily soluble and do not alter the crystallization of the NPs into all-nanoparticle crystals, and (2) one in which the salts are sparingly soluble such that their crystallization competes with the aggregation of the (+)/(−) NPs on the surfaces of the growing crystals, thereby limiting the crystals' growth. We rationalized both types of behaviors using a continuum description of the salt/NP systems and taking into account the interplay between salt solubility and Debye NP-NP screening effects (*i.e.*, crystallization of salt changes the screening length in the system and facilitates cooperative adsorption of the (+)/(−) NPs onto the growing salt microcrystals, for details, see ref. 24). The systems we study here illustrate that this description is only the first-order approximation and the mixtures of co-crystallizing (+)/(−) NPs and inorganic salts can produce a richer repertoire of nano- and microstructures whose morphologies depend strongly on the specific nature of salts used.

We consider $NiSO_4$ and $CuSO_4$ salts which share the same anion and their cations have the same valence. In this way, the Debye screening effects imparted by a given concentration of either of these salts are identical. We note that if it were predominantly the screening effects that control crystallization/assembly in the system, one could expect similar structures formed in the presence of $NiSO_4$ or $CuSO_4$. This, however, is not what we observed in experiments.

In the absence or at very low concentrations of Cu^{2+} or Ni^{2+} salts, the oppositely charged NPs formed regularly faceted, all-nanoparticle crystals such as those illustrated in Fig. 1 (*arrow to the left*) and described in our previous publications.[9,25,26] The effects of the salt on NP organization became significant only at higher concentrations and depended on the salt's nature.

For instance, for $NiSO_4$ at typical nanoparticle concentrations $C_{NP} \sim 0.33$ mM (n_{NP} = 1 μmol) and salt concentrations around C_{NiSO4} = 0.05 mM (n_{NiSO4} = 0.15 μmol), the NPs formed supraspherical (SS) aggregates (Fig. 1, *top right* panel and Fig. 2a,b). These supraspheres had average diameters ~250 nm and were relatively monodisperse (standard deviation in size ~20%). The compositional EDS scans over these structures evidenced clear gold and sulfur peaks but the concentrations of nickel were below the detection limit (Fig. 2b). The mechanism of formation

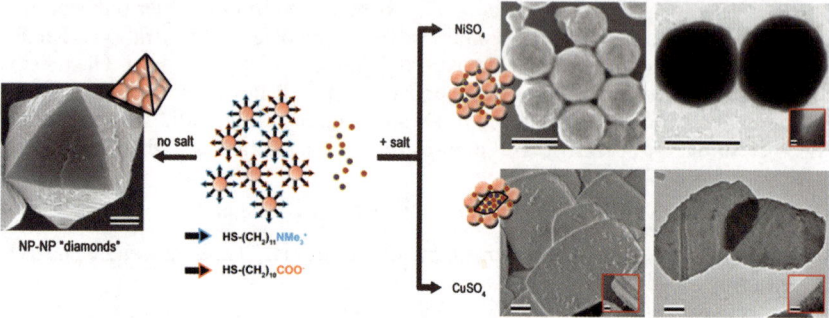

Fig. 1 Self-assembly of oppositely charged nanoparticles in the absence and in the presence of salts. In the absence or at low concentrations of either Ni^{2+} ($C_{NiSO_4} < 0.05$ mM) or Cu^{2+} ($C_{CuSO_4} < 0.1$ mM) salts (left), the positively and negatively charged nanoparticles nucleate and grow into all-nanoparticle crystals. At higher salt concentrations (right), the NPs form supraspheres (with $NiSO_4$, $C_{NiSO_4} > 0.05$ mM) in which the MUA particles are "glued" by the salt cations, or plates (with $Cu(OH)_2/CuSO_4$ $C_{CuSO_4} > 0.1$ mM) in which a nanoscopic needle-like crystals of $Cu(OH)_2$ act as templates onto which the NPs adsorb. Scale bars = 250 nm in all images, (150 nm in SEM inset and 25 nm in both TEM insets).

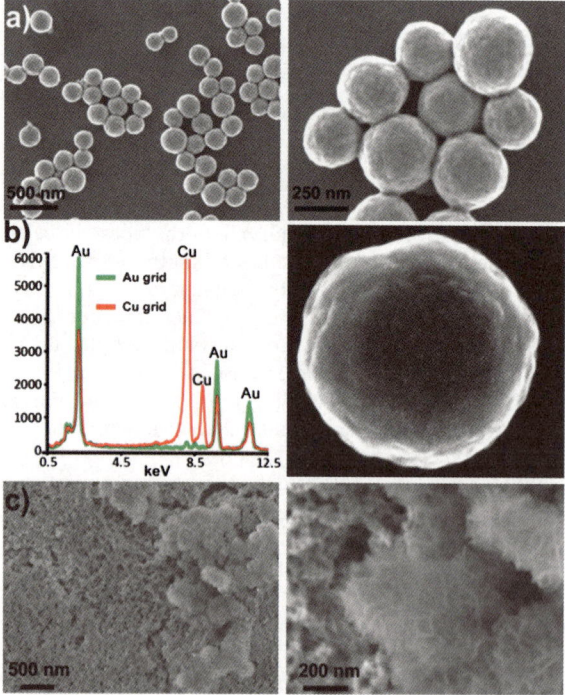

Fig. 2 (a) SEM images of supraspherical assemblies formed from the NPs in the presence of $NiSO_4$ (here, using $C_{NP} \sim 0.33$ mM, $n_{NP} = 1$ µmol, salt concentrations $C_{NiSO_4} = 0.05$ mM, $n_{NiSO_4} = 0.15$ µmol); (b) EDS spectra of a single suprasphere on copper and gold grids reveal the presence of gold (from nanoparticles) and not of any detectable amount of nickel. (c) Typical images of nickel hydroxide aggregates on top of structureless NP precipitates. Such structures form at high concentrations of $NiSO_4$ ($C_{NiSO_4} > 0.15$ mM).

of similar supraspherical aggregates was studied in our earlier papers on nanoparticle assembly mediated by organic ligands.[27,28] Similar to these works, the NPs in our current system can aggregate due to the bridging of the carboxylate groups of the MUA ligands by the Ni^{2+} cations. The energy of such bridges is on the order of few kT and, combined with the van der Waals attractions between NPs' metal cores, can overcome the electrostatic NP–NP repulsions.[15,22,29] The supraspheres then form most likely[27,28] by a nucleation-and-growth mechanism, in which the free MUA and TMA NPs initially nucleate into small, thermodynamically stable (unless smaller than a critical size) clusters that subsequently grow by the addition of single NPs until all NPs available are used. We note that this scenario is expected to be operative only at relatively weak net-attractive interactions. At higher salt concentrations, where the NP–NP attractions become stronger (due to screening effects), the NPs precipitate/flocculate into structureless aggregates. This is illustrated in Fig. 2c which shows NP precipitate with some microcrystals of nickel sulfate formed from an excess of salt in the last stages of water evaporation[30] (the "branched" structure of NiSO$_4$ microcrystals is indicative of diffusional limitations during the process).

The assemblies formed in the presence of copper sulfate (n_{CuSO4} = 0.6–2.0 μmol) were markedly different from those observed in NiSO$_4$ experiments (note: at higher salt concentrations, the NPs gradually precipitated). At n_{CuSO4} = 2.0 μmol, these structures were relatively monodisperse thin plates (up to ~2 μm wide, ~4 μm long and 100–150 nm thick, Fig. 3a). Decreasing the amount of copper salt to n_{CuSO4} = 1.4 μmol led to the formation of square, submicron plates (~0.5 μm × ~0.5 μm × 50 nm, Fig. 3b). Upon further decrease of the n_{CuSO4} to 1.0 μmol, needle-like structures formed (~200 nm × 50 nm × 20 nm, Fig. 3c) and at n_{CuSO4} = 0.6 μmol, the assemblies had a morphology of needles/sheets only ~10 nm thick

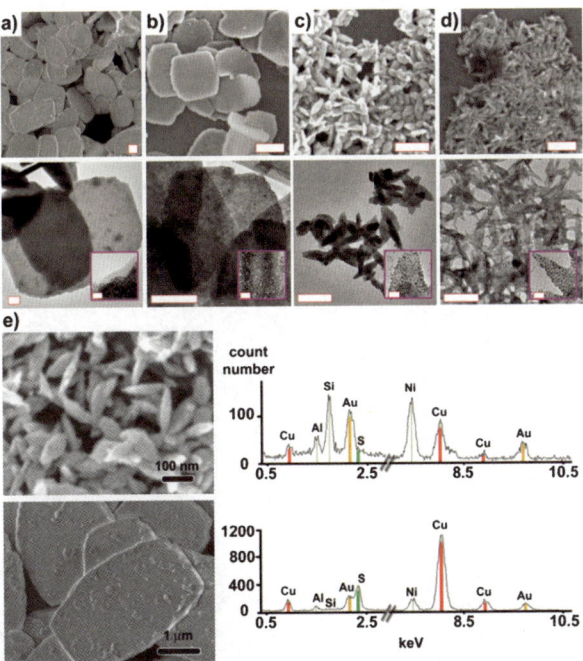

Fig. 3 (a–d) SEM (top row) and TEM (bottom row) images showing size- and shape dependent self-assembly of oppositely charged nanoparticles mediated by Cu^{2+}. The assemblies (a) – (d) correspond to the amounts of CuSO$_4$ 2.0, 1.4, 1.0 and 0.6 μmol, respectively. All scalebars correspond to 500 nm in the SEM images, 200 nm in the TEM images, and 25 nm in the insets. (e) EDS spectra of smaller "needles" (top) from Fig. 3d, and larger "plates" (bottom) depicted in Fig. 3a, grown from copper sulfate and oppositely charged NPs. The lower ratio of intensities of the S:Au peaks in the top image and the higher ratio of the same peaks in the bottom image suggest that small "needles" are composed predominantly of NP-coated copper hydroxide, while large plates contain extra NP layers "glued" together by copper sulfate.

(typical dimensions, ~200 nm × 30 nm × 10 nm, Fig. 3d). The compositional scans of these structures (Fig. 3e) feature prominent peaks corresponding to copper (from salt) and gold (from NPs). Interestingly, in the case of thin needles or sheets (Fig. 3c,d), the sulfur peak is weak compared to the gold peaks, while for the large microplates (Fig. 3a, b), the opposite is true.

The marked differences in the morphologies of structures formed using $NiSO_4$ and $CuSO_4$ can be rationalized by the difference in solubilities of the nickel and copper salts. Specifically, $NiSO_4$ is highly soluble ($K_{SP} = 25$ M^2) and does not precipitate/crystallize during self-assembly of the NPs into supraspheres—as discussed above, the major role of the salt in the system is to form bridges between MUA particles. The mechanism of the self-assembly becomes more complex with $CuSO_4$. While this salt is, in itself, readily soluble, an equilibrium between $CuSO_4$ and poorly soluble $Cu(OH)_2$ is established at basic pH values such as those we used here. The key phenomenon during our SA experiments is therefore that copper hydroxide can crystallize first to form small "seeds" onto which the NPs adsorb. To support this hypothesis, we performed a series of experiments where the amount of TMAOH was kept constant, $n_{base} = 1.6$ μmol, while the amounts of copper sulfate varied, $n_{CuSO4} = 0.6$, 1.0, 1.4, and 2 μmol. At the lowest concentration of copper sulfate, all salt reacted with an excess of base and formed copper hydroxide, which crystallized in the form of thin sheets covered with only a thin film of the NPs (similar to Fig. 3d and Fig. 4a). The EDS spectra evidenced that the ratio of intensities of the sulfur and gold peaks was low. Importantly, its value agreed with the expected ratio of sulfur in the thiolate SAMs to the gold in the NP cores—in other words, the only source of sulfur peaks was from the SAMs and there was no excess sulfur present from $CuSO_4$. When concentration of $CuSO_4$ was increased, not all of it reacted with the base and the proportion of unreacted $CuSO_4$ to the $Cu(OH)_2$ "seeds" increased. As a result, more Cu^{2+} cations remained in solution and were able to bridge[29] more NPs adsorbing onto the $Cu(OH)_2$ seeds (with the SO_4^{2-} counterions contributing to the increased intensity of the EDS sulfur peaks). Consequently,

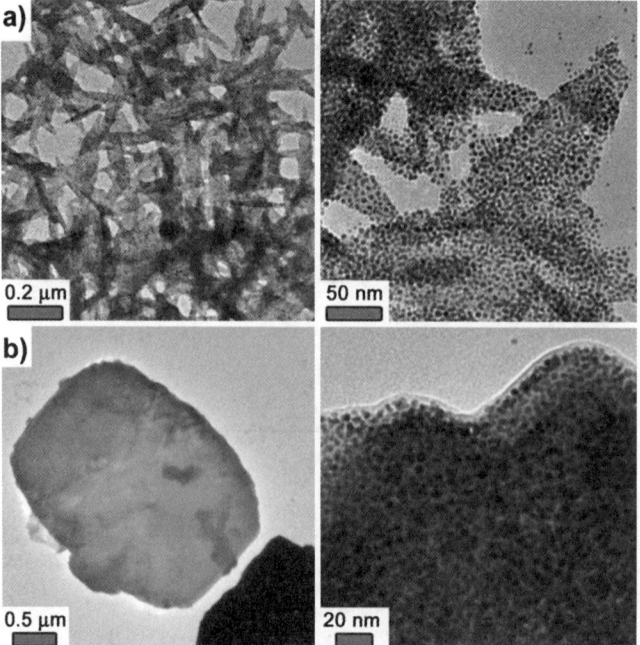

Fig. 4 TEM images of $Cu(OH)_2$ structures coated with (+)/(−) nanoparticles: (a) thin sheets coated with NP monolayers; (b) nanoplates coated with multiple layers of NPs.

the assemblies grew into larger nanoplates covered with multiple layers of nanoparticles. The formation of NP multilayers was confirmed by TEM imaging (Fig. 3a and Fig. 4b) and the presence of copper sulfate on the copper hydroxide "seeds" was evidenced by an increased intensity of sulfur and copper peaks relative to gold, as measured by the EDS, Fig. 3e.

It is interesting to compare and contrast our NP/Cu^{2+} structures with copper hydroxide[31] and copper oxide[32] microstructures prepared by other groups via surfactant-assisted, shape-dependent syntheses. In particular, copper hydroxide microplates with shapes similar to those we obtained were recently reported[31] via a procedure involving copper chloride in the presence of bis-(amidoethyl-carbamoylethyl) octadecylamine as a surfactant and ammonium chloride as a shape modifying agent. Interestingly, when this experiment was performed in the absence of surfactant, spherical $Cu(OH)_2$ nanoparticles (~10 nm in diameter) were obtained, which then grew into irregularly-shaped particles several hundreds of nanometers large. In this context, our current results suggest that oppositely charged nanoparticles can act analogously to organic surfactants by stabilizing crystalline nuclei and modulating the growth of microcrystals. The key difference, however, between small-molecule surfactants and our NPs is that the latter benefit from the polyvalent nature of interactions between the crystals and the multiple ligands immobilized onto the NPs—if these ligands (or their structural analogues) are not tethered onto nanoparticles and are free in solution, they are not capable of regulating the growth of crystals studied here or in our previous paper.[24] This is so because adsorption of individual surfactants onto a surface of a growing microcrystal is entropically unfavorable, with penalty for the translational degrees of freedom of each adsorbed species approximately $kT\ln(Ad/V)$, where A is the total area available to the adsorbate, d is the size of the adsorbate, and V is the solution volume. Consequently, interactions between the adsorbing molecules and the crystal surface must be sufficiently strong to yield an overall favorable/negative free energy of adsorption. In contrast, the entropic penalty is minimal for ligands already preorganized on the NPs. Even though the MUA or TMA ligands are, individually, not necessarily strong and specific binders to salt crystals, their large numbers on the NPs act in concert and add up to appreciable free energies of adsorption onto various surfaces (not only crystals but also glasses, polymers or even metals, for details, see ref. 18 and 33). This process is further aided by the electrostatic attractions between the (+) and the (−) NPs.[34]

Naturally, it would be desirable to extend the NP-surfactant methods to systems that are not necessarily based on charge–charge interactions and to organic microcrystals. In doing so, the attractive NP(+)/NP(−) interactions need to be replaced by other types of interactions capable of harboring the NPs onto the crystals. While the search for such systems is only in its preliminary stages, we have been able to demonstrate the proof-of-the-concept experiments using nanoparticles covered with 1,6-difluoro-4-mercaptophenol (DFMP) ligands (Fig. 5a) and donor–acceptor organic crystals composed of 1,5-bis[2-(2-hydroxyethoxy)ethoxy]naphthalene (donor, D) and pyromellitic diimide (acceptor, A) (Fig. 5b, details on the synthesis and properties of such crystals will be published separately). These molecules form D–A stacks along the crystal, exposing EG-chains and imide groups on the crystal's surface. We designed the SAMs on the NPs in such a way that adsorption of particles onto the D–A crystals would occur through the formation of hydrogen bonds between phenol moieties on the NPs' surfaces and both the EG units and the carbonyl oxygens of the donor and acceptor molecules, respectively. At the same time, the relatively small thickness of the DFMP SAMs on the NP surfaces (~0.6 nm) was beneficial since it translated into strong attractive interactions between the particles (mostly due to van der Waals attractions between the Au cores[15])—this allowed for dense packing of NPs on the surfaces of the organic crystals. The images in Fig. 5c and 5d illustrate the NP coatings formed on both macroscopic as well as microscopic donor–acceptor crystals.

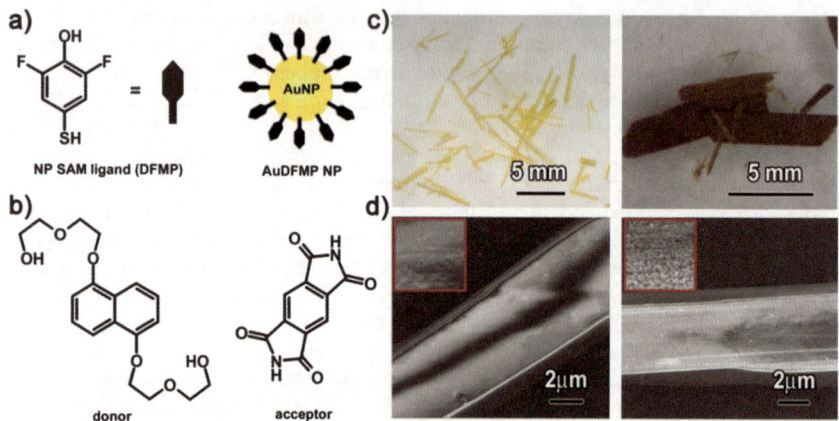

Fig. 5 Structures of (a) 2,6-difluoro-4-mercaptophenol, DFMP, used to form SAMs on AuNPs. (b) Molecules crystallizing into donor–acceptor organic crystals. (c) Optical and (d) SEM images of D–A crystals before (left) and after (right) coating them with AuDFMP NPs.

In summary, we demonstrated a cation-selective self-assembly of oppositely charged nanoparticles as well as first experiments with NP coated organic crystals. One unique feature of these systems is that the ligands on the NPs can interact with the crystals polyvalently—in this way, even weakly binding ligands can turn the NPs into strongly binding crystal-growth modifiers. In the case of nESA, these effects are supplemented by cation-bridging interactions which further enrich the repertoire of composite structures that can be prepared by co-crystallizing salts and nanoparticles. One appealing avenue for future research might be to combine the catalytic properties of the crystalline nanoseeds (e.g., $Cu(OH)_2$ catalyzes oxidation of phenols by hydrogen peroxide[31]) with the selective substrate permeability[35] of the NPs coating these seeds.

References

1 D. L. Feldheim and C. D. Keating, *Chem. Soc. Rev.*, 1998, **27**, 1–12.
2 S. J. Oldenburg, R. D. Averitt, S. L. Westcott and N. J. Halas, *Chem. Phys. Lett.*, 1998, **288**, 243–247.
3 S. H. Sun, *Adv. Mater.*, 2006, **18**, 393–403.
4 H. Zeng, J. Li, J. P. Liu, Z. L. Wang and S. H. Sun, *Nature*, 2002, **420**, 395–398.
5 J. H. Fendler, *Chem. Mater.*, 2001, **13**, 3196–3210.
6 L. S. Zhong, J. S. Hu, H. P. Liang, A. M. Cao, W. G. Song and L. J. Wan, *Adv. Mater.*, 2006, **18**, 2426.
7 J. M. Perez, L. Josephson, T. O'Loughlin, D. Hogemann and R. Weissleder, *Nature Biotechnology*, 2002, **20**, 816–820.
8 A. M. Cao, J. S. Hu, H. P. Liang and L. J. Wan, *Angew. Chem., Int. Ed.*, 2005, **44**, 4391–4395.
9 A. M. Kalsin, M. Fialkowski, M. Paszewski, S. K. Smoukov, K. J. M. Bishop and B. A. Grzybowski, *Science*, 2006, **312**, 420–424.
10 Z. Y. Tang, Z. L. Zhang, Y. Wang, S. C. Glotzer and N. A. Kotov, *Science*, 2006, **314**, 274–278.
11 S. Srivastava, A. Santos, K. Critchley, K. S. Kim, P. Podsiadlo, K. Sun, J. Lee, C. L. Xu, G. D. Lilly, S. C. Glotzer and N. A. Kotov, *Science*, 2010, **327**, 1355–1359.
12 Y. H. Wei, K. J. M. Bishop, J. Kim, S. Soh and B. A. Grzybowski, *Angew. Chem., Int. Ed.*, 2009, **48**, 9477–9480.
13 M. A. Olson, A. Coskun, R. Klajn, L. Fang, S. K. Dey, K. P. Browne, B. A. Grzybowski and J. F. Stoddart, *Nano Lett.*, 2009, **9**, 3185–3190.
14 K. Liu, Z. H. Nie, N. N. Zhao, W. Li, M. Rubinstein and E. Kumacheva, *Science*, 2010, **329**, 197–200.
15 K. J. M. Bishop, C. E. Wilmer, S. Soh and B. A. Grzybowski, *Small*, 2009, **5**, 1600–1630.

16 A. M. Kalsin, B. Kowalczyk, P. Wesson, M. Paszewski and B. A. Grzybowski, *J. Am. Chem. Soc.*, 2007, **129**, 6664.
17 A. M. Kalsin, B. Kowalczyk, S. K. Smoukov, R. Klajn and B. A. Grzybowski, *J. Am. Chem. Soc.*, 2006, **128**, 15046–15047.
18 S. K. Smoukov, K. J. M. Bishop, B. Kowalczyk, A. M. Kalsin and B. A. Grzybowski, *J. Am. Chem. Soc.*, 2007, **129**, 15623–15630.
19 D. A. Walker, K. P. Browne, B. Kowalczyk and B. A. Grzybowski, *Angew. Chem., Int. Ed.*, 2010, **49**, 6760–6763.
20 D. A. Walker, C. E. Wilmer, B. Kowalczyk, K. J. M. Bishop and B. A. Grzybowski, *Nano Lett.*, 2010, **10**, 2275–2280.
21 I. Lagzi, B. Kowalczyk, D. W. Wang and B. A. Grzybowski, *Angew. Chem., Int. Ed.*, 2010, **49**, 8616–8619.
22 D. A. Walker, B. Kowalczyk, M. O. de la Cruz and B. A. Grzybowski, *Nanoscale*, 2011, **3**, 1316–1344.
23 D. Witt, R. Klajn, P. Barski and B. A. Grzybowski, *Curr. Org. Chem.*, 2004, **8**, 1763–1797.
24 B. Kowalczyk, K. J. M. Bishop, I. Lagzi, D. W. Wang, Y. H. Wei, S. B. Han and B. A. Grzybowski, *Nat. Mater.*, 2012, **11**, 227–232.
25 A. M. Kalsin and B. A. Grzybowski, *Nano Lett.*, 2007, **7**, 1018–1021.
26 B. Kowalczyk, A. M. Kalsin, R. Orlik, K. J. M. Bishop, A. Z. Patashinskii, A. Mitus and B. A. Grzybowski, *Chem.–Eur. J.*, 2009, **15**, 2032–2035.
27 R. Klajn, K. J. M. Bishop, M. Fialkowski, M. Paszewski, C. J. Campbell, T. P. Gray and B. A. Grzybowski, *Science*, 2007, **316**, 261–264.
28 R. Klajn, K. J. M. Bishop and B. A. Grzybowski, *Proc. Natl. Acad. Sci. U. S. A.*, 2007, **104**, 10305–10309.
29 D. W. Wang, B. Tejerina, I. Lagzi, B. Kowalczyk and B. A. Grzybowski, *ACS Nano*, 2011, **5**, 530–536.
30 When the pH of NP-salt mixture was raised above ~8 by addition of TMAOH flower-like $Ni(OH)_2$ microcrystals and NPs formed separately and precipitated (see Fig. 2c). Since the solubility of $Ni(OH)_2$ is 2–3 orders of magnitude higher than that of copper hydroxide (pK_{SO} $Ni(OH)_2$ = 17.2, pK_{SO} $Cu(OH)_2$ = 19.7), nickel salts can be stable in solution for longer times during crystallization, screening the interactions between NPs and causing random precipitation of NPs. The excess of salt precipitates toward the end of the evaporation in the form of "wavy" flower-like microstructures.
31 G. H. Lin, W. F. Jia, W. S. Lu and L. Jiang, *J. Colloid Interface Sci.*, 2011, **353**, 392–397.
32 X. D. Liang, L. Gao, S. W. Yang and J. Sun, *Adv. Mater.*, 2009, **21**, 2068–2071.
33 S. Huda, S. K. Smoukov, H. Nakanishi, B. Kowalczyk, K. Bishop and B. A. Grzybowski, *ACS Appl. Mater. Interfaces*, 2010, **2**, 1206–1210.
34 K. V. Tretiakov, K. J. M. Bishop, B. Kowalczyk, A. Jaiswal, M. A. Poggi and B. A. Grzybowski, *J. Phys. Chem. A*, 2009, **113**, 3799–3803.
35 B. Kowalczyk, I. Lagzi and B. A. Grzybowski, *Nanoscale*, 2010, **2**, 2366–2369.

Real-space studies of the structure and dynamics of self-assembled colloidal clusters

Rebecca W. Perry,[a] Guangnan Meng,[b] Thomas G. Dimiduk,[b] Jerome Fung[b] and Vinothan N. Manoharan*[ab]

Received 2nd April 2012, Accepted 6th June 2012
DOI: 10.1039/c2fd20061a

The energetics and assembly pathways of small clusters may yield insights into processes occurring at the earliest stages of nucleation. We use a model system consisting of micrometer-sized, spherical colloidal particles to study the structure and dynamics of small clusters, where the number of particles is small ($N \leq 10$). The particles interact through a short-range depletion attraction with a depth of a few $k_\text{B}T$. We describe two methods to form colloidal clusters, one based on isolating the particles in microwells and another based on directly assembling clusters in the gas phase using optical tweezers. We use the first technique to obtain ensemble-averaged probabilities of cluster structures as a function of N. These experiments show that clusters with symmetries compatible with crystalline order are rarely formed under equilibrium conditions. We use the second technique to study the dynamics of the clusters, and in particular how they transition between free-energy minima. To monitor the clusters we use a fast three-dimensional imaging technique, digital holographic microscopy, that can resolve the positions of each particle in the cluster with 30–45 nm precision on millisecond timescales. The real-space measurements allow us to obtain estimates for the lifetimes of the energy minima and the transition states. It is not yet clear whether the observed dynamics are relevant for small nuclei, which may not have sufficient time to transition between states before other particles or clusters attach to them. However, the measurements do provide some glimpses into how systems containing a small number of particles traverse their free-energy landscape.

1 Introduction

A nucleus growing in a bulk fluid must overcome a number of challenges to become a crystal. The most well-known of these is its high surface area-to-volume ratio, which makes it prone to melting or evaporating back into the fluid. Rarely do nuclei grow to the critical size at which they are no longer unstable. A more subtle challenge arises from the structure of the nucleus, which may differ from that of the final crystal. In this case the nucleus must rearrange in order to become a bulk crystallite, and it must do so on a timescale smaller than that at which new particles attach. If the dynamics of rearrangement are slow, as might happen in a deeply quenched system, growth leads to metastable, disordered structures.[1,2]

These challenges illustrate the complex coupling between energetics, structure, and dynamics that makes nucleation a difficult process to study experimentally. Colloidal systems offer several advantages over molecular systems for such studies: the interparticle energies can be controlled using model attractive interactions such

[a] Harvard School of Engineering and Applied Sciences, Harvard University, Cambridge MA 02138, USA. E-mail: vnm@seas.harvard.edu; Fax: +1 617 495 0416; Tel: +1 617 495 3893
[b] Department of Physics, Harvard University, Cambridge MA 02138, USA

as the depletion force; the structure of the suspension can be studied in real-space, at the single-particle level, using optical or confocal microscopy;[3–5] and the dynamics can be made slow enough to allow the growth of nuclei to be studied in detail.[6,7] But even in colloids it is difficult to observe the embryonic stages of nucleation, when the nuclei are clusters rather than crystallites, and successful nucleation may hinge on a structural transition. The main source of difficulty is the disparity between the rate of cluster formation and the rate of rearrangement, which can differ by orders of magnitude. This makes it nearly impossible to find a cluster—the formation of which is a rare event that can occur anywhere in the bulk—and simultaneously observe its structural transitions. Furthermore, common three-dimensional (3D) microscopy techniques are not fast enough to image the rearrangements of a cluster on timescales short compared to the rotational and translational diffusion time of a nucleus. Thus only the late stages of growth have been investigated in 3D colloidal systems, and the early stages remain elusive.

Here we describe a different approach to addressing these challenges: we study the structure and dynamics of the clusters themselves. To avoid the problem of finding a cluster in the bulk fluid, we localize its assembly in either lithographically-prepared microwells that contain only a small number ($N \approx 10$) of colloidal particles or by using optical tweezers to collect several particles from a dilute gas phase (Fig. 1). We also use a fast 3D imaging technique, holographic microscopy, to capture the structural rearrangements of these colloidal clusters on short timescales.

These experiments do not directly probe nucleation, since the clusters are in a state of artificial isolation:[8] they are either walled off from the bulk fluid or placed in a suspension too dilute to favor growth. Nonetheless, the experiments provide information critical to understanding nucleation and growth, such as the rearrangement timescales and probabilities of obtaining clusters with symmetries that differ from the bulk. A previous article[9] by our group examined the energy landscape and equilibrium probabilities for small clusters ($N < 10$) in detail. Here we expand on these results by presenting (a) the chemical techniques required to control the interparticle interactions and assemble colloidal clusters; (b) a new method to image transition states and rearrangement dynamics of clusters in 3D; and (c) data on the structure and dynamics of such systems for different types of depletion interactions. Although much remains to be done to relate this type of data to bulk nucleation experiments, the results show that all three of the aspects fundamental to nucleation—energetics, structure, and dynamics—can be measured in detail through an approach combining synthesis, fabrication, and modern optical techniques.

2 Background

The central theoretical concept behind our experimental study is the free-energy landscape, a multidimensional surface characterizing the free energy of a system of N particles as a function of all of their configurational degrees of freedom. Understanding the landscape entails mapping out the minima, which represent stable clusters, and the transition pathways between them. Recent theoretical work has shown that the minima of the landscape can be enumerated exhaustively for a small number of hard spherical particles interacting through a short-range attraction. "Short" means that the width of the potential well is much smaller than the radius of the spheres. This limit permits a geometrical solution to the problem of enumerating the minima: the stable clusters must be rigid, or isostatic, sphere packings where the number of contacts, or "bonds," is at least $3N − 6$. To a first approximation, the potential energy of such clusters is proportional to the total number of bonds. Geometrical solutions have enumerated all possible clusters and their energies up to $N = 11$.[10–14] The transition pathways are now beginning to be enumerated through a combination of theory and simulation.[15]

Creating and observing clusters with such short-range interactions in an experimental system requires careful design. We work with dilute colloids to obtain the

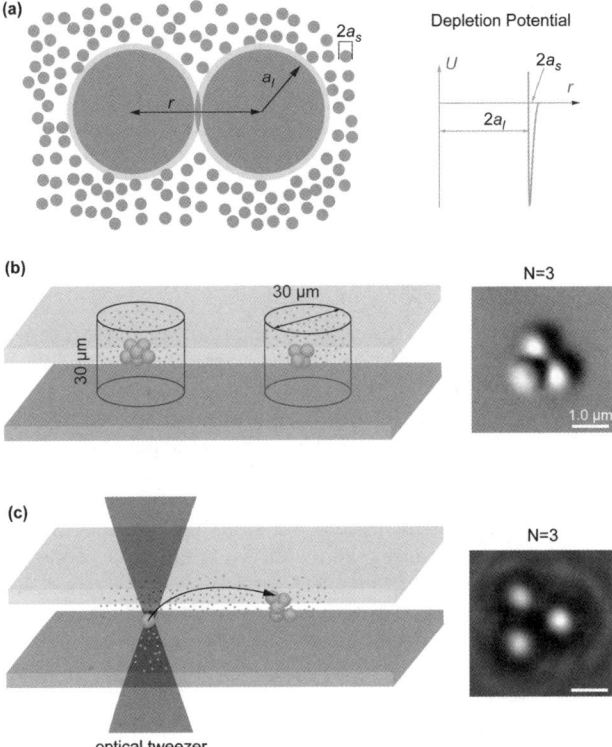

Fig. 1 Experimental systems. (a) Two large spherical particles feel a mutual attraction when they come within a small sphere (depletant) diameter of each other. The width of the depletion interaction potential is much smaller than the large sphere size. (b) To self-assemble clusters of spheres, we deposit small volumes of dilute colloidal suspension into microwells. Within 24 h, clusters form. The image at the right is an optical micrograph (differential interference contrast) of a three particle cluster. (c) A second method of assembling clusters uses an optical tweezer to bring several particles together. Once the desired number of particles is reached, the trap is turned off. The image at the right is an optical micrograph (bright field) of a three particle cluster. Cartoons of clusters in (b) and (c) are not drawn to scale. Micrograph scale bars, 1.0 μm.

clearest possible images of clusters. To favor aggregation in such systems, the attractive interaction between colloidal particles has to be several $k_B T$ deep. At the same time, the binding between colloidal particles has to be reversible. If the particles become stuck together by a strong attractive potential such as the van der Waals interaction, the cluster will not be able to rearrange on experimental timescales.

We therefore use a depletion attraction, a weak entropic interaction in a binary colloidal mixture, as the driving force to assemble colloidal particles into clusters. As Fig. 1 (a) shows, the larger particles experience an effective attraction because the entropy of the smaller spheres, or "depletants," is maximized when the excluded volumes of the larger spheres overlap. The depletion interaction between two large spheres can be modeled by Asakura-Oosawa theory:[16,17]

$$U_{AO}(r) = -k_B T \frac{\pi}{6} \rho_s (2a_s + 2a_l - r)^2 \left(2a_s + 2a_l + \frac{r}{2}\right) \quad (1)$$

where r is the center-to-center distance between the two large spheres, a_s and a_l are the radii of small and large spheres, and ρ_s is the number density of small spheres in the solvent. The range of U_{AO} is approximately the diameter of the depletants, $r < 2a_l + 2a_s$; the minimum of a purely depletion potential occurs at contact, $r = 2a_l$:

$$U_{AO}(r = 2a_l) = -k_B T \rho_s \frac{\pi(2a_s)^3}{6}\left(1 + \frac{3a_l}{2a_s}\right) \approx -(k_B T) 2\pi \rho_s a_s^2 a_l \text{ when } a_s \ll a_l. \quad (2)$$

From eqn (2), when two types of colloidal particles, 1.0 μm particles ($a_l = 500$ nm) and 100 nm depletants ($a_s = 50$ nm, $\phi_s = \rho_s \pi(2a_s)^3/6 = 20\%$), are mixed together, the attractive potential between large particles has a well depth of about $3k_B T$ at contact and a range of 100 nm. At very small separations, the van der Waals force might cause the large particles to stick irreversibly to one another, but this can be prevented by using particles with an electrostatic double layer. The range of the electrostatic repulsive barrier can be tuned through the salt concentration.

The Asakura-Oosawa model for the depletion potential assumes that the small particles reach equilibrium instantaneously as the large particles move, whereas in reality the depletion potential takes some time to saturate due to the finite diffusivity of the depletants. Using theoretical results from Vliegenthart and van der Schoot,[18] we estimate that for the depletants used in our study the potential saturates on timescales orders of magnitude smaller than the diffusion timescale of the large particles and the observed rearrangement timescales for our clusters (section 4.3). Thus the kinetics of the depletants should not significantly affect the dynamics of the clusters. However, this approximation may break down for larger depletants or smaller particles.

3 Experimental

3.1 Colloidal system

Our system consists of negatively-charged polystyrene (PS) microspheres, approximately 1 μm in diameter, and either poly(N-isopropylacrylamide) (PNIPAM) particles or sodium dodecyl sulfate (SDS) micelles as depletants. The sizes of the depletants are chosen so that the range of the depletion attraction is less than 10% of the diameter of the PS particles, so that the attraction is strictly pairwise additive.[19] Whereas the micelles are self-assembled in solution, the PNIPAM particles are synthesized beforehand and added to the suspension. In both systems the depletant scatters negligibly, allowing us to obtain clear images of the clusters through microscopy or scattering. The refractive index of PNIPAM closely matches that of our solvent, water, so the PNIPAM particles are optically transparent in aqueous solution. Furthermore, the strength of the depletion interaction can be easily controlled in both systems simply by modifying the concentration of depletants.

Using PNIPAM spheres allows us to tune the strength and range of the depletion interaction *in situ*. The PNIPAM depletants shrink by 50% in diameter when they are heated above their lower critical solution temperature, around 30° C. This results in a reduction in the magnitude of the interaction strength by a factor of four, according to eqn (2).

We use precipitation polymerization[20,21] to synthesize 80 nm PNIPAM hydrogel particles. The reactor includes a 250 ml three-necked round bottom flask, a magnetic stirrer, a reflux condenser and a nitrogen gas inlet. We dissolve 2.0 g N'-isopropylacrylamide (NIPAM, monomer, 99%, Acros Organics), 0.1 g N,N'-methylenebisacrylamide (BIS, crosslinker, 99%, Promega), and 0.18 g (≈ 6 mM) sodium dodecyl sulfate (SDS, 99%, EMD Chemicals) in 93 ml deionised (DI) water (Milli-Q synthesis grade, Millipore) under gentle stirring. The solution is then heated to 70° C and bubbled with nitrogen for 1 h. To start the polymerization, we inject 40 mg potassium persulfate (KPS, initiator, 99%, Acros Organics) dissolved in

5 ml DI water. During the reaction, the solution is stirred with a magnetic stirrer at 300 rpm and bubbled with nitrogen. After 4 h, the reaction is stopped by cooling the reactor down to room temperature, and the PNIPAM product is collected. To remove unreacted monomer, initiator, and surfactant molecules from the solution, we dialyze the PNIPAM product against DI water for seven days, exchanging DI water every 24 h.

The hydrodynamic radius of the PNIPAM particles is 80 nm at 20° C and 46 nm at 40° C with a lower critical solution temperature around 33° C, as measured by dynamic light scattering (Zetasizer Nano ZS, Malvern Instruments) and shown in Fig. 2. The polydispersity of the particles is less than 5% at all measured temperatures. The weight concentration of particles in the stock solution is 2.13% w/w, as measured by thermogravimetric analysis (TGA, Q5000IR, TA Instruments).

For the microwell experiments described below, we prepare a suspension of 1.0-μm-diameter sulfate latex PS particles (Batch# 2090,1, Invitrogen Molecular Probes, polydispersity (standard deviation in particle diameter) 3%) and 80 nm PNIPAM hydrogel particles in water. For fluorescence microscopy, we use 1.0 μm sulfate fluorescent latex PS particles (FluoSpheres sulfate microspheres, 1.0 μm, red fluorescent (580/605), Invitrogen, polydispersity 5%). The volume fraction of PS particles is 10^{-5}, and the concentration of PNIPAM is 1.0% w/w (estimated volume fraction $\phi_s \approx 0.25$ at 20° C). 15 mM NaCl are added to screen the long-ranged electrostatic repulsion between the PS particles. 0.1% w/w Pluronic P123 (BASF) surfactant is also added to stabilize the PS particles in the salt solution. This procedure ensures that the depletion attraction between PS particles induced by PNIPAM can be reversed by either diluting the PNIPAM particles or by increasing the temperature, thereby shrinking the PNIPAM particle size.

For bulk experiments, we load the PS/PNIPAM suspension (with PS volume fraction 4×10^{-3}) directly into sandwiched glass cover slips through capillary action. The cover slips are separated by 40 μm thick Mylar®A spacers (DuPont Teijin Films) to provide the same thickness across the samples. The edges of the glass cell are sealed with optical glue (NOA-61, Norland Products Inc.) to prevent evaporation.

For the optical-tweezer-assisted assembly method, we prepare a suspension of 1.3-μm-diameter PS particles (Batch #1279,1, Invitrogen Molecular Probes, Surfactant-Free White Sulfate Latex, polydispersity 2.7%), DI water, 5 mM NaCl (EMD, assay (dry basis) 99.0%), and 40 mM SDS (Sigma Aldrich, 99.0%). The volume fraction of

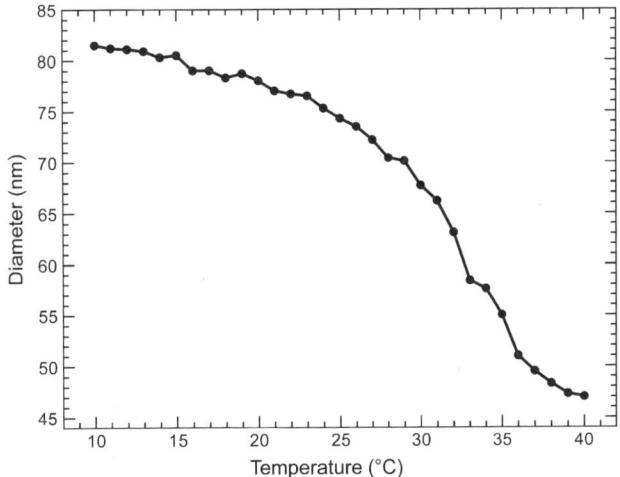

Fig. 2 Temperature dependence of the diameter of the PNIPAM hydrogel depletants.

PS particles is 10^{-6}, on the order of one particle per 100 μm cube. The concentration of SDS is well above the critical micelle concentration, so the surfactant molecules assemble into micelles that act as depletants.[4] Following the analysis of Iracki et al.,[22] we estimate the width of the SDS induced depletion potential, which includes a factor proportional to the Debye length in addition to the physical size of a micelle, to be approximately 30 nm. In terms of the width of the depletion potential, the effective micelle radius, a_s, is 15 nm. Samples are prepared in cells consisting of a glass slide (25 × 76 mm, VWR) and a No. 1 cover slip (22 × 22 mm, VWR). The slide and cover slip are rinsed with DI water and dried with nitrogen before use. We use UV curing epoxy (NOA-61, Norland Products Inc.) to secure 100 μm thick strips of Mylar®A (DuPont Teijin Films) as spacers between the slide and cover slip. After using capillary action to fill the sample chamber with suspension, we seal the sample cell with epoxy (Devcon 5 Minute Epoxy) to prevent evaporation.

3.2 Formation of clusters

We prepare clusters either by letting them self-assemble in lithographically patterned microwells or by bringing particles together in a dilute suspension using an optical tweezer.

3.2.1 Microwell method.
Colloidal clusters can be assembled under equilibrium conditions in microwells, as shown in Fig. 1(b). Since the purpose of the microwells is to isolate a set of particles, and not to confine them, we work at small PS volume fractions such that the volume of each well is 10^5 times the volume of the particles. This ensures that formation of clusters is driven by the attraction between particles rather than by a confinement effect. The solution conditions are the same for every microwell in the plate. Because each plate has tens of thousands of microwells, a single plate yields enough samples to determine ensemble probabilities of cluster structures at small N.

3.2.1.1 Microwell fabrication. Microwell array plates are fabricated using soft lithography.[23] We use standard photolithography procedures to make a master mold of SU-8 with the microwell pattern on the wafer. We first design a photomask pattern using AutoCAD (Autodesk Inc.). The pattern (20 × 20 mm) has an array of circles 30 μm in diameter with a pitch of 60 μm on a square lattice. The pattern is printed on a photomask transparency at 20,000 dpi resolution by CAD/Art Services, Inc. We then spin coat SU-8 photoresist (SU-8 3035, MicroChem Corp.) onto a silicon wafer (University Wafer) at 3000 rpm, setting the thickness of the SU-8 layer at 35 μm. The microwells are made by replica molding on the SU-8 master. We prepare a pre-gel solution by dissolving 10% w/w acrylamide (monomer, 99%, Promega), 0.5% w/w N,N'-methylenebisacrylamide (crosslinker, 99%, Promega), 0.5% w/w allylamine (copolymer, 98%, Alfa Aesar) and 0.1% w/w DAROCUR 1173 (photoinitiator, Ciba Specialty Chemicals Inc.) in DI water. The pre-gel solution is poured onto the SU-8 master mold and covered by a silanized cover slip (see below), which later becomes the bottom "window" of the microwell, through which the clusters can be viewed using an inverted microscope. The solution is placed 10 cm from an UV lamp (B-100YP, UVP) for 10 min to polymerize the hydrogel. The polymerized microwell plate is carefully separated from the SU-8 master mold, rinsed with DI water, and stored in DI water.

3.2.1.2 Microwell functionalization. Because the depletion attraction causes particles not only to stick to one another, but also to the walls of the microwells, we attach similar PNIPAM particles to the microwell walls and glass surfaces that bound the wells (Fig. 3). This matches the roughness of the surface to the scale of the depletants, which has been shown to minimize the depletion interaction between large particles and surfaces.[24] We synthesize a separate batch of poly(N'-isopropylacrylamide-*co*-acrylic acid) (PNIPAM-*co*-AAc) hydrogel particles for this purpose. The PNIPAM-*co*-AAc hydrogel particles are synthesized using the same procedure

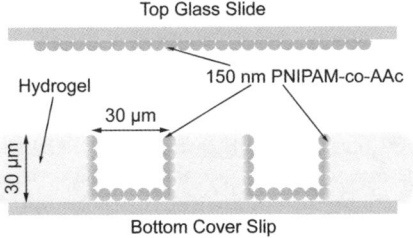

Fig. 3 Schematic of microwells and glass slides, shown from the side. All surfaces are saturated with 150 nm PNIPAM-*co*-AAc particles (not drawn to scale) to suppress the depletion attraction between the PS particles and the surfaces.

as the 80 nm PNIPAM particles, except that we add 200 mg acrylic acid (99%, Sigma) to the reacting solution. The hydrodynamic diameter of these PNIPAM-*co*-AAc particles is 150 nm at 20° C and 60 nm at 40° C. These particles are attached to the microwell boundaries and glass surfaces using silane chemistry. First the glass surfaces, either precleaned No. 1 cover slips (24 × 30 mm, VWR) or precleaned glass slides (25 × 75 mm, VWR), are silanized. Cover slips are silanized in 1.0% w/w 3-methacryloxypropyltrimethoxysilane (98%, Sigma) in anhydrous ethanol solution for 24 h at room temperature. Glass slides are immersed in 1.0% w/w (3-Aminopropyl)triethoxysilane (98%, Sigma) in anhydrous ethanol solution for 24 h at room temperature. Then the cover slips and glass slides are rinsed with anhydrous ethanol and blow-dried with compressed dry air. The silanization is completed by leaving the cover slips and glass slides in an oven at 110° C for 1 h.

The cover slips form the bottom windows of the microwell plates, while the glass slides are coated with PNIPAM-*co*-AAc particles and used to cover the tops of the wells. To coat the microwells and silanized slides with particles, we immerse them in a dialyzed colloidal suspension of 150 nm PNIPAM-*co*-AAc particles for 24 h at room temperature. The amine groups on the surfaces of the glass slide and acrylamide hydrogel microwells slowly bind with the carboxylic acid groups in the PNIPAM-*co*-AAc particles. Afterwards the PNIPAM-*co*-AAc hydrogel particles are irreversibly adsorbed onto the surfaces. After this surface treatment, we are able to form 3D colloidal clusters of PS spheres in the middle of the microwells. Without the surface treatment, PS spheres form two-dimensional (2D) crystallites on the boundaries of the microwells.

3.2.1.3 Sample preparation. Once the microwells are prepared and functionalized, we load them with the PS/PNIPAM suspension described in section 3.1. The hydrogel microwell plate and glass slides are first rinsed with about 100 μL PS/PNIPAM suspension at least five times, so that the hydrogel plate has the same ionic and surfactant concentration as the suspension. After the last rinse, the wells are filled with the suspension, and the microwell plate and glass slide are sealed with epoxy (Devcon 5 Minute Epoxy) around the edges of the cover slip. We find that the number of particles per well is randomly distributed with a mean of about ten. Before putting the sample on the optical microscope for observation and counting, we wait 24 h for the system to reach equilibrium at 22.0 ± 1.0° C. Because the hydrogel microwells tend to deform ten days after sample preparation, the observations and data collection are done within seven days of fabrication, and the sample is discarded afterward.

3.2.2 Assisted assembly of clusters by optical tweezers. In the optical tweezer method, we start with a slide of dilute colloidal suspension and assemble a cluster one particle at a time while observing the system with an optical microscope. The microscope is equipped with an optical trap formed by an 830 nm laser (Sanyo DL-8142-201, with Thorlabs TCM1000T temperature controller and LD1255 current controller) focused through a 60X, 1.2 NA Plan Apo water immersion objective (Nikon). To build a cluster, we start by bringing two PS particles into the optical trap where they form a depletion bond. Then we add particles one-by-one to the cluster until we reach the desired N, as illustrated in Fig. 1(c). Starting with individual particles ensures that none of the particles in the clusters are previously bonded or fused irreversibly. Once each particle is attached to the cluster by at least one bond, we turn the optical tweezer off. At this point, the cluster can explore its configurational space independent of any external potential.

3.3 Optical methods for observation

3.3.1 Optical microscopy. We use an inverted optical microscope (Eclipse TE-2000, Nikon Corp.) equipped with 40X dry (NA = 0.9) and 100X oil-immersion (NA = 1.4) objectives, Nomarski differential interference contrast, and epifluorescence to observe the structures of the colloidal clusters in microwells and the bulk phase behavior. A thermally insulated temperature controlled microscope stage (HSC-60, Instec Inc., with ±0.1° C temperature stability) controls the temperature of the sample during observation. The images and videos are recorded by digital cameras (2560 × 1920, Digital Sight DS-5Mc, Nikon Corp. for still images; and 720 × 720, 40 frames per second, EO-0312C, Edmund Optics for movies) onto a personal computer. For the colloidal clusters, we scan sequentially through the microwells and record videos of clusters in each before analyzing the data.

We resolve the 3D structures of colloidal clusters by scanning through the recorded videos frame by frame. Although the microscope captures a 2D image with narrow depth of field, the rotational motion of the clusters over time allows us to see all of the particles. We map the nearest neighbors for each particle by looking at the 2D image and following it as the structure rotates in 3D space. We then compare this data to the contact matrices or computer renderings of different finite sphere packings identified in theoretical work.[10]

All of the micrographs shown in this paper have been subjected to linear post-processing (brightness and contrast adjustments) to maximize clarity.

3.3.2 Digital holographic microscopy. To quantitatively image the 3D dynamics of the clusters, we use digital holographic microscopy, a fast 3D imaging technique. Our apparatus consists of a Nikon Eclipse TE2000 inverted microscope modified to use a 660 nm laser (Opnext HL6545MG with Stanford Research Systems LDC501 laser diode current and temperature controller) for illumination, as shown in Fig. 4(a). Two lenses expand and shape the laser beam so that a broad plane wave illuminates the sample as shown in Fig. 4(a) and (b). The typical laser power is around 50 mW. The light then scatters from a colloidal cluster in the sample cell (Fig. 4(b)). The interference pattern of the scattered light and transmitted beam is imaged by a 60X, 1.2 NA Plan Apo water immersion objective (Nikon) and magnified by a tube lens before being recorded on a Photon Focus MVD-1024E-160-CL-12 monochrome CMOS camera. In contrast to bright field microscopy techniques, the objective is intentionally defocused so that the focal plane lies 20 to 40 μm downstream of the object of interest. This allows us to better resolve the fringes in the interference pattern. We record the interference patterns at a rate of 100 frames per second with an exposure time of 15 μs for each frame. The images from the camera are sent through CameraLink cable to an EPIX PIXCI E4 frame grabber in a desktop personal computer, where they are recorded to disk.

As illustrated in Fig. 4(b), each 256 × 256-pixel interference pattern (or hologram) represents the scattering from all objects in a sample volume of approximately 30 × 30 × 130 μm, centered above the objective. To remove the effects of scattering from irregularities on the slide or optics, we record a background image with no spheres in the field of view, normalize both it and the hologram images to have a mean value of one, and divide the holograms by the background. The normalization procedure

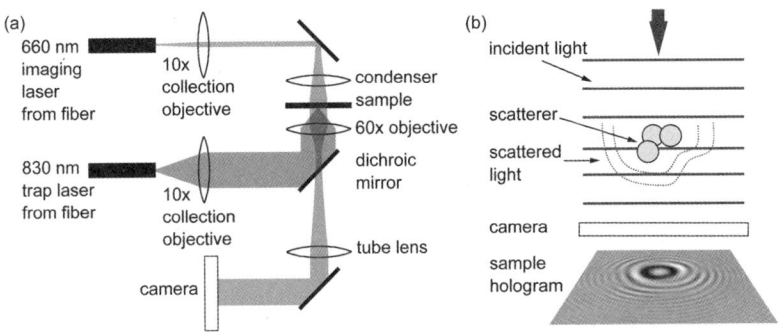

Fig. 4 Digital holographic microscopy measures the 3D positions of each particle in a cluster. (a) Schematic of apparatus. A 660 nm imaging laser illuminates the sample with a plane wave. A counter-propagating 830 nm laser creates an optical trap which is used to assemble the cluster and turned off during a dynamical measurement. (b) Diagram of hologram formation. A portion of the incident light scatters from a cluster of particles and interferes with the transmitted beam, producing a hologram that is captured by the camera.

allows us to compare our data to calculated holograms, which we also normalize to one. Background division removes irrelevant features from the data, making the interference fringes clearer.

Once the background is removed, we fit an exact scattering model to the holograms to determine the positions of the particles, following a technique originally developed for single spheres by Lee et al.[25,26] and later extended to multiple spheres by Fung et al.[27,28] We use a full multisphere scattering code, SCSMFO, that accounts for interference between the scattered waves, near-field coupling, and multiple scattering.[29] This allows us to correctly fit clusters with particles separated by less than a wavelength. We use the Levenberg-Marquardt algorithm to minimize the sum of the squared residuals between a recorded hologram and a hologram calculated from the scattering model. We fit our data using the open-source software package Holopy (http://www.launchpad.net/holopy), developed in our research group.

In our procedure, we assume nothing about the cluster geometry; instead, we fit for all $3N$ particle coordinates, plus an intensity scaling factor that accounts for variations in laser power from frame to frame. To reduce the number of free parameters in the fit, we assume a uniform particle size and refractive index. We use the particle diameter given by the manufacturer of the colloids, 1.3 μm, and a refractive index of 1.585[30] along with a small but nonzero imaginary part of the refractive index, $0.0001i$, to ensure that the scattering calculations converge. Once we determine the coordinates of all of the spheres by fitting, we classify the geometry of the cluster through 3D visualization or by calculating its second moment.

With $3N + 1$ parameters, the minimization problem is computationally complex. The algorithm will not converge to the actual particle positions unless we choose an initial guess for the particle positions that is close to the actual particle locations. We use two methods to generate initial guesses. The most convenient method is to use the particle positions found for the preceding or subsequent frame. This method works well at the high frame rates of our experiments, which ensure that the particles do not move far between frames. But in some cases, such as the first frame of a data series, we must guess the particle positions without any prior information. Thus we use a second method in which we determine approximate particle positions from a numerical reconstruction of a hologram.[31] Although near-field effects prevent reconstructions from providing accurate positions of particles spaced less than a wavelength apart,[32] reconstructing a hologram of a lone cluster still produces an image that resembles a bright field micrograph of the cluster. By reconstructing to various planes within the sample volume, we find a plane in which the particles are approximately in focus. From this image we can estimate the relative positions and connectivity of particles in the cluster, as described in section 3.3.1. This procedure generally yields a sufficiently precise initial guess for our fitting algorithm to converge. The resulting coordinates are then used to initialize the fit for the next frame.

To prevent the Levenberg-Marquardt algorithm from getting trapped in local minima, we allow small overlaps between particles. The algorithm either does not converge or converges to poor solutions when we impose a hard no-overlap condition. Instead, we allow the algorithm to place particles in positions that overlap up to 100 nm without any penalty. Allowing slight overlaps likely helps the fitter to avoid local minima that are due to "jammed" states, in which the most direct way for a particle to move to its true position is through another particle. Although the SCSMFO scattering calculations are not strictly defined for overlapping spheres, they nonetheless converge to within our desired numerical accuracy. When the algorithm attempts to place the spheres in positions with greater than 100 nm overlap, we calculate holograms of particles with reduced diameters such that the overlaps are entirely removed. Using holograms that assume the particles are smaller than their true size leads to a larger value of the objective function, effectively penalizing configurations with large overlaps. Allowing overlaps may also compensate for our assumption that the spheres are all exactly the same size.

4 Results and discussion

4.1 Interactions

If the interactions between particles are irreversible, kinetics rather than thermodynamics will govern the structures of the clusters that assemble. Because our goal is to understand the statistical mechanics and dynamics of clusters near equilibrium conditions, we first demonstrate that the interactions between clusters are reversible and well-controlled, a necessary prerequisite to further studies.

We first examine the bulk phase behavior of PS particles at a volume fraction of 4×10^{-3} and a constant temperature of 20° C. At low concentrations of PNIPAM particles, 0.6% w/w and smaller, the PS particles remain dispersed in a gas phase, and no aggregates or crystallites form even after two weeks. As the concentration of PNIPAM increases above 0.7% w/w, we observe quasi-2D crystals forming on the glass substrates, as shown in the optical micrographs in Fig. 5. The formation of quasi-2D crystals is likely due to the depletion attraction between PS particles and the planar surface, which, unlike our microwell devices, is not treated with a layer of PNIPAM-*co*-AAc particles. When the surface is treated to prevent binding between the PS and the glass, we observe gelation in the bulk at a concentration of 1% w/w PNIPAM. The concentration dependence of the bulk phase behavior confirms that the PS particles attract one another through the depletion forces induced by the PNIPAM particles, and that the interaction can be tuned by changing the concentration of PNIPAM depletants.

At a constant concentration of PNIPAM particles, 0.8% w/w, we find that varying the temperature from 20 to 26° C causes the crystals to sublimate, as shown in Fig. 6. The process is reversible: after the sample is cooled to room temperature, the crystals reform. The temperature dependence is due to the change in depletant diameter on approaching the lower critical solution temperature (LCST) of the PNIPAM polymer. As shown in eqn (2), the magnitude of the depletion potential depends quadratically on the size of the depletants, $U \propto a_s^2$ for constant number density ρ_s. Since the PNIPAM hydrogel particles change their sizes from $a_s = 40$ nm at 20° C to 20 nm at 40° C, the corresponding depletion potential decreases by about a factor of four over the same range. The quadratic dependence of the depletion potential on the PNIPAM size means that a relatively small change in the depletant diameter can have a large effect on the potential and can easily shift the system out of the gas-solid coexistence regime.

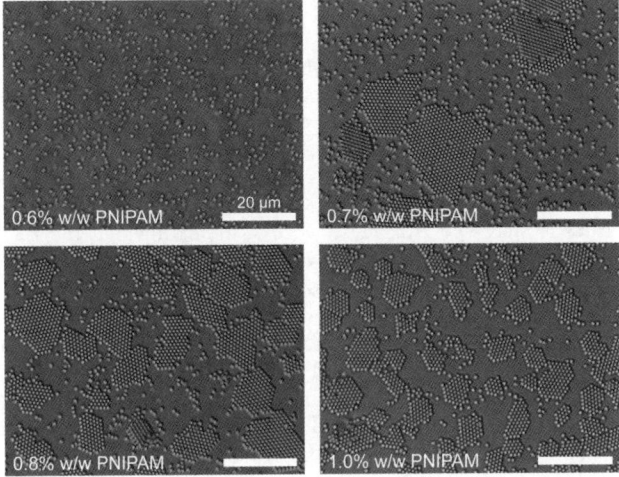

Fig. 5 Optical micrographs of 1.0 μm PS colloidal particles mixed with varying concentrations of PNIPAM particles (not visible under optical microscopy). All samples are at 20° C. PS particles form a gas phase at low PNIPAM concentration and a crystalline phase at higher PNIPAM concentration. Scale bar, 20 μm.

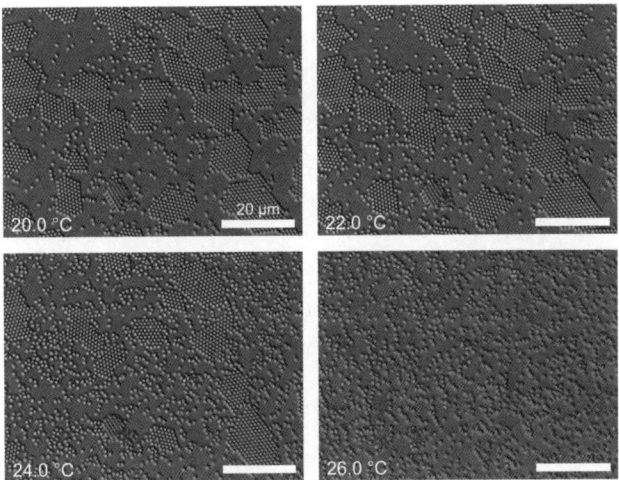

Fig. 6 Optical micrographs of a mixture of 1.0 μm PS colloidal particles and 80 nm PNIPAM (0.8% w/w) particles at different temperatures. The PS particles form a crystal phase at low temperature and a gas at higher temperature. Scale bar, 20 μm.

We also observe that the transition temperature increases with the concentration of PNIPAM hydrogel particles, in qualitative agreement with eqn (2): higher concentrations of PNIPAM increase the depletion depth, placing the system deeper into the two-phase regime, so that a larger decrease in the depletant diameter is necessary to force sublimation. Similar sublimation behavior has been observed in other systems in which the depletant size varies with temperature.[4] These bulk phase behavior results show that the attraction between PS particles can be controlled over a range of a few k_BT by changing either the concentration of PNIPAM or the temperature.

The key to achieving this kind of reversible interaction is control over the electrostatic repulsion between the PS and PNIPAM particles. At low salt concentrations,

5 mM NaCl, we find that the PS particles remain dispersed as singlets even at 1.4% w/w PNIPAM and 10^{-2} volume fraction PS particles. At high salt concentration, 100 mM NaCl, we observe irreversible aggregation of the PS spheres.

We also confirm that the interactions between PS particles are reversible and well-controlled when they are placed in the microwells, where we use a lower PS volume fraction. We find that at a PNIPAM concentration of 1.0% w/w and a PS volume fraction of 10^{-5}, the PS particles form clusters in the middle of the microwells, with no particles stuck to the walls. As shown in Fig. 7, the clusters sublimate if the temperature is increased from 25 to 30° C, indicating that the PS particles in the colloidal clusters are not trapped by van der Waals forces.

Forming clusters in the microwells requires a delicate balance of the PNIPAM and PS concentrations. We choose the PS concentration to obtain the desired average number of particles per well, which is set by the microwell dimensions. Because the PS concentration is low, the PNIPAM concentration must be made high enough to overcome the tendency of the system to sublimate. Indeed, if we reduce the PNIPAM concentration slightly, to a value of 0.9% w/w, clusters no longer form. But if the PNIPAM concentration is too high, the probabilities of formation of particular structures are biased, as shown in Fig. 8. Here we plot the probability of finding a particular cluster structure, the octahedron ($N = 6$), as a function of PNIPAM concentration. The probability of forming an octahedron decreases systematically with the PNIPAM concentration (note that the error bars, calculated using the Wilson score interval method,[33] represent 95% confidence intervals rather than standard errors on the mean). One possible source of this bias could be the variation in the depletion potential as a function of depletant concentration: a previous study showed a secondary repulsive barrier in the depletion potential at higher depletant concentrations.[34] The other possibility is that the formation probabilities become kinetically dominated at higher PNIPAM concentration, which corresponds to a deeper depletion well. The conditions we ultimately choose—1.0% w/w PNIPAM, 10^{-5} volume fraction of PS particles, and 15 mM NaCl—manage to satisfy all constraints to ensure equilibrium assembly conditions.

For the dynamics experiments shown in section 4.3, we demonstrate an alternative method of making clusters that works directly in the gas phase and does not require delicately balancing all concentrations. In these experiments we use SDS micelles instead of PNIPAM particles as the depletant, and we assemble clusters using an optical tweezer. We work at very low PS concentration, volume fraction of 10^{-6},

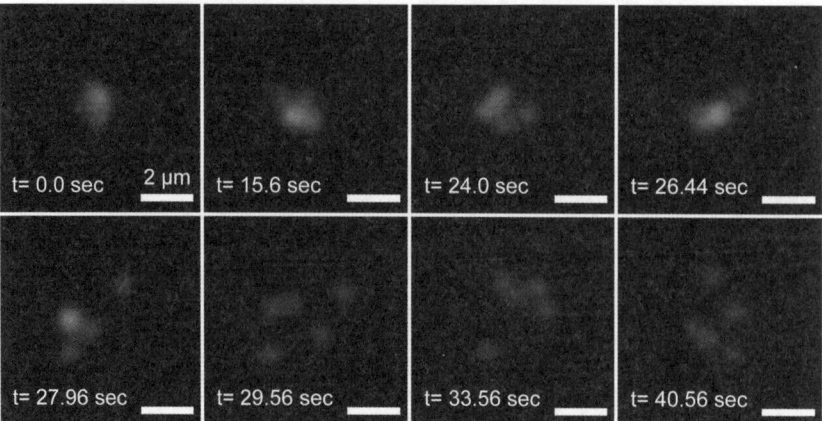

Fig. 7 Optical video microscopy snapshots of a triangular dipyramidal ($N = 5$) colloidal cluster during a sublimation transition as the temperature increases from 20 to 30 °C. Scale bar, 2.0 μm.

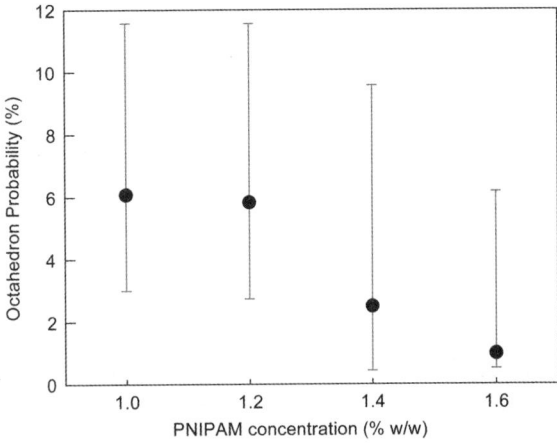

Fig. 8 Probability of observing a 6-particle octahedral cluster (black circular dots) as a function of PNIPAM depletant concentration. Error bars represent the lower and upper limits of the 95% confidence interval, as determined by the Wilson score interval method.[33]

and use an SDS concentration that is sufficiently large to induce an attraction, but not large enough to cause phase separation. We test for reversible interactions by building dimers and measuring how long it takes for them to break apart. To check that the bond angles can change, we build small clusters such as bent, two-bond trimers and look for fluctuations in geometry over time. We find that 40 mM SDS and 5 mM NaCl allow us to assemble pairs of particles that remain bound for tens of seconds after the optical tweezer is turned off. This timescale is long enough to observe structural transitions in larger clusters. Under the same conditions the bond angle in a trimer can fluctuate from 180° to 60°, corresponding to a rigid triangle. In contrast, we do not see bonds break in systems with no salt (40–50 mM SDS, 0 mM NaCl) or too much SDS (250 mM SDS, 5 mM NaCl). In such systems, rigid clusters form and are stable for more than a few minutes.

4.2 Structures

4.2.1 Structures of small colloidal clusters.

The clusters that assemble in the microwells take on a variety of morphologies, depending on N. In general we find that the number of structures at each N increases rapidly with N for $N > 6$. For each $N < 6$ we observe only one structure. We observe dimers for $N = 2$, triangles for $N = 3$, tetrahedra for $N = 4$, and triangular dipyramids for $N = 5$. Following the convention in Hoy et al.,[13] we refer to these structures as "Barlow packings," since all of them are subsets of either a face-centered cubic (FCC) or a hexagonally close-packed (HCP) lattice.

At $N = 6$, we observe two structures, an octahedron (point group O_h) and a "polytetrahedron" (point group C_{2v}), which is a triangular dipyramid capped with a third tetrahedron. Optical micrographs and computer renderings in Fig. 9 show the structure of these two clusters. Whereas the octahedron is a Barlow packing, the polytetrahedron is incompatible with a close-packed lattice. This is the smallest N at which a non-Barlow packing occurs.

Most of the structures at $N = 7$ are non-Barlow packings. We observe at least five different structures, as shown in Fig. 9. In one case, we are not able to determine from the optical micrographs whether the symmetry is C_{2v} or D_{5h} (a pentagonal dipyramid). For our 1 μm particles, these two structures differ only in the location of a small gap of approximately 50 nm. Two of the other structures are chiral enantiomers, both of which we observe in the measurements. Of all of these six clusters, only one, the capped octahedron with symmetry group C_{3v}, is a Barlow packing.

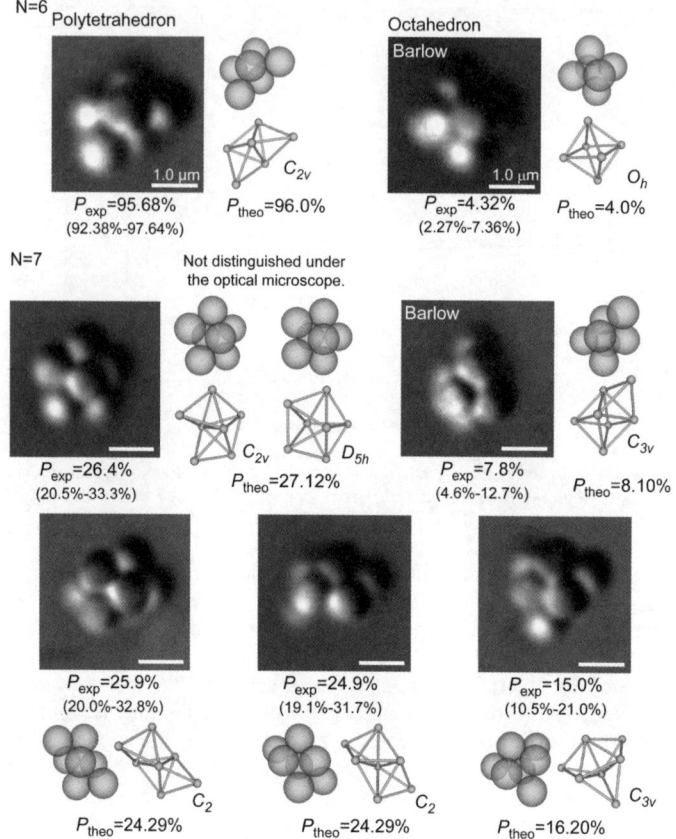

Fig. 9 Optical micrographs and renderings of colloidal clusters for $N = 6,7$, with point groups indicated in Schönflies notation. The measured (with the lower and upper limits of the 95% confidence interval determined by the Wilson score interval method[33]) and calculated probabilities are listed below each structure. Annotations above renderings indicate the clusters that cannot be distinguished under bright field microscopy. Structures that are compatible with crystalline lattices are marked with "Barlow". Scale bar, 1.0 μm.

At $N = 8$, we observe at least eight different structures; again, in one case we cannot determine the symmetry, which could take on at least six possible point groups (Fig. 10). All of the six possible structures are variants on the pentagonal dipyramid motif seen at $N = 7$. Only two of the observed structures are Barlow packings, and both of these are derivatives of an octahedron.

All of the structures that we observe in the experiments correspond to mechanically-stable packings of hard spheres with infinitesimally short-ranged attractions. The set of all such structures up to $N = 9$ was enumerated by Arkus and coworkers.[10] This enumeration was later extended to $N = 10$ by both Arkus and coworkers[11] and Hoy and O'Hern[14] and recently to $N = 11$ by Hoy and coworkers.[13] The enumerated packings correspond to the the minima of the potential-energy landscape as a function of N. Interestingly, up to $N = 9$ all of these idealized packings are degenerate: they contain the same number of contacts between spheres and hence the same potential energy. The theoretical packings are shown in the renderings in Fig. 9 and 10.

Three of the possible structures at $N = 8$ are not observed in any of the approximately 1000 microwells we examine. These structures are annotated as "$P_{\text{exp}} = 0\%$" in Fig. 10. One of them, the gyroelongated square dipyramid (point group D_{3d}) corresponds to a Barlow packing. It is also a derivative of an octahedron.

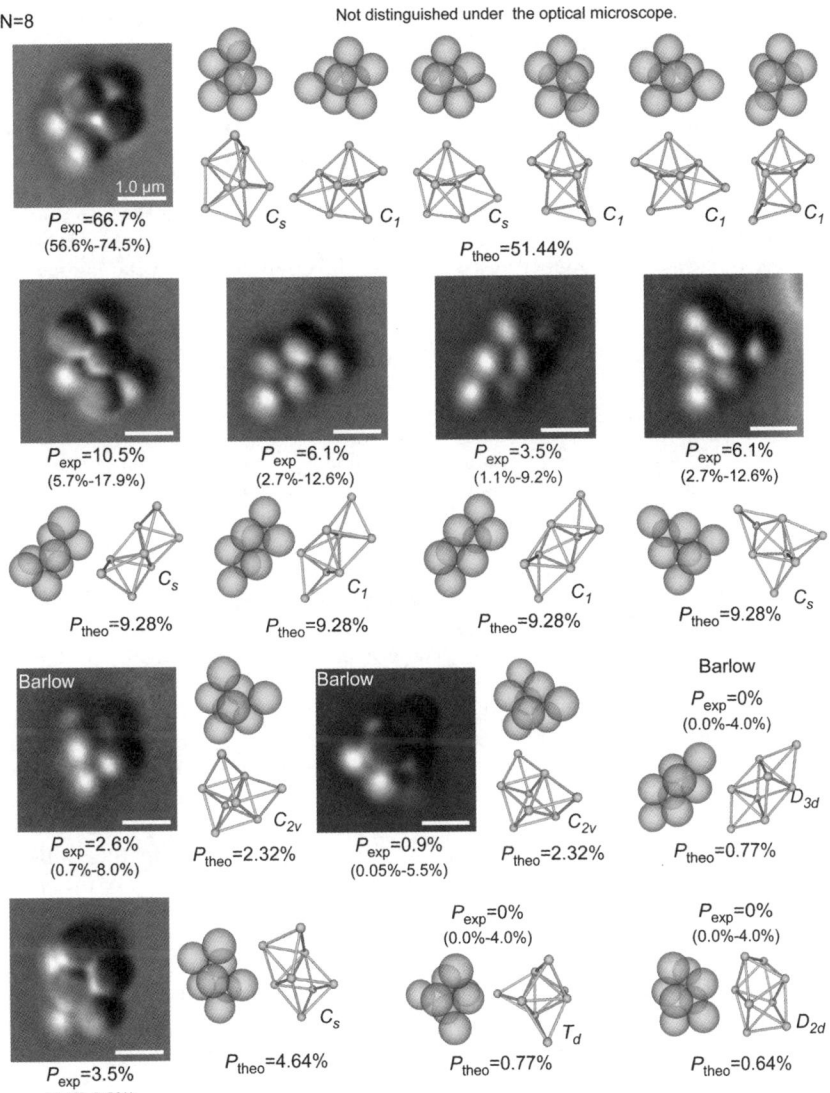

Fig. 10 Optical micrographs and renderings of colloidal clusters for $N = 8$, with point groups indicated in Schönflies notation. The measured (with the lower and upper limits of the 95% confidence interval determined by the Wilson score interval method[33]) and calculated probabilities are listed below each structure. Annotations above renderings indicate the clusters that cannot be distinguished under bright field microscopy. Structures that are compatible with crystalline lattices are marked with "Barlow". Scale bar, 1.0 μm.

4.2.2 Probabilities and free energies. We measure the free energy of each of the cluster structures simply by counting the number of occurrences of each cluster on the microwell plate. If the clusters are in equilibrium, the distribution of cluster structures should follow the Boltzmann distribution, $F_i \propto -k_B T \ln (P_i)$, where P_i is the probability of observing structure i. For example, at $N = 6$, we observe about 4% octahedra and 96% polytetrahedra, implying that the free energy of a polytetrahedron is about $3k_B T$ lower than that of an octahedron. This difference can be attributed only to entropy, since the two structures have the same number of contacts between particles – or "bonds" – and hence the same potential energy. The measured probabilities for each structure are shown in Fig. 9 and 10. Where it is not possible to determine the particular symmetry group of a cluster from the micrographs, we add together the probabilities of all possible structures.

As we showed in previous work,[9] all the measured probabilities agree well with theoretical calculations for the rotational and vibrational entropies. Both sets of probabilities are shown in Fig. 9 and 10. The dominant contribution to the free energy comes from the rotational entropy: structures with higher rotational symmetry are much less likely to form than less-symmetric structures. This is because the symmetry number of a structure is inversely related to the number of permutations of particles that do not change the structure.[15] Each permutation corresponds to a different pathway to the same structure, and in equilibrium, all such pathways are equally probable.[8] We note that at $N = 8$, the three structures we do not observe have high symmetry, and thus low probability ($P_{theo} < 1.0\%$) of formation.

In terms of nucleation, the most striking feature of the results up to $N = 8$ is the low probability of forming a structure compatible with a close-packed lattice. The total probability of all possible Barlow packings is about 4% for $N = 6$, 8% for $N = 7$, and 5% for $N = 8$. The most likely structures are the least symmetric ones, which in general correspond to packings based on a polytetrahedral motif.[35] Hoy et al.[13] found similar probabilities in their theoretical study.

The situation becomes more complicated when there are more than 9 particles in a cluster: at $N = 9$, clusters with soft modes first appear, and at $N = 10$ clusters with greater than $3N - 6$ bonds can form. These structures, many of which are Barlow packings, occur frequently in the experiments, as shown in Fig. 11. This result is qualitatively in agreement with theory: vibrational entropy associated with soft modes stabilizes the non-rigid clusters, while the potential energy associated with the extra bond stabilizes the clusters with $3N - 5$ bonds. Quantitative agreement is more difficult to obtain, since an accurate theoretical calculation of the free energy of the non-rigid clusters requires detailed knowledge of the pair potential. This is because the soft modes dominate the vibrational entropy, and the amplitude of these modes depends on the curvature of the potential near its minimum. Since the probabilities of the non-rigid clusters are non-negligible, any error will also affect a calculation of the probability of forming a cluster with extra bonds.

If at larger N there is a similar correlation between Barlow packings and extra bonds or soft modes—as we expect there might be, since structures with extra bonds and soft modes tend to contain both octahedral and tetrahedral subunits,[11] a necessary precondition for an FCC or HCP substructure—then there could be a significant implication for nucleation in similar kinds of short-range attractive systems: the probability of forming a Barlow cluster would depend not only on the potential depth, but also on the curvature of the potential, or its spring constant.

4.3 Dynamics

The microwell experiments highlight the low probabilities of forming Barlow packings at low N. Even for $N \geq 9$ the probabilities do not exceed 25%, although, as we have noted, these results may depend on the details of the potential. Connecting these results to nucleation barriers in bulk systems requires understanding the internal dynamics of the clusters. As Crocker noted[8] about Meng et al.'s original experiments,[9] our "clusters can equilibrate at leisure in complete isolation, whereas the clusters in an unbounded fluid are continuously bombarded by and grow by absorbing smaller clusters..., all of which may frustrate the equilibration of internal modes." While growth has been well characterized,[36] little is known about the second process, internal equilibration. The rate-limiting step for equilibration is rearrangement between cluster structures, an activated process that requires breaking at least one bond. Here we examine the dynamics of rearrangements in real space.

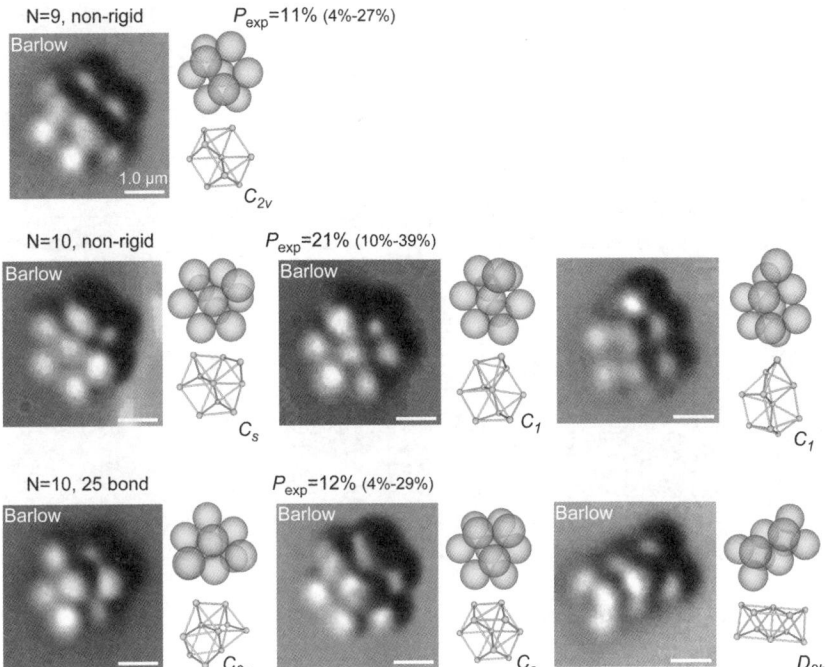

Fig. 11 Optical micrographs and renderings of special colloidal clusters for $N = 9$ and $N = 10$, with point groups indicated in Schönflies notation. The measured (with the lower and upper limits of the 95% confidence interval determined by the Wilson score interval method[33]) probabilities are listed above the categories. Structures that are compatible with crystalline lattices are marked with "Barlow". Scale bar, 1.0 µm.

4.3.1 Dynamics of transitions under bright field microscopy. We find that our clusters can and do transition between different structures after formation. For the PNIPAM system, we find that a six particle cluster changes its structure from a polytetrahedron to an octahedron and back every few minutes to tens of minutes. A typical transition as viewed through optical microscopy is shown in Fig. 12. The transition itself occurs on a timescale of seconds. We observe similar transitions at larger N.

The short timescale of the transition makes it difficult to determine the structure of the transition state. When a cluster is in an energy minimum, we can infer the relative positions of all the particles because the rotational Brownian motion of the cluster eventually brings all the particles within view. In contrast, the lifetime of a transition state is significantly shorter than the timescale of rotational motion, so we can only obtain qualitative data on the transition-state structures. For example, the micrographs in Fig. 12 appear to show that one of the twelve bonds breaks, and a new bond forms between different particles, but we cannot confirm this without quantitative measurements of the 3D positions of all six particles, accurate to 100 nm or better. Further complicating measurements of the dynamics is the long lifetime of the minima relative to that of transition states. Transitions are therefore rare events, and capturing just one of them may require recording tens of thousands of frames.

We therefore use a different experimental technique and, at the same time, modify our system to make it possible to study the dynamics of the clusters. To image the clusters we use holographic microscopy instead of optical microscopy. Holographic techniques can resolve the positions of all the particles in a cluster with at least

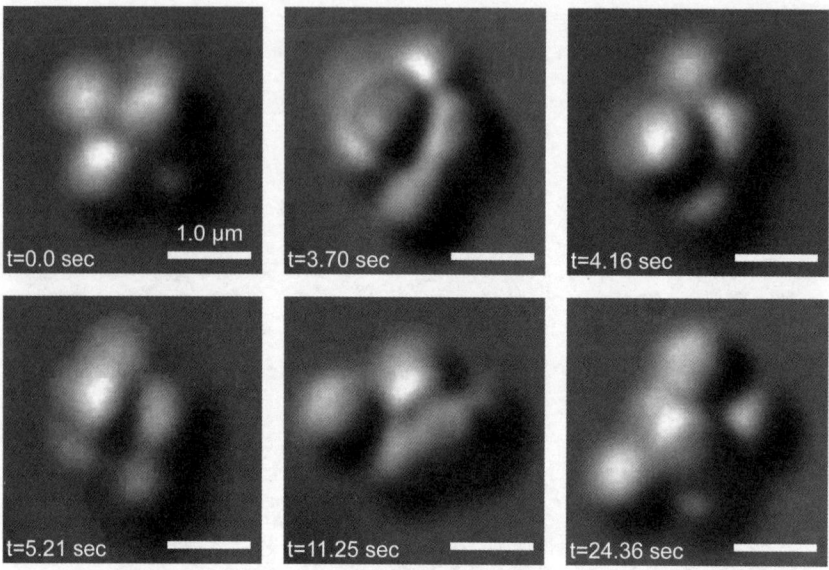

Fig. 12 A transition between an octahedral and polytetrahedral cluster in a microwell, captured using optical microscopy. Scale bar, 1.0 μm.

100 nm precision and 10 ms temporal resolution. We also change the depletant from PNIPAM particles to SDS micelles, and we assemble clusters directly in the gas phase using an optical tweezer. Although the clusters obtained in this way are thermodynamically unstable after the tweezer is turned off, they survive long enough to allow us to study transitions, as noted in section 4.1. Also, the rate of transitions is higher than in the PNIPAM-microwell system. The reasons for this are not clear, but the simplest explanation may be that the potential well is not as deep. In the microwell system we must use a deep potential well to force the particles to aggregate at low concentration. In experiments where we manually concentrate particles using an optical tweezer, we are free to tune the depletant concentration to optimize the kinetics. SDS micelles are more convenient than PNIPAM particles for this purpose because they are much simpler to make and mix. They also lead to a similar "sticky" depletion potential, in which the range of the attraction is much smaller than the diameter of the PS particles.

4.3.2 Validation of holographic microscopy technique. Because holographic microscopy has not previously been used to study dynamics of clusters larger than two particles, we first show that our fitting method yields realistic and accurate particle positions. Since a hologram is a 2D encoding of a 3D system and not simply a projection, we cannot verify the calculated particle coordinates by overlaying them on top of a real-space image, as one might do in standard particle tracking techniques based on optical microscopy.[37] Instead, we verify the calculated coordinates by numerically comparing measured holograms to ones obtained by fitting a scattering solution to the data ("best-fit holograms"). We also compare numerical reconstructions of the measured and best-fit holograms.

An example of the results obtained from our method is shown in Fig. 13(a) for one of the more complicated holograms to fit, one taken of a six-particle cluster that has formed an octahedron. Qualitatively, the data and the hologram calculated from the fit appear identical: the interference rings are in the same locations, and the deviations from a circular symmetry are in the same places. Quantitatively, the model fits the data well. The mean of the squared residuals across all pixels, $\chi^2 \approx 4 \times 10^{-4}$, is within a factor of 10 of the noise floor for the measured holograms, 5×10^{-5}. This corresponds to an uncertainty in the particle positions of 30–45 nm in x, y, and z, consistent with previous findings.[27]

To further verify the accuracy of the fit, we reconstruct both the measured and best-fit hologram, as shown in Fig. 13(b). To generate these reconstructions we numerically propagate light through the hologram to the midpoint of the cluster, as determined by the fit. Although the reconstructions do not account for coupling between the scattered fields of the particles, they nonetheless reveal an approximate image of the cluster. The cluster structure and orientation suggested by the reconstructions agree well with those computed from the fit, as shown in Fig. 13(c).

4.3.3 Dynamics of small clusters.

Having demonstrated that fitting exact scattering solutions to holograms reveals accurate cluster structures, we now examine the measurements of cluster dynamics obtained from time-series of holograms.

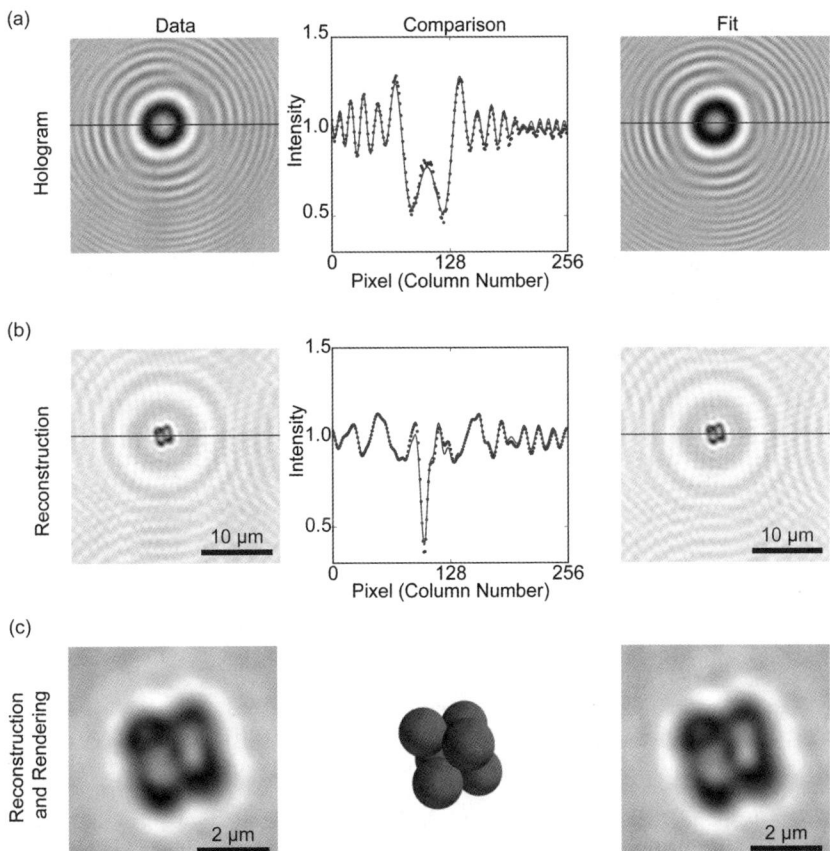

Fig. 13 Fitting holographic microscopy data to exact scattering models reveals the locations of all particles in 3D. (a) A frame of raw holographic data for a 6-particle cluster (left) and the hologram calculated from the best-fit positions of the particles (right). Middle plot shows a comparison between the intensities of the two holograms along the linear cross-sections shown in the images. Dotted line corresponds to the measured hologram and solid line to the best-fit hologram. (b) Holographic reconstructions of the raw data (left) and of the best-fit hologram (right). As above, middle plot shows a comparison between intensities of the two images across a linear cross-section. Scale bars, 10 μm. (c) Close-ups of the two reconstructions along with a rendering of the octahedral cluster generated from the fitted particle locations, showing that the fit agrees qualitatively with the reconstructed images. Scale bars, 2 μm.

We fit for all $3N$ particle coordinates as a function of time, but for simplicity we characterize the cluster structure by an order parameter M_2, the second moment of the mass distribution:[38]

$$M_2 = a_{\text{eff}}^{-2} \sum_{i=1}^{N} |r_i - r_0|^2 \qquad (3)$$

r_i is the location of the ith sphere, r_0 the center of mass of the cluster, and a_{eff} the effective radius of the particles, or half the distance from the center of one particle, across the depletion zone, to the center of a neighboring particle. We take $a_{\text{eff}} \approx a_l + a_s/2$ and $a_s \approx 15$ nm for an SDS micelle.

The variation in cluster structure with time, as characterized by M_2 and real-space renderings of the cluster coordinates, is shown in Fig. 14 for 2-, 3-, and 4-particle clusters. Although such clusters have only a single free-energy minimum, as discussed in section 4.2.1, they show transitions between rigid and non-rigid states as well as rotational and translational Brownian motion. We see a dimer ($N = 2$) break apart, a trimer ($N = 3$) assemble itself into a rigid triangle, and a tetramer ($N = 4$) transition from a tetrahedron to a planar diamond and back to a tetrahedron. Interestingly, the tetrameric transition is an inversion: labeling the particles shows that the handedness of the tetrahedron changes from the beginning to the end of the measured trajectory (see color renderings in Fig. 14). Similar types of tetrahedron-diamond-tetrahedron transitions may occur in larger clusters, where they could represent a mechanism for isomerization between different polytetrahedral configurations.

The data show that the lifetime of a non-rigid state is on the order of seconds. To understand this lifetime we estimate the timescale for the tetrahedron-diamond-tetrahedron inversion. In this transition, one of the end particles must traverse an arc length of approximately 110°. Neglecting translations of the center of mass and global rotations, the path length this end particle must travel is $\frac{110}{180}\pi\sqrt{3}a_l = 2.2\,\mu\text{m}$, since it is $\sqrt{3}a_l$ from the rotation axis. Using the diffusion coefficient for a single particle $D = 3.1 \times 10^{-13}$ m^2 s^{-1}, which we measure in a separate experiment by holographically tracking an unbound particle, we estimate that the rearrangement should occur in about 7 s, which is close to the lifetime we observe. The agreement between the calculation and data shows that the lifetime of the non-rigid state is likely diffusion-limited, and that there are no significant hydrodynamic corrections to the diffusion time for bond rotation.

4.3.4 Dynamics of a transition between two free energy minima.
As described in section 4.2.1, a six particle cluster is the smallest cluster that can transition between two rigid energy minima: an octahedron, which is a Barlow packing, and a polytetrahedron, which is not. Using holographic microscopy, we observe a six particle cluster form a polytetrahedron and transition to an octahedron. The results, summarized in Fig. 15, contain far more detail than can be obtained from the bright field micrographs in Fig. 12.

The ball-and-stick renderings of Fig. 15 show the bonds that form and break during the transition. Initially, there are only 10 bonds between the six particles. Four of the particles, shown in gray, are bound in a rigid tetrahedron. Shortly after $t = 2$ s, the particle labeled in blue bonds to the tetrahedron to form a trigonal dipyramid. Then an additional particle, shown in red, bonds to the dipyramid to complete the formation of a polytetrahedron at around $t = 3$ s. Just before $t = 6$ s, a bond breaks, and the cluster rapidly transitions to an octahedron, which persists until the end of the data set.

The observed timescale for this transition, which transforms the cluster from a structure inconsistent with crystallinity to a Barlow packing, is close to the timescale expected from single-particle diffusion. This transition requires two particles to

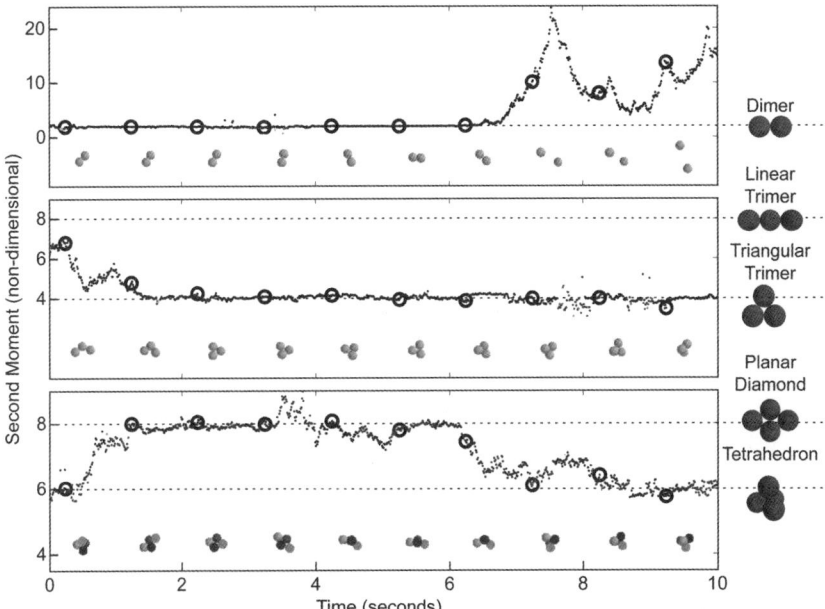

Fig. 14 Cluster dynamics as determined by holographic microscopy for 2-, 3-, and 4-particle clusters. Plots show the second moment M_2 (eqn (3)) as a function of time. The intensity of the data points indicates the relative value of the goodness-of-fit parameter χ^2 (black represents the lowest χ^2). Dotted lines show M_2 for the reference geometries at right (dimer = 2, linear trimer = 8, triangle = 4, planar diamond = 8, tetrahedron = 6). Renderings within each plot show the cluster configurations corresponding to the nearest circled data points.

move from 3 1/3 radii apart to 2 radii apart in order to form a bond. For simplicity, we consider only the time it takes one particle to diffuse a linear distance of $4a_1/3$. From the measured single particle diffusivity, this should take approximately one second, consistent with our observations.

We can estimate the rate of transitions from the observed structure lifetime, which is also on the order of 1 s. We assume that it is equally likely for any of the 12 bonds in the polytetrahedron to break. Only one of these breakages can lead to the formation of an octahedron; the other 11 will result in tetrahedron-diamond-tetrahedron transitions that do not change the structure. Thus we expect that transitions from polytetrahedra to octahedra should in general happen on timescales of tens of seconds; presumably we were fortunate to be able to capture, in our short data set, the breaking of the one bond that would allow an octahedron to form. For comparison we estimate the growth rate as a function of the volume fraction, assuming diffusion-limited conditions. At 10^{-6} volume fraction, new particles arrive at the cluster every hour, at 10^{-4} every few minutes, and at 10^{-2} every few seconds, which is comparable to the time between structural transitions.

The $N = 6$ cluster is the smallest system in which growth can lead to two different outcomes: Barlow packing or polytetrahedral order. A new particle that attaches to an octahedron produces another Barlow packing, while one that attaches to a polytetrahedron produces an $N = 7$ polytetrahedron. At higher volume fractions, when the growth rate is comparable to the transition rate, we might expect that the system has a greater tendency to develop polytetrahedral order, which is incompatible with crystal nucleation. Given the low free energy of the $N = 6$ polytetrahedron relative to that of the octahedron, the prospects for successful nucleation of a crystal from an $N = 6$ embryo seem bleak. However, at higher volume fraction the initial clusters that form may be much larger than six particles, so the $N = 6$ case may not in general

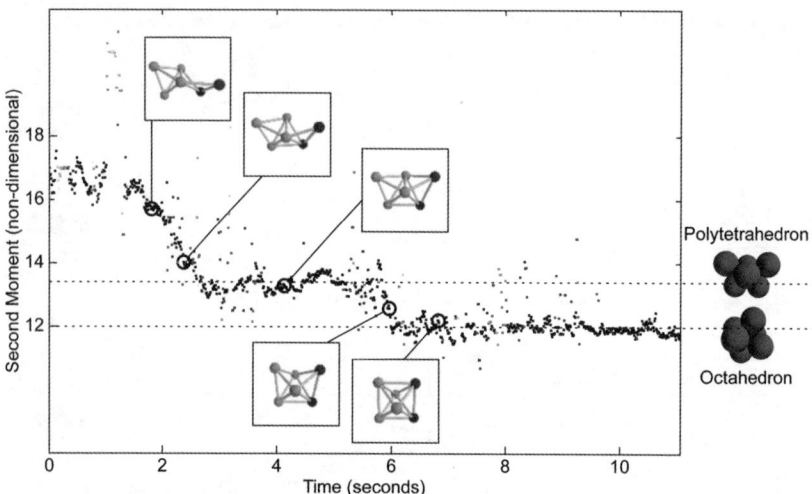

Fig. 15 Transition from polytetrahedron to octahedron in a 6-particle cluster, as measured by holographic microscopy. Plot shows evolution of the second moment M_2 as a function of time, and ball-and-stick insets show the cluster geometry. The insets are oriented to clearly show the cluster structure and do not represent the actual spatial orientation of the clusters. Dashed lines indicate the second moments of the polytetrahedron $\left(13\frac{11}{27}\right)$ and the octahedron (12).

represent a nucleation "bottleneck." Further studies of transitions in larger systems, where extra bonds and soft modes are possible, are necessary to more rigorously relate the cluster dynamics to nucleation probabilities.

5 Conclusions

The work we have shown here represents the first steps toward understanding nucleation through analysis of the thermodynamics and dynamics of colloidal clusters. Much remains to be done on both the experimental and theoretical fronts, particularly for larger clusters. Also, although we have measured transition rates for a few small clusters, we need much more data on both small and large clusters to obtain statistically significant estimates of the transition state lifetimes and transition pathways, which are the key elements missing from the free-energy landscape model of short-range attractive spheres.[9,15] Such studies will require measurements of the interaction potential: as we have shown, the rearrangement timescales for systems with different depletants can vary by orders of magnitude, presumably because of differences in the well depth and width; also, the probabilities of obtaining clusters with soft modes, many of which are Barlow packings, depend on the curvature of the potential and not just the well depth. To measure these features of the potential we must be able to resolve the separation between two colloidal particles to nanometer-scale precision. This is a difficult task, but recent advances in imaging[27] and optical-tweezer-based measurements[39] show that it is possible.

Although the connection to nucleation barriers remains tenuous at this stage, our work demonstrates that the study of colloidal clusters stands to reveal new insights into processes that are key to understanding nucleation, including the formation of clusters and their structural transitions. Modern experimental techniques such as soft lithography and holographic microscopy make it possible to measure all the thermodynamic and dynamical information about a cluster, including its structure, free energy, and fluctuations about free-energy minima. We know of no other experimental system that can be probed in such detail. The main goal for future experiments is to systematically explore the dynamics as a function of N and to obtain, from that wealth of detail, a more complete model of the free-energy landscape that governs nucleation.

Acknowledgements

We thank Jesse Collins, Miranda Holmes-Cerfon, Zorana Zeravcic, Natalie Arkus, Michael Brenner, and Robert Hoy for helpful discussions. Rebecca W. Perry and Thomas G. Dimiduk acknowledge the support of National Science Foundation (NSF) Graduate Research Fellowships. This work was funded by the NSF through CAREER grant no. CBET-0747625 and through the Harvard MRSEC, grant no. DMR-0820484. Computations were performed on the Odyssey cluster, managed by the Harvard FAS Sciences Division Research Computing Group. Microwell device fabrication was performed in part at the Center for Nanoscale Systems (CNS), a member of the National Nanotechnology Infrastructure Network (NNIN), which is supported by the NSF under award no. ECS-0335765. CNS is part of Harvard University.

References

1 V. J. Anderson and H. N. W. Lekkerkerker, *Nature*, 2002, **416**, 811–815.
2 C. P. Royall, S. R. Williams, T. Ohtsuka and H. Tanaka, *Nat. Mater.*, 2008, **7**, 556–561.
3 W. K. Kegel and A. van Blaaderen, *Science*, 2000, **287**, 290–293.
4 J. R. Savage, D. W. Blair, A. J. Levine, R. A. Guyer and A. D. Dinsmore, *Science*, 2006, **314**, 795–798.
5 P. J. Lu, E. Zaccarelli, F. Ciulla, A. B. Schofield, F. Sciortino and D. A. Weitz, *Nature*, 2008, **453**, 499–503.
6 U. Gasser, E. R. Weeks, A. Schofield, P. N. Pusey and D. A. Weitz, *Science*, 2001, **292**, 258–262.
7 J. R. Savage and A. D. Dinsmore, *Phys. Rev. Lett.*, 2009, **102**, 198302.
8 J. C. Crocker, *Science*, 2010, **327**, 535–536.
9 G. Meng, N. Arkus, M. P. Brenner and V. N. Manoharan, *Science*, 2010, **327**, 560–563.
10 N. Arkus, V. N. Manoharan and M. P. Brenner, *Phys. Rev. Lett.*, 2009, **103**, 118303.
11 N. Arkus, V. N. Manoharan and M. P. Brenner, *SIAM Journal on Discrete Mathematics*, 2011, **25**, 1860–1901.
12 D. J. Wales, *ChemPhysChem*, 2010, **11**, 2491–2494.
13 R. S. Hoy, J. Harwayne-Gidansky and C. S. O'Hern, *Phys. Rev. E: Stat., Nonlinear, Soft Matter Phys.*, 2012, **85**, 051403.
14 R. S. Hoy and C. S. O'Hern, *Phys. Rev. Lett.*, 2010, **105**, 068001.
15 F. Calvo, J. P. K. Doye and D. J. Wales, *Nanoscale*, 2012, **4**, 1085–1100.
16 S. Asakura and F. Oosawa, *J. Chem. Phys.*, 1954, **22**, 1255–1256.
17 A. Vrij, *Pure Appl. Chem.*, 1976, **48**, 471–483.
18 G. A. Vliegenthart and P. van der Schoot, *Europhys. Lett.*, 2003, **62**, 600–606.
19 T. Biben, P. Bladon and D. Frenkel, *J. Phys.: Condens. Matter*, 1996, **8**, 10799–10821.
20 X. Wu, R. H. Pelton, A. E. Hamielec, D. R. Woods and W. Mcphee, *Colloid Polym. Sci.*, 1994, **272**, 467–477.
21 M. Andersson and S. L. Maunu, *J. Polym. Sci., Part B: Polym. Phys.*, 2006, **44**, 3305–3314.
22 T. D. Iracki, D. J. Beltran-Villegas, S. L. Eichmann and M. A. Bevan, *Langmuir*, 2010, **26**, 18710–18717.
23 D. B. Weibel, W. R. DiLuzio and G. M. Whitesides, *Nat. Rev. Microbiol.*, 2007, **5**, 209–218.
24 A. D. Dinsmore, A. G. Yodh and D. J. Pine, *Nature*, 1996, **383**, 239–242.
25 S. Lee, Y. Roichman, G. Yi, S. Kim, S. Yang, A. van Blaaderen, P. van Oostrum and D. G. Grier, *Opt. Express*, 2007, **15**, 18275–18282.
26 S. Lee and D. G. Grier, *Opt. Express*, 2007, **15**, 1505–1512.
27 J. Fung, K. E. Martin, R. W. Perry, D. M. Kaz, R. McGorty and V. N. Manoharan, *Opt. Express*, 2011, **19**, 8051–8065.
28 J. Fung, R. W. Perry, T. G. Dimiduk and V. N. Manoharan, *J. Quant. Spectrosc. Radiat. Transfer*, 2012, DOI: 10.1016/j.jqsrt.2012.06.007.
29 D. W. Mackowski and M. I. Mishchenko, *J. Opt. Soc. Am. A*, 1996, **13**, 2266–2278.
30 S. N. Kasarova, N. G. Sultanova, C. D. Ivanov and I. D. Nikolov, *Opt. Mater.*, 2007, **29**, 1481–1490.
31 J. Sheng, E. Malkiel and J. Katz, *Appl. Opt.*, 2006, **45**, 3893–3901.

32 L. Dixon, F. C. Cheong and D. G. Grier, *Opt. Express*, 2011, **19**, 16410–16417.
33 R. G. Newcombe, *Stat. Med.*, 1998, **17**, 857–872.
34 J. C. Crocker, J. A. Matteo, A. D. Dinsmore and A. G. Yodh, *Phys. Rev. Lett.*, 1999, **82**, 4352–4355.
35 D. R. Nelson and F. Spaepen, *Solid State Phys.*, 1989, **42**, 1–90.
36 T. A. Witten and L. M. Sander, *Phys. Rev. Lett.*, 1981, **47**, 1400–1403.
37 J. C. Crocker and D. G. Grier, *J. Colloid Interface Sci.*, 1996, **179**, 298–310.
38 N. Sloane, R. Hardin, T. Duff and J. Conway, *Discrete Comput. Geom.*, 1995, **14**, 237–259.
39 W. B. Rogers and J. C. Crocker, *Proc. Natl. Acad. Sci. U. S. A.*, 2011, **108**, 15687–15692.

PAPER

Aggregation of ferrihydrite nanoparticles in aqueous systems

Virany M. Yuwono, Nathan D. Burrows, Jennifer A. Soltis, Tram Anh Do and R. Lee Penn*

Received 6th June 2012, Accepted 8th June 2012
DOI: 10.1039/c2fd20115a

Crystal growth by non-classical mechanisms, such as oriented aggregation, frequently involves an aggregation step. The aggregation of nanoparticles is sensitive to solution variables like ionic strength and pH, as well as the presence and concentration of other chemical species. Aggregation is a critical first step during the early stages of oriented aggregation. Time-resolved, cryogenic transmission electron microscopy was employed to characterize the degree of aggregation, the reversibility of aggregation, and the influence of additives on aggregation in aqueous suspensions. In this work, freshly synthesized ferrihydrite nanoparticles in aqueous suspension were employed as the model system. These nanoparticles are largely aggregated even with a solution pH several pH units away from the point of zero net proton charge (PZNPC) and a very low ionic strength ($<10^{-4}$ M). Reversibility of aggregation was observed to be time-dependent. Finally, chemical additives dramatically change the evolution of aggregation state with strongly coordinating ligands strongly suppressing aggregation, even after aging at an elevated temperature.

1 Introduction

Control over nanocrystal shape, size, and size distribution remains a long-standing goal in materials synthesis. Non-classical crystal growth mechanisms, which can provide routes for size and shape control, often involve an aggregative step involving intermediate nanoparticles that can be structurally distinct from the final crystal structure.[1–11] Recent work by Yuwono et al. demonstrated that aggregation precedes oriented aggregation, a crystal growth mechanism involving the formation of mesocrystals.[8] Mesocrystals are herein defined as secondary objects composed of primary crystallites in crystallographic registry but lacking direct crystallite–crystallite contact.[6,8]

The kinetics of oriented aggregation is intimately tied to the kinetics and nature of aggregation. In the oriented aggregation mechanism of crystal growth, the reversible formation of a particle–particle complex was proposed.[5] The primary particles composing the particle–particle complex can reorient so as to achieve crystallographic registry. The resulting object, termed a mesocrystal, can either irreversibly convert to an oriented aggregate, with all species previously present in the spaces separating the primary particles either removed or incorporated into the crystal, or disaggregate. Recent work by Burrows et al. demonstrates that the initially formed fractal-like clusters of particles are precursors to mesocrystals.[11] The goal of this work is to directly characterize aggregation as a function of

Department of Chemistry, University of Minnesota, 207 Pleasant St. SE, Minneapolis, MN 55455, United States. E-mail: rleepenn@umn.edu; Fax: +1 612 626 7541; Tel: +1 612 626 4890 changing solution conditions with special attention paid to the reversibility of aggregation.

This study examines the reversibility of aggregation as a function of time after perturbations to the solution have been made. In general, aggregation becomes progressively less reversible with time spent at a pH somewhat near to the point of zero net proton charge (PZNPC), which is the pH at which the particles have no net surface charge with only H^+ and OH^- as the ions contributing charge. In addition, this study examines the effect of adding organic molecules to suspensions of nanoparticles in order to track immediate changes in aggregation as well as after aging at elevated temperature. In general, the organic additives employed here, PEG (600 MW) and alizarin, inhibit aggregation. In this work, ferrihydrite $(Fe_{8.2}O_{8.5}(OH)_{7.4} \cdot 3H_2O)$[12] nanoparticles were selected for their environmental relevance[8,13,15–17] and because, under the right conditions, they phase transform to the more stable phase goethite,[13,14] which can subsequently grow by oriented aggregation.[14–17]

2 Experimental procedures

2.1 Ferrihydrite synthesis

Six-line ferrihydrite nanoparticles were synthesized by the controlled hydrolysis of a homogeneous, aqueous Fe^{3+} solution following methods that have been previously published.[11,14–17] Prior to use, all glassware and plasticware were acid washed with 4 M nitric acid followed by rinsing in distilled water and Milli-Q water (Millipore Milli-Q system, 18.2 MΩ cm^{-1} resistance), three times each. Two batches of ferrihydrite were prepared: Fh_{66} and Fh_{45}. Equal volumes of 0.40 M $Fe(NO_3)_3$ (Fisher Scientific) and 0.48 M $NaHCO_3$ (Sigma-Aldrich) were prepared using Milli-Q water. The solutions were heated in a water bath until the temperature reached 66 °C (Fh_{66}) or 45 °C (Fh_{45}). The sodium bicarbonate solution was added to the ferric nitrate solution using a peristaltic pump, with approximately 30 cm of the tubing submerged in the water bath so as to heat the base solution to the reaction temperature. The delivery of the base solution occurred over approximately ten (Fh_{66}) or twenty (Fh_{45}) minutes under continuous stirring using a magnetic stir bar. The resulting nanoparticle suspension was immediately cooled to room temperature by submerging the capped reaction vessel in an ice bath. The resulting suspension was then microwave-annealed by heating in a 950 W microwave (Samsung) in half-minute intervals until boiling. The suspension was cooled by plunging the capped bottle into an ice bath. The suspension was dialyzed (Spectra Por dialysis tubing, MWCO 2000) at 4 °C against Milli-Q water for 3 days with the dialysis water changed a minimum of three times per day. The pH at this time was 3.9 (Fh_{66}) and the pH of the 10× diluted suspension was 3.63 (Fh_{45}). Cryogenic transmission electron microscopy (cryo-TEM) samples (section 2.2) were prepared from the as-dialyzed suspensions.

2.2 Cryo-TEM sample preparation

A small volume of the sample suspension (typically 3 µL) was applied to a transmission electron microscopy (TEM) copper sample grid with a lacy carbon support film purchased from SPI Supplies (200 mesh Cu grid). The grid was then blotted with filter paper using a Vitrobot (FEI Mark IV) under 100% relative humidity for two seconds to create a thin film of suspension. The grid was then plunged into liquid ethane and quickly transferred to storage under liquid nitrogen. All sample manipulations were performed under liquid nitrogen at cryogenic temperatures. The samples were analysed using low-dose imaging conditions with an FEI F30 TEM operated at 300 keV.

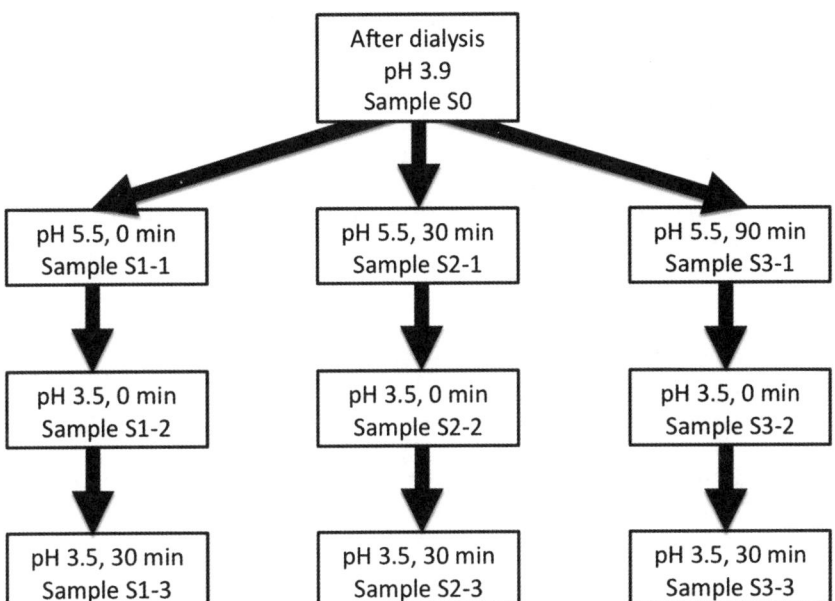

Scheme 1 Flow diagram of pH effects studies on aggregation states of freshly prepared ferrihydrite suspensions.

2.3 Characterizing the reversibility of aggregation

The steps for the aggregation experiments are illustrated in Scheme 1. First, a cryo-TEM sample of the as-dialyzed, Fh_{66} suspension was prepared (S0). Then, three 100 ml Nalgene® bottles containing the Fh_{66} suspension were prepared at room temperature. The pH adjustments for each of these three samples were the same, but the incubation time at the higher pH was varied. First, the pH of each suspension was adjusted to 5.5 ± 0.06 using 0.1 M NaOH (Mallinckrodt). For the suspension in the first bottle, a 3 µL sample was taken for cryo-TEM sample preparation (S1-1) as soon as the pH stabilized. Then, the pH was immediately adjusted to 3.5 ± 0.06 using 0.1 M HNO_3 (Macron Chemicals or Mallinckrodt), and cryo-TEM samples were prepared at time 0–1 min (S1-2) and 30 ± 1 min (S1-3). The second and third suspensions were incubated at pH 5.5 for 30 ± 1 (S2-1) and 90 ± 1 min (S3-1), respectively. Then, the pH was adjusted to 3.5 and cryo-TEM samples prepared at the same time points as previously described (S2-2, S2-3, S3-2 and S3-3). Constant stirring was maintained using a magnetic stir bar and stir plate (Fisher Scientific Isotemp) set at medium speed during the experiments.

2.4 Characterizing the influence of additives on aggregation

The aggregation state of ferrihydrite particles in the presence of organic additives was also characterized. Alizarin (Acros Organics), which is a dihydroxyanthraquinone isomer, and polyethylene glycol (PEG, 600 MW; Alfa Aesar) were added to samples of the Fh_{45} suspension and the samples were diluted so that the ferrihydrite loading was one-tenth of the initial loading. The final alizarin concentration was 0.1 mM and final PEG concentration was 0.1 mM hydroxyl groups per litre. A third sample was prepared as a control sample by diluting Fh_{45} 10-fold using only MilliQ water. The pH of each sample was adjusted, with continuous stirring, to pH 4.00–4.03 with NaOH. Cryo-TEM samples were prepared 4–6 h after the addition of organic species and after 117 h aging at 80 °C. Cryo-TEM samples of Fh-0, PEG-0, and Aliz-0 were prepared after a stable aggregate size was observed using dynamic light scattering. Sample labels are denoted by additive (or lack thereof) and time aged at 80 °C: Fh-0 and Fh-117; Aliz-0 and Aliz-117; and PEG-0 and PEG-117.

Samples for X-ray diffraction (XRD) were prepared at 117 h by applying a droplet of aged suspension to quartz slides. Each drop was allowed to dry before an additional droplet was added. XRD patterns were collected with a PANalytical X'Pert Pro MPD theta–theta diffractometer with a cobalt source and an X'Celerator detector over the range of 20–80° 2θ. The experimental results were compared with the reference powder diffraction files (PDFs) for 6-line ferrihydrite (#29-712) and goethite (#29-713). No evidence of other iron oxides was observed in the XRD patterns.

2.5 Image analysis

A minimum of ten images per sample was collected and analysed. Image processing and analysis were conducted using a semi-automated image processing and analysis scheme developed by Burrows *et al.* using ImageJ® (versions 1.43u–1.46o).[11] ImageJ® is an open-source image processing and analysis program written in Java by Wayne Rasband at the U.S. National Institutes of Health.[18-20] The result of this scheme is an image overlay with primary particles outlined and their positions identified.

The "particle analysis" tool was used to identify the locations of as many particles as possible, limiting the allowed particle area/size so no background was included as primary particles. For images in which particles were missed, the overlay was flattened and the resulting image file was opened in Photoshop®, where a coloured dot matching the overlay was manually applied to each primary particle not included in the computer-generated overlay. The resulting RGB image was opened with ImageJ® and split into three channels. An image from a single colour channel was selected and converted to 8 bit. The semi-automated image processing and analysis scheme was then repeated, which produced an updated file containing particle positions.

Assignment of particles to clusters was accomplished using the Graph plugin, in which the criterion for inclusion of a primary particle in a cluster is some maximum separation distance.[21] This parameter is adjusted until the automated result appeared visually consistent with the experimental image. In the cases where goethite rods appeared, these crystals were accounted for by measuring the size of the crystals and calculating the number of primary particles required to produce the crystal. These values were added to the particles per aggregate count for the clusters in which the rods reside.

3 Results and discussion

3.1 pH effects on the aggregation state of ferrihydrite nanoparticles

Aggregation is far more extensive in the as-dialyzed suspension than expected despite a suspension pH of several pH units lower than the isoelectric point and a very low ionic strength ($\ll 10^{-4}$ mM salt). Fig. 1 shows cryo-TEM micrographs of the vitrified sample prepared using the freshly dialyzed suspension of ferrihydrite nanoparticles at pH 3.9 (S0). At this pH, the nanoparticles in the suspension were present mainly as small clusters and isolated primary particles. The average number of primary particles per aggregate or cluster was 4.9, with 79.1% of the particles residing in clusters. The point of zero net proton charge (PZNPC) for similarly prepared ferrihydrite nanoparticles was experimentally determined to be 8.8, which is nearly five pH units away from the pH of the suspension here.[22] This degree of aggregation far surpassed our expectation and was reproducibly observed.

When the pH was raised to 5.5, which is "only" three pH units away from the PZNPC, the nanoparticles formed large, fractal aggregates, as shown in the

Fig. 1 Freshly synthesized ferrihydrite nanoparticles after dialysis at pH 3.9 (S0).

representative cryo-TEM image of S1-1 (Fig. 2a). The average number of primary particles per aggregate was 30, with nearly all the particles residing in aggregates (96.6%). Even though large aggregates were prominently observed in the images, the average number of particles per aggregate was much smaller because of a wide aggregate size distribution, with nearly two-thirds of the primary particles residing in clusters composed of 100 or more primary particles ($n \geq 100$).

Decreasing the suspension pH to 3.5 resulted in substantial disaggregation. Results from the image analysis of cryo-TEM images of S1-2 (Fig. 2b, prepared immediately after the pH adjustment) demonstrate a dramatic decrease in the average number of particles per aggregate and an approximately 25% decrease in the fraction of particles residing in clusters or aggregates (Table 1). Cryo-TEM images of samples prepared after 30 min of aging at pH 3.5 (S1-3, Fig. 2c) demonstrated little difference in aggregation state between S1-2 and S-1-3. Thus, we conclude that disaggregation was fast after the pH decrease.

Sample S2-1 was prepared after aging for 30 min at pH 5.5. The cryo-TEM micrographs of S2-1 show that the particles formed extensive 3-D lace-like structures (Fig. 2d). Image analysis revealed a high number average per aggregate (2×10^2) and nearly all primary particles residing in aggregates. The number average is much higher than in S1-1, indicating that aggregates increase in size with time at pH 5.5.

Decreasing the pH of the above sample to 3.5 resulted in a substantial decrease in the size of the aggregates but only a modest decrease in the fraction of particles

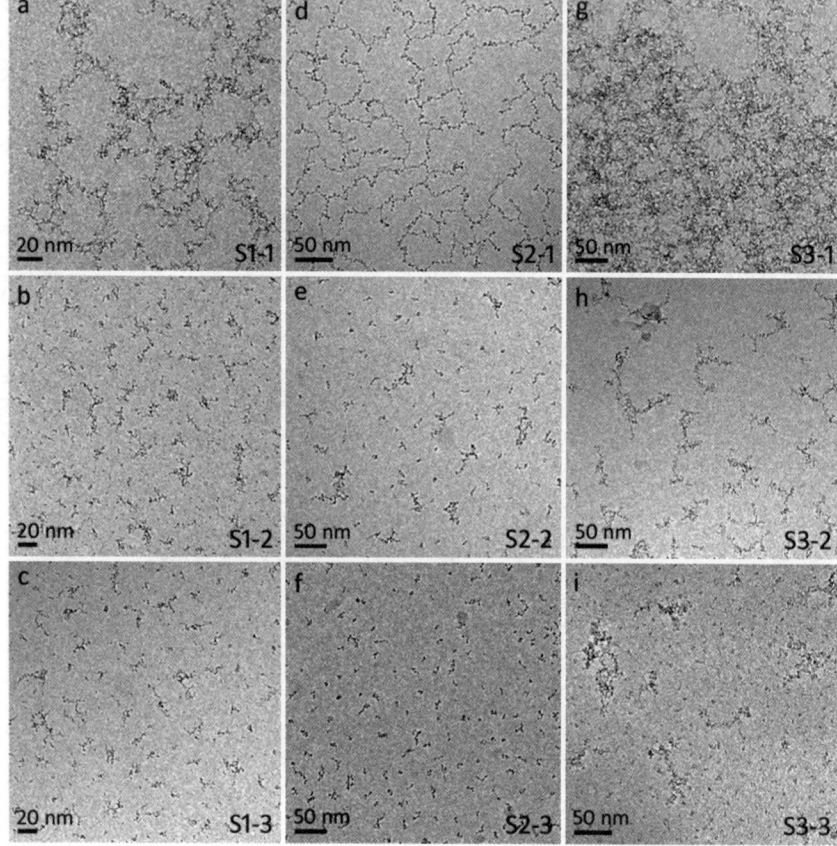

Fig. 2 Representative cryo-TEM micrographs of samples (a) S1-1, (b) S1-2, (c) S1-3, (d) S2-1, (e) S2-2, (f) S2-3, (g) S3-1, (h) S3-2, and (i) S3-3.

Table 1 Results of cryo-TEM image analysis tracking the average number of particles per aggregate and percentage of particles in aggregates as a function of pH and incubation times

	S0	S1-1	S1-2	S1-3	S2-1	S2-2	S2-3	S3-1	S3-2	S3-3
Average #/aggregate	4.9	30	4.7	4.0	220^a	5.2	4.5	103^a	52	23
% In aggregates	79.1	96.6	74.5	75.0	99.7	91.5	84.0	99.8	97.9	95.9
% In aggregates ($n \geq 100$)		1.1	66.4	0.6	0	92.6	5.5	0.4	98.9	66.0	42.1

^{a} The average number of particles per aggregate reported is a minimum estimate as many of the largest aggregates spanned more than the field of view.

residing in aggregates (S2-2, Fig. 2e). After 30 min aging at pH 3.5, the average number of primary particles per aggregate decreased slightly (S2-3, Fig. 2f), but the fraction of particles residing in aggregates dropped quite significantly. While these results were consistent with the S1 series, with fast and immediate disaggregation observed upon the drop in pH, the continued drop with aging may suggest a second type of aggregation evolves with aging at the higher pH.

The final series was sampled from the suspension aged at pH 5.5 for 90 min before decreasing the pH to 3.5. Cryo-TEM analysis of S3-1 revealed extensive aggregation of the nanoparticles (Fig. 2g). The average number of particles per aggregate was around 1000, and almost all of the particles resided in aggregates (99.8%). This number average is the minimum estimate of the aggregate size since many of the aggregates extended beyond the field of view. Decreasing the pH to 3.5 produced the same effect as in the S1 and S2 series. With immediate sampling after the pH decrease (S3-2, Fig. 2h), a dramatic drop in the number of primary particles per aggregate was observed. As with the S2 series, the number of particles per aggregate continued to decrease with aging at the lower pH (S3-3, Fig. 2i), but the overall degree of disaggregation was less extensive than observed in the S1 and S2 series. These results demonstrate that the aggregation is progressively less reversible as a function of aging time at the higher pH.

These results (summarized in Table 1) show that aggregation is substantial even at a pH that is far from the PZNPC (about five pH units below). When the pH of the suspension is increased to within about three pH units of the PZNPC, aggregation immediately and dramatically increases and then slowly continues to increase with time at that higher pH. In general, the disaggregation of the particles after decreasing the pH can be described as a two-step process, a fast disaggregation step followed by a slower disaggregation step. In addition, the aggregation state was progressively less reversible as a function of time spent at the higher pH. This observation is evident when comparing the average number of particles per aggregate and the percentage of particles residing in aggregates for S1-3, S2-3, and S3-3. One possible explanation for the decreased reversibility of aggregation is that during aggregation, water molecules at the interfaces became increasingly ordered as they are coordinated with surface sites and with other water molecules.[22-27] Such ordering would reduce the mobility of molecules and ions near the surface and in the spaces between particles residing in a single aggregate. In our experiments, this could translate to slower and inhibited disaggregation.

3.2 Effects of organic additives on the aggregation state of ferrihydrite nanoparticles

The addition of the organic species alizarin to a suspension of ferrihydrite was found to initially have little to no effect on aggregation state (Fig. 3). In the case of PEG (600 MW) addition, however, a modest change in aggregation state was observed. In PEG-0, the number of particles residing in aggregates decreased while the average number of particles per aggregate slightly increased, which seems to be the result of a modest increase in the $3 \leq n \leq 10$ aggregates (Table 2) as compared to Fh-0.

After aging for 117 h at 80 °C, substantive differences between samples were observed (Fig. 4, Table 2). Representative cryo-TEM images of the aged samples, as well as Fh-0, are shown in Fig. 5. Fh-117 (Fig. 5b) was composed of aggregates containing about one thousand primary particles per aggregate, with no isolated primary particles observed and 99.7% residing in clusters of 25 particles or more ($n \geq 25$; Table 2). In addition, the expected product of crystal growth by oriented aggregation was observed.[8,11,14-16] These crystals were accounted for by measuring the size of the crystals and calculating the number of primary particles required to produce the crystal. No significant difference in the size and shape of the crystals was observed between crystals produced in Fh-117 *versus* PEG-117.

In contrast, the addition of alizarin strongly suppressed aggregation, as can be clearly discerned by the observation that aggregates in Aliz-117 were composed of about three particles on average (Fig. 5c). In addition, only a small fraction of particles (4.8%) resided in aggregates composed of more than 25 primary particles. Finally, no products of crystal growth by oriented aggregation were observed. Interestingly, the alizarin addition also suppressed the phase transformation from ferrihydrite to goethite, as observed *via* XRD. These two observations, the inhibition

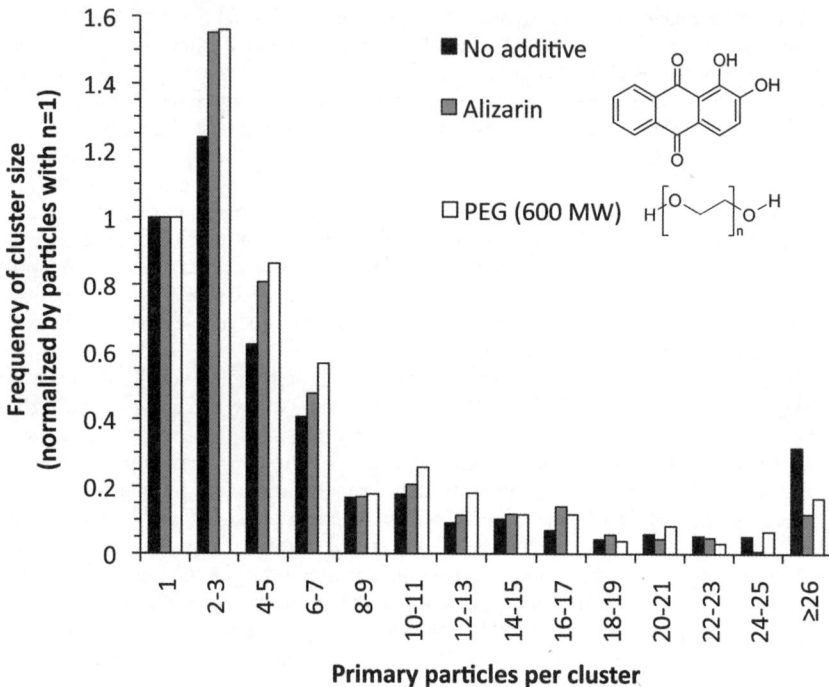

Fig. 3 Bar graph depicting the aggregation state of the untreated sample (black) and the samples treated with alizarin and PEG (600 MW) prior to aging. The frequency is normalized to the number of particles observed in the unaggregated state ($n = 1$). The total number of particles counted was 17 352 (Fh-0), 16 965 (Aliz-0), and 15 915 (PEG-0).

Table 2 Results of cryo-TEM image analysis tracking the average number of particles per aggregate and percentage of particles in aggregates as a function of organic additive and time aged at 80 °C

	Fh-0 $t = 0$ h	Aliz-0 $t = 0$ h	PEG-0 $t = 0$ h	Fh-117 $t = 117$ h	Aliz-117 $t = 117$ h	PEG-117 $t = 117$ h
Average #/aggregate	2.5	2.5	3.4	10^{3a}	3.1	11.9
% In aggregates	77.8	80.2	64.5	100	86.3	97.4
% In aggregates ($n \geq 25$)	7.4	4.7	3.4	99.7	4.8	67.8
% In aggregates ($n \geq 100$)	0	0	0	97.4	0	60.4
# Particles counted	17 352	16 531	15 969	34 196	13 101	22 632

[a] The average number of particles per aggregate reported is a minimum estimate as many of the largest aggregates spanned more than the field of view.

of aggregation as well as the inhibition of phase transformation, are consistent with strong binding of alizarin to the iron oxide surface resulting in stabilization of the ferrihydrite structure.

The aggregation state of PEG-117 was intermediate between the Fh-117 and Aliz-117 (Fig. 4, Fig. 5d, Table 2). In general, far fewer of the aggregates contain more than one hundred primary particles as compared with Fh-117, but the average

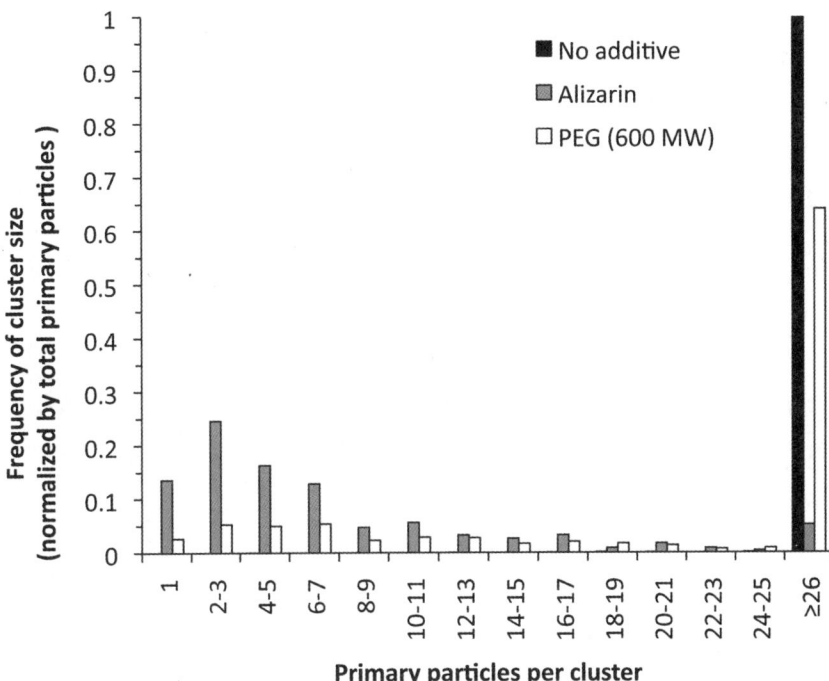

Fig. 4 Bar graph depicting the aggregation state of the untreated sample (black) and the samples treated with alizarin and PEG (600 MW) after 117 h aging at 80 °C and pH 4. The frequency is normalized to the total number of primary particles observed in any state. The total number of particles counted was 32 922 (Fh-117), 13 122 (Aliz-117), and 15 566 (PEG-117).

number of primary particles per aggregate remained quite small (*ca.* 12 particles/aggregate). In addition, the expected product of crystal growth by oriented aggregation was observed.[8,11,14–16] As with the untreated sample, these particles were accounted for by calculating the number of primary particles required to produce each crystal. No significant difference in size or shape of crystals in Fh-117 and PEG-117 was observed.

These results demonstrate that the addition of surface-active species to a suspension of ferrihydrite nanoparticles affects the aggregation state and the eventual transformation to goethite. Adsorbed molecules must be removed from the surface of the primary particles prior to transformation from mesocrystal to oriented aggregate or those species will otherwise be incorporated into the crystal product.[5,16] Therefore, the presence of adsorbed molecules is hypothesized to affect oriented aggregation by two possible mechanisms: (1) affecting particle–particle interactions through steric hindrance and/or changes in surface charge, among other things, and (2) slowing the transformation to single crystal from the mesocrystal state. These results provide important insights to understand the effects of additives on crystal growth by oriented aggregation.

Conclusions

These data provide important information about the early stages of crystal growth by oriented aggregation. Previous work has demonstrated that fractal aggregates precede the formation of mesocrystals and oriented aggregates in the non-classical crystal growth mechanism of oriented aggregation. Furthermore, it has been

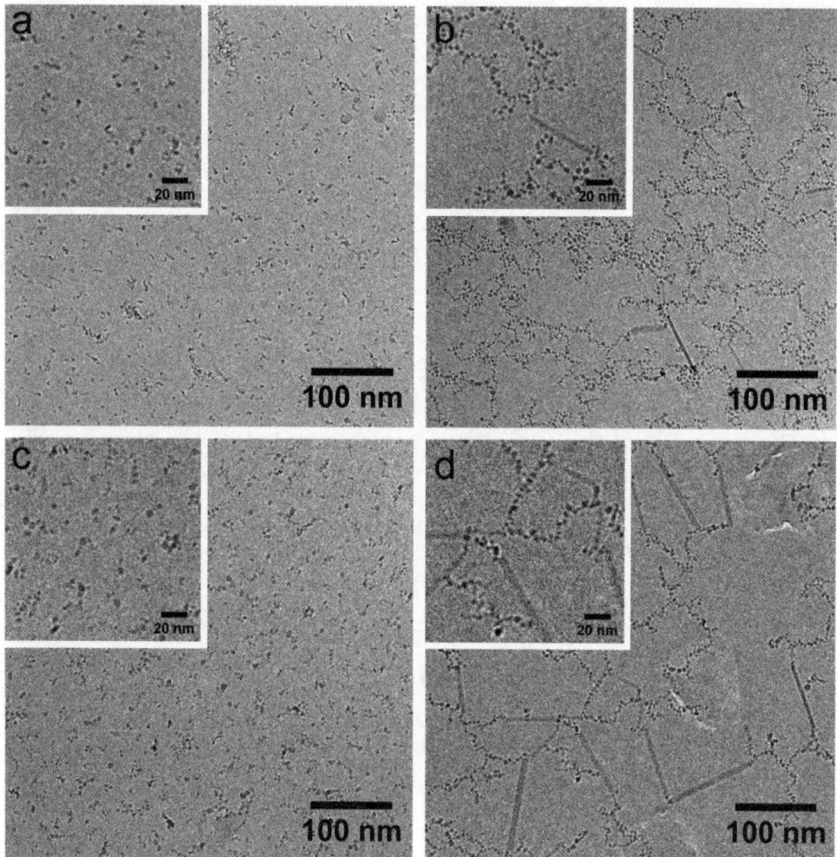

Fig. 5 Cryo-TEM images of Fh$_{45}$ prior to aging (a) and aged for 117 h at 80 °C with (b) no additive, (c) 0.1 mM alizarin, and (d) 600 MW PEG (0.1 mM hydroxyl groups per litre).

demonstrated that primary particles rearrange within such fractal-like aggregates prior to the production of oriented aggregates.[11] The data presented here provide important insights into how primary particles aggregate and disaggregate in response to aqueous solution modifications, including pH changes and the addition of adsorbates. In general, aggregation becomes progressively less reversible when the pH is somewhat close to the PZNPC. The addition of PEG (600 MW) moderately inhibited aggregation while alizarin almost completely inhibited aggregation during aging at elevated temperatures. Thus, the addition of appropriate additives could yield substantive control over crystal growth by oriented aggregation. Finally, cryo-TEM provides a direct method for characterization of aggregation in suspension.

Acknowledgements

We thank the University of Minnesota, National Science Foundation (NSF-0957696 and 1012193) and the Nanostructural Materials and Processes Program (at the University of Minnesota) for financial support. Parts of this work were carried out in the Characterization Facility, University of Minnesota, a member of the NSF-funded Materials Research Facilities Network (www.mrfn.org) *via* the MRSEC program.

References

1 A. Dey, G. de With and N. A. J. M. Sommerdijk, *Chem. Soc. Rev.*, 2010, **39**, 397.
2 A. Baynton, T. Radomirovic, M. I. Ogden, C. L. Raston, W. R. Richmonda and F. Jones, *CrystEngComm*, 2011, **13**, 10–112.
3 X.-L. Fang, C. Chen, M.-S. Jin, Q. Kuang, Z.-X. Xie, S.-Y. Xie, R.-B. Huang and L.-S. Zheng, *J. Mater. Chem.*, 2009, **19**, 6154.
4 R. L. Penn and J. F. Banfield, *Science*, 1998, **281**, 969–971.
5 R. L. Penn, *J. Phys. Chem. B*, 2004, **108**, 12707–12712.
6 R.-Q. Song and H. Cölfen, *Adv. Mater.*, 2010, **22**, 1301–1330.
7 M. Niederberger and H. Cölfen, *Phys. Chem. Chem. Phys.*, 2006, **8**, 3271.
8 V. Yuwono, N. D. Burrows, J. A. Soltis and R. L. Penn, *J. Am. Chem. Soc.*, 2010, **132**, 2163–2165.
9 J.-I. Park, Y.-W. Jun, J.-S. Choi and J. Cheon, *Chem. Commun.*, 2007, 5001–5003.
10 Z. Liu, X. D. Wen, X. L. Wu, Y. J. Gao, H. T. Chen, J. Zhu and P. K. Chu, *J. Am. Chem. Soc.*, 2009, **131**, 9405–9412.
11 N. D. Burrows, C. R. H. Hale and R. L. Penn, *Cryst. Growth Des.*, 2012, in press.
12 F. M. Michel, V. Barron, J. Torrent, M. P. Morales, C. J. Serna, J. F. Boily, Q. Liu, A. Ambrosini, A. C. Cismasu and G. E. Brown, *Proc. Natl. Acad. Sci. U. S. A.*, 2010, **107**, 2787–2792.
13 R. M. Cornell and U. Schwertmann, *The iron oxides: structures, properties, reactions, occurrences and uses*, Wiley-VCH, Weinheim, 2nd edn, 2003.
14 D. J. Burleson and R. L. Penn, *Langmuir*, 2006, **22**, 402–409.
15 R. L. Penn, J. Erbs and D. Gulliver, *J. Cryst. Growth*, 2006, **293**, 1–4.
16 R. L. Penn, K. Tanaka and J. Erbs, *J. Cryst. Growth*, 2007, **309**, 97–102.
17 U. Schwertmann and R. M. Cornell, *Iron oxides in the laboratory: preparation and characterization*, Wiley-VCH, Weinheim, 2000, 2nd edn.
18 M. Abramoff, P. Magelhaes and S. Ram, *Biophotonics Inter.*, 2004, **11**, 36–42.
19 W. S. Rasband, *ImageJ*, U.S. National Institutes of Health, Bethesda, Maryland, USA, 1997–2011.
20 T. J. Collins, *BioTechniques*, 2007, **43**, 25–30.
21 B. Tupper, *Graph*, 2010. http://imagej.nih.gov/ij/plugins/graph/index.html.
22 J. J. Erbs, T. S. Berquó, B. C. Reinsch, G. V. Lowry, S. K. Banerjee and R. L. Penn, *Geochim. Cosmochim. Acta*, 2010, **74**, 3382–3395.
23 S. K. Ghose, G. A. Waychunas, T. P. Trainor and P. J. Eng, *Geochim. Cosmochim. Acta*, 2010, **74**, 1943–1953.
24 Z. Zhang, P. Fenter, L. Cheng, N. C. Sturchio, M. J. Bedzyk, M. Předota, A. Bandura, J. D. Kubicki, S. N. Lvov, P. T. Cummings, A. A. Chialvo, M. K. Ridley, P. Bénézeth, L. Anovitz, D. A. Palmer, M. L. Machesky and D. J. Wesolowski, *Langmuir*, 2004, **20**, 4954–4969.
25 S. Kerisit, D. J. Cooke, D. Spagnoli and S. C. Parker, *J. Mater. Chem.*, 2005, **15**, 1454.
26 D. Spagnoli, J. P. Allen and S. C. Parker, *Langmuir*, 2011, **27**, 1821–1829.
27 A. Navrotsky, *Int. J. Quantum Chem.*, 2009, **109**, 2647–2657.

Biomimetic type morphologies of calcium carbonate grown in absence of additives

Jens-Petter Andreassen,* Ralf Beck and Margrethe Nergaard

Received 25th March 2012, Accepted 6th June 2012
DOI: 10.1039/c2fd20056b

This report demonstrates how typical particle morphologies documented in biomimetic mineralization studies of calcium carbonate will precipitate also from solutions without adding modulating additives, at supersaturation levels below the solubility level of amorphous calcium carbonate (ACC). In the literature, hexagonal plates and flower structures of vaterite, as well as dumbbell structures of aragonite are explained by non-classical aggregation mechanisms from precursor crystals, assisted and stabilized by biomolecules, ions and templates, or by transformation from ACC. By performing experiments at both depleting and constant low supersaturation ratios for a range of temperatures, we show that the vaterite morphology changes from hexagonal monocrystalline plates expressing the basal (001) faces, to dendritic flower shapes and finally to spherulites, as a function of increasing supersaturation. Aragonite goes through a similar transition from monocrystalline elongated structures to polycrystalline dumbbells and spherical structures. We conclude that the key to understand the shape development is quantification of the activity based supersaturation and the realization that calcium carbonate forms along classical crystallization pathways. The higher number of crystals required for aggregation based growth is not favoured at these low supersaturation values. Dislocation and surface nucleation driven crystal growth is responsible for faceted morphologies at moderate supersaturation, whereas dendritic and spherulitic growth patterns appear due to interface instability at higher driving forces.

Introduction

The action of biomolecular, organic and inorganic additives in modulating the solid phase structure and appearance of calcium carbonate has been a matter of great interest in biomimetic mineralization studies, as well as for industrial applications.[1] Calcium carbonate displays a fascinating array of morphologies when precipitated in both natural and synthetic environments. Observations of complex shapes for the three anhydrous polymorphs of calcium carbonate are frequent in the field of biomimetic mineralization,[2] usually explained as a consequence of specific interactions between biomolecules and crystals at different length scales. Certain morphological observations are repeatedly reported in the literature for a range of crystalline materials independent of the nature of the solid precipitating and on the type of additives. Such typical shapes are spherical particles,[3,4] dumbbell[5,6] or peanut structures[7,8] and flower structures.[9,10] The non-equilibrium shape expression, and the fact that some of these structures are polycrystalline, has led to non-classical views on the formation mechanisms, characterised by assembly of precursor crystals, by both

---

Norwegian University of Science and Technology, Department of Chemical Engineering, 7491 Trondheim, Norway. E-mail: jensp@chemeng.ntnu.no; ralfb@chemeng.ntnu.no; margrethe. nergaard@ntnu.no

directional and non-directional aggregation mechanisms.[11,12] For calcium carbonate, such mechanisms are usually suggested when in the presence of molecules which play an essential role in the hierarchical organisation of the mineral structures.[13-15] The amorphous calcium carbonate (ACC) phase has been suggested to act as a template[16] and polymorph dictating precursor[17] for some of these crystalline calcium carbonate particles.

In contrast to the aggregation based non-classical models, similar polycrystalline particle morphologies are also explained by the classical crystallization paradigm by morphology patterns appearing at higher driving forces resulting in kinetic roughening and interface instability. At sufficiently low supersaturation crystal growth is driven by step propagation originating from dislocations,[18] and the surface morphology can be altered by additives interacting with these steps.[19] It has also been shown that the overall crystal morphology is a direct consequence of the different additive's ability to influence the shape of the dislocation patterns.[20] A higher driving force favours two dimensional nucleation on the crystal surface,[21] which readily results in hopper crystals.[22] The size of the two-dimensional nucleus is reduced by increasing the supersaturation, leading to a rough, highly kinked surface and to a switch from surface integration to kinetics controlled by the diffusion of growth units from the surrounding fluid bulk.[23] Depending on the diffusion field around the particle and the crystal geometry, certain parts of the crystal surface, like corners and edges, will experience higher supersaturation values and thereby grow faster, leading to morphological instability and to dendritic growth of monocrystalline structures.[24] Spherulitic growth is observed as non-crystallographic branching due to non-epitaxial surface nucleation at yet higher supersaturation, as reported frequently in crystallization of polymers and minerals from viscous melts, as well as for minerals grown from solution.[25] The particle morphology will in this case depend on the mechanism of spherulitic growth. Non-epitaxial growth-front nucleation from a central nucleus, results in isotropic type 1 spherulites, while type 2 spherulites start from an elongated precursor crystal by branching on the two fast growing tips of the structure, leading to dumbbell intermediates, and finally ending up as fully curled polycrystalline spheres.

The interpretation of the role of additives in crystallization processes will thus depend on the understanding of crystal growth; is it driven by aggregation of precursor crystals or by attachment of ions, and how will these mechanisms determine the morphology of the resulting crystalline particles? The chemical potential difference, expressed by the solution supersaturation, has a pivotal role in explaining morphology shifts in classical crystallization by monomer addition, whereas the role of the supersaturation in the non-classical paradigm is less clear. It seems reasonable to assume that an aggregation based mechanism relies on a high nucleation rate and hence a higher supersaturation, due to the high number of nano-sized precursor units required to build the final particles.[26]

Biomimetic mineralization studies are frequently performed by experimental protocols where the resulting driving force is difficult to quantify because it relies on time-dependent absorption or desorption of carbon dioxide in different geometrical configurations. Generation of supersaturation is quite often accomplished by simultaneous in-diffusion of carbon dioxide and ammonia, caused by decomposition of solid ammonium carbonate, to a solution placed in a desiccator containing calcium ions and the additive, at a rate determined by a certain number of holes pinched in the parafilm covering the solutions.[27,13] Alternatively, supersaturated conditions can be initiated by destabilizing a calcium bicarbonate solution by outgassing of carbon dioxide over time by the so-called Kitano-method.[28,16] Crystallization of calcium carbonate by these methods in the absence of additives represents a reference or a blank for biomimetic studies since the product is the typical rhombohedral morphology of calcite when supersaturation builds up over time at ambient temperatures. The outcome of adding additives or introducing template structures to this system is often a departure from the reference, resulting in an alteration of the calcite morphology or a polymorphic shift to vaterite or aragonite when calcite growth is inhibited. This kinetic stabilization opens up a parameter window for the other polymorphs, at low temperatures and supersaturations, where these structures are not normally detected. Typical resulting morphologies of vaterite are monocrystalline hexagonal plate particles in the presence of a N-trimethylammonium derivative of hydroxyethyl cellulose[29] or florets with hexagonal symmetry under Langmuir monolayers.[30] The hexagonal particles are claimed to be the product of nano-aggregation whereas the florets are said to form by dendritic growth. In these studies stabilization of the basal (001) face by an additive is essential to develop these morphologies. In another recent study of hexagonal vaterite[31] performed by mixing 10 mM calcium chloride and 10 mM ammonium carbonate, the hexagonal morphology is directly related to the presence of NH_4^+ and the particles are templated by the initially formed ACC through a type of solid state transformation. Lens and rosette shaped vaterite has also been produced[32] by gas diffusion experiments in the presence of ammonium ions and their formation is explained by stabilization of [001] nanosheets that subsequently undergo nano-aggregation to produce stacked monocrystalline hexagonal super-plates. The morphology of aragonite observed in experiments in the presence of various additives at room temperature, are typically dumbbell structures. Similar superstructures are identified at higher temperature by decomposition of urea in calcium and additive containing solutions. The interpretation of the shape information leads to formulation of different advanced aggregation mechanisms assisted by the varying additives in question, and explained at a consequence of specific interactions with these molecules, like hexamethylenetetramine,[33] soluble starch,[34] poly(N-vinyl pyrrolidine),[35] trisodium citrate.[36]

In the present work, we emphasize the importance of the thermodynamic driving force for morphology development by focusing our attention at similar vaterite and aragonite particles in aqueous solutions at ambient (20–40 °C) temperatures. We demonstrate how the same morphologies will develop in the absence of any additives and without any precursor ACC, by simply controlling the activity based solution supersaturation at moderate levels at different temperatures.

Experimental

Crystallization experiments

In order to elucidate the effect of supersaturation and temperature on the particle morphology in additive free experiments, three different protocols were applied. Spontaneous precipitation experiments (Table 1) from the mixture of low concentration Na_2CO_3 and $CaCl_2$ solutions resulting in high pH ∼ 10, were performed in a 1 litre stirred (500 rpm) vessel. The reactor volume was sealed and the gas head volume was minimized to avoid absorption of CO_2 during the experiments. Spontaneous experiments were also performed at pH ∼ 7 by mixing a carbonate–bicarbonate solution, made by equilibrating solutions of different concentrations of NaOH with carbonate dioxide until the pH was stable, with a similarly equilibrated solution of $CaCl_2$, and by continuous addition of CO_2 as the reaction proceeds. Knowledge of the alkalinity (NaOH added), the partial pressure of CO_2 (∼1 bar), the free calcium concentration and activity coefficients facilitate the calculation of the supersaturation, without concern for desorption of CO_2 during the experiments. Due to the lower carbonate activity of this system (Table 2), the calcium concentration was set to a higher value to reach comparable initial supersaturation values relative to the high pH experiments. The carbonate activity is replenished during precipitation in this case, due to exchange with the CO_2 gas phase as long as the alkalinity is not significantly depleted. As such, these experiments work in a semi-batch fashion until all the calcium is consumed, giving a more moderate drop in supersaturation. The activity coefficients will have a lower value in the moderate pH experiments due

Table 1 Experimental conditions for spontaneous precipitation at high pH: S_a and S_v refers to the activity based supersaturation (eqn (2)) for aragonite and vaterite. The temperature is given in °C and concentrations as mol kg^{-1} solvent

T	S_a	S_v	pH	c, CaCl$_2$	c, freeCa^{2+}	a, Ca^{2+}	c, Na$_2$CO$_3$	c, freeCO$_3^{2-}$	a, CO$_3^{2-}$	$\gamma_\pm$
30	8.4	5.2	10.36	1.6×10^{-3}	1.0×10^{-3}	7.4×10^{-4}	1.6×10^{-3}	5.5×10^{-4}	4.0×10^{-4}	0.74
40	3.8	2.4	9.94	5.8×10^{-4}	4.6×10^{-4}	3.6×10^{-4}	5.8×10^{-4}	1.8×10^{-4}	1.4×10^{-4}	0.79
40	4.6	2.9	9.99	7.3×10^{-4}	5.5×10^{-4}	4.2×10^{-4}	7.3×10^{-4}	2.3×10^{-4}	1.8×10^{-4}	0.77
40	5.3	3.4	10.03	8.7×10^{-4}	6.3×10^{-4}	5.2×10^{-4}	8.7×10^{-4}	2.4×10^{-4}	1.9×10^{-4}	0.82
40	5.9	3.8	10.06	1.0×10^{-3}	7.0×10^{-4}	5.2×10^{-4}	1.0×10^{-3}	3.3×10^{-4}	2.4×10^{-4}	0.74
40	7.0	4.5	10.10	1.2×10^{-3}	8.3×10^{-4}	6.0×10^{-4}	1.2×10^{-3}	4.1×10^{-4}	3.0×10^{-4}	0.72
40	10.3	6.5	10.20	2.2×10^{-3}	1.2×10^{-3}	8.2×10^{-4}	2.2×10^{-3}	7.0×10^{-4}	4.7×10^{-4}	0.66

to higher ionic strength as a result of the alkalinity manipulation (Table 2). Experimental details regarding these setups have been explained in previous reports.[37,38] The onset of precipitation was determined by a daily calibrated pH-meter (Mettler-Toledo SevenMulti) and samples were extracted at different time intervals after the first change in the pH value, as a result of calcium carbonate precipitation.

The third group of experiments were performed in a constant composition setup where similar initial supersaturated solutions with high pH (~10) were prepared as the initial working solution. Sodium chloride was added to the working solution in order to maintain a constant ionic strength upon addition of reactants form the burettes.[39] At the onset of precipitation the *in situ* pH-probe (Synrode, Metrohm) triggers the addition of CaCl$_2$ and Na$_2$CO$_3$ solutions from two burettes to maintain the solution pH as constant, by means of the Metrohm titrator "902 Titrando" and the software "Tiamo 2.2". In all experiments, the burette concentrations of CaCl$_2$ and Na$_2$CO$_3$ were 0.128 mol kg^{-1} solvent. The constant composition approach was pioneered by Nancollas and co-workers and for details on the underlying theory we refer to these papers.[39] The constancy of the pH during the experiments guarantees a constant supersaturation, as long as the alkalinity, total calcium and the ionic strength is maintained constant, and CO$_2$ ingress is avoided.[40] Some constant composition experiments were also performed by seeding premade vaterite spherulites into the working solution of different initial supersaturations. The vaterite seed particles were manufactured as explained previously.[26,41]

Liquid samples from the experiments were extracted at different time intervals and filtered through Millipore membrane filters (0.22 μm) and washed with ethanol, dried overnight (60 °C) and investigated by scanning electron microscopy, SEM (Hitachi N-3400 and Zeiss Ultra) after sputtering with gold. Powder-XRD patterns were recorded on a Bruker D8-Focus to confirm the presence of different polymorphs (data not shown).

Solution speciation and supersaturation

The driving force for precipitation of calcium carbonate is given by the difference between the chemical potential in solution, and the solid phases, $\Delta\mu$. The relationship between the driving force, $\Delta\mu$, the activity product of calcium and carbonate ions and the thermodynamic solubility product, K_{sp} is given by eqn (1):

$$\frac{\Delta\mu}{RT} = \ln\frac{a_{Ca^{2+}} \cdot a_{CO_3^{2-}}}{K_{sp}} \tag{1}$$

where R is the universal gas constant and T is the absolute temperature. The square root of the activity product of calcium and carbonate ions divided by the thermodynamic solubility product is defined as the supersaturation ratio S:

Table 2 Experimental conditions for spontaneous precipitation at moderate pH. S_a and S_v refers to the activity based supersaturation (eqn (2)) for aragonite and vaterite. The temperature is given in °C and concentrations as mol kg^{-1} solvent

T	S_a	S_v	pH	c, CaCl$_2$	c, freeCa^{2+}	a, Ca^{2+}	c, NaOH	c, freeCO$_3^{2-}$	a, CO$_3^{2-}$	$\gamma_\pm$
40	3.8	2.4	6.90	5.0 × 10^{-3}	4.9 × 10^{-3}	1.3 × 10^{-3}	1.2 × 10^{-1}	1.4 × 10^{-4}	3.8 × 10^{-5}	0.27
40	4.2	2.7	6.97	5.0 × 10^{-3}	4.9 × 10^{-3}	1.2 × 10^{-3}	1.4 × 10^{-1}	2.1 × 10^{-4}	5.2 × 10^{-5}	0.25
40	4.6	2.9	7.03	5.0 × 10^{-3}	4.8 × 10^{-3}	1.1 × 10^{-3}	1.7 × 10^{-1}	2.9 × 10^{-4}	6.7 × 10^{-5}	0.23
40	5.0	3.1	7.08	5.0 × 10^{-3}	4.8 × 10^{-3}	1.1 × 10^{-3}	1.9 × 10^{-1}	3.8 × 10^{-4}	8.4 × 10^{-5}	0.22
40	5.3	3.4	7.12	5.0 × 10^{-3}	4.8 × 10^{-3}	9.8 × 10^{-4}	2.1 × 10^{-1}	5.0 × 10^{-4}	1.0 × 10^{-4}	0.21
40	6.0	3.8	7.21	5.0 × 10^{-3}	4.7 × 10^{-3}	8.6 × 10^{-4}	2.7 × 10^{-1}	8.1 × 10^{-4}	1.5 × 10^{-4}	0.18
40	7.0	4.5	7.71	5.0 × 10^{-3}	4.6 × 10^{-3}	7.1 × 10^{-4}	3.6 × 10^{-1}	1.6 × 10^{-3}	2.5 × 10^{-4}	0.15
40	8.1	5.2	7.44	5.0 × 10^{-3}	4.4 × 10^{-3}	5.7 × 10^{-4}	5.0 × 10^{-1}	3.2 × 10^{-3}	4.1 × 10^{-4}	0.13

$$S = \sqrt{\frac{a_{Ca^{2+}} \cdot a_{CO_3^{2-}}}{K_{sp}}} \qquad (2)$$

The activity of calcium and carbonate ions is defined as the product of the mean activity coefficient, $\gamma_\pm$ and the free concentration of the respective species. The concentration of the ions is reduced in solution (free concentration) due to association with other solution constituents. In this work all equilibria concerning the species, CaHCO$_3^+$, HCO$_3^-$, CO$_3^{2-}$, H$^+$, OH$^-$, H$_2$O, CO$_2$(aq), CO$_2$(g), Ca^{2+} as well as the aqueous complex of calcium carbonate, CaCO$_3^0$(aq) have been taken into consideration. MultiScale 7.0 was used to calculate the free ion concentrations and the activities, utilizing the Pitzer[42] model. Thermodynamic solubility products for the different polymorphs of calcium carbonate were adopted from the work of Plummer and Busenberg.[43] It has recently been proposed that the association of calcium and carbonate ions in solution is not as ion pairs but as clusters of a higher number of ions in clusters, but we still regard the equilibrium values reported in the literature to be valid, irrespective of the nature of the association.[44] Activity based supersaturation values are essential for validating growth mechanisms of crystals[45] due to the non-ideal behaviour, even at very low thermodynamic driving forces (ionic strength). Since mechanistic studies within the field of biomimetic mineralization quite often are performed on the basis of concentration only, we have reported both the initial concentration, the free concentrations, as well as the resulting activities of calcium and carbonate ions (Table 1 and 2) for the experiments performed at high and moderate pH, respectively, limited to the experiments that we report results for in this paper. These initial supersaturation conditions, expressed as the ionic activity product (IAP)along with the IAP-values for the constant composition experiments, are reported in Fig. 1, to show their position relative to the two thermodynamic K_{SP}s for ACC reported in literature[46,47] and the K_{SP}-values for the three polymorphs.[43]

Results and discussion

Spontaneous batch experiments

All three polymorphs were detected in the experiments performed by spontaneous precipitation (Table 1 and 2). Most experiments were performed well below the solubility of ACC as shown in Fig. 1. The content of calcite increases with time due to the inherent solution assisted transformation by dissolution of the other polymorphs[48] as the solution becomes saturated or undersaturated. It is thus more challenging to study the morphological development of vaterite and aragonite in the absence of additives that can stabilize them kinetically by inhibiting calcite. This was solved by ending the experiments before the dissolution process was initiated, in order to investigate their morphology response to the initial driving force and temperature. For the constant composition experiments this is not a problem since the solution is kept supersaturated with respect to all the polymorphs throughout the experiment.

The reduction of pH with time illustrates the particle formation rate and the de-supersaturation of the solutions as shown below for experiments performed at pH ~ 10 and pH ~ 7 in Fig. 2 and 3, respectively. For the experiments at low initial supersaturation values there is a considerable induction time, whereas experiments at S_a-values approaching the solubility of ACC precipitate instantaneously. The reduced nucleation rate at lower supersaturations is also evident from the smaller slope of the pH-curve due to fewer growing crystals.

The fact that aragonite is detected even at the lowest temperatures studied, although in small amounts, tells us that generalized rules and polymorph diagrams as a function of temperature is virtually impossible to construct due to the simultaneous large effect of the initial supersaturation.[49] We find that hexagonal platelet morphologies of vaterite expressing their (001) faces are formed in the whole temperature range, but that low supersaturation conditions are required. This means that there is no need for additives in order to stabilize this morphology as often claimed in the literature.[31,32] The morphology development of these particles as a function of increasing the initial supersaturation is shown in Fig. 3. At higher driving forces the particles become thicker and lens shaped, *i.e.* the growth rate of the (001) face increases more rapidly than the growth rate of the six prism faces. We see a direct analogy to the formation of snow crystals in this respect. Snow crystals form as hexagonal plates at supersaturation below the kinetic roughening transition value of the driving force. The thickness of the plates is a matter of the relative growth rates of the prism and basal facets.[24,50] The thickness development of the vaterite plates is limited due to the available supersaturation during the experiment, but for an initial supersaturation of $S_v = 6.5$ the plates are substantially thicker with a layered morphology indicating that surface instability is initiated. At $S_v = 8.7$ at 30 °C (Fig. 4a) the thickness of the plates increased even further and tilting of the layers towards the *c*-axis closes the particles around the poles to form spheres and

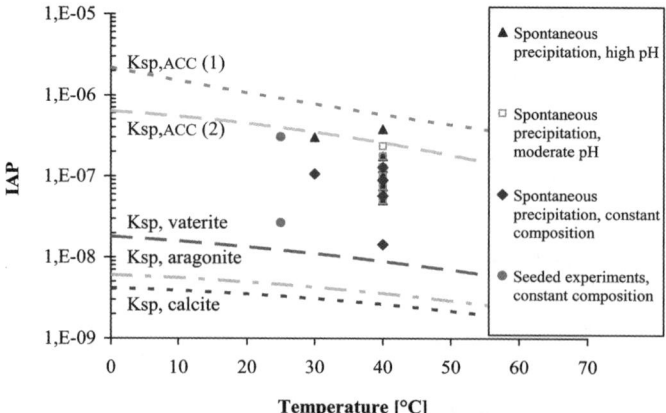

Fig. 1 The initial ionic activity product (IAP) in all reported experiments. The constant composition data points will stay at the same position during the experiments resulting in constant supersaturation, whereas the points for the spontaneous de-supersaturation experiments at high and moderate pH will, given enough time, move down and pass the solubility lines of vaterite and aragonite before ending up at the calcite solubility line. The solubility products, $K_{SP,ACC}(1)$ and $K_{SP,ACC}(2)$ were adopted from Clarkson[47] and Brecevic and Nielsen,[46] respectively, and for the crystalline polymorphs from Plummer and Busenberg.[43]

this could very well be the mechanism for the formation of spherulitic seed particles of vaterite at this temperature. The vaterite seed spherulites (Fig. 4b) are produced at high initial concentrations and have grown in a constant supersaturation environment dictated by the dissolution of the resulting amorphous compound (corresponding to $S_v = 5.7$, Fig. 1).[26] Fig. 3 also reveals important information about the development of aragonite as the initial supersaturation is increased in additive free experiments. Both star shaped type 1 spherulites and elongated needle shaped (Fig. 3a, upper left corner) particles are identified. The needles start to branch into dumbbell structures at $S_a = 6.0$ and $S_a = 7.0$ (Table 1 and Fig. 3b and c) and finally end up as fully grown type 2 spherulites at $S_a = 10.3$ (Fig. 3d).

The experiments at moderate pH are performed at a carbonate ion activity about ten times lower than at high pH, hence the calcium activity is increased accordingly to start at a comparable supersaturation (Table 2). The lower activity coefficients compared to high pH are a result of higher ionic strength originating from the added sodium hydroxide required to produce the level of carbonate activity. The de-supersaturation is shown as a decline in pH in Fig. 5. As seen from the figure and as explained in the experimental section, precipitation occurs at a lower de-supersaturation rate as compared to high pH. Due to similar values of the initial supersaturation, however, the induction times are more or less comparable, but the time-dependent evolution of the vaterite morphology is different (Fig. 6).

Floret-type particles are probably produced as a result of roughening of the prism faces leading to progressive dendritic growth perpendicular to the basal face as a function of time. It is interesting to note that similar morphologies are produced in experiments based on the decomposition of urea, which work in a similar semi-batch fashion; by temperature- and time-dependent release of carbonate ions into a calcium containing solution.[51] From the micrographs in both Fig. 3 and Fig. 6 it seems evident that growth on the basal faces is dominated by surface nucleation. The surface roughness is probably a result of polynuclear dominated growth and insufficient lateral growth of these islands to fill in the voids, leading to protrusions on the surface.[21] These details in the surface are often interpreted as nanoparticles originating from the solution, which integrate in the surface by a mechanism that guarantees monocrystallinity (mesocrystals). However, the enormous number of nanoparticles for such a mechanism would require homogeneous bulk nucleation at a very high supersaturation.[26] It is also inherently difficult to produce a high number of crystalline nanoparticles of calcium carbonate since the favoured product at higher supersaturation is nano-particles of ACC. However, in the present study

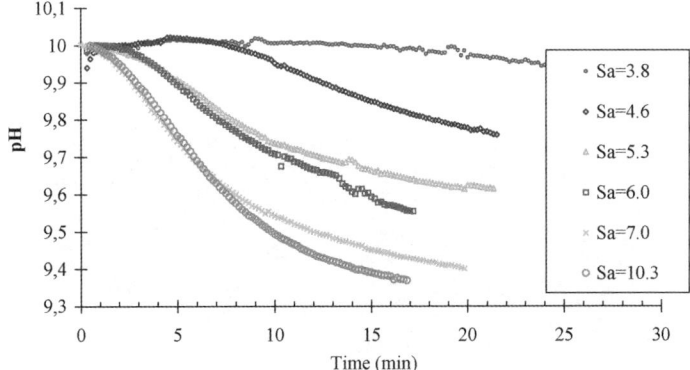

Fig. 2 pH-development with time for different supersaturations in spontaneous precipitation experiments at high pH and 40 °C. The starting points are shifted from the respective initial pH-values (Table 1) to pH = 10 for easier comparison. Samples withdrawn during the experiment are seen as disturbances ($S_a = 3.8$, $S_a = 5.3$ and $S_a = 5.0$) in the pH-curve.

Fig. 3 Development of vaterite plates and type 2-spherulites of aragonite as a result of increased supersaturation in spontaneous precipitation experiments at high pH and 40 °C. Time refers to minutes after experiment start and S_v is the initial supersaturation. From upper left to lower right: (a) $S_v = 2.9$ after 20min, (b) $S_v = 3.8$ after 16 min, (c) $S_v = 4.5$ after 16 min, and (d) $S_v = 6.5$ after 16 min. The magnification is 3200× and the total scale bar 10μm.

the solutions are not supersaturated with respect to ACC. For systems that do not produce amorphous compounds as easily, like $BaSO_4$, the production of crystalline nanoparticles is just a matter of increasing the supersaturation sufficiently. The resulting nanoparticles do not however, show any specific tendency to aggregate.[52] As seen in Fig. 6d, increasing the supersaturation to $S_a = 8.1$ results in instability also along the c-axis, causing dendritic-like patterns also in that direction. No electron diffraction studies have been performed in this work and consequently it is not possible to determine the shift in growth mechanism leading from monocrystalline (dendritic) to polycrystalline spherulitic particles. The monocrystallinity of lens-shaped and floret morphologies has been demonstrated previously[30] and the seed particles (Fig. 4b) have been shown to be polycrystalline.[26] Detailed investigations of this shift will be a matter of future investigations.

Once precipitation starts in these batch systems, the supersaturation will drop according to the relative abundance of polymorphs, as well as their respective growth rates; two aspects which are mutually dependent. Differences in the de-supersaturation rate might affect the ongoing nucleation of the different polymorphs, the

Fig. 4 (a) Spherical vaterite precipitated in a high pH-experiment at the conditions of $S_a = 8.4$ and 30 °C, 13 min after experiment start. The magnification is 5000× and the total scale bar 10 μm. (b) Seed crystals of vaterite precipitated from initial $S_v \sim 80$, resulting in $S_v = 5.7$ during the transformation of ACC. The magnification is and 6400× and the scale bar is 2 μm.

supersaturation they experience and hence their morphology. In order to verify that the aragonite spherulite type 2 branching and vaterite development is truly an effect of changes in the supersaturation level, constant composition experiments were designed and performed at supersaturation values comparable to the initial values of the de-supersaturation experiments.

Four different experiments were performed at 40 °C with different initial supersaturation values in the working solutions (Fig. 7). Titration of calcium and carbonate from the burettes starts when carbonate ions disappear from the solution due to nucleation and growth of calcium carbonate, leading to a reduction in pH. Titration sets in later when the initial supersaturation level is lowered, reflecting the induction times seen in Fig. 2. At $S_a = 2$ nothing happens within 35 min as can be seen from the constancy of the pH. If the temperature is lowered to 30 °C, the onset of titration is delayed as a result of slower precipitation kinetics which can be seen from the two added volume curves at $S_a = 5$. The pH is kept constant also during the titration at $S_a = 5$ and 30 °C, but at some point the feed-back control breaks down as a consequence of the exponential increase in particle surface area that increases. Similar behaviour is seen for all these experiments, as a result of aragonite switching from faceted to spherulitic growth.

This development of aragonite can be seen from the micrographs in Fig. 8. The experimental conditions are similar to those leading to Fig. 3(a) and the vaterite morphology is comparable. Instability is initiated on the fast growing tips of aragonite after 5 min (Fig. 8a) and ten minutes later the type 2 spherulitic curling pattern is well developed. The curling leads to complete spherical structures after 25 min (data not shown).

By starting at a higher initial supersaturation, equal to the level applied for Fig. 3(b), it can be shown how sustained supersaturation at this level promotes far more spherulitic curling leading to sheaf of wheat particles and type 1 structures already after three minutes (Fig. 9a), which branch substantially during the next five minutes (Fig. 9b). At this higher level of supersaturation the particles will develop into fully grown spherulites much faster than at $S_a = 5$, but the experiments are difficult to run further due to the high demand for titrants caused by the large surface area of aragonite.

We found that the morphology results by reducing the temperature to 30 °C, were fairly similar. In this case the onset of precipitation was delayed, but once the

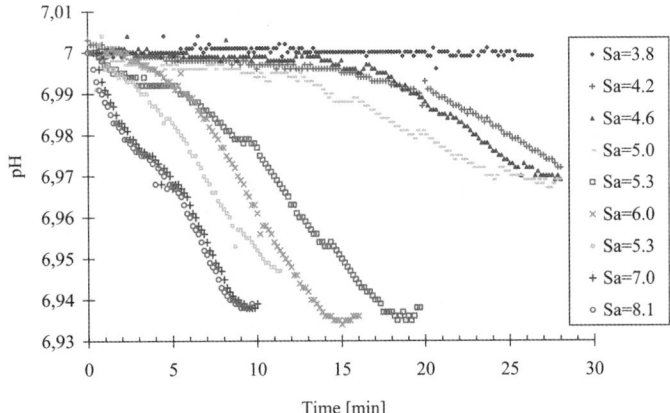

Fig. 5 pH-development with time for different supersaturations in spontaneous precipitation experiments at moderate pH and 40 °C. The starting points are shifted from the respective initial pH-values (Table 1) to pH = 7. Disturbances in the pH-curve (e.g. after 5, 10 and 15 min for $S_a = 5.3$) correspond to withdrawn samples.

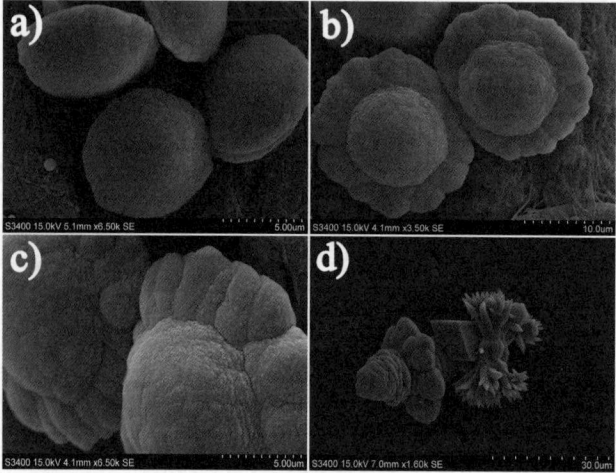

Fig. 6 Time development of vaterite plates at an inital supersaturation of $S_v = 3.4$ ($S_a = 5.3$) at moderate pH: (a) after 5 min, scale bar = 5 μm, (b) after 11 min, scale bar = 10 μm, (c) after 16 min, scale bar = 5 μm. The effect of increasing the initial supersaturation to $S_v = 5.2$ ($S_a = 8.1$) is shown in (d), scale bar = 30 μm.

branching of aragonite starts the rate of titration escalates. The samples shown in the micrographs in Fig. 10 were collected five minutes apart. Compared to the experiment at $S_a = 5$ and 40 °C, the repeating branching unit size is reduced and as a consequence, the spherulites develop a denser morphology. This is in line with our previous investigations of the branching frequency at higher temperatures, based on depleting supersaturation in batch experiments.[38]

The behaviour of these constant composition experiments, based on initial spontaneous precipitation, was dominated by the polycrystalline growth of aragonite. In order to test the growth behaviour of vaterite at constant supersaturation, working solutions were prepared and immediately seeded with polycrystalline vaterite spherulites.[26,41] This was done to further investigate the shift from monocrystalline to polycrystalline growth regimes as a result of supersaturation variations, in agreement with classical crystallization predictions. Seeding the solutions allows us to investigate vaterite at conditions where it will not precipitate spontaneously in sufficient amounts. The vaterite seed spherulites (Fig. 4b) have grown in a constant supersaturated environment, dictated by the dissolution of the resulting amorphous compound (corresponding to $S_v = 5.7$). Hence, we wanted to study the difference in morphology when these seed particles are grown at a similarly high and a much lower supersaturation in the constant composition set-up at 25 °C, as illustrated in Fig. 1. It is challenging to run the set-up at high supersaturation due to simultaneous spontaneous nucleation in the solution, causing some calcite to nucleate. However, it was possible to run the experiment at $S_v = 5$ at 25 °C and the result is shown in Fig. 11. The large size of the vaterite crystals (Fig. 11a) compared to the seeds (Fig. 4b, same magnification) only 7 min after seeding, demonstrates the high growth rate of vaterite at these conditions. The grain sizes in the grown layers on the particles is in the <50 nanometre range and the grains seem to be more or less randomly oriented probably reproducing the polycrystalline features of the underlying seed spherulite (Fig. 11b).

When the supersaturation was reduced to the low value of $S_v = 1.5$ the final size of the vaterite particles was reduced due to a much smaller growth rate, as can be seen by comparing Fig. 11 (a) and Fig. 12(a), at the same magnification. However, growth for one hour at these conditions caused the surface grain size

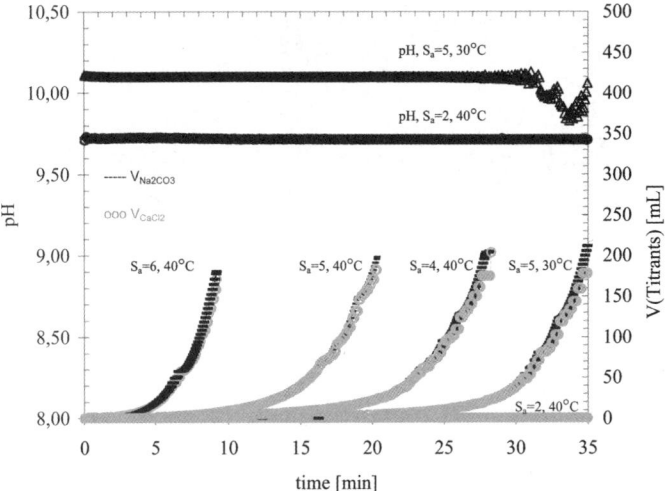

Fig. 7 Volume of CaCl$_2$ solution (from burette 1) and Na$_2$CO$_3$ solution (from burette 2) that is titrated into the crystallizer in order to keep the supersaturation level constant by means of maintaining solution pH. The pH-curves for $S_a = 5$ at 30 °C and $S_a = 2$ at 40 °C are shown for comparison.

to increase and re-established some of the symmetry of the monocrystalline vaterite. The grains are growing out from the equator of the particle and there are poles developing (Fig. 12b), and the whole of the particle is starting to resemble the hexagonal stacked plate morphology seen in the spontaneous experiments (Fig. 3d and 4a).

Increasing the temperature from 25 °C to 60 °C at a comparably low supersaturation ($S_v = 1.3$) gives substantially larger particles as can be expected from a classical crystallization point of view, as seen by comparison at the same magnification in Fig. 12 (a) and 13(b). The morphology is less compact and dominated by hexagonal prismatic crystals oriented perpendicular to the pole-axis. The hexagonal plates within each particle are competing for the supersaturation, the number of plates is going down with time and we speculate that the structures will end up as one large hexagonal plate given enough time at these conditions.

The switch from polycrystalline to monocrystalline growth is in accordance with our previous findings where a simpler semi-batch approach was used.[41] Vaterite grows according to a polycrystalline growth mode at high initial supersaturation

Fig. 8 Aragonite particles and vaterite plates grown at a constant supersaturation ratio of $S_a = 5$ at 40 °C. The depicted crystals have been withdrawn from the reactor, washed and dried after (a) 5 min and (b) 15 min. The magnification is 1600× and the scale bar is (a) 10 μm and (b) 3 μm.

Fig. 9 Aragonite particles and vaterite plates grown at 40 °C at a constant supersaturation ratio of (a) $S_a = 4$ after 3 min and (b) $S_a = 6$ after 8 min. The magnification is 1600× and the scale bar is 10 μm in both cases.

and according to a monocrystalline growth mode at low initial supersaturation. These results show how combinations of monocrystalline and polycrystalline growth habits can be alternated by seeding with polycrystalline spherulites of vaterite. The resulting hexagonal monocrystalline platelets that grow from the nanopatterned spherulite surface vary in size with the supersaturation. It produces a pattern which is different on the spherulite poles compared to the equators, in correspondence with the underlying symmetry produced by either a type 2 spherulitic growth mechanism or the stacked layered plate development, illustrated by Fig. 4. For higher values of supersaturation, the surface units are again in the nano-size range, not as a consequence of bulk nucleation and subsequent transport of such units to the surface by assembly, but by crystal growth in the classical sense, and due to growth front nucleation characteristic of spherulitic growth. Results of phase field modelling[25] cohere with the observed results. High supersaturation gives rise to a high rate of growth front nucleation, in turn leading to extended branching, small crystal grains and compact crystal morphology. We are currently studying particles at alternating high and low supersaturation in constant composition experiments to demonstrate this point further (work in progress). This will produce particles with cross-sectional morphology variations as often found in nature due to shifting environmental conditions.[22]

Implications for biomimetic mineralization studies and for the precursor aggregation based paradigm

Numerous studies during the last decade are explaining complex morphologies as a result of additive assisted aggregation of precursor crystals. Aggregation-based mechanisms imply that a high number of crystalline precursor particles are nucleated in the solution and are collectively dictating the shape of the resulting polycrystalline or monocrystalline structures. Additives are claimed to assert a

Fig. 10 Aragonite particles, plates of vaterite and rhombohedra of calcite grown at 30 °C a constant supersaturation ratio of $S_a = 5$ after (a) 30 min and (b) 35 min. The magnification is 1600× and the scale bar 10 μm in both cases.

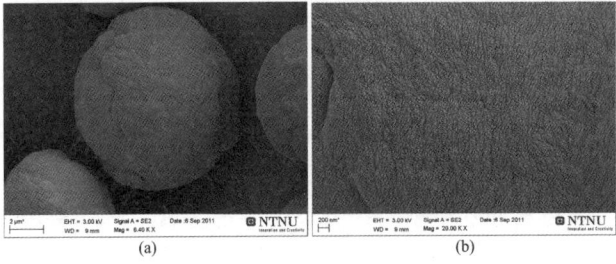

Fig. 11 Vaterite particles grown from vaterite seed crystals (Fig. 4b) at a constant supersaturation ratio of $S_v = 5$ at 25 °C after 7 min. The magnification is 6400× and the scale bar is 2 μm in (a) and 20 000× and 200 nm in (b).

specific controlling function in this assembly process. The present work suggests that this is not the case. The same morphologies can be produced without any additives and the thermodynamic driving force applied is too low for the high nucleation rates required to produce a sufficient number of nanoparticles, to support such a mechanism. The framework of "classical crystallization" is sufficient for understanding our results. Classical crystallization mechanisms are often associated with faceted mono-crystalline particles only, but this is a distorted simplification. Dendritic growth patterns are well studied as a consequence of increasing supersaturation leading to Mullins–Sekerka instability.[53] Spherulitic polycrystalline growth is also well studied, but mostly in other disciplines than solution crystallization, like in melt crystallization of polymers where spherulitic growth is a major field of interest.[54] Classical crystallization relies on quantification of the thermodynamic driving force and that should also apply to proposed revisions of these concepts. The new hypotheses within aggregation based growth originate from studies where the supersaturation is inaccessible. Only initial concentrations are reported and in the case of calcium carbonate, the carbonate concentration is often varying over time depending on the experimental design. Activity measures are not provided even though the solutions are non-ideal due to significant ionic strength.[32] We hence recommend experiments that fix the activity of both calcium and carbonate prior to solids formation as proposed in this study, rather than gas diffusion experiments, in order to study the effect of additives on the formation mechanism and morphology development of calcium carbonate. In order to evaluate the effects of additives on the growth of the metastable polymorphs these should ideally be seeded to a system of constant supersaturation in order to keep above the solubility of these polymorphs at all times, to prevent transformation.

Fig. 12 Vaterite particles grown from vaterite seed crystals (Fig. 4b) at a constant supersaturation ratio of $S_v = 1.5$ at 25 °C after 60 min. The magnification is 6400× and the scale bar is 1 μm in (a) and 20 000× and 200 nm in (b).

Fig. 13 Vaterite crystals grown from vaterite seed crystals (Fig. 4b) at a constant supersaturation ratio of $S_v = 1.3$ at 60 °C. The depicted crystals have been arrested after 30 min of growth. The magnification is 1600× and the scale bar is 10 μm in (a) and 6400× and 2 μm in (b).

Conclusion

In this report we show that morphologies like hexagonal plates, florets, dumbbells and spheres, produced by biomimetic mineralization studies in presence of a wide range of different additives, can be produced without any additives. Since the mechanism of particle formation is explained by these additives' role in an assembly process of precursor crystals, we also conclude that the true formation mechanism is different. Based on studies of spontaneous and seeded constant composition studies, performed at quantifiable activity based supersaturation values below the solubility of ACC, we propose that calcium carbonate forms along a classic pathway. Hexagonal plate particles of vaterite are produced at low supersaturation by a mechanism of surface integration of ions. The particles develop into florets by dendritic growth and to spheres by spherulitic growth. Similarly, aragonite dumbbells are produced when non-crystallographic branching of the initial needle structures occurs at intermediate driving force. Over time or by increasing the supersaturation, polycrystalline spherulites develop.

References

1 K. Sangwal, *Additives and crystallization processes: from fundamentals to applications*, Wiley, Chichester, 2007.
2 N. Sommerdijk and G. de With, *Chem. Rev.*, 2008, **108**, 4499–4550.
3 K. J. Kim, *J. Cryst. Growth*, 2000, **208**, 569–578.
4 J. P. Andreassen, E. M. Flaten, R. Beck and A. E. Lewis, *Chem. Eng. Res. Des.*, 2010, **88**, 1163–1168.
5 O. Prymak, V. Sokolova, T. Peitsch and M. Epple, *Cryst. Growth Des.*, 2006, **6**, 498–506.
6 S. Busch, H. Dolhaine, A. DuChesne, S. Heinz, O. Hochrein, F. Laeri, O. Podebrad, U. Vietze, T. Weiland and R. Kniep, *Eur. J. Inorg. Chem.*, 1999, 1643–1653.
7 F. C. Meldrum and S. T. Hyde, *J. Cryst. Growth*, 2001, **231**, 544–558.
8 N. Sasaki, Y. Murakami, D. Shindo and T. Sugimoto, *J. Colloid Interface Sci.*, 1999, **213**, 121–125.
9 C. F. Mu, Q. Z. Yao, X. F. Qu, G. T. Zhou, M. L. Li and S. Q. Fu, *Colloids Surf., A*, 2010, **371**, 14–21.
10 J. X. Fang, H. J. You, P. Kong, Y. Yi, X. P. Song and B. J. Ding, *Cryst. Growth Des.*, 2007, **7**, 864–867.
11 E. Matijevic, *Chem. Mater.*, 1993, **5**, 412–426.
12 M. Ocana, R. Rodriguezclemente and C. J. Serna, *Adv. Mater.*, 1995, **7**, 212–216.
13 A. N. Kulak, P. Iddon, Y. T. Li, S. P. Armes, H. Colfen, O. Paris, R. M. Wilson and F. C. Meldrum, *J. Am. Chem. Soc.*, 2007, **129**, 3729–3736.
14 R. Q. Song and H. Colfen, *Adv. Mater.*, 2010, **22**, 1301–1330.
15 A. W. Xu, Y. R. Ma and H. Colfen, *J. Mater. Chem.*, 2007, **17**, 415–449.
16 E. M. Pouget, P. H. H. Bomans, J. Goos, P. M. Frederik, G. de With and N. Sommerdijk, *Science*, 2009, **323**, 1455–1458.
17 D. Gebauer, P. N. Gunawidjaja, J. Y. P. Ko, Z. Bacsik, B. Aziz, L. J. Liu, Y. F. Hu, L. Bergstrom, C. W. Tai, T. K. Sham, M. Eden and N. Hedin, *Angew. Chem., Int. Ed.*, 2010, **49**, 8889–8891.

18 H. H. Teng, P. M. Dove, C. A. Orme and J. J. De Yoreo, *Science*, 1998, **282**, 724–727.
19 K. Delak, J. Giocondi, C. Orme and J. S. Evans, *Cryst. Growth Des.*, 2008, **8**, 4481–4486.
20 C. A. Orme, A. Noy, A. Wierzbicki, M. T. McBride, M. Grantham, H. H. Teng, P. M. Dove and J. J. DeYoreo, *Nature*, 2001, **411**, 775–779.
21 R. F. Xiao, J. I. D. Alexander and F. Rosenberger, *Phys. Rev. A: At., Mol., Opt. Phys.*, 1991, **43**, 2977–2992.
22 I. Sunagawa, *Bulletin De Mineralogie*, 1981, **104**, 81–87.
23 P. Meakin and B. Jamtveit, *Proc. R. Soc. London, Ser. A*, 2010, **466**, 659–694.
24 E. Yokoyama and T. Kuroda, *Phys. Rev. A: At., Mol., Opt. Phys.*, 1990, **41**, 2038–2049.
25 L. Granasy, T. Pusztai, G. Tegze, J. A. Warren and J. F. Douglas, *Physical Review E*, 2005, 72.
26 J. P. Andreassen, *J. Cryst. Growth*, 2005, **274**, 256–264.
27 A. W. Xu, W. F. Dong, M. Antonietti and H. Colfen, *Adv. Funct. Mater.*, 2008, **18**, 1307–1313.
28 J. M. Didymus, P. Oliver, S. Mann, A. L. Devries, P. V. Hauschka and P. Westbroek, *J. Chem. Soc., Faraday Trans.*, 1993, **89**, 2891–2900.
29 A. W. Xu, M. Antonietti, H. Colfen and Y. P. Fang, *Adv. Funct. Mater.*, 2006, **16**, 903–908.
30 B. R. Heywood, S. Rajam and S. Mann, *J. Chem. Soc., Faraday Trans.*, 1991, **87**, 735–743.
31 E. M. Pouget, P. H. H. Bomans, A. Dey, P. M. Frederik, G. de With and N. Sommerdijk, *J. Am. Chem. Soc.*, 2010, **132**, 11560–11565.
32 N. Gehrke, H. Colfen, N. Pinna, M. Antonietti and N. Nassif, *Cryst. Growth Des.*, 2005, **5**, 1317–1319.
33 L. Chen, F. Z. Huang, S. K. Li, Y. H. Shen, A. J. Xie, J. Pan, Y. P. Zhang and Y. Cai, *J. Solid State Chem.*, 2011, **184**, 2825–2833.
34 J. H. Xiang, H. Q. Cao, J. H. Warner and A. A. R. Watt, *Cryst. Growth Des.*, 2008, **8**, 4583–4588.
35 H. X. Guo, Z. P. Qin, P. Qian, P. Yu, S. P. Cui and W. Wang, *Adv. Powder Technol.*, 2011, **22**, 777–783.
36 X. Y. Xu, Q. Y. Lai, Y. Zhao, Y. J. Hao, H. M. Zeng and L. Wang, *Cryst. Res. Technol.*, 2010, **45**, 712–716.
37 E. M. Flaten, M. Seiersten and J. P. Andreassen, *J. Cryst. Growth*, 2010, **312**, 953–960.
38 R. Beck, E. Flaten and J. P. Andreassen, *Chem. Eng. Technol.*, 2011, **34**, 631–638.
39 T. F. Kazmierczak, M. B. Tomson and G. H. Nancollas, *J. Phys. Chem.*, 1982, **86**, 103–107.
40 R. Beck, M. Seiersten and J.-P. Andreassen, *In preparation*, 2012.
41 R. Beck and J. P. Andreassen, *Cryst. Growth Des.*, 2010, **10**, 2934–2947.
42 K. S. Pitzer, *J. Phys. Chem.*, 1973, **77**, 268–277.
43 L. N. Plummer and E. Busenberg, *Geochim. Cosmochim. Acta*, 1982, **46**, 1011–1040.
44 D. Gebauer and H. Colfen, *Nano Today*, 2011, **6**, 564–584.
45 H. H. Teng, P. M. Dove and J. J. De Yoreo, *Geochim. Cosmochim. Acta*, 2000, **64**, 2255–2266.
46 L. Brecevic and A. E. Nielsen, *J. Cryst. Growth*, 1989, **98**, 504–510.
47 J. R. Clarkson, T. J. Price and C. J. Adams, *J. Chem. Soc., Faraday Trans.*, 1992, **88**, 243–249.
48 T. Ogino, T. Suzuki and K. Sawada, *Geochim. Cosmochim. Acta*, 1987, **51**, 2757–2767.
49 J. Kawano, N. Shimobayashi, A. Miyake and M. Kitamura, *Journal of Physics-Condensed Matter*, 2009, 21.
50 K. G. Libbrecht, *Rep. Prog. Phys.*, 2005, **68**, 855–895.
51 L. F. Wang, I. Sondi and E. Matijevic, *J. Colloid Interface Sci.*, 1999, **218**, 545–553.
52 A. Petrova, W. Hintz and J. Tomas, *Chem. Ing. Tech.*, 2008, **80**, 359–363.
53 W. W. Mullins and R. F. Sekerka, *J. Appl. Phys.*, 1963, **34**, 323.
54 J. H. Magill, *J. Mater. Sci.*, 2001, **36**, 3143–3164.

PAPER

Computer simulation of soft matter at the growth front of a hard-matter phase: incorporation of polymers, formation of transient pits and growth arrest†

Richard P. Sear*

Received 16th January 2012, Accepted 9th May 2012
DOI: 10.1039/c2fd20044a

Biominerals are typically composites of hard matter such as calcite, and soft matter such as proteins. There is currently considerable interest in how the soft matter component is incorporated into the hard matter component. This would typically be a protein that does not fold up into a single rigid domain but is closer to a simple polymer, being incorporated into a growing inorganic crystal in aqueous solution. Here I use computer simulation to study a very simple (2D lattice gas) model of a growing phase and a polymer. This allows me to study the microscopic dynamics of incorporation or rejection of a single polymer by the growing phase. It also allows me to look at how high concentrations of absorbing polymer can both arrest crystal growth, and change the shape of crystals. I find that the incorporation of a single polymer into the growing phase is due to slow dynamics of the polymer at the growth front. These slow dynamics are then unable to keep up with the advancing growth front. This is an intrinsically far-from-equilibrium process and so occurs even when incorporation is thermodynamically highly unfavourable. During the incorporation process, large polymers create large and deep, but transient, pits in the growth front.

1 Introduction

Our bones and teeth are composites of soft matter and hard matter. The soft matter is proteins such as collagen and the hard matter is hydroxyapatite crystals. Bones and teeth are just two examples of a huge range of biominerals that are composites of an inorganic crystal such as hydroxyapatite or calcite, and proteins.[1] Soft and hard matter are very different. Soft matter is as the name suggests rather soft, the effective modulus for compressing a polymer in solution may be kPa. By contrast crystals like hydroxyapatite and calcite have GPa moduli, see Table 1. Soft matter also has characteristic length-scales that can be tens of nanometres or even larger. Inorganic crystals have lattice constants of a few Ångstroms. Many interactions in soft matter are relatively weak, on the order of the thermal energy kT or less, while the ions in highly insoluble inorganic crystals are held in their lattices by strong, many kT, interactions.

These very large differences in properties between soft and hard matter are expected to be reflected in near total immiscibility of soft and hard matter at equilibrium. However, biominerals are formed far from equilibrium and remain there.

Department of Physics, University of Surrey, Guildford, Surrey, GU2 7XH, United Kingdom. E-mail: r.sear@surrey.ac.uk; Tel: +44(0)1483 686793

† Electronic supplementary information (ESI) available: 6 movies of the dynamics are available. See DOI: 10.1039/c2fd20044a

Table 1 Comparison of the typical values for relevant properties of soft and hard matter. Soft matter has larger length scales than hard matter and so lower free-energy densities (and hence moduli). The soft matter object considered is a linear polymer, with radius of gyration R_G. In order to obtain specific numbers I take $R_G = 10$ nm. The hard matter is a crystal with a unit cell of order a across, and with a Young's modulus E; E is approximately 100 GPa for calcite.[2] One of the sides of the unit cell of calcite is 0.5 nm across. I give three hard-matter free-energy densities. The first is the elastic modulus (top of the three rows). The second is the free-energy driving force per unit volume for crystallisation (middle row), and the third is the free energy cost per unit volume of creating a cavity of radius R_G (bottom row). S is the supersaturation of the solution and γ_X is the crystal–solution interfacial tension. I take a supersaturation $S \approx kT$, and an interfacial tension $\gamma_X \approx kT/\text{Å}^2 \approx 100$ mJ m^{-2}

	Soft matter (polymer)	Hard matter (crystal)
Lengthscale	$R_G \approx 10$ nm	$a \approx 0.5$ nm
Free energy cost of volume change	$kT/R_G^3 \approx kT/(10\text{nm})^3$ ≈ 1 kPa	$E \approx 100$ GPa (elastic compression) $S/a^3 \approx 10$ MPa (crystallisation) $\gamma_X/R_G \approx (kT/\text{Å}^2)/10$ nm ≈ 10 MPa (cavity)

The soft matter is incorporated into the growing crystal far from equilibrium, and then is trapped. Motivated by this, here I use computer simulation to study the dynamics of a simple model polymer at the growth front of a growing hard matter phase. I am interested in how the dynamics of the interaction of hard and soft matter, are affected by their very large differences in properties. Note the soft matter dynamics here are those at the growth front. Although soft matter is expected to be almost totally immiscible with hard matter, soft matter often strongly adsorbs on surfaces. I am also interested in what features of the polymer control whether or not it is incorporated into the growing crystal.

Any incorporation of polymers or other large soft-matter objects such as micelles, into inorganic crystals will only occur due to out-of-equilibrium processes. So to understand incorporation it is essential to understand the dynamics of the process. Here, I use as simple as model as possible: a two-dimensional lattice model. The use of simulation allows me to track every detail of the microscopic kinetics so for the first time we can observe exactly how soft matter is incorporated into a growing hard phase. Observing this detail is not possible in experiments.

In brief, I find that the large size of the polymer results in slow dynamics, and that these slow dynamics are what causes single polymers to be included in the growing phase. The slow dynamics drive the system far from equilibrium and produces a far from equilibrium hard–soft matter composite. By contrast a small weakly interacting oligomer may rapidly reach local equilibrium at the growth front and then it can 'surf' along on the growth front instead of being incorporated into the growing crystal. When the polymer concentration is high many polymers can bind to and hence cover the growth front. Then the slow dynamics of the polymer can also greatly slow down the rate at which a growth front advances.

As far as I am aware, simulations have not previously been performed on the dynamics of the incorporation of polymers into crystals. Although Muthukumar has developed a theory for the effect of polymer adsorption on crystal anisotropy.[3] There is, however, a significant experimental literature on a diverse range of soft matter objects inside crystals. Examples are naturally occurring proteins,[4-7] polysaccharide polymer gels,[8-12] and copolymer micelles.[13] The crystalline phase in most studies is calcite, but it is a general phenomenon, other crystals can incorporate soft matter objects.[5,6] Crystals can even incorporate colloidal particles.[14]

There are extensive results on the effect of polymers in solution on crystallisation. Both homopolymers[10,15,16] and copolymers[12] can bind to growing crystal surfaces, and this binding affects the growth of the crystal. By changing the growth rates of different crystal faces by different amounts, polymers can change the shapes of crystals. Biological polymers, for example peptides[17-20] and alginates,[21] have been shown in *in vitro* experiments to also do this. Indeed this is not restricted to ionic crystals, there are peptides that bind to the surface of ice and inhibit the growth of ice crystals.[17]

It is clear that biominerals such as our bones, sea-urchin spines, *etc.*, all have incorporated proteins, and these natural polymers act to control crystallisation.[1,22] Note that in general in experiments (but not here) it can be difficult to distinguish between the effect a polymer may be having on the nucleation stage of crystallisation and its effect on growth. However it is clear that synthetic polymers and proteins can affect crystal growth. They can do so in a way that is specific to one of a substance's polymorphs and to specific faces of a polymorph.

There are also extensive results on the adsorption and desorption of polymers on static, *i.e.*, not growing, solid surfaces, including flat crystalline surfaces like mica.[23,24] Simply because polymers are large objects they usually adsorb very strongly on solid surfaces, *i.e.*, adsorb irreversibly. Indeed, unless the polymer solution is very dilute the polymer molecules adsorb in such numbers that the solid surface is completely covered with polymer. The physics behind this is simple, if the free energy change per monomer is only, for example, 0.5 kT, then the free energy of adsorbing a polymer 100 monomers long is 50 kT—more than enough to drive irreversible adsorption.

For surfaces in contact with highly dilute polymer solutions, isolated single polymers can adsorb, typically irreversibly, *via* a process with rather complex, often far from equilibrium, dynamics.[23,25-28] As adsorption is so strong, desorption of a single polymer is typically studied by pulling the polymer off the surface using an AFM tip. This requires a sustained force of typically tens of pN or more.[23] The lifetime of a single polymer at a surface will often be too large to measure in experiment. A single polymer on a surface, although it may never desorb, can diffuse on this surface. This diffusion has been studied, and found to be much slower than in the bulk of the solution.[29-31]

At all but high dilutions, a static solid surface will acquire a covering layer of adsorbed polymer. The polymers in this layer may be dynamic,[26] *i.e.*, the polymers will slowly turn over, but coverage will be complete. The expectation would be that such a layer would greatly inhibit crystal growth, so all but very low polymer concentrations should inhibit the growth of any crystal whose surfaces attract the polymers.

The rest of this paper is organised as follows. The next section describes our results for a single polymer interacting with a growth front. The third section contains our results for the effects on growth of a high concentration of polymers at the growth front. The fourth section presents our conclusions, and the final section gives details of the model and of the simulation method. There is supplementary information† in the form of 6 movies that illustrate the dynamics we study here. As the processes we study here are intrinsically dynamic, they are best understood by seeing the dynamics in these movies.

2 Results for a single polymer

In this section we will consider a single polymer interacting with a growth front. This will be the case when the polymer solution is dilute. We are interested in what polymer properties determine whether the polymer is incorporated in the growing phase, or is pushed ahead of the growth front. We will vary the length of the polymer, the interaction between the polymer and the growing phase, and the speed of the polymer dynamics.

Our model for the growing crystal, the hard matter part of our system, is a simple two-dimensional lattice gas. It is the lattice gas that can be mapped onto the famous two-dimensional Ising model[32] solved by Onsager.[33] See section 5 for details of our model. The model has a vapour and a condensed phase. The vapour is our model of the solution and the condensed phase is our model of the crystal. We work at low temperature where the vapour is dilute and in the condensed phase essentially every lattice site is occupied by a particle. This can be seen in simulation snapshots such as those in Fig. 1 where the condensed phase is solid red because lattice sites filled by particles are shown in red, while the vapour is solid white as the density of particles is very low there. I work at constant supersaturation, *i.e.*, the difference in chemical potential between the condensed and the vapour phase, S, is constant. The temperature is also constant, and so the ratio of the energy of attraction of two neighbouring lattice-gas particles, ε, and the thermal energy is constant and the same for all simulation runs at $\varepsilon/kT = 9$. We mostly work at $S/kT = 0.1$ but some runs are at a larger supersaturation of $S/kT = 0.2$. Also, simulations are done in a square simulation box of n_L by n_L sites.

The polymer is a standard linear lattice homopolymer with L monomers. Here we study L between 4 and 256. Each monomer occupies one lattice site, and is shown in green in our simulation snapshots. The polymer is flexible and its dynamics are *via* individual monomer moves. See section 5 for details of the model polymer. Like the Ising model for the hard matter phase, our model polymer is quite standard.[34] The interaction between a monomer and lattice-gas particles is ε_M. We consider values of ε_M in the range $0 < \varepsilon_M < \varepsilon$. Because $\varepsilon_M > 0$, the polymer adsorbs on the surface of the growth front—this reduces the interfacial energy of this interface. However, because $\varepsilon_M < \varepsilon$ there is an energy cost for incorporation of the polymer into the condensed phase, as then particle–particle bonds are replaced by weaker particle–monomer bonds. So the polymer is essentially a surfactant as it prefers the interface and reduces the interfacial tension.

The model also has an additional parameter, s_p. This controls the relative speeds of the dynamics of the polymer, and of the dynamics of the condensed-phase. For large values of s_p the monomers move rapidly relative to motion of a layer of the condensed phase, while for small s_p the monomers move slowly. Changing s_p does not change the equilibrium behaviour, so changing s_p allows us to distinguish between equilibrium and intrinsically non-equilibrium behaviour. See the Model section, section 5, for details on s_p, but essentially s_p is the ratio between the rate of moves of a monomer and rate at which an attempt is made to add or remove a lattice-gas particle.

2.1 Growth mechanism of the condensed phase

We start our calculations by performing simulations of growth without a polymer. Movie 1† shows growth of a few layers in a small ($n_L = 30$) system, and Fig. 1 shows two snapshots from that movie. The position of the growth front of a larger ($n_L = 400$) system without polymer is shown as the black dashed curve in Fig. 2. From Fig. 2 we see that the growth rate is approximately 2×10^{-3} sites cycle^{-1}, for $n_L = 400$ at $\varepsilon/kT = 9$ and $S/kT = 0.1$.

Fig. 1 Two snapshots of growth in a small system $n_L = 30$ across. These show the growth mechanism of the condensed phase in this model. Empty lattice sites are white, sites filled with particles are red (grey in greyscale image). Snapshot A) shows two complete layers plus the start of a third layer on top of the second complete layer. The third layer is very incomplete, it consists of just 4 particles. In this 2D model the nucleus is a single particle so these 4 particles are over the barrier and will grow. Snapshot B) shows the system 10 cycles later when the third layer has grown so that it is now approximately half complete. 40 cycles later this layer will be complete. The dynamics of this process are shown in movie 1.† The system is at $\varepsilon/kT = 9$ and $S/kT = 0.2$.

From Movie 1† and Fig. 1 we see that the growth front advances single layer by single layer. Each layer first nucleates (in our two-dimensional system the nucleus is a single particle) and then grows laterally. The growth mechanism is nucleation and growth.

Nucleation is an activated process. This is apparent in Movie 1† where many cycles pass before each layer nucleates. Once this single particle is there, growth to the left and to the right does not involve a barrier as the energy cost of adding particles to the left or right of an existing particle is zero. However, growth of the number of particles in an incomplete layer is not monotonic, the number of particles in an incomplete layer undergoes a biased random walk, biased to increase the number of particles by the supersaturation. After a time the layer will either be complete or have dissolved. Then after a waiting time, the next layer will nucleate, and so on. At this low temperature there is a wide range of values of ε/kT and h/kT over which the growth mechanism is as described here. The mechanism is the two-dimensional version of the conventional crystal growth mechanism where successive layers each form by nucleation followed by growth in the area of the layer.[35]

2.2 A polymer at the growth front

Now let us consider a polymer at the growth front. Movie 2† and Fig. 3 show a system with a single short $L = 10$ polymer. The monomers of this polymer attract the lattice-gas particles 30% as strongly as the particles bind to each other: $\varepsilon_M/\varepsilon = 0.3$. Note that the polymer binds to the surface effectively irreversibly, even though

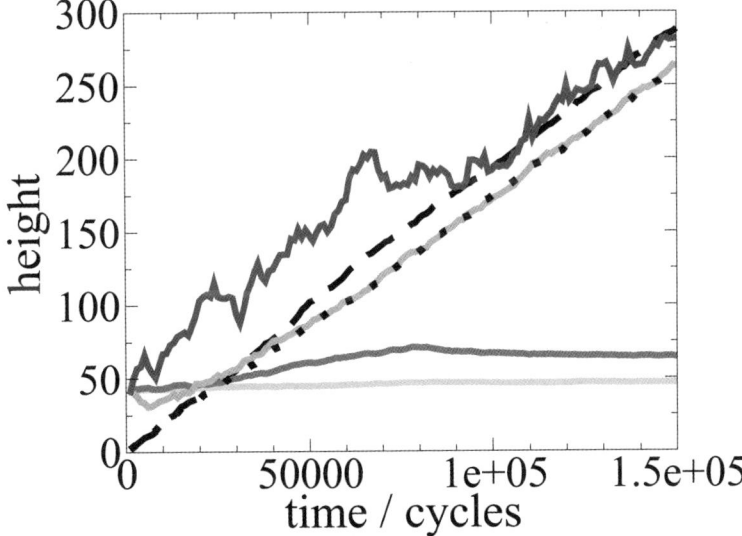

Fig. 2 Plots of growth fronts (black curves), and the centres of mass of polymers (coloured curves), both as a function of time. The length of the polymer is $L = 100$. The growth front is defined as being the height of the top layer that is more than half occupied by lattice-gas particles, in units of the lattice spacing. The black dashed curve is the growth front in the absence of polymer, while the black dotted curve is the growth front in the presence of polymer with $s_p = 10$. Note that the two growth fronts are almost parallel. The growth fronts for other values of s_p are also very similar and so are not shown for clarity. The centres of mass of polymers with $s_p = 0.1, 1, 10$ and 100 are shown as cyan, violet, green and red curves, respectively. In greyscale these four curves can be distinguished as at large times, $s_p = 0.1$ is the bottom curve, then the $s_p = 1, 10, 100$ curves are in this order from bottom to top. When the polymer is incorporated into the condensed phase its centre of mass becomes almost static (see the cyan and violet curves), while if it surfs along on the interface its centre of mass tracks the growth front (see the red and green curves). The parameter values are $\varepsilon/kT = 9$, $\varepsilon_M/\varepsilon = 0.1$ and $S/kT = 0.1$, for all polymers, and the box size is $n_L = 400$.

it is short and the attraction between its monomers and the particles is only 30% as strong that between the particles of the condensed phase. The total interaction when it lies flat on the surface is $10\varepsilon_M = 27kT$—enough for essentially irreversible binding. Thus all but very short and weakly interacting polymers will bind to the growth front and essentially never unbind.

Thus for polymers binding to a growth front, the lifetime of binding is much larger than the timescale of adding a layer. This is true even for short polymers of only $L = 10$ monomers, and the lifetime of binding will increase exponentially with L. Note that it is clear from Movie 2† that individual monomers detach and reattach but these small movements of one or a few monomers do not change the position of the centre of mass of the polymer significantly. Thus motion of the polymer as a whole is very slow in comparison to the rate of advance of the growth front, and so the growth front tends to advance past the polymer.

In Fig. 2 we see that the polymer has essentially no effect on the rate of advance of the growth front. There the dashed black curve is the growth front without polymer, while the dotted black curve is the growth front with a single polymer of length $L = 100$. These two curves are almost parallel, which tells us that the growth front is advancing at almost the same rate with and without polymer. This is also true for shorter and for longer polymers, and for all polymer speeds, s_p (data not shown).

So, we expect that in experiment the growth rate will not be affected by concentrations of polymer that are low enough that polymer covers only a small fraction of the growth front. This is in systems where growth proceeds *via* nucleation and growth of new layers. When the polymers cover only a small fraction of the total area, they do not affect nucleation, and layers can simply grow around them. (Systems that grow *via* a single spiral defect may be different.)

As a single polymer does not slow growth there are really only two possible things it can do: be incorporated into the growing phase, or 'surf' along on the growth front. See Movie 3† for a polymer being incorporated into a growth front, and Movie 4† for a polymer surfing along on a growth front. For $\varepsilon_M/\varepsilon < 1$, surfing along on the growth front will be energetically favoured. See section 5.3.

2.3 Incorporation of the polymer into the condensed phase

We have performed a number of simulations in which we varied both L and $\varepsilon_M/\varepsilon$. Each simulation was of 10^5 cycles. This is long enough for the growth front to move approximately 175 lattice sites. At the end of each run we checked to see if the polymer had been incorporated into the growing phase, or was surfing along at the growth front.

The results are shown in Fig. 4. For attractions between monomers and particles $\varepsilon_M \geq 0.3\varepsilon$, polymers of all lengths are incorporated. Note that for $\varepsilon_M < \varepsilon$, incorporation is energetically unfavourable. Therefore, here incorporation must be due to out of equilibrium kinetics. Also, it is interesting to note that the strength of attraction needed to produce incorporation is relatively insensitive to polymer length. Stronger attractions are needed to force the incorporation of shorter polymers but

Fig. 3 A snapshot of a growth front with a polymer of length $n_L = 10$ bound to it. The system is $L = 100$ across, but only a part 50 sites wide is shown. Empty lattice sites are white, sites filled with particles are red (dark grey in greyscale), and the monomers of the polymer are green (lighter grey). The monomers at the two ends of the polymer are distinguished by being a darker green than the rest of the monomers. The dynamics of this process are shown in Movie 2.† The system is at $\varepsilon/kT = 9$, $S/kT = 0.1$, and $\varepsilon_M/\varepsilon = 0.3$.

the increase in $\varepsilon_M/\varepsilon$ is small and it is only for very short polymers. This is perhaps unsurprising as for all except very small values of ε_M binding of the polymer to the growth front is irreversible for all except very small values of L. For most values of ε_M a bound polymer essentially never unbinds even if it is 10 monomers long.

Finally, it should be noted that Fig. 4 is for runs of length 10^5 cycles, a time long enough for the growth front to advance by approximately 175 sites. If longer runs are done then eventually all polymers are trapped by the growing phase, even ones that surf on the front for shorter runs. Thus the green squares in Fig. 4 do not indicate polymers that are never incorporated, just those that are incorporated very inefficiently. Here, inefficiently means after surfing on the growth front for at least hundreds of sites of growth, and in most cases much more.

As $\varepsilon_M/\varepsilon < 1$, it is energetically unfavourable for a polymer to be incorporated into the condensed phase. Thus we suspect that incorporation is due to the system being far from equilibrium. We can test this. Within a simulation model such as ours, it is straightforward to change the speed of the dynamics, while keeping the equilibrium behaviour completely unchanged. If this affects incorporation behaviour then it must be a non-equilibrium effect. We vary the parameter s_p, that controls the relative rates of attempts to move a monomer, and to occupy/fill a lattice site. We do so while keeping ε/kT, $\varepsilon_M/\varepsilon$ and S/kT all constant, so only the dynamics change. As s_p increases, the dynamics of the polymer speed up relative to the dynamics of the growth front.

The results are shown in Fig. 2. The different coloured solid curves in this figure are for $L = 100$ polymers that are identical except for the speed of their dynamics. The cyan curve is the centre of mass of the slowest polymer and the red curve is the centre of mass of the fastest polymer. Clearly, the two slower polymers (cyan and violet curves) are incorporated into the growth front, while the faster polymers (green and red curves) move sufficiently fast to surf along on the growth front. Movie 3† shows a polymer being incorporated, while Movie 4† shows a polymer surfing along on the growth front and not being incorporated into the growing phase. The same trend is also seen both for shorter and for longer polymers (data not shown), *i.e.*, decreasing s_p slows the polymer and promotes incorporation. In all cases if the dynamics are slow enough the polymer is incorporated.

This supports our hypothesis that polymers are incorporated not because this is the equilibrium state but because the dynamics of motion of the polymer absorbed

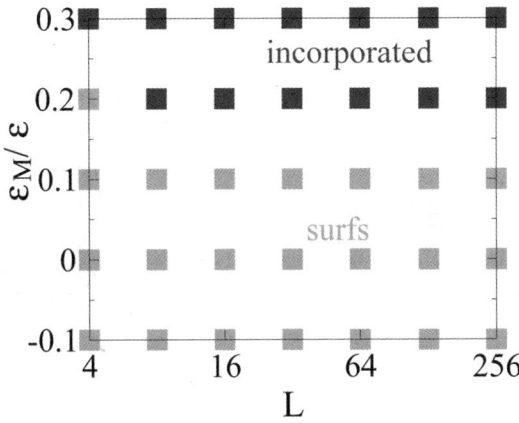

Fig. 4 Plot of the polymer lengths, L, and interaction parameter $\varepsilon_M/\varepsilon$ values for which the polymer is incorporated into the growing condensed phase (red squares, or dark grey in greyscale), or surfs along the growth front (green squares, or lighter grey in greyscale). Incorporation is determined for a single polymer at the end of a simulation run of 10^5 cycles. The system is at $\varepsilon/kT = 9$, $S/kT = 0.1$, $s_p = 10$ and the box size is $n_L = 400$.

on the surface is too slow to keep up with the growth front and so the system is forced into the highly non-equilibrium state in which the polymer is incorporated into the condensed phase. In other words, incorporation is a kinetic not an equilibrium effect. If this is general, it implies that for a polymer, micelle, or other soft matter object, to be incorporated into a growing crystal, all that is needed is to slow the kinetics of the soft matter at the growth front, until they cannot keep up with the growth front.

2.4 Pit formation

Movie 3† and Fig. 5 show a growth front advancing on a long, $L = 256$, polymer. Note that as the growth front grows around the polymer a transient pit is created in the front. This is created by the front advancing forward but not growing sideways over the polymer until the front is a considerable distance past the polymer.

As can be seen in Movie 3† and Fig. 5 the polymer absorbs along a part of the growth front around 100 sites across, blocking growth of this width of the growth front. However, over the rest of the growth front, growth continues. This moves the growth front past the polymer, causing a pit to form. The absorbed polymer is at the bottom of this pit. However, as the depth of the pit increases, the two sides of the pits become growth fronts for lateral growth, *i.e.*, for growth at 90° to the growth front. The lateral growth of these fronts will eventually cause the pit walls to move together, closing off the pit and irreversibly incorporating the polymer. Because of this lateral growth the pit is only ever transient, ultimately the polymer will always be enclosed on all sides.

3 Results for a high density of polymers: growth arrest

The results of the previous section were all for a single polymer at a growth front that is much larger than the polymer. The single polymer cannot inhibit growth along

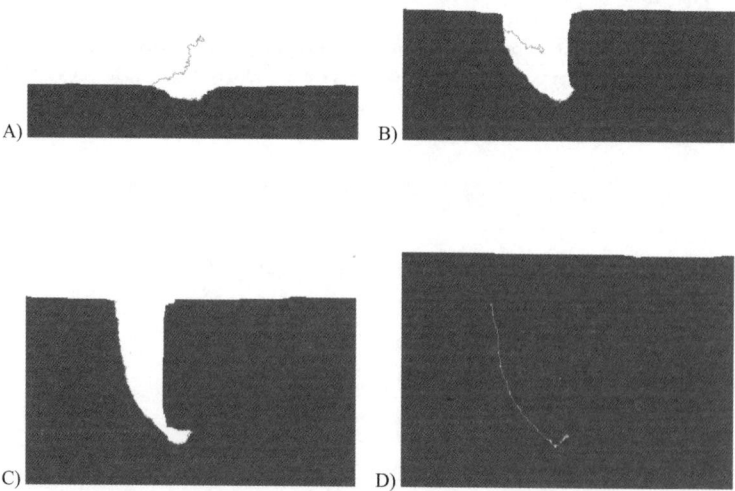

Fig. 5 Four snapshots of growth and incorporation of an $L = 256$ polymer. Empty lattice sites are white, sites filled with particles are red (dark grey in greyscale) and the polymer is green (lighter grey). Snapshot B) is 53 000 cycles after snapshot A), snapshot C) is another 38 000 cycles later, and snapshot D) is 32 000 cycles after snapshot C). Snapshot A) shows the initial binding of the polymer to the surface, and the start of growth around the polymer. Snapshots B) and C) show the polymer at the bottom of a pit that has formed around it as the growth front grows past the polymer. Snapshot D) shows the polymer in its final position incorporated in the condensed phase after the growth front has passed it by. The dynamics of this process are shown in Movie 3.† The parameter values are $\varepsilon/kT = 9$, $S/kT = 0.1$, $\varepsilon_M/\varepsilon = 0.3$, and $s_p = 10$.

this front and so is either pushed ahead of the growth front or incorporated into the growing phase. This behaviour will be observed when the polymer is very dilute in solution so that polymer molecules never cover more than a very small fraction of the growth front. In this section we consider what happens when the concentration of polymer is large so that polymers cover a significant fraction of the growth front. We find that the polymers can then greatly inhibit, and even almost completely arrest growth. See Movie 5† for almost completely arrested growth, and the snapshot in Fig. 6 for an almost arrested growth front covered with polymer.

In Movie 5† we see growth that is very slow due to adsorbed polymer. Although the growth front is almost completely covered with the polymers, the bond between a monomer and a lattice-gas particle is relatively weak, $\varepsilon_M = 2.7kT$. Thus, occasionally monomers lift off from the surface and the condensed phase can grow by a few more lattice-gas particles. This allows the growth front to very slowly advance. Larger values of ε_M greatly slow this.

Studies of a range of values of L and ε_M show that all but short polymers with small values of ε_M adsorb essentially irreversibly on to the growth front. Then, at sufficiently high concentrations these polymers cover the growth front and greatly slow growth. Very short weakly interacting polymers do not bind to the growth front and so do not slow growth.

3.1 Anisotropic growth due to anisotropic binding

When polymers bind to growth fronts along one axis but not to fronts along another axis, they can make growth rates anisotropic, and so produce a domain of the condensed phase that is anisotropic. In the absence of polymer or when polymer binds isotropically, symmetry forces the growing condensed-phase domains of our model to be square. Within our model it is straightforward to make binding on horizontal and vertical growth fronts different. We simply make the polymer–lattice-gas particle interaction dependent on direction. For details see section 5.2.

The result is shown in Movie 6† and in Fig. 7. The polymers only bind to the horizontal growth fronts, not to the vertical growth fronts. Thus growth along the horizontal direction is largely arrested while vertical growth is almost unaffected by the polymer and so is much faster. The result is the highly anisotropic domain

Fig. 6 Snapshot of a simulation where growth is partially arrested by short, $L = 10$, absorbing polymers. There is a high density of polymers: a total of 225 polymers in a square box $n_L = 400$ sites across. Only a section of 200×100 sites is shown in the snapshot. The interaction between a monomer and a lattice-gas particle is $\varepsilon_M/\varepsilon = 0.3$. The supersaturation $S/kT = 0.1$, $s_P = 10$, and $\varepsilon/kT = 9$. See Movie 5† for the growth dynamics; this is the final snapshot from that movie. It is at the end of a simulation run of 10^6 cycles. The lattice-gas particles are red (dark grey in greyscale) and the polymers are green (lighter grey). To make it easier to see where one polymer ends, and another begins, monomers at the two ends of the polymers are darker green than the others.

Fig. 7 Snapshot of a simulation of anisotropic growth of the condensed phase, due to anisotropic binding of the polymer. The simulation is of a box $n_L = 500$ sites across with 700 short ($L = 10$) polymers that adsorb onto the vertical growth fronts but are repelled by the horizontal growth fronts. The interaction between a monomer and lattice-gas particle is $\varepsilon_M/\varepsilon = 0.3$ along the horizontal axis but is $\varepsilon_M/\varepsilon = -0.3$ along the vertical axis. The supersaturation $S/kT = 0.1$, $s_p = 10$ and $\varepsilon/kT = 9$. The snapshot is after 10^5 cycles. See Movie 6† for the growth dynamics. The lattice-gas particles are red (dark grey in greyscale) and the polymers are green (lighter grey).

of the condensed phase shown in Fig. 7. Without polymers the condensed-phase domain would be square.

It is well known that polymers can influence the shape or morphology of crystals, and this is presumed to be as a result of the polymer binding more strongly to some crystal faces than others, therefore affecting the growth rates of different faces by different amounts.[5,12,19,36] In our systems, high concentrations of polymers can almost completely arrest growth on the faces they bind to. The result is highly anisotropic, needle-like domains of the condensed phase.

4 Conclusion

Here we have used computer simulation to study polymers at the growth front of a growing phase. Our model is only a very crude model of a crystal such as calcite growing in the presence of polymers, polymer micelles or other soft matter objects. However, we may hope that our generic predictions will hold for most systems of crystals and polymers. For example, we found in our simple model that due to their large size, polymers have slow dynamics, which results in the polymer at the growth interface being very far from equilibrium. In contrast the rapid dynamics of small oligomers allow them to equilibrate at the interface. These slow dynamics of the polymer at the interface then mean that growth occurs around it and the polymer is incorporated into the phase, in a process that is intrinsically very far from equilibrium. This prediction looks rather generic as it just relies on slow dynamics, and real three-dimensional polymers on surfaces have such slow dynamics.[23,25–31] So although there are many differences between our model and experiments, because the effect is so generic, we expect that in both cases incorporation can be driven by the slow dynamics of the soft matter at the growth front failing to keep up with the advance of this front. This may even be true in complex biomineralisation systems.

We considered both dilute polymer solutions and more concentrated ones. The most important findings of our simulations can be summarised as:

1. A single polymer has almost no effect on the velocity of the growth front of the condensed-matter phase.

2. Even a very weak attraction between the monomers of the polymer and the growth front is enough to drive incorporation of the polymer into the growing phase. This incorporation therefore occurs even when it is thermodynamically highly unfavourable. It is an intrinsically far-from-equilibrium process that relies on the slow dynamics of the polymer at the growth front interface.

3. Larger polymers create transient pits as the growth front grows around them. As the growth front passes the polymer, the growth rate is almost as fast as without the polymer. However, its sideways growth is very slow until the pit is quite deep, and this creates a steep-sided but transient pit in the growth front.

4. When the polymer concentration is high enough for the polymers to mostly cover the growth front, then polymers dramatically slow growth.

5. When the polymer concentration is high enough to slow growth, and the polymers only bind to growth fronts along one axis, then the growth becomes highly anisotropic, and so the size of the domains that form changes from square to needle-like.

These results are all for homopolymers. Similar effects are seen for co-polymers (results not shown). If one of the blocks of a co-polymer attracts the growing phase, the polymer will tend stick to the growth front and be incorporated into the growing phase. Sufficiently high concentrations of co-polymers can also arrest growth.

Future experimental work could usefully study the behaviour of polymers as a function of their molecular weight. As the details of the monomer–surface interaction are not known for any polymer–crystal pair, interpreting data on polymers with different chemistries is difficult. However, varying the molecular weight keeps this interaction constant and so would allow a much 'cleaner' interpretation of the data, and so a more powerful experiment. Our results suggest that small oligomers would be excluded from the crystal but sufficiently large polymers would be included. This should be true providing the monomers of which they are made attract the crystal surface relatively weakly.

Diffusion of adsorbed single polymers at a surface can be studied in experiment, and the diffusion constant, D_S, measured.[29–31] If this could be done on a growing crystal surface, then the prediction is as follows. If the growth steps on the growth front move a distance equal to the polymer radius, R_G, in less time than the polymer takes to diffuse this distance, R_G^2/D_S, then these steps will grow past and around a polymer. If the rate of advance of the growth front is also so fast that it creates and deepens a pit faster than the polymer can diffuse out of it, then the polymer will be incorporated into the growing crystal.

5 Simulation method and model

The model is based on a simple two-dimensional lattice gas on a square lattice. Without the polymer, the model is the lattice gas that is equivalent to the two-dimensional Ising model studied by Onsager.[32] Then each lattice site is either empty, or occupied by a single particle. Simulations are performed at constant chemical potential μ. I write this as $\mu = \mu_{co} + S$, where μ_{co} is the chemical potential at coexistence, and S is the supersaturation, *i.e.*, the chemical potential driving force for growth. The lattice-gas particles attract each other *via* a nearest-neighbour interaction such that for every pair of particles on neighbouring lattice sites, the energy changes by $-\varepsilon$. Each site has four nearest neighbours. We work at fixed temperature T, and chemical potential μ. Thus the number of particles (but not the number of polymers) varies. The correct Boltzmann weight of a state is $\exp(-U/kT + N\mu/kT)$. Here U is the total energy and N is the total number of lattice gas particles.

In the absence of polymer, the dynamics are Glauber dynamics, which are the standard implementation of the Metropolis Monte Carlo rule with non-conserved numbers of lattice-gas particles.[32,37] With these dynamics a lattice site is chosen at random, and the site is flipped from empty to occupied if it is empty, or from occupied to empty if it is occupied. If this increases the Boltzmann weight of the state the move is accepted, if it decreases the Boltzmann weight, the move is accepted with probability equal to the ratio of the new and old Boltzmann weights.

We use the Glauber algorithm because as it fixes the chemical potential it allows a 'clean' study of growth rates, without the complication of movement of the growth front changing the concentration profile near the growth front, and so changing the growth rate. Also, the growth rate is slow, see Fig. 2. This is because it occurs via nucleation and growth, as can be seen in Movie 1.† Then as the rate is determined largely by the nucleation barrier, it depends only weakly on diffusion. Thus, as the growth rate depends weakly on whether the microscopic dynamics are Glauber or diffusive, and as the results of Glauber dynamics are more straightforward to interpret, we use these dynamics.

A single homopolymer chain is modelled as a linear sequence of L identical monomers, each of which occupies a single lattice site. Successive monomers along the polymer chain are constrained by bonds that limit the pair of successive monomers to be on either nearest-neighbour or next nearest-neighbour sites, i.e., with monomer m on site (i_m, j_m), and monomer $m + 1$ on site (i_{m+1}, j_{m+1}), the bond enforces the constraint $(i_{m+1} - i_m)^2 + (j_{m+1} - j_m)^2 \leq 2$. Each site has 8 such neighbours. Neighbouring monomers do not interact except that a maximum of one monomer can occupy a site, i.e., the monomer–monomer interaction is just excluded volume. Therefore the polymer is in the "good solvent" regime, which is appropriate for modelling polymers that are highly water soluble. It also means that in the vapour phase of the lattice-gas model (our model for the aqueous solution), the polymer chain is swollen:[38] the radius of gyration $R_G \simeq L^{3/4}$. A monomer attracts a neighbouring lattice-gas particle with an attraction of strength ε_M, i.e., for every neighbouring pair of a monomer and a particle the energy changes by $-\varepsilon_M$.

In the presence of a polymer, each move is either: 1) an attempt to move a single monomer; 2) an attempt to remove a lattice-gas particle from a filled site or add a lattice-gas particle to an empty site. Each move is with probability f_p, an attempt to move a monomer, and with probability $1 - f_p$, an attempt to remove/add a particle. For a given simulation the number of monomers, M_p, is fixed, as is the number of sites that are not fixed, M_{FL}. Then the probability $f_p = s_p M_p/(s_p M_p + M_{FL} - M_p)$. Here s_p is the ratio between the rate at which monomers move, and the rate at which sites fill or empty with lattice-gas particles. For almost all of our simulations we fix $s_p = 10$, but we do vary s_p in Fig. 2. Except for that figure $s_p = 10$ always. This is fast enough to allow short weakly interacting polymers to be highly dynamic on the growth front, although strong interactions dramatically slow the dynamics of course. A cycle is defined as $(s_p - 1)M_p + M_{FL}$ attempted moves, so that in one cycle we make one attempt on average to flip each of the $M_{FL} - M_p$ lattice sites not occupied by a monomer. We also make an average of s_p attempts to move each of the M_p monomers.

The dynamics of the polymer are as follows. When a monomer is selected, an attempt is made to move it to one of its 8 nearest or next-nearest neighbouring sites, selected at random. If the site selected to move the monomer is occupied, the move is rejected. If the site is vacant then the move is accepted if the move results in an increase in the Boltzmann weight. If it results in a decrease in the weight it is accepted with probability equal to the ratio of the new and old weights.

5.1 Growth front simulations

These are done in a simulation box with the bottom row of sites fixed in the occupied state and the top row fixed in the empty state, so here $M_{FL} = n_L(n_L - 2)$. There are periodic boundary conditions along the horizontal direction. See Fig. 1, 3, 5 and 6 for snapshots from simulations of this type. One or more polymers are then introduced into the simulation. If it is one polymer then the polymer is placed with its lowest monomer a few sites above the bottom of the simulation box. When many polymers are required they are placed in a regular array in the simulation box. In all cases the polymers are simulated independently before placing in the simulation box to equilibrate their internal conformation. Then the simulation is run on and the condensed phase grows from the bottom row of sites fixed to be occupied, see for example Movie 1.† The position of the growth front, required for the results of Fig. 2, is defined as being the top row of the lattice with more than $n_L/2$ lattice-gas particles in it, *i.e.*, the top layer that is more than half complete.

5.2 Anisotropic growth simulations

These are done in a simulation box $n_L = 500$ sites across. There are periodic boundary conditions along the horizontal and vertical axes. See Fig. 7 for a snapshot from a simulation of this type. We start with a square domain of the condensed phase 161 by 161 sites across surrounded by 700 polymers. This domain is significantly larger than the critical nucleus, ensuring that it will grow not shrink. Then the system is first partially equilibrated by running a simulation in which the lattice-gas particles are fixed but the polymers are moved for 10^6 attempted moves per monomer. During this time polymers adsorb on the immobilised domain. Then once this is done the simulation is begun and the condensed-phase domain begins to grow.

For these simulations (and only these simulations) the interaction between a monomer and a lattice-gas particle (only) is made anisotropic. The interaction between a monomer and lattice-gas particle is $\varepsilon_M/\varepsilon = 0.3$ along the horizontal axis, but is $\varepsilon_M/\varepsilon = -0.3$ along the vertical axis. Thus polymers bind to vertical growth fronts but do not bind to horizontal growth fronts, creating anisotropic growth. Of course if bonding is the same along both axes growth along both axes occurs at the same rate, with or without polymer.

5.3 Thermodynamic equilibrium

Here we will briefly consider the energetics and equilibrium behaviour of our system. For simplicity, let us consider the polymer in a completely extended configuration. If we bring a polymer in this configuration from the vapour phase to lie flat on the growth front, the energy change is $-L\varepsilon_M$. For $\varepsilon_M > 0$, this is negative and it becomes very large for large L. However if we move the extended polymer from the vapour into the bulk of the condensed phase the energy change is $(2L + 2)(\varepsilon - \varepsilon_M)$. This is positive if $\varepsilon_M < \varepsilon$, and so is then unfavourable.

Here we study the regime where the polymer weakly binds to the condensed phase, *i.e.*, where $0 < \varepsilon_M < \varepsilon$. Then at equilibrium, the polymer will be adsorbed at the interface and *not* as an inclusion in the condensed phase. The polymer is essentially a surfactant that has a minimum energy when at the interface. This implies that in this parameter range any incorporation in the growing condensed phase can only be due to non-equilibrium effects.

Acknowledgements

The simulations were inspired by experimental results of Kang Rae Cho working with Jim de Yoreo at the Laurence Berkeley National Laboratory, and Fiona Meldrum (University of Leeds).

References

1. S. Mann, *Biomineralization*, Oxford University Press, Oxford, 2001.
2. M. F. Ashby, L. J. Gibson, U. Wegst and R. Olive, *Proc. R. Soc. London, Ser. A*, 1995, **450**, 123.
3. M. Muthukumar, *J. Chem. Phys.*, 2009, **130**, 161101.
4. A. Herman, L. Addadi and S. Weiner, *Nature*, 1988, **331**, 546.
5. F. C. Meldrum and H. Cölfen, *Chem. Rev.*, 2008, **108**, 4332.
6. E. Bonucci, *J. Bone Miner. Metab.*, 2009, **27**, 255–264.
7. C. Gilow, E. Zolotoyabko, O. Paris, P. Fratzl and B. Aichmayer, *Cryst. Growth Des.*, 2011, **11**, 2054.
8. L. A. Estroff, L. Addadi, S. Weiner and A. D. Hamilton, *Org. Biomol. Chem.*, 2004, **2**, 137.
9. H. Li, H. L. Xin, D. A. Muller and L. A. Estroff, *Science*, 2009, **326**, 1244.
10. H. Li and L. A. Estroff, *Adv. Mater.*, 2009, **21**, 470.
11. H. Li, Y. Fujiki, K. Sada and L. A. Estroff, *CrystEngComm*, 2011, **13**, 1060.
12. Z. Deng, G. J. M. Habraken, M. Peeters, A. Heise, G. de With and N. A. J. M. Sommerdijk, *Soft Matter*, 2011, **7**, 9685.
13. Y. Y. Kim, K. Ganesan, P. Yang, S. Kulak, A. N. Borukhin, S. Pechook, L. Ribeiro, K. R., S. J. Eichhorn, S. P. Armes, B. Pokroy and F. C. Meldrum, *Nat. Mater.*, 2011, **10**, 890.
14. Y.-Y. Kim, L. Ribeiro, F. Maillot, O. Ward, S. J. Eichhorn and F. C. Meldrum, *Adv. Mater.*, 2010, **22**, 2082.
15. Y. Peng, A.-W. Xu, B. Deng, M. Antonietti and H. Cölfen, *J. Phys. Chem. B*, 2006, **110**, 2988.
16. R. Kim, C. Kim, S. Lee, J. Kim and I. W. Kim, *Cryst. Growth Des.*, 2009, **9**, 4584.
17. A. L. D. Vries and T. J. Price, *Philos. Trans. R. Soc. London, Ser. B*, 1984, **304**, 575.
18. M. M. Tomczak, M. K. Gupta, L. F. Drummy, S. M. Rozenzhak and R. R. Naik, *Acta Biomater.*, 2009, **5**, 876.
19. R. W. Friddle, M. L. Weaver, S. R. Qiu, A. Wierzbicki, W. H. Casey and J. J. De Yoreo, *Proc. Natl. Acad. Sci. U. S. A.*, 2010, **107**, 11.
20. M.-K. Liang, O. Deschaume, S. V. Patwardhan and C. C. Perry, *J. Mater. Chem.*, 2011, **21**, 80.
21. L. Lakshtanov, N. Bovet and S. Stipp, *Geochim. Cosmochim. Acta*, 2011, **75**, 3945–3955.
22. S. Weiner and L. Addadi, *Annu. Rev. Mater. Res.*, 2011, **41**, 21.
23. M. Seitz, C. Friedsam, W. Jöstl, T. Hugel and H. E. Gaub, *ChemPhysChem*, 2003, **4**, 986.
24. A. P. Gunning, A. R. Kirby, A. R. MacKie, P. Kroon, G. Williamson and V. J. Morris, *J. Microsc.*, 2004, **216**, 52.
25. D. Panja, G. T. Barkema and A. B. Kolomeisky, *J. Phys.: Condens. Matter*, 2009, **21**, 242101.
26. H. E. Johnson and S. Granick, *Science*, 1992, **255**, 966.
27. J. F. Douglas, H. M. Schneider, P. Frantz, R. Lipman and S. Granick, *J. Phys.: Condens. Matter*, 1997, **9**, 7699.
28. K. Konstadinidis, S. Prager and M. Tirrell, *J. Chem. Phys.*, 1992, **97**, 7777.
29. S. A. Sukhishvili, Y. Chen, J. D. Müller, E. Gratton, K. S. Schweizer and S. Granick, *Nature*, 2000, 406.
30. S. A. Sukhishvili, Y. Chen, J. D. Müller, E. Gratton, K. S. Schweizer and S. Granick, *Macromolecules*, 2002, **35**, 1776.
31. J. S. S. Wong, L. Hong, S. C. Bae and S. Granick, *Macromolecules*, 2011, **44**, 3073–3076.
32. D. Chandler, *Introduction to modern statistical mechanics*, Oxford University Press, New York, 1987.
33. L. Onsager, *Phys. Rev.*, 1944, **65**, 117.
34. A. D. Sokal, in *Monte Carlo and molecular dynamics simulations in polymer science*, ed. K. Binder, Oxford University Press, New York, 1995.
35. J. W. Mullin, *Crystallization*, Butterworth Heinemann, Oxford, 2001.
36. S.-H. Yu and H. Cölfen, *J. Mater. Chem.*, 2004, **14**, 2124.
37. K. Binder and D. W. Heermann, *Monte Carlo simulation in statistical physics: an introduction*, Springer Verlag, Heidelberg, 2010.
38. P.-G. de Gennes, *Scaling concepts in polymer physics*, Cornell University Press, Ithaca, New York, 1979.

General discussion

Dr Zhang opened the discussion of the paper by Professor Alfons van Blaaderen: Due to the surface coating the silica rods carry charges, the question is: how do the interactions (electrostatic repulsion or additive induced depletion attraction) affect the resulting phase behavior, in particular, the liquid crystal phase, when compared with a pure hard rod system?

Professor Van Blaaderen responded: Surface coatings on silica can indeed modify surface charges, but also 'pure' silica can already acquire charge (*e.g.*, by dissociation of surface silanol groups). Unfortunately, these kinds of equilibria and mechanisms in non watery solvents are far less well established. However, if the Debye–Hückel screening length is made/kept sufficiently thin as compared to the particle size, the colloidal interactions become more and more steep and will start to look more and more like a hard particle. In the results presented in this paper the double layers are sufficiently thin that a comparison with systems (computer simulations mostly) is warranted. As mentioned in the paper, a 'simple' rescaling does not, however, seem trivial. This is why our rescaling does not completely (meaning quantitatively) follow the hard rod results, although qualitatively it is quite similar. If the double layer is made much larger than in the present paper, results not discussed, we find strong differences with hard rod results, even to the extreme case where we find a plastic crystal of the rods. In such a system there is 3D long-ranged positional order of the centers of gravity of the rods (on a body centered lattice), but completely free rotations of the particles. (This is the counterpart of a liquid crystal where there is long ranged orientational order of some kind, but no positional order.)

Dr Zhang remarked: From the SAXS profiles along the gravity, one can see some small oscillations at the low q region, are they corresponding to the periodicity of the layers in the smectic phase?

Professor Van Blaaderen responded: Indeed, as mentioned in the text, these oscillations are due to the periodicity of the layers in the smectic phase, for the most ordered samples even six oscillations are visible.

Professor Kato said: The silica rods behave like liquid crystalline molecules that show nematic and smectic liquid crystal phases. For molecular-based liquid crystals, we use some techniques to obtain macroscopically ordered alignment. Can you achieve macroscopic alignment of the silica rods using external stimuli such as electric fields?

Professor Van Blaaderen replied: It is even an important part of our motivation to develop this new colloidal model system to try to investigate quantitatively on a single particle level the kind of questions raised here. For instance, for molecular systems that form liquid crystals it is quite hard to separate geometric effects from (more specific) chemical effects on structured surfaces that are often used to align such liquid crystal phases. We have indeed started to make structured walls and investigate how the surface corrugations influence the alignment of our liquid crystal phases. Also electric field effects are easily studied and are also very beneficial for obtaining much better longer-range order with fewer defects. Preliminary results can be found in the PhD thesis of Anke Kuijk available for download at the following website: www.colloid.nl: Fluorescent colloidal silica rods—synthesis and phase behavior, Anke Kuijk, PhD Thesis, Utrecht University, 2012.

Professor Kato asked: For molecular-based nematic liquid crystals, response times for electric fields are in the range of milliseconds. How long do the silica rods take time to respond to electric fields?

Professor Van Blaaderen answered: Preliminary results on electric field induced response time show that individual rods' response times can be quite fast, even on the tens of millisecond time scales. In the case of the response of a nematic phase, the response becomes much slower (seconds or parts of seconds).

Professor De Yoreo remarked: I would have expected systems of longer rods to have a higher defect density because it is more difficult for them to reorient. The data show the opposite is true. Why?

Professor Van Blaaderen responded: We agree that the rotational dynamics of the smaller rods is indeed faster than that of the longer rods. Nevertheless, the amount of defects was indeed less for these systems. We speculate that this may be caused by the fact that the thermodynamic driving force to form the liquid crystal phases was partially responsible for this observation. As it is experimentally difficult to determine how close a system has stayed to its equilibrium concentration of defects, it may be hard to guess how much the dynamics of the particles still played a role and to what extent.

Mr Campbell commented: Given that your silica rods are bullet-shaped, did you observe any additional alignment effects arising from this asymmetry?

Professor Van Blaaderen replied: We indeed asked ourselves the same question and were able to look into this quite quantitatively for some of the fluorescently labelled systems. According to the labeling procedure, the fluorescent intensity is going down from the start of rod growth (the part which has the round tip) towards the end of the synthesis (the 'flat' part of our 'bullets'). For these systems we could check for each individual particle in the nematic and smectic phases whether there was orientation. No preferred orientation was found. More qualitatively, this was also our observation in the SEM images we took after drying some of the smectic and crystalline samples. It is interesting to remark though that in experiments in which we grew a small amount of polymethylmethacrylate onto the rods (using them as seeds) we did find specificity: the organic *pmma* only grew on the flat end of the rods. The mechanism behind this intriguing phenomenon is under investigation.

Mr Khan asked: Is the creation of these rods/bullets froma sol–gel or precipitation method of manufacture?

Professor Van Blaaderen responded: As mentioned in our paper, the synthesis of our rods is described in more detail in ref. 16. We developed this method which very briefly consists of the growth of silica from the hydrolysis and condensation of a tetraalkoxysilane in water droplets containing the polymer polyvinylpyrrolidone and ammonia formed in pentanol. Because the initial silica particle is adsorbed to the water pentanol interface, subsequent growth from the water phase makes the resulting structure more and more anisotropic.

Mrs Virone opened the discussion of the paper by Professor Bartosz A. Grzybowski: Are nano-particles acting as templates for the nucleating phase?
Can the nano-particles influence the system like inducing polymorphic transformation (or more in genral: could the nano-particles be used by any means as a controlling tool to steer the process nucleation–growth–polymorphism)? If yes, how?

Can the nanoparticles form an eutectic system (at a certain concentration and temperature) with the nucleating phase?

Professor Grzybowski replied: No, the nanoparticles do not nucleate the inorganic crystals. The exact argument is outlined in our *Nature Materials* paper.[1]

Therein, we decribe crystal growth by a model which assumes that (i) NP adsorption onto the microcrystals is described by a Langmuir-type kinetic model and (ii) the rate of crystal growth is proportional to the free crystal surface area. This model reproduces the experimentally observed scaling of the crystal size, $d \sim 1/\chi$ (where $\chi =$ CNP/Csalt).

Importantly, an alternative scenario, in which the NPs would nucleate the growth of the microcrystals is unlikely—assuming the number of nucleated crystals scales with the number of NPs, one would then expect the scaling $d \sim (1/\chi)^{1/3}$, which is not observed experimentally.

As to the rest of the question: we have not yet tried the control of polymorph, although this is an excellent idea.

1. Bartlomiej Kowalczyk, Kyle J. M. Bishop, Istvan Lagzi, Dawei Wang, Yanhu Wei, Shuangbing Han and Bartosz A. Grzybowski, *Nat. Mater.*, 2012, **11**, 227–232.

Dr Sommerdijk addressed Professor Grzybowski: It is suggested that the two populations of oppositely charged nanoparticles jointly deposit on the surface of a developing salt crystal that nucleated in solution. I find it difficult to understand that the crystallization of the salts would not start with the heterogeneous nucleation on the surface of the nanoparticles that stronly interact with the developing crystal.

Professor Grzybowski responded: The nanoparticles do not nucleate the growth of inorganic crystals. The exact argument is outlined in our *Nature Materials* paper[1] where we decribe crystal growth by a model which assumes that (i) NP adsorption onto the microcrystals is described by a Langmuir-type kinetic model and (ii) the rate of crystal growth is proportional to the free crystal surface area. This model reproduces the experimentally observed scaling of the crystal size, $d \sim 1/\chi$ (where χ = CNP/Csalt). Importantly, an alternative scenario, in which the NPs would nucleate the growth of the microcrystals is unlikely—assuming the number of nucleated crystals scales with the number of NPs, one would then expect the scaling $d \sim (1/-)^{1/3}$, which is not observed experimentally.

1. B. Kowalczyk, I. lagzi, K. J. M. Bishop, D. Wang and B. A. Grzybowski, Charged nanoparticles as supramolecular surfactants for controlling the growth and stability of inorganic microcrystals, *Nat. Mater.*, 2012, **11**, 227–232.

Professor Van Blaaderen said: Regarding the role of spherical particles on the nucleation of crystals: a nice model study on the hard-sphere particle system has been done by Daan Frenkel and his group[1] using computer simulations. They found that inclusions with a certain radius of curvature and above acted as 'nucleation catalysts' *i.e.* the spherical hard particles nucleated on the spherical inclusion, but also fell off after some growth. Larger spherical inclusions also reduced the nucleation barrier but they remained covered by the crystallites that they spawned. A nice experimental verification of these simulations is provided in ref. 2.

1. A. Cacciuto, S. Auer and D. Frenkel, Onset of heterogeneous crystal nucleation in colloidal suspensions, *Nature*, 2004, **428**(6981), 404–406.
2. V. W. A. de Villeneuve, R. P. A. Dullens, D. G. A. L. Aarts, *et al.*, Colloidal hard-sphere crystal growth frustrated by large spherical impurities, *Science*, 2005, **309**(5738), 1231–1233.

Dr Sear asked: Are your systems of nanoparticles adsorbed onto the crystal/solution interface in some way the crystalline analogue of Pickering emulsions? These are emulsions formed when nanoparticles (or other size particles) adsorb onto the interface between two fluid phases (*e.g.*, oil and water). Although Pickering emulsions are fluid systems and you have a crystal, you show that the nanoparticles are behaving as surfactants, and this is exactly the role of nanoparticles in Pickering emulsions.

If so, could you steal some ideas from Pickering emulsions, which are well studied systems, and use them in your systems?

Professor Grzybowski replied: This is certainly an interesting analogy though I am not sure how far it can carry us in designing the NP/crystal systems. What we are focusing on with our NPs is to combine electrostatics with more specific interactions (H-bonding, donor–acceptor) to use our method to control the growth of organic microcrystals—one avenue of current research is enantioselection (growth of crystals made of one stereoisomer).

Professor De Yoreo remarked: Doesn't decrease in size with increase in nanoparticle-to-salt concentration imply an impurity effect rather than a surfactant effect?

Professor Grzybowski responded: I am not sure I understand the question correctly. The effect is akin to the growth of nanoparticles controlled by surfactants (*e.g.*, thiolates on Au). The more thiols per fixed amound of Au salt precursor, the smaller the AuNPs that grow. In our present system, the more nanoparticle "surfactants", the smaller the inorganic crystals onto which these nanoparticles adsorb. A more detailed theoretical model can be found at B. Kowalczyk, I. lagzi, K.J.M. Bishop, D. Wang and B.A. Grzybowski, Charged nanoparticles as supramolecular surfactants for controlling the growth and stability of inorganic microcrystals, *Nat. Mater.*, 2012, **11**, 227–232.

Mr Khan said: After coating the material, do you find different properties in the particles' ability to flow/assemble amongst themselves. There are techniques that use coatings of other materials on colloids for anti-caking/flow aids to reduce the agglomeration of particulates. Do you find that you have different powder properties because of this coating on the crystals?

Professor Grzybowski replied: This is a very interestign point—we have done only preliminary studies on how the coatings affect the stability of the coated microcrystals. What we observe in these experiments is that the (+) (−) NP coatings on different crystals can to some degree fuse and bind the crystals—it would thus appear they are acting as inter-crystalline "glue".

Dr Sear asked: Are the nanoparticles incorporated into the growing ionic crystals? If so, could the reduction in ionic crystal size you observe in the presence of nanoparticles, be due to incorporated nanoparticles straining the lattice of the growing ionic crystal and so slowing or arresting growth by that mechanism?

Professor Grzybowski answered: We have done numerous TEM and HRTEM on these crystals—even in very small crystallites, which are "transparent" to the probing bean (*i.e.*, we can "see through") we do not see any nanoparticles included in the crystals.

Dr Staniland said: You state that the difference in self-assembled nanoparticles you obtain are "not dependent on the salt's specific nature" but only on the salt's solubility. I want to know how we can discount the salts nature. In your study you use Cu^{2+} and Ni^{2+}. They can be quite different in nature. As we know from the paper by Dr Dove, they mentioned Cu^{2+} inhibits $CaPO_4$ crystallisation whereas Ni^{2+} doesn't.

Professor Grzybowski responded: The salts we use both have the same anions and divalent cations—both Cu^{2+} and Ni^{2+} can bridge carboxylic acid groups on the nanoparticles. Since we are getting diametrically different crystals with these salts, the effect cannot be attributed to a simple valence of the cations. What the salts differ in is solubility products and this is the variable that controls the progress of crystallization.

Mr Davis asked: Is it possible to remove the nanoparticle coatings without affecting the crystal within? Would it be possible to form these coatings on calcium carbonate crystals *in situ* in a flow cell?

Professor Grzybowski replied: We can remove the coatings by (i) crosslinking the nanoparticles and (ii) dissolving the crystalline core. This procedure produces "empty NP boxes". On the other hand, it is not possible—at least from now—to peel off the NP skin without destroying the crystals. As to the flow cell, yes, probably possible, although the kinetics of crystal growth is slow (hours), so the flow rates would have to be very low.

For more info, please refer to B. Kowalczyk, I. lagzi, K. J. M. Bishop, D. Wang and B. A. Grzybowski, Charged nanoparticles as supramolecular surfactants for controlling the growth and stability of inorganic microcrystals, *Nat. Mater.*, 2012, **11**, 227–232.

Dr Christenson opened the discussion of the paper by Professor Vinothan Manoharan: Would it be possible to get open clusters, or chains of particles if the attractive interaction were weaker? A chain-like structure should have a larger entropy to compensate for the loss in attraction.

Professor Manoharan answered: When we weaken the attraction by reducing the concentration of the depletant particles, we find that the clusters evaporate rather than form chains. This result shows that for a small enough potential well depth, the translational entropy of the particles is the dominant contribution to the free energy. At larger depths, the potential energy dominates. Thus I would expect a chain to be an unstable structure regardless of the depth of the potential well: the increase in vibrational entropy comes at the cost of a reduction in both the translational entropy and the potential energy. However, chains might become metastable or stable states when the interaction is longer-ranged or anisotropic.

Dr Christenson asked: Could you please comment on the relevance of your work to the nucleation of atomic and molecular systems?

Professor Manoharan responded: We cannot infer anything directly about nucleation, as our study examines only isolated, finite-sized clusters. Insofar as these clusters represent possible structures that may occur during the embryonic stages of nucleation, our results might be used to estimate nucleation barriers associated with structural transitions. But more generally, the results highlight the importance of entropy in determining the most probable cluster structures.

In atomic and molecular systems, the thermodynamics of cluster formation are more complicated than in the colloidal system we study. The interactions are longer-ranged and can be anisotropic, non-pairwise additive, or both. Extensive theoretical work on model atomic systems such as Lennard-Jones or Morse clusters has shown that structures with high symmetry are favored at low temperature—exactly the opposite of what we find in our colloidal system. This is a consequence of the range of the potential, which for atomic and molecular systems extends far enough for two particles to interact even if they are not on adjacent vertices of a cluster. However, at higher temperatures the potential energy must compete with entropic effects such as the permutational contribution we discuss in our study.

These entropic effects, which tend to favor lower-symmetry structures, should not be ignored in theoretical studies that attempt to relate the structure of nuclei to the most favorable cluster structures.

Professor Gilbert opened the discussion of the paper by Professor R. Lee Penn: The beautiful mesocrystals you observe at the cryo-TEM appear to be twinned occasionally. Can you explain how this is possible, given that aggregated nanoparticles are not continuous, space-filling single crystals?

Professor Penn replied: Oriented aggregation requires structural accord only at the interface between particles.[1] During the discussion, several additional points were made regarding the frequency of twinning. The twinned mesocrystals are much less common than the untwinned mesocrystals. Thus, we can conclude that the formation of the twinned crystals is energetically less favorable than the formation of the untwinned crystals. The frequency of twinned mesocrystals and crystals in the system featured in this paper is low, right around 1% by number. The final oriented aggregates are composed of around one hundred primary crystallites.[2] Thus, the frequency of twinned interfaces is estimated to be less than 10^{-4}. From that, we can conclude that the energetic difference between forming an interface between particles with parallel alignment *versus* twinned alignment is fairly large.

1. R. L. Penn and J. F. Banfield, Oriented attachment and growth, twinning, polytypism, and formation of metastable phases: Insights from nanocrystalline TiO_2, *Am. Mineralogist*, 1998, **83**, 1077–1082.
2. R. Lee Penn, J. Erbs, and D. Gulliver, Controlled growth of α-FeOOH nanorods by exploiting oriented aggregation, *J. Cryst. Growth*, 2006, **293**, 1–4.

Professor De Yoreo said: The images shown do not reveal that the particles are co-aligned. What is the evidence that they are, do they aggregate before coming into alignment and what is the driving force for this initial aggregation event?

Professor Penn answered: The images presented were from the early stages of aging, before many oriented aggregates have formed. In images of samples from later times, the sizes and shapes of the aggregates match the sizes and shapes of the final crystals, even when there is separation between the primary crystallites. High-resolution imaging supports this conclusion as well.[1]

Our current hypothesis is that the particles come together and form a reversible particle–particle complex that is analogous to an outersphere complex. Then, the rearrangement of the particles occurs while they are associated with one another *via* this complex state.[2,3]

1. R. L. Penn, *J. Phys. Chem. B*, 2004, **108**, 12707–12712.
2. V. M. Yuwono, N. D. Burrows, J. A. Soltis and R. L. Penn, *J. Am. Chem. Soc.*, 2010, **132**, 2163–2165.
3. N. D. Burrows, C. R. H. Hale and R. L. Penn, *Cryst. Growth Des.*, 2012, in press. DOI 10.1021/cg3004849.

Professor Manoharan remarked: Did you measure the fractal dimension of the aggregates? This could provide some clues about the interaction between the nanoparticles, and would provide a basis for quantitative comparison to theory or simulation.

Professor Penn replied: We do have some measurements of fractal dimension. For the cryo-TEM images of the dialyzed suspension and the suspension with PEG added, the fractal dimensions measured from cryo-TEM images range from 1.3 to 1.5. The fractal dimension is quite sensitive to the number density of particles in a particular image, with lower values for images with less dense particle coverage.

For images of suspensions adjusted to higher pH, fractal dimensions are around 1.6. These results are most consistent with diffusion limited aggregation (*i.e.*, high sticking probabilities).

Dr Christenson asked: What is the charge of the ferrihydrite particles? This should not be too difficult to measure.

Professor Penn replied: Previous measurements at pH 4 show that the *ca*. 5 nm nanoparticles are positively charged, as would be expected, with zeta potentials ranging from 40 to 75 mV.[1] The surface charge density can be estimated (equations in ref. 1), and our calculated surface charge densities range from approximately 4×10^{-20} to 8×10^{-20} C nm^{-2}, which is an elementary charge density of 0.3 to 0.5 e nm^{-2}. The particles in the initial suspensions are most likely more charged while the particles adjusted to pH 5.5 are most likely less charged. Unfortunately, we do not yet have zeta potential measurements for the suspensions prepared with PEG nor alizarin.

1. N. D. Burrows, C. R. H. Hale and R. L. Penn, *Cryst. Growth Des.*, 2012, in press, DOI 10.1021/cg3004849.

Ms Asenath-Smith commented: In your results, you show a trend towards greater aggregation at higher pH as well as a decreasing number of aggregates per particle with decreasing pH. Have you considered the pH dependence of the solubility of iron oxide in terms of your results?

Professor Penn replied: Yes, we have considered solubility, and we performed these experiments such that, for the most part, the iron(III) solubility is low. We have tracked primary particle size as a function of aging at elevated temperature and at various pHs, and coarsening is observed to increase predictably when the iron(III) solubility is higher.[1]

1. Two-Step Growth of Goethite from Ferrihydrite, D. J. Burleson and R. L. Penn, *Langmuir*, 2006, **22**, 402–409.

Dr Christenson remarked: What is the average size of the particles and how far apart are they, *i.e.* how much water remains between them? Van der Waals forces should hold these particles together provided any water film is thinner than a few layers of molecules.

Professor Penn answered: Previous work has shown that the initial size of the ferrihydrite nanoparticles depends on the temperature at which the forced hydrolysis was performed. In this work, the forced hydrolysis was performed at either 45 or 66 °C. Based on the previous work, we estimate the particle sizes are approximately 5.5 and 7.0 nm in length, respectively.[1,2]

Many of the images of the aggregates and mesocrystals are taken at modest underfocus so as to enhance contrast. Using images closer to zero defocus, the average separation distance ranges from appearing to be in direct contact to about one-half nanometer. For regions in which the primary particles appear to have space between them, the distance separating them seems quite consistent, with over 250 measurements yielding a separation distance of 0.5 ($\pm$0.1) nm. We estimate that, assuming ice-like water, there could be 3–4 layers of water residing in the space between the crystallites. This could be interpreted as consistent with work looking at hydrated surfaces, results from which have shown that structured water resides on the crystal surfaces.[3–6]

In the end, this is a difficult measurement to make from the cryo-TEM images because these structures are three dimensional objects, and the arrangements of the primary particles within the secondary mesocrystal are most likely not within a single plane.

1. R. L. Penn, J. J. Erbs and D. M. Gulliver, *J. Cryst. Growth*, 2006, **293**, 1–4.
2. N. D. Burrows, C. R. H. Hale and R. L. Penn, *Cryst. Growth Des.*, 2012, in Ppress, DOI 10.1021/cg3004849.
3. S. K. Ghose, G. A. Waychunas, T. P. Trainor and P. J. Eng, *Geochim. Cosmochim. Acta*, 2010, **74**, 1943–1953.
4. S. Kerisit, D. J. Cooke, D. Spagnoli and S. C. Parker, *J. Mater. Chem.*, 2005, **15**, 1454.
5. D. Spagnoli, B. Gilbert, G. A. Waychunas and J. F. Banfield, *Geochim. Cosmochim Acta*, 2009, **73**, 4023–4033.
6. Z. Zhang, P. Fenter, L. Cheng, N. C. Sturchio, M. J. Bedzyk, M. Predota, A. Bandura, J. D. Kubicki, S. N. Lvov, P. T. Cummings, A. A. Chialvo, M. K. Ridley, P. Benezeth, L. Anovitz, D. A. Palmer, M. L. Machesky and D. J. Wesolowski, *Langmuir*, 2004, **20**, 4954.

Dr Sommerdijk asked: Prof Penn showed cryo-transmission electron micrographs (cryoTEM) illustrating different stages of aggregation-based crystallization of ferrihydrite nanopartices.[1] When discussing the distance between the mineral particles visible in the cryoTEM images it is important to note that the presented images are recorded using phase contrast, applying significant underfocus values. Recording TEM images at underfocus causes Fresnel fringes, *i.e.* the formation of a white ring surrounding the darker object. The underfocus value determines the intensity of this ring and thereby the directly interpretable resolution.[2] From this it can be inferred that the distance between the nanoparticles is smaller than what appears from the images, but for a real estimation of this distance recordings closer to focus are required.

1. V. Yuwono, N. Burrows, J. Soltis and R. L. Penn, *Faraday Dicsuss.*, 2012, **159**, DOI:c2fd20115a.
2. H. Friedrich, P. M. Frederik, G. de With and N. A. J. M. Sommerdijk, *Angew. Chem., Int. Ed.*, 2010, **49**, 7850.

Professor Penn answered: This is absolutely correct. In our response to the question of Hugo Christenson, we address this issue directly. In attempting to measure the separation distance between primary nanoparticles, we used images collected using a focus condition much closer to zero defocus. However, we still face the challenge that these objects are three dimensional, and the interfaces between the primary particles may not be parallel to the beam. High-resolution tomography would be better suited for such measurements.

Professor Cölfen opened the discussion of the paper by Dr Jens-Petter Andreassen: Why is the hexagonal vaterite (001) face found under the low supersaturation conditions without additives? (001) is a highly charged face only exposing Ca^{2+} or CO_3^{2-} ions. This should reconstitute to a lower energy face. What stabilizes the (001) face? Could ions in solution like Na^+ or Cl^- do this?

Dr Andreassen answered: We performed our investigation without additives, apart from the counter ions of calcium and carbonate, Cl^- and Na^+. We cannot say from this study to which extent different ionic environments will affect the expression of the (001) face. We found that the supersaturation is the key parameter in this respect. At lower supersaturation the growth rate of the six prism faces is higher than the (001) face, resulting in hexagonal plates, in analogy to what is also seen for ice crystals.

Professor Van Blaaderen remarked: Both directional and non-directional aggregation mechanisms are considered non-classical crystal nucleation and growth mechanisms. Processes like directional and non-directional aggregation are strongly affected by ionic strength and the amount of charge on aggregation units. Control over the activity based solution supersaturation levels, even at moderate levels can still significantly change the ionic strength of the solution and/or the particle charge on growing crystals of amorphous aggregates. Why are therefore these 'non-classical' mechanisms completely excluded in an experimental study in which supersaturation levels are varied to arrive at different crystal morphologies? Especially, in view of the fact that in other papers it is claimed that equilibrium amorphous aggregates can exist at even low levels of supersaturation?

Dr Andreassen responded: I agree that aggregation processes would be highly affected by ionic strength within the predictions set by the DLVO theory, provided that there are units there, in sufficient numbers that can actually aggregate (see other answers regarding this). However we find the same morphologies in our high and low pH-series. The low pH series has a substantially higher ionic strength (Table 2) than the high pH case. The difference in ionic strength within each series, due to different supersaturation levels is small compared to the difference between the series. So we do not exclude "non-classical" mechanisms—we simply point to the fact that without any additives we find the same morphologies that are said to be made by an additive directed (and stabilized) assembly process of prenucleated crystals. In that case the proposed mechanism cannot be correct—at least the role of the additives is not interpreted correctly. In addition, I would say that an aggregation mechanism is unlikely due to the fact that the high numbers of crystals needed are not produced at these low driving forces and the high number of aggregating units is not observed. The equilibrium solubility of amorphous aggregates is not relevant for this discussion unless you claim that the crystalline particles are formed by solid-state transformation of ACC. It has been shown several times in the literature that this is not the case.

Miss Barber asked: Explain set up and samples preparation.
Why do you think you can't observe ACC prior to vaterite for the cases shown? What's the duration to observe evolution throughtout those vaterite crystals?
MultiScale is used for solutions prediction—clarify the Na_2CO_3 addition to the solution.

Professor Andreassen responded: For explanations of set-up and sample preparation I refer to the paper and references therein. We did not observe ACC but we also did not look for it. If we do precipitation at much higher supersaturations, we find ACC in the SEM, but if the amount is small it will transform rapidly while handling the solution samples. Whether ACC is present at these low supersaturation values depends on the solubility of ACC and the literature values are differing. I am not sure what is asked for regarding clarification of the sodium carbonate addition, in relation to MultiScale. I have to refer to the description given in the paper.

Dr Gebauer asked: I absolutely agree with the point of view that dislocation- and nucleation-driven crystal growth leads to faceted crystal morphologies, whereas at high levels of supersaturation, dendritic and finally spherulitic patterns can be observed.[1] These are classical growth models that can be rationalized to proceed ion-by-ion, as opposed to the particle-mediated crystallization pathways reported in recent literature. That said, do you think that it is always possible to conclude the active pathway without a doubt based upon the observation of the outcome of a crystallization process? Is it always the classical growth mode if we observe one of these distinct morphologies, or do we have to consider every individual case in detail? Is it really necessary that all constituent nanoparticles are nucleated in parallel when it comes to particle-mediated crystallization? By principle, the constituent particles could also nucleate sequentially, while the crystal could grow particle-by-particle. Do you see the possibility that ripening processes (where more stable particles must form at very low levels of supersaturation) could follow particle-mediated pathways, especially when organic additives are present that stabilize nanoparticles?

1. J.-P. Andreassen, R. Beck and M. Nergaard, *Faraday Discuss.*, 2012, **159**, DOI: 10.1039/c2fd20056b.

Professor Andreassen answered: In my view the classical growth paradigm (if we include dendritic and spherulitic growth) has the capacity to explain the morphologies observed for crystalline systems. It relates shifts in crystal growth mechanisms (and morphology) to the supersaturation (and temperature) and as such it should not be necessary to consider every individual case. A revision of the classical crystallization paradigm is only necessary if it fails to explain the morphologies that we observe, and that the shifts in mechanisms (and morphologies) do not follow changes in the supersaturation as postulated. To observe only the outcome of a crystallization process is not sufficient for inferring something about the process—it should also be supported by supersaturation monitoring and sampling during the process. Polycrystalline particles discovered at the end of a crystallization process, based on diffusion methods, for instance, are not saying anything conclusive about their formation mechanism. A particle mediated/aggregation-based mechanism should be able to account for the number of primary particles required, and also to show that these high particle numbers are found in the solution during particle formation and how the numbers change due to the alleged aggregation process. The concept of aggregation-based growth should also explain where the high numbers of particles arise from. Is the supersaturation high enough to promote extensive nucleation of nanoparticles throughout the duration of growth? Sequential nucleation seems unlikely since over time there will be enough surface area to consume the supersaturation that builds up. Regarding stabilization of nano-particles by additives, and their role in aggregation of nanoparticles to produce aggregates, I would expect the additives to actually stabilize the particles by preventing their assembly into larger particles.

Professor Andreassen further commented: As a response to your comment regarding agreement with the conclusions in our paper, lack of proof that aggregation is not responsible for the morphologies we find in absence of additives. This study was performed at low increments in the initial supersaturation levels, and at these low driving forces we do not observe the high number of precursor crystals that would be required to account for such an aggregation mechanism. We do however observe many of the morphologies that in literature are claimed to be formed by assembly (with additives). We have not probed the solutions by high resolution techniques but we think that the high number (1) of 10–20 nm crystals required should be quite easy to detect if they were there, and we think that the burden of proof is on those who suggest a revision of the classical growth paradigm and that such revisions should be based on quantifiable supersaturation, which is the main underlying parameter in the classical understanding of crystal growth. That aggregation of precursor crystals is an unlikely growth mechanism is something we have discussed in previous reports.[1,2]

1. J. P. Andreassen, *J. Cryst. Growth*, 2005, **274**, 256–264.
2. R. Beck and J. P. Andreassen, *Cryst. Growth Des.*, 2010, **10**, 2934–2947.

Professor De Yoreo asked: What is the growth mechanism that leads to these hierarchical structures without particle aggregation, is it surface-induced nucleation or something else and are their certain orientations of the nuclei selected that represent low interfacial energy (like twin planes)?

Professor Andreassen replied: The complex structures caused by spherulitic growth are usually explained by non-epitaxial surface nucleation[1,2] at the growth front, and selection criteria are not considered as far as I know.[1,2]

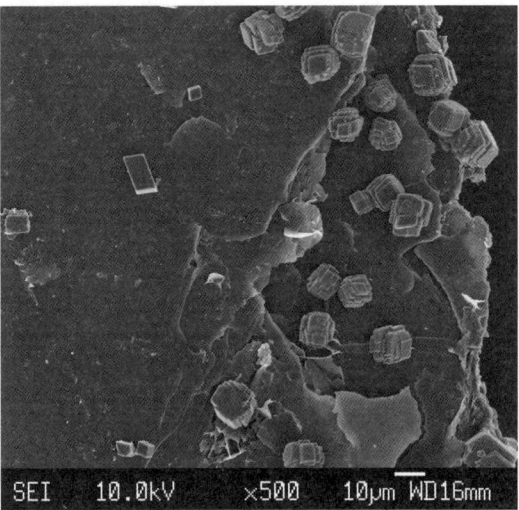

Fig. 1 Calcite crystals grown on the frayed edges of mica surfaces from 100 mM CaCl$_2$ solution using the ammonia diffusion method. The configuration of the cell is such that the largest supersaturation is near the edges of the mica sheets.

1. L. Granasy, T. Pusztai, T. Borzsonyi, J. A. Warren and J. F. Douglas, *Nat. Mater.*, 2004, **3**, 645.
2. L. Granasy, T. Pusztai, G. Tegze, J. A. Warren and J. F. Douglas, *Phys. Rev. E*, 2005, **72**, 011605.

Mrs Nergaard commented: We demonstrate that there is a strong link between supersaturation and the resulting morphology of the precipitated calcium carbonate. For the application of crystallization, particle morphologies can be of great importance, and such morphology predictions are of great interest. In classical theory, the underlying tool to describe a system's behavior is thermodynamics. Proper quantification of the numbers of nanoparticles nucleated as well as aggregation rates at different conditions, could give rise to a similar tool within non-classical crystallization, but this is so far not seen.

Dr Christenson remarked: There is very little mention of the evolution of calcite morphology with supersaturation in your paper. In this context I recall calcite morphologies such as those in the attached image (Fig. 1), obtained in connection with recent studies of CaCO$_3$ crystal growth on mica.[1] Am I correct in assuming that the morphology of the crystals near the frayed edges of the mica surfaces may be the result of extensive surface nucleation at high supersaturations (crystal growth from 100 mM CaCl$_2$ with the ammonia diffusion method and no additives)?

1. C. J. Stephens, Y. Mouhamad, F. C. Meldrum and H. K. Christenson, *Cryst. Growth Des.*, 2010, **10**, 734–738.

Dr Andreassen replied: We did not observe much calcite in the experiments reported in the paper. The calcite abundance increases as the temperature is lowered or when given time to transform from the initially precipitated vaterite/aragonite mixture. The morphologies that we then observe are the classical calcite rhombohedra. However we have previously shown that calcite spherulites precipitate at low temperatures, in absence of additives.[1] To me it also looks like the calcite morphologies in the attached image are the result of surface nucleation. On some of the calcite faces 2D nucleated islands are visible. 2D nucleation could of course also be the source of steps also at lower supersaturation, but invisible due to small step heights. Extensive step bunching and incomplete step propagation might explain why this is evident in the picture. I am not sure if you observe these morphologies all over the mica surface or only at the frayed edges, but step bunching is affected both by the level and distribution of supersaturation over the (crystal) surface, and by diffusion and fluid flow.[2]

1. R. Beck and J. P. Andreassen, *Cryst. Growth Des.*, 2010, **10**, 2934–2947.
2. A. A. Chernov, L. N. Rashkovich and P. G. Vekilov, *J. Cryst. Growth*, 2005, **275**, 1–18.

Dr Gebauer asked: In the introduction of the *Faraday Discussions* paper,[1] you indicate that the "pre-structured" precursor phase of amorphous calcium carbonate (ACC) described in our work[2] would dictate the polymorphism of the crystalline state. However, as detailed in the original publication,[2] the different proto-structures in ACC do not lead to polymorph selection; they appear to be one out of many factors that influence crystallization. The notion of proto-calcite and proto-vaterite ACC was introduced based upon evidence that was obtained by means of solid-state-NMR and IR spectroscopy as well as Ca K-edge EXAFS analyses.

1. J.-P. Andreassen, R. Beck and M. Nergaard, *Faraday Discuss.*, 2012, **159**, DOI: 10.1039/C2FD20056B.
2. D. Gebauer, P. N. Gunawidjaja, J. Y. P. Ko, Z. Bacsik, B. Aziz, L. Liu, Y. Hu, L. Bergström, C.-W. Tai, T.-K. Sham, M. Edén, N. Hedin, *Angew. Chem. Int. Ed.*, 2010, **49**, 8889–8891.

Dr Andreassen responded: Thank you for the clarification and I apologize for the incorrect statement when referring to your paper.

Mr Khan opened the discussion of the paper by Dr Richard P. Sear: This model looks interesting and I wonder if this could be analogous to many of the crystal growth inhibitors that are in saliva which reduce remineralisation of denitin and enamel?

Interestingly the one and many polymer model on crystal growth, could this explain how enamel lesions (white spots) occur if there is a build up of material (hard or soft). I would also ask that in the model, would the polymer trap ions/ACP material below it if inversely as you have shown that it stops growth intially from crystals growing by blocking material above it?

Dr Sear responded: Good question. I am not sure that I know enough about these inhibitors in saliva to answer it. If they are intrinsically disordered proteins, then they should look roughly like a polymer in solution. Then my model might be a reasonable first attempt at modelling the behaviour of these inhibitors in saliva on nm and larger lengthscales. If there is some quantitative data available, *e.g.*, growth rate as a function of concentration, then this could be compared with the results of simulation to test this idea.

As for trapping ions, I have not studied dissolution in my model, but it would be easy to do. Presumably, high polymer concentrations would inhibit dissolution as well as growth, and that could be studied quantitatively in the model. It may even be possible to do something like cycle the model through crystallisation and demineralisation cycles to see if polymers could build up in a part of the model crystal and act as a weak spot where a lesion could start, but that is speculative. Thanks for these thought-provoking questions.

Dr Christenson commented: In the simulation which has some snapshots in Fig. 5 of your paper the growing crystal appears to first surround the polymer molecule leaving a void around it, which later fills in. How is this possible, and would it not be likely that in a real system incorporation of molecules into a crystal lattice would sometimes lead to empty pockets around these molecules?

Dr Sear answered: It is possible because I use what are called Glauber dynamics (see the final section of the paper), which here basically have the effect of fixing the supersaturation all along the condensed-phase/solution interface. The advantage of this is that supersaturation there is constant and so I can look at growth rates such as in Fig. 2, but a disadvantage is the phenomenon you ask about. Growth is possible even of a front that is not connected to the bulk, and so a void can fill in even if there is no diffusion path from the void to the bulk. This is not physical.

As a computational physicist, I am not perhaps the best person to ask about real systems. However, I could change the dynamics in my model such that the particles have to diffuse from the solution to the growth front. Then if this diffusion was sufficiently fast, the growth mechanism of nucleation and growth of the layers would still be rate limiting. I would then expect that voids like the one in Movie 3 should still form, but would then be unable to fill and remain as an empty pockets.

So the current model clearly under predicts empty pockets, and so it would be useful to study pockets both in experiment and in models with improved dynamics. Thank you for a stimulating question.

Professor De Yoreo asked: What happens if you differentiate between binding to steps and terraces and what is the impact of supersaturation?

Dr Sear replied: I have not studied a model in which the polymer binds preferentially to steps, although it certainly could be done, I don't know how I would put in preferential bonding to a terrace as it would just be a flat surface in my model.

At low polymer concentrations I would guess that, in my model at least, preferential binding to steps would have little effect, so long as binding to the surface was strong the polymer would be incorporated. At higher polymer concentrations then binding to steps would clearly slow forward growth of the interface as it would slow the growth of a layer, which occurs at steps. However, under the conditions I work it is really nucleation that is rate limiting, so it is possible that binding to steps would just slow growth of the condensed phase. But additional simulations would have to be done to test this.

Varying supersaturation over a wide range has little effect on polymer incorporation. Of course reducing the supersaturation reduces growth speed but except for all but very short polymers the polymer dynamics are so slow that reducing the growth rate has little effect.

Professor Penn remarked: Is it possible to simulate oriented aggregation in the presence of additives like the polymer molecules used in your simulation? Specifically, it'd be interesting to test whether a molecule present within the space between the two primary particles would be incorporated or removed prior to conversion to an oriented aggregate.

Dr Sear responded: The short answer is that it would be a demanding calculation, but yes it could be done. A relatively simple model would need to be used to keep the computational cost reasonable. Simulation would allow access to the microscopic details of the dynamics, and to the structure of the resulting (*meso*)crystal, including molecule incorporation. These details are inaccessible in experiment and so here simulation would presumably be very useful in understanding the system. So I would say it is a good idea.

A slightly longer answer is that my model is a lattice model and so unsuited to studying oriented aggregation—the domains I study cannot rotate, at least not by angles other than 90°. So an off lattice model would be needed. There has already been simulation work on the sintering of nanoparticles without additives.[1] Adding a simple model oligomer to a model such as this and studying its effect on two nanoparticles coming together and sintering could be studied. An explicit model of a metal nanoparticle in explicit water is not doable, in my opinion.

1. L. J. Lewis, P. Jensen and J-L. Barrat, *Phys. Rev. B*, 1997, **56**, 2248.

Dr Toroz said: The question with regard to this work is on how large the size of the polymer should be: if there is any post critical size that has been monitored (*i.e.* number of residues in a peptide *etc.*).

Dr Sear answered: The effects I observe, for example the incorporation of the polymer, are mostly insensitive to the size of the polymer, provided it is large enough that the polymer/surface interaction is many kT strong. Then the binding between the polymer and growing surface is effectively irreversible, and increasing it further by making the polymer larger, has little effect. The size at which the binding becomes many kT depends on the monomer/surface interaction strength of course, but unless it is quite weak, a 10-mer is approximately as likely to be incorporated as a 100-mer, and even smaller oligomers are incorporated.

Dr Kiley remarked: Interactions such as hydrogen bonding are very important in understanding solvent–mineral interactions. An excellent example of this is the effect of water/ethanol mixtures on calcium carbonate systems. These interactions are strongly dependent upon atomistic details and are therefore difficult to represent in a coarse-grained system. An alternative is an adaptive resolution approach, in which the representation of the solvent molecule is changed dynamically to provide both computational speed-up and chemical accuracy at the same time.

PAPER

A metastable liquid precursor phase of calcium carbonate and its interactions with polyaspartate†

Mark A. Bewernitz,[a] Denis Gebauer,[b] Joanna Long,[c] Helmut Cölfen[d] and Laurie B. Gower[*e]

Received 23rd April 2012, Accepted 6th June 2012
DOI: 10.1039/c2fd20080e

Invertebrate organisms that use calcium carbonate extensively in the formation of their hard tissues have the ability to deposit biominerals with control over crystal size, shape, orientation, phase, texture, and location. It has been proposed by our group that charged polyelectrolytes, like acidic proteins, may be employed by organisms to direct crystal growth through an intermediate liquid phase in a process called the polymer-induced liquid-precursor (PILP) process. Recently, it has been proposed that calcium carbonate crystallization, even in the absence of any additives, follows a non-classical, multi-step crystallization process by first associating into a liquid precursor phase before transition into solid amorphous calcium carbonate (ACC) and eventually crystalline calcium carbonates. In order to determine if the PILP process involves the promotion, or stabilization, of a naturally occurring liquid precursor to ACC, we have analyzed the formation of saturated and supersaturated calcium carbonate–bicarbonate solutions using Ca^{2+} ion selective electrodes, pH electrodes, isothermal titration calorimetry, nanoparticle tracking analysis, ^{13}C T_2 relaxation measurements, and ^{13}C PFG-STE diffusion NMR measurements. These studies provide evidence that, in the absences of additives, and at near neutral pH (emulating the conditions of biomineralization and biomimetic model systems), a condensed phase of liquid-like droplets of calcium carbonate forms at a critical concentration, where it is stabilized intrinsically by bicarbonate ions. In experiments with polymer additive, the data suggests that the polymer is kinetically stabilizing this liquid condensed phase in a distinct and pronounced fashion during the so called PILP process. Verification of this precursor phase and the stabilization that polymer additives provide during the PILP process sheds new light on the mechanism through which biological organisms can exercise such control over deposited $CaCO_3$ biominerals, and on the potential means to generate *in vitro* mineral products with features that resemble biominerals seen in nature.

[a]*Department of Biomedical Engineering, University of Florida, Gainesville, FL, 32611-64000, USA. E-mail: bewerni1@ufl.edu; Tel: +1-352-846-3337*
[b]*Department of Chemistry, Physical Chemistry, University of Konstanz, Universitätsstrasse 10, Box 714, D-78457 Konstanz, Germany. E-mail: Denis.Gebauer@uni-konstanz.dez; Fax: +49 7531 883139; Tel: +49 7531 882169*
[c]*Department of Biochemistry and Molecular Biology, University of Florida, Gainesville, FL, USA*
[d]*Department of Chemistry, Physical Chemistry, University of Konstanz, Universitätsstrasse 10, D-78457 Konstanz, Germany. E-mail: Helmut.Coelfen@uni-konstanz.de; Fax: +49 7531 883139; Tel: +49 7531 884063*
[e]*Department of Materials Science and Engineering, University of Florida, Gainesville, FL, 32611-6400, USA. E-mail: lgowe@mse.ufl.edu; Tel: +1-352-846-3336*

† Electronic supplementary information (ESI) available. See DOI: 10.1039/c2fd20080e

Introduction

Calcium carbonate is one of the most abundant biominerals on Earth.[1-4] Calcium carbonate is very important in many industrial processes such as in fillers for papermaking and other composites, in carbon dioxide storage for environmental concerns,[5] and in biomedical applications, such as drug delivery,[6] due to its low cost and biocompatibility.[7] From an engineering standpoint, understanding the biological formation and regulation of calcium carbonate biominerals is very intriguing because biologically-derived calcium carbonate exhibits an incredible array of crystals with complex and non-faceted morphologies. In nature, it is usually found as the mineral component in skeletal structures in invertebrates, such as in the spine of sea urchins[8] and in the nacre of molluscs, but plays a role in the otoliths and bones of vertebrates as well.[9,10] The additional control over mineral product displayed by biological organisms, as opposed to *in vitro* techniques, is believed to be due to cellular manipulation of an amorphous calcium carbonate (ACC) precursor phase which may act as a reactive intermediate in generating complex functional materials.[11,12] This may be due to the fact that ACC is not subject to constraints of crystal lattice energies and can be "moulded" into an endless array of non-equilibrium morphologies.[3]

Manipulation of the ACC precursor may be achieved by templating and altering environmental influences to direct the formation of crystalline calcium carbonate product, which may be calcite, aragonite, and even vaterite [13] (although the latter case often serves as a crystalline intermediate).[14-16] It has been demonstrated with *in vitro* model systems that the crystalline morphology can be directed by the presence of negatively charged polyelectrolytes (that are believed to emulate the acidic proteins of biominerals) during the precipitation process, such as is proposed by the polymer-induced liquid-precursor (PILP) process, where a pseudomorphic transformation of a fluidic ACC precursor can lead to crystals with non-equilibrium morphologies.[17,18]

During the PILP process, $CaCO_3$ appears to nucleate through a multistep-step process where the polymer (charged, anionic polyelectrolyte) associates with Ca^{2+} and CO_3^{2-} ions to form an intermediate liquid phase prior to solid nucleation. This is why the PILP process is so successful in generating "molten" morphologies and films because the precursor nanodroplets are believed to meld together to form smooth mineral products that lack the facets found in crystals grown by the classical crystallization process. It is difficult, however, to theoretically explain how a true liquid metastable phase can be induced by the presence of a polymer while assuming that, in the absence of polymer additives, the nucleation follows the classical view of crystallization.

According to the classical view, the formation of $CaCO_3$ nucleation proceeds *via* a nucleation and growth mechanism where the basic constituents of the crystal (ions, molecules, or atoms, depending on the material) remain soluble until a critical concentration is achieved, allowing for a critical nucleus to be formed. Recently, it has been found that $CaCO_3$ nucleation occurs through a non-classical process, where a stable prenucleation solute interaction, called a prenucleation cluster (PNC) forms. The PNC formation has been reported at pHs ranging from 9.0 to 10.0 and has been reported to be stable with respect to the initial ions in solution, prior to the formation of a metastable solid nucleus of ACC.[19] Gebauer *et al.* demonstrated that Ca^{2+} and CO_3^{2-} ions associate into clusters of ions (having a typical diameter on the order of *ca.* 2–3 nm) in undersaturated, saturated and supersaturated conditions.[20] The presence of PNCs has been corroborated by cryo-TEM experiments in solutions saturated with respect to calcite.[21] Evidence has been obtained in the form of computer simulations that suggest that $CaCO_3$ PNCs form as stable, rapidly exchanging, branched "polymers" of alternating Ca^{2+} and CO_3^{2-} ions.[22] This dynamically ordered liquid-like oxyanion polymer (DOLLOP) model for the formation of PNCs proposed by Gale et al. further supports the notion that PNCs are non-classical by nature.

Other studies have shown evidence of a multi-step model in addition to that of the PNCs proposed by Gebauer, where amorphous calcium carbonate might be formed through a liquid–liquid phase separation from the bulk solution in the form of droplets. It has been proposed for some time now that a liquid–liquid phase transition might be induced by the presence of additives, such as polymers, to form droplets that can grow from tens of nanometres to as large as microns in diameter.[17] However, there has been considerable evidence recently that $CaCO_3$ transitions through very large liquid precursor droplets in the absence of any additives.[23–25] By analyzing an acoustically levitated droplet of saturated $CaCO_3$ (with respect to calcite) at a pH of 6.3, Wolf et al.[25] demonstrated that a liquid-phase of $CaCO_3$ precursor droplets forms in supersaturated conditions, which arises from an increase in pH upon outgassing of CO_2. The liquid precursor phase was observed to form with droplet diameters of up to several hundred nanometres. Rieger et al.[23] quenched and freeze captured a solution after inducing precipitation by rapidly mixing $CaCl_2$(aq) and Na_2CO_3(aq) (presumably at a pH in the upper 10s) and used cryo-TEM to visualize droplets of $CaCO_3$ precursor up to 2 μm in diameter. Faatz et al.[24] used a CO_2 outgassing technique at pH 7 to demonstrate that ACC with diameters of hundreds of nm can be obtained through a suspected liquid–liquid phase separation prior to solidification into the amorphous state. Evidence of multi-step mechanisms is not limited only to $CaCO_3$ mineralization, but can occur in other metal carbonates,[24–27] in calcium oxalate systems[28] phosphate-coordinated systems,[29–33] as well as in organics, such as biopolymers,[34,35] amino acids[36–39] and organic pigments.[40]

In these previous works by Wolf, Rieger, and Faatz, the detected liquid precursor droplets in their experiments suggests that the liquid phase is metastable with respect to the phases that occur at later stages, and that the structural transitions toward a solid, ACC-like form follows a downhill energetic sequence, similar to the dehydration and subsequent crystallization of ACC.[41]

Our group has primarily focused on the polymer-induced liquid-precursor (PILP) process and its relevance to biomineralization. In this case, the liquid precursor phase exists long enough to be manipulated into non-equilibrium morphologies, the hallmark of biominerals. Thus, it is highly desirable to understand what the mechanistic role of the polymer is in this process in light of these other new findings of various precursor phases. For example, rather than the polymer interacting with ions, as we had originally assumed, one must now consider if the polymer might actually be interacting with any of these other species that have been detected in $CaCO_3$ solutions.[23–25] Thus, the focus of this paper can be expressed in two parts: 1) without polymer additives, can the $CaCO_3$ liquid precursor be detected at a more neutral and biologically friendly pH of 8.5? 2) Does PILP form through polymer interaction with a pre-existing liquid precursor, with PNCs (as previously proposed),[42] or by some other interaction?

Given this focus, we carried out studies of the early stages of precipitation in a fashion similar to Gebauer et al., except that we conducted our studies at pH 8.5 and allowed the ever increasing supersaturation to evolve by using punctuated injections of aqueous $CaCl_2$ into bicarbonate buffer, allowing for time between injections to allow for system equilibration. At various points in the evolution, we analyzed the state of the solution using a Ca^{2+} ion selective electrode, isothermal titration calorimetry (ITC), nanoparticle tracking analysis (NTA) light scattering, analytical ultra-centrifugation (AUC) and carbon specific nuclear magnetic resonance (^{13}C NMR) spectroscopic techniques such as Carr–Purcell–Meibloom–Gill (CPMG) T_2 measurement and pulsed field gradient stimulated echo (PFG-STE) diffusion-NMR. It should be noted that we did not modulate the pH during the supersaturation evolution as did Gebauer et al. The PILP-generating technique used in our lab that we wished to model does not maintain a constant pH. Rather, we allowed the pH to evolve with the titration and monitored it throughout the titration. Using these methods at a moderate pH of ~8.5, we discovered a new phase transition of a liquid condensed phase (LCP) that occurs at a critical concentration of bound calcium. We then used the same techniques with a polymer (polyaspartic acid–sodium salt) present to determine if the formation of the polymer-induced liquid-precursor (PILP) phase is indeed a polymer-stabilized LCP phenomenon.

Materials and methods

Generation of super-saturated solution in the absence of polymer additive

Using a micropipette, 20 mM sodium carbonate (Fisher) was titrated into a 20 mM sodium bicarbonate (Fisher) solution to generate a 20 mM, pH 8.5 carbonate buffer solution. Calcium chloride (Fisher) solution with a concentration of 4.5, 6, or 10 mM was titrated into 29 ml of carbonate buffer, which was stirred at 100 rpm using a magnetic stir bar. The volumes of titration were 200 μl each except for the first injection which was only 40 μl to account for infinite dilution phenomenon. Titrations were injected at an approximate rate of 20 μl s^{-1} immediately over the rotating stir bar to ensure adequate mixing. The titrations were made in a punctuated fashion, 2 min of constant pH, and [Ca^{2+}]$_{Free}$ measurements were acquired before injecting more CaCl$_2$(aq). The solutions were prepared with nanopure water and all were filtered using a 0.22 μm Millipore syringe filter prior to any titrations. Each of these titrations was conducted in triplicate and the error expressed in the results is plus or minus two standard deviations.

Generation of super-saturated solution in the presence of polymer additive

Using a micropipette, 300 mM sodium carbonate was titrated into a 300 mM sodium bicarbonate solution to generate a 300 mM, pH 8.5 carbonate buffer solution. This solution was then titrated into 29 ml of a stirred 10 mM CaCl$_2$ solution which may or may not contain 20 μg ml^{-1} polyaspartic acid sodium salt (monodisperse, Alamanda Polymers), depending on the experiment. The volumes of titration were 200 μl each except for the first injection which was only 40 μl, matching the titration conditions for the non-additive experiments. Titrations were injected at an approximate rate of 10–20 μl s^{-1}, always over the rotating stir bar to ensure adequate mixing and to minimize the formation of strong concentration gradients that might affect the experiments. The titrations were made in a punctuated fashion; 2 min of constant pH, and [Ca^{2+}]$_{Free}$ measurements were acquired before injecting more titrant. The injections were made using disposable micropipette tips that were disposed of after each injection and replaced prior to a new injection. Each experiment was conducted in triplicate, including the control, and the error shown is plus or minus two standard deviations.

Ca^{2+} electrode and pH electrode measurements

The free Ca^{2+} concentration, [Ca^{2+}]$_{Free}$, in the titrated solution was obtained using a Ca^{2+} ion selective electrode (Radiometer Analytical, ISE-K-Ca, E11M006) in conjunction with a reference electrode (Radiometer Analytical E21M009). A calibration standard curve for calculating free Ca^{2+} concentration was generated by titrating the experiment-appropriate concentration of CaCl$_2$(aq) into nanopure water which had been brought to pH 8.5 by the addition of trace amounts of NaOH (Fisher) (aq). The pH evolution of the titration was obtained using a standard pH electrode. Both pH and [Ca^{2+}]$_{Free}$ values had to remain constant for at least 2 min of mixing before adding another titration injection to verify that the solution equilibrated and that solid nucleation had not yet occurred. It is important to note that we are not quantitatively accounting for CO$_2$ net diffusion out of the solutions during our experiments.

Isothermal titration calorimetry (ITC) of phase transition

All measurements were made using a VP ITC (MicroCal™). Carbonate–bicarbonate buffer and $CaCl_2$(aq) was generated as described above. The injection of $CaCl_2$(aq) into the reservoir carbonate–bicarbonate buffer was maintained at exactly the same ratio as the titration experiments to give exact punctuated enthalpies that correspond to the titration experiment. The experiments were conducted at a controlled temperature of 298 Kelvin. The rest time between injections was adjusted to allow for complete thermodynamic equilibration between the reaction vessel and the reference vessel. A stir speed of 180 rpm was used for all the experiments. A reference power of 2 μcal s^{-1} was used due to the very subtle, endothermic nature of the reaction. The solutions were not degassed due to the fraction of CO_2 that is in equilibrium within the carbonate–bicarbonate buffer at pH 8.5. No bubbling phenomena were observed during the course of the ITC experimentation. Control enthalpy profiles of $CaCl_2$(aq) injections into water, water injections into bicarbonate buffer, and water injections into water were subtracted from the raw data to normalize the data.

Nanoparticle tracking analysis (NTA) of emergent phase

The number count and the hydrodynamic radius of the emergent phase droplets were analyzed using the NTA light scattering technique. Samples were analyzed using an LM20 analyzer (Nanosight™) and the data was processed using an NTA analytical software suite (Nanosight™). Samples for analysis were prepared as described in the "Generation of super-saturated solution" section. 0.3 ml of sample was used in each analysis.

Analytical ultracentrifugation (AUC)

AUC was performed on an Optima XL-I (Beckman-Coulter, Palo Alto, CA) using the Rayleigh interference optics at 60 000 rpm and 25 °C. The experiments were performed in 12 mm titanium double sector cells (Nanolytics, Potsdam, Germany). All experiments were evaluated using the software SEDFIT applying Lamm equation modelling for 1–4 noninteracting species to determine sedimentation and diffusion coefficients as well as concentration of up to 4 species in a solution.[43,44]

NMR, PFG-STE, spin–spin (T_2) relaxation time measurement

All NMR experiments were conducted on a Bruker Avance DRX 500 MHz vertical bore system using a *xyz* gradient TXI probe with a 1H and 2H interior coil, ^{13}C and ^{15}N exterior coil, and *xyz* gradients. All carbonate–bicarbonate buffer solutions were generated as described above except using 100% ^{13}C-enriched sodium carbonate and sodium bicarbonate ingredients (Cambridge Isotopes) to enhance signal/noise. All experiments were conducted at 298 Kelvin. Deuterium oxide was used to obtain a lock at a volume fraction of 2.5% of the total sample. Data was processed using TOPSPIN™ software and MATLAB™ derived software when deconvolution of overlapping spectral peaks was required. The T_2 relaxation times of the various species in solution were obtained using a Carr–Purcell–Meibloom–Gill (CPMG) sequence with increasing tau (τ) times of 40, 120, 200, 400, 600, 800, 1000, 2000, 3000, and 4000 msec. PFG-STE ^{13}C diffusion experiments were conducted using the variation of the Bruker stegs1s pulse sequence based upon the pulsed-field gradient spin echo (PFG-SE) technique for measuring diffusion. ^{1}H was decoupled from ^{13}C for the entirety of the pulse sequence. We used 1.8 s diffusion times and 1 msec gradient pulse times. All processing was zero-filled twice and was done with 0.3 Hz line-broadening to allow for characterization of NMR spectral features. We used a gradient with strength of 50 g cm^{-1} for the gradient pulses which were varied equidistantly between 2% and 95% to generate 16 1-D slices for analysis. The PFG-STE pulse sequence used, as well as many of the relevant variables chosen, is shown in Fig. S.1† of the supplementary information.

Results

Metastable liquid condensed phase (LCP)

The binding behavior of Ca^{2+} ion as aqueous $CaCl_2$ was titrated at various concentrations into 20 mM carbonate buffer (pH 8.5) is shown in Fig. 1a. Even though care was taken to minimize temporary concentration gradients when introducing $CaCl_2(aq)$ to the bicarbonate buffer, there still appear to be some kinetic effects due to the injection. This is evident by the slight difference in binding fraction between the various concentrations of $CaCl_2(aq)$ injectant. The bound fraction (slope of the displayed line) would be expected to be very similar if the system were at true thermodynamic equilibrium. Therefore, interpretation of the data should be approached from a qualitative standpoint. To additionally demonstrate the kinetic binding effects that can occur due to imperfect addition of ion to counterion, and to eliminate the possibility that the character of the Ca^{2+} binding profiles are due to nucleation of mineral at the nozzle (micropipette tip), an additional titration was performed with the nozzle just above the solution. The results, shown in Fig. S.2† of the supplementary information, demonstrate that the qualitative character of the Ca^{2+} binding profile is conserved and not due to nucleation at the nozzle even if kinetic binding effects are enhanced. Additionally, we did not quantitatively

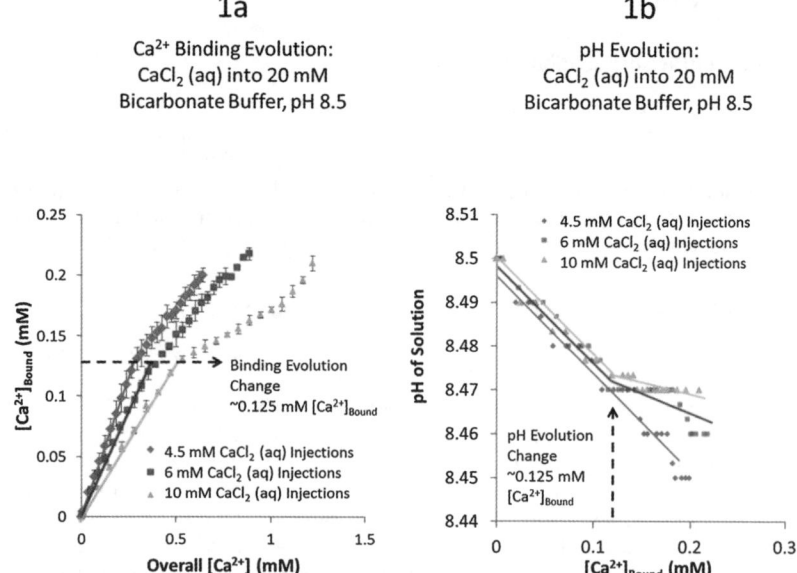

Fig. 1 Left (1a) the evolution of bound Ca^{2+} vs. overall titrated $[Ca^{2+}]$ as $CaCl_2(aq)$ is titrated into 20 mM carbonate buffer, pH 8.5. Initially, the binding profile is linear which is expected for prenucleation cluster formation. However, a discontinuity in Ca^{2+} binding is observed at a value of ~0.125 mM $[Ca^{2+}]_{Bound}$, which is evidence of a phase transition. The evolution was done in triplicate and results were averaged. Right (1b) the pH evolution of the pH 8.5 20 mM bicarbonate buffer system due to the injection of $CaCl_2(aq)$. The pH decreased due to the consumption of CO_3^{2-} to form ion pair–PNCs until a discontinuity occurs at a consistently measured $[Ca^{2+}]_{Bound}$ of ~0.125 mM, further suggesting a phase transition at the same bound calcium concentration. The change in pH evolution in the upward (basic) direction suggests that a larger fraction of bicarbonates and/or a reduction of carbonates are participating in Ca^{2+} binding. Lines were included as an aid.

account for carbon dioxide gassing-out that would occur with this solution over time. This can lead to changes in binding affinity and pH over time. For these reasons, we interpreted the data qualitatively.

Initially, during the titration, the fraction of calcium that binds is constant with each additional injection, which is expected for the formation of calcium carbonate prenucleation clusters.[20] However, at a critical bound calcium concentration (~0.125 mM in this case), the constant Ca^{2+} binding affinity decreases yielding a new linear binding affinity slope, suggesting a change in the types of products forming. Note—this change in binding behaviour occurs prior to solid nucleus formation, which would be evidenced by the inability to achieve apparent equilibrium due to decreasing pH (massive carbonate binding) and decreasing free calcium concentration (massive binding of Ca^{2+} to form solid mineral). Interestingly, at the point of the transition, the $[Ca^{2+}]_{Free}[CO_3^{2-}]_{Free}$ ion product is different for all three $CaCl_2$(aq) titration concentrations (0.4, 0.7, 0.1 mm^2 for 4.5, 6, and 10 mM injections, respectively). Solid nucleation and growth leading to precipitation eventually occurred at concentrations of bound Ca^{2+} between 0.2 and 0.22 mM.

The pH evolution of the solution was monitored during the titration and the results are shown in Fig. 1b. The pH of the solution was decreasing initially, as would be expected, prior to the phase transition due to the sequestering of CO_3^{2-} carbonate ions as predicted to occur during PNC formation. The pH evolution flattened considerably at the same bound Ca^{2+} concentration (~0.125 mM) as the change in binding affinity was observed, indicating that the Ca^{2+} binding affinity to bicarbonate has increased relative to carbonate. As nucleation is a singular event, the discontinuity in the pH development and/or the change in Ca^{2+} binding development strongly points toward the nucleation of a second phase. Because the pH development becomes flatter, less carbonate, or more bicarbonate ions are binding after nucleation of this phase, suggesting that the bicarbonate ion is participating in the emergent phase. The data displayed in Fig. 1 suggest that there is a critical bound calcium concentration that leads to a phase transition which involves the bicarbonate ion.

In order to gain further insight into the thermodynamics of this phase transition, isothermal titration calorimetry (ITC) was conducted using the same punctuated titration method as was used for the Ca^{2+} binding and pH development experiments to measure the enthalpy change upon nucleation of the emergent phase. All three titration concentrations were tested, and as can be seen in Fig. 2, a first-order phase transition, identified by the apparent enthalpic discontinuities, is occurring at similar conditions as identified in the titration experiments, further corroborating the notion that a phase transition has occurred. The phase transition occurred 3 injections earlier for the 4.5 mM injection and 2 injections early for the 6 mM injection in the ITC experiment as compared to the desktop titrations, possibly due to the more controlled environment of the ITC reaction chamber. The enthalpic discontinuity is positive, demonstrating that the driving force of the phase transition is entropic, which is often a sign of liberation of hydration waters.

Nanoparticle tracking analysis (NTA) was used to examine the size of the emergent phase species and to estimate the amount of species in solution after the phase transformation. Samples were analyzed prior to, at, and after the observed phase transition (the 7th, 10th, and 13th injection, respectively) for the case of the 6 mM $CaCl_2$ titrant. The results of the analysis are shown in Fig. 3. Prior to the detected phase transition (7th injection), no phase was detectable since pre-nucleation clusters of ca. 2–3 nm in diameter are well below the 20 nm diameter threshold for the NTA method and therefore there isn't any data for this condition to present. At the detected phase transition (the 10th injection), large species emerged with a diameter size distribution of ~60 nm. Additional $CaCl_2$ injections (13th injection) resulted in an increase in size of up to 100 nm diameter, suggesting that increasing Ca^{2+} concentration leads to growth and/or coalescence of the emergent phase. This strongly suggests that the emergent phase is stable with respect to the solution

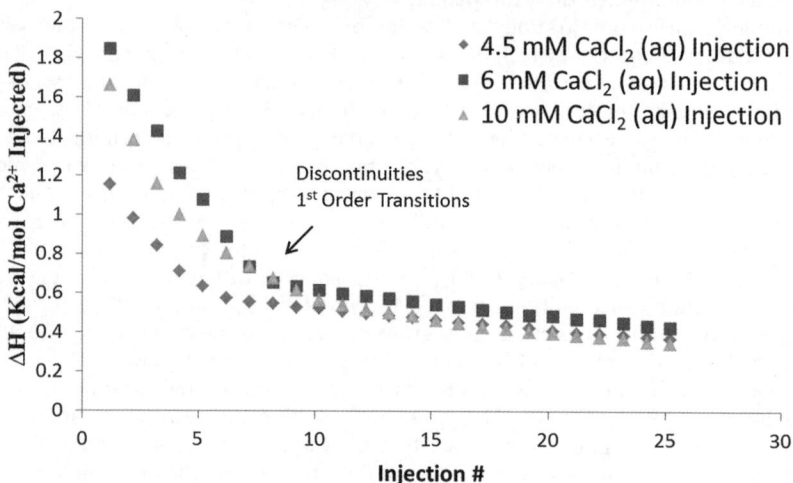

Fig. 2 The enthalpy of reaction during the titration of $CaCl_2$(aq) into 20 mM carbonate buffer, pH 8.5. The data shows that there is an endothermic phase transition occurring, indicating that the phase transition is entropically driven and probably due to liberation of hydration waters.

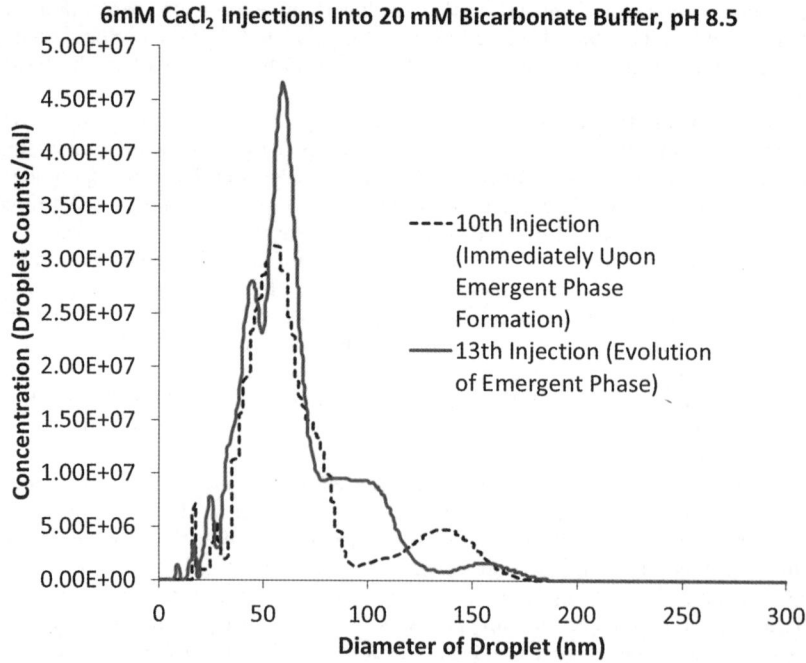

Fig. 3 The emergent phase size distribution, according to nanoparticle tracking analysis (NTA), of the solution at and after the phase transition (injection 10 and 13, respectively for a 6 mM $CaCl_2$(aq) injection). Prior to the phase transition no species were detected. At the phase transition (10th injection), droplets with a distributed diameter averaging ~60 nm emerge. The addition of more Ca^{2+} to the solution yields more detectable phase emerging at 60–70 nm diameter and they seem to grow larger as well.

species, because if it were metastable, it could only occur in the form of microscopic fluctuations in the solution and would not grow with further Ca^{2+} injections.

A video of the raw data used to track the scattering phase is shown in the supplementary information Fig. S.3.† In the video one can see that the droplets showed somewhat unusual behaviour, almost as if the species were warping and adopting a temporary aspect ratio. The species were not observed to coalesce or split permanently; they came close to doing so and then re-established the former interfaces prior to the interaction. This behavior seems consistent with the Wolf *et al.* observation of $CaCO_3$ precursor droplets having emulsion-like behavior with a resistance to coalescence due to weak electrostatic stabilization.[45] According to the NTA results, the emergent species is very dilute (on the order of 10^{-12} species liter^{-1}), which suggests that the aggregation/coalescence of particles/droplets that yield the emergent phase arises due to weak interactions. Note—this phase would be difficult to detect using standard dynamic light scattering techniques, which require a much higher particle/droplet count and a known refractive index.

Analytical ultracentrifugation (AUC) was used to detect the species present in solution for the various points of $[Ca^{2+}]$ evolution for the case of the 6 mM $CaCl_2$(aq) injection. The results are presented in Fig. 4. Prior to the emergent phase formation, ions/ion pairs and PNCs are present as indicated by the squares and circles, respectively. At the expected point of phase emergence, larger species suddenly emerge (D & E in Fig. 4). After a few more injections, only ions/ion pairs are detected. This result seems counterintuitive after a phase transformation, but if the volumes of the phases are considered, it becomes understandable. As determined by NTA analysis, the LCP droplets are very dilute and represent a very small fraction of the overall volume. Literature reports on emulsion systems demonstrate coalescence with subsequent demixing allowing for a determination of the respective phase volumina.[46] However, the small volume fraction of this newly formed liquid phase with respect to the mother solution, would lead to an undetectable phase boundary and therefore, only the ions and prenucleation clusters in the major liquid phase can be seen. Due to the apparent liquid nature of this emergent phase as inferred from the NTA and AUC analyses, we refer to these as liquid condensed phase (LCP) droplets.

Nuclear magnetic resonance was used to further analyse the constituents involved in the LCP in comparison to the solvated carbonate–bicarbonate species in the bulk

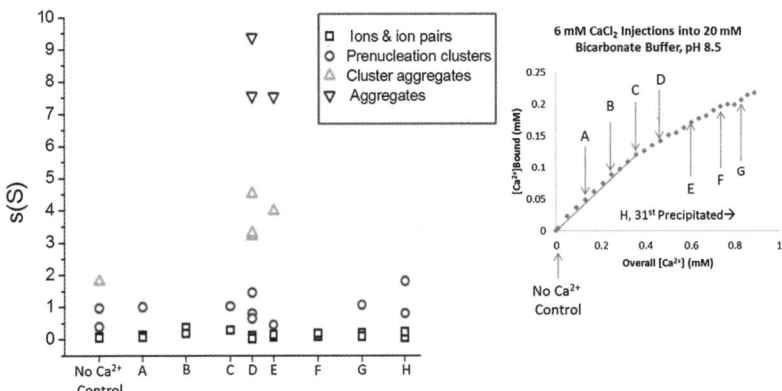

Fig. 4 Sedimentation coefficients (25 °C) for the detected species in solution by AUC for the injections indicated in the inset. The emergence of LCP occurs at point C. It is evident that larger species are present for D & E, just after the phase transition. The control consists of just carbonate buffer. PNCs are also seen in the control because they can form with sodium carbonate species as well.

solution. Specifically, Carr–Purcell–Meiboom–Gill (CPMG) spin echo and pulse field gradient stimulated echo (PFG-STE) diffusion NMR techniques were employed. CPMG measures the T_2 relaxations, which investigates the immediate surrounding chemical environment of the species. PFG-STE measures the self-diffusion of the carbonate–bicarbonate species. Self-diffusion is the diffusion coefficient of a species in the absence of any chemical potential gradient but for our situation it is synonymous to the translational diffusion coefficient. Firstly, it should be noted that a 1D spectrum of the system after the phase transition (17 injections of 6 mM $CaCl_2$(aq) titrant shown in Fig. 5) does not indicate the presence of solid precipitation products at the $CaCO_3$ or $Ca(HCO_3)_2$ predicted chemical shifts. Solids would result as a broad NMR spectrum with a carbonate peak at ~170 ppm or a bicarbonate peak at ~160 ppm, which are not observed. However, if the carbonate–bicarbonate is behaving as a solvated solute or a dynamic liquid, then they can interconvert through proton exchange and the result is an intermediate NMR spectrum, with one peak in-between the two chemical shifts due to rapid (de)protonation interconversion between the two states.[47,48] Thus, NMR provides additional independent evidence of the liquid character of the nucleated LCP phase. The emergent LCP phase is further evident by the distortion of the resultant peak resulting in an asymmetric broadening and "bulging" of the peak in the up-field direction (bicarbonate direction) (Fig. 6). There is a slight widening seen in the 7th injection prior to the phase transition as compared to the buffer control but it is minor and very symmetric, suggesting the presence of only one phase. The asymmetric upfield direction bulging of the solution with 17 injections of $CaCl_2$(aq) suggests that this portion of ions is favouring bicarbonate in the bicarbonate–carbonate interconversion to a greater extent than the bulk solution and that there may be an additional phase in the system. This is consistent with the pH measurement shown in Fig. 1b, which also suggests a favoring of bicarbonate binding (or a less pronounced carbonate binding) after the phase transition. We do not consider this emergent asymmetric bulging to be due to the aggregation of PNCs alone, because of its bicarbonate-favored directionality. Also, the dynamics of PNCs is reported as being on the order of pico and nano seconds, which would be far too short to resolve from the bulk solution over time-averaged data of 5.5 s for these NMR experiments.

The broadened peak was deconvoluted for analysis using Gaussian distributions to model overlapping peaks, resulting in a bulk solution peak (main peak) which resembles in width and chemical shift the carbonate buffer control, and an emergent

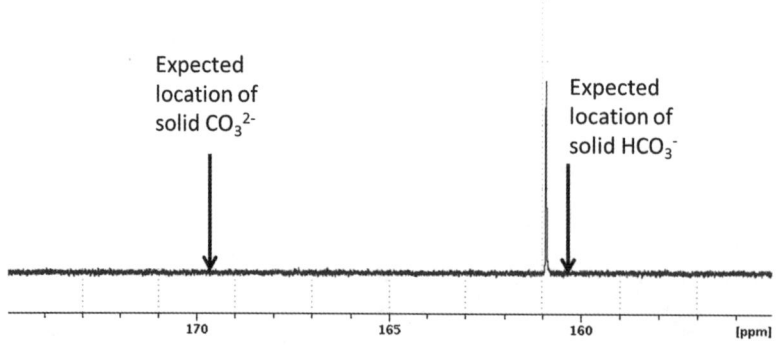

Fig. 5 A 1-D ^{13}C spectrum of the solution after the phase transition (17th injection of 6 mM $CaCl_2$(aq) into 20 mM carbonate buffer, pH 8.5). There are no detectable peaks at the solid carbonate or bicarbonate chemical shifts (marked with arrows). The effect of the nucleated liquid precursor phase is seen as a widening of the peak because it is still behaving as a (non-solid) phase.

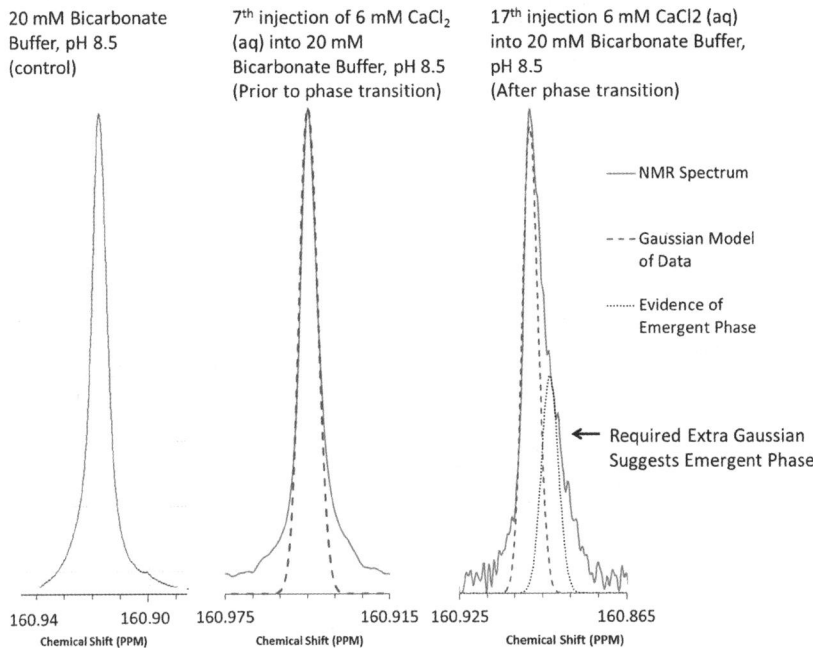

Fig. 6 (Left) 1-D NMR ^{13}C spectrum of 20 mM bicarbonate buffer, pH 8.5 (control). (Center) First slice of the ^{13}C NMR T_2 measurement of the 20 mM bicarbonate buffer with 7 injections of 6 mM CaCl$_2$(aq). (Right) First Slice of the ^{13}C NMR T_2 measurement of the 20 mM bicarbonate buffer with 17 injections of 6 mM CaCl$_2$(aq) (after phase emergence). Both the control (left) and the solution prior to phase transition (center) symmetrical Gaussian distributions of signal. After the phase transition (right), the evidence of an emergent phase manifests as an asymmetric bulge shifted upfield from the bulk solution peak which requires an additional Gaussian peak to model. The data was deconvoluted in order to attribute T_2 relaxation measurements and ^{13}C NMR diffusion measurements to each modeled portion of the peak.

peak (small peak) which is suspected to be the emergent liquid condensed precursor phase (see Fig. 6). To see an example of our Gaussian modelling of overlapping peaks, see Fig. S.4† of the supplementary information. Measuring the T_2 relaxation and diffusion properties of the carbonates and bicarbonates in the buffer solution by means of NMR is an excellent way to establish that there are distinctly different species present after the phase transition. The T_2 relaxations of both deconvoluted peaks were determined by using the CPMG technique, which measures the reduction in peak intensity (M/M$_0$) *vs.* the change in the tau delay according to the following equation:

$$ln\left(\frac{M}{M_0}\right) = \frac{\tau}{-T_2} \quad (1)$$

The result is a linear relationship between the natural log of the signal attenuation and the tau delay with a slope of $1/-T_2$ which can be found by fitting the data. The results, shown in Fig. 7(top), show that the average T_2 relaxation of 0.9 s for the ions associated with the LCP is much faster than the average bulk solution ions T_2 relaxation of 1.5 s, suggesting that the emergent ions expressing bicarbonate-bias are tumbling at a different rate and/or are in a different chemical environment than the ions making up the bulk solution over several-second time averages. The bulk solution T_2 relaxation of 1.5 s is identical to the T_2 relaxation that was obtained for pure carbonate buffer suggesting that the carbonate–bicarbonate ions making

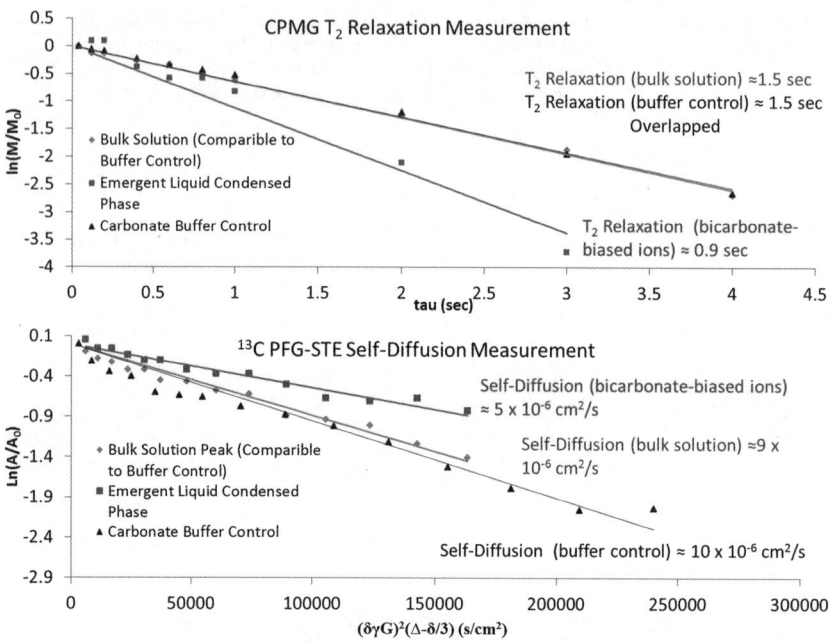

Fig. 7 The results of the CPMG T_2 relaxation measurement (top) and the ^{13}C PFG-STE self-diffusion measurement (bottom) of the deconvoluted NMR peaks. The CPMG plot (top) is plotting the reduction in peak intensity (left side of eqn (1)) vs. the tau delay (right side of eqn (1)). The result is a linear relationship with a slope of $-1/T_2$. The diffusion measurement plot (bottom) plots the degree of attenuation (left side of eqn (2)) vs. the right side of the equation, where the slope of the resulting line is $-D$ (negative diffusion constant). The bicarbonate-biased ions (bulge) have a shorter T_2 relaxation time and a slower diffusion, suggesting that a fraction of the carbonate–bicarbonate ions are slowed in rotation and in diffusion, presumably due to interactions with Ca^{2+} and the emergent LCP. The self-diffusion of the bulk solution carbonates–bicarbonates is roughly the same as the self-diffusion of carbonate buffer alone ($\sim 10 \times 10^{-6}$ cm^2 s^{-1}).

up the bulk solution are tumbling and/or are in a chemical environment similar to bicarbonate buffer control (control) (shown in Fig. S.5† of the supplementary information). T_2 relaxations of solvated, ^{13}C-enriched, carbonate–bicarbonate solutions like the ones discussed in this paper are primarily due to chemical shift anisotopy (CSA). Chemical shift anisotropy occurs if the electron shielding around the dipole is anisotropic, as is the case of asymmetric carbonate and bicarbonate ions. When CSA is present, the extent to which the dipole interacts with an applied magnetic field depends on the rate of molecular tumbling of a dipole. In dilute, non-viscous solution states, the effect of CSA is often averaged-out due to rapid tumbling of molecules. When carbonates/bicarbonates rotations are inhibited by interactions/binding with Ca^{2+}, the reduced tumble rate of molecules enhances the CSA effect and leads to a reduction of T_2 relaxation times. The relaxation rates of the Ca^{2+}-bound/interacting carbonates and bicarbonates are on the order of solution state relaxations. Solids have much shorter T_2 relaxations, on the order of less than a millisecond, and therefore, the detected bicarbonate-biased ions exhibit solute character, which may rely on the internal rapid dynamics of the liquid calcium carbonate precursor phase.

As mentioned above, ^{13}C PFG-STE diffusion NMR is an excellent non-invasive technique for establishing the co-presence of two distinct diffusions within a solution, indicating the presence of an emergent liquid phase. In the case of this study, we assumed isotropic, unbiased diffusion and therefore were concerned only with diffusion rates in the z-direction (measured by the magnetic field gradient created in the z-direction). The nucleus of interest, in this case the ^{13}C nucleus of carbonate–bicarbonate, is subjected to a 90° pulse for an encoding step, followed by another 90° pulse to store the magnetization. After an evolution time (diffusion time), a final 90° pulse is applied after the evolution time to refocus the transverse magnetization for acquisition. A field gradient of amplitude, G and duration, δ, is applied during the first and third 90° pulses to apply and remove, respectively, the spatial encoding to the ^{13}C nuclei. This ensures that the extent of refocusing of the applied 90° encoding pulse is dependent on the degree of deviation from the initial position that was spatially-encoded by the gradient. Translational movement of the dipole during the diffusion time results in an attenuation of the intensity of signal (A/A$_0$) which is related to D (the self-diffusion constant) through the following expression:

$$ln\left(\frac{A}{A_0}\right) = -D(\delta\gamma G)^2\left(\Delta - \frac{\Delta}{3}\right) \quad (2)$$

where G is the gradient strength, γ is the gyromagnetic ratio of ^{13}C, δ is the duration of the gradient application, and Δ is the diffusion time. By plotting the left side of the equation *vs.* the right side with stepwise increases in the gradient strength, G, we generated a line with slope −D (negative self-diffusion constant), as seen in Fig. 7(bottom). This figure compares the diffusions of the emergent LCP (obtained by 17 injections of 6 mM CaCl$_2$(aq) into 20 mM bicarbonate buffer, pH 8.5) with the mother bulk solution and a bicarbonate buffer control. This required the deconvolution of NMR spectra in a similar fashion as described for the T2 determination above. A description of the method and the attenuation of signal are shown in Fig. S.6† of the supplementary information. The results indicate two distinct diffusions exist in the LCP-containing solution, a slower one of 5×10^{-6} cm^2 s^{-1}, which is attributed to bicarbonate-biased ions contributing to the emergent phase, and a faster one of 9×10^{-6} cm^2 s^{-1}, which is that of the bicarbonates–carbonates in the bulk solution. The self-diffusion of the bulk solution peak is very similar to the diffusion we obtained from our control of carbonate buffer alone (10×10^{-6} cm^2 s^{-1}), and is in close accordance with literature values for the self-diffusion of carbonate buffer obtained by simulation[49] and by experiment[50] under similar conditions. It should be noted that the slower diffusion attributed to the emergence of a LCP is not that of the large LCP droplets seen using the NTA technique which, given their size of >60 nm, would be orders of magnitudes slower based on Einstein–Stokes diffusion (and not detectable with the gradient used in this experiment), but rather are believed to be that of small Ca^{2+} bound carbonate–bicarbonate ion-pair species that might be constituents of the large metastable droplets. According to the Einstein–Stokes equation (using the kinematic viscosity of water at 25 °C), the bulk solution and buffer control carbonates–bicarbonates have an effective diameter of 0.5 nm. The slower diffusing carbonates–bicarbonates have an effective diameter of 1 nm. The favouring of bicarbonate, the apparent slowing of ^{13}C-based ion rotations, and the roughly doubling of effective diameters suggests that there is a significant amount of calcium bicarbonate monodentate and calcium bicarbonate bidentate ion-pairing phenomenon present, which are being detected over relatively long time averages as compared to the bulk mother solution.

Another interpretation is that the emergent LCP is more viscous than the mother solution. The average effective size of the detected carbonate–bicarbonate may be a singular ion still if the viscosity of the emergent phase were twice that of water, or about 0.018 dyne-sec cm^2. Both interpretations suggest that there is a liquid–liquid separation occurring that is forming a LCP. It is important to stress that the effective diameters calculated are based on diffusion of the ions, which are expected to be an ensemble average due to the dynamic nature of ion-pairs and PNCs and free ions along with their associated waters of hydration. This needs to be considered when comparing these values to the size of PNCs determined by means of cryo-TEM[21] and AUC.[20] However, the average diameter of the Ca bicarbonate-biased ion pairing species in our 2-phase system of 1 nm is relatively consistent with the diameters of 0.75 nm and 2.0 nm obtained by cryo-TEM and AUC, respectively.

The data as a whole suggests that, at a critical bound Ca^{2+} concentration, Ca bicarbonate ion pairs begin to associate due to entropic driving forces (often the liberation of hydration waters) into a sparse, but visible, metastable liquid-condensed phase. Due to evidence of LCP droplet growth during experiment evolution, the phase is believed to be stable with respect to the solution state. The bound Ca bicarbonate species believed to comprise the LCP appear to have a long enough dynamic lifetime to be detectable using solution state NMR, (unlike PNCs), and they appear to be slowly translating in space and slowly tumbling as compared to the free carbonate–bicarbonate in the bulk solution. It is reasonable to assume that the emergent droplets are comprised of bicarbonates given the data, but, considering how dispersed the droplets are, the bulk of the detected bicarbonate-bias species being detected by the NMR must be in solution, possibly exchanging onto and off of the droplets. It is important to point out that the values of T_2 and self-diffusion obtained by NMR are on the scale of solutes in solution, supporting the notion that the LCP is a liquid phase distinct from the mother solution. The values of T_2 and self-diffusion for solids would be many orders of magnitude smaller.

Experiments in the presence of polyaspartate

In the titration and NMR experiments, the method of generating a super-saturated solution with polymer additive differed from that without additives. The reason for this was to mimic our lab's convention of adding carbonate–bicarbonate ion to a solution containing $CaCl_2$ and polymer additive to initiate the PILP process. Also, the polymer additive has such a strong stabilizing effect on the solution that the concentration of bicarbonate buffer needed to be increased to generate a sufficient super-saturation to observe eventual solid nucleation and precipitation.

A Ca^{2+} binding profile, in the presence of PAsp (14 000 g mol^{-1} and 27 000 g mol^{-1} MW), as 300 mM bicarbonate buffer, pH 8.5 was titrated into a 10 mM $CaCl_2$, 20 μg ml^{-1} PAsp solution is shown in Fig. 8. As is the case with Fig. 1, there is presumably some kinetic binding effects occurring in this titration and therefore the Ca^{2+} binding profile should be treated as qualitative. As a control, a Ca^{2+} profile was conducted for a titration in the absence of PAsp for comparison. In this case, one can see an apparent discontinuity in Ca^{2+} concentration at an early stage that is analogous to the one seen in Fig. 1a. As with the discontinuity shown in Fig. 1a, this potential discontinuity suggests that a phase transition has occurred. In order to aid in visualization, lines have been added to the inset of the figure to demonstrate discontinuity.

In the presence of polymer, the first thing that one notices is how much farther the Ca^{2+} profiles extend prior to precipitation relative to the control (bottom curve, with no polymer). In addition to this stabilizing influence, the shapes of the titration curves show an interesting trend. The binding of Ca^{2+} with respect to injections of $CaCl_2$ (aq) appears to display a discontinuity (Fig. 8, inset), suggesting that LCP is forming in the presence of polymer as well, as evidenced by the same type of discontinuity as seen in the non-additive case (Fig. 1a). The LCP-like Ca^{2+} binding behaviour continues for many more injections beyond where the additive-free control would have precipitated (see precipitation indicators in Fig. 8). The continuation of the same type of binding that forms the LCP phase is a strong indication that LCP is being stabilized by the presence of the PAsp polymer additive. Thus the polymer's role in the PILP process may be to stabilize the LCP.

It is relevant to note that Verch *et al.*, in a "fingerprinting" study on the effect of polymer additives, also detected a significant nucleation inhibition and a species after nucleation of $CaCO_3$ in presence of PAsp (MW 6800 g mol^{-1} and MW

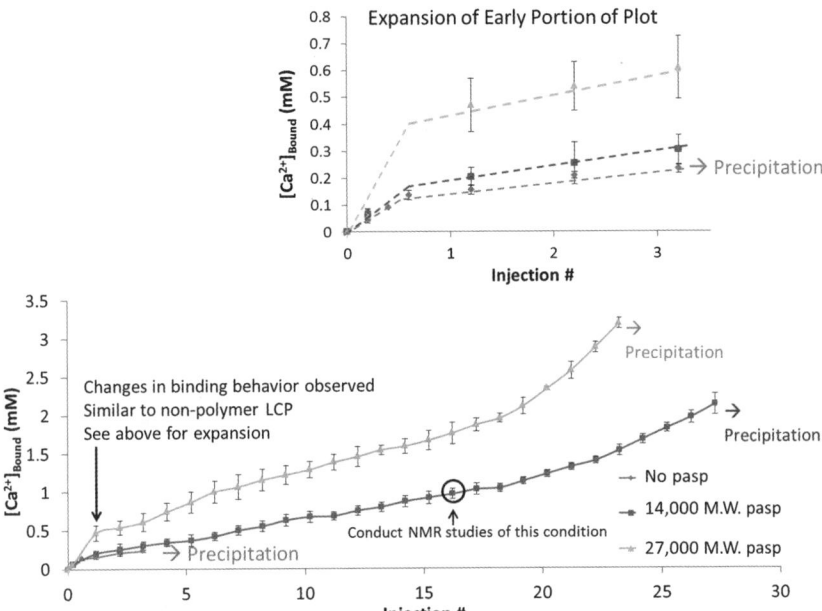

Fig. 8 The [Ca^{2+}] profile of a system where 300 mM bicarbonate buffer, pH 8.5 was injected (40 μl initial, 200 μl for the rest) into 10 mM CaCl$_2$ with 20 μg ml^{-1} PAsp (no PAsp for control). The results qualitatively show evidence of a phase transition and are analogous to the LCP formation in the absence of polymer shown in Fig. 1a. After the presumed LCP formation, the solution is stabilized against solid nucleation and precipitation for many more injections, suggesting that LCP is stabilized due to the interaction of polymer with the LCP phase. This may be the basis for the PILP process. NMR studies to analyze the two phases in solution (bulk solution and phase-separated PILP) were conducted at the location indicated by the circle.

27 000 g mol^{-1}) that is much more soluble than ACC, which they related to a PILP phase (although they worked at a higher pH of 9.75).[51]

A profile of the pH evolution with the addition of 200 μl injections of 300 mM, pH 8.5 bicarbonate buffer into 10 mM CaCl$_2$(aq) containing 20 μg ml^{-1} of PAsp is shown in Fig. 9. It is apparent when comparing the pH evolution of bicarbonate buffer into aqueous PAsp solution (diamonds and squares) to the pH evolution of bicarbonate buffer into pure water (dotted orange line) that PAsp has a dampening effect on the evolution by maintaining a lower pH. This leads to a higher bicarbonate : carbonate ratio at the early points in the titration. This would be a very effective means for directing more bicarbonate to bind with Ca^{2+} if it were present. When Ca^{2+} is present with the PAsp (triangles and crosses) the pH development suggests this is the case. In comparison to the controls in the absence of Ca^{2+}, the Ca^{2+} containing solutions actually do demonstrate additional bicarbonate binding in terms of a bicarbonate : carbonate ratio, until a certain threshold is met. At this point the pH evolution flattens considerably at a much lower pH than the various controls. This supports the assertion that a critical amount of Ca bicarbonate ion pairs accumulates leading to a phase transition (LCP) and a change in Ca^{2+} affinities for bicarbonate and carbonate. The preference to maintain a pH in the range of a lower 8s supports the view that an emergent bicarbonate-rich phase is participating in directing the binding preference for Ca^{2+}. Fig. 9 supports the view that PAsp (and potentially other additives) promote and stabilize the LCP phase and perhaps is the description for the so-called PILP process.

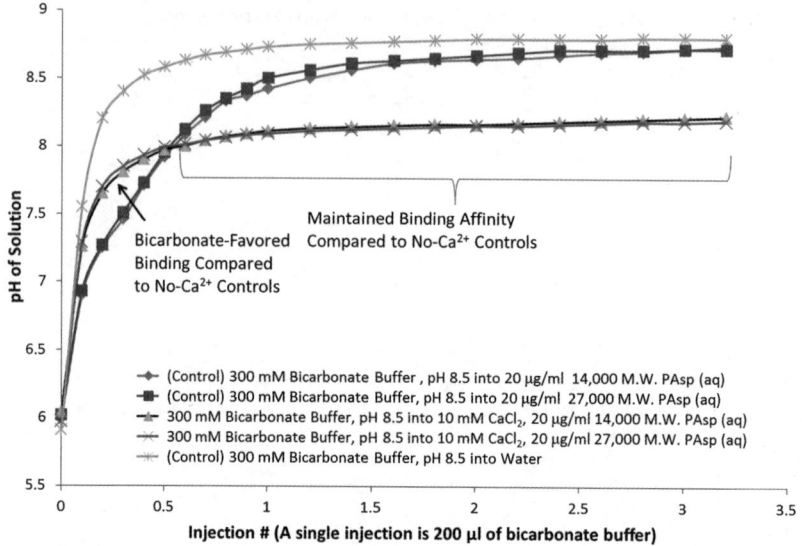

Fig. 9 The pH evolution of a 10 mM CaCl$_2$ solution containing 20 μg ml^{-1} PAsp due to the punctuated titration of 300 mM bicarbonate buffer, pH 8.5. The presence of PAsp has a mitigating effect on the upward evolution of the solution, allowing for more bicarbonates to be present (diamond and squares). When Ca^{2+} is present with the PAsp, the pH evolution increases in the basic direction faster than the evolution with PAsp alone, before leveling-off at a preferred pH. This suggests that when enough bicarbonate has bound, an equilibrium is established, presumably due to an emergent phase, resulting in a consistent, flattened pH evolution. The typical injection is 200 μl, but to characterize the early titration, fractions of 200 μl injection were used.

To analyse the PILP phase which is suspected to be polymer-promoted and stabilized LCP, a 1D NMR spectrum of the system at the condition indicated with a circle in Fig. 8 was obtained and is shown in Fig. 10 and 11. The effect of apparent PILP formation severely distorts the spectral peak. As seen in Fig. 10, there is an emergent peak splitting from the bulk solution peak in the upfield direction, indicating that a significant fraction of the buffer is favoring bicarbonate as compared to the bulk solution over a large multisecond time-average. Fig. 11 shows the time evolution of the two-phase bulk solution-PILP sample. The process required hours to develop but solid nucleation and precipitation did not occur as evidenced by the unchanging area of the peaks. Fig. 11 shows a 1D NMR spectral comparison after 18 h (to attempt to achieve equilibrium) of the one-phase, pH 8.5, bicarbonate buffer and PAsp solution *versus* the multiphase bicarbonate–PAsp–Ca^{2+} solution at the 17th injection. The two-phase system emergent peak (bulge of separation from bulk solution) is shifted more upfield and is much larger than the NMR evidence of LCP precursor phase in the absence of additive, suggesting that polymer is stabilizing and accumulating a bicarbonate-biased ionic interaction, which fits the description of the LCP phase, within a second bulk solution (mother-solution) phase.

We used the same parameters and techniques to measure the T$_2$ relaxation and diffusion properties of the PILP system as described above for the LCP system. The results are shown in Fig. 12. Raw data for the CPMG (control), PFG-STE (control), CPMG (phase-separated PILP solution), and PFG-STE (phase-separated PILP solution) are shown in Fig. S.7–S.10† of the supplementary information, respectively. As with the LCP NMR experiments, the diffusion values obtained for the suspected PILP phase are not being presented as the diffusion of the large droplets. We believe this is the effective diffusion of the ion solute species that collect into, and are in exchange with, the PILP phase droplets. The diffusion

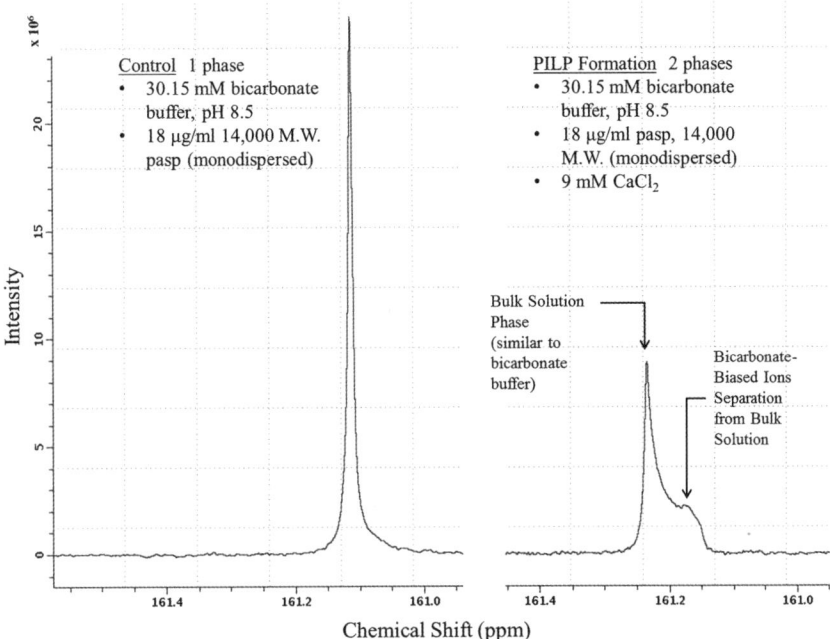

Fig. 10 A comparison between the 1D NMR spectra of bicarbonate buffer with polyaspartic acid sodium salt (left) and the separated bicarbonate-biased ions (right). The presence of PAsp leads to a broad peak which is shifted upfield (bicarbonate weighted direction), distinct from the remaining ions in solution (bulk solution). This behavior is similar to the behavior of the liquid condensed phase except it is enhanced greatly in the presence of polymer, suggesting that the PILP process stabilizes the liquid condensed phase.

measurements were consistent in scale with those obtained for the LCP system. The presence of PAsp in the solution appeared to reduce the carbonate–bicarbonate ion diffusion in general, but the ratios are similar to the LCP experiments and controls. The diffusion of phase separated ions (3×10^{-6} cm^2 s^{-1}) was a little less than half the diffusion of the carbonate–bicarbonate in the bulk solution (7×10^{-6} cm^2 s^{-1}), and the bulk solution diffused at the same rate as the bicarbonate buffer control (both at 7×10^{-6} cm^2 s^{-1}). Using the Einstein–Stokes equation, the effective diameter of the ionic species was 0.7 nm for the bulk solution and control carbonate–bicarbonate ions, and was 1.6 nm for the bicarbonate-biased ions. This is consistent with the slowed ions existing disproportionately as calcium monodentate/bidentate over this time-average. Note—the presence of the PAsp led to an overall slowing of the experimental and control diffusions, leading to the calculation of a larger effective diameter. This may be due to increased viscosity caused by the presence of the polymer. Still, the ratio of diffusions (bicarbonate-biased ion pairs : mother-solution : control) were consistent with the non-polymer case and demonstrate the separation of two distinct equilibria, which are comprised of ions diffusing at different rates. As explained for the non-polymer case, this could be due to singular ions in a more viscous emergent PILP phase. If this were the case, the emergent PILP phase would have a viscosity of 0.02 dyne-sec cm^2, according to the Einstein–Stokes equation.

The T_2 relaxation of the upfield bulge was 1.1 msec, which suggests that the bicarbonate-biased ions are slowly tumbling and rotating in solution. It was almost exactly the same as the bulk solution; both were shorter than the relaxation measured for the polymer–bicarbonate buffer control (1.5 msec). Although a distinction between suspected PILP phase and the bulk solution would have been helpful in

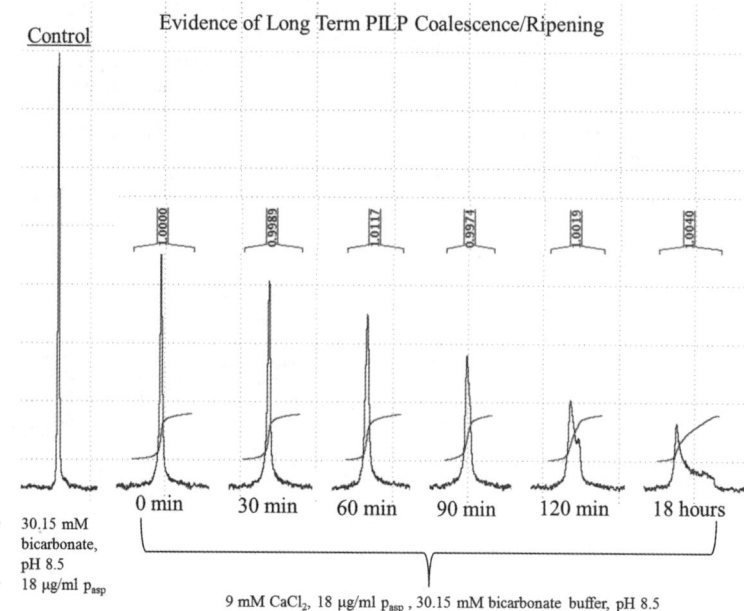

Fig. 11 The time evolution of a 1D NMR spectrum of the phase-separated PILP-containing solution shown in Fig. 10. Initially, it is broader than the control solution (absent of CaCl$_2$) as was observed for the formation of liquid condensed phase. However, in time a more distinct phase separation occurs to yield a large fraction of bicarbonate-bias of the ions distinct from the bulk solution phase, suggesting the possibility of slow ripening and coalescence of LCP phase. The red, bracketed numbers above each spectral peak represent the area under the peak, normalized with respect to the 0 min spectrum. The amount of signal remains constant suggesting that all the ions are still behaving as solution state; therefore, solid nucleation and precipitation has not occurred.

distinguishing between types of phase environments, it is not surprising that they are similar due to the appearance of heavy chemical exchange between the them, as evidenced by the bridging between the suspected PILP peak and the mother solution peak (see Fig. 10).

As a side note, the spectrum of the separated PILP phase shown in Fig. 10 resembles a powder pattern, which is an NMR spectral pattern obtained when there are solid particles present. However, a true powder pattern due to solids in solution would yield a broad spectrum peak with a width of 100 ppm. The width of the bulk solution–PILP peak is only 0.1 ppm. This could mean that there is a slowing in the rotation of the bicarbonate–carbonate ions but that the ion's rotations are still very rapid as expected for a solute in solution.

Discussion

The nucleation pathway of calcium carbonate through a multi-step process, where the first step is the formation of a metastable liquid and the second is the solid nucleation formation (within or outside the metastable liquid) has been at the forefront in recent literature.[3,34,52] Here, we have presented data that supports the first step of this assertion through a combination of techniques: in the absence of polymer additives, the ITC, Ca^{2+} binding profiles, and pH evolution experiments indicate that there is a phase transition. The pH measurements suggest the carbonate–bicarbonate binding becomes bicarbonate-biased, as do NMR peak emergences in the bicarbonate direction suggesting that bicarbonate is playing a role in the emergent phase. The fluidic

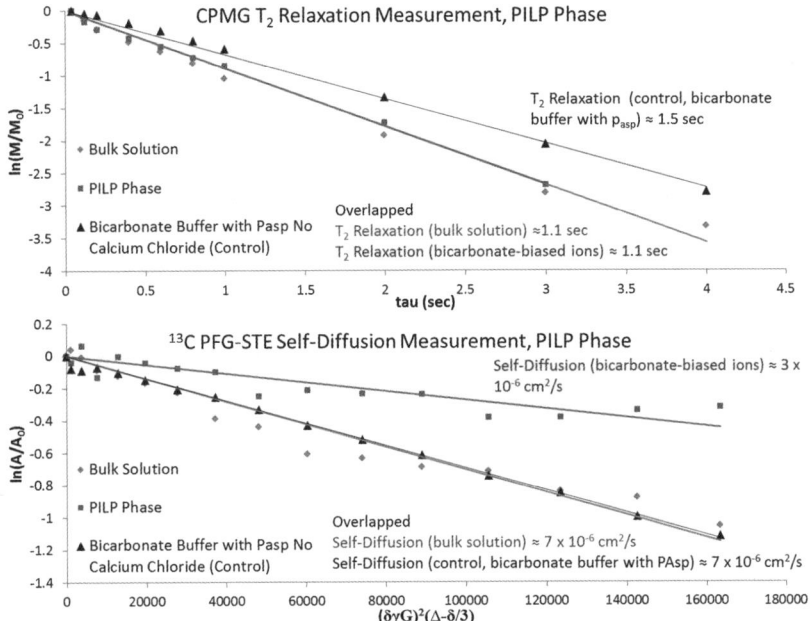

Fig. 12 The results of the CPMG T2 relaxation measurement (top) and the 13C PFG-STE self-diffusion measurement (bottom) of the PILP phase (suspected polymer stabilized LCP), bulk solution, and bicarbonate buffer with polyaspartic acid (control). The CPMG plot (top) is plotting the left side of equation 1 *vs.* the tau on the right side. The result is a linear relationship with a slope of $-1/T_2$. The diffusion measurement plot (bottom) plots the left side of equation 2 *vs.* the right side where the slope of the resulting line is $-D$ (diffusion constant). The bicarbonate-biased ions have the same T_2 relaxation as the bulk solution due to the large amount of chemical exchange, but have a shorter T_2 relaxation than the bicarbonate buffer control, suggesting that rotations of the bicarbonate-biased ions are slowed. The diffusion of the bicarbonate-biased ions is slowed with respect to the bulk solution and the bicarbonate buffer control in a way similar to LCP, suggesting that polymer is kinetically stabilizing the LCP phase.

character of this nucleated phase is suggested in the sedimentation properties of the species in solution and the apparent co-existence of a distinct solution-like liquid phase within the mother solution, as measured by AUC and NMR self-diffusion and relaxation times, respectively.

Gebauer *et al.*[20] and Wolf *et al.*[45] claim that PNCs do not contain significant amounts of the bicarbonate ion. Computer simulations have suggested that DOLLOP PNCs are ubiquitous and exhibit incredibly quick dynamics on the order of picoseconds and nanoseconds.[22] For these reasons, we believe that the nucleation phenomenon we are seeing cannot be described by PNC theory. The apparent bicarbonate-biased constituents we detect accompanying the precursor phase manifest their properties for long enough lifetimes to detect, suggesting that the LCP precursor phase is not forming from carbonate-rich PNCs. We propose that in the LCP phase, there is a bicarbonate-biased Ca^{2+} interaction which is distinct from the Ca^{2+} interaction with carbonates–bicarbonate in the bulk solution over significant time averages. The presence of Ca bicarbonate ion pairing phenomenon or even clustering does not challenge the validity of predominantly Ca carbonate PNCs, like DOLLOP. These two phenomena are not mutually exclusive. At high pHs it is expected that the numerous CO_3^{2-} ions, with their greater binding affinity to Ca^{2+} would dominate. However, at more neutral pHs, like the pH in which this study was conducted, there are relatively few CO_3^{2-} ions as compared to HCO_3^{-} ions and weaker Ca bicarbonate ion-pairing or clustering phenomenon become

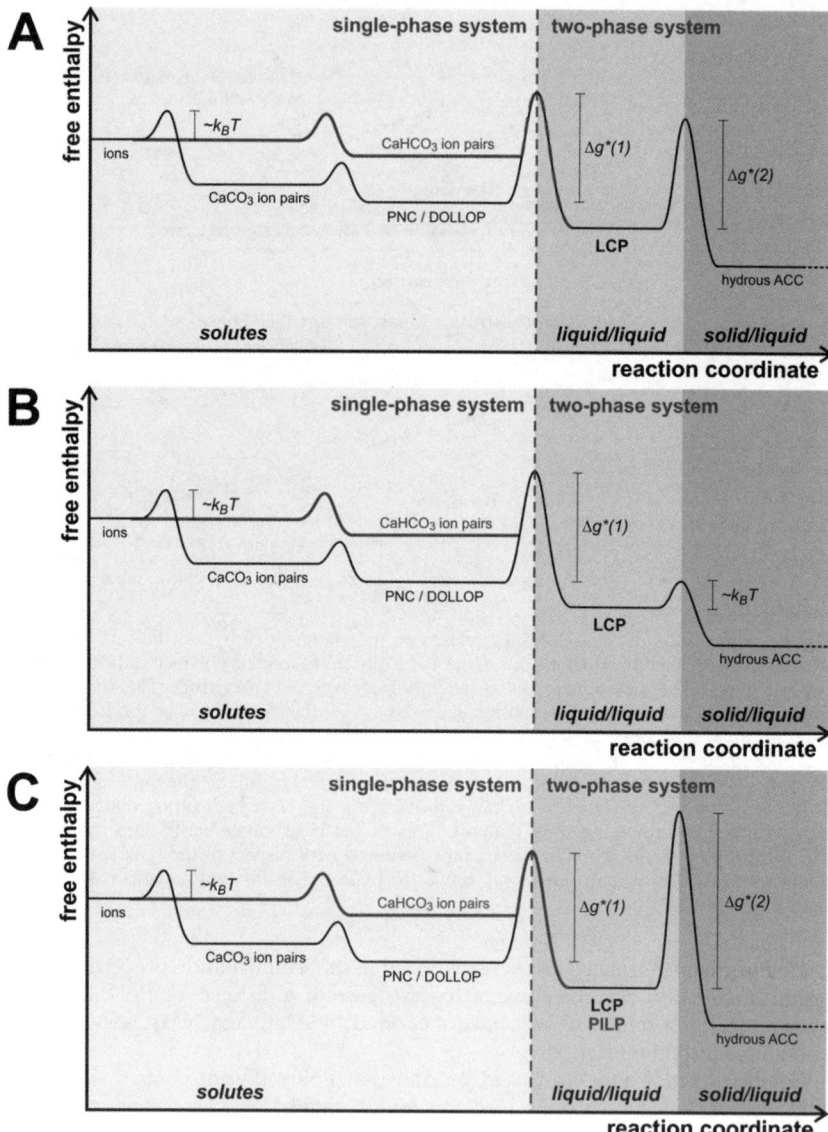

Fig. 13 An overview of the energetics (not to scale) of calcium carbonate precipitation from supersaturated solution, putting LCP into a global context with earlier findings. The diagram has been truncated on the right where the hydrous ACC will ultimately transform into a more stable crystalline phase. In solution, calcium bicarbonate and calcium carbonate ion pairs form, the latter of which can associate into larger species, PNCs including dynamically ordered, liquid-like oxyanion polymers (DOLLOP). The different prenucleation species form virtually spontaneously (thermal energy k_BT), while formation of LCP is associated to a barrier of nucleation $\Delta g^*(1)$. (A) At neutral pHs, the nucleation step involves calcium bicarbonate species, and with it, leads to a bicarbonate-bias in the LCP and an intrinsic kinetic stabilization, $\Delta g^*(2)$, (B) At higher pH levels, calcium bicarbonate ion pairing becomes negligible, carbonate species dominate the nucleation process, and render LCP transient due to a reduced intrinsic stabilization. The barrier $\Delta g^*(2)$, may even vanish. (C) Addition of PAsp may lead to a distinct increase of the barrier $\Delta g^*(2)$ due to a pronounced role of bicarbonate species in the LCP in the presence of polymer. The bicarbonate pathway may be preferred in the presence of polymer, and/or the polymer may stabilize bicarbonate species within LCP.

significant, leading to accumulation and nucleation of a bicarbonate-rich metastable liquid phase like the one that has been proposed in this study.

This is significant because it has been proposed that nucleation of solid $CaCO_3$ can occur through nucleation inside of a metastable liquid phase.[53] The seed nucleus is proposed to form within the droplet and its growth and polymorph is subject to the environment of the metastable droplet, rather than that of the bulk solution. If metastable droplets of bicarbonate biased Ca^{2+} interactions phase-separate and are stabilized by the presence of polymer (PILP process), as is possible with the droplets observed in this study, then this might explain the incredible amount of nucleation inhibition that polymers (particularly negatively charged, aspartic and glutamic acid-rich polymers) give to a solution. An emergent metastable bicarbonate droplet and its smaller constituents would sequester Ca^{2+} and carbonate–bicarbonate ions from the solution, thus mitigating the solid-phase nucleation through PNC non-classical pathways. However, the resulting concentrated metastable droplet would have difficulty forming a seed nucleation because it is bicarbonate-biased and would require the extra step of releasing the H^+ ions prior to organization into a nucleus. We propose that LCP and PILP (PAsp-stabilized LCP) plays a role in the calcium carbonate energetic cascade from supersaturated solution to crystallinity as described in Fig. 13.

Knowing what the PILP phase is may provide insight into various biological system's mechanism of control over morphology, phase, and location of the final biomineral products. Although the presence of PNCs and LCP are interesting with respect to theoretical models of crystallization mechanisms, without the polymer, the final products simply resemble those predicted by classical models. It is the accumulating and stabilizing effect of the polymer that allows for manipulation of the liquid condensed phase into an endless array of non-equilibrium morphologies. On the other hand, without this propensity of the mineral reactants to form this liquid condensed phase, it is not clear that the polymer would be able to "induce" a liquid precursor phase. This work shows the significance of both sides of this interesting mineralization system.

Acknowledgements

We thank the Deutsche Forschungsgemeinschaft and National Science Foundation for financial support of this project within the "Materials World Network to Study Liquid Precursor Formation and Crystallization at Interfaces: Fundamentals towards Applications" (NSF Grant DMR-0710605 and DFG Grant CO 194/5-1, VO 829/4-1). Dirk Haffke (Konstanz) is acknowledged for NTA analysis, Rose Rosenberg, Dirk Haffke and Antje Völkel (Konstanz) for AUC experiments.

References

1 H. A. Lowenstam and S. Weiner, *On Biomineralization*, Oxford University Press, N. Y., 1989.
2 S. Mann, J. Webb and R. J. P. Williams, *Biomineralization: Chemical and Biochemical Perspectives*, VCH Publishers Inc., N. Y., 1989.
3 L. B. Gower, *Chem. Rev.*, 2008, **108**, 4551–4627.
4 F. C. Meldrum and H. Coelfen, *Chem. Rev.*, 2008, **108**, 4332–4432.
5 S.-W. Lee, S.-B. Park, S.-K. Jeong, K.-S. Lim, S.-H. Lee and M. C. Trachtenberg, *Micron*, 2010, **41**, 273–282.
6 Y. Zhao, Y. Lu, Y. Hu, J.-P. Li, L. Dong, L.-N. Lin and S.-H. Yu, *Small*, 2010, **6**, 2436–2442.
7 S. Biradar, P. Ravichandran, R. Gopikrishnan, V. Goornavar, J. C. Hall, V. Ramesh, S. Baluchamy, R. B. Jeffers and G. T. Ramesh, *J. Nanosci. Nanotechnol.*, 2011, **11**, 6868–6874.
8 J. Seto, Y. Ma, S. A. Davis, F. Meldrum, A. Gourrier, Y.-Y. Kim, U. Schilde, M. Sztucki, M. Burghammer, S. Maltsev, C. Jäger and H. Cölfen, *Proc. Natl. Acad. Sci. U. S. A.*, 2012, **109**, 3699–3704.

9 I. Sethmann and G. Woerheide, *Micron*, 2008, **39**, 209–228.
10 D. Ren, Z. Li, Y. Gao and Q. Feng, *Biomed. Mater.*, 2010, **5**, 055009.
11 E. Beniash, L. Addadi and S. Weiner, *J. Struct. Biol.*, 1999, **125**, 50–62.
12 N. Gehrke, N. Nassif, N. Pinna, M. Antonietti, H. S. Gupta and H. Colfen, *Chem. Mater.*, 2005, **17**, 6514–6516.
13 F. C. Meldrum, *Int. Mater. Rev.*, 2003, **48**, 187–224.
14 J. Rieger, T. Frechen, G. Cox, W. Heckmann, C. Schmidt and J. Thieme, *Faraday Discuss.*, 2007, **136**, 265–277.
15 J. Aizenberg, *Bell Labs Tech. J.*, 2005, **10**, 129–141.
16 P. Fratzl, F. D. Fischer, J. Svoboda and J. Aizenberg, *Acta Biomater.*, 2010, **6**, 1001–1005.
17 L. B. Gower and D. J. Odom, *J. Cryst. Growth*, 2000, **210**, 719–734.
18 B. Guillemet, M. Faatz, F. Grohn, G. Wegner and Y. Gnanou, *Langmuir*, 2006, **22**, 1875–1879.
19 D. Gebauer and H. Cölfen, *Nano Today*, 2012, **6**, 564–584.
20 D. Gebauer, A. Volkel and H. Colfen, *Science*, 2008, **322**, 1819–1822.
21 E. M. Pouget, P. H. H. Bomans, J. Goos, P. M. Frederik, G. de With and N. Sommerdijk, *Science*, 2009, **323**, 1455–1458.
22 R. Demichelis, P. Raiteri, J. D. Gale, D. Quigley and D. Gebauer, *Nat. Commun.*, 2011, **2**, 590.
23 J. Rieger, T. Frechen, G. Cox, W. Heckmann, C. Schmidt and J. Thieme, *Faraday Discuss.*, 2007, **136**, 265–277.
24 M. Faatz, F. Grohn and G. Wegner, *Adv. Mater.*, 2004, **16**, 996.
25 S. E. Wolf, L. Mueller, R. Barrea, C. J. Kampf, J. Leiterer, U. Panne, T. Hoffmann, F. Emmerling and W. Tremel, *Nanoscale*, 2011, **3**, 1158–1165.
26 S. J. Homeijer, R. A. Barrett and L. B. Gower, *Cryst. Growth Des.*, 2010, **10**, 1040–1052.
27 S. J. Homeijer, M. J. Olszta, R. A. Barrett and L. B. Gower, *J. Cryst. Growth*, 2008, **310**, 2938–2945.
28 F. F. Amos, L. Dai, R. Kumar, S. R. Khan and L. B. Gower, *Urol. Res.*, 2009, **37**, 11–17.
29 A. Dey, P. H. H. Bomans, F. A. Mueller, J. Will, P. M. Frederik, G. de With and N. A. J. M. Sommerdijk, *Nat. Mater.*, 2010, **9**, 1010–1014.
30 S. S. Jee, R. K. Kasinath, E. DiMasi, Y.-Y. Kim and L. B. Gower, *CrystEngComm*, 2011, **13**, 2077–2083.
31 S.-S. Jee, T. T. Thula and L. B. Gower, *Acta Biomater.*, 2010, **6**, 3676–3686.
32 T. T. Thula, D. E. Rodriguez, M. H. Lee, L. Pendi, J. Podschun and L. B. Gower, *Acta Biomater.*, 2011, **7**, 3158–3169.
33 S. S. Jee, L. Culver, Y. Li, E. P. Douglas and L. B. Gower, *J. Cryst. Growth*, 2010, **312**, 1249–1256.
34 P. G. Vekilov, *Cryst. Growth Des.*, 2010, **10**, 5007–5019.
35 O. Galkin, W. Pan, L. Filobelo, R. E. Hirsch, R. L. Nagel and P. G. Vekilov, *Biophys. J.*, 2007, **93**, 902–913.
36 Y. Jiang, H. Gong, D. Volkmer, L. Gower and H. Coelfen, *Adv. Mater.*, 2011, **23**, 3548.
37 Y. Jiang, L. Gower, D. Volkmer and H. Colfen, *Phys. Chem. Chem. Phys.*, 2012, **14**, 914–919.
38 Y. Jiang, L. Gower, D. Volkmer and H. Cölfen, *Cryst. Growth Des.*, 2011, **11**, 3243–3249.
39 S. Wohlrab, H. Colfen and M. Antonietti, *Angew. Chem., Int. Ed.*, 2005, **44**, 4087–4092.
40 Y. Ma, G. Mehltretter, C. Pluege, N. Rademacher, M. U. Schmidt and H. Coelfen, *Adv. Funct. Mater.*, 2009, **19**, 2095–2101.
41 A. V. Radha, T. Z. Forbes, C. E. Killian, P. U. P. A. Gilbert and A. Navrotsky, *Proceedings of the National Academy of Sciences*, 2010, **107**, 16438–16433.
42 D. Gebauer, A. Verch, H. G. Boerner and H. Coelfen, *Cryst. Growth Des.*, 2009, **9**, 2398–2403.
43 P. Schuck, *Biophys. J.*, 2000, **78**, 1606–1619.
44 P. Schuck, *Biophys. J.*, 1998, **75**, 1503–1512.
45 S. E. W. S. E. Wolf, L. Muller, R. Barrea, C. J. Kampf, J. Leiterer, U. Panne, T. Hoffmann, F. Emmerling and W. Tremel, *Nanoscale*, 2011, **3**, 1158–1165.
46 K. Strenge and A. Seifert, *Prog. Colloid Polym. Sci.*, 1991, **86**, 76–83.
47 H. Nebel, M. Neumann, C. Mayer and M. Epple, *Inorg. Chem.*, 2008, **47**, 7874–7879.
48 F. M. Michel, J. MacDonald, J. Feng, B. L. Phillips, L. Ehm, C. Tarabrella, J. B. Parise and R. J. Reeder, *Chem. Mater.*, 2008, **20**, 4720–4728.
49 R. E. Zeebe, *Geochim. Cosmochim. Acta*, 2011, **75**, 2483–2498.
50 K. Kigoshi and T. Hashitani, *Bull. Chem. Soc. Jpn.*, 1963, **36**, 1372–1372.
51 A. Verch, D. Gebauer, M. Antonietti and H. Colfen, *Phys. Chem. Chem. Phys.*, 2011, **13**, 16811–16820.
52 D. Gebauer and H. Coelfen, *Nano Today*, 2011, **6**, 564–584.
53 P. G. Vekilov, *Cryst. Growth Des.*, 2004, **4**, 671–685.

PAPER

The role of cluster formation and metastable liquid—liquid phase separation in protein crystallization

Fajun Zhang,[*a] Felix Roosen-Runge,[a] Andrea Sauter,[a] Roland Roth,[b] Maximilian W. A. Skoda,[c] Robert M. J. Jacobs,[d] Michael Sztucki[e] and Frank Schreiber[a]

Received 15th February 2012, Accepted 19th March 2012
DOI: 10.1039/c2fd20021j

We discuss the phase behavior and in particular crystallization of a model globular protein (beta-lactoglobulin) in solution in the presence of multivalent electrolytes. It has been shown previously that negatively charged globular proteins at neutral pH in the presence of multivalent counterions undergo a "re-entrant condensation (RC)" phase behavior (Zhang *et al.*, *Phys. Rev. Lett.*, 2008, **101**, 148101), *i.e.* a phase-separated regime occurs in between two critical salt concentrations, $c^* < c^{**}$, giving a metastable liquid–liquid phase separation (LLPS). Crystallization from the condensed regime has been observed to follow different mechanisms. Near c^*, crystals grow following a classic nucleation and growth mechanism; near c^{**}, the crystallization follows a two-step crystallization mechanism, *i.e*, crystal growth follows a metastable LLPS. In this paper, we focus on the two-step crystal growth near c^{**}. SAXS measurements indicate that proteins form clusters in this regime and the cluster size increases approaching c^{**}. Upon lowering the temperature, *in situ* SAXS studies indicate that the clusters can directly form both a dense liquid phase and protein crystals. During the crystal growth, the metastable dense liquid phase is dissolved. Based on our observations, we discuss a nucleation mechanism starting from clusters in the dilute phase from a metastable LLPS. These protein clusters behave as the building blocks for nucleation, while the dense phase acts as a reservoir ensuring constant protein concentration in the dilute phase during crystal growth.

Introduction

Crystallization plays a decisive role in many processes in nature and industry. For example, the pharmaceutical industry requires production of the desired crystal form of the drug molecules which is important for their biofunction and stability.[1] However, it has long been known that the unique features of crystallization, including crystal lattice and polymorphism, particle size, and its distribution, are defined in the nucleation stage. In spite of the huge efforts over the last decades, our understanding of the early stage of crystallization is still limited. A breakthrough over the last decade has revealed new insight into this step: studies have shown that

[a] *Institut für Angewandte Physik, Universität Tübingen, Auf der Morgenstelle 10, 72076 Tübingen, Germany. E-mail: fajun.zhang@uni-tuebingen.de*
[b] *Institut für Theoretische Physik, Universität Erlangen-Nürnberg, Germany*
[c] *STFC, ISIS, Rutherford Appleton Laboratory, Chilton, Didcot, OX11 0OX, UK*
[d] *Department of Chemistry, Chemistry Research Laboratory, University of Oxford, UK*
[e] *European Synchrotron Radiation Facility, 6 rue Jule Horowitz, F-38043 Grenoble, France*

in colloid and protein solutions, the attractive potential is short-ranged as compared to their size, which is crucial for their phase behavior, and a "two-step" mechanism has been proposed to explain the crystallization behavior in these systems under suitable conditions, *i.e.* nucleation events follow a metastable liquid–liquid phase separation (LLPS).[1–3]

In the region of LLPS, however, colloidal systems with spherical isotropic short-range potential will become a gel or arrested phase instead of a dense liquid phase. In other words, the gelation line coincides with the phase separation boundary.[4] Proteins, which differ from conventional colloids by having a non-spherical shape and inhomogeneous charge pattern, have shown liquid–liquid coexistence under suitable conditions[5–17] and gelation at higher volume fraction or lower temperature.[18–21] Near LLPS, recent studies reveal the formation of dense protein clusters. The formation and physical origin of clusters in colloidal and protein solutions have attracted significant attention in soft matter studies during the last few years. For instance, it is a crucial step for understanding the mineralization process.[22] The cluster phenomenon is closely related to the interactions and phase behavior of these systems. For example, counter-balanced interactions have been reported to lead to equilibrium as well as transient clusters in concentrated protein solution.[23–25] Recent studies have revealed another type of cluster phase existing in concentrated protein solutions, such as lysozyme, hemoglobin, *etc.*[26–28] The clusters can be very big with 10^5 to 10^6 molecules. A theoretical model has been proposed to explain the origin of such a long-living cluster phase.[28]

The consequence of the formation of protein clusters (transient or equilibrium) and metastable LLPS is that it changes the kinetic pathway of crystal nucleation significantly. There is increasing evidence that clusters, nanoscale amorphous precipitates, and other more complex precursors in the aqueous phase play an important role in crystallization.[22,29–31] Computer simulations by ten Wolde and Frenkel show that far from the metastable LLPS critical point, the nucleation follows the classical nucleation mechanism. However, when approaching the critical point, the critical nucleus becomes highly disordered, liquid-like droplets which further follow a structural change to eventually become crystalline.[32] This finding inspires a two-step model, *i.e.* nucleation occurs within the dense liquid phase, which corresponds to the separation of order parameters (density and structure) during crystallization. It applies not only to proteins and colloidal systems, but also to small molecular systems.[33,34] However, the role of the protein cluster as well as the dense liquid phase during nucleation and protein crystallization is still not entirely clear.[3]

We have recently studied the phase behavior of globular proteins in solution in the presence of multivalent metal ions. It has been shown that solutions of negatively charged globular proteins at neutral pH in the presence of multivalent counterions undergo a "re-entrant condensation (RC)" phase behavior,[16,17,35–37] *i.e.* a phase-separated regime occurs in between two critical salt concentrations, $c^* < c^{**}$, including a metastable liquid–liquid phase separation (LLPS).[17] Crystallization from the condensed regime follows different mechanisms. Near c^*, crystals grow following a classic nucleation and growth mechanism; near c^{**}, the crystallization occurs from phase-separated protein solutions, indicating a two-step mechanism, *i.e.* crystal growth follows a metastable LLPS.[16]

In this work, we aim to achieve new insights into the role of LLPS and clustering in protein crystallization by presenting a study of the structural evolution during a two-step process. The questions we are interested in are the following: (1) what is the structure of proteins in solution near c^{**}? Are they still in their dimeric state or forming clusters? (2) If clusters are formed, how do the size and structure of the protein clusters depend on the location in the phase diagram? (3) Can the nuclei of crystals be formed *via* cluster–cluster aggregation? (4) What is the relationship between clusters and the LLPS? With these questions in mind, we have performed systematic SAXS measurements on a series of solutions, and the crystal growth has been followed using *in situ* measurements as a function of temperature.

Experimental

Materials

Globular β-lactoglobulin (BLG) from bovine milk (L3908) and yttrium chloride (YCl$_3$) were purchased from Sigma-Aldrich. Solutions were prepared by mixing stock solutions of BLG (67 mg ml^{-1}) and YCl$_3$ (100 mM). The phase diagram (protein concentration c_p vs. salt concentration c_s) was determined at room temperature (~22 °C) by monitoring the optical transmission of a series of protein solutions containing different salt concentrations.[35] The c_p values were determined by UV absorption using an extinction coefficient of 0.96 ml mg^{-1} at a wavelength of 278 nm.[38] Note that the presence of high concentration of buffer (such as HEPES and Tris buffer) can affect the phase behavior and the solubility of yttrium salts. However, with lower buffer concentration (about 5 mM), the effect on the solubility of yttrium salts is negligible. To avoid the effect of other ions, no buffer was used in this work for sample preparation except for specific reference data set in Fig. 2.[16]

Small-angle X-ray scattering (SAXS) measurements

The SAXS measurements were performed at the ESRF (Grenoble, France) on the beamline ID02 with a sample-to-detector distance of 2 m or a combination of 0.85 and 5 m in order to cover a larger q range of 0.04 to 7.8 nm^{-1}.[39] The data were collected by a high sensitivity fiber-optic coupled CCD (FReLoN) detector placed in an evacuated flight tube. The protein solutions were loaded using a flow-through capillary cell (diameter ~2 mm; wall thickness ~10 μm). No variation that would indicate radiation damage of SAXS profiles was observed comparing successive short exposures of cumulated time up to 3 s. For the temperature scan, each temperature was held for 15 min to allow equilibration. The incident and transmitted beam intensities were simultaneously recorded for each SAXS pattern with an exposure of 0.3 s. The 2D data were normalized to an absolute scale and azimuthally averaged to obtain the intensity profiles, and the solvent background was subtracted. For more detailed information on data reduction and q-resolution calibration, see ref. 40.

Results and discussions

Phase behavior of protein solutions in the presence of multivalent counterions

We first briefly summarize the phase diagram of BLG (c_p) as a function of the YCl$_3$ concentration (c_s) at room temperature (Fig. 1) with an extended range of c_p. This diagram provides a guide for optimizing the conditions for protein crystallization. For a given protein concentration c_p, an increase of the salt concentration c_s above a certain threshold (c^*) results in the protein solution becoming turbid and entering a two-phase state. When c_s is increased further (above c^{**}), the protein solution turns clear again. Thus, the two salt concentrations c^* and c^{**} divide the phase diagram into three regimes (Fig. 1). Regimes I and III contain clear protein solution, whereas the protein condenses (or aggregates) in Regime II. The addition of YCl$_3$ leads to a charge inversion on the protein surface, which has been proved using zeta potential measurements.[16,37] The experimental observations on the crystal growth from different regions indicate different growth mechanisms:[16] near c^*, crystal growth follows the classical nucleation theory, i.e. nucleation occurs from homogeneous supersaturated solutions. Near c^{**}, the solution has a transition temperature, T_h. Below T_h, a metastable LLPS occurs before crystallization, apparently resembling the so-called "two-step" nucleation mechanism. In the remaining part of paper, we will focus on the structure of protein clusters and crystallization in Regime III.

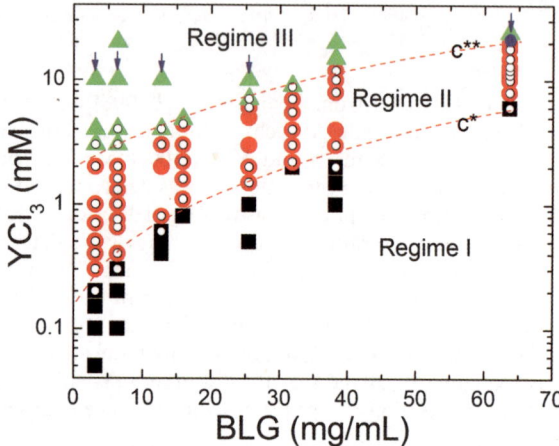

Fig. 1 Extended phase diagram of protein BLG as a function of protein (c_p) and salt, YCl$_3$ (c_s).[16] (a) Phase diagram at room temperature (22 °C). Solid symbols present the sample solutions at different regimes (see text). Small open symbols present the samples solutions where crystallization was observed at 4 °C. The small arrows in Regime III indicate the samples measured by SAXS.

Protein clusters and their structure in Regime III studied by SAXS

We now examine the solution structure of proteins in Regime III using SAXS. By fixing the salt concentration and varying the protein concentration, we approach the phase transition boundary, c^{**}. The selected solutions are shown in Fig. 1 as indicated by vertical arrows. The additional sample with higher protein concentration (67.0 mg mL^{-1}) with 15 mM YCl$_3$ near c^{**} is chosen for a temperature dependent measurement. Fig. 2 shows a series of SAXS profiles of BLG in solution with 10 mM YCl$_3$ and different protein concentrations. The scattering curves have been scaled by their protein concentrations in terms of the absolute intensity.

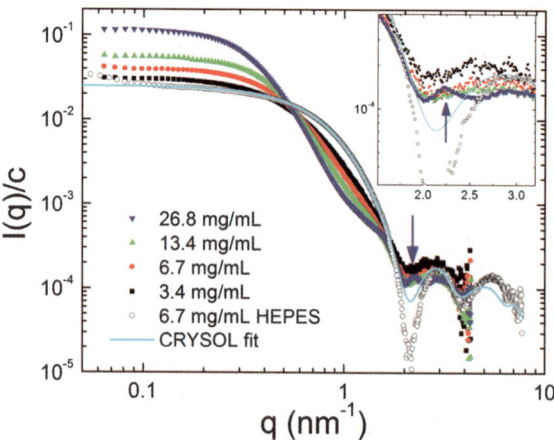

Fig. 2 SAXS profiles of BLG solutions with YCl$_3$ in the re-entrant regime. For a given value of c_s, a clear transition from dimer to cluster was observed with increasing protein concentration. For comparison, the SAXS curve of BLG in HEPES buffer is also shown, which can be well described by a dimer using the crystal structure (PDB code of 1BEB). Inset shows the expanding of the region with the arrow.

For comparison, a SAXS profile for BLG in 20 mM HEPES buffer is also plotted. These data were collected at two sample-to-detector distances (0.85 and 5 m) covering a large q range from 0.03 to 7.8 nm^{-1}. Two scattering minima are clearly visible at $q = 2.2$ and 4.2 nm^{-1}, respectively. In HEPES buffer (pH 7.0), BLG is dissolved as dimers,[41–43] which is further confirmed by fitting the SAXS data using CRYSOL[44] with its crystal structure (PDB code 1BEB). The radius of gyration, R_g, of the BLG dimer obtained by the fitting procedure is 2.33 nm, which is in good agreement with those reported in the literature.[45] For solutions with 10 mM YCl$_3$ and increasing BLG concentration, the scattering curves change smoothly. At 3.4 mg mL^{-1} protein, well below the phase boundary, the scattering intensity in the low q region ($q < 0.5$ nm^{-1}) is higher as compared to the dimeric state, whereas in the intermediate q region (0.5 nm^{-1} < q < 1.4 nm^{-1}), the scattering intensity is lower. With increasing the protein concentration, this trend becomes more significant and two crossing points are observed at $q = 0.5$ and 1.4 nm^{-1}, respectively. With 26.8 mg mL^{-1} protein, a kink is well-developed in the q region between the crossing points, indicating the formation of clusters. Basic structural parameters of these clusters, such as R_g, and the normalized forward intensity, $I(0)$, are obtained from a Guinier analysis.[46] As shown in Table 1, R_g increases when approaching the phase boundary. $I(0)$ is related to the molecular weight of the cluster, which can be used to estimate the number of dimers within the clusters. Simply dividing the normalized $I(0)$ by 0.0248 (the value obtained from the form factor), the numbers are estimated as 1.3, 1.7, 2.7, 4.5, respectively. These results are listed in Table 1. A new maximum appears at $q = 2.2$ nm^{-1} for $c_p > 20$ mg mL^{-1} as indicated by the arrow in Fig. 2 (expanded in the figure). This maximum, which corresponds to a center-to-center distance of $d = 2\pi/q = 2.82$ nm, is very close to the monomer–monomer distance within a dimer (3.08 nm) (calculated from the crystal structure: 1BEB). Thus, the appearance of this peak is attributed to the nearest neighbour correlation within clusters, indicating the liquid-like structure within the clusters.

The cluster phase observed in our system is due to the balance of electrostatic repulsion and the attraction.[47] We have demonstrated that in the re-entrant regime, the effective surface charge of proteins is inverted from negative to positive, which provides the long range repulsion.[35,37] On the other hand, the bridging effect of the yttrium cations provides the short range attraction. Although other interactions, such as van der Waals attraction may also contribute to the overall attractive potential, we assume that the bridging effect dominates, which is consistent with the crystallography study.[16] Clustering for qualitatively comparable interactions has been observed also for lysozyme and colloids in solution.[25]

Since the bridging effect seems to contribute anisotropic attractions, not only the crystal structure, but also the structure of these clusters should reflect the anisotropic interaction between the proteins being bridged by counterions. Indeed, the SAXS profiles of the cluster phases can be reasonably fitted by typical cluster structures created based on the crystal structure determined in our previous work

Table 1 Structural parameters obtained by Guinier analysis on the SAXS data

Sample	$I(0)/c$/cm^{-1}	R_g/nm	Number of dimer in cluster[a]
BLG 6.7 HEPES	0.0248	2.33	Monodisperse dimer
BLG 3.4 YCl$_3$ 10 mM	0.0321	3.08	1.3
BLG 6.7 YCl$_3$ 10 mM	0.0432	3.36	1.7
BLG 13.4 YCl$_3$ 10 mM	0.0663	3.67	2.7
BLG 27.0 YCl$_3$ 10 mM	0.1126	4.90	4.5

[a] These values are calculated by dividing $I(0)$ by 0.0248, $I(0)$ of the pure dimer solution.

(PDB 3PH5)[16] with 2, 3 and 4 dimers, as shown in Fig. 3. The clusters are created as follows: we first choose one dimer from the lattice and set its center of mass to be the origin. Then the positions of the other dimers within the clusters are chosen based on the shortest center-to-center distance to the origin. The experimental SAXS profile of BLG 13.4 mg mL^{-1} with 10 mM YCl$_3$ is used to compare with these cluster structures using CRYSOL.[44] The best fitting comes from the cluster with 3 dimers (consistent with Table 1) which reproduces the R_g and $I(0)$ quantitatively, and other features of the SAXS profile qualitatively, such as the position of the kink, the shoulder for dimer, the minimum and maximum from the form factor as indicated by arrows in Fig. 3. Of course, this simple structural model of the cluster cannot reproduce all the details since the size and structure of the clusters will have a distribution. In spite of these difficulties, it is plausible to assume that the protein clusters have a "pre-crystalline" structure (Fig. 3) similar to those created from the crystal structure.

Having elaborated on the formation of small clusters with dimers as the principle building blocks from Fig. 2 and 3, we now review the full hierarchic structure of protein assembly in solution. Fig. 4 shows the SAXS profile of a sample close to the phase boundary with BLG 67 mg mL^{-1} with 15 mM YCl$_3$. A pronounced maximum at $q = 2.2$ nm^{-1} is clearly visible. As mentioned above, the scattering peak at $q = 2.2$ nm^{-1} is attributed to the monomer–monomer (M–M) correlation within clusters, which cannot be reproduced from the crystal structures in Fig. 3. The shoulder at $q \sim 1.8$ nm^{-1} corresponds to the dimer scattering, and the shoulder at $q \sim 0.3$ nm^{-1} corresponds to the form factor of the protein cluster. The number of dimers within the clusters is estimated to be around 3–4 for this sample since the intensities continuously increase in the low q region and both the forward intensity and the radius of gyration cannot be precisely determined by the Guinier analysis. The continuous increase indicates that the clusters may further build up higher level structures with a larger length scale, which makes the M–M correlation more significant.

Crystal growth followed by *in situ* SAXS

We now present the results of crystal growth directly from the cluster solution by an *in situ* SAXS observation. Fig. 5 shows a series of SAXS profiles for BLG

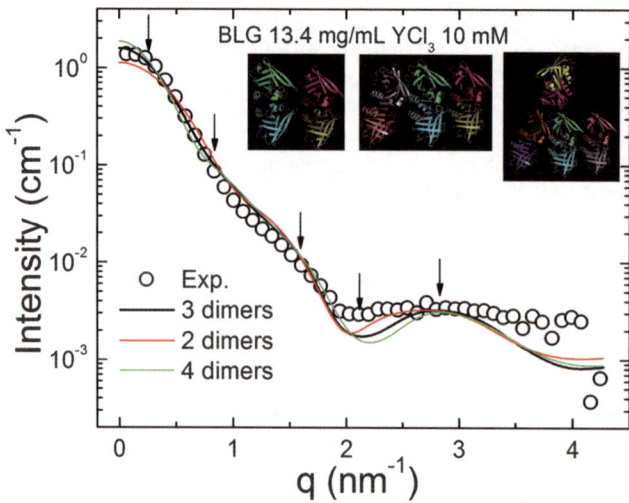

Fig. 3 Experimental SAXS data of protein clusters compared to those created using the crystal structure (PDB code of 3PH5). Only 10% of experimental data are plotted for clarity. The inset shows the structure of protein clusters created by the crystal structure with 2, 3 and 4 dimers.

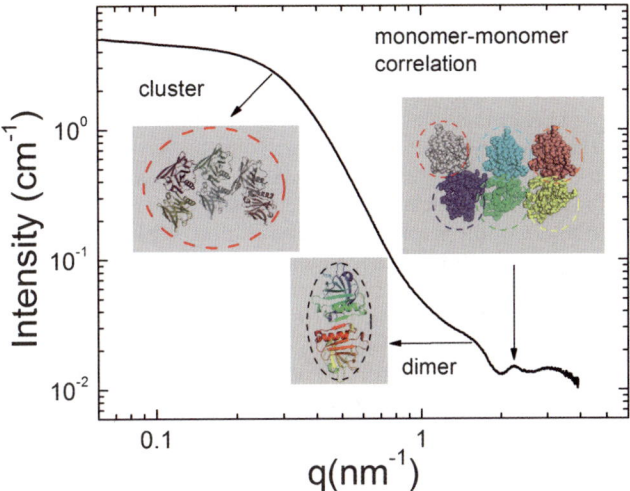

Fig. 4 Structural hierarchy of BLG solutions in re-entrant regime revealed by SAXS. The peak and shoulders in the SAXS curve at q equals $q = 2.2$, 1.8 and 0.3 nm^{-1} correspond to the monomer–monomer correlation, the form factor of a dimer, and the cluster, respectively.

67 mg mL^{-1} with YCl$_3$ 15 mM. The transition temperature, T_h, is about 22 °C. At 25 °C (after preparation), the SAXS curve is the same as shown in Fig. 4. The sample is firstly quenched to 10 °C, then heated up stepwise to 22 °C, and then cooled down stepwise to 10 °C. The increment of heating and cooling is 1 °C and the time for equilibration is 15 min. After quenching to 10 °C, the scattering intensity in the low q region (region (1) in Fig. 5) increases as compared to that at 25 °C, indicating the formation of larger clusters. The solution becomes turbid. In the meantime, the M–M correlation peak (region (3) in Fig. 5) becomes weaker. In the intermediate q region (region (2) in Fig. 5), where the scattering is mainly contributed by the form factor of a dimer, no changes could be observed. The subsequent stepwise heating

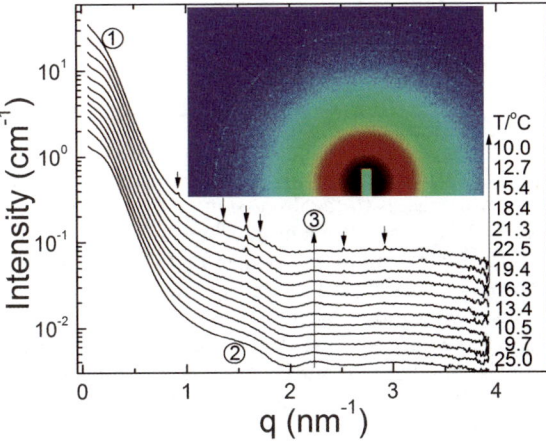

Fig. 5 SAXS curves at different temperatures during cooling. Crystallization occurs below 25 °C. The intensity of the maximum at $q = 2.2$ nm^{-1} decreases with lowering temperature, and the low q intensity increases steadily. Bragg peaks appearing in the intermediate q range have been indexed using the crystal structure (see text). The curves are shifted upward for clarify. The inset shows the 2D scattering pattern at 10 °C.

shows the inverse trend, *i.e.* with increasing temperature, the intensity in the low q region decreases, and the M–M correlation peak becomes stronger. The solution becomes clear again. This observation suggests that the formation of larger clusters is a reversible process, which is consistent with our previous measurement on T_h.[16] So far, no Bragg peaks could be observed. Upon further lowering the temperature, one observes the following significant changes: (i) Bragg peaks appear, indicating the crystallization of proteins, indicating that protein crystals can grow directly from the clusters. The inset 2D pattern of the final state shows sharp circles indicating the isotropic distribution of the orientation in the solution. We have recently determined the structure of crystals grown under similar conditions in the phase diagram. All crystals have an orthorhombic structure with space group $P2_12_12_1$ and contain one dimer in their asymmetric unit (PDB code 3PH5 and 3PH6), which can be used to index the Bragg peaks in the SAXS profiles.[16] For example, the first three Bragg peaks at $q = 0.924$ nm^{-1}, $q = 1.348$ nm^{-1} and $q = 1.571$ nm^{-1} can be assigned as (002), (012) and (100) of the crystal structure, respectively. This means that although M–M correlation is visible in the cluster phase, the dimer represents the building block of the crystals. (ii) In the low q region, the scattering intensity increases with decreasing temperature, indicating the formation of larger objects, *i.e.* clusters and crystals. (iii) No change is observed in region (2) except the appearance of Bragg peaks. (iv) the M–M correlation peak at $q = 2.2$ nm^{-1} continuously decreases in intensity as the temperature decreases. The temperature dependence of the M–M correlation peak may be related to the flexibility of the monomers within the cluster.

The crystal growth of similar samples has been followed using an optical microscope.[16] It shows that after quenching the solution below T_h, the solution becomes turbid instantly. After some time (hours) the dense liquid phase appears, then crystal growth appears to start from the dilute phase. Over time, the crystal growth consumes the material and leads to the dissolution of the droplets. Our observations of crystal growth in the presence of YCl$_3$ suggest that the most common way is the growth from the dilute phase directly leading to the coexistence of a dense liquid phase and crystals. From the results of the SAXS measurements, we now know that even in the dilute phase, proteins form clusters (Fig. 2, 3) and crystals grow directly from such protein clusters (Fig. 5). These observations suggest that the protein clusters undergo two competing pathways during cooling, *i.e.* either forming a dense liquid phase or reordering into a crystal, which can modify the pathways of crystal growth as to be discussed below.

Discussion

Pathway of the two-step crystallization

We first discuss the possible pathway of protein crystallization during a two-step growth procedure, which leads to the discussion of the role of protein clusters and a metastable LLPS. A two-step mechanism following a metastable LLPS suggests that the nucleation occurs within the dense liquid phase instead of the dilute phase.[3,32,48] Because the surface free energy at the interface between the crystal and the solution is significantly higher than at the interface between the crystal and the dense liquid, the barrier for nucleation of crystals from the solution would be much higher. This would lead to much slower nucleation of crystals directly from the solution than inside the clusters. While indeed some experimental observations are consistent with this prediction, *e.g.* for glucose isomerease and hemoglobin,[10,49] other observations however, indicate that crystallization prefers to start in the dilute phase.[50] While this has often been attributed to the high viscosity or gelation of the dense liquid phase, a clear understanding of their exact role is still missing.

In our system, we observe in addition to the metastable LLPS a non-negligible effect of clustering. Although the mechanism of clustering observed here may not be the same as in the concentrated protein solutions,[26–28] it can be assumed to play a similar role on protein crystallization. The protein clusters observed show interesting features with respect to crystallization, which make them suitable as building blocks for crystal formation. Firstly, the clusters have a pre-crystalline structure, since the cation binding sites represent specific interaction patches. These imprint a favorable structure, as implied by the SAXS scattering curves of the clusters being simulated well using the clusters created from the crystal structure. Secondly, clusters seem to have an internal flexibility, as evidenced by the pronounced M–M correlation peak in larger clusters. This flexibility enables local reorientation within the clusters, rendering the pre-crystalline structure of clusters to be an ideal building block for crystallization, since the enthalpy cost of nucleation *via* local reorientation is much lower than for a hypothetical nucleation directly from a dimer in solution. Thirdly, the cluster size and its distribution vary throughout the phase diagram; in particular, the clusters grow when the solution conditions approach the coexistence region. Large pre-crystalline clusters can serve as the precursor (stabilized nuclei) for further crystal growth, as recently observed in various systems.[22,29–31]

Inspired by our observations, we propose an alternative pathway of protein crystallization in Fig. 6. The monodisperse dimeric proteins in solution form clusters upon adding YCl_3 due to the bridging effect of cations (Fig. 6a), as indicated by SAXS measurements (Fig. 2). Upon lowering the temperature across the phase boundary (below T_h), LLPS takes place, leaving a dense phase and a dilute phase with clusters (Fig. 6b). In solution, clusters can move freely and restructure themselves internally. Furthermore, clusters can assemble to larger clusters with a pre-crystalline structure. As a next step, internal reorientation turns the clusters to stable precursor nuclei (Fig. 6c). Because the dense liquid phase is metastable with respect to the crystalline phase, the crystals grow and consume the protein molecules from the dense liquid phase which dissolves and disappears (Fig. 6d). This pathway differs from the previous view of the two-step mechanism involving a metastable LLPS by the phases where nucleation starts. While the previous view predicts the crystals to be nucleated within the dense liquid phase, our observations suggest that both condensed phases can arise from the protein clusters in the protein-poor phase.

The role of clusters and dense liquid phase in protein crystallization

The possible pathway of protein crystallization discussed above (Fig. 6) leads to a further discussion of the role of the protein clusters and the metastable dense liquid phase during the two-step crystallization procedure. In our system, both the dense liquid phase and crystals grow directly from the protein clusters (Fig. 6): we thus speculate that while the protein clusters enhance the nucleation, the dense liquid phase may play a role in optimizing the conditions for crystallization. Some clues

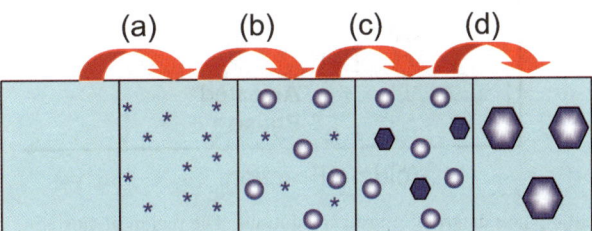

Fig. 6 Schematic illustration of the phase transitions (LLPS and crystallization) from protein clusters (see text for details). (a) Upon addition of YCl_3, protein clusters formed *via* ion bridging. (b) Lowering temperature to $T < T_h$ and crossing the phase boundary, the clusters first aggregate with a liquid-like structure. (c) After the induction time protein crystals grow directly from the clusters. (d) With time increasing, while the crystals grow steadily, the dense liquid phase dissolves and disappears.

arising from the detailed studies on lysozyme in solution support this explanation: Pan *et al.* have shown that the protein clusters exist between the solubility line and the LLPS in the phase diagram,[23] which coincides with the optimum conditions for protein crystallization predicted using the second virial coefficient as predictor by Vliegenthart and Lekkerkerker.[51] Our observations from *in situ* SAXS (Fig. 5) indicate that nucleation is possible from protein clusters, and microscope observations indicate that crystals grow from the dilute solution instead of the dense liquid phase. Employing this picture, the metastable LLPS occurs as a helpful side effect at these thermodynamic conditions, since the LLPS indicates suitable interaction conditions and ensures that the dilute protein solutions stay at a defined concentration. Furthermore, the dense liquid phase potentially acts as a reservoir feeding the dilute phase with protein molecules and thus stabilizing the thermodynamic conditions during the ongoing crystal growth.

Connection of two-step crystallization to the phase diagram

In addition to the two-step nucleation mechanism,[3,32,48] we have discussed an alternative pathway of two-step crystallization *via* cluster precursors. Considering the complexity of protein phase behavior, we expect both to be relevant under suitable conditions, which we discuss with the schematic phase diagram in Fig. 7. Vliegenthart and Lekkerkerker[51] elaborated on earlier findings of George and Wilson[52] to provide a criterion for optimum crystallization conditions based on the reduced second virial coefficient. Besides the metastable critical point enhancing nucleation *via* critical density fluctuations,[32] the identified optimum conditions in terms of the reduced second virial coefficient correspond to a temperature window, *i.e.* both the dilute and the dense phase (indicated with grey shading). Considering the lower nucleation barrier in the dense phase, the optimum conditions for protein crystallization should be represented by the dense solution as predicted in the two-step nucleation mechanism,[3,32,48] although in practice, such condition often leads to polycrystallites instead of single crystals, which are needed for the crystallographic study.

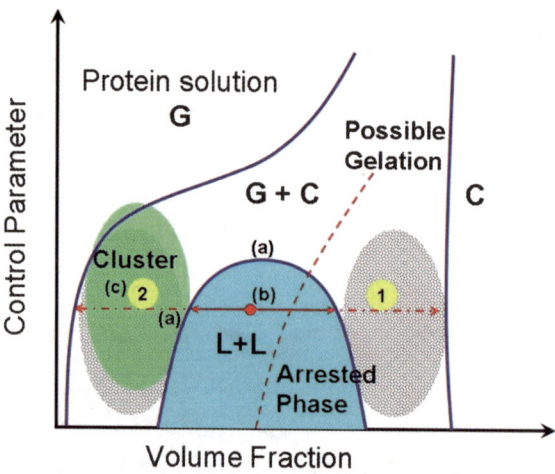

Fig. 7 Schematic phase diagram of protein solutions. The region of clustering[35] is marked in green and the grey shaded area corresponds to the optimum condition for crystallization.[46] Two-step nucleation mechanism predicts that protein solutions located within LLPS region (red point) undergo a LLPS in the first step and then nucleation occurs within the dense liquid phase (along pathway (1)). An alternative pathway proposed in this work follows pathway (2): after LLPS, protein clusters in the dilute phase with the optimized thermodynamic conditions for crystallization can initialize crystal growth with a reduced energy barrier. Note that the exact conditions or the location of the gelation line will vary with the specific system.

These conditions, however, are not necessarily accessible for all systems, since arrested phase behaviour may render the high-concentration coexistent phase a gel instead of an equilibrium solution.[4,19–21,53] While for isotropic interaction potentials, a metastable LLPS is generally arrested,[4,20] anisotropic and patchy spheres have been shown to phase separate completely into a metastable "gas" and "empty liquid" phase.[21,53] At least for the arrested systems, however, protein clusters in the dilute phase could step in as precursors, providing an alternative (even dominate) way of crystallization. Interestingly, the cluster phase is expected to overlap with the optimum crystallization conditions based on the second virial coefficient. The proposed mechanism *via* cluster precursors could thus explain the frequent observation of crystal growing in the dilute phase.

Finally, since there appear to be different views in the literature regarding the terminology of the "two-step" crystallization process, we would like to comment on the different scenarios. In protein solutions, depending on the position in the phase diagram (labeled in Fig. 7 as a, b and c), at least three scenarios for the intermediate state could be realized: (a) a transient density fluctuation or cluster near the critical point or near the G–L binodal, this scenario is supported by simulations and theoretical studies;[32,54,55] (b) a metastable co-existence of liquid phases, which has been observed in several protein systems;[10,49] and (c) protein clusters as shown in our system and other proteins in solution.[27,28] In real protein solutions, these mechanisms might also occur at the same time, rendering a deeper understanding of protein crystallization an interesting challenge for future research.

Conclusions and outlook

In summary, our SAXS study of the structure and crystallization from bovine β-lactoglobulin (BLG) solutions in the presence of YCl_3 leads to the following conclusions. First, near the phase boundary, the balance between the bridging effect of yttrium cations and effective charge inversion of proteins leads to the formation of small protein clusters which have a pre-crystalline structure. Second, *in situ* SAXS measurements on the crystallization together with previous observation by optical microscopy indicate that these protein clusters can form both a dense liquid phase and crystals. The crystals occur later compared to the dense liquid phase due to the higher energy barrier of nucleation. This procedure leads to the competing coexistence of these two structures. Third, due to the metastable character of the dense liquid phase, it re-dissolves during the ongoing crystal growth and disappears in the post-crystal growth stage. The observed two-step crystal growth procedure suggests that clusters play the main mechanistic role for nucleation, while LLPS allows the system to access and stabilize suitable conditions.

Acknowledgements

We acknowledge the useful discussion with Dr Zocher (IFIB, Universität Tübingen) and Dr T. Narayanan (ESRF, Grenoble, France). We acknowledge financial support from Deutsche Forschungsgemeinschaft (DFG) and beamtime allocation at ESRF, Grenoble, France.

References

1 J. D. Gunton, A. Shiryayev and D. L. Pagan, *Protein Condensation—kinetic pathways to crystallization and disease*, Cambridge University Press, New York, 2007.
2 P. G. Vekilov, *Cryst. Growth Des.*, 2004, **4**, 671–685.
3 P. G. Vekilov, *Nanoscale*, 2010, **2**, 2346–2357.
4 P. J. Lu, E. Zaccarelli, F. Ciulla, A. B. Schofield, F. Sciortino and D. A. Weitz, *Nature*, 2008, **453**, 499–503.
5 M. Delaye, J. I. Clark and G. B. Benedek, *Biochem. Biophys. Res. Commun.*, 1981, **100**, 908–914.

6 M. L. Broide, C. R. Berland, J. Pande, O. Ogun and G. B. Benedek, *Proc. Natl. Acad. Sci. U. S. A.*, 1991, **88**, 5660–5664.
7 M. Muschol and F. Rosenberger, *J. Chem. Phys.*, 1997, **107**, 1953–1962.
8 O. Galkin and P. G. Vekilov, *Proc. Natl. Acad. Sci. U. S. A.*, 2000, **97**, 6277–6281.
9 O. Annunziata, N. Asherie, A. Lomakin, J. Pande, O. Ogun and G. B. Benedek, *Proc. Natl. Acad. Sci. U. S. A.*, 2002, **99**, 14165–14170.
10 O. Galkin, K. Chen, R. L. Nagel, R. E. Hirsch and P. G. Vekilov, *Proc. Natl. Acad. Sci. U. S. A.*, 2002, **99**, 8479–8483.
11 S. Grouazel, J. Perez, J. P. Astier, F. Bonneté and S. Veesler, *Acta Crystallogr., Sect. D: Biol. Crystallogr.*, 2002, **58**, 1560–1563.
12 O. Annunziata, O. Ogun and G. B. Benedek, *Proc. Natl. Acad. Sci. U. S. A.*, 2003, **100**, 970–974.
13 Q. Chen, P. G. Vekilov, R. L. Nagel and R. E. Hirsch, *Biophys. J.*, 2004, **86**, 1702–1712.
14 N. Dorsaz, G. M. Thurston, A. Stradner, P. Schurtenberger and G. Foffi, *Soft Matter*, 2011, **7**, 1763–1776.
15 Y. Wang, A. Lomakin, R. F. Latypov and G. B. Benedek, *Proc. Natl. Acad. Sci. U. S. A.*, 2011, **108**, 16606–16611.
16 F. Zhang, G. Zocher, A. Sauter, T. Stehle and F. Schreiber, *J. Appl. Crystallogr.*, 2011, **44**, 755–762.
17 F. Zhang, R. Roth, M. Wolf, F. Roosen-Runge, M. W. A. Skoda, R. M. J. Jacobs, M. Sztucki and F. Schreiber, *Soft Matter*, 2012, **8**, 1313–1316.
18 F. Cardinaux, T. Gibaud, A. Stradner and P. Schurtenberger, *Phys. Rev. Lett.*, 2007, **99**, 118301.
19 T. Gibaud, F. Cardinaux, J. Bergenholtz, A. Stradner and P. Schurtenberger, *Soft Matter*, 2011, **7**, 857–860.
20 T. Gibaud and P. Schurtenberger, *J. Phys.: Condens. Matter*, 2009, **21**, 322201.
21 F. Sciortino and E. Zaccarelli, *Curr. Opin. Solid State Mater. Sci.*, 2011, **15**, 246–253.
22 A. Navrotsky, *Proc. Natl. Acad. Sci. U. S. A.*, 2004, **101**, 12096–12101.
23 Y. Liu, E. Fratini, P. Baglioni, W. R. Chen and S. H. Chen, *Phys. Rev. Lett.*, 2005, **95**, 118102.
24 A. Shukla, E. Mylonas, E. Di Cola, S. Finet, P. Timmins, T. Narayanan and D. I. Svergun, *Proc. Natl. Acad. Sci. U. S. A.*, 2008, **105**, 5075–5080.
25 A. Stradner, H. Sedgwick, F. Cardinaux, W. C. K. Poon, S. U. Egelhaaf and P. Schurtenberger, *Nature*, 2004, **432**, 492–495.
26 O. Gliko, W. Pan, P. Katsonis, N. Neumaier, O. Galkin, S. Weinkauf and P. G. Vekilov, *J. Phys. Chem. B*, 2007, **111**, 3106–3114.
27 W. Pan, O. Galkin, L. Filobelo, R. L. Nagel and P. G. Vekilov, *Biophys. J.*, 2007, **92**, 267–277.
28 W. Pan, P. G. Vekilov and V. Lubchenko, *J. Phys. Chem. B*, 2010, **114**, 7620–7630.
29 J. F. Banfield, S. A. Welch, H. Zhang, T. T. Ebert and R. Lee Penn, *Science*, 2000, **289**, 751–754.
30 G. Furrer, B. L. Phillips, K.-U. Ulrich, R. Pöthig and W. H. Casey, *Science*, 2002, **297**, 2245–2247.
31 S. Mintova, N. H. Olson, V. Valtchev and T. Bein, *Science*, 1999, **283**, 958–960.
32 P. R. ten Wolde and D. Frenkel, *Science*, 1997, **277**, 1975–1978.
33 D. Erdemir, A. Y. Lee and A. S. Myerson, *Acc. Chem. Res.*, 2009, **42**, 621–629.
34 P. E. Bonnett, K. J. Carpenter, S. Dawson and R. J. Davey, *Chem. Commun.*, 2003, 698–699.
35 F. Zhang, M. W. A. Skoda, R. M. J. Jacobs, S. Zorn, R. A. Martin, C. M. Martin, G. F. Clark, S. Weggler, A. Hildebrandt, O. Kohlbacher and F. Schreiber, *Phys. Rev. Lett.*, 2008, **101**, 148101.
36 L. Ianeselli, F. Zhang, M. W. A. Skoda, R. M. J. Jacobs, R. A. Martin, S. Callow, S. Prévost and F. Schreiber, *J. Phys. Chem. B*, 2010, **114**, 3776–3783.
37 F. Zhang, S. Weggler, M. Ziller, L. Ianeselli, B. S. Heck, A. Hildebrandt, O. Kohlbacher, M. W. A. Skoda, R. M. J. Jacobs and F. Schreiber, *Proteins: Struct., Funct., Bioinf.*, 2010, **78**, 3450–3457.
38 *Handbook of Biochemistry: Selected Data for Molecular Biology*, ed. H. A. Sober, The Chemical Rubber Co., Cleveland, OH, 1970.
39 M. Sztucki, T. Narayanan, G. Belina, A. Moussaïd, F. Pignon and H. Hoekstra, *Phys. Rev. E: Stat., Nonlinear, Soft Matter Phys.*, 2006, **74**, 051504.
40 T. Narayanan, in *Soft Matter: Characterization*, ed. R. Borsali and R. Pecora, Springer, Berlin-Heidelberg, 2008, vol. II, ch. 17, pp. 899–952.
41 U. M. Elofsson, M. A. Paulsson and T. Arnebrant, *Langmuir*, 1997, **13**, 1695–1700.
42 M. Gottschalk, H. Nilsson, H. Roos and B. Halle, *Protein Sci.*, 2003, **12**, 2404–2411.

43 K. Vogtt, N. Javid, E. Alvarez, J. Sefcik and M.-C. Bellissent-Funel, *Soft Matter*, 2011, **7**, 3906–3914.
44 D. I. Svergun, C. Barberato and M. H. J. Koch, *J. Appl. Crystallogr.*, 1995, **28**, 768–773.
45 C. Moitzi, L. Donato, C. Schmitt, L. Bovetto, G. Gillies and A. Stradner, *Food Hydrocolloids*, 2011, **25**, 1766–1774.
46 O. Glatter and O. Kratky, *Small angle X-ray scattering*, Academic Press London, 1982.
47 J. Groenewold and W. K. Kegel, *J. Phys. Chem. B*, 2001, **105**, 11702–11709.
48 V. Talanquer and D. W. Oxtoby, *J. Chem. Phys.*, 1998, **109**, 223–227.
49 D. Vivares, E. W. Kaler and A. M. Lenhoff, *Acta Crystallogr., Sect. D: Biol. Crystallogr.*, 2005, **61**, 819–825.
50 Y. Liu, X. Wang and C. B. Ching, *Cryst. Growth Des.*, 2010, **10**, 548–558.
51 G. A. Vliegenthart and H. N. W. Lekkerkerker, *J. Chem. Phys.*, 2000, **112**, 5364–5469.
52 A. George and W. W. Wilson, *Acta Crystallogr., Sect. D: Biol. Crystallogr.*, 1994, **50**, 361–365.
53 E. Bianchi, J. Largo, P. Tartaglia, E. Zaccarelli and F. Sciortino, *Phys. Rev. Lett.*, 2006, **97**, 168301.
54 D. W. Oxtoby, *Acc. Chem. Res.*, 1998, **31**, 91–97.
55 J. F. Lutsko, *Adv. Chem. Phys.*, 2012, **151**, arXiv:1104.3413v1101.

PAPER

Polymer-induced liquid precursor (PILP) phases of calcium carbonate formed in the presence of synthetic acidic polypeptides—relevance to biomineralization

Anna S. Schenk,[a] Harshal Zope,[b] Yi-Yeoun Kim,[a] Alexander Kros,[b] Nico A. J. M. Sommerdijk[c] and Fiona C. Meldrum[*a]

Received 5th April 2012, Accepted 6th June 2012
DOI: 10.1039/c2fd20063e

Polymer-induced liquid precursor (PILP) phases of calcium carbonate have attracted significant interest due to possible applications in materials synthesis, and their resemblance to intermediates seen in biogenic mineralisation processes. Further, these PILP phases have been formed *in vitro* using polyelectrolytes such as poly(aspartic acid) which bears many structural parallels to the highly acidic biomacromolecules that are associated with biogenic calcium carbonate. This article describes experiments which investigate how the composition of acidic polypeptides determines their ability to form PILP phases of $CaCO_3$, and therefore whether it is feasible that the acidic biomacromolecules extracted from $CaCO_3$ biominerals could also function in this way. A series of random copoly(amino acid)s constructed from 80–20%, 50–50% and 20–80% aspartic acid and serine residues were synthesised and their effect on $CaCO_3$ precipitation was determined. A strong correlation between the composition and function of the polypeptide was observed. Only the polypeptide containing 80% aspartic acid residues (Asp80%–Ser20%) induced the formation of continuous $CaCO_3$ films, which provide a fingerprint of an intermediary PILP phase, while addition of Mg^{2+} also facilitated the formation of expanded film-like structures with the polypeptide Asp50%–Ser50%. In contrast, the weakly-acidic polypeptide Asp20%–Ser80% had only a minor effect on the crystal morphologies and also failed to aid infiltration of $CaCO_3$ into small pores. These results therefore demonstrate that counter-ion induced phase separation of highly acidic biomacromolecules proteins appears to be entirely feasible based upon their composition, but that evidence for the operation of this mineralisation mechanism *in vivo* is still required.

1 Introduction

Nature has developed a range of diverse and often elaborate strategies for the fabrication of biominerals with unique structures and properties. These are typically based on both soluble and insoluble organic matrices, which enable organisms to precisely control the size,[1] shape,[2] organisation,[3] and even polymorph[4] of the mineral

---

[a]*School of Chemistry, University of Leeds, Woodhouse Lane, Leeds, LS2 9JT, UK. E-mail: F.Meldrum@leeds.ac.uk*
[b]*Leiden Institute of Chemistry, Soft Matter Chemistry, Einsteinweg 55, 2333 CC Leiden, The Netherlands*
[c]*Laboratory of Materials and Interface Chemistry, Eindhoven University of Technology, P. O. Box 513, 5600 MB Eindhoven, The Netherlands*

phase. Investigation of biomineralisation strategies has also revealed a further key feature - that biominerals often form *via* an amorphous precursor phase[5,6] rather than by the immediate precipitation of the crystalline phase of the mature biomineral. This mechanism offers many potential advantages, such as providing a facile route to moulding the biomineral into its final desired form[7] and the provision of high concentrations of reagents, which gives rise to rapid growth rates. Again, organic macromolecules have been shown to play a role in the stabilisation of these transient phases.[8]

Inspired by the intricate structures observed in biominerals, synthetic strategies have been developed based on those which operate during biomineralisation. The precipitation of minerals in the presence of soluble organic additives is one concept that has proven highly successful in the regulation of crystal morphology and nanostructure.[9] Of particular note in this regard is the formation of so-called "polymer-induced liquid precursor" (PILP) phases. These have been best-studied for calcium carbonate due to their apparent resemblance to biogenic amorphous calcium carbonate (ACC)[10] and possible applications in materials synthesis.

Gower *et al.* were the first to observe that the precipitation of calcium carbonate in the presence of μg mL^{-1} concentrations of the highly acidic polyelectrolyte poly(aspartic acid) (PAsp) leads to the deposition of precipitates with unusual, non-crystallographic morphologies such as films or fibres.[11] The reaction was performed by exposing an aqueous solution of $CaCl_2$ and the polymer to ammonium carbonate vapour,[10] which leads to the formation of non-crystalline droplets within the solution. This was attributed to the phase separation of an amorphous, liquid-like material consisting of a highly hydrated $Ca^{2+}/PAsp/CO_3^{2-}$ complex.[12] If the droplets are stable for sufficiently long periods of time, they can also coalesce to form an amorphous thin film on a substrate, before ultimately crystallising.[10]

Due to its fluid-like nature, the calcium carbonate PILP phase can be moulded into non-crystallographic shapes[13,14] and can infiltrate into small volumes.[10,15] Indeed, Kim *et al.* recently demonstrated that a $CaCO_3$ PILP phase prepared in the presence of poly(aspartic acid) can penetrate into the nanometer-diameter pores of polycarbonate track etch membranes. The data they presented also suggested that the infiltration of the pores is driven by capillary action. Calcite nano-rods with very high aspect ratios of up to 100 were obtained in this process.[16]

While poly(aspartic acid) and poly(acrylic acid)[17] were for many years the only synthetic polyelectrolytes known to support calcium carbonate PILP formation, it is now recognised that many polyelectrolytes can generate thin films of calcium carbonate *via* PILP phases. These include DNA,[18] ovalbulmin,[19] a calcification-associated peptide,[20] and a short synthetic polypeptide (PhosSer-Ser-Asp)$_3$ (repetitive units of phosphoserine, serine and aspartic acid), which resembles the sequence of the acidic domains in phosphophoryn, a bioprotein associated with dentine mineralization.[21] As a further example from biology, Marsh also reported the self-association of the protein phosphophoryn in the presence of calcium or magnesium ions.[22] A recent study using the soluble polyamine poly(allylamine hydrochloride) (PAH) also showed that $CaCO_3$ PILP formation is not restricted to negatively charged polyelectrolytes. PAH carries a net positive charge at pH values lower than 10, but is highly effective at generating thin films and fibres analogous to those observed with PAsp.[23]

PILP phases have also been observed for systems other than calcium carbonate.[24–27] Microspheres of DL-glutamic acid (Glu) with hierarchical structures were generated *via* a PILP phase in the presence of the oppositely charged polyelectrolyte poly(ethyleneimine) (PEI),[28] and different water/alcohol solvent mixtures could be used to optimise the reaction products. A detailed investigation of the complex phase behaviour shown by this quaternary system (Glu, PEI, water, ethanol) revealed that the existence region of the liquid precursor phase is limited to only a small area in the phase diagram, while stable coacervates are formed over a wide range of reactant concentrations.[29] It has also been suggested that calcium phosphate PILP phases can be formed in the presence of PAsp. Here, infiltration of the mineral into a collagen matrix was taken as evidence for this mechanism, rather than the observation of droplets or expanded regions of thin films.[24]

Therefore, while liquid-like precursor phases are undoubtedly of great value to bio-inspired materials synthesis (for example in the fabrication of patterned films[30]), it remains an open question as to whether a "PILP" mechanism actually operates during the *in vivo* formation of biominerals. In this paper, we explore whether it is feasible that acidic polypeptides with compositions comparable to those isolated from calcium carbonate biominerals could generate PILP-type phases of $CaCO_3$. Taking a systematic approach, we here use random copoly(amino acid)s containing aspartic acid and serine residues in a heterogeneous distribution along the polymer chain. The proportion of charged aspartate monomers is varied between 20 and 80%. The biopolymers associated with calcium carbonate biominerals are typically highly acidic, and often contain up to 50% negatively charged amino acid residues.[31] Indeed, Caspartin, a water-soluble protein extracted from calcitic prisms of the mussel *Pinna noblis*, contains 69% aspartic acid (Asp) and 8% glutamic acid (Glu).[32]

Our study investigates how the composition of our random copoly(aspartic acid and serine amino acids) determines their effect on calcium carbonate precipitation, and in particular whether they induce the formation of a mineral precursor phase with liquid-like characteristics. Our investigations focus on the formation of fingerprint morphologies such as calcium carbonate thin films and fibres, where the occurrence of such unusual structures, in combination with the ability to infiltrate into small volumes, such as the pores in track etch membranes, can be taken as a strong indicator for the formation of a polymer-induced liquid precursor (PILP) phase. In order to appropriately consider the biological environment, where calcium carbonate based tissues often form in a Mg-rich environment, we also investigate how addition of this divalent ion affects PILP formation. This study offers new insight into the relationship between polypeptide composition and phase separation during mineralisation, and provides the basis for discussion of the relevance of PILP mechanisms to biomineralisation processes.

2 Experimental

Calcium carbonate was precipitated in the presence of synthesised random copoly(amino acid)s with target compositions of Asp20%–Ser80%, Asp50%–Ser50%, Asp80%–Ser20%, with the aim of investigating the relationship between the composition of the polypeptide and its ability to support the formation of a polymer induced liquid precursor (PILP) phase. The generation of a PILP phase was estimated from the morphologies of the product calcium carbonate precipitates, the ability of the mineral to infiltrate small volumes (here, the pores of track etched membranes), and from analysis of polymer/Ca^{2+} solutions using dynamic light scattering.

2.1 Chemicals

Fmoc-L-Asp(tBu)-OH was purchased from IRIS Biotech, GmbH, Germany, while Fmoc-L-Ser-OH, $CaCl_2 \cdot 2H_2O$, $MgCl_2 \cdot 6H_2O$ and $(NH_4)_2CO_3$ were obtained from Sigma-Aldrich. All reagents were used without further purification. Deionized Millipore water was used in the preparation of the crystallisation solutions.

2.2 Synthesis and characterization of the acidic polypeptides

The desired random polypeptide libraries (Asp20%–Ser80%, Asp50%–Ser50%, Asp80%–Ser20% of length 24) were prepared using the solid phase synthesis method. The scale of the synthesis was 0.1 mmol, following the Fmoc strategy and using standard Fmoc-derivatized amino acids. Briefly, the synthesis was performed on a fully automated parallel peptide synthesizer Syro I (Multisyntech, Witten, Germany) using rink-Amide resin (Iris biotech, substitution 0.53 mmol g^{-1}) as a solid support. The desired mix of amino acid solutions in DMF (LiCl, 1 g L^{-1}) was prepared and activation of the amino acids was achieved using HCTU–DIEA (1 : 2). Fmoc deprotection was carried out using a 40% (v/v) piperidine solution in dimethylformamide. All couplings were performed for 45 min at room temperature, while deprotections were carried out for 3 min in 40% piperidine, followed by 12 min in 20% piperidine. A gentle flow of N$_2$ was maintained throughout the synthesis. The peptides were then cleaved from the resin using a trifluoroacetic acid (TFA) : triisopropylsilane (TIS):H$_2$O (95 : 2.5 : 2.5, v/v/v) mixture for 3 h. The resulting solutions were precipitated in cold diethyl ether (dry) and centrifuged at 400 rpm for 10 min. The pellets were thoroughly washed 3–4 times in cold diethyl ether (dry) and were vacuum-dried to obtain powdered co-poly(amino acid)s, which were employed in the crystallisation experiments.

The random copoly(amino acid)s were characterised using mass spectrometry (MS) and infra-red (IR) spectroscopy. Mass spectroscopy was performed by direct injection of 2 μl of a 2 μM solution of a polypeptide in water/acetonitrile (1/1, v/v) and 0.1% formic acid into a mass spectrometer (Thermo Finnigan LTQ Orbitrap) equipped with an electrospray ion source in positive mode (source voltage 3.5 kV, gas flow 10, capillary temperature 275 °C) with resolution $R = 60.000$ at $m/z = 400$ (mass range = 150–2000 or 150–4000). Dioctylpthalate ($m/z = 391.28428$) was used as a "lock mass". IR was carried out on solid samples using a Perkin Elmer ATR-IR spectrometer.

The phase separation behaviour of the different polypeptides in the presence of Ca^{2+} cations was investigated using Dynamic light scattering (DLS). A Malvern Zetasizer Nano ZS (Malvern Instruments, UK) was used to determine the sizes of the peptide aggregates in aqueous solution (pH = 10, adjusted with NaOH). Typically, 10–20 repetitive measurements were recorded for each specimen at an internal temperature of 25 °C (after 2 min equilibration time). The solutions were filtered with a 450 nm syringe filter system prior to measurements.

2.3 Precipitation of calcium carbonate in the presence of acidic polypeptides

2.3.1 Bulk crystallization. Calcium carbonate was precipitated in the presence of random copoly(amino acid)s using the ammonium carbonate vapour diffusion method. Reactant solutions containing CaCl$_2$·2H$_2$O (10 mM) and varying concentrations of Asp20%–Ser80%, Asp50%–Ser50% or Asp80%–Ser20% (5 μg mL^{-1}, 10 μg mL^{-1}, 50 μg mL^{-1}, 100 μg mL^{-1}) were prepared in Petri dishes (10 mL total solution volume). In order to facilitate film formation, MgCl$_2$·6H$_2$O (30 mM) was added to some solutions.

Glass cover slides, which had been cleaned in Piranha solution (75% H$_2$SO$_4$ and 25% H$_2$O$_2$), were used as substrates for the precipitation of the mineral. The glass slides were placed upright in the Petri dishes containing the reaction solutions, which were then sealed with Parafilm that was pierced with 5 needle holes. Two additional Petri dishes, each filled with 5 g ammonium carbonate, were also covered with Parafilm (punched with 5 needle holes) and were placed at the bottom of a desiccator to generate a slow release of CO$_2$ into the sealed chamber. After the crystallisation reaction had been allowed to proceed for reaction times ranging from 2 h to 3 days, the glass substrates were removed from the solutions and were rinsed first with Millipore water and then ethanol, before allowing them to dry at room temperature.

2.3.2 Infiltration into the pores of track etched membranes. Calcium carbonate was precipitated in the pores of track-etched (TE) membranes (pore diameter 50 nm, Isopore, Millipore) according to previously published protocols.[16,33] Briefly, the membranes were plasma-cleaned and were then placed in glass vials containing aqueous solutions of CaCl$_2$·2H$_2$O (10 mM) mixed with one of the copolypeptides Asp20%–Ser80%, Asp50%–Ser50% or Asp80%–Ser20% (50 μg mL^{-1}). The reactant solutions were then sealed and left overnight to enable a complete filling of the pores. Subsequently, the glass vials were placed in a desiccator and exposed to ammonium carbonate vapor. The reactions were allowed to proceed for 24 h, after which time the membranes were removed from the solutions and were rinsed with ethanol. The majority of crystalline material attached to the surfaces of the membranes was then scraped off using a glass cover slide and the membranes were then transferred into microcentrifuge tubes (Eppendorf) where they were dissolved in dichloromethane (DCM). After centrifugation, the solvent was exchanged for fresh and the dissolution/centrifugation cycle was repeated two more times. The precipitates were then washed with methanol and ethanol, and the intra-membrane particles isolated in this way were finally re-dispersed in ethanol and pipetted onto glass slides or TEM grids for further characterisation.

2.4 Characterization of the calcium carbonate precipitates

The calcium carbonate precipitates were characterised using a range of techniques including light microscopy, scanning electron microscopy (SEM), transmission electron microscopy (TEM) and Raman microscopy. Light microscopy and polarized optical microscopy (POM) were performed using a Nikon Eclipse LV 100 microscope operated in the reflected light mode, giving information on the overall morphologies and crystallinity of the calcium carbonate precipitates. The morphologies and surface textures of the mineral films and crystals were also further investigated with SEM using a LEO 1530 Gemini FEG-SEM instrument operated at an acceleration voltage of 3 kV. All samples were mounted on SEM stubs with adhesive conducting pads and coated with a Pt/Pd alloy. Raman microscopy using a Renishaw Raman 2000 System instrument was performed for polymorph identification, and 10–20 randomly selected crystals were typically analysed per sample. Nanorods precipitated in the pores of track etch membranes were further supported on lacy carbon grids and investigated by TEM. Electron microscopy and diffraction were carried out on a Tecnai TF20 FEGTEM instrument equipped with a Gatan Orius SC600A CCD camera operated at 200 kV.

3 Results

The influence of three random copoly(amino acid)s of compositions Asp20%–Ser80%, Asp50%–Ser50% and Asp80%–Ser20% on the precipitation of calcium carbonate was studied with the purpose of investigating the relationship between the proportion of acidic groups in the polypeptide and its ability to support the formation of a polymer induced liquid precursor (PILP) mineral phase. The copoly(amino acid)s were prepared with a total of 24 amino acids and their composition provided a large spread in Asp contents (Fig. 1A). While the serine side groups represent neutral residues on the peptide chain, the aspartic acid residues (pK_a 3.9) are completely deprotonated at pH 9.5 – the pH where calcium carbonate precipitation takes place in the ammonium carbonate diffusion technique – and therefore exhibit a negative charge. Infra-red spectra of the copoly(amino acid)s (Fig. 1B) revealed an increase in the absorptions at 1200 cm^{-1} and 1650 cm^{-1} when the aspartic acid content was raised from 20 to 80%. The vibration bands appearing at these spectral positions can be assigned to the stretching modes of carboxylic acid C–O and C=O bonds, respectively.

Calcium carbonate precipitation in bulk and at the air–water interface

To study the general effects of the selected random copoly(amino acid)s on calcium carbonate precipitation, each polymer was mixed with an aqueous solution of calcium chloride to give final concentrations of [Ca^{2+}] = 10 mM and [peptide] = 50 µg mL^{-1} and these solutions were then exposed to ammonium carbonate vapor in a sealed desiccator. On completion of the reaction, the products deposited on glass

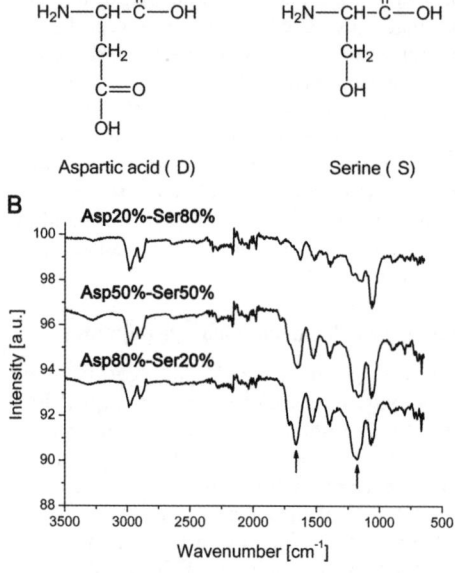

Fig. 1 Random copoly(amino acid)s with different compositions. (A) The polypeptides are composed of aspartic acid (Asp, D) and serine (Ser, S) in varying proportions. (B) IR spectra of Asp20%–Ser80% ($D_{20}S_{80}$), Asp50%–Ser50% ($D_{50}S_{50}$) and Asp80%–Ser20% ($D_{80}S_{20}$). With increasing aspartic acid content, stronger absorption is observed for the vibration bands appearing at ~1200 cm^{-1} and 1650 cm^{-1} (black arrows).

substrates, which had been placed upright in the reactant solutions, were characterized by polarized optical microscopy and scanning electron microscopy.

Representative images of the $CaCO_3$ precipitates formed are shown in Fig. 2. As a general trend, film-like structures and micrometer-sized crystalline patches were

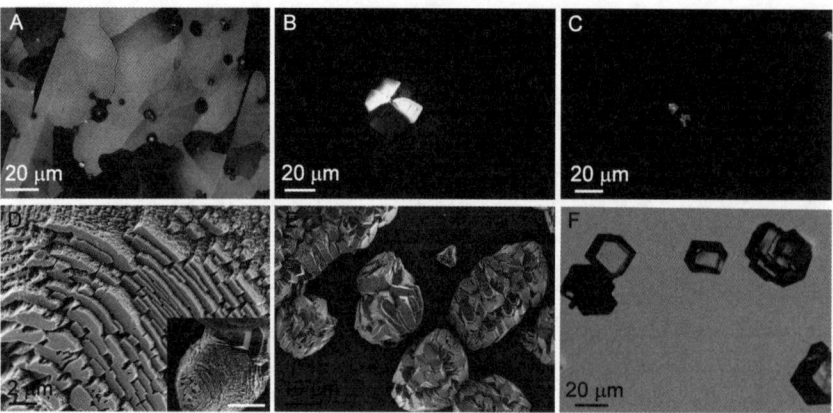

Fig. 2 Bulk crystallization of calcium carbonate in the presence of different random copoly(amino acid)s (additive concentrations of 50 μg mL^{-1}). (A–C) Polarized light microscopy images of films formed at the air water interface where they were precipitated in the presence of (A) Asp80%–S20% (image recorded after annealing at 300 °C for 2 h), (B) Asp50%–Ser50% and (C) Asp20%–Ser80%. Electron micrographs of crystals formed in bulk in the presence of Asp80%–Ser20% and Asp50%–Ser50% are shown in (D) and (E), respectively. The inset in (D) gives an overview of the overall morphology of the crystal (scale bar = 10 μm). (F) Light microscopy image of rhombohedral calcite crystals grown in the presence of Asp20%–Ser80%.

usually observed in regions on the glass substrates which had been in contact with the air–water interface (Fig. 2A–C), while blocky calcium carbonate crystals formed in the bulk solution (Fig. 2D–F). Looking at the contrasting behaviour of the different polymers, Asp80%–Ser20% induced the formation of expanded calcium carbonate thin films at the air water interface after 5 h reaction time (Fig. 2A). However, the films obtained only became visible between crossed polarizers upon annealing at 300 °C for 2 h. This indicates that the calcium carbonate films formed in the presence of Asp80%–Ser20% had been amorphous before crystallization was induced at elevated temperatures. The resulting crystalline films were composed of both single-crystalline and polycrystalline patches and were reminiscent of films prepared in the presence of poly(aspartic acid).[12] Investigation of the crystals formed in the bulk solution in the same reaction solution using Raman microscopy revealed that they were calcite, while SEM showed that they were elongated, with rounded, roughened surfaces (Fig. 2D).

In contrast, when the copoly(amino acid)s Asp50%–Ser50% and Asp20%–Ser80% were added to the reaction solutions, individual patches rather than expanded fused films were found at the air–water interface (Fig. 2B and C). Asp50%–Ser50% induced the precipitation of relatively large, sheet-like patches, which were up to 20 μm in diameter but did not fuse into a continuous film. In the case of Asp20%–Ser80%, very few crystalline patches formed, and those which did were extremely small. Annealing failed to increase the quantity of crystalline material, which indicates that the precipitates were already fully crystalline. Investigation of the particles formed in bulk solution showed that those precipitated in the presence of Asp50%–Ser50% (Fig. 2E) were calcite (as confirmed by Raman microscopy) with roundish morphologies. Their surface features were coarser and more facetted than those observed for Asp80%–Ser20%. Considering then the effect of Asp20%–Ser80%, the bulk crystals took the form of partially inter-grown calcite rhombohedra with smooth surfaces (Fig. 2F). Together, these results show that an increase in the charged aspartic acid residues on the peptide chain leads to greater modifications of the morphologies of the calcium carbonate precipitates. The Asp-rich copoly(amino acid) Asp80%–Ser20% also appeared to stabilise an amorphous calcium carbonate phase for significant periods of time.

The structures of the crystalline films and patches were also further investigated using SEM (Fig. 3). The expanded films formed by the addition of Asp80%–Ser20% were rather smooth (Fig. 3A), and only at high magnifications (Fig. 3B, inset) could porosity in the order of tens of nanometers be observed. The edges of the film fragments were rough and rounded, non-facetted mineral granules measuring approximately 100 nm in size (white arrows) and granule aggregates could be seen in these regions. These observations are consistent with a model put forward by Gower et al., which suggested that crystalline films formed in the presence of poly(aspartic acid) result from PILP droplet coalescence and subsequent solidification.[12]

In a sample produced with the Asp-deficient polypeptide Asp20%–Ser80% a crystalline patch deposited in the region covered by the air/water interface was found to be overgrown with co-oriented crystals (Fig. 3C). This observation points to the underlying patch being single-crystal in nature, such that it directs the growth of newly formed crystals along just one specific crystallographic axis.

Further insight into the efficiency of the three copoly(amino acid)s as crystal growth modifiers was gained by varying the concentrations of the organic additives while keeping the calcium concentration constant at 10 mM. The peptide containing the highest proportion of aspartic acid residues (Asp80%–Ser20%) still induced the formation of thin, partially crystalline patches of calcium carbonate at concentrations as low as 5 μg mL^{-1} (Fig. 4A). However, it is notable that these patches are polycrystalline in structure, as compared with the single crystal domains produced at higher concentrations of this peptide, and shown in Fig. 2A. the higher concentration of polymer therefore appears to sufficiently stabilise the amorphous precursor phase such that transformation to single crystals can occur. Similarly,

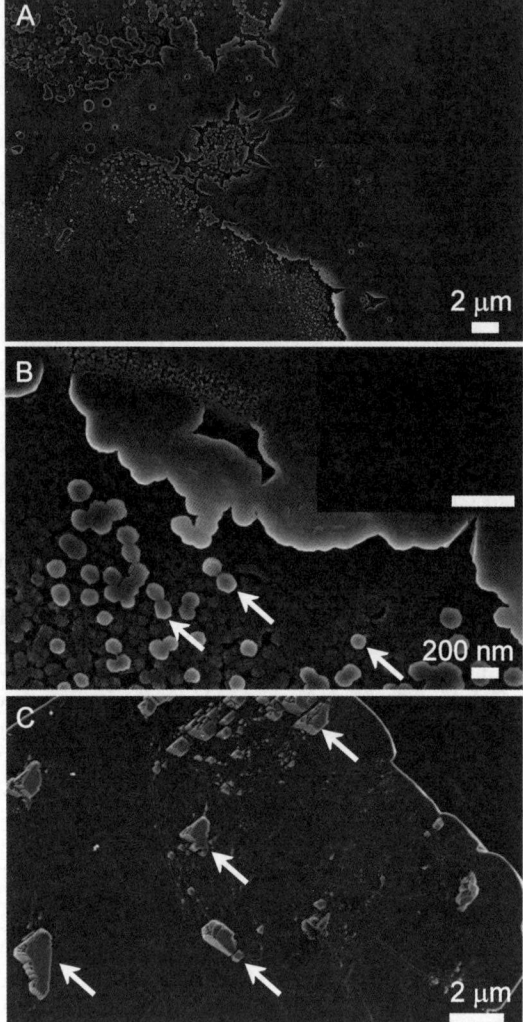

Fig. 3 Electron micrographs of films and patches. Expanded calcium carbonate films with rough edges are observed in a sample prepared in the presence of Asp80%–Ser20% (A) A higher magnification of the edge-region is presented in panel (B). The white arrows point to non-facetted granules of mineral. The inset shows the porosity of the film (scale bar = 200 nm). (C) Co-oriented crystals (indicated by white arrows) grow out of the surface of a crystalline patch found at the air–water interface in a calcium carbonate sample precipitated in the presence of Asp20%–Ser80%.

while the numbers and sizes of the crystalline patches deposited at the air–water interface also increased when the concentration of Asp50%–Ser50% and Asp20%–Ser80% was increased to 100 µg mL^{-1}, (Fig. 4B and C), expanded film-like structures were never observed. Instead, the calcite crystals formed in the bulk solution simply appeared more spherical as the concentration of the additive Asp50%–Ser50% was increased from 50 µg mL^{-1} to 100 µg mL^{-1} (Fig. 4B, inset). These results clearly demonstrate that the effect of the poly(amino acid)s on calcium carbonate precipitation is not simply due to the absolute number of functional aspartic acids units present in the reactant solution, but that the composition of the peptide molecules is also crucial.

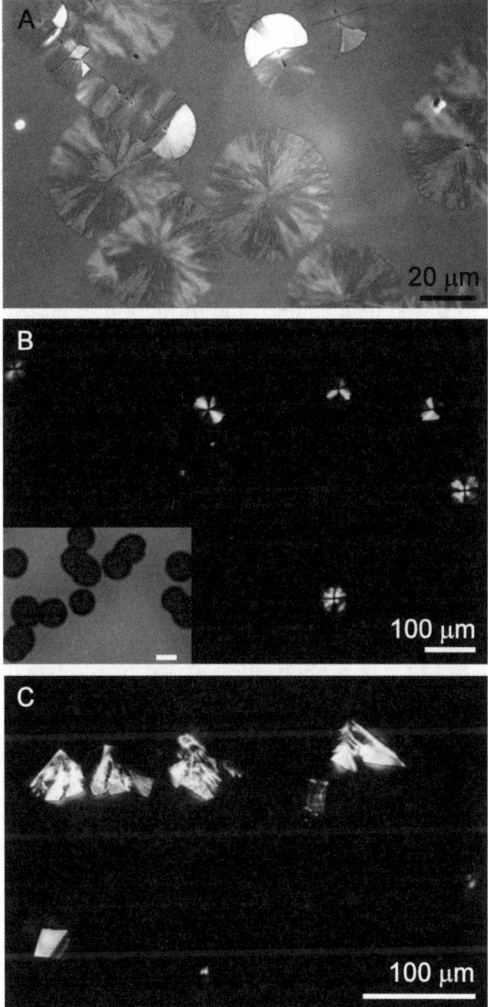

Fig. 4 Variation of the peptide concentration. (A) Thin crystalline patches are formed at the air–water interface in a sample prepared in the presence of a low amount (5 μg mL^{-1}) of Asp80%–Ser20%. (B) A higher concentration (100 μg mL^{-1}) of Asp50%–Ser50% leads to an increased number of individual crystalline patches at the air water-interface, but does not induce the formation of expanded films. The inset shows spherical particles, which were formed at the bottom of the glass substrate (scale bar = 100 μm). Relatively thick crystalline patches were observed at the air–water interface in a sample precipitated with 100 μg mL^{-1} Asp20%–Ser80% (C).

The effect of magnesium on CaCO$_3$ precipitation in the presence of the copoly(amino acid)s

Biogenic calcium carbonate produced by marine organisms is often formed in the presence of high amounts of magnesium ions, which are known to play an important role in the stabilisation of biogenic amorphous calcium carbonate (ACC).[34] Magnesium also has a similar effect on synthetic ACC,[35] where it inhibits the amorphous to crystalline transition.[36] Indeed, Cheng et al. have demonstrated that addition of magnesium ions to CaCO$_3$ PILP (prepared using poly(aspartic acid)) can not only result in the formation of expanded films of calcium carbonate,[37] but that the amount of polymer required to induce the formation of thin films is reduced by a factor of 10. The development of continuous sheets of calcium carbonate in the presence of Mg^{2+} has also been reported for a polymer-free system,[35] but in this case much higher concentrations of reactants ($[Ca^{2+}]$ = 60 mM) were used and thin featureless films were only obtained at a very high $Mg^{2+}:Ca^{2+}$ ratio of 10 : 1. In the study presented here, we investigated whether addition of Mg^{2+} ions can lead to the formation of thin films of calcium carbonate, even in the presence of the polypeptides which did not promote effective film formation in its absence. The $Mg^{2+}:Ca^{2+}$ ratio used in these experiments was 3 : 1.

As described above, the polypeptide Asp80%–Ser20% could induce the deposition of calcium carbonate films even in the absence of magnesium. However, addition of Mg^{2+} ions to this system enabled the formation of large expanded calcium carbonate films in the presence of very low concentrations of polymer ([Asp80%–Ser20%] = 10 μg mL^{-1}). When the sample was removed from the crystallization solution after 1 day, the edge of a film was visible at the air/water interface after drying (Fig. 5A). Under crossed polarizers, however, the area covered by the film was dark, which is indicative of an amorphous structure (Fig. 5B). Crystallinity (appearing as bright spots) was initially only observed for small spherical particles (~10 μm in diameter) which were deposited on the surface of the film and directly on the glass substrate. These crystalline particles were similar in appearance to the solidified PILP droplets that Cheng et al. observed on a magnesium-bearing calcium carbonate film produced in the presence of 2 μg mL^{-1} poly(aspartic acid) at a $Mg^{2+}:Ca^{2+}$ ratio of 4 : 1.[37] Raman microscopy confirmed that the spherical particles were calcite.

Annealing of these 1 day amorphous films at 300 °C resulted in their crystallisation and the formation of polycrystalline domains (Fig. 5C). Fine cracks were observed in these heated films due to the rapid release of water from the highly hydrated precursor phase. The same amorphous sample was also left in air for 7 days before being examined by polarized light microscopy. Even after this time no crystalline film was observed, and only after annealing at 300 °C for 30 min did a polycrystalline calcium carbonate film covering the glass substrate become apparent. Films formed after 2 days incubation in the reactant solution were also examined, but again no crystalline film was observed directly after drying. Leaving this sample in air for 7 days did, however, result in crystallisation, and fused polycrystalline patches measuring approximately 100 μm in diameter were observed in POM (Fig. 5D). Our investigations therefore confirm that magnesium ions can exert a stabilizing effect on ACC and can delay the crystallisation of calcium carbonate.

The effect of magnesium ions on the precipitation of calcium carbonate in the presence of Asp50%–Ser50% and Asp20%–Ser80% was also investigated. Notably, the polypeptide Asp50%–Ser50% (20 μg mL^{-1}), which had failed to support effective film-formation in Mg-free solutions, also induced the formation of expanded precursor films when Mg^{2+} was present in sufficiently high quantities. These became crystalline upon annealing. Addition of magnesium never induced the formation of such expanded films in the presence of the Asp-poor copoly(amino acid) Asp20%–Ser80% (20 μg mL^{-1}), and only an increased number of small crystalline patches at the air/water interface, accompanied by dumbbell-shaped particles, was observed.

In conclusion, Mg^{2+} ions have a strong effect on the precipitation of calcium carbonate from solutions containing Asp-rich copoly(amino acid)s. The presence of Mg^{2+} ions in the reaction solutions appears to render the precise acidic amino acid content in the peptide chain less crucial. At magnesium levels of $Mg^{2+}:Ca^{2+}$ = 3 : 1 – which are still lower than the $Mg^{2+}:Ca^{2+}$ ratio found in modern seas,[38] - expanded mineral films could be achieved with a polypeptide containing only 50% of the functional, Ca^{2+}-binding units (Asp50%–Ser50%). However, a higher concentration of this peptide was needed (20 μg mL^{-1}) to create the same effect as observed with 10 μg mL^{-1} Asp80%–Ser20%.

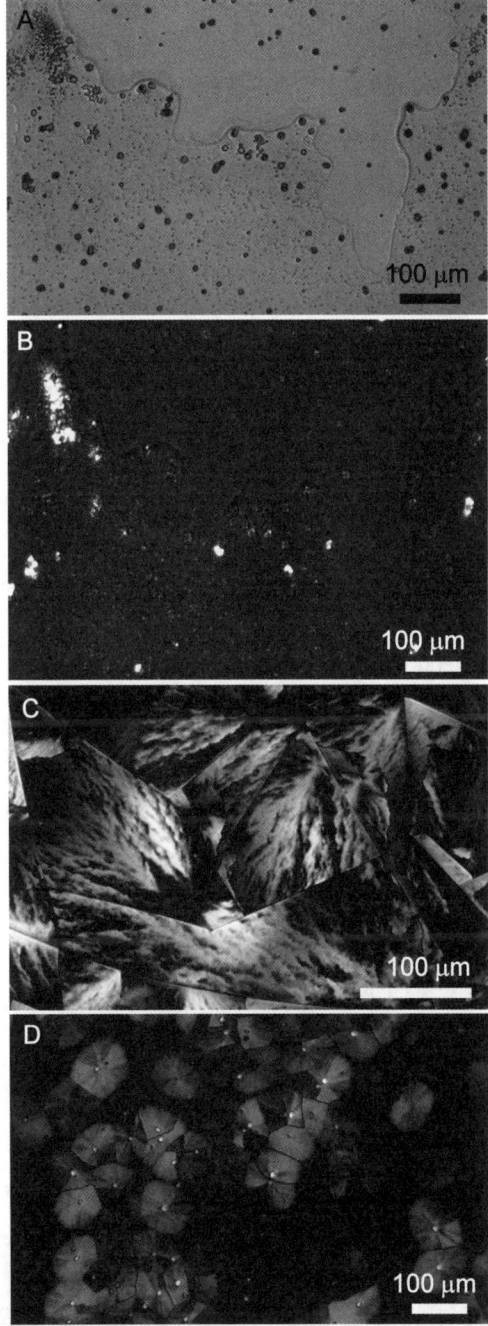

Fig. 5 The effect of Mg^{2+} addition. (A) An expanded film-like structure accompanied by a number of small, roundish particles were visible in a calcium carbonate sample precipitated in the presence of Mg^{2+} (30 mM) and Asp80%–Ser20% (50 μg mL^{-1}) after 1 day reaction time. Under crossed polarizers (B) the particles appeared bright, whereas the film did not show any signs of crystallinity. After annealing at 300 °C for 2 h an expanded film consisting of polycrystalline domains became visible in POM (C). Crystalline patches were also seen in a sample removed from the reaction after 2 days, dried and left in air for 1 week (D).

Precipitation in confinement

An important feature of the polymer-induced liquid precursor phase is its ability to infiltrate into small volumes. Kim *et al.* recently precipitated calcium carbonate *via* a calcium carbonate PILP phase in polycarbonate track-etch membranes with rod-shaped pores measuring 200 nm or 50 nm in diameter.[16] The liquid-like characteristics of the PILP phase (in this case generated with poly(acrylic acid)) enabled a filling of this confined reaction space by capillary action. Subsequent crystallization of the precursor yielded single crystal calcite nanorods with extremely high aspect ratios of up to 100.

Here, we use the infiltration into small membrane pores as an indicator for the formation of a liquid precursor phase. Plasma-treated polycarbonate membranes with 50 nm pores were immersed in aqueous solutions containing $CaCl_2$ (10 mM) and one of the copoly(amino acid)s ([peptide] = 50 µg mL^{-1}). These solutions were then exposed to ammonium carbonate vapour for 1 day. After dissolution of the membrane, the crystalline material isolated was examined using scanning electron microscopy.

No rod-shaped, intra-membrane particles were isolated either in control experiments carried out in the absence of polymer, or when the $CaCO_3$ was precipitated in the presence of Asp20%–Ser80%. In contrast, when the two polypeptides with aspartic acid contents of 50% and higher (Asp80%–Ser20% or Asp50%–Ser50%) were added to the reaction solution, mineral nanorods up to 10 µm in length were observed (Fig. 6A and B). This provides good evidence that in these cases, the crystallization proceeds *via* a liquid precursor phase. Substantial filling of the membrane pores could be achieved, particularly with the peptide comprising 80% charged residues. The larger mineral particles and polymeric material seen in Fig. 6 are due to incomplete dissolution of the membrane.

The nanorods were further examined by TEM (Fig. 7) revealing a dense mineralization of the rod-shaped material precipitated in the presence of the most acidic peptide Asp80%–Ser20% (Fig. 7A and B). Moreover, an electron diffraction pattern characteristic of a single-crystal (Fig. 7C) and continuous lattice fringes (Fig. 7D) were observed for this sample. When Asp50%–Ser50% was added to the reactant solution, the nanorods formed within the membrane pores appeared to be composed of small granules (Fig. 7E and F) and showed diffraction corresponding to a polycrystal (inset in Fig. 7F).

The phase behaviour of the polypeptide additives

In the well-characterized PAsp/$CaCO_3$ system, the highly negatively charged poly-(aspartate) is believed to sequester Ca^{2+} ions from the surrounding solution. This leads to a liquid–liquid phase separation yielding isotropic droplets (∼2–4 µm) of a hydrated Ca^{2+}/PAsp/CO_3^{2-} complex when carbonate ions are present.[12] In support

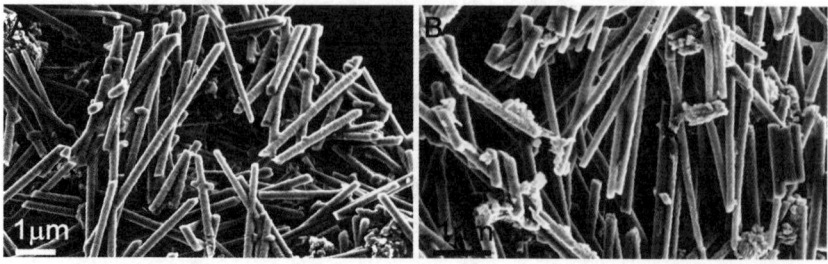

Fig. 6 Infiltration into track etch membrane pores. SEM micrographs of calcium carbonate rods resembling the shape of the template pores are shown. The samples were produced in the presence of Asp80%–Ser20% (A) and Asp50%–Ser50% (B), respectively.

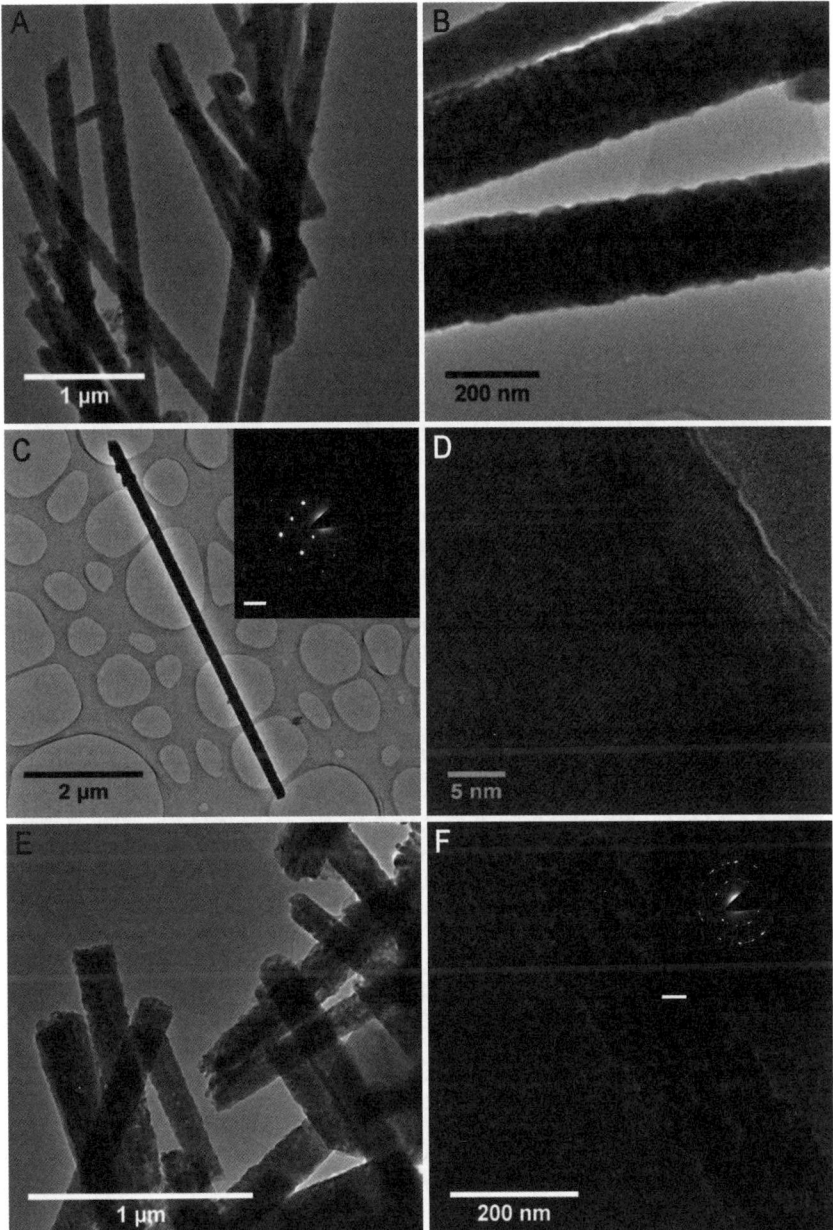

Fig. 7 Transmission electron microscopy of nanorods isolated from track etch membranes (pore diameter = 50 nm). (A–D) Calcium carbonate nanorods prepared in the presence of Asp80%–Ser20%. The rods appear densely mineralized (A and B). Selected area diffraction (inset in C) recorded for a representative rod (C) yields diffraction spots indicative of a single crystal (scalebar inset = 3 nm^{-1}). Continuous lattice fringes are observed over a wide area (D). Those nanorods prepared in the presence of Asp50%–Ser50% appear more granular (E and F) and give rise to an electron diffraction pattern corresponding to a polycrystal (inset in (F), scale bar inset = 3 nm^{-1}).

of this concept, Gower and Odom have reported an increased turbidity developing in the reactant solution after the critical induction time (>2 h after the ammonium carbonate diffusion was started).[12]

Poly(acrylic acid), which provides another example of a highly carboxylated polymer, is also known to induce the formation of a liquid-like precursor during calcium carbonate precipitation,[10,17] and phase separation in the presence of Ca^{2+} ions is well described for this negatively charged polyelectrolyte.[39,40] Recently, calcium carbonate films and fibres were also observed on precipitation of this mineral in the presence of the positively charged polymer poly(allylamine hydrochloride) (PAH). This was again attributed to a microphase separation phenomenon rather than to a specific interaction between the polymer and the growing crystal,[23] where the polyelectrolyte now underwent microphase separation in the presence of oppositely charged carbonate ions.[41] Hydrated Ca^{2+}/PAH/CO_3^{2-} droplets were formed, which grew, coalesced and deposited on the substrate.

In order to obtain information on whether the copoly(amino acid)s investigated here could induce a phase separation, dynamic light scattering (DLS) was used to determine whether aggregates form in aqueous solutions of the copoly(amino acid)s in the presence and absence of calcium cations. In order to collect a sufficiently intense signal to allow reliable fitting of the data, it was necessary to prepare highly concentrated peptide solutions ([peptide] = 100 µg mL^{-1}). These concentrations are 2 times higher than those chosen for the bulk crystallization experiments, so a concentration effect on the dimensions of the agglomerates cannot be excluded. Despite this limitation, the results of the light scattering study provide valuable information about the influence of the counter ion (here Ca^{2+}) on the behaviour of the polymers in solution.

All three copoly(amino acid)s formed agglomerates in the submicrometer size range (Fig. 8). When Ca^{2+} ions were added to an aqueous solution of Asp80%–Ser20%, the average size of the polymer aggregates increased from 70 nm to ~250 nm (Fig. 8A). The same trend was observed for both of the other polymers. Asp50%–Ser50% formed agglomerates with sizes between 400 and 1000 nm in the presence of calcium, as compared with 20 nm in its absence (Fig. 8B). Similarly, the peptide with the lowest amount of aspartic acid side groups, Asp20%–Ser80%, showed a substantial increase in the aggregate size upon addition of Ca^{2+} ions (~60 nm in the absence compared to ~1000 nm in the presence of calcium, Fig. 8C). Since a 450 nm syringe filter was used for transferring the peptide solutions into the cuvettes used for measuring, it can be assumed that the larger aggregates must have developed by aggregation of smaller droplets.

Samples were also drawn from the reactant solutions ([Ca^{2+}] = 10 mM, [peptide] = 50 µg mL^{-1}) after they had been exposed to ammonium carbonate vapour for 2 h and investigated with optical microscopy. At this time, the majority of the precipitates formed in the presence of Asp20%–Ser80% had already crystallized to calcite rhombohedra. In contrast, very little crystalline material was observed in the Asp80%–Ser20%/calcium carbonate system, although considerable quantities of non-crystalline precipitates, which were similar in appearance to the PILP droplets reported by Gower and Odom for poly(aspartic acid)/calcium carbonate after short reaction times, were observed.[12] This provides a further indication that the driving force for rapid crystallization was reduced in the presence of the Asp-enriched peptide.

4 Discussion

These experiments investigate whether it is feasible that acidic polypeptides, with compositions comparable to those of the biogenic macromolecules associated with $CaCO_3$ biominerals, can induce the formation of a so-called "polymer-induced liquid precursor" (PILP) phase. Indeed, the biomacromolecules extracted from biogenic calcite are characteristically highly acidic and often comprise at least 50 mol% acidic amino acids (aspartic acid and glutamic acid).[31,32] Our results show that all three copoly(amino acid)s considered can influence the precipitation of calcium carbonate under certain reaction conditions but that the magnitude of

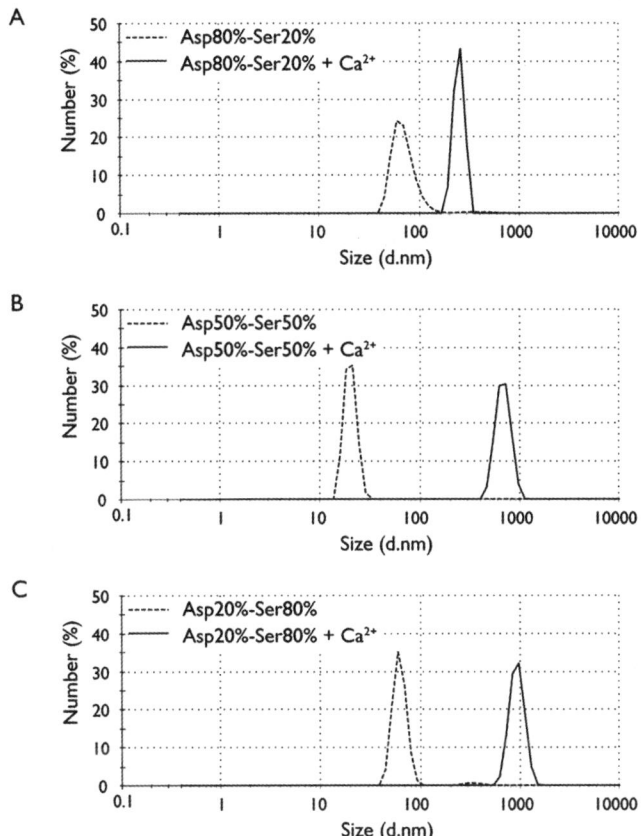

Fig. 8 Dynamic light scattering of copoly(amino acid)s. Curves plotting the size distribution of the peptide agglomerates by intensity are presented for Asp80%–S20% (A), Asp50%–Ser50% (B) and Asp20%–Ser80% (C).

the effects observed, and whether the resulting structures were clearly indicative of a liquid precursor phase (fingerprint morphologies), depends strongly on their compositions. While the most acidic polypeptide Asp80%–Ser20%, which is enriched in aspartate residues, induced the formation of expanded film-like structures even at relatively low concentrations, Asp50%–Ser50% and Asp20%–Ser80% merely led to the precipitation of a few sheet-like patches at the air/water interface. It should be noted however, that infiltration into the small pores of track etched membranes was achieved even in the absence of magnesium ions, when calcium carbonate was precipitated in the presence of the synthetic peptide containing 50% aspartate residues.

The effect of Mg^{2+} ions on the ability of the polypeptides to generate a PILP phase was also studied. The presence of significant Mg concentrations, as occurs in biological environments, can have a significant effect on the calcification process, as indicated by the ultimate incorporation of Mg in many calcite biominerals. Addition of excess magnesium ions to the reaction solutions here facilitated the deposition of expanded precursor films, even when only half of the side groups of the peptide chain were charged (Asp50%–Ser50%).

It therefore seems possible that those biopolymers with the highest reported densities of acidic residues may undergo microphase separation. Further, this process is aided by magnesium ions, such that their presence during biogenic calcification could facilitate the phase separation of macromolecules that carry 50% negatively charged side groups. Biomacromolecules comprising only around 20% acidic residues are also frequently classified as "highly acidic", but in the light of the above results it appears unlikely that these are sufficiently acidic to direct calcium carbonate precipitation *via* a PILP process. Only very minor effects on the morphology of the deposited mineral were observed when Asp20%–Ser80% was added to the reactant solution.

In general, phase separation of macromolecules is accompanied by conformational changes, and therefore requires a high degree of flexibility. This could be unfavourable if a specific stereochemical interaction is required between the highly acidic protein and certain crystal faces. In biology, the aggregation of proteins and protein fragments is often associated with disease and physiological anomalies.[42–44] Indeed, it is known to play a key role in diseases such as Alzheimers and Parkinsons, in which trivalent metal ions have been implicated.[44] This is supported by recent studies which have demonstrated that proteins can display re-entrant behaviour in the presence of trivalent metal counter ions.[44,45] The physiological function of many soluble proteins therefore appears to depend on their ability to retain their conformation and remain soluble under a wide range of solution conditions.

Running counter to this line of argument, recent studies have revealed that some proteins extracted from biogenic $CaCO_3$ (*e.g.* from the prismatic layer of mollusc shells) are functional even in the unfolded (disordered) state.[46] As reviewed by Evans, the terminal sequences in the polypeptides associated with the growth of the nacreous and the prismatic layer in mollusc shells commonly adopt an unfolded, structurally labile conformation in solution. Therefore, it was suggested that the stabilization of the internal protein structure is only achieved by the interaction with ion clusters or mineral surfaces.[47] This concept is in keeping with the observation that the aragonitic nacre layer in the marine bivalves *Mercenaria mercenaria* and *Crassostrea gigas* form *via* an amorphous precursor phase.[48] Unfolded proteins could have specifically evolved for the efficient adaptation to disordered surfaces and/or for the transport of ion clusters to such surfaces.[47]

It is also now quite well-established that mineralization of the diatom cell wall, a silica-based material with an intricate microstructure, is based on microphase separation process. Silaffins, which are highly phosphorylated, polyamine functionalized zwitterionic biomacromolecules, together with long chain polyamines, are proposed to play a key role in this micropatterning process.[49,50] In the presence of silicic acid these mocleules can undergo microphase separation, which leads to the precipitation of spherical microdroplets. Silica precipitation then occurs around these droplets giving rise to elaborate honeycomb structures.[49,51]

Although we have established here that the most acidic macromolecules associated with $CaCO_3$ biominerals are capable of inducing formation of a PILP phase, the question remains as to whether this mechanism could actually operate *in vivo*. The best characterised example of the formation of biogenic calcite *via* an amorphous precursor phase is undoubtedly that of sea urchin larval spicules.[52,53] Beniash *et al.* reported that the amorphous calcium carbonate (ACC), which acts as a precursor for spicule growth in sea urchin embryos, is present as well-defined vesicle-bound granules.[52] These granules are then transferred into the syncytium, the membrane-bound compartment in which the spicule forms. The membrane is tightly bound to the forming spicule and the amorphous to crystalline transition occurs in the absence of bulk liquid. It has also been recently observed that *in vivo* bone-lining cells concentrate a non-crystalline calcium phosphate phase within intracellular vesicles.[6]

Thus while some features of these processes – such as the ability to concentrate an amorphous mineral precursor phase within a confined volume – appear to closely reconcile with the experimental observations of PILP, others do not. Experimentally, PILP droplets are highly unstable and rapidly aggregate/coalesce either in solution or on a substrate. These characteristics therefore do not seem to correlate with the stability observed for the membrane-bound amorphous granules. *In vitro* studies have also shown that it is extremely difficult to scale-up PILP processes to give larger quantities of the product materials. Indeed, very dilute solutions are required to give the characteristic fingerprint morphologies. This was also observed in the in-depth study of the Glutamic acid/PEI system, where PILP was only formed within a tiny area of the phase diagram.[29]

In conclusion, our work clearly demonstrates that counter-ion induced phase separation of highly acidic proteins would appear to be entirely feasible, in the light of their chemical compositions and structures. However, it still remains to build a bridge between these observations and those made of calcification *via* amorphous precursor phases *in vivo*.

Acknowledgements

This work was supported by the Engineering and Physical Sciences Research Council [grant number EP/I001514/1] (AS and FM). This programme grant funds the Materials in Biology (MIB) consortium. We further thank the EPSRC for financial support *via* grant EP/H005374/1 (YYK and FM). H.Z. and A. K. acknowledge the support of the European Research Council (ERC) *via* an ERC starting grant.

References

1 P. Fratzl, O. Paris, K. Klaushofer and W. J. Landis, *J. Clin. Invest.*, 1996, **97**, 396–402.
2 J. Aizenberg, A. Tkachenko, S. Weiner, L. Addadi and G. Hendler, *Nature*, 2001, **412**, 819–822.
3 A. Berman, J. Hanson, L. Leiserowitz, T. F. Koetzle, S. Weiner and L. Addadi, *Science*, 1993, **259**, 776–779.
4 G. Falini, S. Albeck, S. Weiner and L. Addadi, *Science*, 1996, **271**, 67–69.
5 Y. Politi, T. Arad, E. Klein, S. Weiner and L. Addadi, *Science*, 2004, **306**, 1161–1164.
6 J. Mahamid, A. Sharir, D. Gur, E. Zelzer, L. Addadi and S. Weiner, *J. Struct. Biol.*, 2011, **174**, 527–535.
7 S. Weiner, I. Sagi and L. Addadi, *Science*, 2005, **309**, 1027–1028.
8 S. Raz, P. C. Hamilton, F. H. Wilt, S. Weiner and L. Addadi, *Adv. Funct. Mater.*, 2003, **13**, 480–486.
9 F. C. Meldrum and H. Coelfen, *Chem. Rev.*, 2008, **108**, 4332–4432.
10 L. B. Gower, *Chem. Rev.*, 2008, **108**, 4551–4627.
11 L. A. Gower and D. A. Tirrell, *J. Cryst. Growth*, 1998, **191**, 153–160.
12 L. B. Gower and D. J. Odom, *J. Cryst. Growth*, 2000, **210**, 719–734.
13 X. G. Cheng and L. B. Gower, *Biotechnol. Prog.*, 2006, **22**, 141–149.
14 N. Gehrke, N. Nassif, N. Pinna, M. Antonietti, H. S. Gupta and H. Colfen, *Chem. Mater.*, 2005, **17**, 6514–6516.
15 M. J. Olszta, E. P. Douglas and L. B. Gower, *Calcif. Tissue Int.*, 2003, **72**, 583–591.
16 Y.-Y. Kim, N. B. J. Hetherington, E. H. Noel, R. Kröger, J. M. Charnock, H. K. Christenson and F. C. Meldrum, *Angew. Chem., Int. Ed.*, 2011, **50**, 12572–12577.
17 X. R. Xu, J. T. Han and K. Cho, *Chem. Mater.*, 2004, **16**, 1740–1746.
18 N. Sommerdijk, E. N. M. van Leeuwen, M. R. J. Vos and J. A. Jansen, *CrystEngComm*, 2007, **9**, 1209–1214.
19 S. E. Wolf, J. Leiterer, V. Pipich, R. Barrea, F. Emrnerling and W. Tremel, *J. Am. Chem. Soc.*, 2011, **133**, 12642–12649.
20 A. Sugawara, T. Nishimura, Y. Yamamoto, H. Inoue, H. Nagasawa and T. Kato, *Angew. Chem., Int. Ed.*, 2006, **45**, 2876–2879.
21 L. J. Dai, X. G. Cheng and L. B. Gower, *Chem. Mater.*, 2008, **20**, 6917–6928.
22 M. E. Marsh, *Biochemistry*, 1989, **28**, 339–345.
23 B. Cantaert, Y.-Y. Kim, H. Ludwig, F. Nudelman, N. A. J. M. Sommerdijk and F. C. Meldrum, *Adv. Funct. Mater.*, 2012, **22**, 907–915.
24 M. J. Olszta, X. Cheng, S. S. Jee, R. Kumar, Y.-Y. Kim, M. J. Kaufman, E. P. Douglas and L. B. Gower, *Mater. Sci. Eng., R*, 2007, **58**, 77–116.
25 S. J. Homeijer, M. J. Olszta, R. A. Barrett and L. B. Gower, *J. Cryst. Growth*, 2008, **310**, 2938–2945.
26 S. J. Homeijer, R. A. Barrett and L. B. Gower, *Cryst. Growth Des.*, 2010, **10**, 1040–1052.
27 S. Wohlrab, H. Colfen and M. Antonietti, *Angew. Chem., Int. Ed.*, 2005, **44**, 4087–4092.

28 Y. Jiang, L. Gower, D. Volkmer and H. Coelfen, *Cryst. Growth Des.*, 2011, **11**, 3243–3249.
29 Y. Jiang, L. Gower, D. Volkmer and H. Coelfen, *Phys. Chem. Chem. Phys.*, 2012, **14**, 914–919.
30 Y.-Y. Kim, E. P. Douglas and L. B. Gower, *Langmuir*, 2007, **23**, 4862–4870.
31 S. Weiner, *Am. Zoologist*, 1984, **24**, 945–951.
32 F. Marin, R. Amons, N. Guichard, M. Stigter, A. Hecker, G. Luquet, P. Layrolle, G. Alcaraz, C. Riondet and P. Westbroek, *J. Biol. Chem.*, 2005, **280**, 33895–33908.
33 E. Loste, R. J. Park, J. Warren and F. C. Meldrum, *Adv. Funct. Mater.*, 2004, **14**, 1211–1220.
34 Y. Politi, D. R. Batchelor, P. Zaslansky, B. F. Chmelka, J. C. Weaver, I. Sagi, S. Weiner and L. Addadi, *Chem. Mater.*, 2010, **22**, 161–166.
35 E. Loste, R. M. Wilson, R. Seshadri and F. C. Meldrum, *J. Cryst. Growth*, 2003, **254**, 206–218.
36 K. J. Davis, P. M. Dove and J. J. De Yoreo, *Science*, 2000, **290**, 1134–1137.
37 X. G. Cheng, P. L. Varona, M. J. Olszta and L. B. Gower, *J. Cryst. Growth*, 2007, **307**, 395–404.
38 S. M. Stanley, *Palaeogeogr., Palaeoclimatol., Palaeoecol*, 2006, **232**, 214–236.
39 R. Schweins and K. Huber, *Eur. Phys. J. E: Soft Matter Biol. Phys.*, 2001, **5**, 117–126.
40 M. A. V. Axelos, M. M. Mestdagh and J. Francois, *Macromolecules*, 1994, **27**, 6594–6602.
41 H. Daiguji, E. Matsuoka and S. Muto, *Soft Matter*, 2010, **6**, 1892–1897.
42 F. Horkay, P. J. Basser, A. M. Hecht and E. Geissler, *Phys. Rev. Lett.*, 2008, **101**, 4.
43 R. Piazza, *Curr. Opin. Colloid Interface Sci.*, 2004, **8**, 515–522.
44 F. J. Zhang, S. Weggler, M. J. Ziller, L. Ianeselli, B. S. Heck, A. Hildebrandt, O. Kohlbacher, M. W. A. Skoda, R. M. J. Jacobs and F. Schreiber, *Proteins: Struct., Funct., Bioinf.*, 2010, **78**, 3450–3457.
45 F. Zhang, M. W. A. Skoda, R. M. J. Jacobs, S. Zorn, R. A. Martin, C. M. Martin, G. F. Clark, S. Weggler, A. Hildebrandt, O. Kohlbacher and F. Schreiber, *Phys. Rev. Lett.*, 2008, 101.
46 K. Delak, S. Collino and J. S. Evans, *Biochemistry*, 2009, **48**, 3669–3677.
47 J. S. Evans, *Chem. Rev.*, 2008, **108**, 4455–4462.
48 I. M. Weiss, N. Tuross, L. Addadi and S. Weiner, *J. Exp. Zool.*, 2002, **293**, 478–491.
49 N. Kroger, S. Lorenz, E. Brunner and M. Sumper, *Science*, 2002, **298**, 584–586.
50 M. Sumper and E. Brunner, *Adv. Funct. Mater.*, 2006, **16**, 17–26.
51 E. Brunner, K. Lutz and M. Sumper, *Phys. Chem. Chem. Phys.*, 2004, **6**, 854–857.
52 E. Beniash, L. Addadi and S. Weiner, *J. Struct. Biol.*, 1999, **125**, 50–62.
53 E. Beniash, J. Aizenberg, L. Addadi and S. Weiner, *Proc. R. Soc. London, Ser. B*, 1997, **264**, 461–465.

PAPER

Precipitation of ACC in liposomes—a model for biomineralization in confined volumes†

Chantel C. Tester,[a] Ching-Hsuan Wu,[a] Steven Weigand[b] and Derk Joester*[a]

Received 30th April 2012, Accepted 6th June 2012
DOI: 10.1039/c2fd20088k

Biomineralizing organisms frequently precipitate minerals in small phospholipid bilayer-delineated compartments. We have established an *in vitro* model system to investigate the effect of confinement in attoliter to femtoliter volumes on the precipitation of calcium carbonate. In particular, we analyze the growth and stabilization of liposome-encapsulated amorphous calcium carbonate (ACC) nanoparticles using a combination of *in situ* techniques, cryo-transmission electron microscopy (Cryo-TEM), and small angle X-ray scattering (SAXS). Herein, we discuss ACC nanoparticle growth rate as a function of liposome size, carbon dioxide flux across the liposome membrane, pH, and osmotic pressure. Based on these experiments, we argue that the stabilization of ACC nanoparticles in liposomes is a consequence of a low nucleation rate (high activation barrier) of crystalline polymorphs of calcium carbonate.

Introduction

Biological control over mineral nucleation and growth results in crystal shapes that differ dramatically from those formed by bulk precipitation from solution.[1] For example, sea urchin embryo primary mesenchyme cells (PMC) deposit single crystalline calcite in the form of smooth spicules that do not exhibit the typical {104} faces. Remarkably, PMC can change the crystallographic growth direction in response to external factors (Fig. 1). This "bottom-up" process results in an intricately shaped single crystal with smoothly curving surfaces.

As in many other biomineralizing organisms, precipitation in PMC occurs in an environment delimited by phospholipid bilayer membranes that act as selective diffusion barriers.[2] Compartmentalization is integral to cellular control over local supersaturation, the location of the mineral, and its chemical composition. In the case of the sea urchin embryo, formation of the crystalline mineral is preceeded by deposition of amorphous calcium carbonate (ACC) precursor phases.[3] This is but one example for the widespread use of amorphous phases in biomineralization.[1,4]

Amorphous precursors are more soluble and less dense than crystalline polymorphs, facilitating their storage and transport within the cell. In principle, amorphous-to-crystalline transformation can occur by dissolution-reprecipitation or

---

[a]*Department of Materials Science and Engineering, Northwestern University, 2220 Campus Drive, Evanston, IL 60208, USA. E-mail: d-joester@northwestern.edu; Fax: +847 491 7820; Tel: +847 491 7443*
[b]*Northwestern University/DND-CAT, APS/ANL Sector 5, Building 432/A002, 9700 S. Cass Ave., Argonne, IL 60439*

† Electronic Supplementary Information (ESI) available: Liposome size distributions, UV-Vis spectra and results of an osmotic pressure study. See DOI: 10.1039/c2fd20088k/

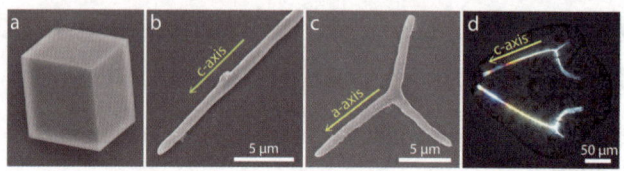

Fig. 1 (a) Synthetic calcite single crystal displaying characteristic rhombohedral structure and {104} facets. Depending on the concentration of a growth factor, *Strongylocentrotus purpuratus* primary mesenchyme cells deposit single crystals of calcite ($CaCO_3$) in the form of cylindrical (b) or triradiate (c) spicules. The cylindrical spicule is elongated parallel to the calcite *c*-axis. The triradiate grows parallel to the three *a*-directions. (d) Regulation of the crystal growth direction during development results in intricately shaped single crystals in this early pluteus stage *S. purpuratus* embryo.

solid-to-solid transformation. In either case, the amount of water liberated during crystallization is significantly reduced. This may be an important aspect of maintaining a high growth speed while minimizing the formation of pores.[2a] Colloids of synthetic amorphous precursor nanoparticles readily infiltrate organic scaffolds and wet surfaces.[5] By shaping the confining phospholipid membrane, cells could thus control the shape of the isotropic precursor. In the sea urchin embryo, one controlled nucleation event at the onset of mineralization—the deposition of a calcite rhombohedron—is sufficient to control the transformation of amorphous precursors that are added during growth. It is currently not clear how the amorphous precursors are synthesized and added to the growing biomineral.

Many biominerals show characteristic features at length scales on the order of 30–50 nm. These have been suggested to be the consequence of a synthetic mechanism that involves deposition of colloidal particles of similar sizes.[6] In addition, there are now several examples of echinoderms,[2b] mollusks,[7] and vertebrates,[4d,8] where amorphous precursors such as ACC or amorphous calcium phosphate are present in small phospholipid vesicles. Secretory vesicles play an important role in the directed transport of proteins. It is conceivable that delivery of both proteinaceous and mineral building blocks occurs by directed transport and fusion of such vesicles. This would require that the amorphous cargo is somehow stabilized.

The stabilization of amorphous precursors has frequently been ascribed to organic matrix macromolecules and inorganic additives present at the mineralization site.[9] However, the amount of these impurities can be very small. For example, in the sea urchin embryo spicule, less than 0.5wt% organic matter is present. At the same time, the surface area-to-volume ratio in attoliter to femtoliter size vesicles is very large. The surface chemistry of the phospholipid bilayer and its selective permeability for certain species, *e.g.* water and gases, may thus play an important role. Finally, confinement influences the kinetics of phase transformations, as evident in the large freezing point depression of small water droplets.[10]

Liposomes, *i.e.* small unilamellar phospholipid bilayer vesicles, are a versatile platform for modelling intracellular, biologically controlled mineralization. Variation of liposome diameter, lipid composition, and the composition of the interior and exterior solution allow systematic investigation of the role of confinement, surface chemistry, and soluble additives. Following seminal work on precipitation of iron oxide,[11] liposomes have been investigated primarily as models of matrix vesicles in bone mineralization.[12]

We have previously reported that nanoparticles of ACC can be synthesized inside phosphatidylcholine (**1**) liposomes.[13] Particle size can be controlled by choosing the concentration of Ca^{2+} and the liposome diameter. ACC particles greater than 200 nm can be synthesized in liposomes of 1 μm diameter and remain stable for much longer than in bulk solution. Herein, we investigate the growth of ACC nanoparticles over time using a combination of *in situ* techniques in an attempt to find an explanation for this unexpected stability.

Materials and methods

Materials

Soy phosphatidylcholine (PC, Avanti Polar Lipids, Alabaster, AL). Calcium chloride dihydrate (BDH Chemicals, Westchester, PA). Ammonium carbonate, Sodium chloride, Dichloromethane, Chloroform (Mallinckrodt Baker, Phillipsburg, NJ). Phenol red (Sigma Aldrich, St. Louis, MO). SephadexTM G-25, Nucleopore track-etched membranes (GE Healthcare Biosciences AB, Uppsala, Sweden). Unless otherwise noted, all solutions where prepared in ultra-pure water ($\rho = 18.2$ MΩ cm) prepared with a Barnstead NanoDiamond UF + UV purification unit.

Liposome preparation

Lipid films were prepared by rotary evaporation of a solution of PC (20 mg) in CH_2Cl_2 (10 mL) at 40 °C and 700 mbar. All subsequent steps were performed at room temperature. The lipid film was rehydrated with 5 mL of 1 M or 0.1 M $CaCl_2$ for at least 30 min. The resulting vesicle suspension was extruded eleven times through a polycarbonate track-etched membrane (pore size 0.1 or 1.0 μm) using an Avanti Mini-Extruder (Avanti Polar Lipids, Alabaster, AL). Liposomes were subjected to size exclusion chromatography (Sephadex G-25, eluent: 1.5 M NaCl). Dynamic light scattering (Malvern Zetasizer nano) was used to confirm the size of the liposomes and to identify the fractions with the highest liposome concentration (viscosity 1.0383 cP, refractive index 1.345). Fractions were pooled, stored at 4 °C, and used within 48 h.

CaCO$_3$ precipitation

Precipitation was initiated by gently mixing suspensions of $CaCl_2$-containing liposomes in 1.5 M NaCl with $(NH_4)_2CO_3$. Two concentrations of $(NH_4)_2CO_3$ were used: 1) 900 μL of liposome suspension were mixed with 100 μL $(NH_4)_2CO_3$ (1M) for a final concentration of 100 mM $(NH_4)_2CO_3$; 2) 800 μL of liposome suspension were mixed with 200 μL $(NH_4)_2CO_3$ (1.5 M) for a final concentration of 300 mM $(NH_4)_2CO_3$.

Influence of pH on precipitation

Liposomes were prepared as described above, but lipid films were rehydrated with 5 mL of a 1 M $CaCl_2$ containing 50 μM phenol red. Precipitation was initiated by mixing 800 μL of liposome suspension with 200 μL of 1.5 M $(NH_4)_2CO_3$. At the end of the reaction, the pH of the liposome interior was further increased by mixing 900 μL of the liposome-stabilized ACC suspension with 100 μL of 3wt% NH_4OH.

Cryo-TEM

Copper grids coated with a lacy carbon support film (Ted Pella, Redding, CA) were rinsed with chloroform and plasma treated (E.A. Fischione Plasma Cleaner) for 12 s. Liposome suspensions were plunge frozen in liquid ethane using an automated vitrification robot (FEI Vitrobot™ Mark IV) at a given time after the addition of $(NH_4)_2CO_3$ solution. The liposome suspension (5 μL) was pipetted onto a grid within the vitrobot enclosure (25 °C and 100% rel. humidity). Excess liquid was removed by automatic blotting (2 blots for 1 s each) before plunging into liquid ethane.

Imaging of liposomes was performed at 120 kV, on a field emission gun-equipped Jeol 1230 TEM employing a Gatan cryo-holder operating at approximately −180 °C. For optimum contrast of both inorganic precipitates and the phospholipid bilayer membrane, underfocusing was used for balancing mass-thickness and phase contrast.[14] Images were recorded using a Hamamatsu ORCA side mounted CCD camera (1024 × 1024 pixels). The size of the liposomes and encapsulated precipitates were measured using ImageJ.[15]

Influence of osmotic pressure on precipitation

Precipitation was initiated by gently mixing 800 µL of liposome suspension with 200 µL of 1) 0.5 M $(NH_4)_2CO_3$ for a final osmolarity of 2.7 Osmkg^{-1}; 2) 0.5 M $(NH_4)_2CO_3$ plus 1.5 M NaCl for a final osmolarity of 3.3 Osmkg^{-1}. The size of the resulting precipitates was measured by Cryo-TEM of samples plunge frozen 20 h from the addition of $(NH_4)_2CO_3$.

Small- and wide angle X-ray scattering

Simultaneous SAXS/WAXS was conducted at beamline 5-ID-D at the Advanced Photon Source (APS), using the BioSAXS setup. SAXS and WAXD patterns were recorded in parallel using a MARUSA, Inc. (SAXS) and Roper Scientific (WAXS) CCD camera system. The sample-to-detector distances were 0.199 m and 8.599 m for the wide- and small angle scattering detectors, calibrated using lanthanum hexaboride powder and silver behenate standards, respectively. The wavelength of radiation was set to 0.95372 Å (13 keV), resulting in q-ranges of 0.0016–0.1 Å^{-1} and 0.45–3.7 Å^{-1}. To minimize radiation damage, measurements were made while continuously flowing the liposome suspension at a rate of 4 µL s^{-1} through a quartz capillary flow cell (ID 0.15 cm, wall thickness 30 µm). Data was acquired over the course of 11 h with an exposure time of 3 s. The resulting isotropic scattering patterns were azimuthally averaged using FIT2D,[16] and corrected for varying incident beam intensity and sample transmission using a $CdWO_4$ crystal-coupled diode in the beam stop. The data was then normalized to an absolute scale using the measured transmission, sample thickness, and a scale factor calibrated from water. The final scattering curves were calculated as the average of three subsequent exposures of the same sample. When calculating the scattering intensity of liposome-encapsulated nanoparticles, the intensity of the liposome suspension at $t = 0$ was subtracted.

SAXS data analysis

SAXS data analysis was performed using the IRENA 2 package for Igor Pro.[17] Particle diameters were calculated using a unified Guinier exponential/power law.[18]

Results and discussion

Liposome preparation

Calcium-loaded liposomes were prepared by rehydration of a dry film of phosphatidylcholine (1) with an aqueous $CaCl_2$ solution (1.0 M). At high ionic strength, it is more challenging to produce liposomes with a narrow size distribution. When a polydisperse suspension of multilamellar vesicles was extruded through a track-etched membrane with 100 nm pores, the resulting unilamellar liposomes were 100 ± 25 nm in diameter. The polydispersity of liposomes produced by extrusion through larger pores was generally higher and the average diameter fell below the pore diameter. Herein, we refer to liposome suspensions by the nominal size of the track-etched pores of the membrane used during extrusion (e.g. 100 nm or 1 µm). For full characterization of the liposome suspensions by DLS, please see ESI†. Following extrusion, unencapsulated $CaCl_2$ was removed by size exclusion chromatography with an isosmotic aqueous NaCl solution (1.5 M) as mobile phase. The concentration of enecapsulated $CaCl_2$ is assumed to be equal to that of the rehydrating solution.

1

Precipitation of liposome-encapsulated ACC nanoparticles

Precipitation of ACC nanoparticles inside liposomes was induced by adding a concentrated solution of $(NH_4)_2CO_3$ to the liposome suspension as described previously.[13] The amorphous nature of the particles was confirmed by *in situ* wide angle X-ray scattering (WAXS) and selected area diffraction in Cryo-TEM (data not shown). Liposome-encapsulated ACC nanoparticles with sizes between 15 and 200 nm did not crystallize for at least 20 h.[13] ACC prepared from supersaturated solutions usually contains approximately one equivalent of water $(CaCO_3 \cdot H_2O)$.[19] We assume that this is the case for the liposome-encapsulated ACC nanoparticles as well.

Growth kinetics

The growth of ACC nanoparticles over time was followed using Cryo-TEM and *in situ* small angle X-ray scattering. To arrest growth, liposome suspensions were plunge-frozen in liquid ethane. Cryo-TEM of frozen-hydrated samples reveals that within 15 min from addition of ammonium carbonate, exactly one nanoparticle of ACC forms in each liposome. Precipitates remain approximately spherical over the course of the reaction. This allows the radius of the particles to be extracted in a straightforward manner using SAXS (see below). The size of the nanoparticle scales with that of the encapsulating liposome ($R^2 = 0.7$).[13] The distribution of particle sizes is thus dependent on the polydispersity of the liposome diameter (ESI).† While Cryo-TEM imaging allows very accurate determination of particle shapes and sizes, throughput is relatively slow.

We therefore investigated the kinetics of growth under a range of conditions using *in situ* SAXS. SAXS patterns were recorded at increasing time intervals from the addition of ammonium carbonate. Patterns were azimuthally averaged and plotted as the scattering intensity *versus* the scattering vector, $q = 4\pi \sin\theta \, \lambda^{-1}$ (Fig. 2).

In general, the scattered intensity $I(q)$ can be written as

$$I(q) = |A(q)|^2 = |\int_v \rho(r) e^{iqr} dr|^2 \tag{Eq 1}$$

where $A(q)$ is the normalized amplitude of scattering, and $\rho(r)$ the electron density of the scattering particle. From this expression, the increase in scattered intensity with time can be attributed to the increase in volume of the growing ACC nanoparticles.

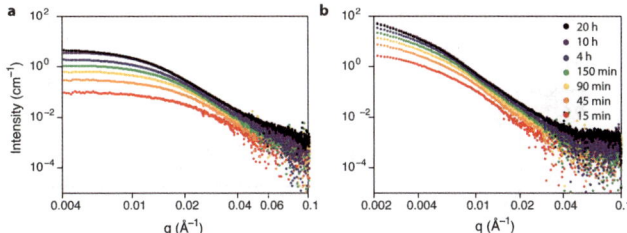

Fig. 2 Small angle X-ray scattering intensity recorded at increasing time from addition of aqueous $(NH_4)_2CO_3$ to (a) 100 nm and (b) 1 μm liposomes loaded with 1M $CaCl_2$. Scattering from the liposome suspension at $t = 0$ has been subtracted. The increase in intensity over time as well as the shift in the Guinier shoulder towards smaller q with increasing time is indicative of the growth of encapsulated ACC nanoparticles.

In the limit of small q, the scattering function simplifies to the Guinier approximation:

$$I(q) = \rho_0^2 v^2 e^{\frac{1}{3}q^2 R_g^2} \qquad \text{(Eq 2)}$$

Using $r^2 = (5/3)R_g^2$ for spherical particles, average particle radii were extracted at each time point. Comparison between the particle sizes measured by Cryo-TEM and SAXS indicates that the values measured by SAXS are consistently greater than those measured by Cryo-TEM by a factor of 1.3 (data not shown). This discrepancy is a consequence of the bias created by the greater scattering intensity of large particles in SAXS.

Plots of particle volume vs. time indicate that nanoparticle growth is slow over the first 20 h, with maxima between 1 nm^{3}s^{-1} in 100 nm liposomes at 100 mM external ammonium carbonate concentration (pH = 8.5–8.6) and 350 nm^{3}s^{-1} in 1 μm liposomes at 300 mM $(NH_4)_2CO_3$ (Fig. 3, Table 1). Based on the initial growth rate of the particles and assuming that the density and molecular weight of ACC is approximately equal to that of calcium carbonate monohydrate ($CaCO_3 \cdot H_2O$, 2.49 g cm^3, 118 g mol^{-1}), we calculate the net flux of carbonate integrated over the surface of the liposome ($J \cdot A$) and the steady state flux (J).

Depending on the combination of liposome diameter and ammonium carbonate concentration, the net influx of carbonate can be varied by more than two orders of magnitude, from ~10 ions s^{-1} to more than 4000 ions s^{-1}. As expected, the flux is only dependent on the ammonium carbonate concentration. While the flux, at a given external ammonium carbonate concentration, seems to be slightly higher in larger liposomes, given the error of the measurements and estimates, we believe that this is not statistically significant. Even at the highest ammonium carbonate

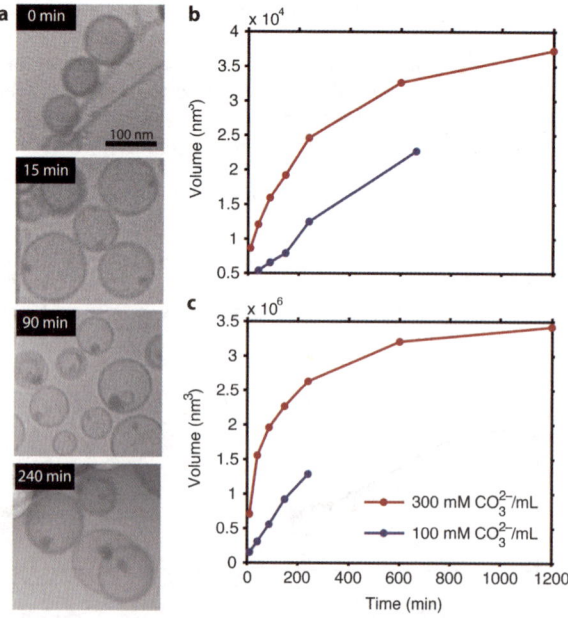

Fig. 3 (a) Cryo-TEM images of liposome suspensions plunge frozen before and at increasing time from addition of 100 mM $(NH_4)_2CO_3$. Exactly one nearly spherical nanoparticle of ACC precipitates inside each liposome. Growth was followed by SAXS in (b) 100 nm and (c) 1 μm liposomes at 100 mM (blue line) and 300 mM (red line) ammonium carbonate concentration in the outside medium.

Table 1 Flux of carbonate across the phospholipid membrane, estimated from ACC nanoparticle sizes measured by SAXS. $\dot{V}$: initial ACC nanoparticle growth rate, $J \cdot A$: integrated carbonate flux, A: exterior liposome surface area, J: carbonate flux

$r_i/\mu m$	$[(NH_4)_2CO_3]/mM$	$\dot{V}/nm^3\ s^{-1}$	$J \cdot A/CO_3^{2-}\ s^{-1}$	$A/\mu m^2$	$J/CO_3^{2-}\ \mu m^{-2}\ s^{-1}$
0.047	100	0.8 ± 0.1	10 ± 2	0.028	360 ± 72
0.497	100	98 ± 8	1,200 ± 100	3.103	386 ± 32
0.047	300	2.6 ± 0.8	30 ± 10	0.028	1080 ± 360
0.497	300	346 ± 76	4,000 ± 1000	3.103	1289 ± 320

concentration, the flux we observe in liposomes is more than 4 orders of magnitude smaller than the flux of carbonate necessary to sustain the growth of sea urchin embryo spicules (diameter ~1.2 μm, growth rate ~10 μm h^{-1}, $\dot{V} \approx 3.14 \times 10^6$ nm^3 s^{-1}, $J \approx 2.3 \times 10^7$ CO$_3^{2-}$ μm^{-2} s^{-1}).[20] For comparison, time-resolved SAXS studies of bulk CaCO$_3$ precipitation was so rapid that it required the use of a stopped-flow device to capture the formation of ACC. Particles with a diameter of 32 nm form within 0.5 s after mixing of CaCl$_2$ (9 mM) and Na$_2$CO$_3$ (9 mM) solutions ($\dot{V} \approx 34\,000$ nm^{3}s^{-1}), more than 100 times faster than the fastest rate we observed in liposomes, but ~60 times slower than during spiculogenesis.[21]

At 100 mM ammonium carbonate, the growth rate is constant over at least 4 h. Increasing the ammonium carbonate concentration by a factor of 3 at constant liposome diameter leads to a proportional increase in the initial particle growth rate. However, the growth rate decreases over time and the system appears to reach equilibrium at about 20 h. Based on the expected calcium concentration inside the liposome and the particle size at 20 h, the reaction does not go to completion (70 ± 10% of the expected final particle diameter). Several possibilities for this, including pH changes and osmotic pressure build up are discussed below.

The relatively slow growth rate and its dependence on the concentration of (NH$_4$)$_2$CO$_3$ indicate that precipitation is limited by the flux of carbonate (CO$_3^{2-}$) across the membrane. As the phospholipid bilayer is a diffusion barrier to ions, carbonate transport into the liposome must occur by the diffusion of CO$_2$ (Scheme 1). This is possible because carbonate is in equilibrium with bicarbonate and carbonic acid. The latter dehydrates to carbon dioxide, which dissolves in and diffuses across the membrane.

$$CO_2 + H_2O \rightleftharpoons H_2CO_3 \xrightleftharpoons{H_2O} HCO_3^- + H_3O^+ \xrightleftharpoons{H_2O} CO_3^{2-} + 2H_3O^+ \quad (Eq\ 3)$$

Inside the liposome, carbonate is in equilibrium with carbonic acid that is formed by the reverse process, i.e. hydration of carbon dioxide. Given the high calcium concentration and low solubility product of ACC, the equilibrium is shifted towards carbonate formation inside the liposome. Protons liberated in the rehydration of CO$_2$ on the inside of the liposome are buffered by the co-diffusion of ammonia.

The permeability ($P = k_p D/\Delta x$) of the liposome membrane for CO$_2$ is approximately 0.2 cm s^{-1} ($k_p \approx 1.7$, $D \approx 5 \times 10^{-8}$ cm^{2}s^{-1}, $\Delta x = 6$ nm) and approximately 0.13 cm s^{-1} for NH$_3$.[22] At this high CO$_2$ permeability, the carbonate flux is not limited by diffusion of CO$_2$ across the membrane, but by the slow kinetics of dehydration of H$_2$CO$_3$ ($k = 3.7 \times 10^{-2}$ s^{-1}). This is why, in cells, this reaction is typically catalysed by carbonic anhydrase, an enzyme with extremely fast turnover ($k_{cat} = 10^4$–10^6 s^{-1}).[22a]

Early growth of ACC nanoparticles in liposomes is thus limited only by the availability of carbonate ions. Consequently, the initial growth rate is dependent only on carbon dioxide flux and liposome surface area. Under the assumption that the flux of

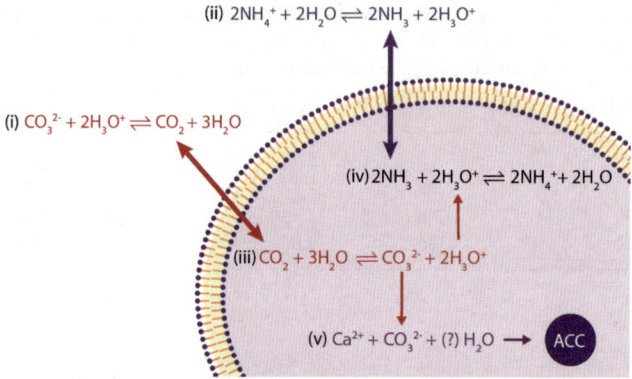

Scheme 1 Precipitation of ACC on the inside of Ca^{2+}-loaded liposomes. On the outside of the liposome, carbon dioxide (i) and ammonia (ii) are present in equilibrium with carbonate and ammonium ions in the ammonium carbonate solution. The neutral species diffuse across the membrane. (iii) Hydration of CO_2 to CO_3^{2-} releases protons that lower the pH inside the liposomes. Co-diffusion of ammonia (iv), driven by its conversion to NH_4^+, buffers the pH on the inside of liposome. Finally, the coupled equilibria are drained by precipitation of ACC (v) from encapsulated calcium and carbonate formed in (iii).

carbon dioxide across the membrane is constant, the number of carbon dioxide molecules diffusing across the membrane is

$$\dot{n} = 4\pi r_L^2 J \qquad (Eq\ 4)$$

where r_L is the radius of liposome and J is the flux of carbon dioxide. This steady state assumption is reasonable as long as the ammonium carbonate concentration on the outside and calcium concentration on the inside are not significantly depleted and the pH on either side does not change significantly.

All of the carbon dioxide that enters the liposome must be incorporated into the growing ACC nanoparticle as carbonate. The rate at which ions are added to the amorphous phase is then,

$$\dot{n} = \frac{\dot{V}\rho}{M_w} \qquad (Eq\ 5)$$

where V and ρ are the particle volume and density, respectively. Under the assumption that the volume is zero at $t = 0$ and the density is time independent,

$$\dot{V} = \frac{4\pi r_L^2 J M_w t}{\rho} \qquad (Eq\ 6)$$

Hence, the particle grows linearly with time at a rate dependent on the flux of carbon dioxide and the surface area of the liposome. The flux is proportional to the carbon dioxide concentration gradient, which is approximately equal to the concentration of ammonium carbonate. This is in good agreement with the observed growth rate during the first four hours at 100 mM external ammonium carbonate concentration (Fig. 3). At higher ammonium concentration, a linear growth regime is observed at shorter times only.

Influence of pH and osmotic pressure

Progressive attenuation of the growth rate with time could be a consequence of depleting Ca^{2+} and changing pH and/or osmolarity inside the liposome. Based on the nanoparticle size, reactions proceed to less than 50% conversion of the available calcium. The final Ca^{2+} concentration is thus greater than 500 mM, and Ca depletion is unlikely to play a role.

Hydration of CO_2 and formation of CO_3^{2-} in the liposome will increase the H_3O^+ concentration (Scheme 1). The resulting drop in pH shifts the equilibrium (Eq (3)) to the left until the CO_2 concentration inside and outside the liposome are equal. With no concentration gradient to drive transport of CO_2 into the liposome, particle growth must stop. This can be demonstrated by replacing $(NH_4)_2CO_3$ with Na_2CO_3 (300 mM) in the outside solution. In the absence of ammonia to buffer the pH of the liposome interior, no ACC nanoparticles form within 20 h after addition of Na_2CO_3 (Fig. 4a,b).

We tracked the pH inside the liposome over the course of ACC precipitation using phenol red (2) co-encapsulated in the liposome (Fig. 4c). Phenol red is a membrane impermeable pH indicator that is yellow when protonated and pink when deprotonated, with a visual transition interval from pH = 6.8–8.2. Upon addition of $(NH_4)_2CO_3$ to the liposome suspension, an immediate colour change from yellow to pink was observed by eye and UV-Vis spectrometry (ESI†). This corresponds to a rapid pH increase from the slightly acidic $CaCl_2$ solution (pH = 5.5) and indicates that ammonia diffuses into the liposome. Over the course of the reaction (20 h), the colour of the liposome suspension slowly reverts to yellow, consistent with acidification of the liposome interior. Addition of a 0.3 wt% solution of ammonium hydroxide (NH_4OH) at 20 h to increase the amount of free ammonia resulted in an immediate colour change to pink. ACC nanoparticles in liposomes treated with NH_4OH at 20 h showed an 18% increase in diameter compared to particles in untreated liposomes, as observed by Cryo-TEM (data not shown). We conclude from these experiments that acidification of the liposome interior resulting from hydration of CO_2 limits the growth of ACC nanoparticles in liposomes. This implies that the flux of ammonia across the liposome membrane falls to zero before the pH and ammonium ion concentration gradients are dissipated.

Given the basic character of ammonia ($pK_a(NH_4^+) = 9.25$), we expect that the equilibrium is far on the side of the ammonium ion. The net reaction inside the liposome is:

$$(CO_2)_{aq} + (x + 1)H_2O + (Ca^{2+})_{aq} + 2(NH_3)_{aq} \rightarrow 2(NH_4^+)_{aq} + (CaCO_3 \cdot xH_2O)_s \quad (Eq\ 7)$$

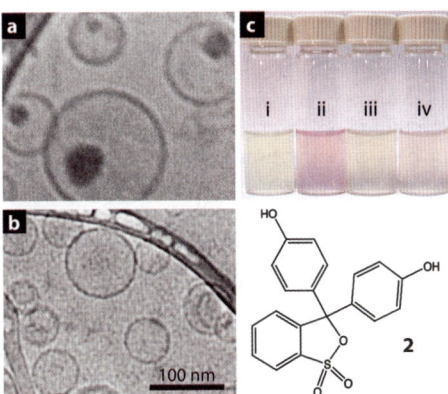

Fig. 4 (a) Cryo-TEM image of liposome suspensions (300 mM $(NH_4)_2CO_3$, 20 h) show encapsulated ACC nanoparticles with a diameter of 29 ± 7 nm. (b) No precipitation is observed when $(NH_4)_2CO_3$ is replaced with the same concentration of Na_2CO_3. (c) Suspensions of liposomes loaded with 1 M $CaCl_2$ and 50 μM phenol red (i) before and (ii) immediately after addition of $(NH_4)_2CO_3$. De-protonation of phenol red results in an immediate colour change from yellow to pink. (iii) Over time, the pH inside the liposome drops. (iv) At 20 h, the pH is raised and the reaction restarted by addition of 0.3 wt% NH_4OH.

Since the neutral species are free to diffuse across the membrane, only Ca^{2+} and NH_4^+ influence the osmolarity inside the liposome. With two ammonium ions produced for every calcium ion consumed, the osmolarity inside the liposome must increase from 3.0 Osm kg^{-1} to 4.0 Osm kg^{-1} over the course of ACC precipitation. At the same time, the osmotic pressure of the outside solution stays nearly constant. The osmotic pressure differential is expected to oppose the flux of ammonia and carbon dioxide across the membrane. A reduced flux of ammonia would lead to slow acidification of the liposome interior, further reducing the carbon dioxide concentration gradient and ultimately shutting down ACC nanoparticle growth altogether.

Based on this analysis, we predicted that the final ACC nanoparticle size would depend on the osmotic pressure difference between the inside and outside solution. Raising the osmolarity of the outside solution should reduce the buildup of osmotic pressure and thus allow growth to continue longer. A decrease in the osmolarity of the outside solution should have the opposite effect. However, when we compared the ACC particle size 20 h after addition of ammonium carbonate between liposomes that had been suspended in solutions with an osmolarity of 2.7 Osm kg^{-1} and 3.3 Osm kg^{-1}, we found no difference. This may be due to the relatively small differences in osmolarity that we could access experimentally. A detailed investigation of the effect of the osmotic pressure difference is ongoing.

ACC nanoparticle growth and stabilization

Precipitation of ACC requires that the solution inside the liposome is supersaturated with respect to ACC ($pK_{sp} = 5.97$ at 20 °C, $\sigma_{ACC} = \ln [Ca^{2+}][CO_3^{2-}] - pK_{sp}$).[23] In 100 nm liposomes ($V \approx 0.5$ aL) loaded with 1 M $CaCl_2$, just one carbonate ion is sufficient to do so ($\sigma_{ACC} = 3$). Particle growth rates indicate that, independent of the liposome diameter, the solubility limit of ACC is exceeded within the first second after addition of ammonium carbonate. At a similar degree of supersaturation in bulk, ACC is precipitated, but crystallizes within seconds unless stabilizing additives are employed.[9] For example, ACC nanoparticles precipitate within 30 s after mixing of Na_2CO_3 and $CaCl_2$ solutions ($\sigma_{ACC} = 75$) and completely transformed into a mixture of calcite, vaterite, and aragonite within an hour.[24]

We postulate that the extended stability of liposome-encapsulated ACC can be attributed to confinement in small volumes, here ~0.5 aL in 100 nm liposomes and ~0.5 fL in 1 μm liposomes.[13] Liposome encapsulated nanoparticles cannot grow beyond a size controlled by the number of encapsulated calcium ions. Under the conditions described herein, we expect that particles grow to at most one third of the diameter of the liposome. The size of such particles makes them metastable with respect to the crystalline polymorphs.[25] However, a critical nucleus is required for crystallization. In bulk precipitation, even a very small nucleation current—in the extreme case, just one nucleation event in the entire volume—is sufficient for completion of the phase transformation. In liposomes, where mass transfer between particles is shut down, Ostwald ripening does not occur. Consequently, a small nucleation current would lead to a small number of encapsulated crystalline particles.

The Meldrum group reported prolonged ACC stability due to confinement in picoliter droplets.[26] ACC initially precipitates, then transforms into a single crystal within 5 to 30 min. In bulk solution crystals form within 2 min. Under the assumption that only one nucleus forms in a 5 pL droplet over the course of 30 min ($I = 1.1 \times 10^{11}$ $m^{-3}s^{-1}$), a nucleation event would be expected in only 0.42% of 1 μm liposomes ($V \approx 0.5$ fL) over 20 h. In 100 nm liposomes ($V \approx 0.5$ aL), the fraction would be three orders of magnitude lower still. In comparison, the lower bound for the steady state nucleation rate of ACC can be estimated based on the observation that exactly one ACC particle has formed in 100 nm liposomes after 15 min. The value ($I = 2.55 \times 10^{18}$ $m^{-3}s^{-1}$) is at least nine orders of magnitude larger than the nucleation rate of calcite.

While these estimates are rough, a low effective nucleation rate for calcite is sufficient to explain the stabilization of ACC nanoparticles confined in liposomes. In fact, it may be possible to quantify the nucleation rate by determining the frequency of crystalline particles in liposomes. This is not a trivial undertaking in small liposomes. However, liposomes or droplets with volumes in the 5–50 fL range (5–10 μm diameter) may be ideally suited for this purpose.

Conclusions

In summary, we show that the growth of stable ACC nanoparticles is controlled by the rate of carbon dioxide transport into calcium-loaded liposomes. For sustained growth, it is critical to prevent acidification of the liposome. Co-diffusion of ammonia is an effective way to do so. However, growth stops significantly before all Ca^{2+} is consumed as a consequence of slow acidification of the liposome. This is likely the consequence of a slow osmotic pressure build up in the liposome that prevents influx of ammonia. The closest equivalent to this precipitation process in liposomes is biomineralization in foraminifera.[27] Calcium uptake occurs primarily by endocytosis of seawater into vacuoles. Mitochondria near the calcification site create a high local CO_2 concentration in the cytosol. This results in a carbon dioxide flux across the membrane of the vacuole. Cells actively maintain a pH gradient from the vacuole (high pH) to the relatively acidic cytosol.

The growth rate of ACC nanoparticles may be conveniently controlled over several orders of magnitude by adjusting carbonate concentration and liposome surface area. However, the precipitation is slow relative to bulk precipitation. Biological mineralization is frequently cyclic with large differences in mineralization rates. While the liposome model system captures slow growth well, modelling fast growth would require carbonic anhydrase as a catalyst.

ACC nanoparticles exhibit extended stability without the aid of additives when confined within small volumes. We believe this is due to a much lower nucleation rate and higher activation barrier for crystalline nuclei compared to nuclei of the amorphous phase. Note that the removal of water during crystallization of hydrated ACC is not hindered in liposome-encapsulated ACC nanoparticles, as the phospholipid membrane is permeable to water.

Biomacromolecules, for instance highly acidic proteins, that are occluded in biominerals may thus not be as important for the direct stabilization of ACC as previously assumed. However, their role may still be critical in the processing of amorphous precursors, for instance in modulating local supersaturation, and stabilizing, or flocculating ACC colloids. Finally, macromolecules may be responsible for controlling the speed at which amorphous-to-crystalline transformations occur once a nucleus has formed.

We anticipate that liposome-stabilized ACC will provide a convenient platform to determine the nucleation rate of the crystalline polymorphs of calcium carbonate and systematically investigate the influence of membrane chemistry, particle size, and co-encapsulated additives on nucleation, growth, and phase transformations.

Acknowledgements

This work was in part supported by the NSF (Grants No. DMR-0805313 and DMR-1106208), by the MRSEC program of the National Science Foundation (DMR-1121262) at the Materials Research Center of Northwestern University, and by the International Institute for Nanotechnology at Northwestern University. SAXS/WAXS was performed at DND-CAT at the APS. Use of the APS was supported by the US Dept. of Energy (BES), under Contract No. DE-AC02-06CH11357. Cryo-TEM was performed at NU BIF and UV-visible spectroscopy was performed at the Keck Biophysics Facility. We thank Peter Voorhees for discussion.

References

1 H. A. Lowenstam, S. Weiner, *On Biomineralization*, Oxford University Press, New York, 1989.
2 (a) S. Weiner and L. Addadi, *Annu. Rev. Mater. Res.*, 2011, **41**, 21–40; (b) E. Beniash, L. Addadi and S. Weiner, *J. Struct. Biol.*, 1999, **125**, 50–62.
3 (a) Y. Politi, Y. Levi-Kalisman, S. Raz, F. Wilt, L. Addadi, S. Weiner and I. Sagi, *Adv. Funct. Mater.*, 2006, **16**, 1289–1298; (b) Y. Politi, R. A. Metzler, M. Abrecht, B. Gilbert, F. H. Wilt, I. Sagi, L. Addadi, S. Weiner and P. Gilbert, *Proc. Natl. Acad. Sci. U. S. A.*, 2008, **105**, 20045–20045.
4 (a) E. Baeuerlein, 2nd ed., Wiley-VCh, Weinheim, 2004; (b) E. Beniash, R. A. Metzler, R. S. K. Lam and P. Gilbert, *J. Struct. Biol.*, 2009, **166**, 133–143; (c) H. Hashimoto, S. Yokoyama, H. Asaoka, Y. Kusano, Y. Ikeda, M. Seno, J. Takada, T. Fujii, M. Nakanishi and R. Murakami, *J. Magn. Magn. Mater.*, 2008, **320**, 2310–2310; (d) J. Mahamid, B. Aichmayer, E. Shimoni, R. Ziblat, C. H. Li, S. Siegel, O. Paris, P. Fratzl, S. Weiner and L. Addadi, *Proc. Natl. Acad. Sci. U. S. A.*, 2010, **107**, 6316–6321.
5 (a) E. Loste, R. J. Park, J. Warren and F. C. Meldrum, *Adv. Funct. Mater.*, 2004, **14**, 1211–1220; (b) F. Nudelman, K. Pieterse, A. George, P. H. H. Bomans, H. Friedrich, L. J. Brylka, P. A. J. Hilbers, G. de With and N. Sommerdijk, *Nat. Mater.*, 2010, **9**, 1004–1009.
6 A. Baronnet, J. P. Cuif, Y. Dauphin, F. Farre and J. Nouet, *Mineral. Mag.*, 2008, **72**, 617–626.
7 D. E. Jacob, R. Wirth, A. L. Soldati, U. Wehrmeister and A. Schreiber, *J. Struct. Biol.*, 2011, **173**, 241–249.
8 J. Mahamid, A. Sharir, D. Gur, E. Zelzer, L. Addadi and S. Weiner, *J. Struct. Biol.*, 2011, **174**, 527–535.
9 L. Addadi, S. Raz and S. Weiner, *Adv. Mater.*, 2003, **15**, 959–970.
10 H. R. Pruppacher, *Microphysics of clouds and precipitation/by Hans R. Pruppacher and James D. Klett*, D. Reidel Pub. Co, Dordrecht, Holland; Boston:, 1978.
11 S. Mann, J. P. Hannington and R. J. P. Williams, *Nature*, 1986, **324**, 565–567.
12 (a) N. R. Blandford, G. R. Sauer, B. R. Genge, L. N. Y. Wu and R. E. Wuthier, *J. Inorg. Biochem.*, 2003, **94**, 14–27; (b) E. D. Eanes, *Anat. Rec.*, 1989, **224**, 220–225; (c) B. R. Genge, L. N. Y. Wu and R. E. Wuthier, *Anal. Biochem.*, 2007, **367**, 159–166.
13 C. C. Tester, R. E. Brock, C.-H. Wu, M. R. Krejci, S. Weigand and D. Joester, *CrystEngComm*, 2011, **13**, 3975–3978.
14 H. Cui, T. K. Hodgdon, E. W. Kaler, L. Abezgauz, D. Danino, M. Lubovsky, Y. Talmon and D. J. Pochan, *Soft Matter*, 2007, **3**, 945–955.
15 M. D. Abramoff, P. J. Magalhaes and S. J. Ram, *Biophotonics international*, 2004, **11**, 36–42.
16 (a) A. P. Hammersley, in *ESRF Internal Report*, 1997; (b) A. P. Hammersley, S. O. Svensson, A. Thompson, H. Graafsma, A. Kvick and J. P. Moy, *Rev. Sci. Instrum.*, 1995, **66**, 2729–2733.
17 J. Ilavsky and P. R. Jemian, *J. Appl. Crystallogr.*, 2009, **42**, 347–353.
18 G. Beaucage, *J. Appl. Crystallogr.*, 1995, **28**, 717–728.
19 A. L. Goodwin, F. M. Michel, B. L. Phillips, D. A. Keen, M. T. Dove and R. J. Reeder, *Chem. Mater.*, 2010, **22**, 3197–3205.
20 K. A. Guss and C. A. Ettensohn, *Development*, 1997, **124**, 1899–1908.
21 J. Bolze, B. Peng, N. Dingenouts, P. Panine, T. Narayanan and M. Ballauff, *Langmuir*, 2002, **18**, 8364–8369.
22 (a) J. Gutknecht, M. A. Bisson and F. C. Tosteson, *J. Gen. Physiol.*, 1988, **69**, 779–794; (b) A. Walter and J. Gutknecht, *J. Membr. Biol.*, 1986, **90**, 207–217.
23 J. R. Clarkson, T. J. Price and C. J. Adams, *J. Chem. Soc., Faraday Trans.*, 1992, **88**, 243–249.
24 D. Pontoni, J. Bolze, N. Dingenouts, T. Narayanan and M. Ballauff, *J. Phys. Chem. B*, 2003, **107**, 5123–5125.
25 C. J. Stephens, S. F. Ladden, F. C. Meldrum, H. K. Christenson, Adv. Funct. Mater., **20**, pp. 2108–2115.
26 C. J. Stephens, Y. Y. Kim, S. D. Evans, F. C. Meldrum and H. K. Christenson, *J. Am. Chem. Soc.*, 2011, **133**, 5210–5213.
27 S. Bentov, C. Brownlee and J. Erez, *Proc. Natl. Acad. Sci. U. S. A.*, 2009, **106**, 21500–21504.

The role of the amorphous phase on the biomimetic mineralization of collagen

Fabio Nudelman,[a] Paul H. H. Bomans,[a] Anne George,[b] Gijsbertus de With[a] and Nico A. J. M. Sommerdijk*[a]

Received 3rd April 2012, Accepted 1st June 2012
DOI: 10.1039/c2fd20062g

Bone is a hierarchically structured composite material whose basic building block is the mineralized collagen fibril, where the collagen is the scaffold into which the hydroxyapatite (HA) crystals nucleate and grow. Understanding the mechanisms of hydroxyapatite formation inside the collagen is key to unravelling osteogenesis. In this work, we employed a biomimetic *in vitro* mineralization system to investigate the role of the amorphous precursor calcium phosphate phase in the mineralization of collagen. We observed that the rate of collagen mineralization is highly dependent on the concentration of polyaspartic acid, an inhibitor of hydroxyapatite nucleation and inducer of intrafibrillar mineralization. The lower the concentration of the polymer, the faster the mineralization and crystallization. Addition of the non-collagenous protein C-DMP1, a nucleator of hydroxyapatite, substantially accelerates mineral infiltration as well as HA nucleation. We have also demonstrated that Cu ions interfere with the mineralization process first by inhibiting the entry of the calcium phosphate into the collagen, and secondly by stabilizing the ACP, such that it does not convert into HA. Interestingly, under these conditions mineralization happens preferentially in the overlap regions of the collagen fibril. Our results show that the interactions between the amorphous precursor phase and the collagen fibril play an important role in the control over mineralization.

1. Introduction

Bone is a hierarchically structured composite material whose basic building block is the mineralized collagen fibril, where the collagen is the scaffold into which the hydroxyapatite (HA) crystals nucleate and grow.[1,2] The triple-helical collagen molecules are packed in a quasi-hexagonal manner and are axially organized into parallel arrays, with their ends separated by a 40 nm gap and neighbouring molecules staggered by 67 nm.[3–5] The fibril, thus, has a periodic cross-striated structure where a densely packed 27 nm-long region—the overlap zone—alternates with the less dense 40 nm-long gap zone.

Understanding the mechanisms of HA nucleation and growth inside the collagen lies at the core of unravelling bone formation, and therefore has been widely investigated, both in *in vivo* and *in vitro* systems. Early studies on bone formation have been able to establish the spatial and crystallographic relationship between collagen and the mineral. Conventional transmission electron microscopy (TEM) studies,

[a]*Laboratory of Materials and Interface Chemistry and Soft Matter CryoTEM Unit, Eindhoven University of Technology, P.O. Box 513, 5600 MB Eindhoven, The Netherlands. E-mail: n.sommerdijk@tue.nl; Fax: + 31 40 244 5619; Tel: +31 40 247 5870*
[b]*Department of Oral Biology, University of Illinois, Chicago, USA. E-mail: anneg@uic.edu; Fax: + 1 312 996 6044; Tel: +1 312 413 0738*

combined with X-ray and electron diffraction have shown that the HA crystals nucleate and grow within 40 nm gaps present in the collagen fibril, and have their c-axis aligned parallel to the long axis of collagen.[1,2,6–11] The collagen fibrils, however, cannot induce HA mineralization on their own, and require acidic non-collagenous proteins (NCPs) to promote HA formation.

One of the major issues in bone formation has been to characterize the pathways through which the calcium and phosphate ions are translocated to the mineral deposition site within the collagen. Two main modes have been proposed: 1) the crystals are actively nucleated from solution by the NCPs associated to the collagen gap zones, without intervention of cellular processes;[2,12] and 2) matrix vesicles that bud from the plasma membrane accumulate ions extracellularly through their molecular composition.[13,14] Recent research has demonstrated that osteoblasts concentrate within intracellular vesicles calcium phosphate in a non-crystalline, disordered phase, composed of amorphous calcium phosphate (ACP).[15,16] This precursor phase is then delivered to the collagen at the bone growth front, infiltrates into the fibrils, and then crystallizes into HA. Although the presence of an amorphous precursor to HA in bone mineralization has only recently been identified, this strategy has been well known for several years, in particular in calcium carbonate-based biomineralization in invertebrates. In these organisms, amorphous calcium carbonate forms the first mineral deposit that subsequently transforms either into calcite or aragonite.[17–20] The importance of using amorphous phases as precursors is that these materials are isotropic and easily moldable, and thus allow the overcoming of directional restrictions of crystals when building mineralized structures.[21] Moreover, this strategy allows the transport of large enough quantities of the mineral to the growth front that are necessary to build mineralized structures.

One issue, however, that has been overlooked is whether the amorphous precursor can play an active role in helping to direct the mineralization process itself by interacting with the pre-formed organic matrix template. In *in vitro* studies on collagen mineralization, we have demonstrated that the charge interaction between the collagen and the ACP drives the entry of the mineral into the fibril. Subsequently, the charged residues of collagen form nucleation sites that mediate the transformation of ACP into oriented HA.[22] Thus, parameters such as net charge, particle size and chemistry of the ACP particles are important in mediating the interaction with collagen and hence the mineralization process.

Here, we employed cryo-transmission electron microscopy (cryoTEM) and cryo-electron tomography (cryoET) to further investigate the role of the ACP in the mineralization of collagen. We employed a biomimetic mineralization system, where the NCPs were replaced by polyaspartic acid (pAsp), which inhibits HA formation in solution and stabilizes the ACP, directing it into the collagen.[23,24] In the first step of the work, we investigated the influence of the concentration of pAsp on the mineralization. Furthermore, we accidentally found that Cu ions derived from Cu cryoTEM grids inhibit HA formation, stabilizing the ACP for long periods of time. We then studied how the mineralization of collagen proceeded under these conditions, both in presence of pAsp and the C-terminal fragment of the NCP dentin matrix protein 1 (C-DMP1), which is known to induce HA nucleation.[25–27] We show that the stabilization of the ACP phase here realized both by different concentrations of pAsp and by the presence of Cu ions is an important aspect in controlling mineralization.

2. Experimental

2.1 CryoTEM and cryoET

Sample vitrification was performed using an automated vitrification robot (FEI Vitrobot™ Mark III) for plunging in liquid ethane.[28] CryoTEM Au or Cu grids, R2/2 Quantifoil Jena grids (Quantifoil Micro Tools GmbH) were surface plasma treated for 40 s using a Cressington 208 carbon coater prior to use. For 2D imaging and tomography, samples were studied on the TU/e CryoTitan (FEI, www.cryotem.nl), equipped with a field emission gun (FEG) operating at 300 kV and a post column Gatan energy filter (GIF). Images were recorded using a post-GIF 2k × 2k Gatan CCD camera. Low-dose selected area electron diffraction (LDSAED) and energy-dispersive X-ray analysis (EDX) were performed on a FEI Tecnai 20 (Type Sphera) equipped with a LaB_6 filament operating at 200 kV and a Gatan cryo holder operating at −170 °C was used. The images were recorded using a 1k × 1k Gatan CCD camera and Cryo-EDX spectra were recorded on an EDAX detector in cryo-STEM mode, spot size 3.

The alignment and 3-D reconstructions of the data sets were performed in IMOD or Inspect3D v.3.0 (FEI Company). For the segmentation and visualization of the 3D volume, Amira 4.1.0 (Mercury Computer Systems) was used. Image analysis was performed using Gatan Digital Micrograph and NIH ImageJ.

2.2 Assembly of collagen on TEM grids:

Type I collagen extract from horse tendon was kindly provided by Dr Giuseppe Falini (Department of Chemistry, University of Bologna, Italy) and prepared as described elsewhere.[29] Briefly, 1 g extract was mixed overnight with 10 ml aqueous acetic acid (50 mM, pH 2.5), centrifuged at 5000 rpm for 10 min and the supernatant was collected and stored at 4 °C. At pH 2.5 the collagen fibrils are disassembled and remain in solution.

CryoTEM grids were laid on a 15 µl drop of collagen solution for 10 s. The excess of collagen solution was manually blotted and the grids were transferred to a 15 µl drop of Hepes buffer (10 mM, pH 7.4) containing NaCl (150 mM), for 30 min.[30] The increase in pH to physiological levels triggers collagen assembly into fibrils and its subsequent precipitation.[29,31] This procedure was performed inside a glove box,[28] where temperature and humidity are controlled at 22 °C and 100% relative humidity. The grids were then transferred to the mineralization solution (see below).

2.3 Mineralization experiments:

Collagen mineralization in presence of polyaspartic acid (pAsp) was achieved by incubating the collagen-adsorbed cryoTEM grids in Hepes buffer (10 mH, pH 7.4, Sigma) containing $CaCl_2$ (2.7 mM, Merck), K_2HPO_4 (1.35 mM, Merck) and polyaspartic acid (10 µg ml^{-1}, molecular weight 2000–11 000 Da, Sigma) at 37 °C, as described elsewhere.[22,24]

Collagen mineralization in presence of C-DMP1 was achieved by incubating collagen-adsorbed cryoTEM grids in Hepes buffer (10 mM, pH 7.4, Sigma) containing $CaCl_2$ (2.7 mM, Merck), K_2HPO_4 (1.35 mM, Merck) and C-DMP1 (15 µg ml^{-1}, see below) + polyaspartic acid (1.5 µg ml^{-1} −10 µg ml^{-1}, molecular weight 2000–11000 Da, Sigma) at 37 °C. In the control experiments collagen was mineralized without any additives.

After mineralization, grids were washed with MilliQ water for 10 min. The excess water was removed by manual blotting and the grids were transferred to the automated vitrification robot. Air dried samples were prepared by washing the grids in MilliQ water for 10 min; manual blotting with filter paper to remove the excess of water; dehydrating the samples in 100% ethanol for 5 min and air drying.

2.4 Staining of collagen with uranyl acetate

Staining of collagen with uranyl acetate was performed on some samples immediately before vitrification of the grids in liquid ethane, following the procedure described by Traub et al.,[10] with modifications. After removal of the samples from the mineralization solution and subsequent washing with MilliQ water, the grids were incubated in 0.5% uranyl acetate in MilliQ water for 15 s, then washed with

MilliQ water for 1 min, manually blotted and transferred to the automated vitrification robot.

3. Results

In order to investigate the role of the amorphous phase in the mineralization of collagen, we first investigated the effect of pAsp concentration on the infiltration of ACP into the fibrils and its subsequent transformation into oriented hydroxyapatite (HA) crystals. When pAsp was present at a concentration of 1.5 µg ml^{-1} in the mineralization solution, the collagen fibrils were completely impregnated with amorphous calcium phosphate (ACP) after 24 h of reaction (Fig. 1A). At this stage, the ACP had already started to convert into oriented HA crystals, as shown by the presence of needle-like, dense objects within the amorphous phase (Fig. 1A, arrows), as reported.[22] Higher concentrations of pAsp resulted in considerably slower mineralization rates. Between 3 and 10 µg ml^{-1} we could only observe aggregates of ACP infiltrating into the collagen through the region at the border between the gap and overlap regions (Fig. 1B–D).[22] These results show that by increasing the amount of pAsp in the solution and hence making the ACP phase more stable, the rate of infiltration of the ACP into the collagen significantly slows down. At a concentration

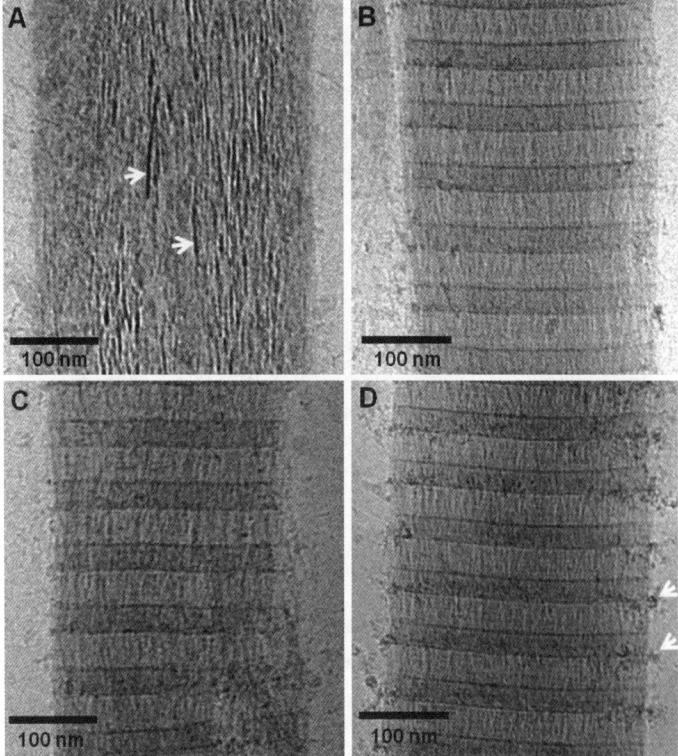

Fig. 1 CryoTEM images of collagen mineralized with calcium phosphate for 24 h in presence of different concentrations of pAsp, on cryoTEM grids made of Au. (A) 1.5 µg ml^{-1} of pAsp. White arrows: apatite crystals nucleating and growing within the amorphous calcium phosphate phase. (B) 3 µg ml^{-1} of pAsp. (C) 6 µg ml^{-1} of pAsp. (D) 10 µg ml^{-1} of pAsp. White arrows: amorphous calcium phosphate particles infiltrating into the collagen. Panel D was adapted from Nudelman, F. *et al.*,[22] *Nature Mater.*, 2010, **9**, 1004–1009. Reprinted with permission from Macmillan Publishers Ltd. Nature Materials, www.nature.com/nmat, Copyright 2010.

of 10 μg ml⁻¹ of pAsp, 72 h were necessary for the complete mineralization of collagen with HA crystals (Fig. 2). In this case, infiltration of the fibril with ACP occurred within 24–48 h (Fig. 2A and B), followed by the crystallization of ACP into HA (Fig. 2C). Interestingly, ACP and oriented HA crystals were found only on well organized regions of the collagen (Fig. 2D, black circle area). On poorly organized areas, very little mineral was present, consisting mostly of few HA crystals randomly oriented (Fig. 2D, white circle area and white dashed circle). These observations suggest a role for the collagen not only in directing the infiltration of the mineral phase, but also in templating the orientation and growth of the HA crystals. In addition, with the progression of the mineralization, the collagen fibrils expanded in the direction perpendicular to their long axis, becoming deformed by the developing mineral (Fig. 2E). The cross-sectional area of the fibril at the point with the

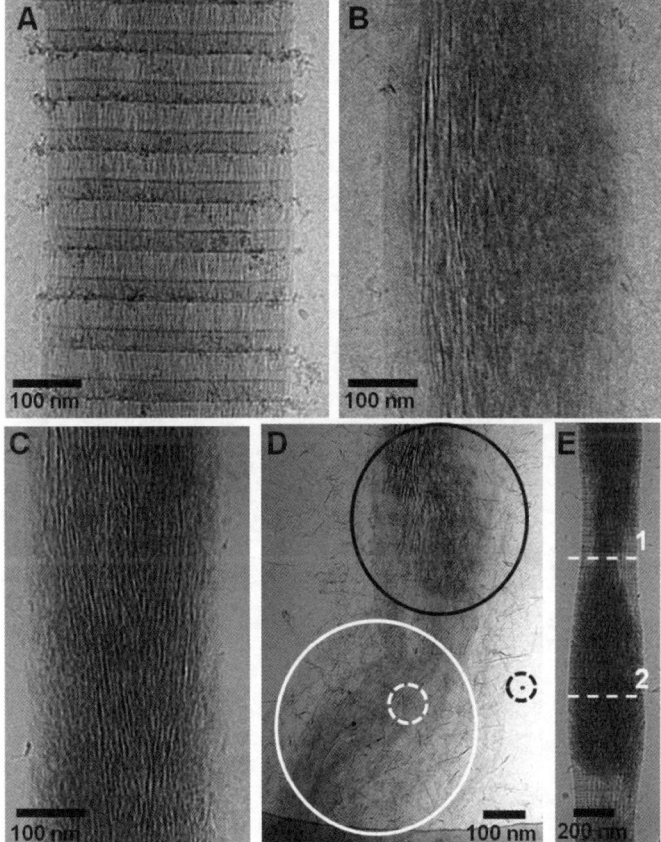

Fig. 2 CryoTEM images of collagen at different stages of mineralization in presence of 10 μg ml⁻¹ of pAsp, on cryoTEM grids made of Au. (A) Mineralization for 24 h. (B) Mineralization for 48 h. (C) Mineralization for 72 h. (D) Collagen fibril containing two regions, one well organized (black circle area) and one poorly organized (black circle area). The well organized area is partially mineralized with amorphous calcium phosphate and apatite, while the poorly organized one contains only few, randomly oriented apatite crystals (white dashed circle). Dashed black circle: 10 nm gold marker. (E) A partially mineralized collagen fibril, where deformation caused by the presence of the mineral can be observed. Dashed-lines: Region with least (dashed-line 1) and most amount of mineral (dashed-line 2) from where the cross-sectional area of the fibril was calculated. Panel E was adapted from Nudelman, F. *et al.*,[22] *Nature Mater.*, 2010, **9**, 1004–1009. Reprinted with permission from Macmillan Publishers Ltd. Nature Materials, www.nature.com/nmat, Copyright 2010.

least amount of mineral (dashed-line 1) increased from 53 000 nm² to 102 000 nm² at the point with the highest mineral content (dashed-line 2). Although in both cases the collagen contained a mixture of ACP and HA, the expansion of the fibril is proportional to the amount of calcium phosphate that is inside. Therefore, it is likely that it was the infiltration of the ACP into the collagen that caused the fibril to expand, rather than the transformation of ACP into HA. Regarding the presence of HA crystals randomly oriented on poorly ordered regions of collagen (Fig. 2D), we must point out that these images are 2-D projections of a 3-D object. Therefore, we cannot distinguish whether the crystals are actually present inside or on the surface of the collagen.

All the above experiments were performed with collagen absorbed on TEM grids that were made of Au. We accidentally found that when grids made of Cu were used instead, the mineralization of collagen changed drastically. Using 10 μg ml^{-1} of pAsp, after 24 h of reaction there was substantially fewer ACP aggregates infiltrating the collagen when compared to experiments performed on Au grids (Fig. 3A). After 48 h, most of the mineral was associated to the overlap region (Fig. 3B), becoming denser after 72 h but still with no signs of crystallinity (Fig. 3C and D). Cryo-energy dispersive X-ray spectroscopy (cryoEDX) confirmed that the precipitates found on the collagen were indeed composed of calcium phosphate (Fig. 3E). Our results

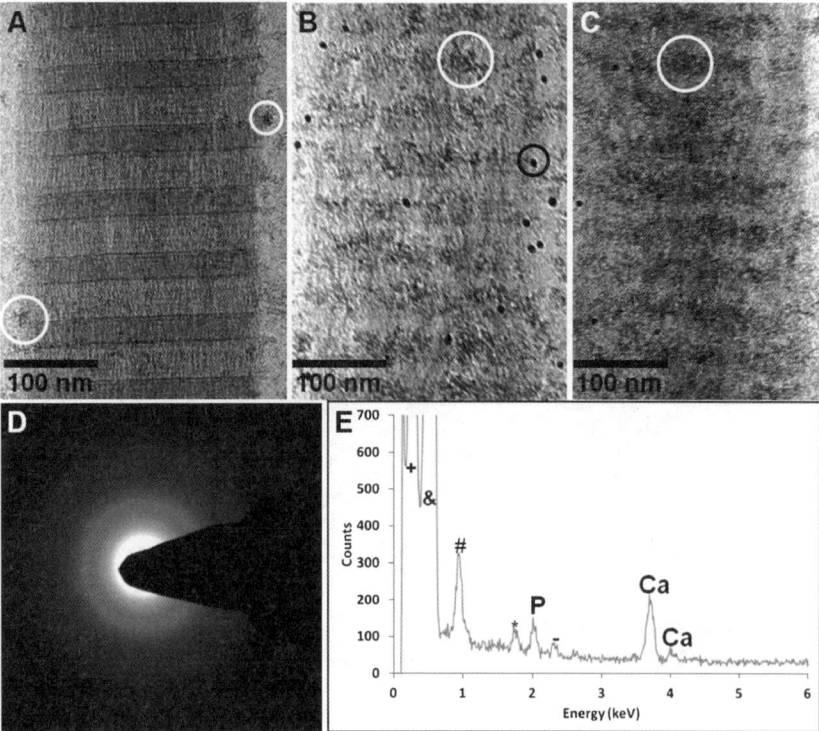

Fig. 3 CryoTEM images of collagen at different stages of mineralization in presence of 10 μg ml^{-1} of pAsp, on cryoTEM grids made of Cu. (A) Mineralization for 24 h. White circle: amorphous calcium phosphate particles associated to the collagen. (B) Mineralization for 48 h. White circle: amorphous calcium phosphate particles associated to the overlap region of collagen. Black circle: 10 nm gold markers. (C) Mineralization for 72 h. White circle: amorphous calcium phosphate particles associated to the overlap region of collagen. (D) Low-dose selected-area electron diffraction of C, showing that the mineral is still amorphous. (E) Cryo-energy dispersive X-ray spectroscopy (cryoEDX) measurement of C, confirming that the precipitates are indeed composed of calcium phosphate.

indicate that Cu ions were being released from the grid into the solution, and were interfering with the crystallization process. Indeed, using atomic absorption spectroscopy, we found that after 72 h of reaction Cu ions were present in the solution at a concentration of 5 μg ml^{-1}, and after 1 week their concentration increased to 11 μg ml^{-1}.

In order to verify whether ACP was still entering the collagen or was accumulating only on the surface of the fibril, we performed cryo-electron tomography (cryoET)[32] on samples that were mineralized for 48 h (Fig. 4). Indeed, cryoET showed the presence of mineral particles inside the collagen, although in smaller amounts than found when Au grids were used (Fig. 2). In fact, while in experiments performed without Cu (*i. e.* on Au grids) the ACP phase was found as a continuous matrix permeating the collagen, in the presence of Cu the calcium phosphate was found as discrete particles with a broad size distribution, ranging from 10 to 60 nm in size. Further analysis of the reconstruction data confirmed that most of the mineral particles were found indeed in the overlap region, while the gap zone was the least mineralized area (Fig. 4B–D).

In the next step, we combined chemical staining with cryoTEM[22] to map if the infiltration of the ACP into the fibrils occurs through the same mechanism when Cu grids are used as compared to experiments done with Au grids. Uranyl acetate is a staining agent that binds to the charged amino acids of collagen, increasing the local mass density, thus generating a pattern of bands that mark the location of clusters of positively and negatively charged amino acids.[33,34] This staining has

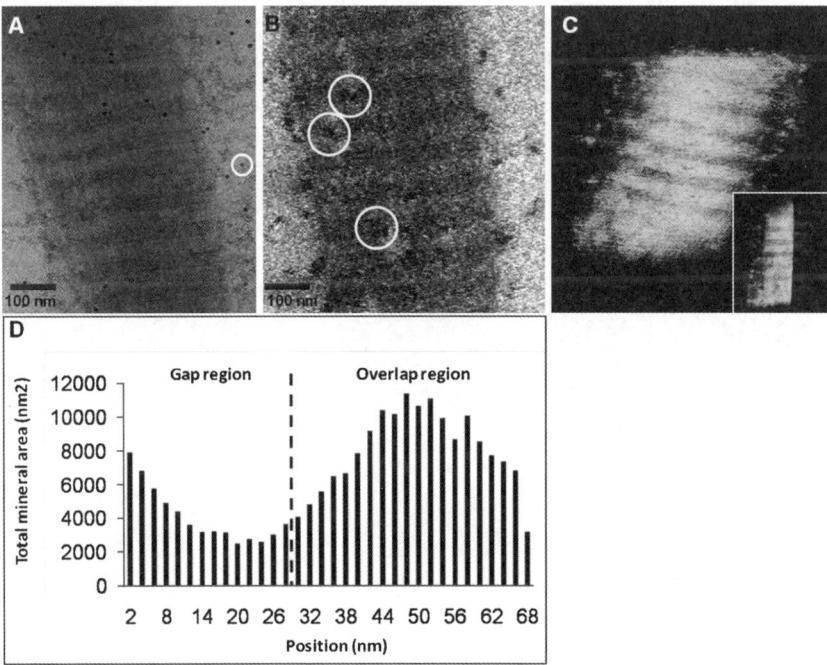

Fig. 4 Cryo-electron tomography of a collagen fibril mineralized for 48 h in the presence of 10 μg ml^{-1} of pAsp, on a Cu grid. (A) Two-dimensional cryoTEM image. (B) Slice from a section of the reconstructed 3-dimensional volume, showing calcium phosphate particles inside the collagen (white circles). (C) Computer-generated 3-dimensional visualization of the reconstruction, where the collagen fibril is depicted in white and the calcium phosphate precipitates in red. Inset: only half of the collagen fibril is shown, revealing the calcium phosphate precipitates inside the fibril, predominantly in the overlap region. (D) Graph showing the distribution of calcium phosphate within a 67 nm repeat.

been widely used to visualize collagen in conventional (reviewed in[33]) and in cryoTEM,[10,22] and the resulting staining pattern can be directly correlated to the amino acid sequence of collagen and to the crystal structure. When the mineralization of collagen was performed on Au grids, we observed the infiltration of the mineral into the fibrils through the a-band region, as previously reported (Fig. 5A and B, black arrows).[22] This region corresponds to the C-terminal end of the collagen molecules, and has been described as being highly positively charged and thus the most favourable for the negatively charged pAsp–mineral complex to interact with. When the experiments were performed on Cu grids, we still observed the ACP associated with the same areas of the collagen, namely the a-band region (Fig. 5C, black arrows). Analysis of the mass–density profile of the non-mineralized (Fig. 5D) and mineralized on Au and Cu grids (Fig. 5E and F, respectively) confirmed the location of the mineral phase on the collagen. The peaks corresponding to the a-bands significantly increased in intensity in both cases when compared to the non-mineralized collagen. The peaks corresponding to the other staining bands, on the other hand, did not change, suggesting little or no interaction between the ACP and the other regions of the collagen fibril, even in the presence of Cu ions.

We investigated the stability of the amorphous phase by letting the mineralization reaction proceed for 4 weeks (Fig. 6A and B). At this stage, the collagen fibrils were completely calcified, with mineral present both inside and on the surface of the collagen. However, even after such long incubation time the mineral phase consisted predominantly of ACP, as shown by LDSAED (Fig. 6A, inset). The amorphous phase did not convert into HA even when the samples were air-dried (Fig. 6B, inset). We performed control experiments where no additives were present in the mineralization solution. After 72 h, we observed the precipitation of few HA crystals in the solution and on the surface of the collagen (not shown). When incubated for 1 week and air dried, the samples contained large amounts of HA crystals, covering the whole surface of the collagen fibrils and of the carbon support film on the cryoTEM grid (Fig. 6C). In this case, LDSAED demonstrated that the mineral is composed of

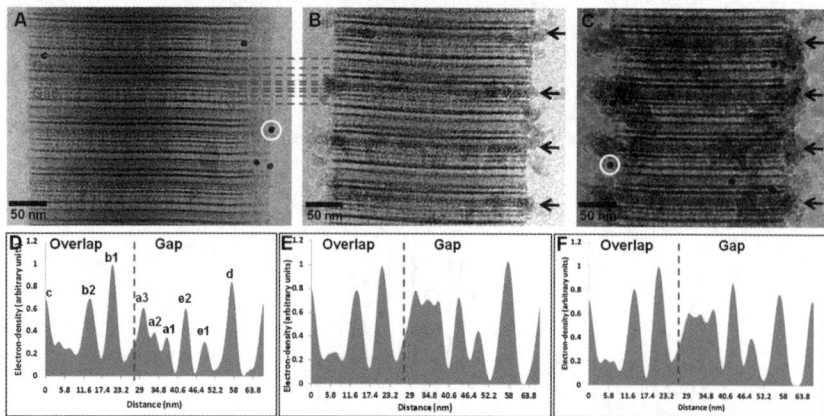

Fig. 5 Uranyl acetate map of collagen during the early stages of mineralization. (A) Non-mineralized collagen. White circle: 10 nm gold marker. (B) Collagen mineralized for 24 h in presence of 10 μg ml^{-1} of pAsp, on Au grids. (C) Collagen mineralized for 24 h in presence of 10 μg ml^{-1} of pAsp, on Cu grids. White circle: 10 nm gold marker. (D) Intensity profile of A, non-mineralized collagen. Dashed line: border between gap and overlap regions (C-terminus). (E) Intensity profile of B, collagen mineralized for 24 h on Au grids. Dashed line: border between gap and overlap regions (C-terminus). (F) Intensity profile of C, collagen mineralized for 24 h on Cu grids. Dashed line: border between gap and overlap regions (C-terminus). Panels A and D were adapted from Nudelman, F. et al.,[22] Nature Mater., 2010, **9**, 1004–1009. Reprinted with permission from Macmillan Publishers Ltd. Nature Materials, www.nature.com/nmat, Copyright 2010.

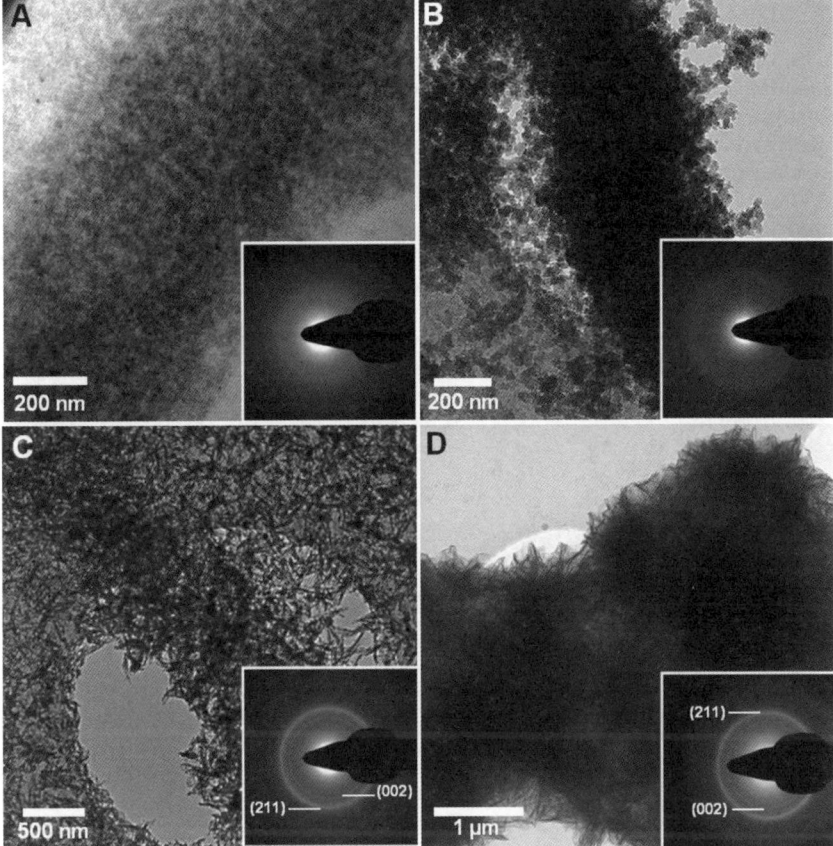

Fig. 6 (A) CryoTEM image of collagen mineralized for 4 weeks in presence of 10 μg ml^{-1} of pAsp on Cu grids. The calcium phosphate is still amorphous, as shown by the LDSAED (inset). (B) TEM image of dried collagen mineralized for 4 weeks in presence of 10 μg ml^{-1} of pAsp on Cu grids. Even after freeze-drying, the mineral shows no signs if crystallinity, as shown by the LDSAED (inset). (C) TEM image of dried collagen mineralized for 2 weeks without additives, on a Cu grid. The surface of the collagen and of the carbon support film on the grid are completely covered with apatite crystals. Inset: LDSAED, showing the (002) and (211) reflections of apatite. (D) TEM image of dried HA crystals precipitated in presence of 10 μg ml^{-1} of pAsp on a Cu grid after 2 weeks of reaction, without collagen. Inset: LDSAED, showing the (002) and (211) reflections of apatite.

HA, as shown by the (002) and (211) reflections. Interestingly, when pAsp was present but collagen was absent, we could also obtain HA on air-dried samples (Fig. 6D).

With the aim of understanding how the Cu ions could be influencing mineralization, we performed experiments in the presence of the C-terminal fragment of the dentin matrix protein 1 (C-DMP1). This protein is an acidic non-collagenous protein that is expressed during the initial stages of bone and dentin mineralization.[25,26] *In vitro* studies have shown that in the presence of collagen, it promoted HA nucleation on the surface of the fibrils.[27] Moreover, the C-terminal fragment of this protein was identified as the domain containing HA nucleation sites. Furthermore, C-DMP1 was demonstrated to induce the biomimetic mineralization of collagen when used in combination with pAsp.[22] Thus, we decided to investigate whether this protein could induce the formation of HA crystals inside the collagen also in the presence of Cu ions. In control experiments performed on Au grids, C-DMP1 was used at a concentration of 15 μg ml^{-1} and in the absence of pAsp. Under these conditions, we observed that after 24 h this protein induced the formation HA crystals mainly on the surface of the collagen fibrils (Fig. 7A). When C-DMP1 was used in combination with 10 µg ml^{-1} of pAsp, also on Au grids, the collagen fibrils were mineralized with oriented HA crystals, formed through an ACP precursor, as described (Fig. 7B). Interestingly, when the experiments were performed on Cu grids, C-DMP1 lost its ability to nucleate HA. In the absence of pAsp, globular structures consisting of calcium phosphate formed on the surface of the collagen (Fig. 7C, black arrows). Although LDSAED still needs to be performed, it is very likely that these structures were composed of ACP. When used in combination with pAsp on Cu grids, the fibrils were mineralized with ACP that formed predominantly on the overlap region, in a similar way as when pAsp was used alone (Fig. 7D). However, the presence of C-DMP1 significantly accelerated mineralization, such that after 24 h the fibrils contained large amounts of ACP when compared to mineralization with only pAsp. These results indicate that the Cu ions are interacting with the protein and changing its activity. Further studies need to be conducted to clarify the effect on the Cu ions on the protein.

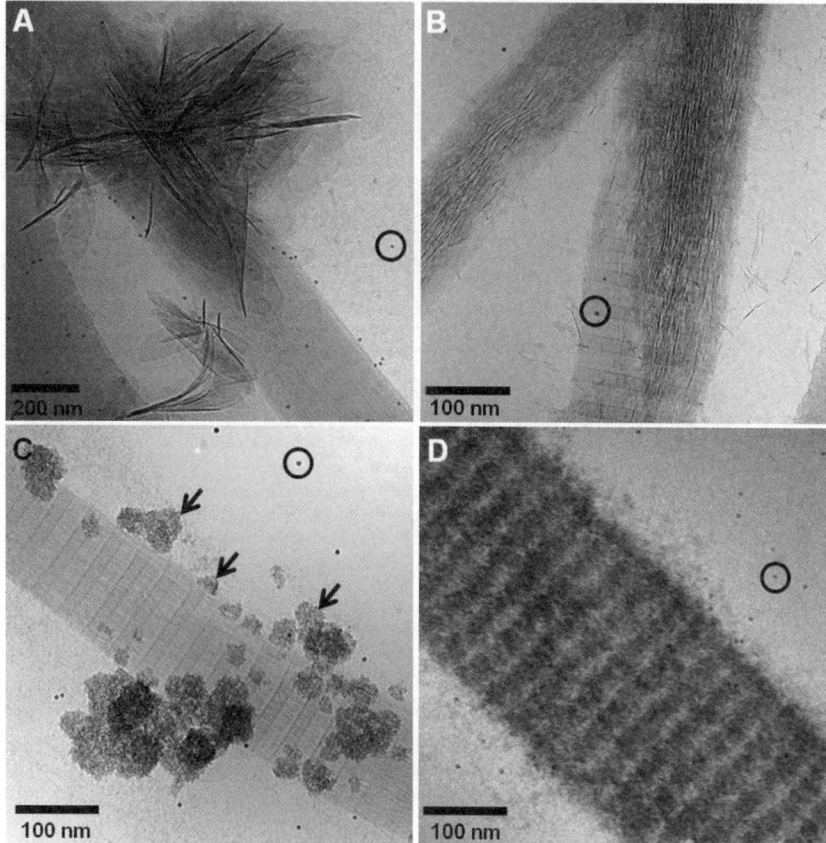

Fig. 7 CryoTEM images of collagen mineralized in presence of 15 µg ml^{-1} of C-DMP1 on Au and Cu grids. (A) Collagen mineralized for 24 h in presence of 15 µg ml^{-1} of C-DMP1 on Au grids. Black circle: 10 nm gold markers. (B) Collagen mineralized for 24 h in presence of 15 µg ml^{-1} of C-DMP1 and 10 µg ml^{-1} of pAsp on Au grids. Black circle: 10 nm gold markers. (C) Collagen mineralized for 24 h in presence of 15 µg ml^{-1} of C-DMP1 on Cu grids. Black arrows: globular structures of calcium phosphate on the surface of collagen. Black circle: 10 nm gold markers. (D) Collagen mineralized for 24 h in presence of 15 µg ml^{-1} of C-DMP1 and 10 µg ml^{-1} of pAsp on Cu grids. Black circle: 10 nm gold markers.

4. Discussion

4.1 Effect of the concentration of pAsp

We have shown that the rate of collagen mineralization is highly dependent on the concentration of the HA nucleation inhibitor, namely pAsp. At low concentration of pAsp (1.5 μg ml^{-1}, Fig. 1A), the infiltration of the calcium phosphate into the collagen and its subsequent transformation into HA happens within 24 h. At higher concentration of pAsp (above 3 μg ml^{-1}, Fig. 1B–D), this process is substantially slowed down. Interestingly, when pAsp at higher concentrations is combined with C-DMP1, an HA nucleator, the mineralization of collagen is substantially accelerated, reaching the same rate as with pAsp at low concentration. Considering that the main function of pAsp is to transiently stabilize ACP and allow it to enter the collagen,[24,35] these results suggest that there must be a balance between the stabilization of ACP in solution and the promotion of HA nucleation. Interestingly, we observed that the size of the ACP aggregates infiltrating into the collagen increased at higher concentrations of pAsp (Fig. 1B–D). Previously, we have observed that pAsp at 10 μg ml^{-1} stabilizes the formation of loosely packed assemblies of calcium phosphate complexes[36] that are 1 nm in size.[22] These assemblies further aggregate, forming larger and denser structures that are similar to the ones observed entering the collagen fibril (Fig. 1D). It is likely that the size and the stability of these aggregates of calcium phosphate particles are dependent on the concentration of pAsp. At lower concentrations of pAsp, these aggregates will tend to be smaller and less stable, while at higher concentrations they will grow larger in size and become more stable. We note, however, that at higher concentration of pAsp (10 μg ml^{-1}) the calcium phosphate precipitates from the solution within 24 h, while infiltration of the ACP into the collagen is still ongoing and takes 24–48 h. Thus, the speed of the mineralization reaction is not dependent on the stabilization of the calcium phosphate into the solution, but on aggregate size. It is possible that at 1.5 μg ml^{-1} of pAsp, the assemblies of calcium phosphate complexes will be small enough to quickly diffuse through the collagen fibril, while at higher concentrations of the polymer (above 3 μg ml^{-1}), the diffusion of the particles will take substantially longer and thus the mineralization will be slower.

4.2 The role of the 3-D structure of collagen in controlling mineralization

It is noteworthy that intrafibrillar mineralization occurred only when the collagen fibril was well organized, and not when it was poorly ordered (Fig. 2D). This result is significant since it highlights that the 3-D structure of collagen is essential for its function in controlling mineralization. Since the fibril is composed of individual collagen molecules organized in a parallel, staggered array, this organization results in the proper alignment of the positively charged amino acids of neighbouring molecules into domains that in turn are able to interact with the ACP and mediate its infiltration into the fibril. Furthermore, the lateral and axial organizations of the collagen molecules also creates HA nucleation sites by organizing the charged amino acids of adjacent molecules into a 3-D structural template that induces oriented HA nucleation.[37–39] Thus, on disorganized regions of the fibril such structural template is absent and diffusion of ACP into the collagen and its transformation into oriented HA crystals cannot occur. These findings bear important consequences for studies that use single collagen molecules to investigate mineralization.[38]

4.3 Expansion of collagen during mineralization

The expansion of the fibril during mineralization shows that, contrary to what has previously been assumed, availability of space does not limit crystal growth. The liquid-crystalline nature of collagen allows enough flexibility so that the molecules can be pushed aside during mineralization.[6] In addition, during HA formation the intermolecular spacing between the collagen molecules in fact decreases, leading to regions with a close-packed arrangement of molecules.[6] Moreover, it must be considered (1) that the increase in the cross-sectional area of the collagen is proportional to the amount of mineral inside the fibril; and (2) that mineralization starts with filling the intermolecular spaces with ACP. It is likely, then, that it was during the infiltration stage that the molecules were pushed aside and the fibril expanded. We further propose that the crystallization of ACP into HA will cause the shrinkage of the fibrils due to extrusion of water and the decrease of the intermolecular spacing, as observed in the turkey tendon.[6]

4.4 The effect of Cu ions on collagen mineralization

We have also demonstrated that Cu ions interfere with the mineralization process first by inhibiting the entry of the calcium phosphate into the collagen, and second by stabilizing the ACP, such that it does not convert into HA. Interestingly, HA crystals still form either in the absence of additives, or in the presence of pAsp but without the collagen fibril. In other words, the Cu ions are preventing crystal formation inside the collagen, but not in solution. Although Cu ions are known to inhibit the transformation of ACP into HA,[40] our observations mean that stabilization of the ACP by the Cu in our experiments occurs not only by the direct interaction of the ion with the calcium phosphate. Cu must also be complexing with the pAsp, the C-DMP1, and possibly with the collagen as well, and therefore affecting their function, both in mediating the infiltration of the ACP into collagen and subsequent HA nucleation.

At present, we do not have further information on the possible mechanism through which Cu ions interfere with the mineralization process. Regarding the infiltration of the ACP into the collagen, we speculate that Cu may associate to the sites in the collagen through which the calcium phosphate goes in, and therefore slowing the diffusion of the ACP into the fibril. Alternatively, it is also possible that in the presence of Cu, larger complexes of ACP are formed in solution which cannot easily diffuse into the collagen.[41] Nevertheless, the mineral infiltration site remains the same, namely the C-terminal end of the collagen molecule. It is interesting to note that when the mineralization experiments were performed employing both pAsp and C-DMP1, mineral infiltration into the collagen was significantly faster (Fig. 7D), either in the presence of Cu ions (using Cu grids) or in their absence (using Au grids). The fast infiltration of ACP into the collagen, without nucleation of HA, means that one of the functions of C-DMP1 is to help drive the ACP into the fibril. More experiments on the activity of the C-DMP1 need to be performed to understand its precise mechanism of action during collagen mineralization.

Considering the inhibition of the transformation of ACP into HA inside the collagen, we hypothesize that the Cu ions may bind to the nucleation sites in the collagen,[37,39] thus inhibiting their activity. Interestingly, in the presence of Cu ions the C-DMP1 also lost the ability to nucleate HA, leading to the formation of globular mineral structures instead, which are likely composed of ACP (Fig. 7C). Furthermore, Cu ions may also induce protein aggregation, which could explain the formation of globular mineral structures of calcium phosphate and C-DMP1.

It is interesting to note that when Cu ions are present, the ACP accumulates inside the collagen mainly in the overlap region of the fibril, both induced by pAsp alone (Fig. 3 and 4) and by pAsp combined with C-DMP1 (Fig. 7D). *In situ* TEM observations of HA formation on the mineralizing turkey tendon have demonstrated that it is the gap region of collagen the site where the HA crystals nucleate and grow.[2,8] More specifically, the region in the gap zone that corresponds to the e-band of uranyl acetate staining, has been implicated as the site where the HA crystals nucleate and grow.[10] In this respect, we also note that also the biomimetic mineralization system that we employ lacks this specificity, since HA crystals form everywhere within a 67 nm repeat. Although collagen has nucleation sites both in the gap and overlap regions,[37,39] it is possible that part of the function of the NCPs is to target HA formation to specific sites. It must be pointed out, however, that there are studies that described HA nucleation and growth occurring concomitantly in the gap and overlap regions in the mineralizing turkey tendon,[6,42–44] which is in line with our observations.

4.5 Outlook

The present work raises several questions on the possible interactions between the ACP and the collagen. Nevertheless, our results show that the interactions between the amorphous precursor phase and the collagen fibril play an important role in the control over mineralization. Further experiments need to be conducted to fully understand the importance of the particle size and stability of the ACP for its infiltration into the collagen. In addition, the stabilization of ACP by Cu ions is also an interesting topic that demands further research to unravel how this ion interacts with the collagen fibrils, the pAsp and the C-DMP1, and thus affecting the mineralization process. Understanding these mechanisms can lead to further insights on the role of the amorphous precursor phase on collagen mineralization and hence on bone formation.

Acknowledgements

We thank G. Falini (University of Bologna, Italy) for kindly providing the horse tendon collagen; Laura J. Brylka (Eindhoven University of Technology) for help with the initial mineralization experiments; J. van Roosmalen (Eindhoven University of Technology, The Netherlands) for help with the tomography reconstructions. Supported by the Dutch Science Foundation, NWO, The Netherlands; by the European Community (FP6, project code NMP4-CT-2006-033277 TEM-PLANT) and by NIH Grant DE 11657 (A.G).

References

1 S. Weiner and H. D. Wagner, *Annu. Rev. Mater. Sci.*, 1998, **28**, 271–298.
2 M. J. Glimcher and H. Muir, *Philos. Trans. R. Soc. London, Ser. B*, 1984, **304**, 479–508.
3 A. J. Hodge and J. A. Petruska, in *Aspects of Protein Structure*, ed. G. N. Ramachandran, Academic Press, New York, 1963, pp. 289–300.
4 A. Miller, *Philos. Trans. R. Soc. London, Ser. B*, 1984, **304**, 455–477.
5 J. P. R. O. Orgel, T. C. Irving, A. Miller and T. J. Wess, *Proc. Natl. Acad. Sci. U. S. A.*, 2006, **103**, 9001–9005.
6 P. Fratzl, N. Fratzl-Zelman and K. Klaushofer, *Biophys. J.*, 1993, **64**, 260–266.
7 D. J. S. Hulmes, T. J. Wess, D. J. Prockop and P. Fratzl, *Biophys. J.*, 1995, **68**, 1661–1670.
8 W. J. Landis, M. J. Song, A. Leith, L. Mcewen and B. F. Mcewen, *J. Struct. Biol.*, 1993, **110**, 39–54.
9 W. Traub, T. Arad and S. Weiner, *Proc. Natl. Acad. Sci. U. S. A.*, 1989, **86**, 9822–9826.
10 W. Traub, T. Arad and S. Weiner, *Matrix - Collagen and Related Research*, 1992, **12**, 251–255.
11 W. Traub, T. Arad and S. Weiner, *Connect. Tissue Res.*, 1992, **28**, 99–111.
12 A. Veis and A. Perry, *Biochemistry*, 1967, **6**, 2409.
13 S. Y. Ali, S. W. Sajdera and H. C. Anderson, *Proc. Natl. Acad. Sci. U. S. A.*, 1970, **67**, 1513.
14 H. C. Anderson, R. Garimella and S. E. Tague, *Front. Biosci.*, 2005, **10**, 822–837.
15 J. Mahamid, B. Aichmayer, E. Shimoni, R. Ziblat, C. Li, S. Siegel, O. Paris, P. Fratzl, S. Weiner and L. Addadi, *Proc. Natl. Acad. Sci. U. S. A.*, 2010, **107**, 6316–6321.
16 J. Mahamid, A. Sharir, L. Addadi and S. Weiner, *Proc. Natl. Acad. Sci. U. S. A.*, 2008, **105**, 12748–12753.
17 L. Addadi, Y. Politi, F. Nudelman and S. Weiner, *39th Course of the International School of Crystallography on Engineering of Crystalline Materials Properties*, Erice, Italy, 2007.
18 E. Beniash, J. Aizenberg, L. Addadi and S. Weiner, *Proc. R. Soc. London, Ser. B*, 1997, **264**, 461–465.
19 Y. Politi, T. Arad, E. Klein, S. Weiner and L. Addadi, *Science*, 2004, **306**, 1161–1164.
20 I. M. Weiss, N. Tuross, L. Addadi and S. Weiner, *J. Exp. Zool.*, 2002, **293**, 478–491.

21 L. Addadi, S. Raz and S. Weiner, *Adv. Mater.*, 2003, **15**, 959–970.
22 F. Nudelman, K. Pieterse, A. Georgè, P. H. H. Bomans, H. Friedrich, L. J. Brylka, P. A. J. Hilbers, G. de With and N. A. J. M. Sommerdijk, *Nat. Mater.*, 2010, **9**, 1004–1009.
23 A. S. Deshpande and E. Beniash, *Cryst. Growth Des.*, 2008, **8**, 3084–3090.
24 M. J. Olszta, X. G. Cheng, S. S. Jee, R. Kumar, Y. Y. Kim, M. J. Kaufman, E. P. Douglas and L. B. Gower, *Mater. Sci. Eng., R*, 2007, **58**, 77–116.
25 A. George and A. Veis, *Chem. Rev.*, 2008, **108**, 4670–4693.
26 G. He, T. Dahl, A. Veis and A. George, *Nat. Mater.*, 2003, **2**, 552–558.
27 G. He and A. George, *J. Biol. Chem.*, 2004, **279**, 11649–11656.
28 M. R. Vos, P. H. H. Bomans, P. M. Frederik and N. A. J. M. Sommerdijk, *Ultramicroscopy*, 2008, **108**, 1478–1483.
29 A. Tampieri, G. Celotti, E. Landi, M. Sandri, N. Roveri and G. Falini, *J. Biomed. Mater. Res.*, 2003, **67A**, 618–625.
30 H. M. H. F. Sanders, M. Iafisco, E. M. Pouget, P. H. H. Bomans, F. Nudelman, G. Falini, G. de With, M. Merkx, G. J. Strijkers, K. Nicolay and N. A. J. M. Sommerdijk, *Chem. Commun.*, 2011, **47**, 1503–1505.
31 F. Z. Jiang, H. Horber, J. Howard and D. J. Muller, *J. Struct. Biol.*, 2004, **148**, 268–278.
32 F. Nudelman, G. de With and N. A. J. M. Sommerdijk, *Soft Matter*, 2011, **7**, 17–24.
33 J. A. Chapman, M. Tzaphlidou, K. M. Meek and K. E. Kadler, *Electron Microsc. Rev.*, 1990, **3**, 143–182.
34 A. J. Hodge and F. O. Schmitt, *Proc. Natl. Acad. Sci. U. S. A.*, 1960, **46**, 186–197.
35 L. B. Gower, *Chem. Rev.*, 2008, **108**, 4551–4627.
36 A. S. Posner and F. Betts, *Acc. Chem. Res.*, 1975, **8**, 273–281.
37 E. P. Katz and S. Li, *J. Mol. Biol.*, 1973, **80**, 1–15.
38 A. Kawska, O. Hochrein, A. Brickmann, R. Kniep and D. Zahn, *Angew. Chem., Int. Ed.*, 2008, **47**, 4982–4985.
39 W. J. Landis and F. H. Silver, *Cells Tissues Organs*, 2009, **189**, 20–24.
40 Y. Okamoto and S. Hidaka, *J. Biomed. Mater. Res.*, 1994, **28**, 1403–1410.
41 D. Toroian, J. E. Lim and P. A. Price, *J. Biol. Chem.*, 2007, **282**, 22437–22447.
42 M. E. Maitland and A. L. Arsenault, *Calcif. Tissue Int.*, 1991, **48**, 341–352.
43 A. L. Arsenault, *Calcif. Tissue Int.*, 1988, **43**, 202–212.
44 A. L. Arsenault, *Calcif. Tissue Int.*, 1991, **48**, 56–62.

… PAPER

Revisiting geochemical controls on patterns of carbonate deposition through the lens of multiple pathways to mineralization†

D. Wang,[a] L. M. Hamm,[a] A. J. Giuffre,[a] T. Echigo,[a] J. Donald Rimstidt,[a] J. J. De Yoreo,[b] J. Grotzinger[c] and P. M. Dove[*a]

Received 19th April 2012, Accepted 17th May 2012
DOI: 10.1039/c2fd20077e

The carbonate sedimentary record contains diverse compositions and textures that reflect the evolution of oceans and atmospheres through geological time. Efforts to reconstruct paleoenvironmental conditions from these deposits continue to be hindered by the need for process-based models that can explain observed shifts in carbonate chemistry and form. Traditional interpretations assume minerals precipitate and grow by classical ion-by-ion addition processes but are unable to reconcile a number of unusual features contained in Proterozoic carbonates. The realization that diverse organisms produce high Mg carbonate skeletal structures by non-classical pathways involving amorphous intermediates raises the question of whether similar processes are also active in sedimentary environments. This study examines the hypothesis that non-classical pathways to mineralization are the physical basis for some of the carbonate morphologies and compositions observed in natural and laboratory settings. We designed experiments with a series of different solution Mg : Ca ratios and saturation environments to investigate the effects on carbonate phase, Mg content, and morphology. Our observations of diverse carbonate mineral compositions and textures suggest geochemical conditions bias the mineralization pathway by a systematic relationship to Mg : Ca ratio and the abundance of carbonate ions. Environments with low Mg levels produce calcite crystallites with 0–12 mol% $MgCO_3$. In contrast, the *combination* of high initial Mg : Ca and rapidly increasing saturation opens a non-classical pathway that begins with extensive precipitation of an amorphous calcium carbonate (ACC). This phase slowly transforms to aggregates of very high Mg calcite nanoparticles whose structures and compositions are similar to natural disordered dolomites. The non-classical pathways are favored when the local environment contains sufficient Mg to inhibit calcite growth through increased solubility—a thermodynamic factor, and achieves saturation with respect to ACC on a timescale that is shorter than the rate of aragonite nucleation—a kinetic factor. Aragonite is produced when Mg levels are high but saturation is insufficient for ACC precipitation. The findings provide a physical basis for anecdotal claims that the interplay of kinetic and thermodynamic factors underlies patterns of carbonate precipitation and suggest the need to expand traditional interpretations of geological carbonate formation to include non-classical pathways to mineralization.

[a] *Department of Geosciences, Virginia Tech, Blacksburg, VA 24061, USA*
[b] *Molecular Foundry, Lawrence Berkeley National Laboratory, Berkeley, CA 94720, USA*
[c] *Division of Geological and Planetary Sciences, California Institute of Technology, Pasadena, CA 91125, USA*

† Electronic supplementary information (ESI) available. See DOI: 10.1039/c2fd20077e

1 Introduction

The precipitation of carbonate minerals to form sediments and sedimentary rocks provides our most comprehensive proxy record of global environmental change over the course of Earth history. Of particular interest are the ancient carbonate rocks of the Proterozoic. Reaching backward in time more than two billion years, this unique interval of Earth history marks the evolution of our planet from an inorganic to a biological world. The era is characterized by deposition of extensive carbonate sediments with unusual textures and mineral chemistries. Many of the features are rarely seen again in the geologic record and are distinct from the biological calcification of skeletal structures that dominated carbonate precipitation in the Phanerozoic.[1] These shifts in carbonate phases, compositions, and textures are broadly agreed to reflect important changes in the interactions between physical and biological processes.[2] However, we have only an incomplete picture of the mechanistic basis for the diverse patterns of carbonate mineral precipitation recorded in these deposits. Therefore, the mineralization processes that regulate distributions of trace elements and isotopes within carbonate structures—the proxies of greatest paleoenvironmental interest—remain crucial targets of research.

Calcite and aragonite are the most important polymorphs of calcium carbonate ($CaCO_3$) that form in modern seawater.[3-6] Also of considerable interest is dolomite, a Mg-rich carbonate ($CaMg(CO_3)_2$) that is widely found in ancient sediments but today seen only in restricted marine environments.[7] While equilibrium thermodynamics is thought to have the principal control on geological mineral precipitation, kinetic factors are also known to play important roles in the formation of these different phases. For example, the long-standing "dolomite problem" refers to observations that dolomite only very stubbornly precipitates at low temperature despite being significantly oversaturated in seawater as compared to calcite and aragonite.[7] Similarly, seawater is less supersaturated with respect to aragonite than to calcite, yet aragonite tends to precipitate first from seawater due to kinetic inhibition of calcite.[4,5,8]

Until recently, most interpretations of how carbonate minerals form in sedimentary environments were based upon the assumption that mineralization occurs by the ion-by-ion attachment of solutes to the mineral surfaces by the classical terrace–ledge–kink (TLK) model of crystal growth.[9,10,11] While TLK theory has provided the foundation for designing and interpreting many studies of calcite, a long-standing enigma is the inability of this growth process to produce low temperature carbonates with more than 10–12 mol% Mg.[3,12,13] Moreover, *in situ* observations confirm that calcite growth by classical step propagation becomes fully inhibited at Mg levels far below those measured in seawater, thus precluding the possibility that very high Mg carbonates, such as protodolomite, can form by this pathway at low temperature.[8,14,15]

The biomineralization community has faced a similar dilemma regarding the processes by which organisms exert species-specific controls on the Mg content of carbonate skeletons. For example, the coccoliths of calcifying algae contain approximately 0–3 mol% $MgCO_3$ (low Mg calcite, LMg calcite)[16] while coralline red algae contain 5–15 mol% $MgCO_3$ (high Mg calcite, HMg calcite).[17] In contrast, certain echinoderm species readily produce complex calcified biominerals with as much as 30 mol% $MgCO_3$[18] while the teeth of sea urchins can contain up to 50 mol%.[19] It is now recognized that these very high Mg calcite (VHMg calcite) structures form by a two-stage mineralization process that begins with the accumulation of amorphous calcium carbonate (ACC) in a localized environment[20-23] followed by its transformation to 5–100 nm nanoparticles. The resulting crystals can have significant crystallographic co-alignment at the nanoscale to produce the complex superstructures we know as skeletal biominerals.[24-27]

Studies of synthetic inorganic materials show similar processes can produce crystalline aggregates of co-aligned nanoparticles—often referred to as mesocrystals—with a wide variety of mineral chemistries.[28,29] Details vary but the conditions required for mesocrystal formation are generally characterized by 1) high supersaturation conditions that favor extensive nucleation; 2) the presence of inhibitors that limit crystallite growth; and 3) quiescent environments that allow interaction potentials of particle surfaces to favor aggregation along preferred orientations. Studies also demonstrate that organic and some polymeric species can mediate crystal–crystal interactions during mesocrystal formation.[28–30]

Emerging concepts from biomineral and synthetic materials studies of the amorphous-to-mesocrystal pathway raise the question of whether a similar process could be the missing link to interpreting the origins of some high and very high Mg carbonates. The influence of magnesium is of particular interest because the ratio of Mg : Ca in seawater has varied over geologic history[4] and Mg slows the growth rate of carbonate minerals. However, few experiments have investigated this influence while monitoring the pathway to mineral formation. Two recent studies show the Mg content of ACC can attain (and exceed) the composition range of VHMg calcite and dolomite.[31,32] Those studies, however, cannot provide insights into the nature of the final crystalline products. Here we tested the hypothesis that the interplay of Mg : Ca ratio and the abundance of carbonate ion regulate the pathway to carbonate formation. To investigate this idea, we designed two calcification environments that varied the abundance of carbonate ion and Mg : Ca of the solution. By monitoring the products that formed, we found composition and morphological evidence for three pathways to calcite formation. The relations also suggest a plausible explanation for the formation of calcite *versus* aragonite at low temperatures.

2 Experimental

The saturation state of a solution with respect to ACC and calcite is determined by the relations

$$K_{sp,calcite} = (a_{Ca}{}^{2+}\ a_{CO_3}{}^{2-})/a_{CaCO_3,calcite} \quad (1)$$

$$K_{sp,ACC} = (a_{Ca}{}^{2+}\ a_{CO_3}{}^{2-})/a_{CaCO_3,ACC} \quad (2)$$

where a_i is the chemical activity of an ion or solid and the solubility products are $K_{sp,calcite} = 10^{-8.43}$ and $K_{sp,ACC} = 10^{-6.39}$,[33] respectively. Thus, the reported K_{sp} of pure ACC is approximately 100× greater than calcite. If the ACC solubility value is correct, an ACC-saturated solution with equivalent activities of Ca^{2+} or CO_3^{2-} contains approximately 10× more Ca^{2+} and 10× more CO_3^{2-} than a calcite-saturated solution.

2.1 Mineralization

Solutions were prepared using reagents of $CaCl_2 \cdot 2H_2O$ (Sigma Aldrich, 99+%) and $MgCl_2 \cdot 6H_2O$ (Sigma Alrich, 99%). Most experiments were conducted with Ca^{2+} concentrations of 10 mM and Mg levels were set to obtain starting molar ratios of Mg : Ca = 0.0 to 6.0. Additional experiments were conducted at $[Ca^{2+}]$ = 25 mM and, again, Mg levels were set accordingly.

Supersaturation was achieved using two established methods. For the ammonium carbonate diffusion method, 10 g of $(NH_4)_2CO_3$ (Sigma Aldrich) was dissolved in 20 mL distilled deionized water to provide the carbonate source.[31,34–37] After adding the salt solutions described above to six replicate vessels that contained silica glass substrates or carboxylated surfaces prepared by standard methods (Wallace *et al.*, 2009) using MUA on gold. The solutions were exposed to the ammonium carbonate source and incubated in closed vessels at 40 °C and 100% humidity. The ammonium carbonate solutions were removed after 1 day and the precipitates were removed and washed with anhydrous ethanol after 3 days.

A second method used a batch approach to mix solutions of $CaCl_2 \cdot 2H_2O$, $MgCl_2 \cdot 6H_2O$, and $NaHCO_3$ to produce reactants with lower carbonate levels and pH = 8.5. Initial Ca^{2+} concentrations were 10 mM. To keep the initial saturation state equivalent for all batch experiments, $[HCO_3^-]$ varied from 12 to 30 mM depending on solution Mg : Ca, which ranged from 0.0 to 6.0. Solutions were mixed in 3 mL wells containing silica glass substrates for precipitate collection. After 6 days at 25 °C, the substrates were removed and washed with anhydrous ethanol.

For both methods, parallel experiments were conducted to monitor pH and decreases in the solution Mg^{2+} and Ca^{2+} concentrations by sampling 2 mL from the replicates while also while observing the onset and progression of mineralization using an optical microscope. Solutions for chemical analysis were filtered through 0.2 μm nylon membranes (Whatman) and analyzed for Mg and Ca by inductively coupled plasma atomic emission spectrometry (Spectro ARCOS SOP).

2.2 Preparation of Coorong dolomite

Disordered dolomites from the Coorong Lakes, South Australia were sonicated in 30% peroxide for two hours, centrifuged for 10 min, and washed with ethanol. The samples were dispersed on substrates for analysis.

2.3 X-Ray diffraction (XRD)

For each treatment, three replicate samples were analyzed on a PANalytical X'PERT PRO with Cu–Kα_1 radiation operated at 45 kV and 40 mA from 20°–60° 2θ at scan speeds of 0.02° sec^{-1}. $MgCO_3$ compositions were determined using a linear correlation[39] of peak position to composition. All sample (104) peaks were corrected by the silicon (111) peak at 28.439° 2θ using a powdered silicon internal standard (NIST 640D), deposited by ethanol dispersion. Noting that structural considerations such as lattice disorder are convoluted into (104) peak position, especially in a disordered material, a recent study of calcites (0–50 mol% $MgCO_3$) found measurements using a linear correlation represent a lower bound on $MgCO_3$ content.[40]

2.4 Electron microscopy

SEM images were collected using a FEI Helios 600 Nanolab operated at 5 kV accelerating voltage. TEM samples were prepared by established focused ion beam (FIB) methods[41] on a FEI Helios 600 Nanolab. TEM selective area electron diffraction (SAED) patterns and dark field images were collected on a Philips EM420 operated at 120 kV.

3 Results

3.1 Morphology

For the ammonium carbonate diffusion method, visible precipitation began in the Mg-free experiments that contained the lower concentration of calcium (10 mM) with the formation of euhedral calcite crystallites (Fig. 1A). At the higher calcium level (25 mM), ACC was observed for the first few minutes of the experiment. For both conditions, any visible ACC underwent rapid dissolution to become a source of ions for the growing crystallites. The resulting crystallites exhibited a rhombohedral morphology, without evidence of polycrystals, indicating growth by step propagation. Calcite crystallites also formed, albeit more slowly, from solutions with an initial Mg : Ca level of 1.0 to produce the elongated 'birdseed' morphology (Fig. 1B) previously reported for Mg-bearing calcites.[36,37,42]

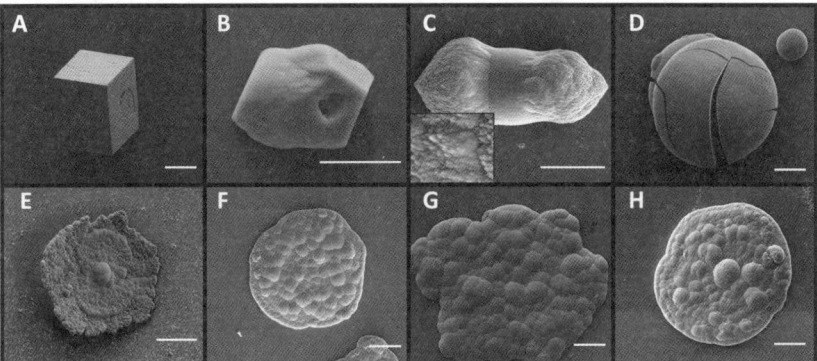

Fig. 1 Systematic changes in calcite morphology indicate shifts in the predominant mineralization processes as step growth becomes progressively inhibited by Mg concentrations in solution. (**A**) Mg : Ca = 0. Crystallites grow by a classical step growth process and exhibit the rhombohedral morphology typical of the pure calcite system; (**B**) Mg : Ca = 1. Elongated birdseed crystallites that also express (104) facets; (**C**) Mg : Ca = 2.0. Crystallites that form at higher Mg levels are elongated assemblages of small calcite particles (inset); (**D, E**) onset of transition at Mg : Ca = 3.0 produces a mixture of morphologies with high Mg calcite spherules and aggregates of very high Mg calcite nanoparticles, respectively. (**F**) Mg : Ca = 4.0; (**G**) Mg : Ca = 5.0; (**H**) Mg : Ca = 6.0. For all Mg : Ca levels, films of ACC transform into similar low-relief aggregates of very high Mg calcite nanoparticles aggregates that exhibit a botryoidal texture. All scale bars = 20 microns.

A significant morphological change was seen in the calcite crystallites that form from solutions that began with a Mg : Ca ratio of 2.0 to 3.0. After the extensive formation of ACC particles on the surface, calcite crystallites developed with dumbbell and spherulitic morphologies, respectively (Fig. 1CDE). The structures are regular but lack evidence of faceting seen at the lower Mg : Ca solutions. The dumbbell structures had the appearance of being comprised of smaller particles with like morphology (inset Fig. 1C). Raman spectroscopy confirmed that all of the crystallites were calcite. At the higher Mg : Ca = 3.0, the spherulitic calcites are accompanied by a small proportion of disc-shaped aggregates that marked the beginning of a transition to a different mineralization process (Fig. 1E).

When the initial Mg : Ca levels were ≥4.0, a third type of calcite morphology was observed (Fig. 1FGH). Precipitation began with the accumulation of high Mg ACC onto the substrate as a continuous sheet, without evidence of crystallites (Fig. 2A). The ACC layer subsequently transformed to circular, low relief aggregates of calcite nanoparticles (Fig. 2BC). As aggregates developed, the surrounding ACC was depleted (Fig. 2BC) and the calcite developed with a distinctive botyroidal (anisotropic and radial) texture without macroscopic evidence of external crystallographic control (Fig. 2D). TEM dark field images, however, showed significant co-alignment of the constituent nanoparticles within the aggregates (Fig. S1†). The morphology of these calcites was independent of solution Mg : Ca level (Fig. 1FGH) but their rate of transformation, significantly slowed with increasing Mg : Ca ratio of the initial solution. For example, solutions with a Mg : Ca ratio of 4.0 and 6.0 underwent the ACC to calcite transformation in approximately 1 day and 3 days, respectively. Allowing the experiments to continue after all of the ACC had transformed to calcite, we observed the formation of late stage aragonite. ACC and aragonite were not observed to co-exist in any of the experiments.

3.2 Composition

Measurements of the corresponding precipitate compositions showed the Mg content of ACC was linearly dependent upon solution Mg : Ca ratio (Fig. 3A) as

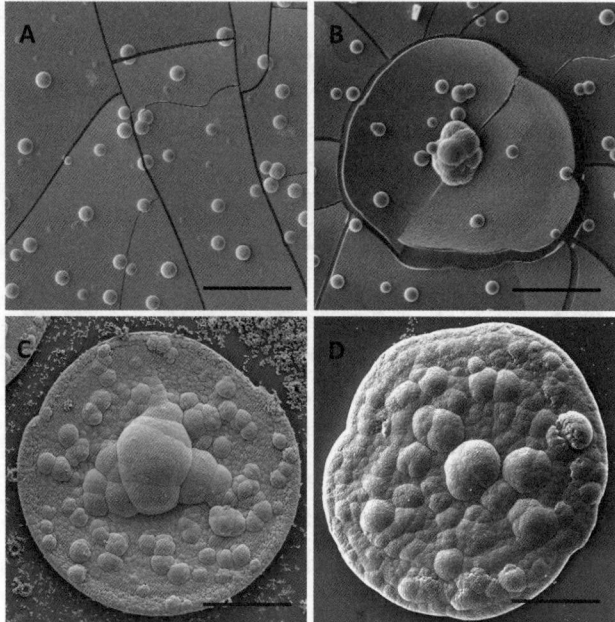

Fig. 2 Under extreme conditions, ACC forms as a sheet that transforms into very high Mg calcite. All scale bars = 25 microns. (**A**) Typical morphology after 5 h for initial Mg : Ca of 5.0 shows a continuous film of ACC. (**B**) After 12 h, the very high Mg calcite begins to form as circular discs that develop from the ACC film. (**C**) At an intermediate stage (24 h), these regions evolve into low relief aggregates of very high Mg calcite aggregates. The surrounding ACC is depleted and appears as discrete particles that reveal the underlying glass substrate. (**D**) Late stage (3 days) very high Mg calcite aggregates.

reported previously.[31,32] Comparisons of the calcite compositions and the fractionation patterns of the calcites that transformed from this metastable intermediate provided independent evidence for multiple mineralization pathways. Field (**1**) shows that when solutions have low Mg : Ca levels, the ACC and euhedral calcites that formed have the same Mg content, and thus, the same fractionation within experimental errors. Our measurements of 0–12 mol % $MgCO_3$ in the calcite crystallites are consistent with reports for classical step growth.[13]

At the higher Mg : Ca levels that correspond to Field (**2**), the ACC transformed to HMg calcites of 10–20 mol% $MgCO_3$. Note that these levels of Mg are lower than what was measured for its amorphous precursor. Thus, the transformation process reduced the Mg content in the calcite by up to 10%. Qualitative observations showed the amount and size of crystallites decreased in parallel with an increasing Mg content of the calcite that formed. This is consistent with measurements showing that Mg incorporation slows the rate of calcite growth by the step propagation process.[8] The change in kinetics is a consequence of the higher solubility of the Mg–calcite solid solution products ($Ca_{1-x}Mg_xCO_3$), a thermodynamic effect.[8]

At Mg : Ca ratios ≥ 3, Field (**3**) shows that the highest Mg ACC transformed to a VHMg calcite that contained 35–40 mol % $MgCO_3$. The compositions exhibit a plateau in Mg content without evidence of fractionation with the composition of the solution or its amorphous precursor. While our estimates of the higher Mg contents may contain as much as 2% error (*e.g.* experimental), the amounts of Mg in the crystallites are nonetheless >2× greater than measurements for the crystallites in Field (**2**). Significantly, values of up to 40 mol% $MgCO_3$ are within the composition range of disordered dolomites.[43]

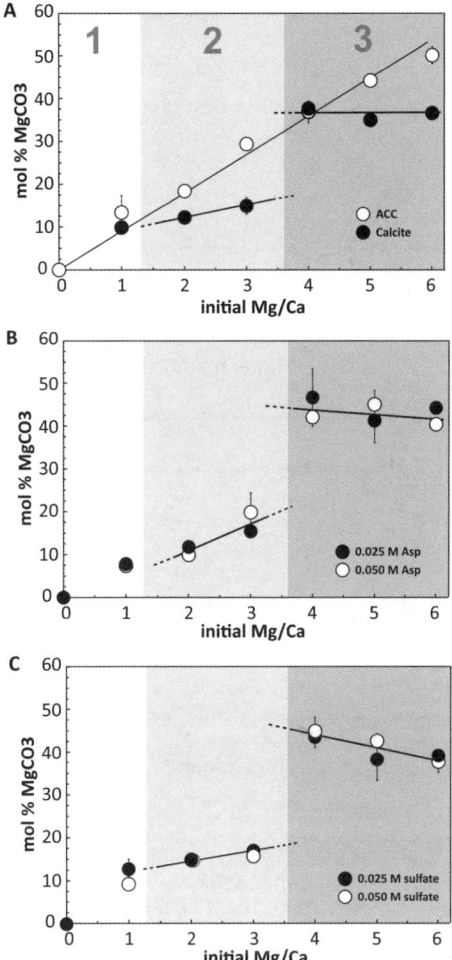

Fig. 3 Pathway to mineralization influences calcite composition. (**A**) The composition of ACC exhibits a linear dependence upon initial Mg : Ca ratios as reported previously.[31,32] The composition of calcites that transform from the ACC shift in composition according to Mg : Ca levels: for solutions with low Mg : Ca levels (**1**), the ACC transformation produces calcites with similar compositions. Measurements of 0–12 mol % $MgCO_3$ are consistent with studies for classical step growth. At higher Mg : Ca levels (**2**), Mg is excluded during the ACC transformation to produce calcite with a lower Mg content than its amorphous precursor. The calcites that form from ACC at these highest Mg : Ca levels (**3**) exhibit a plateau in Mg content. For the conditions of these experiments, these VHMg calcites contain approximately 35–40 mol % $MgCO_3$. Calcites that transform from ACC in solutions that contain two concentrations (0.025 M and 0.05 M) of (**B**) the composition of calcites that form from ACC in the presence of aspartate; and (**C**) sulfate solutions show similar compositions. Precipitation onto carboxyl-functionalized substrates gives similar trends (Fig. S3†)

Measurements of chemical composition changes during the mineralization process showed the evolution of total solution Mg, Ca and Mg : Ca ratio over the course of experiments conducted at initial solution Mg : Ca = 2.0 and 4.0 (Fig. 4AB). A maximum pH of 10 was reached during the course of the experiments. Recall that these conditions produced HMg calcite crystallites and VHMg calcite nanoparticle aggregates, respectively. As seen in Fig. 4AB, sharp increases occurred in solution Mg : Ca ratios during the first few hours until becoming approximately constant (Fig. 4AB).

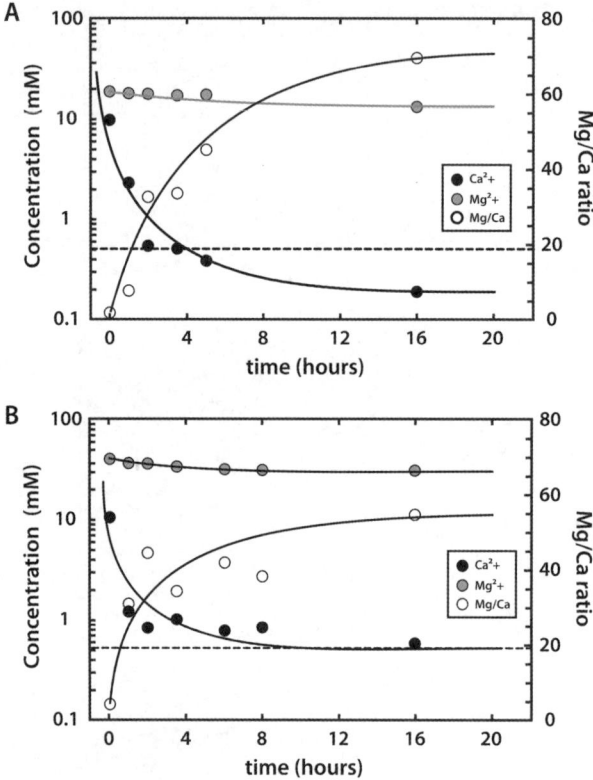

Fig. 4 Measurements of solution chemistry during precipitation for experiments conducted at (A) initial Mg : Ca = 2 and (B) intial Mg : Ca = 4. The data show minimal changes in total Mg compared to the large decrease in calcium concentration. Thus, both solutions acquire very high Mg : Ca levels as precipitation proceeds. For Mg : Ca solutions = 2, the rapid decrease in Ca^{2+} concentration to approximately 0.5 mM is indicative of very early precipitation of Mg calcite. In contrast, solutions with initial Mg : Ca = 4 show that total Ca concentration is maintained at the higher value of 0.8 mM for 16 h, indicating the inhibition of calcite formation and the stability of ACC. Dashed line indicates stoichiometric solubility of ACC at 40 °C.[33]

For experiments at Mg : Ca ≤ 3.0, Fig. 3ABC shows the Mg content of crystallites that formed is dependent upon initial solution composition. This suggests, in Field (2), Mg levels are set by the initial, or very early solution composition (*e.g.* Fig. 4A). In contrast, in Field (3), experiments conducted at Mg : Ca = 4.0 showed Mg content is not dependent upon initial solution composition. In this case, the nanoparticle calcite is not transformed from ACC until several hours after the start of the experiment when Mg : Ca levels approached the long asymptote of the solution conditions (Mg : Ca = 30–40) at the time of transformation (Fig. 4B). This suggests the 'compositional gap' (Fig. 3ABC) between Fields (2) and (3) is an artifact of changing solution conditions and delay of calcite formation in Field (3).

Additional experiments investigated the influence of organic and inorganic constituents that are commonly found in biological and sedimentary environments on mineralization by these processes. The acidic aspartate (Asp, $C_4H_6NO_4^-$) and other carboxylated biomolecules are implicated in promoting calcification and Mg uptake[22,31] while studies have debated the role of sulfate (SO_4^{2-}) as both an inhibiting[44] and catalyzing agent[45] in the formation of high Mg carbonates. For the conditions of these experiments, we find that neither of these molecules significantly influenced composition (Fig. 3B and C, respectively) or morphology (Fig. S2†).

Because the single Asp residue has small effects on mineralization compared to larger carboxylated macromolecules[15,22] found in eutrophic environments, further studies of biomolecule effects are needed. Parallel experiments conducted on the carboxyl-terminated substrates (Fig. S3†) produced the same composition and texture trends as seen in the additive-free experiments.

4 Discussion

Observations of distinctive patterns in calcite morphology (Fig. 1) and composition (Fig. 3) suggest an interplay between Mg content—an inhibitor to step growth—and the supersaturation—which has a primary control on rates of nucleation and step growth. Low levels of Mg in solution allow crystallites of calcite to nucleate and grow by the classical process, including those solutions with saturation states that exceed ACC solubility. As Mg is introduced to solution and the growing calcite incorporate progressively higher levels of this ion, the rate of growth slows.[8,14] Calcite compositions reflect the initial Mg : Ca conditions and may be related by a fractionation coefficient, shown in Fields (**1**) and (**2**) of Fig. 3A. This interpretation is consistent with studies that estimate a Mg : Ca ratio of 1.0 to 2.0 as the threshold for inhibiting calcite growth[8,14,15,46] and the transition to aragonite precipitation.[4,47]

In contrast, the pathway to producing VHMg calcite is favored when Mg levels are sufficient to inhibit classical growth *and* levels of carbonate ion accumulate rapidly to high values. If the saturation state reaches ACC solubility, the negligible free energy barrier to ACC formation allows for rapid precipitation of very high Mg ACC. This amorphous phase is thermodynamically unstable, and transforms to particles of VHMg calcite. However, subsequent growth of these particles is severely inhibited in the high Mg environments and the final products exhibit mesocrystal features as aggregates of nanoparticles with spherulitic and botryoidal textures.

4.1 Conceptual model

Combining results from this work and insights from previous studies suggests a process-based conceptual model for how geochemical conditions can bias pathways different types of calcite. The schematic diagram (Fig. 5) suggests a systematic, albeit qualitative, relationship between changes in the Mg : Ca conditions and saturation state through the supply of carbonate ion to produce calcite through three types of processes. In Field (**1**), low Mg environments combined with lower carbonate levels to produce calcite by the classical ion-by-ion addition process [8,10,14,15] with compositions that reflect fractionation trends.

A second pathway is favored by the combination of moderate to high carbonate levels and higher Mg : Ca conditions (Field **2**). Here, ACC is the initial precipitate and it subsequently transforms to HMg calcite with Mg contents up to 20 mol% $MgCO_3$. This composition range is reported for the skeletal structures of diverse calcifying organisms[17,22,48–50] and also seen in some sedimentary environments.[1,51]

Field (**3**) represents extreme settings that have high Mg conditions and a *rapid* increase in carbonate ion concentration to reach very high saturation states. The *combination* of these conditions is required to produce the very high Mg ACC that subsequently transforms to VHMg calcite (Field **3**). At first glance, one might expect the highest Mg levels to favor aragonite, but we suspect the initial appearance of ACC and then calcite is a kinetic effect. That is, the rapid climb to high carbonate levels effectively 'leapfrogs' the aragonite field by reaching ACC saturation quickly. This favors precipitation of very high Mg ACC that (slowly) transforms by extensive nucleation to VHMg calcite. Subsequent growth of this calcite is severely inhibited by the high Mg levels to limit growth beyond the nanoparticle size range.

Finally, the conditions associated with the three qualitative fields in Fig. 5 suggest a geochemical window in the upper left portion of the diagram where calcite does not readily precipitate. This environment is associated with solutions that have high Mg

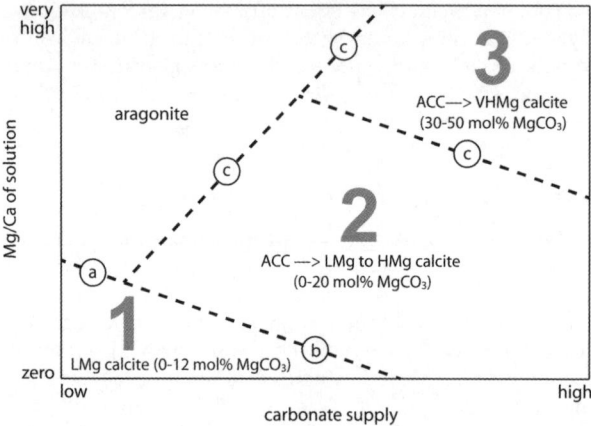

Fig. 5 Conceptual model for how the interplay of Mg : Ca ratio and carbonate supply bias pathways to CaCO$_3$ mineralization. When initial Mg : Ca ratio and saturation are low, calcite crystallites nucleate and grow by classical processes (Field **1**). As Mg : Ca of the solution is increased, the growth rate of calcite slows to zero.[8,14,42] There are two consequences of this inhibition: higher Mg : Ca levels in combination with a low supersaturation in carbonate level favor aragonite precipitation. However, higher Mg : Ca levels in combination with higher saturations promotes the rapid formation of a Mg–ACC that subsequently transforms to LMg to HMg calcites (Field **2**) or, in the case of extreme environments, VHMg calcite (Field **3**). The dashed lines denote the highly schematic nature of this illustration and serve as a reminder that the influence of geochemical factors such as inhibitors and pH are not considered here. **Annotations on the figure:** (a)[5,6] (b)[77] (c) *this study.*

levels but the carbonate supply is lower, or increased slowly to a level that is below ACC solubility. Previous studies indicate this region should favor aragonite because the calcite to aragonite transition is believed to occur at Mg : Ca ratios of approximately 0.9 to 2.0 in the laboratory[5,6] and seawater,[4] respectively.

To test this idea, we conducted additional experiments using a batch method with initial Mg : Ca ratios = 0.0 to 6.0 and carbonate levels that were insufficient to precipitate ACC.[33] Indeed, initial Mg : Ca > 2 produce 100% aragonite without evidence of calcite crystallites or nanoparticle aggregates (Fig. 6AB).

The diagram and experimental evidence, however, raise the point that multiple pathways to mineralization can emerge from a combination of kinetic and thermodynamic factors. Conditions that favor HMg and VHMg calcite *versus* aragonite provide good examples. Here, the local environment must 1) contain sufficient Mg to inhibit calcite growth through increased solubility—a thermodynamic factor; and 2) achieve a saturation with respect to ACC on a timescale that is shorter than the rate of aragonite nucleation—a kinetic factor. These insights provide a physical basis for the long-standing idea that the interplay between kinetic and thermodynamic factors govern patterns of carbonate precipitation in sedimentary environments.[13]

The conceptual model in Fig. 5 is consistent with previous field[52–54] and laboratory studies[5,6] that noted the importance of carbonate ion concentration as a primary control on the appearance of calcite *versus* aragonite. However, the construct is admittedly highly simplified and qualitative. The diagram does not account for the effects of minor and trace impurities on mineralization. For example, one would expect that Sr^{2+}, large macromolecules or SO$_4^{2-}$ could influence ACC formation or the stability of calcite *versus* aragonite (see later discussion). One could also ask why this conceptual model is different from the construct reported by Given and Wilkinson in 1985.[55] At that time, the scientific community had not recognized a low temperature pathway to classical growth of VHMg calcite. Moreover, it was not known that some aragonite ooids are likely deposited as ACC layers that transform to VHMg calcite (see later discussion) and are eventually altered to aragonite.

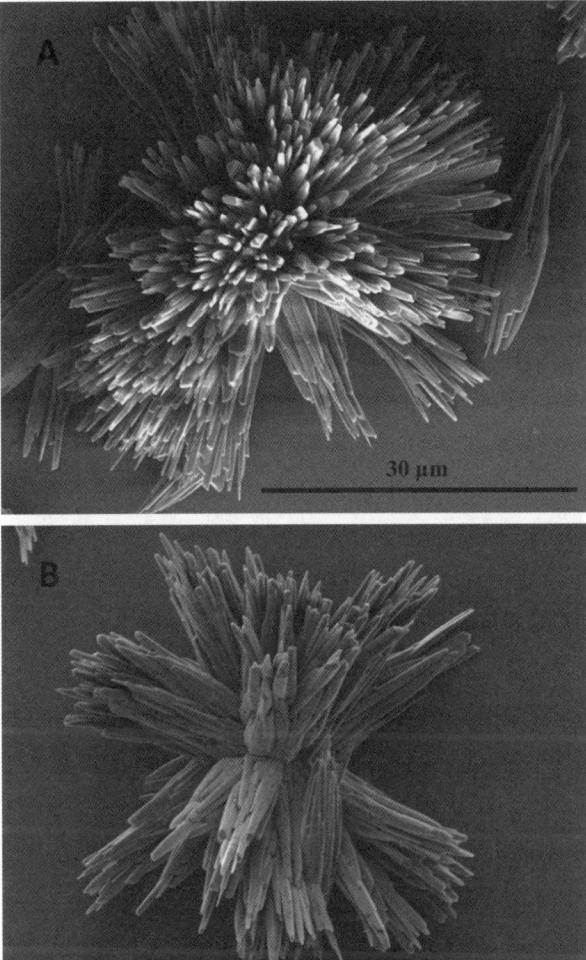

Fig. 6 Aragonite nucleation and growth is favored by the combination of high Mg : Ca levels and a slow saturation of the solution to values that are below ACC solubility.[33] ACC and aragonite are not observed to co-exist under any of the conditions investigated in this study. (**A**) Mg : Ca = 2.5 and (**B**) Mg : Ca = 4.5. Scale bar = 30 microns.

5. Implications

Insights from this study suggest the need to expand the traditional picture of how calcium carbonate minerals are formed to include the possibility of non-classical mineralization processes. Multiple pathways to carbonate precipitation may better explain some of the diverse patterns of carbonate deposition in the sediment record. Many questions are unanswered and need careful consideration to determine the scope of their importance.

5.1 ACC as precursor to disordered dolomite

The ability of the ACC pathway to produce VHMg calcite provides a possible explanation for the source of very high Mg carbonate needed to make disordered dolomites at low temperatures.[56,57] While disordered carbonates are proposed by a number of studies as possible precursors to ordered dolomite,[58,59] the study shows the process by which calcite with very high Mg concentrations can precipitate. For a similar non-classical process to be generally applicable to the widespread occurrence of dolomite, the mechanism proposed here must also be able to operate over long periods of time in order to explain the volumes that are seen in the rock record. We suggest that diffusion boundary layers in pore systems or microbe–water interfaces are involved, where local, micron-scale environments can rapidly reach ACC saturation. Our findings support the idea that the precipitation of high and very high Mg calcite is favored in locally isolated or transport limited environments that allow Mg : Ca ratios to climb while also supplying high carbonate levels. This interpretation is consistent with that of Hardie,[7] who proposed that both of these conditions are essential for dolomite formation and extensive field evidence that nearly all Proterozoic carbonates began as calcite or aragonite. Our findings also support anecdotal evidence that ACC formation is a key step *en route* to dolomite.[56,58]

One field test of this concept is found in modern sedimentary dolomites from the Coorong Lakes, South Australia[60] that form in shallow evaporative waters with high Mg : Ca ratios of 5–50 and very high supersaturations.[61] The disordered dolomites that occur in this geologically young sedimentary environment may provide an 'evolutionary link' to ordered dolomites in the rock record. Indeed, Coorong carbonates are 1–3 μm particles that express pseudo facets, yet are composed of co-oriented 50–200 nm particles that are consistent with the dolomite structure (Fig. 7A). SAED analysis shows these pseudofacets likely arise from partial ordering and co-alignment of the particles[43] (Fig. 7A inset). These features are comparable to the VHMg calcite produced in this study for sulfate-free (Fig. 7B) and sulfate-bearing (Fig. 7C) experiments that both exhibit some particle co-alignment (Fig. 7AB insets).

If disordered dolomites are indeed the products of an ACC to VHMg calcite conversion, then the rock record should contain evidence for their primary roots. Observations from this study suggest three plausible markers for origins in VHMg calcite that could be preserved in geological carbonates: first, some ancient carbonates may preserve relic structures of VHMg calcite as spherulites. For example, Proterozoic dolomites from Namibia show spherical structures[62] that are similar to the VHMg calcite aggregates reported herein. Other spherules are seen in controversial reports of 'nannobacteria' that could be relic VHMg calcite.[63] Dolomite films or coatings that line the pore spaces of supratidal carbonates[64] could also result from repeated evaporations of sea water to form sheets of VHMg calcite similar to the low relief aggregates in this experimental study. Another possibility is carbonate mud, whose primary composition could conceivably have been ACC such as we hypothesize for the Coorong sediments. To explain the replacement of preexisting grains is, however, a more difficult challenge.

5.2 Other environments

Carbonate mineralization *via* ACC may also occur during the precipitation of ooids. As small coated grains, often <2 mm,[1] the origins of these spherical structures in carbonate platform environments have been debated for more than a century.[65–67] Efforts to interpret ooid origins may have been complicated by observations of the aragonite that forms at later stages and because ACC formation was not recognized at the time. Some studies have also argued for a direct biological origin.[68] However, a recent analysis of individual layers that comprise natural ooids from the Bahamian Archipelago suggests each 'overgrowth' begins as a veneer of ACC that subsequently transforms to VHMg calcite.[69] This interpretation suggests ACC saturation can be readily achieved within a local microenvironment, likely as an indirect consequence of microbial activity to produce locally high levels of carbonate ions.

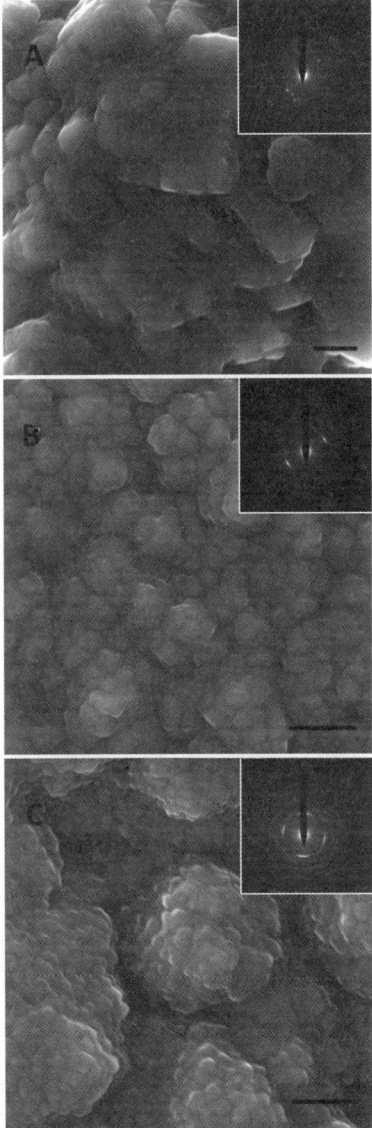

Fig. 7 SEM images of VHMg calcites produced in this study are similar to modern dolomitic carbonates. Natural and laboratory samples are comprised of aggregates of nanoparticles that possess some degree of interparticle co-alignment. All scale bars = 200 nm. (**A**) Modern dolomite from Coorong Lakes, Australia is comprised of 1–3 μM particles that express macro-facets consistent with termination. These particles possess partial ordering into 100–150 nm subdomains (inset). Individual particles are highly co-aligned along two predominant orientations along the [452] zone axis as seen in SAED spot patterns. (**B**) Calcites synthesized from control solutions at Mg : Ca = 5 are comprised of 58 ± 12 nm particles (35 mol% $MgCO_3$). (inset) Arcs in SAED patterns from synthetic samples indicate partial co-alignment of particles. (**C**) Sulfate solutions with initial Mg : Ca = 5 produce smaller 33 ± 6.8 nm calcite nanoparticles (45 mol% $MgCO_3$).

It appears likely that the formation of ACC in open marine environments of the modern ocean should be ruled out since both Mg : Ca and oversaturation are thought to be buffered to moderate values. Thus, settings that allow local fluids to evolve, such as in pore spaces associated with diagenetic environments or intercellular settings, may be most promising. In both of these situations, both Mg : Ca ratios and saturation state can become elevated as carbonate or gypsum ($CaSO_4$) precipitation occurs. The correlation of such environments with dolomite precipitation has been long recognized.[70,71] Nevertheless, the mechanism proposed herein has an additional requirements: first, the reported solubility of ACC argues that ion concentrations be maintained or oscillate to high values relative to calcite saturation (*i.e.*, >10×). Second, aragonite formation must be prevented. We show this can be achieved under conditions that supply a high carbonate level to favor the higher energy precipitation of ACC.

The botryoidal aggregates of calcite produced in this study bear a striking similarity to the larger scale (1–2 mm) carbonate botryoids that are seen in the Upper Cretaceous cements of many Mediterranean localities.[72] These very high Mg calcites exhibit radial morphologies that were originally interpreted as aragonite. Low-relief botryoidal or "mound" structures are also noted as an intermediate morphology during the hydrothermal synthesis (218 °C) of dolomite.[58] However, the length scale of the individual nanocrystals (lateral dimensions of 10–200 nm) is consistent with the observations of this study. Though anecdotal, these observations suggest the origins of botryoidal carbonates with radial or botryoidal fabrics should be revisited. Almost nothing is known about how the ACC to calcite pathway influences the uptake of trace and minor elements beyond the marked changes in Mg fractionation reported herein. Finally, the influence of this pathway on isotopic signals is also unknown but synthetic disordered dolomites exhibit oxygen isotopes that are offset to significantly lower values than expected.[56]

5.3 Biologically mediated carbonate formation

Microbial activity in sediments is widely recognized in carbonate mineralization. Generally, it is assumed that the extracellular polymeric matrix associated with biological processes directly modifies carbonate nucleation and growth.[62] Alternatively, it has been shown microbial metabolic rates also directly control rates of carbonate mineralization.[73,74] This relationship is implicitly contained in Fig. 5 through carbonate supply and the connection of this key variable to the many environmental parameters that control microbial metabolic rates (nutrient availability,[75] temperature,[76] metabolic process, *etc.*) and geochemical controls of cation concentrations. Their interactions, thus, have a strong influence on the pathway to carbonate mineralization.

In conclusion, the findings suggest the need to revisit the traditional picture of carbonate formation to consider non-classical pathways to mineralization and the consequences of interplays between kinetic and thermodynamic factors.[7,13] A good example is seen in our observations that the very high Mg calcites are favored by a local environment that: 1) contains sufficient Mg to inhibit crystallite growth through increased solubility—a thermodynamic factor; and 2) achieves a saturation with respect to ACC on a timescale that is shorter than the rate of aragonite nucleation—a kinetic factor. For example, the intimate associations of calcite and aragonite that are observed in some cements[55] may be explained as the interplay of small fluctuations in the local geochemical environment. By revisiting the geochemical conditions that bias the energy landscape of mineralization toward different amorphous and crystalline phases, an updated view of relationships between mineralization patterns and environment may emerge.

Acknowledgements

The research was supported by awards to PMD from the US Deptartment of Energy (DOE BES-FG02-00ER15112) and the National Science Foundation (NSF OCE-1061763). DW thanks the Mineralogical Society of America for the Grant for

Student Research that also supported this work. LMH thanks the NSF for support through the Graduate Research Fellowship Program. This work was also supported by the Office of Science, Office of Basic Energy Sciences, Division of Chemical Sciences, Geosciences, and Biosciences, of the U.S. Department of Energy under Contract No. DE-AC02-05CH11231. The opinions, findings, and conclusions or recommendations expressed in this material are those of the authors and do not necessarily reflect the views of the NSF or the DOE. We thank JF Read for thoughtful insights into geological carbonates.

References

1 M. E. Tucker and V. P. Wright, *Carbonate sedimentology*, Blackwell Scientific Publications, London, 1990.
2 A. H. Knoll and K. Swett, *Am. J. Sci.*, 1990, **290A**, 104–132.
3 J. W. Morse, R. S. Arvidson and A. Luttge, *Chem. Rev.*, 2007, **107**, 342–381.
4 S. M. Stanley and L. A. Hardie, *Palaeogeogr., Palaeoclimatol., Palaeoecol*, 1998, **144**, 3–19.
5 J. W. Morse, Q. W. Wang and M. Y. Tsio, *Geology*, 1997, **25**, 85–87.
6 J. Lee and J. W. Morse, *Geology*, 2010, **38**, 115–118.
7 L. A. Hardie, *Journal of Sedimentary Petrology*, 1987, **57**, 166–183.
8 K. J. Davis, P. M. Dove and J. J. De Yoreo, *Science*, 2000, **290**, 1134–1137.
9 W. K. Burton, N. Cabrera and F. C. Frank, *Philos. Trans. R. Soc. London, Ser. A*, 1951, **243**, 299–358.
10 A. A. Chernov and E. I. Givargizov, *Modern crystallography: crystal growth*, Springer-Verlag, Berlin, 1984.
11 J. J. De Yoreo, L. A. Zepeda-Ruiz, R. W. Friddle, S. R. Qiu, L. E. Wasylenki, A. A. Chernov, G. H. Gilmer and P. M. Dove, *Cryst. Growth Des.*, 2009, **9**, 5135–5144.
12 L. S. Land, *Aquat. Geochem.*, 1998, **4**, 361–368.
13 R. L. Folk and L. S. Land, *Bulletin—American Association of Petroleum Geologists*, 1975, **59**, 60–68.
14 L. E. Wasylenki, P. M. Dove and J. J. De Yoreo, *Geochim. Cosmochim. Acta*, 2005, **69**, 4227–4236.
15 A. E. Stephenson, J. J. DeYoreo, L. Wu, K. J. Wu, J. Hoyer and P. M. Dove, *Science*, 2008, **322**, 724–727.
16 H. M. Stoll, J. R. Encinar, J. I. G. Alonso, Y. Rosenthal, I. Probert and C. Klaas, *Geochem., Geophys., Geosyst.*, 2001, **2**, 14.
17 K. E. Chave, *J. Geol.*, 1954, **62**, 266–283.
18 J. N. Weber, *Am. J. Sci.*, 1969, **267**, 537–566.
19 J. S. Robach, S. R. Stock and A. Veis, *J. Struct. Biol.*, 2006, **155**, 87–95.
20 Y. Ma, S. Weiner and L. Addadi, *Adv. Funct. Mater.*, 2007, **17**, 2693–2700.
21 E. Beniash, J. Aizenberg, L. Addadi and S. Weiner, *Proc. R. Soc. London, Ser. B*, 1997, **264**, 461–465.
22 S. Raz, S. Weiner and L. Addadi, *Adv. Mater.*, 2000, **12**, 38–42.
23 C. E. Killian, R. A. Metzler, Y. U. T. Gong, I. C. Olson, J. Aizenberg, Y. Politi, F. H. Wilt, A. Scholl, A. Young, A. Doran, M. Kunz, N. Tamura, S. N. Coppersmith and P. Gilbert, *J. Am. Chem. Soc.*, 2009, **131**, 18404–18409.
24 Y. Oaki, A. Kotachi, T. Miura and H. Imai, *Adv. Funct. Mater.*, 2006, **16**, 1633–1639.
25 I. Sethmann, R. Hinrichs, G. Worheide and A. Putnis, *J. Inorg. Biochem.*, 2006, **100**, 88–96.
26 D. Vielzeuf, N. Floquet, D. Chatain, F. Bonnete, D. Ferry, J. Garrabou and E. M. Stolper, *Am. Mineral.*, 2010, **95**, 242–248.
27 Y. Oaki and H. Imai, *Small*, 2006, **2**, 66–70.
28 R. Q. Song and H. Colfen, *Adv. Mater.*, 2010, **22**, 1301–1330.
29 H. Colfen and M. Antonietti, *Mesocrystals and nonclassical crystallization*, John Wiley & Sons, Ltd, Chichester, 2008.
30 A. N. Kulak, P. Iddon, Y. T. Li, S. P. Armes, H. Colfen, O. Paris, R. M. Wilson and F. C. Meldrum, *J. Am. Chem. Soc.*, 2007, **129**, 3729–3736.
31 D. B. Wang, A. F. Wallace, J. J. De Yoreo and P. M. Dove, *Proc. Natl. Acad. Sci. U. S. A.*, 2009, **106**, 21511–21516.
32 A. V. Radha, A. Fernandez-Martinez, Y. Huc, Y.-S. Jun, G. A. Waychunas and A. Navrotsky, *Geochim. Cosmochim. Acta*, 2012, **90**, 83–95.
33 L. Brecevic and A. E. Nielsen, *J. Cryst. Growth*, 1989, **98**, 504–510.
34 T. Y. J. Han and J. Aizenberg, *Chem. Mater.*, 2008, **20**, 1064–1068.
35 G. E. Henderson, B. J. Murray and K. M. McGrath, *J. Cryst. Growth*, 2008, **310**, 4190–4198.

36 Y. J. Han and J. Aizenberg, *J. Am. Chem. Soc.*, 2003, **125**, 4032–4033.
37 Y. J. Han, L. M. Wysocki, M. S. Thanawala, T. Siegrist and J. Aizenberg, *Angew. Chem., Int. Ed.*, 2005, **44**, 2386–2390.
38 A. F. Wallace, J. J. De Yoreo and P. M. Dove, *J. Am. Chem. Soc.*, 2009, **131**, 5244–5250.
39 D. N. Lumsden, *Journal of Sedimentary Petrology*, 1979, **49**, 429–436.
40 F. Zhang, H. Xu, H. Konishi and E. E. Roden, *Am. Mineral.*, 2010, **95**, 1650–1656.
41 H. Y. Li, H. L. Xin, D. A. Muller and L. A. Estroff, *Science*, 2009, **326**, 1244–1247.
42 K. J. Davis, P. M. Dove, L. E. Wasylenki and J. J. De Yoreo, *Am. Mineral.*, 2004, **89**, 714–720.
43 M. R. Rosen, D. E. Miser and J. K. Warren, *Sedimentology*, 1988, **35**, 105–122.
44 P. A. Baker and M. Kastner, *Science*, 1981, **213**, 214–216.
45 P. V. Brady, J. L. Krumhansl and H. W. Papenguth, *Geochim. Cosmochim. Acta*, 1996, **60**, 727–731.
46 A. E. Stephenson, J. L. Hunter, N. Han, J. J. DeYoreo and P. M. Dove, *Geochim. Cosmochim. Acta*, 2011, **75**, 4340–4350.
47 S. M. Stanley, J. B. Ries and L. A. Hardie, *Proc. Natl. Acad. Sci. U. S. A.*, 2002, **99**, 15323–15326.
48 S. Gayathri, R. Lakshminarayanan, J. C. Weaver, D. E. Morse, R. M. Kini and S. Valiyaveettil, *Chem.–Eur. J.*, 2007, **13**, 3262–3268.
49 A. Becker, A. Ziegler and M. Epple, *Dalton Trans.*, 2005, 1814–1820.
50 J. B. Ries, *Geology*, 2004, **32**, 981–984.
51 A. Spadafora, E. Perri, J. A. McKenzie and C. Vasconcelos, *Sedimentology*, 2010, **57**, 27–40.
52 P. A. Sandberg, *Nature*, 1983, **305**, 19–22.
53 R. L. Folk, *Journal of Sedimentary Petrology*, 1974, **44**, 30–39.
54 B. H. Wilkinson, *Geology*, 1979, **7**, 524–527.
55 R. K. Given and B. H. Wilkinson, *Journal of Sedimentary Research*, 1985, **55**, 109–119.
56 M. Schmidt, S. Xeflide, R. Botz and S. Mann, *Geochim. Cosmochim. Acta*, 2005, **69**, 4665–4674.
57 I. J. Kelleher and S. A. T. Redfern, *Mol. Simul.*, 2002, **28**, 557–572.
58 S. E. Kaczmarek and D. F. Sibley, *J. Sediment. Res.*, 2007, **77**, 424–432.
59 M. J. Malone, P. A. Baker and S. J. Burns, *Geochim. Cosmochim. Acta*, 1996, **60**, 2189–2207.
60 H. C. W. Skinner, *Am. J. Sci.*, 1963, **261**, 449–472.
61 D. T. Wright and D. Wacey, *Sedimentology*, 2005, **52**, 987–1008.
62 M. Sanchez-Roman, C. Vasconcelos, T. Schmid, M. Dittrich, J. A. McKenzie, R. Zenobi and M. A. Rivadeneyra, *Geology*, 2008, **36**, 879–882.
63 J. Martel and J. D. E. Young, *Proc. Natl. Acad. Sci. U. S. A.*, 2008, **105**, 5549–5554.
64 Z. Lasemi, M. R. Boardman and P. A. Sandberg, *Journal of Sedimentary Research*, 1989, **59**, 249–257.
65 H. C. Sorby, *Proceedings of the Geological Society of London*, 1879, **35**, 56–95.
66 P. J. Davies, B. Bubela and J. Ferguson, *Sedimentology*, 1978, **25**, 703–729.
67 S. J. Gaffey, *Journal of Sedimentary Petrology*, 1983, **53**, 193–208.
68 R. L. Folk and F. Leo Lynch, *Sedimentology*, 2001, **48**, 215–229.
69 S. M. A. Duguid, T. K. Kyser, N. P. James and E. C. Rankey, *J. Sediment. Res.*, 2010, **80**, 236–251.
70 I. P. Montañez, *Proc. Natl. Acad. Sci. U. S. A.*, 2002, **99**, 15852–15854.
71 J. E. Adams and M. L. Rhodes, *AAPG Bulletin*, 1960, **44**, 1912–1920.
72 D. J. Ross, *Journal of Sedimentary Research*, 1991, **61**, 349–353.
73 C. Glunk, C. Dupraz, O. Braissant, K. L. Gallagher, E. P. Verrecchia and P. T. Visscher, *Sedimentology*, 2011, **58**, 720–736.
74 C. Dupraz, R. P. Reid, O. Braissant, A. W. Decho, R. S. Norman and P. T. Visscher, *Earth-Sci. Rev.*, 2009, **96**, 141–162.
75 P. A. del Giorgio and J. J. Cole, *Annu. Rev. Ecol. Syst.*, 1998, **29**, 503–541.
76 P. B. Price and T. Sowers, *Proc. Natl. Acad. Sci. U. S. A.*, 2004, **101**, 4631–4636.
77 Y. Nishino, Y. Oaki and H. Imai, *Cryst. Growth Des.*, 2008, **9**, 223–226.

General discussion

Dr Gebauer opened the discussion of the paper by Dr Laurie B. Gower: During the discussions, Dr Sommerdijk implied that at low pH-levels of around pH 8.5, associations between calcium and carbonate before nucleation would become unimportant, and that any pre-nucleation species formed with calcium should be bicarbonate-based. In other words, it was implied that pre-nucleation clusters based on calcium-carbonate interactions did not exist, at least at this pH level, and that the NMR results obtained for the proposed liquid condensed phase (LCP) showed that pre-nucleation species had been bicarbonate-based. These points are not sustainable.

First of all, it is important to note that LCP forms upon a first-order transition after some time of addition of calcium into bicarbonate solution, that is, LCP represents a nucleated second phase. This is evidenced by means of calcium measurements, pH titration as well as isothermal titration calorimetry.[1] Hence, LCP does not represent pre-nucleation clusters, which, on the other hand, form spontaneously in solution already before nucleation of LCP. We can use the ion-pairing constants of calcium with carbonate and bicarbonate to assess the level of ionic interactions in the solutions before nucleation. While there is considerable disparity in the literature regarding the exact values of these ion pairing constants,[2] there is a consistent trend with the calcium-carbonate interaction being 100-fold stronger than the calcium-bicarbonate interaction in terms of equilibrium constants. On average, reports give $K(CaCO_3^0) = \sim 1000$ M^{-1} and $K(CaHCO_3) = \sim 10$ M^{-1}.[3,4] In bi/carbonate buffer pH 8.5, the population is *ca.* 1–2% carbonate and 98% bicarbonate, and taking into account the ion association constants given above, calcium-carbonate interactions will still be significant, leading to pre-nucleation cluster formation,[5] while calcium-bicarbonate interactions start to become important. However, as shown by means of computer simulations, the bicarbonate ion forms a rather unstable link with calcium in pre-nucleation clusters, and these species consequently cannot contain significant amounts of bicarbonate even at rather low pH-levels.[6] The reason why pre-nucleation clusters cannot be observed by means of solution NMR is addressed in detail in my reply to another question (asked by Mike Nielsen, below, with regard to the same paper). In brief, the solutions investigated here by means of NMR contain *ca.* 20 mM bicarbonate, ~ 0.2 mM carbonate and ~ 0.3 mM calcium. Taking the calcium-carbonate interactions leading to pre-nucleation cluster formation into account, it can be inferred that only $\sim 0.3\%$ of carbonate species are bound in pre-nucleation clusters (which is still a rather large fraction of *relevant* ions), rendering their detection by means of effects on chemical shifts virtually impossible. The NMR signal prior to nucleation of LCP therefore reflects the buffer composition, which is essentially 98% bicarbonate as expected.

That said, NMR proves that LCP is even richer in bicarbonate than the starting solution, however, strictly following the argumentation above, this does not mean that calcium-carbonate interactions within LCP did not lead to pre-nucleation cluster formation. Pre-nucleation clusters in LCP do also form, as calcium-carbonate interactions are much stronger than calcium-bicarbonate interactions. Without doubt, the LCP phase is richer in bicarbonate than the mother solution, but this does not reflect the composition of pre-nucleation clusters at all. NMR shows that LCP constitutes a second phase with liquid character which exists next to the starting solution. The same kind of interactions occur within this phase, while it remains as yet unknown how the calcium ions are distributed among the two different liquid phases. Likely, though, LCP is more concentrated in ionic species in general if we consider the viscosities.[1]

1. M. A. Bewernitz, D. Gebauer, J. Long, H. Cölfen and L.B. Gower, *Faraday Dicsuss.*, 2012, **159**, DOI: 10.1039/c2fd20080e.
2. J.-Y. Gal, J.-C. Bollinger, H. Tolosa and N. Gache, *Talanta*, 1996, **43**, 1497–1509.
3. L. N. Plummer and E. Busenberg, *Geochim. Cosmochim. Acta*, 1982, **46**, 1011–1040.
4. E. W. Moore and H. J. Verine, *J. Am. Physiol.*, 1981, **241**, G182–G190.
5. D. Gebauer, A. Völkel, H. Cölfen, *Science*, 2008, **322**, 1819–1822.
6. R. Demichelis, P. Raiteri, J. D. Gale, D. Quigley and D. Gebauer, *Nat. Commun.*, 2011, **2**, 590.

Dr Sommerdijk replied: It is important to note first that it has not been established in which type of cluster or complexes the Ca, CO_3^{2-} or HCO_3^- ions are bound. Furthermore, for structures containing interactions of the type CO_3–Ca–CO_3–Ca or CO_3–Ca–HCO_3, as suggested for pre-nucleation clusters and DOLLOP, there are no experimental data giving the strength of the Ca-CO_3 or Ca–HCO_3 bonds in the species involved.

However, using the experimentally derived ion-pair constants for the known $CaCO_3$ or $CaHCO_3^-$ complexes,[1] one can calculate that for a solution containing 1–2 % carbonate and 98–99% bicarbonate, 25–50% of these complexes are present as $CaHCO_3^-$. This implies that although the ion-pair binding constant for $CaCO_3$ is 100× larger than $CaHCO_3^-$, at the conditions specified (pH = 8.5, a pH value which I would not describe as low) the Ca–HCO_3 bond can compete with the Ca–CO_3 bond. Therefore a significant proportion of the pre-nucleation species must contain bicarbonate bound calcium, alongside calcium bound to carbonate. This latter type of calcium may consist of purely calcium carbonate ion pairs and pre-nucleation clusters, but equally well complexes, or clusters, consisting of calcium bound to both carbonate and bicarbonate may exist at this point. It would not be unreasonable to assume that such mixed carbonate species—which are also observed in lower pH simulations[2]—would be the starting point for the nucleation of the—indeed second—LCP phase.

1. E. W. Moore and H. J. Verine, *J. Am. Physiol.*, 1981, **241**, G182–G190.
2. R. Demichelis, P. Raiteri, J. D. Gale, D. Quigley and D. Gebauer, *Nat. Commun.*, 2011, **2**, 590.

Dr Gebauer remarked: In his reply, Dr Sommerdijk states that the equilibrium constants of calcium–carbonate interactions within pre-nucleation clusters had not been determined experimentally yet. This is not quite correct, as there are experimental values for calcium carbonate interactions based upon a multiple-binding equilibrium.[1] These agree virtually quantitatively with the respective parameters obtained independently by means of computer simulations.[2] Ion pair formation (to which Dr Sommerdijk refers as "known complexes") cannot be separated from the formation of clusters (higher "complexes") when evaluating the concentrations of free ion. Thus, literature values for ion pairing constants do contain the contribution from ionic interactions within pre-nucleation clusters. The ion pairing constants from the literature consequently provide a good estimate of the interactions within the clusters themselves.

That said, here is a major misunderstanding: There is no doubt that at low pH-values, calcium will be bound to bicarbonate based on $K(CaHCO_3^{+;0}) = \sim 10$ M^{-1}. However, it is crucial to note that the free enthalpy associated to this equilibrium constant is $\Delta G = -kT \ln 10 = \sim -2kT$. This energy relates to the binding strength of calcium to bicarbonate, and is independent of pH (*i.e.* independent of the bicarbonate level). The same calculation can give a measure for the binding strength between calcium and carbonate with $K(CaCO_3^0) = \sim 1000$ M^{-1}.

Clearly, the calcium–bicarbonate bond cannot compete with the calcium-carbonate bond at any pH value, contrary to Dr Sommerdijk's statement. The absolute change from $\sim 2kT$ to $\sim 7kT$ in ΔG considering bicarbonate and carbonate binding to calcium, respectively, is all that matters here.

It is actually the essence of the law of mass action: The calcium–bicarbonate interaction strength in solution is essentially the same as thermal energy, and in order for calcium-bicarbonate pairing to become significant, the bicarbonate concentration has to be increased distinctly, thereby forcing the equilibrium into the direction of ion pairs. But the calcium–bicarbonate interactions remain weak independent of pH, and bonds can be easily broken by thermal energy. The respective probabilities can be evaluated thoroughly within concepts of statistical thermodynamics.

Consequently, solid calcium bicarbonate does not exist and cannot be precipitated from aqueous solution at any "reasonable" conditions, to the best of my knowledge. You always get $CaCO_3(s)$. The ionic calcium–carbonate interactions in solution, in the range of few kT, are just strong enough to allow for the formation of branched or unbranched chains, or rings in solution, which are highly dynamic and behave similar to a liquid (a structural form called DOLLOP).[2] As also pointed out in ref. 2, stronger ionic interactions would lead to a very different binding behavior of free ions in the pre-nucleation stage, and the findings indicate that DOLLOP is the structural form representing pre-nucleation clusters. When bicarbonate binds to these species, however, it forms an unstable link and, handwavingly said, acts as a chain terminator. This means that at any low pH value (be it pH = 5, 7 or 8.5), calcium bicarbonate ion pairs are certainly present, however, the interaction strength between these ions is too low to allow for bicarbonate playing a role in larger clusters or ion associates. As pointed out in ref. 2, the calcium–bicarbonate interactions are actually unimportant at experimental conditions when it comes to associations beyond simple ion pairs, that is, at millimolar concentrations where calcium carbonate solutions typically become supersaturated. Calcium bicarbonate will not play a major role in DOLLOPs of any realistic macroscopic solution ensemble at any pH. If it did, we should be able to obtain solid calcium bicarbonate.

1. D. Gebauer, A. Völkel and H. Cölfen, *Science*, 2008, **322**, 1819–1822.
2. R. Demichelis, P. Raiteri, J. D. Gale, D. Quigley and D. Gebauer, *Nat. Commun.*, 2011, **2**, 590.

Dr Beck addressed Dr Gebauer:

I agree with Dr Gebauer that at pH = 8.5 the solution consists of considerable amounts of carbonate. Let us, for example, consider one solution at 25 °C and 1 bar that was prepared by initially:
- NaCl: 10 mmol L^{-1}
- $NaHCO_3$: 5 mmol L^{-1}
- $CaCl_2$: 5 mmol L^{-1}
- NaOH: 0.31 mmol L^{-1}.

That solution then consists of:
- Na^+: 15.31 mmol L^{-1}
- Cl^-: 20 mmol L^{-1}

Alkalinity: 5.31 mmol L^{-1}.
- Total carbon species C,tot: 5 mmol L^{-1}.
- Total calcium species Ca,tot: 5 mmol L^{-1}

The alkalinity is defined as the sum of all titratable bases.[1]

Using the following activity-based equilibrium constants, the formula activity = gamma*concentration, and using the Davies modification to the Debye–Hückel equation, the pH, activity coefficients, the ion speciation and supersaturation can be calculated iteratively:

autoprotolysis of water:
- $K,w = 1.06 \times 10^{-14}$

CO_3^{2-} and HCO_3^- equilibrium:
- $K,HCO_3^- = 4.69 \times 10^{-11}$.

Equilibrium between HCO_3^-, H^+, dissolved CO_2 and water determined:
- $K,CO_2 = 4.33 \times 10^{-7}$

Complex for $CaCO_3$:
- $K,CaCO_3^0 = 1.63 \times 10^3$

The equilibrium between Ca^{2+} and HCO_3^- is characterized by the following equilibrium constant:
- $K,CaHCO_3^+ = 1.23 \times 10^1$.

Solubility product for calcite:
Ksp,calcite = 3.2×10^{-9}.

The calculations result in pH = 8.5, activity coefficients for monovalent ions of $\gamma_1 = 0.85$ and for divalent ions of $\gamma_2 = 0.52$.

The speciation of the solution constituents is as follows:
- CO_3^{2-}: 0.1 mmol L^{-1}
- CO_2(aq): 0.028 mmol L^{-1}
- HCO_3^-: 4.5 mmol L^{-1} $CaHCO_3^+$: 0.14 mmol L^{-1} $CaCO_3^0$: 0.23 mmol L^{-1}
- Ca^{2+}: 4.7 mmol L^{-1}.

The supersaturation ratio Sc = $(aCa^{2+}*aCO_3^{2-}/Ksp,c)^{0.5}$ with respect to a hypothetically forming calcite phase is then:
- Sc = 6.6

It can be seen that the amount of carbonate is considerable. The fraction of CO_3^{2-} to HCO_3 is 2%.

The question is, how it can be quantified how much of the calculated amount of $CaCO_3^0$ forms as a complex and how much forms as prenucleation clusters? The same question applies to $CaHCO_3^+$ and HCO_3^-: how can it be determined whether the amount of bicarbonate and the amount of hydrogen carbonate actually form as complex and how much forms as prenuclation clusters that actually influence the nucleation of calcium carbonate phases?

1. B. Kaasa, K. Sandengen and T. Oestvold, SPE International Symposium on Oilfield Scale 95075, 2005, 1–13.

Dr Gebauer replied: We are happy to find independent verification of our speciation of the carbonate buffer at pH 8.5.

As detailed in the replies to other questions and comments referring to our paper in the first session, experiments show that calcium carbonate pre-nucleation clusters contain most of the calcium that is bound in the prenucleation stage. The clusters are more stable than ion pairs of calcium carbonate. The specific strength of interactions would lead to a size distribution reminiscent of the outcome of polycondensation reactions. This suggests that the prenucleation clusters have been concealed by the ion pair concept.

In brief, why should the relatively strong interaction between calcium and carbonate ions be limited, and not allow for the formation of polymers beyond ion pairs? The only reason would be energetic costs due to the generation of surface, as classically assumed. But when the ions maintain their hydration as in DOLLOP, this argument does not quite hold anymore.

For calcium bicarbonate, the interactions are too weak to allow for "polymerizations" at relevant conditions.

Mr Nielsen enquired: How do you reconcile Laurie Gower's statement that the lifetime of the pre-nucleation clusters is too short to detect with NMR, with the idea that there is a stable population of pre-nucleation clusters constituting a significant percentage of the calcium in the solution? If there is a population of clusters thermodynamically stable with respect to ion species in solution, that constitutes a distribution of sizes among which any given cluster interconverts, should that not be detectable with NMR?

Dr Gebauer replied: The kinetic interpretation of thermodynamic equilibrium considers the equilibrium constant K of formation of a certain species as the quotient of the respective rates of formation and decomposition. This means that a highly dynamic system can be associated with a value of $K > 1$ (*i.e.* thermodynamically stable; $\Delta G = -RT \ln K$), because the ratio of the respective rate constants is important. The notion that pre-nucleation clusters of calcium carbonate[1] are thermodynamically stable with respect to ions in the single-phase solution system has recently been corrobortated by means of computer simulations.[2] The work also indicates that the clusters are highly dynamic, and can change conformation on timescales in the regime of hundreds of picoseconds,[2] suggesting that the clusters can also very quickly exchange ions with the solution leading to rather short cluster lifetimes. Still, the clusters represent a constant population.

As pointed out by Laurie Gower during the discussions, and as also stated in our paper,[3] the dynamics of the clusters are quick as compared to NMR timescales and thus no separate NMR resonance for pre-nucleation clusters can be observed. The NMR signal represents an ensemble average of all species present in solution, which contain ^{13}C. If we compare the NMR signal from pure bicarbonate buffer with that obtained in presence of calcium ions, prior to nucleation of LCP, (Fig. 6, left and center spectrum),[3] there is no obvious shift, however, a slight broadening is apparent that may or may not be due to the presence of pre-nucleation clusters.

That said, it is most important to note the relative quantities of ^{13}C species in the system. The solution, which was investigated by means of NMR in the present study (referred to above), was prepared by adding 7 successive injections of 6 mM calcium solution (each injection 200 μl) into 29 mL of 20 mM bicarbonate buffer, pH 8.5.[3] That is, the system investigated by means of NMR here contained roughly 0.3 mM calcium ions as opposed to essentially 20 mM bicarbonate and *ca.* 0.2 mM carbonate. At pH 9.00, *ca.* 30% of calcium is bound in pre-nucleation clusters at 10 mM carbonate buffer concentration.[1] Even less calcium would be bound at pH 8.5, but the carbonate buffer concentration utilized in the study was 20 mM. We can assume that in the present case, roughly 20% of the calcium ions in solution would be bound in pre-nucleation clusters at any time (*i.e.* ~0.06 mM). Considering that pre-nucleation clusters consist of equal amounts of calcium and carbonate,[1] this means that *ca.* 0.06 mM carbonate would be bound in pre-nucleation clusters too. This is only 0.3% of the entire population containing ^{13}C, as there is a great excess of bi/carbonate over calcium. The problem here is that we do not exactly know the individual chemical shift of carbonate in pre-nucleation clusters as compared to free carbonate. Due to the solute character of pre-nucleation clusters (the ions virtually retain their solvation characteristics in DOLLOP),[2] a minor shift as compared to free carbonate ions is expected, which in combination with our considerations regarding relative quantities ultimatley suggests that also only a minor, if detectable, effect of pre-nucleation cluster formation can be seen in solution NMR in principle.

1. D. Gebauer, A. Völkel and H. Cölfen, *Science*, 2008, **322**, 1819–1822.
2. R. Demichelis, P. Raiteri, J. D. Gale, D. Quigley and D. Gebauer, *Nat. Commun.*, 2011, **2**, 590.
3. M. A. Bewernitz, D. Gebauer, J. Long, H. Cölfen and L. B. Gower, *Faraday Dicsuss.*, 2012, **159**, DOI: 10.1039/c2fd20080e.

Dr Beck asked Dr Gebauer: Is it possible that the observation that the measured effective free calcium concentration is lower than the total calcium concentration[1,2] can be caused by complex formation and interactions with other constituents (reducing the calcium activity)?

1. D. Gebauer, A. Völkel and H. Cölfen, *Science*, 2008, **322**, 1819–1822.
2. M. A. Bewernitz, D. Gebauer, J. Long, H. Cölfen and L. B. Gower, *Faraday Dicsuss.*, 2012, **159**, DOI: 10.1039/c2fd20080e.

Dr Gebauer answered: No, this is not the case. Please refer for details to our reply to Dr Beck's question regarding our paper in the first session.

Dr Verch addressed Dr Gower and Mr Bewernitz: This comment is related to the pH and calcium ion-measurements shown in Fig. 1 in your paper.

Both recorded graphs flatten at a certain point, which indicates the appearance of a new phase. A flattening of the pH-curve is a sign of a reduced carbonate ion binding, as you say in your paper. However, I have difficulties interpreting this observation as an indication of a new bicarbonate-rich phase, especially when considering the development of free calcium ions. These graphs are, as stated before, also flattened, which means that less calcium ions must be bound. The changes of the pH and calcium curves, you have shown, seem to be very similar in their intensities. I find it difficult to draw any conclusions about the composition of the newly formed phase from the presented data, especially when taking into account that the pH measurement can be quite erroneous if the carbon dioxide diffusion into the reaction vessel is not controlled. I would interpret these data as the transformation into a new phase with a lower formation/equilibrium constant.

Could the authors elaborate on the experiment and conclusions drawn in more detail?

Mr Bewernitz replied: The data regarding the pH and bound Ca^{2+} evolution is difficult to interpret. In addition, the diffusion of carbon dioxide was not controlled in these experiments even though control experiments show that the effect of carbon dioxide diffusion on the pH evolution is linear (see supplement). We disagree with the questioner, however, that this indicates that we cannot draw ANY conclusions from the data.

I must correct you on a misconception present in your question. The pH and bound Ca^{2+} evolutions were not interpreted as direct evidence of a bicarbonate-rich phase emergence. We used the NMR primarily as evidence that the emergent liquid phase is bicarbonate-rich as compared to the mother solution. We used the pH and bound Ca^{2+} evolutions primarily as evidence that a phase transition of sorts is occurring as they are both state variables and a discontinuity of their evolution is evidence that a change of state has occurred. As mentioned above, the control experiments, which include water injections into bicarbonate buffer and the bicarbonate buffer exposed to air, both have similar pH evolutions and are both linear. Therefore, we make the qualitative argument that the discontinuity is due to the interaction of ions and not due to the out-gassing of CO_2.

We agree with the questioner that, without controlling for CO_2 diffusion, we shouldn't draw many quantitative conclusions. This is the main reason why we treated the data as qualitative (and explicitly stated so in the paper). I want to correct you regarding the pH measurements themselves. The unaccounted-for out-gassing of CO_2 does not make the pH measurements "erroneous". The pH value is the value within the solution and it is a truly measured value. It does not matter whichever pathway the solution took to achieve this condition. The emergent phase appears at the flattening pH behavior, and was therefore included as the conditions present when observing the phase change system behavior. When further research is conducted that controls for the diffusion of CO_2, we would not be surprised to see a slight change in the situation around the phase transition.

Qualitatively, the flattening of the pH behavior is just as important to consider as the discontinuity. In a dynamic system, any binding or phase-forming events that consume bicarbonate or carbonate is going to be energetically taxing on the system because the bicarbonate buffer has to do work to supply the appropriate ratio of reactants (carbonate and bicarbonate) and adjust further from its initial equilibrium. A flattening of the pH evolution would occur when the carbonate and bicarbonate consumption matches the carbonate and bicarbonate ratios of the mother solution bicarbonate buffer, leading to a minimization of the energetic "tax" on the buffer to adjust. Where does this occur? Fig. 1 shows this occurring at a pH of $\sim$8.47 and Fig. 9 shows this occurring at a pH of $\sim$8.15. The products due to PNC and LCP formation would require a consumption of bicarbonate and carbonate of roughly

98 : 1 or larger, respectively, to maintain these pH's. However, since CO_2 gassing-out is also contributing to this pH flattening and that aspect is not being controlled-for, the true ratio would still be expected to favor bicarbonate but at a smaller, unknown, ratio. For this reason, we qualitatively reported that the pH evolution may suggest that there is a bicarbonate product forming but utilized NMR to firmly come to the conclusion.

With regard to your final interpretation that the Ca^{2+} binding profile suggests there is a phase transition (rather than an emergence) into a new phase: This is a possible scenario and we considered it until the Analytical UltraCentrifugation (AUC), Nanoparticle Tracking Analysis (NTA), and NMR data detected a distinct liquid phase. It is possible that more than one thing is occurring, a phase emergence in addition to other phase transitions (we mention this possibility in the paper as well) but the data suggest that at least one thing, a phase emergence, is occurring.

Speculation: To further address your comment about the decreasing binding affinity for Ca^{2+} after the phase emergence: You are correct that this is a bit puzzling considering that the emergence of an energetically favorable Ca^{2+} binding phase in addition to the Ca^{2+} binding affinity in the PNC's would be expected to lead to an *increase* in overall Ca^{2+} binding affinity, not a decrease. However, consider that LCP formation sequesters many bicarbonates into a new distinct phase. This would require the mother solution phase to adjust by converting some carbonates into bicarbonates to replace those lost in the sequestering. This would have the effect of making the $Ca-CO_3$ ion pairing less favorable, releasing Ca^{2+}. In addition, the carbonate/bicarbonate concentration in the mother solution phase will be reduced, also negatively affecting the $Ca-CO_3$ ion pairing. This might explain why there is a reduction of Ca^{2+} binding affinity after the emergence of LCP, and may also explain the flattening of the pH evolution of the mother solution phase shown in Fig. 1, as discussed above. It is important to note that it is not known whether the LCP contains a stoichometric ratio of 1 calcium ion for every 2 bicarbonate ions. Being a liquid–liquid phase separation we can assume that there is a partition of ion species between phases including Na^+ and Cl^- as well as the all important Ca^{2+}, HCO_3^- and CO_3^{2-} ions (and Mg^{2+} and K^+ if we are more biomimetic with our conditions). Other than the reasonable assumption that the phases are charge balanced overall, it is not known how the ions are partitioned. It is quite possible, given the reduction of bound Ca^{2+} affinity in the mother solution, that the LCP contains fewer Ca^{2+} ions than would be expected for a 1 Ca^{2+} : 2 HCO_3^- ratio with the difference of charge made up with Na^+ and Cl^- partitioning. This would lead to more free Ca^{2+} in the mother solution phase and reduction in the detected Ca^{2+} binding affinity as seen in Fig. 1. This is speculation and we decided to interpret only what appeared more concrete; that qualitatively, a phase transition occurs as evidenced by the discontinuous concentration profile.

Wolf *et al.* have demonstrated the presence of a liquid phase that formed in bicarbonate buffer conditions which he called liquid amorphous calcium carbonate (LACC).[1] This phase may be similar or the same as the liquid phase we detect in our paper, being due to Wolf *et al.* working with an even more neutral conditions than we did where bicarbonate could play a distinct role. However, due to the unknown ionic composition of the bicarbonate-rich phase we detected, we chose to call it liquid condensed phase (LCP). If the bicarbonate-rich phase consists primarily of bicarbonates then it is not calcium carbonate in a chemical sense, but would be more appropriately called liquid calcium bicarbonate. If the liquid phase has less Ca^{2+} than would stoichiometrically match the bicarbonates (a possibility described above) then it wouldn't chemically be calcium bicarbonate but rather a liquid condensed phase (LCP) of ions. Finally, as has been mentioned in Dr Gower's response to Mr Ihli's question below, we cannot be certain at this stage, that solid ACC nucleation occurs through LCP or in spite of it (even though PILP, presumably composed of LCP, is a precursor). Therefore we did not feel comfortable declaring it a precursor phase without more research. For these reasons, we chose LCP as the most conservative nomenclature.

1. S. E. Wolf, J. Leiterer, V. Pipich, R. Barrea, F. Emmerling and W. Tremel, *J. Am. Chem. Soc.*, 2011, **133**, 12642–12649.

Dr Wolf asked: The effect, which you describe, that the polymer actually *reduces* the final pH of the solution, is really remarkable. Based on our observation with the Kitano system under levitated conditions,[1] we believe that the liquid ACC densifies in the course of time and thus looses water of hydration. But what is the source to bind such an amount of water? It seems to us that the pH is a crucial parameter for the liquid/liquid phase separation as it shifts the equilibrium of carbonate to bicarbonate towards bicarbonate. This prolongs the lifetime of the liquid ACC: We are typically working at a low pH of 6.3 employing the Kitano method and under such conditions, we are capable of quite easily observing the liquid ACC, even in absence of polymer. But at higher pH, like reported by Rieger in the Faraday Discussion 136 on Crystal Growth[2] or in the contribution of Faatz *et al.*,[3] the lifetime is drastically reduced. Thus, the pH-reducing effect of polymer you reported is really of particular importance.

1. S. E. Wolf, *et al.*, *J. Am. Chem. Soc.*, 2008, **130**(37),12342–7.
2. J. Rieger, *et al.*, *Faraday Discussions*, 2007, **136**, 265–277.
3. M. Faatz, *et al.*, *Adv. Mater.*, 2004, **16**(12), 996–1000.

Dr Gower replied: Yes, we agree, pH may be very important. It would be nice to do further studies at different pHs. Regarding the source to bind such an amount of water, it seems that it is not so much as binding the water, but rather that the ions are condensing to a new phase that happens to have a substantial amount of water still present, but a little less than the surrounding mother solution. In addition, the carbon dioxide may play an ancillary role in the formation of an additional liquid phase. It must partition into both liquid phases during a liquid–liquid phase separation just like all of the ions. If the emergent liquid phase is neutral or acidic with respect to the mother solution (as is the case with LCP), then the amount of dissolved molecular carbon dioxide may be considerable. Its participation in the carbonate : bicarbonate : carbon dioxide buffer system would mean that its consumption/uptake from the mother solution would have an effect on the pH and *vice versa*. We imagine it to be difficult to sequester carbon dioxide into neutral LCP droplets at high pH's (>9) when it is such a dilute minor component of the buffer system at those conditions. This may be the reason why liquid–liquid phase separation was so easily observed at even lower pH values than in our system, as demonstrated by Wolf *et al.*, whose experiments are conducted at relatively low pH's where carbon dioxide is prevalent and possibly facilitating its partitioning. Perhaps in addition to many possible roles, polymer plays a pH-reducing role in the solution. This speculation doesn't really get to the question of lifetime kinetics but we agree that the pH and the polymer additive's effect on it does seem to be of particular importance.

Mr Ihli addressed Dr Gower and Mr Bewernitz: If I understood the paper correctly, the observation of a liquid condensed phase (LCP)—a precursor to amorphous calcium carbonate (ACC)—polymer assisted or not, was made at near neutral pH levels "majorly". Further, the resulting LCP was shown to dominantly contain bicarbonate ions. Assuming the limited pH range at which LCP were observed in this case, couldn't the presence of bicarbonate ions in solution be seen as a prerequisite to LCP formation and ultimately ACC? Now ACC is also known to form at higher pH levels[1] and thus in the absence of any significant amounts of bicarbonate ions in solution. Doesn't this fact then directly imply that at least two possible pathways exist for the formation of ACC and that in turn solubility values, resulting supersaturation levels *etc.* should be evaluated in hindsight to this, *i.e.* not just temperature and concentration but also carbonate species initially involved in the formation process? In that sense how should the recent finding of "positive PILP"[2] be evaluated, requiring a higher pH level (~9.5) for effective LCP associated structure formation in terms of LCP composition and subsequent crystallization pathway.

1. N. Koga, Y. Nakagoe and H. Tanaka, *Thermochim. Acta*, 1998, **318**, 239.
2. B. Cantaert, Y.-Y. Kim, H. Ludwig, F. Nudelman, N. A. J. M. Sommerdijk and F. C. Meldrum, *Adv. Funct. Mater.*, 2012, **22**, 907.

Dr Gower answered: Yes, I agree with the assessment that this could mean that ACC can form *via* different processes. We note, however, that we have not done studies to determine if the LCP solidifies and forms an ACC phase. However, Wolf *et al.*[1] have shown that a liquid amorphous calcium carbonate (LACC) phase forms under their conditions of neutral pH, and it transforms into solid ACC which then crystallizes into calcite. They measured a large Avrami coefficient which indicates that the transformation rate to calcite is high, and they suggest that secondary nucleation occurs on amorphous calcium carbonate particles, with a dissolution assisted crystallization mechanism.

Because the evidence suggests that PILP forms via accumulation of an LCP type of phase, and we know that PILP droplets do solidify into ACC, that is a possibility that LCP may also form an ACC phase. But it is possible that the polymer could be involved, or it may be that the bicarbonates have transferred into carbonate groups at that point that it has become ACC. Likewise, studies by Cantaert *et al.*[2] of "positive PILP" show the ACC phase does form within the PILP phase formed by positively charged polymers It is not clear what exact species are present in all of these systems, and particularly when they are transforming to ACC.

This query brings up a good point—if the ACC (or any polymorph) of calcium carbonate originates in the LCP phase, then many of our saturation values for calcium carbonate polymorph nucleation (particularly ACC) may have to be re-evaluated. The critical condition (saturation, pH temp, *etc.*) in which solid ACC nucleation occurs would be the condition inside the LCP droplet, not the global conditions of the mother-solution phase. However, the exact conditions within the LCP droplets (and PILP phases) are currently unknown (other than they appear to be bicarbonate-rich at the conditions utilized in the paper). In addition, we have not examined higher pH values to determine if there is a LCP, and if it still has bicarbonate bias or if other species are more prevalent. More work is needed in this area.

1. S. E. Wolf, J. Leiterer, M. Kappl, F. Emmerling, W. Tremel, Early Homogenous Amorphous Precursor Stages of Calcium Carbonate and Subsequent Crystal Growth in Levitated Droplets. *J. Am. Chem. Soc.*, 2008, **130**, 12342–12347.
2. B. Cantaert *et al.*, Think Positive: Phase Separation Enables a Positively Charged Additive to Induce Dramatic Changes in Calcium Carbonate Morphology. *Adv. Funct. Mater.*, 2012, **22**(5), 907–915.

Professor Meldrum asked: In your paper you state that in the absence of polymer, the quantity of liquid condensed phase (LCP) formed is very small. What relevance does the LCP phase have on the formation of ACC in the absence of polymer? Is the LCP a necessary precursor to the formation of ACC when polymer is not present?

Mr Bewernitz responded: See Dr Gower's response above.

Professor De Yoreo commented: What do we know for certain about the dynamics of liquid precursor particle dehydration, subsequent particle–particle interactions and eventual crystallization?

Dr Gower responded: Unfortunately, we have not done any quantitative analysis of the particle dehydration rate. What we know is mostly based on observations that if we have a shallow dish, the films are more uniform and less granular, as though the precursor phase has a better flow. With deep vials, the particles grow partially fuse into large aggregates. In the light scattering studies of $CaCO_3$ PILP formation,[2] the droplets seemed to grow by coalescence, as judging from the growth rate. Also, it seems as though the droplets rapidly solidify when they adsorb to a surface, but this is based on simple observations of scratching the PILP formed films. In contrast, when a large drop of PILP phase had accumulated underneath an air bubble in our first PILP report,[1] it remained a liquid and did not solidify for weeks (except for a large spherulite floating in it). This may be due to the large thickness of the 100 μ drop, which could not allow exclusion of polymer or water, or achieve stoichiometric balance (particularly if bicarbonates are present). This is why I believe many biominerals seem to deposit the precursor phase (ACC or PILP) layer-by-layer, as opposed to simply filling a large preformed vesicle. And of course the dehydration dynamics are probably very strongly impacted by the polymer composition and molecular weight, as well as concentration. For example, in the CaP PILP phase, the higher molecular weight polymer seems to stabilize smaller droplets, while lower MW polymers allow for more growth/agglomeration of the droplets, and this latter phase doesn't lead to as much intrafibrillar mineralization. I assume that Mother Nature has evolved more specialized proteins to optimize such parameters (or stabilizes the phase inside of vesicles).

Regarding crystallization, based on the discussion at the meeting, I think you are referring to whether we know for sure that the preformed ACC phase does not dissolve and recrystallize, *versus* the proposed pseudo-solid-state transformation. I say pseudo because there is presumably a fair amount of hydration waters present, which can allow for localized diffusion and reorganization of the ions. So I don't consider it pure solid state transformation (although Gilbert's group have apparently shown that an anhydrous ACC can crystallize in biominerals,[3] which I found to be quite surprising). From *in situ* optical microscope studies,[1] one can see that the crystallization (birefringence) slowly propagates across the film. There does not appear to be any dissolution as this is occurring. Of course this is a micron scale observation, so I cannot say what is happening at the nanoscale. But there are cases where the films dissolve and recrystallize, and in those cases you tend to get microfaceting, or large three-dimensional growth of more classical crystallites, so it is obvious when this is occurring. Plus, the amorphous films will crystallize after being removed from the solution, so there is not much surrounding water available for them to dissolve. They do however crystallize more slowly when in an air environment (which I again attribute to the difficulty in exclusion of polymer, water, or protons if it has bicarbonate excess).

1. L. B. Gower and D. J. Odom, Deposition of calcium carbonate films by a polymer-induced liquid-precursor (PILP) process, *J. Cryst. Growth*, 2000, **210**(4), 719–734.
2. E. DiMasi, T. Liu, M. J. Olszta and L. B. Gower, Laser light scattering studies of a polymer-induced liquid-precursor (PILP) process for mineralization, in: *Biological and Bio-Inspired Materials and Devices*, ed. K. H. Sandhage, S. Yang, T. Douglas, A. R. Parker and E. Di-Masi, Warrendale, PA, Mat. Res. Society Proceedings, 2005. p. K10.16.11–K10.16.17.
3. A. V. Radha, T. Z. Forbes, C. E. Killian, PUPA Gilbert and A. Navrotsky, Transformation and crystallization energetics of synthetic and biogenic amorphous calcium carbonate, *Proc. Natl. Acad. Sci. U. S. A.*, 2010, **107**(38), 16438–16443.

Dr Christenson remarked: A useful test that indicates a liquid-like character is whether or not a substance is imbibed in a capillary, as we showed in a recent study of PILP in track-etch membranes.[1] The same test also shows whether or not an object such as an acicular crystal is hollow.

1. Y.-Y. Kim, N. B. J. Hetherington, E. H. Noel, R. Kröger, J. M. Charnock, H. K. Christenson and F. C. Meldrum, *Angew. Chem., Int. Ed.*, 2011, **50**, 12572–12577.

Dr Wolf remarked: In the recent years, we learned that polymers have a vast multitude of effects during crystallization, and you showed in your contribution that even the pH is strongly affected by a small amount of polymer. We showed in our contribution herein and earlier[1] that—once the liquid ACC phase is formed—the polymer affects the emulsion as well in terms of depletion stabilisation and destabilisation, among other effects like reducing the solution's supersaturation. Can you shortly outline your current opinion of which are the major and important effects a polymer excerts during a PILP process?

1. S. E. Wolf *et al.*, *J. Am. Chem. Soc.*, 2011, **133**(32), 12642–12649.

Dr Gower responded: I have always considered the polymer's primary role as the ability to sequester a highly concentrated pool of ions, yet while inhibiting the crystallization reaction within this presumably highly supersaturated local pool of ions. Based on our report here, while there are already localized regions of enriched pools of ions, even without the polymer (LCP), the amount of this phase seems to be rather small. The polymer sequesters far more ions (much longer calcium binding titration curve shown in Fig. 8) to form much more of the phase. Plus, the phase does not tend to nucleate calcite crystals *via* ion-by-ion growth, but rather first solidifies into an amorphous solid phase prior to crystal nucleation which propagates throughout the amorphous phase. This pseudomorphic transformation is key to forming non-equilibrium morphologies, which of course I always point out is the hallmark of biominerals. Based on the findings that this phase is enriched in bicarbonate species, this may play a role in the inhibitory activity of the phase which is slow to nucleate. It is also possible that the polymer could poison any nuclei that form. I believe that the polymer is somehow responsible for holding onto excess hydration waters, which then creates the liquid-like character of the PILP phase. In our prior report (L. Dai, E. P. Douglas, L. B. Gower, Compositional analysis of a polymer-induced liquid-precursor (PILP) amorphous $CaCO_3$ phase, *J. Non-Cryst. Solids*, 2008, **354**, 1845–1854), we did find that there was a correlation between the exclusion of the polymer and the release of water from the precursor phase. Or, perhaps the extra hydration is related to the bicarbonate species involved in the LCP phase, which the polymer carries on further. Lastly, as your papers have proposed, the polymer may play a role in stabilizing or destabilizing the droplets.

Professor Meldrum asked: In bulk solution, PILP phases of calcium carbonate form as a low concentration phase. How do you envisage that PILP could form *in vivo*? How can biology capture these droplets (as the "granules" which have been observed in biological systems) to transfer them to the confined volume in which mineralization takes place?

In your paper you also state that PILP/ amorphous precursor phases enable organisms to make the unusual morphologies characteristic of many biominerals. Why do you state this given that it has often been shown[1–5] that an amorphous precursor phase is not required for morphological control? Forming a crystal within a mould is quite sufficient.

1. N. B. J. Hetherington, A. N. Kulak, Y-Y. Kim, E. H. Noel, D. Snoswell, M. Butler and F. C. Meldrum, "Porous Single Crystals of Calcite from Colloidal Crystal Templates: ACC is Not Required for Nanoscale Templating", *Adv. Func. Mater.*, 2011, **21**(5), 948–954.
2. A. S. Finnemore, M. R. J. Scherer, R. Langford, S. Mahajan, S. Ludwigs, F. C. Meldrum and U. Steiner, "Nanostructured Calcite Single Crystals with Gyroid Morphologies", *Adv. Mater.*, 2009, **21**(38–39), 3928–3932.
3. B. Wucher, W. Yue, A. N. Kulak and F. C. Meldrum, "Designer Crystals: Single Crystals with Complex Morphologies", *Chem. Mater.*, 2007, **19**(5), 1111–1119.
4. W. Yue, A. N. Kulak and F. C. Meldrum, "Growth of Single Crystals in Structured Templates" *J. Mater. Chem.*, 2006, **16**(4), 408–416.
5. R. J. Park and F. C. Meldrum, "Shape-Constraint as a Route to Calcite Single Crystals with Complex Morphologies", *J. Mater. Chem.*, 2004, **14**, 2291–2296.

Dr Gower answered: Regarding the first question—how does biology capture these droplets and transfer them to confined volume?—I have not yet put a lot of thought into this aspect because of my limited knowledge of cellular mechanics. What I do know is that many of the biominerals are formed by fusion of small incoming vesicles with a larger compartment (the mineral deposition vesicle), within which the biomineral is formed. It seems likely that these vesicles contain a PILP-like phase enriched with a high concentration of ions (or amorphous phase), because it is hard to imagine the vesicles simply containing a dilute solution of ions would be able to achieve the kinetics seen for biomineral formation. I understand that the perplexing issue which motivates this question is that the PILP droplets make up a very small volume of the bulk liquid (in the *in vitro* system), so a vesicle could not simply contain a supersaturated solution with polymer, because only a small amount of PILP would be formed in the vesicle. Perhaps ions could be continuously pumped into a vesicle that contains the polymer to build up a larger content of PILP phase, where the vesicle expands as it is slowly built up. It could also be possible that the PILP droplets are formed first and lipids naturally assemble around them due to surface activity to enclose the droplet in a vesicle. This seems like an excellent topic for Derk Joester to address with the vesicle system he is developing, and I would enjoy collaborating with his group in trying to find a way to form PILP phase in vesicles.

Regarding the second question—that an amorphous precursor phase is not required for morphological control: This is true as *in vitro* studies in Meldrum's group have elegantly shown that crystals with unusual non-crystallographic shapes can be grown in confined spaces. However, there is clear evidence that biominerals do not form by the classical crystallization process because most of the molded biominerals examined to date have a colloidal/nanogranular texture,[1] which does not occur in crystals grown by the classical process. In contrast, crystals formed by the PILP process produce the same dense colloidal/nanogranular texture.[1,2]

Another point is that the Meldrum Group has grown crystals in a hard template, and not a soft vesicle, which may or may not constrain the shape of a growing crystal. I am not aware of any vesicle type studies which show that the crystals entirely fill the vesicle and take on its shape.

In addition, it is worth noting that not all biominerals are formed in confined compartments, yet can have non-equilibrium morphologies. This is most evident in the nacreous tablets of mollusks, which are not grown in a confined space, at least not according to some reports (Nakahara's detailed TEM study on nacre formation shows that sheets of organic matrix are stacked on top of one another, and only expand upward as the mineral tablet forms and pushes the organic sheet upward to form what appears to be, after the fact, a compartment[1,3]). The ability of the PILP process to form nacre-like tablets of the same thickness and morphology, without a compartment, seems to be more than just coincidence, and is likely relevant to the process of natural selection of a valuable feature with respect to evolution (see *Chem. Rev.* paper for more detailed discussion[1]). As a final note, it is not just that we happened to mimic one particular feature of a biomineral, which could be considered coincidence, but it is the sheer number of features which can be emulated with the PILP process that builds a strong argument in favor of our hypothesis that the PILP process plays a fundamental role in biomineralization.[1] What are the odds of all of these being a coincidence?

(i) Hexagonal calcite tablets with a microdefect pattern in alternate narrower sectors[1,4] → seminacre tablets of calcite in bryozoans (definitely there is no constraining compartment in seminacre) with preferential etching in alternate narrower sectors

(ii) Single-crystalline tablets of aragonite[2] → mollusk nacre

(iii) Fibrous crystals of calcite[5] → sea urchin teeth, and others

(iv) Micromolded crystals with curved smooth surfaces[4,6] → urchin spine, coccoliths, *etc*.

(v) Templating and patterning of crystals on organic templates[7] → locational control found in many biominerals

(vi) Colloidal/nanogranular texture[1,2] → urchin spine, coral, nacre, sponge spicules, dental enamel

(vii) Large and anisotropic lattice strain (we haven't yet measured it, but it is large enough to be seen in polarized light microscopy)[4] → anisotropic lattice distortions in biogenic aragonite[8]

(viii) High-magnesium calcite[9] → corals, urchin teeth, *etc.*

(ix) "Soft" particle coatings[10] → core-shell microcapsules of dinoflagellate cysts

(x) Interpenetrating composites, such as intrafibrillar mineralization of collagen[11,12] → bone and dentin

(xi) Mineral coatings, cement, and dense spherulites with concentric layers[13,14] → kidney stones

1. L. B. Gower, Biomimetic Model Systems for Investigating the Amorphous Precursor Pathway and Its Role in Biomineralization, *Chem. Rev.*, 2008 Nov, **108**(11), 4551–4627.
2. F. F. Amos, D. M. Sharbaugh, D. R. Talham, L. B. Gower, M. Fricke and D. Volkmer, Formation of single-crystalline aragonite tablets/films *via* an amorphous precursor. *Langmuir*, 2007, **23**(4), 1988–1994.
3. H. Nakahara, An electron microscope study of the growing surface of nacre in two gastropod species, Turbo cornutus and Tegula pfeifferi, *Venus*, 1979, **38**, 205–211.
4. L. B. Gower and D. J. Odom, Deposition of calcium carbonate films by a polymer-induced liquid-precursor (PILP) process, *J. Cryst. Growth*, 2000, **210**(4), 719–734.
5. M. J. Olszta, S. Gajjeraman, M. Kaufman and L. B. Gower, Nanofibrous calcite synthesized *via* a solution-precursor-solid mechanism, Chem. Mater., 2004, **16**(12), 2355–2362.
6. X. Cheng and L. B. Gower, Molding Mineral within Microporous Hydrogels by a Polymer-Induced Liquid-Precursor (PILP) Process, *Biotechnol. Progress*, 2006, **22**(1), 141–149.
7. Y. Y. Kim, E. P. Douglas and L. B. Gower, Patterning inorganic ($CaCO_3$) thin films *via* a polymer-induced liquid-precursor process, *Langmuir*, 2007, **23**(9) 4862–4870.
8. B. Pokroy, J. P. Quintana, Ea. N. Caspi, A. Berner and E. Zolotoyabko, Anisotropic lattice distortions in biogenic aragonite, *Nat. Mater.*, 2004, **3**, 900–902.
9. X. G. Cheng, P. L. Varona, M. J. Olszta, L. B. Gower, Biomimetic synthesis of calcite films by a polymer-induced liquid-precursor (PILP) process 1. Influence and incorporation of magnesium, *J. Cryst. Growth*, 2007, **307**, 395–404.
10. V. M. Patel, P. Sheth, A. Kurz, M. Ossenbeck, D. O. Shah and L. B. Gower, Synthesis of Calcium Carbonate Coated Emulsion Droplets for Drug Detoxification, in: *Concentrated Dispersions: Theory, Experiments, and Applications*, ed. B. Markovic and P. Somansundaran, American Chemical Society, Washington, DC, 2002. p. 15–25.
11. M. J. Olszta, X. G. Cheng, S. S. Jee, R. Kumar, Y. Y. Kim, M. J. Kaufman, *et al.*, Bone structure and formation: A new perspective, *Mater. Sci. Eng. R Rep.*, 2007, **58**(3–5), 77–116.
12. A. K. Burwell, T. Thula-Mata, L. B. Gower, S. Habeliz, M. Kurylo, S. P. Ho, *et al.*, Functional Remineralization of Dentin Lesions Using a Polymer-Induced Liquid-Precursor Process, *PLoS One*, 2012, **7**(6), e38852.
13. F. F. Amos, L. Dai, R. Kumar, S. R. Khan and L. B. Gower, Mechanism of formation of concentrically laminated spherules: implication to Randall's plaque and stone formation, *Urol. Res.*, 2009, **37**(1) 11–17.
14. L. B. Gower, F. F. Amos, S. R. Khan, Mineralogical Signatures of Stone Formation Mechanisms, *Urol. Res.*, 2010, **38**(4), 281–292.

Professor Gilbert addressed Dr Gower and Professor Meldrum: In biominerals multiple nucleation sites appear to be avoided at all costs,[1] and this must be done by the animal with some sort of active control. I find it astounding that when the amorphous nanorods isolated from the track etch membranes crystallize, they produce single crystals. What keeps them from having multiple nucleation sites, and hence multiple crystal orientations?

1. C. E. Killian, R. A. Metzler, Y. U. T. Gong, I. C. Olson, J. Aizenberg, Y. Politi, F. H. Wilt, A. Scholl, A. Young, A. Doran, M. Kunz, N. Tamura, S. N. Coppersmith and P. U. P. A. Gilbert, *J. Am. Chem. Soc.*, 2009, 131.

Professor Meldrum replied: Our studies on the formation of calcite nanorods in track etch membrane pores have shown that the vast majority are single crystals of calcite—despite their high aspect ratios, which can be up to 50–100 times when they are formed in membrane pores with small diameters.[1–3] This work, however, only allows us to characterise the calcium carbonate rods after they have been extracted from the membrane pores, and rods which are fully or partially amorphous are very difficult to extract intact due to poor mechanical properties. It is therefore not possible for us to directly study the progress of crystallisation in the membrane pores, only the final crystal product.

We cannot rule out that there are indeed multiple nucleation sites. But if there are, a single site must eventually "win out", possibly through Ostwald ripening mechanisms, to give a single crystal product. Alternatively, if it is the nucleation step that is limiting, and crystallisation proceeds rapidly after nucleation, one could envisage how crystallisation could proceed from a single nucleation site. The former mechanism, however, seems more likely.

1. Y. Y. Kim, N. B. J. Hetherington, E. H. Noel, R. Kröger, J. M. Charnock, H. K. Christenson and F. C. Meldrum, *Angew. Chem., Int. Ed.*, 2011, **50**(52), 12572–12577.
2. E. Loste, R. J. Park, J. Warren and F. C. Meldrum, *Adv. Funct. Mater.*, 2004, **14**(12), 1211–1220.
3. E. Loste and F. C. Meldrum, *Chem. Commun.*, 2001, **10**, 901–902.

Dr Gower answered: I don't necessarily agree that it requires active control by the animal to yield only one nucleation site, since the *in vitro* systems show that only one nucleation occurs naturally quite often. It seems to me that the whole amorphous phase is temporally inhibited, either by entrapped polymer, water, or bicarbonates, but if one nucleation event is able to occur, growth may be rapid enough to avoid the possibility of another nucleation event. Nevertheless, it makes sense that the biological system would add some nucleating template to aid in orientational or polymorph control, as well as the singular nucleation event.

Professor Meldrum asked: In our work investigating the formation of rod-like calcium carbonate crystals in track etch membrane pores[1] we can produce single crystals of calcite with aspect ratios up to 100 times. These exhibit a helical twist, but the long axis of the particle is completely random. We have also examined calcium carbonate fibres generated in the presence of a block copolymer,[2] and poly(-allyl amine) hydrochloride[3] and these also show no preferred orientation wih respect to the fibre axis.

1. Y. Y. Kim, N. B. J. Hetherington, E. H. Noel, R. Kröger, J. M. Charnock, H. K. Christenson and F. C. Meldrum, "Capillarity Creates Single-Crystal Calcite Nanowires from Amorphous Calcium Carbonate", *Angew. Chem., Int. Ed.*, 2011, **50**(52), 12572–12577.
2. Y. Y. Kim, A. N. Kulak, Y. Li, T. Batten, M. Kuball, S. P. Armes and F. C. Meldrum, "Substrate-Directed Formation of Calcium Carbonate Fibres", *J. Mater. Chem.*, 2009, **19**, 387–398.
3. B. Cantaert, Y-Y. Kim, H. Ludwig, F. Nudelman, N. A. J. M. Sommerdijk and F. C. Meldrum, "Think Positive: Phase Separation Enables a Positively Charged Additive to Induce Dramatic Changes in Calcium Carbonate Morphology", *Adv. Funct. Mater.*, 2012, **22**(5), 907–915.

Professor Roberts responded: This is most interesting but it is not clear whether the "randomness" is total or whether there is order in (say) one crystallographic direction, say along the long crystal axis as would be mediated by an in-plane rotation between an assembly of aggregated particles. Would this be feasible? Also, it would be good to know what the current state of play regarding the development of a mechanistic understanding as to how biological interfaces seem to mediate oriented crystal aggregation whereas synthetically this is not yet possible.

Dr Verch remarked: We have done some TEM-experiments on the microstructure of calcite nano-rods grown in track-etch membranes. The growth orientation of our particles appears to be random. When we take diffraction series along these rods the lattices twist slightly from one end to the other. A rolling of the rods could be excluded. High-resolution TEM on these rods shows their single crystallinity without any grain boundaries.

Professor Frenkel said: During the discussion the question arose if multiple nucleation events could take place in a single fibril. It would seem that, knowing the overall nucleation rate and the average time it takes for the crystallisation front to travel along the fibril, this question can be answered by using Poisson statistics.

Dr Christenson responded: This would certainly be possible in principle, but in our work on the track-etch membranes[1] we were unable to get this information.

1. Y.-Y. Kim, N. B. J. Hetherington, E. H. Noel, R. Kröger, J. M. Charnock, H. K. Christenson and F. C. Meldrum, *Angew. Chem., Int. Ed.*, 2011, **50**, 12572–12577.

Professor Roberts commented: The morphology of the polymer modified $CaCO_3$ crystals as presented in your slide presentation was most interesting. Can you provide any information concerning the polymorphic form and crystal orientation of any of the habit-modified crystals? For example, in samples with a high aspect ratio and assuming that they are single orientated crystals, then what is the needle axis? Intuitively one would expect this to be the 3-fold axis for calcite phase but is there *e.g.* any TEM data which quantifies this in any way? Also, do you have any information concerning the micro-structure of the crystals produced and in particular the formation mechanisms involved where precursor droplets aggregate?

Dr Gower responded: There was one set of experiments where some of the fibers that grew atop of calcite rhombs appeared to have an epitaxial relationship to the underlying "seed" rhomb crystal.[1] This was evident by the well aligned fibers extending from the rhomb, although diffraction work was not performed on these fibers. But this was not all that common. In the fibers that we did examine, there did not seem to be a preferred crystallographic orientation (Meldrum's group has apparently examined many fibers and also did not find a preferred orientation). This may seem surprising, but apparently results from the fibers being initially amorphous, and then crystallization propagating along the fiber through the amorphous phase. I speculate that if the amorphous phase nucleates at the base where it is attached to the seed crystal, then there could be some epitaxy with the seed crystal. But most of the images we have show that the seed crystal becomes covered with a lumpy amorphous $CaCO_3$ coating. So I speculate that it depends on whether or not this coating quickly crystallizes with epitaxy to the underlying rhomb, then this epitaxy might also carry through to the fiber. But if this coating remains amorphous for some time, then the nucleation could occur randomly anywhere along the fiber. One particularly interesting observation was when there was a bend in the fiber, yet the crystallographic orientation remained the same. We considered this evidence that the crystallization most likely proceeded across an already formed amorphous fiber.

1. M. J. Olszta, S. Gajjeraman, M. Kaufman and L. B. Gower, Nanofibrous calcite synthesized *via* a solution-precursor-solid mechanism, *Chem. Mater.*, 2004, **16**(12), 2355–2362.

Professor Roberts said: Also, do you have any information concerning the microstructure of the crystals produced and in particular the formation mechanisms involved where precursor droplets aggregate?

Dr Gower replied: The microstructure of the crystals appears to be composed of nanoparticle subunits, which overall are uniform in crystallographic alignment (although we have seen some less uniform domains). We did more extensive analysis of the $BaCO_3$ and $SrCO_3$ fibers, and one can see a colloidal or nanogranular texture, particularly in the $SrCO_3$ system.[2] We had originally proposed a SPS (solution–precursor–solid) mechanism to explain the fiber formation mechanism based on the fact that we observed bobble tips analogous to those seen in VLS (vapor–liquid–solid) and SLS (solution–liquid–solid) systems.[1] But because a bobble tip is not always present, and the more nanogranular texture can be seen in some, we proposed a modified mechanism where we suggest that the PILP nanodroplets (or nanoparticles, depending on the state of the precursor phase), adsorb at points of high energy, which would be the mineral surfaces with high curvature, such as on the lumpy mineral coating, which then leads to an autocatalytic effect for preferential adsorption of droplets at the tip of the protrusion that, with time, will turn into and extend the tip of the fiber in a one-dimensional fashion. If the tip stays as a liquid-like droplet, it could follow the SPS mechanism and lead to more homogeneous fibers. But if the tip solidifies rapidly upon deposition, it would likely lead to the more granular texture. Meldrum and colleagues have proposed an alternative mechanism (using a similar reaction system).[3] They state "Directional aggregation and fibre formation can then be attributed to selective adsorption of copolymer on specific crystal faces, which endows the particles with structural anisotropy." Given that analogous fibers can be formed without the use of a copolymer system, it seems unlikely (to me) that two different mechanisms are operative. In addition, they also formed amorphous fibers, so had to come up with an entirely different mechanism to explain this: "In the current system, the amorphous precursor particles may condense at specific charged sites on the substrate, giving rise to charge anisotropy in the adsorbed particle due to redistribution of the copolymer chains (a schematic diagram of the proposed mechanism is shown in Fig. 11). Polarisation of further precursor particles moving close to the charged tip of the developing fibre then results in their attraction and adsorption to the fibre end". I have a hard time envisioning redistribution of a copolymer on the particles, and find it hard to believe that their similar fibers are formed by different mechanisms. In any case, this is still apparently an open ended question and deserves more study.

1. M. J. Olszta, S. Gajjeraman, M. Kaufman and L. B. Gower, Nanofibrous calcite synthesized *via* a solution-precursor-solid mechanism, *Chem. Mater.*, 2004, **16**(12), 2355–2362.
2. S. J. Homeijer, R. A. Barrett and L. B. Gower, Polymer-Induced Liquid-Precursor (PILP) Process in the Non-Calcium Based Systems of Barium and Strontium Carbonate, *Cryst. Growth Des.*, 2010, **10**(3), 1040–1052.
3. Y. Y. Kim, A. N. Kulak, Y. T. Li, T. Batten, M. Kuball, S. P. Armes, *et al.*, Substrate-directed formation of calcium carbonate fibres, *J. Mater. Chem.*, 2009, **19**(3), 387–398.

Dr Wolf communicated: Reviewing the literature, it becomes clear that the terms "mesocrystal" and "mesocrystallization" have no clear cut definition. I propose that we accept the following definition, which initially was given in Helmut Cölfen's book on nonclassical crystallization: (i) The crystal has to scatter like a single crystal—both in X-ray diffraction and electron diffraction. (ii) On the mesoscale, it must not be a compact single crystal but a colloidal crystal which is built up from individual and crystallographically aligned nanocrystals. The term mesocrystallization should be defined as just the formation of a mesocrystal without implying a distinct pathway of formation, *e.g.* whether the initial particles are already crystalline (*i.e.*, oriented attachment) or still amorphous (*Faraday Discuss.*, 2012, **158**, DOI: 10.1039/c2fd20045g) during attachment. The condition (i) was initially defined as an option but the typcial current use of the term "mesocrystal" implies that it scatters like a single-crystal. With regard to the discussion "what is in fact amorphous" started by Prof. Gilbert, we should consider both X-ray and electron diffraction. Accepting this definition, the space-filling character of biominerals would be a special feature of biominerals and would not exclude them from the class of mesocrystals, as currently discussed in ref. 2.

Would you agree?

1. Helmut Cölfen, and Markus Antonietti, *Mesocrystals and Nonclassical Crystallization*, Wiley-VCH, 2008, p. 96.
2. L. Yang, C. E. Killian, M. Kunz, N. Tamura and P.U.P.A. Gilbert, *Nanoscale*, 2011, **3**, 603–609.
3. J. Seto *et al.*, *Proc. Natl. Acad. Sci. U. S. A.*, 2012, **109**(10), 3699–3704.

Professor Cölfen communicated in reply: I think the definition of a mesocrystal is quite clear (see also comment to question 436 by Prof. Gillbert). The problem is just that some people confuse the definition of a structure with a formation mechanism. Therefore I fully agree with you that the term "mesocrystal" does not imply any definition of its formation pathway.

I just repeat the initial definitions:

R. S. Song and H. Cölfen, *Adv. Mater.*, 2010, **22**, 1301–1330. "The notation "mesocrystal" is an abbreviation for a mesoscopically structured crystal, which is an ordered superstructure of crystals with mesoscopic size (1–1000 nm)."
H. Cölfen and M. Antonietti, *Mesocrystals and Nonclassical Crystallization*, Wiley, 2008, ch. 4.4. "The notation Mesocrystal is an abbreviation for "Mesoscopically structured crystal". We define mesocrystals as colloidal crystals which are built up from individual nanocrystals, which are aligned in a common crystallographic register."
M. Niederberger and H. Cölfen, *Phys. Chem. Chem. Phys.*, 2006, **28**, 3271–3287. "Mesocrystals are colloidal crystals composed of individual nanocrystals that are aligned in a common crystallographic fashion, exhibiting scattering properties similar to a single crystal."

Please note that the newer definitions just require mutual order of the nanocrystals and not only the perfect orientation resulting in a single crystalline diffraction behaviour. Single crystal scattering behaviour together with proven separated nanocrystal superstructure certainly demonstrates a mesocrystal. It can be debated to what degree of mutual nanoparticle disorder the structure is still a mesocrystal. It seems that the transition between a perfectly 3D ordered mesocrystal and unordered polycrystal is continuous.[1]

1. Continuous structural evolution of calcium carbonate particles: A unifying model of copolymer-mediated crystallization: Alex N. Kulak, Peter Iddon, Yuting Li, *et al.*, *J. Am. Chem. Soc.*, 2007, **129**(12), 3729–3736.

Professor Penn addressed Dr Gower and Professor Cölfen: It seems there is quite a bit of confusion with regard to the term "mesocrystal". The term refers specifically to an object and does not imply a mechanism by which the object formed. In fact, mesocrystals seem to be the product of quite a range of growth mechanisms. Recent publications from a number of groups have described the mesocrystal in many different ways. In our work, we define a mesocrystal as an object composed of primary crystallites that are crystallographically aligned with respect to one another but that do not have direct contact with one another. However, there is no implied mechanism by which such an object is formed. Should we consider avoiding the use of this term? The term is compelling because it implies heirarchical structure and order and, once defined, provides an eloquent and descriptive term.

Dr Gower answered: I am not sure that I am the one to be addressing this question since I didn't initiate the mesocrystal terminology, and have rarely used it when describing PILP formed crystals. I was worried people would mistakenly think I was referring to the process of oriented attachment, which is one mechanism that some people seemed to automatically associate with mesocrystals. Given the very small amount of polymer that is in the PILP formed crystals, it didn't seem necessary to me to describe it in any certain way other than that the crystallization proceeds across the amorphous phase through a pseudomorphic transition, and entraps some polymer along the way. Certainly- in the PILP system there appear to be colloidal subunits which could thus be considered the basis of a mesocrytal, but I am pretty sure they are not fully separated by the polymer since the crystallization process proceeds relatively uniformly across the coalesced precursor particles. Although it appears to be a continuous amorphous phase, some memory of the colloidal constituents remain. I think this is why people have a hard time with accepting the concept of the PILP phase being liquid like. Although it is perplexing, the colloidal texture may simply result from polymer that was excluded during the A-to-C transformation. The results of this study (the NMR relaxation and diffusion times) show that the LCP and PILP phases are definitely not solids. In my very first paper on PILP,1 I could see that the phase that had accumulated below the surface of an air bubble flowed easily, like a non-viscous liquid, so I was surprised when people still question this issue. In fact, I have video footage of that flowing liquid, but they didn't have Supplements to journal papers at that time, so it was not published. I recognize that I have gone off topic, but since this was actually a heavily discussed question at the conference (even though no one posted such questions to me after the fact), I thought I would add it in.

1. L. B. Gower, D. J. Odom, Deposition of calcium carbonate films by a polymer-induced liquid-precursor (PILP) process, *J. Cryst. Growth*, 2000, **210**(4), 719–734.

Professor Cölfen responded: I fully agree. The term mesocrystal describes a structural feature or an object and NOT a formation mechanism. Mesocrystals can be formed by different mechanisms apart from nanoparticle aggregation. Some of them are outlined in ref. 1, others will likely be revealed in the future. See also my answer to Prof Gilbert's question below.

1. R.S. Song and H. Cölfen, *Adv. Mater.*, 2010, **22**, 1301–1330.

Professor Gilbert enquired: Aggregation and crystallization: these are two separate and well distinct events. In echinoderm biominerals, which form from two amorphous precursor phases[1] aggregation of amorphous nanoparticles occurs first, space is filled,[2] water is excluded,[3] then crystallinity propagates through the amorphous material random-walking in 3D.[4] If the nanoparticles were crystalline first, and then aggregated by oriented attachment,[5] they would form a *bona fide* mesocrystal. To this day, there is no evidence in the literature of such a mesocrystalline biomineral. The only evidence in nature of a biomineral mesocrystal is in bacteria,[6] but not in echinoderms as claimed by other authors.[7]

1. Y. Politi, R. A. Metzler, M. Abrecht, B. Gilbert, F. H. Wilt, I. Sagi, L. Addadi, S. Weiner and P. U. P. A. Gilbert, *Proc. Natl. Acad. Sci. U. S. A.*, 2008, **105**, 17362.
2. L. Yang, C. E. Killian, M. Kunz, N. Tamura and P. U. P. A. Gilbert, *Nanoscale*, 2011, **3**, 603.
3. Y. U. T. Gong, C. E. Killian, I. C. Olson, N. P. Appathurai, A. L. Amasino, M. C. Martin, L. J. Holt, F. H. Wilt and P. Gilbert, *Proc. Natl. Acad. Sci. U. S. A.*, 2012, **109**, 6088.
4. C. E. Killian, R. A. Metzler, Y. U. T. Gong, I. C. Olson, J. Aizenberg, Y. Politi, F. H. Wilt, A. Scholl, A. Young, A. Doran, M. Kunz, N. Tamura, S. N. Coppersmith and P. U. P. A. Gilbert, *J. Am. Chem. Soc.*, 2009, **131**, 18404.
5. R. L. Penn and J. F. Banfield, *Geochim. Cosmochim. Acta*, 1999, **63**, 1549.
6. J. F. Banfield, S. A. Welch, H. Z. Zhang, T. T. Ebert and R. L. Penn, *Science*, 2000, **289**, 751.
7. J. Seto, Y. R. Ma, S. A. Davis, F. Meldrum, A. Gourrier, Y. Y. Kim, U. Schilde, M. Sztucki, M. Burghammer, S. Maltsev, C. Jager and H. Cölfen, *Proc. Natl. Acad. Sci. U. S. A.*, 2012, **109**, 7126.

Mr Seto responded: As misunderstood and misinterpreted during the Faraday Discussions, a mesocrystal is not indicative of a formation mechanism, rather it is an organization and structural feature of a crystalline material. We state in Seto *et al.*, *Proc. Natl. Acad. Sci. U. S. A.*, 2012 that a mesocrystal is comprised "of a

3D array of isooriented single crystal particles of size 1–1000 nm (mesoscale dimensions)" and in the same work state that crystallization does indeed follow a tortuous path (as cited in the references provided by Gilbert's group). And mesocrystals may form from oriented attachment, but not exclusively by this route. Again, "the term mesocrystal defines the structure of a material rather than its mechanism of formation." This was correctly reiterated by Lee Penn and others at the Faraday Discussion.

Professor Meldrum replied: The answer here depends on the definition of a mesocrystal. If a mesocrystal is defined in terms of its structure—where it behaves as a single crystal but is comprised of iso-oriented subunits—then it is possible for a biomineral to be a mesocrystal.[1] The current definition of a mesocrystal makes no statement about its mechanism of formation—only about the final structure of the crystal product. There is no requirement for aggregation of crystalline blocks to make a mesocrystal—and I would completely agree that this cannot occur in a biological system.

However, there is strong evidence from sea ucrhin, mollusk and sponge systems, that calcium carbonate crystals can retain a memory of the amorphous particles from which they formed. This generates their "mesocrystal" structure.

1. J. Seto, Y. R. Ma, S. A. Davis, F. Meldrum, A. Gourrier, Y. Y. Kim, U. Schilde, M. Sztucki, M. Burghammer, S. Maltsev, C. Jaeger and H. Cölfen, *Proc. Natl. Acad. Sci. U. S. A.*, 2012, **109**(10), 3699–3704.

Dr Gower answered: I agree that there is no evidence of oriented attachment occurring in biologically controlled mineralizations because the evidence thus far shows that most are formed from an accumulation of amorphous precursor. I think the confusion arises from the possibility of multiple mechanisms leading to meso-structured crystals. In fact, the Seto paper that was cited at the end of the question makes a point to emphasize this point to avoid confusion. They state "It is emphasized that the term mesocrystal defines the structure of a material rather than its mechanism of formation. Therefore, while oriented aggregation of crystalline nanoparticles can give rise to either a single crystal or mesocrystal product, in analysis of the structure of sea urchin spines we here present data which suggests that a mesocrystal can also form when a dense array of amorphous nanoparticles crystallizes to give a highly cooriented end-product material."

In contrast, one person (I believe it was Lee Penn) in the session defined mesocrystal as having aligned units that do not touch, so this would not apply to biominerals since the amorphous particles touch and essentially coalesce into a dense unit. I originally avoided using the term mesocrystal for PILP-formed products because I feared some people would automatically assume that it meant oriented attachment. I tend to think of the PILP formed crystals and biominerals as single crystals with entrapped impurities, because there is only a singular nucleation event, and the terminology of co-oriented subunits could lead to confusion as to how they are formed.

Professor De Yoreo responded: The oriented attachment (OA) of the kind I presented was not studied in a biological context and the conclusions of that study have never been extrapolated to such a context. The observation of co-aligned nano-crystals in nature serves as nothing more than motivation to study the mechanisms of OA, and indeed is only one of a number of motivations that have nothing to do with biomienrals (*e.g.*, branched nanowires, nanoparticle superlattices, *etc.*). Having said that, I don't know why one would extrapolate from a miniscule handful of biological examples to the general conclusion that OA does not happen during the formation of biominerals. Maybe it does, maybe it doesn't. It is a big and rich universe out there and, in fact, the production of nanoparticles by microbial systems—where OA has been shown and is very likely—is a far more widespread phenomenon than the production of spines, plates, bones, *etc.* Coupled to that open view is the realization that no one has ever directly observed how a real biomineral forms. Moreover, in the OA experiments presented by myself, the particles are moving in a fluid with the viscosity of very thick honey (1 million times that of bulk water). They jostle about keeping within 10 nm of one another over long periods of time. Are we sure that even in the "space-filled" region of a biomineral growing from nanoparticles of ACC, that the particles are not jostling about prior to consolidation through crystallization? And would we know if a particle attached because it crystallized or crystallized because it attached?

Professor Cölfen answered: This comment is based on the misunderstanding that mesocrystals would be formed exclusively *via* oriented attachment. This is certainly not true. The term Mesocrystal defines a structure and NOT a formation mechanism (for definitions see below). One possibility is oriented attachment, but in ref. 1 this pathway is not described. Instead, the mesocrystal structure of sea urchin spines is revealed by several complementary techniques demonstrating calcite nanocrystals with a thin ACC surface layer, aligned in crystallographic register. This structure is compatible with all above cited results for the formation mechanism so that there is no contradiction at all. It is just very important not to confuse the term Mesocrystal with a formation mechanism *via* aggregation of nanoparticles. The below citations give the definitions of a mesocrystal in the literature. They clearly show that the term "Mesocrystal" has nothing to do with formation *via* oriented attachment nor any other mechanism for nanoparticle alignment and is purely a definition of a hierarchical structure.

In ref. 2 the notation "mesocrystal" is an abbreviation for a mesoscopically structured crystal, which is an ordered superstructure of crystals with mesoscopic size (1–1000 nm).

In ref. 3 the notation "mesocrystal" is an abbreviation for "mesoscopically structured crystal. We define mesocrystals as colloidal crystals which are built up from individual nanocrystals, which are aligned in a common crystallographic register".

From ref. 4: "Mesocrystals are colloidal crystals composed of individual nanocrystals that are aligned in a common crystallographic fashion, exhibiting scattering properties similar to a single crystal."

Whereas it is likely true that no biomineral mesocrystal formed *via* oriented attachment is known so far, several biominerals were reported to have a mesocrystalline structure including corals, nacre, egg shells, echinoderm biominerals and more. One paper summarizing these structures is ref. 5, but there are several other papers with detailed structural characterization of biomineral mesocrystal structures including the above cited Seto *et al.* paper on sea urchin spines.[1]

1. J. Seto, Y. R. Ma, S. A. Davis, F. Meldrum, A. Gourrier, Y. Y. Kim, U. Schilde, M. Sztucki, M. Burghammer, S. Maltsev, C. Jager and H. Colfen, *Proc. Natl. Acad. Sci. U. S. A.*, 2012, **109**.
2. R. S. Song and H. Cölfen, *Adv. Mater.* 2010, **22**, 1301–1330.
3. H. Cölfen and M. Antonietti, *Mesocrystals and Nonclassical Crystallization*, Wiley 2008 Chapter 4.4.
4. M. Niederberger and H. Cölfen, *Phys. Chem. Chem. Phys.*, 2006, **28**, 3271–3287
5. Y. Oaki, A. Kotashi, T. Miura and H. Imai, *Adv. Funct. Mater.*, 2006, **16**, 1633–1639

Dr Gower opened the discussion of the paper by Professor Fiona C. Meldrum: You state that "Experimentally, PILP droplets are highly unstable and rapidly aggregate/coalesce either in solution or on 30 a substrate. These characteristics therefore do not seem to correlate with the stability observed for the membrane-bound amorphous granules." First of all, the fact that biominerals put membranes around the amorphous droplets or particles may in fact be what is providing the stabilization that is needed. For example, the membrane could prevent the exclusion of the polymer, or water, which is needed for the crystallization reaction to occur. Secondly, our *in vitro* model systems admittedly use a very crude mimic to the biological proteins, so it is likely that the proteins are more optimized for increasing the stability of the phase. Lastly, the length scales in biomineral systems are very small, and PILP droplets are stable at these length and time scales. For example, the mantle epithelial cells in mollusk may release droplets that only have to travel a short distance to the forming nacreous tablet.

Dr Schenk replied: I agree that in our study we investigate a simplified *in vitro* model system, which cannot fully resemble biological conditions. We found that based on their chemical composition, it appears to be possible that highly acidic biomacromolecules could indeed undergo phase separation and thus direct $CaCO_3$ precipitation *via* a liquid precursor mechanism. However, in my view, the fact that—at least *in vitro*—PILP is usually formed in very dilute solutions seems to contradict the concept of a PILP mechanism acting in biological mineralization, where dilution would be associated with unrealistically large volumes. Nevertheless, as you mention, we cannot exclude the hypothesis that biological systems could be optimized to overcome these difficulties.

Dr Gower asked: *In vitro* studies have also shown that it is extremely difficult to scale-up PILP processes to give larger quantities of the product materials. Indeed, very dilute solutions are required to give the characteristic fingerprint morphologies. This was also observed in the in-depth study of the Glutamic 35 acid/ PEI system, where "PILP was only formed within a tiny area of the phase diagram.[29]" This was an organic PILP system, which could be very different than $CaCO_3$ mineral PILP system. (It was too difficult to do the phase diagram for the $CaCO_3$ PILP system.) In any case, the biomineralizing organism is continuously producing and adding the vesicles or droplets to the crystal deposition vesicle, so it is not limited by the batch type reactions we use. This brings up an interesting point though, that mollusk nacre forms the same thickness of the tablets that we deposit *in vitro*, so it may be limited by the supersaturation (ion content) of the reaction medium. Therefore, nacre, and many other biominerals, seem to deposit the biomineral layer-by-layer (concentric laminations are very common).[1] This may be a consequence of the limited formation of PILP phase. Alternatively, thin layers may be needed in order to allow the polymer and water to be excluded.

1. L. B. Gower, Biomimetic Model Systems for Investigating the Amorphous Precursor Pathway and Its Role in Biomineralization, *Chem. Rev.*, 2008, **108**(11), 4551–4627.

Dr Schenk responded: Indeed, the calcium carbonate PILP might be different from the organic PILP system, for which the phase diagram was determined by Jiang *et al.*[1] However, the formation of liquid-like precursor phases to calcium carbonate generally seems to require very dilute reactant solutions. In our understanding, such conditions would not be expected to occur in biological systems due to the large volumes required. Therefore, our concern is not so much related to whether a sufficient amount of mineral can be formed by a PILP mechanism acting *in vivo*, but rather to the question how to generate a PILP phase within the vesicles in the first place.

1. Yuan Jiang, Laurie Gower, Dirk Volkmer and Helmut Cölfen, The existence region and composition of a polymer-induced liquid precursor phase for DL-glutamic acid crystals, *Phys. Chem. Chem. Phys.*, 2012, **14**, 914–919.

Dr Sear addressed Dr Schenk and Dr Gower: You both show interesting results for what you call liquid phases, PILP and LCP. For example, you find that PILP can infiltrate narrow pores. Am I correct in thinking that your PILP and LCP are indeed liquid phases, in the sense of having a relatively low viscosity, low enough to see significant flow? If so can you estimate the viscosities?

My impression is that people think that ACC is amorphous not liquid-like, *i.e.*, has a very high or even effectively almost infinite viscosity. So, are your PILP and LCP very different from ACC in having a lower viscosity? If so, does this imply that the ions and water in PILP are much more mobile than in ACC, *i.e.*, have much higher diffusion coefficients than in ACC?

Finally, your polymer concentrations appear to be quite low; I would guess that the volume fraction is perhaps 1% at most in your PILP or LCP. If the polymer volume fraction is indeed that low, how can this dilute polymer strongly affect ion and water dynamics, and the bulk viscosities?

Dr Gower answered: Yes, we believe the PILP droplets are very different from a solid ACC phase. We originally proposed the concept of PILP of being a liquid phase based on evidence that it flows, such as droplets/particles melding together into a continuous film or non-equilibrium shaped crystal.[1] In one circumstance an accumulation of PILP droplets under an air bubble was seen to flow easily, like a non-viscous liquid.[1] However, since it did not ever crystallize, it is not certain that it was representative of the common PILP phase.

In this *Faraday* paper, we included a absolute viscosity estimation in the *Faraday* paper based on some assumptions. We estimated that the viscosity of the LCP and PILP phase was about twice the viscosity of water. See page 15 and page 20 of the paper. This is well within flow of liquids. For a simple reference, the LCP and the PILP have viscosities that are on the order of milk as compared to water. This is very much lower than a solid which is approaching near infinite viscosity. The viscosity will presumably change with time as the particles dehydrate, but in this Faraday study, it seemed to stay low during the duration of the NMR study (a kinetic study was not performed), so it is not clear if this value will be the same as our other studies where the concentration of carbonates changes with time (such as in the vapor diffusion or CO_2 escape methods).

Regarding the amount of polymer, in one study where the PILP phase was centrifuged and collected at different stages,[2] the polymer was found to be present at high levels in the early stages, and was excluded with time. Corresponding to this was a loss in the hydration waters as well. So it may not be possible to fully define a polymer concentration or viscosity unless we are able to make the phase more stable. Indeed, I have always found it to be amazing that such a small amount of polymer could cause such a pronounced affect. But now that we know there is an inherent tendency to produce a LCP under the proper conditions, the influence of the polymer seems more reasonable.

1. L. B. Gower, D. J. Odom, Deposition of calcium carbonate films by a polymer-induced liquid-precursor (PILP) process, *J. Cryst. Growth*, 2000, **210**(4), 719–734.
2. L. Dai, E. P. Douglas, L. B. Gower, Compositional analysis of a polymer-induced liquid-precursor (PILP) amorphous $CaCO_3$ phase, *J. Non-Cryst. Solids*, 2008, **354**, 1845–1854.

Dr Schenk answered: In our study, we have worked with PILP phases prepared in the presence of a series of copoly(amino acid)s being rich in aspartic acid. Unfortunately, direct measurements of the viscosity were not possible, since it turned out to be very difficult to concentrate the liquid-like droplets without crystallization to occur.

In my understanding, PILP can be regarded as a highly hydrated, amorphous phase, which exhibits properties typically attributed to liquids such as the ability to infiltrate into porous templates. Kim *et al.* have recently demonstrated that effective infiltration into the small pores of track-etched membranes can be achieved with a calcium carbonate PILP phase generated by the polyelectrolyte poly(acrylic acid).[1] The authors provided strong evidence for an infiltration of the template driven by capillary action. Thus, the suggested mechanism is reminiscent of the uptake of liquids into pores.

PILP and solid amorphous calcium carbonate (ACC) are indeed usually considered to be two different species and it has been suggested that liquid-like PILP droplets transform into ACC by dehydration before they eventually crystallize.

1. Y.-Y. Kim, N. B. J. Hetherington, E. H. Noel, R. Kröger, J. M. Charnock, H. K. Christenson and F. C. Meldrum, Capillarity Creates Single-Crystal Calcite Nanowires from Amorphous Calcium Carbonate, *Angew. Chem., Int. Ed.*, 2011, **50**, 12572–12577.

Dr Sear opened the discussion of the paper by Dr Fajun Zhang: Do you think that your larger clusters, those that give the shoulder at $q \sim 0.3$ nm^{-1}, are liquid-like (*i.e.*, not gels) and so are in dynamic equilibrium with the surrounding solution? Can you form and dissolve them reversibly? Do you know what sets this lengthscale of $2\pi/0.3 \sim 20$ nm?

Dr Zhang responded: Under the experimental conditions (protein, salt concentration and temperature), these clusters are liquid like, but they cannot merge into macroscopic liquid droplets because of temperature. In our recent study, we have observed that below a critical temperature, these clusters will merge into liquid droplets and the process is reversible. A macroscopic liquid–liquid phase separation has also been observed when using a higher protein concentration.[1] In another study, we have determined a complete phase diagram of HSA in the presence of YCl$_3$.[2] A closed area of LLPS region has been determined. Within the LLPS region, the solutions after preparation are turbid, when examined under a microscope, one can see small (a few micrometers) liquid droplets flowing around and they can merge. The dense and dilute liquid phases can be easily separated by centrifugation.

As for the length scale defined by the $q \sim 0.3$ nm^{-1} around the shoulder in the SAXS profile, this is a simple estimation of the size of the clusters based on the Guinier law,[3] *i.e.* in the low q region ($q < 1/R_g$), where R_g is the radius of gyration of the particle, the scattering intensity follows: $I(q) \propto \exp(-(1/3)q^2 R_g^2)$. Therefore, this q corresponds to $qR_g \sim 1$, or $R_g \sim 3.5$ nm about 3 dimers within the clusters.

1. F. Zhang, *et al.*, Novel approach to controlled protein crystallization through ligandation of yttrium cations, *J. Appl. Cryst.*, 2011, **44**, 755–762.
2. F. Zhang, *et al.*, Charge-controlled metastable liquid–liquid phase separation in protein solutions as a universal pathway towards crystallization, *Soft Matter*, 2012. **8**(5), 1313–1316.
3. O. Glatter and O. Kratky, *Small angle X-ray scattering*, Academic Press, London, 1982.

Dr Nudelman opened the discussion of the paper by Professor Dr Derk Joester: What is the size of the ACC particles that you observe inside the liposomes, and what is the stabilization mechanism that you propose? We found that ACC particles are stable as long as they stay smaller than 70–80 nm (Nudelman *et al.*, *Nanoscale*, 2010, **2**, 2436–2439), meaning that they must grow beyond a minimum size before they can crystallize. Could this mechanism be related to your findings?

Professor Dr Joester communicated in reply: Depending on the size of the liposomes and the progress of the reaction, we observe stable ACC particles 10–200 nm in diameter. Due to the absence of strong membrane-mineral interaction, observed in Cryo-TEM, we believe that the particles are stabilized by the combination of slow kinetics (*i.e.* high transition state energy) and confinement inside the liposomes. In bulk solution, where particles are free to aggregate and grow by Ostwald ripening, in the extreme case, one supercritical nucleus is sufficient to lead to complete phase transformation. Inside the liposomes, confinement by the phospholipid membrane shuts down mass transfer between particles, requiring the formation of a critical nucleus in every single liposome in order to achieve complete crystallization. A low nucleation rate inside the liposomes would therefore explain the extended stability of ACC nanoparticles under confinement. We speculate that unlike most surfaces, the inner liposome membrane is particularly unsuited for heterogeneous nucleation of a crystalline calcium carbonate polymorph. In the bulk system, the most likely mechanism for nucleation is again heterogeneous nucleation at interfaces. The observation of an apparent maximum size for "stable" ACC particles in bulk might be a consequence of the likelihood (*i.e.*, an activation barrier-associated time constant) for a critical nucleus to form at the particle surface, or in its environment. If it is true that there is a liquid condensed phase which appears as a solid ACC particles in CryoTEM, then these nucleation events may also occur inside the liquid. In either case, the observed maximum size may primarily be dependent on the kinetics of ACC particle/droplet growth *vs.* the probability of crystal nucleation. The apparent maximum size for "stable" ACC particles/droplets may thus depend more on supersaturation and mass transport than on an inherent difference in free energy that depends only on the size.

Dr Sear communicated: When you discuss the stability of your ACC confined in liposomes, you assume that the nucleation rate scales linearly with volume. This is a standard assumption, and it is true for homogeneous nucleation. It can also be true for heterogeneous nucleation, but only if the impurities heterogeneous nucleation is occurring on are not too diverse. If the samples contain random impurities whose surfaces are highly variable from one impurity to the next, then the effective nucleation rate may not scale linearly with system size.

When nucleation is dominated by highly variable impurities, defining a nucleation rate is problematic[1]. However, if we are interested in a time to observe nucleation, then we can assess the rate using the median nucleation time of a set of samples, *i.e.*, the time at which half of a set of samples all with the same volume, have nucleated. It is better to use the median than the arithmetic mean here as there can be a very wide spread of values of the nucleation time, which can make determining the arithmetic mean difficult.

In Fig. 1, I have plotted the median nucleation time (in arbitrary units), as a function of the number of impurities, which I take to be proportional to the system size. Nucleation occurs on impurities, and for the impurities I use the clusters of ref. 1. If the nucleation rate were proportional to system size, the median nucleation time would scale as (system size)$^{-1}$. This is not the case here. The dependence of the median nucleation time is, for this case, well fit by a power-law dependence on system size, but the exponent is -1.98, not -1. This is just an example, the exponent will vary; the point is that it is not in general -1.

Thus, for this model system, assuming that the nucleation rate is proportional to the volume dramatically underestimates how fast the median nucleation time increases as the volume decreases. This may be relevant to understanding your very interesting results showing that the small isolated domains of ACC that you can form are very long lived.

1. R. P. Sear, *J. Phys., Condens. Matter*, 2012, **24**, 052205.

Professor Dr Joester communicated in reply: This is a very good point. For the analysis presented in the paper we assume homogenous nucleation. Similar to what has been described for nucleation and growth in small, dispersed metal droplets, we expected that heterogeneous nucleation would only have a very minor influence on the system (see, for example: Turnbull, *J. Applied Phys.*, 1950). We note that we never observe impurity particles in liposomes before the precipitation starts. We estimate that this puts an upper bound of $\sim$5 nm on the size of such particles. Extrapolating from the droplet system described by Meldrum and coworkers (Stephens, *et al.*, *J. Am. Chem. Soc.*, 2011) is somewhat dangerous, as we do know very little about the presence of heterogeneous nucleators in the volume of the droplet or at the air/water and substrate/water interfaces. In any case, our extrapolation likely overestimates the expected nucleation rate inside the liposomes.

If, instead, crystallization is driven by a low number of efficient nucleators inside the liposomes, as suggested by Richard Sear, we dramatically overestimate the nucleation rate expected for liposomes when extrapolating from the Meldrum experiments. As we point out in the manuscript, our estimate is thus a (very) generous upper bound. However, our intention was not to derive accurate rates, but rather to provide an order of magnitude estimate of whether, in principle, we should reasonably expect to find crystalline particles in liposomes by cryoTEM or detect them by *in situ* WAXS. We find that even when overestimating the nucleation rate dramatically, we do not expect to find/detect them under the experimental constraints. A more accurate modeling of the nucleation rate would thus not lead to a different conclusion. However, our prediction that we expect to be able to detect nucleation of crystalline calcium carbonate in 5–10 μm liposomes depends on whether the mechanism of nucleation is homogeneous or heterogeneous.

Professor De Yoreo enquired: Many of the particles seem to have rough surfaces, as if they are made up of other particles, yet you only see one particle in any vesicle. How do you account for the morphology and could you use cryoEM to capture the dynamics of ACC particle formation in action?

Professor Dr Joester communicated in reply: We agree that at certain stages, the ACC particles have very rough surfaces that may result from the aggregation of smaller particles. We also interpret this roughness as an indication that the ACC is solid, not liquid, which would result in coalescence and the formation of a smooth surface. We think that we only observe one particle per liposome for several reasons: 1) the particles diffuse rapidly through the volume of the liposome and will encounter each other frequently. Aggregation may thus be very rapid. 2) Resolution for our cryoTEM and SAXS is on the order of single digit nm, and we can reliably detect particles at about 10 nm or larger. If smaller particles are present, we would not detect them before they join in the large aggregate.

Dr Sommerdijk communicated: In the images some of the calcium carbonate particles show facets eventhough electron diffraction characterizes them as amorphous. This is very similar to observations we made previously[1] where we detected that the development of facets can actually preceed the development of structure as detectable by electron diffraction.

1. E. M. Pouget, P. H. H. Bomans, A. Dey, P. M. Frederik, G. de With and N. A. J. M. Sommerdijk, *J. Am. Chem. Soc.*, 2010, **132**, 11560.

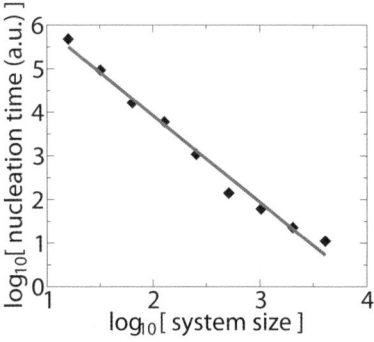

Fig. 1 Log (base 10) of the median time to nucleate (in arbitrary units), as a function of the log (base 10) of the number of impurities in the system (assumed proportional to system size). The system is that studied in ref. 1. For each system size, 16 samples are studied and the median nucleation time of these 16 nucleation times determined, and shown as black diamonds. The red line is a straight-line fit: log[time] = 7.89 − 1.98 × log[system size].

Professor Dr Joester communicated in reply: This is an interesting point. We would like to caution that what appears to have facets by eye, may or may not be faceted in reality—we find that on inspection of these images it is very difficult to differentiate between facets and more or less random angular or flattened surfaces. We also have no indication that these particles diffract electrons. Unlike the bulk system, they do not appear to evolve rapidly. However, we have plans to go back and carry out more SAED experiments of similar particles to compare with the above reference.

Dr Nudelman enquired: From the images that you show, it seems that all the particles are attached to the membrane of the lipossomes. Is it possible that they are forming on the lipid bilayer?

Professor Dr Joester answered: While some particles do appear to be in contact with the lipid bilayer, we believe that this is a transient interaction resulting from the diffusion of the nanoparticle in the confined liposome interior. We base this conclusion on 1) the absence of conformal contact between the precipitate and the membrane and 2) the formation of a single particle, both of which we would not expect if the particles were nucleating on or interacting strongly with the phospholipid membrane. While cryo-electron tomography is needed to fully characterize the extent of membrane mineral interaction, our results suggest the particles form in solution rather than on the lipid bilayer. This argument is further supported by the fact that with minor changes in lipid head group chemistry, as shown during the discussion (unpublished results), we can create liposomes that show strong ACC/lipid interaction, where multiple particles aggregate as conformal patches at the membrane.

Dr Sear asked: Does your method of producing ACC in the confinement of the liposomes, with the time-dependent properties in the liposome such as pH, result in ACC that is different from that produced in bulk?

The reason I ask is that my guess, I don't know whether you agree, is that ACC is not a thermodynamic phase. If it is not, then ACC properties will depend on how it is prepared, and this will include dynamic properties such as the mobilities of water, and of calcium and carbonate ions. Then if these mobilities are much lower in your ACC than in ACC prepared in bulk, this would presumably be a possible explanation for your ACC being longer lived. It would also imply that the ACC properties may depend on liposome size.

Professor Dr Joester responded: From our current data we cannot say whether or not the structure of ACC produced in liposomes is different from ACC produced in bulk solution. However, experimentation by Reeder and coworkers indicates that short-range order and the amount of water in ACC are not strongly dependent on the way in which ACC is prepared. We thus expect that the liposome-stabilized ACC is not significantly different and the liposome-encapsulated ACC is not stabilized because of its structure.

Professor Penn remarked: Is it possible that the objects inside the liposomes were crystalline prior to exposure to the electron beam?

Professor Dr Joester replied: We do not believe that the objects inside the liposomes were crystalline prior to exposure to the electron beam. Due to the fact that ACC is metastable relative the crystalline polymorphs of calcium carbonate (Radha *et al.*, *Proc. Natl. Acad. Sci.*, 2010) our expectation would be the opposite, that exposure to intense radiation could induce crystallization. This has been demonstrated in the characterization of biogenic ACC (Politi *et al.*, *Science*,

2004). In our system, we expect that imaging under cryogenic conditions should limit the rate of any radiation-induced transformation.

Dr Zhang communicated: The SAXS measurements in your paper have a maximum q value of 0.1 Å^{-1} that means the smaller particles with size smaller than 10 nm cannot be precisely determined, and the cryo-TEM images indicate the size of the ACC particles after about 8 h, is around 10 to 20 nm. In this case, the structural information of the early stage of the growth is missing. Are there special reasons for not following the early growth by performing SAXS experiments in a larger q range?

Professor Dr Joester responded: The q-range was selected to follow the growth of the particles over extended periods of time and in liposomes up to 1 μm in diameter. As you suggest, it would very interesting to now combine these results with data recorded at larger q-values to capture early stages of growth.

Professor Roberts remarked: The key problem is that characterisation techniques are needed to determine the crystalline order in ACC and thus assess if this material is truly amorphous (only local atomic ordering with no recognisable 3D ordering) or just nano-crystalline with a small crystallite size. Electron diffraction using a TEM offers an attractive option but electron beam heating and radiation damage can make unambiguous determination challenging particularly when the particles are fully or partially solvated. In principle, X-ray absorption spectroscopy (XAFS) using synchrotron radiation provides an attractive alternate option as the technique allows, in principle, the derivation of a partial radial distribution function centred around the absorbing atomic site. For reasonably heavy absorbing species this can work well and enable determination of both the short range order as well as providing confirmation of crystalline order. The latter can be done through detection of outer shell co-ordinations which would be expected to be absent in a truly amorphous solid,[2] see Fig. 2. However, for lighter atomic species which are associated with the use of absorption spectroscopy around Ca site in CaCO$_3$, the technique can be much less effective as this system lacks near neighbouring atomic sites which have a high enough electron density (backscattering amplitude) to unambiguously identify there really is no long range order. Specifically, a lack of Ca-Ca correlations within the ACC structure is needed. Table 1 reveals that for the Calcite structure these are at about 4.11 Å in inter-atomic distance. This is probably too far for the identification of such a low back scatterer such as Ca to have total confidence in such an assessment. Perhaps a better approach would be to use X-ray absorption spectroscopy centred around absorption at one the lighter elements, such as C, to probe ACC. Table 2 reveals that the near neighbour sites would be much heavier Ca ions which would also be closer at *ca.* 3.25 Å thus offering greater enhancement of back scattering. Concomitantly, this would afford a greater likelihood for a successful characterisation of any short- and longer-range order in CaCO$_3$. Facilities for this kind of measurements are becoming more readily available, *e.g.* using beam line I09 (ref. 3) at the Diamond Light Source in the UK and at other synchrotron radiation centres worldwide. Tables 1 and 2 also provide the corresponding data for aragonite and vaterite phases from which one can easily see the challenge in phase identification not just between the three polymorphs but more importantly any incipient ordering in the ACC which might form a precursor phase for one of these.

References

1. Applications of synchrotron X-radiation to problems in material science, A. R. Gerson, P. J. Halfpenny, S. Pizzini, R. Ristic, K. J. Roberts D. B. Sheen and J. N. Sherwood, in *X-ray Characterisation of Materials*, ed. E. Lifshin, Wiley-VCH Verlag, ISBN 3-527-29657-3, Weinheim, 1999, pp. 105–170.

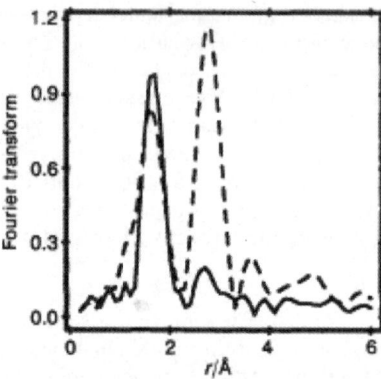

Fig. 2 Fourier transform of Fe K-edge XAFS spectra of nearly amorphous iron oxide stabilized within a polymer matrix (——) showing the lack of inter-atomic coordination beyond the first shell at about 1.8 Å in comparison with crystalline γ-FeOOH (- - - -), after [2] (reproduced with permission of the RSC).

2. Investigation of the local structure around iron dispersed in vinyl chloride vinylidene chloride (VC-VdC) copolymer coatings on mild steel using glancing-angle X-ray absorption spectroscopy, S. Pizzini, K. J. Roberts, I. Dring, P. J. Moreland, R. J. Oldman and J. Robinson, *J. Mater. Chem.*, 1992, **2**, 49–55.
3. http://www.diamond.ac.uk/Home/Beamlines/I09.html

Professor Dr Joester communicated in reply: Thank you for the suggestion. Characterization of the short-range order of the particles by XAFS is in fact an ongoing effort. Due to the low particle concentration in our samples, the need to work under cryogenic conditions to minimize radiation damage, and the low absorption energy of calcium, we have not been able to achieve an adequate signal to noise ratio to characterize the structure by Ca-Kα XAFS.

Professor Kato opened the discussion of the paper by Dr Nico A. J. M. Sommerdijk: What are the role and chemical function of the collagen fibers? Is the specific functional group on the surface of the collagen important? What happens if a simple plastic such as polyethylene having the same morphologies as that of collagen used in this study is employed for the same experiment of the crystallization of hydroxyapatite?

Dr Nudelman replied: The role of the collagen fibrils is to serve as a substrate for hydroxyapatite formation, generating a organic–inorganic composite material with

Table 1 Interatomic distances and coordination numbers (CN) of surrounding nearest neighbours to the calcium atomic site for the three polymorphs of $CaCO_3$. Calcite Ca–Ca distance shown in far left column as being over 4 Å, not ideal for significant backscatter in Ca edge EXAFS spectra. Distances for aragonite and vaterite shown in centre and right hand side columns, respectively

Calcite			Aragonite			Vaterite		
Atom	Distance (Å)	CN	Atom	Distance (Å)	CN	Atom	Distance (Å)	CN
O	2.40	6	O	2.38–2.65	9	O	2.28–2.82	8
C	3.25	6	C	2.94–3.39	5	C	3.00–3.34	6
O	3.50	6	Ca	3.89	2	O	3.88–4.10	2
Ca	4.11	6				Ca	4.13–4.13	6

Table 2 Interatomic distances and coordination numbers (CN) of surrounding nearest neighbours to the carbon atomic site for the three polymorphs of Ca(CO)$_3$. Calcite C–Ca distance 3.25 Å shown in left hand column, greater chance of significant backscatter for C-edge EXAFS spectra. Distances for aragonite and vaterite shown in centre and right hand side respectively

Calcite			Aragonite			Vaterite		
Atom	Distance (Å)	CN	Atom	Distance (Å)	CN	Atom	Distance (Å)	CN
O	1.29	3	O	1.22–1.29	3	O	1.21–1.28	3
Ca	3.25	6	C	2.88	2	Ca	3.00	2
O	3.46–3.76	9	Ca	2.94–2.99	3			

high stiffness and toughness. In this process, collagen controls mineral formation and templates the morphology and sizes of the crystals. Collagen has an active role in two crucial steps in hydroxyapatite formation: first, the positively charged groups that mediate the infiltration of the negatively charged mineral into the fibril. Second, the clusters of charged amino acids (both positively and negatively charged) form nucleation sites that control the transformation of amorphous calcium phosphate into hydroxyapatite.[1] Regarding the plastic substrate, it is possible that if it has active functional groups that interact with the calcium phosphate, it will also be able to mediate hydroxyapatite formation in a fashion similar to collagen. Without the functional groups, however, it may just be an inert scaffold.

1. F. Nudelman *et al.*, *Nat. Mater.*, 2010, **9**, 1004–1009.

Mr Khan asked: In this study, were you able to see an increased de-mineralisation of the collagen when DMP-1 was used to nucleate/increase mineral infiltration of the collagen.

Since your results show that the use of this protein increases infiltration of the collagen, is this a one way process? Would the mineralised collagen release ACP or calcium phosphate ions from crystallised material faster if the material within the collagen was placed into a low mineral solution/demineralised zone such as demineralised dentin?

Dr Nudelman responded: If we leave the mineralized collagen in water or in a low mineral solution, the calcium phosphate will eventually dissolve. However, we never looked at how long it takes, and how the dissolution rate is influenced by the presence of additives such as poly-aspartic acid or DMP1.

Professor Meldrum enquired: Under control conditions, does copper stabilise ACP in bulk solution?

Dr Nudelman said: It is known that copper ions inhibit apatite formation in solution (Y. Okamoto and S. Hidata, *J. Biomed. Mater. Res.*, 1994, **28**, 1403–1410). In our experiments, we observed apatite formation both when poly-aspartic acid was absent from the system, or when poly-aspartic acid was present, but there was no collagen (Nudelman *et al.*, *Faraday Discuss.*, 2012, **159**, Fig. 6C–D). Therefore, the effect of the copper ions is not restricted to the calcium phosphate, but they must also be interacting with the collagen, the poly-aspartic acid and the C-DMP1. The exact mechanism is not clear.

Dr Gower remarked: In this paper, and particularly in you prior *Nature Materials* paper,[1] it appears that the mineral is always found in the interior of the collagen fibrils first. This is particularly evident in the tomographic imaging in that paper, but also seen in Fig. 2e and 7d in this *Faraday* paper. How do you explain that the mineral gets to the interior of the fibrils first? If it was diffusion, the particles would follow a concentration gradient and be more prevalent near the exterior. And particularly since they supposedly have electrostatic interactions with domains in the collagen, why do they keep moving past these domains to the interior?

Secondly, how do such large particles (as shown in Fig. 5B and C, and the supplement of the *Nature* paper with DMP1) diffuse into such tight spaces? And, most impressively, how do particles diffuse beyond the gap regions and preferentially end up in the overlap zones, which has very limited space?

Of course you probably know the point I am trying to make. Does your data support the concept of solid particles diffusing into the interstices of a collagen fibril? Or does it support our proposed mechanism of capillary infiltration of a liquid like phase?[2]

1. F. Nudelman, K. Pieterse, A. George, P. H. H. Bomans, H. Friedrich, L. J. Brylka, *et al.* The role of collagen in bone apatite formation in the presence of hydroxyapatite nucleation inhibitors, *Nat. Mater.*, 2010, **9**(12), 1004–1009.
2. M. J. Olszta, X. G. Cheng, S. S. Jee, R. Kumar, Y. Y. Kim, M. J. Kaufman, *et al.*, Bone structure and formation: A new perspective, *Mater. Sci. Eng. R-Rep.*, 2007, **58**(3–5), 77–116.

Dr Nudelman replied: I agree with your model of capillary infiltration to explain how the mineral gets to the interior of the fibrils. Indeed, if it was simple diffusion, it would be expected that more mineral would be found close to the exterior, and less in the interior. Regarding the tight spaces inside the collagen, however, we must pay attention that it does not seem to be much of an issue for the infiltration of calcium carbonate. The collagen fibril has a liquid-crystalline nature and it is capable of expanding to accomodate the mineral. This is particularly evident in Fig. 2E of this paper, where the infiltration of the amorphous calcium phosphate into the collagen caused the fibril to expand.

Dr Rodríguez Blanco opened the discussion of the paper by Dr Patricia M. Dove: It is well known that, in the absence of foreign ions, ACC breaks down in solution very rapidly (1–2 min at ambient temperature) and transforms to crystalline $CaCO_3$ (Rodríguez-Blanco *et al.*, 2011). It is also accepted that this breakdown involves a dehydration process (Radha *et al.*, 2010; Rodríguez-Blanco *et al.*, 2012). However, in the presence of Mg, this amorphous phase can be stable for much longer time (hours, days). This is interpreted as a consequence of the higher energy needed to dehydrate the Mg ion in comparison to the Ca ion (di Tommaso and de Leeuw, 2010; Moomaw and Maguire, 2008). Mg located within the porous structure of ACC (Goodwin *et al.*, 2010) retards the dehydration and breakdown of this amorphous phase. Furthermore, from experimental data it is known that ACC stability can be longer/shorter depending on the presence and concentration of other common ions (Sr, SO4, PO4, *etc.*) in solution. However, there is lack of information on the activation energies required to dehydrate these ions. This lack of information is quite surprising, considering that these ions are present in seawater and they have a strong influence on the crystallization pathways of $CaCO_3$ from ACC. Is there any information available about the hydration shells of common ions in seawater which can be helpful for the geochemical community? Does anybody know if there is any comparative data, activation energies or general studies on this topic?

References

Di Tommaso, and De Leeuw, *Phys. Chem. Chem. Phys.*, 2010, **12**, 894–901.
Moomaw and Maguire, *Physiology (Bethesda)*, 2008, **23**, 275–285.
Radha *et al.*, *Proc. Natl. Acad. Sci.*, 2010, **107**, 16438–16443.
Rodríguez-Blanco *et al.*, *Nanoscale*, 2011, **3**, 265–271.
Rodríguez-Blanco *et al.*, *J. Alloys Compd.*, 2012, DOI:10.1016/j.jallcom.2011.11.057.
Goodwin *et al.*, *Chem. Mater.*, 2010, **22**, 3197–3205.

Dr Dove replied: The geochemical community is very interested in the kinetics and thermodynamics of ion hydration but there are many areas where data are sparse or absent. A number of references related to IIA cations are called out in the recent paper by Hamm *et al.*, *J. Phys. Chem. B*, 2012, **114**, 10488–10495). While it is agreed that Mg^{2+} prolongs the lifetime of ACC, we found no evidence that moderate levels of SO_4 have an affect on the time to ACC transformation or the composition of the calcite that forms (Fig. 3C).

Professor Roberts communicated: The key question to me is whether the ACC formed is truly amorphous, that is to say, it has strong local ordering, *i.e.* 1st co-ordination sphere, as in a glassy state, but no longer range 3D order as one would see in a crystalline solid. Thus if ACC is amorphous, then why is aragonite not formed also from ACC? Doesn't this observation suggest that ACC has a degree of 3D order and one which perhaps displays a structure rather close to that of calcite?

Dr Dove communicated in reply: It is intriguing that, for the conditions of our experiments, we do not observe ACC transforming into aragonite, regardless of the magnesium level. It provides evidence that magnesium does not always determine aragonite/calcite polymorph selection—an old concept that is finally being tested. Getting back to your question, an understanding of ACC structure is a central need for moving this field forward with mechanistic and not speculative interpretations. Some degree of short-range order certainly seems possible for ACC but perhaps its preference for transforming into calcite is a consequence of 1) the predominant ion coordination(s) and 2) the low probability of forming differently coordinated material. As you know, the coordination number about the calcium atom is distinctively different in calcite and aragonite with six and nine oxygens, respectively. More generally, insights into the structure of amorphous materials and their evolution into crystalline products are needed to inform this gray zone between solution and crystal for a wide variety of chemical systems. One other aspect of the ACC-to-crystal transformation to consider is that this process still required nucleation, which brings with it the usual set of thermodynamic and kinetic barriers. Because calcite is more stable, the thermodynamic driver is larger (and the barrier smaller). Moreover, we simply do not understand enough the chemical reaction pathways, bond-breaking and-making, *etc.* to say anything about the relative magnitudes of the kinetic barriers to forming on polymorph or another, even if ACC is truly amorphous in the sense you describe and does not possess any 3D order.

Dr Sand remarked: With 50% v/v alcohol aragonite can be the first phase to form from ACC and 98–100% aragonite can be obtained using an appropriate shaking speed. Details in K. K. Sand, J. D. Rodriques-Blanco, E. Makovicky, L. Benning and S. L. S. Stipp, Crystallization of $CaCO_3$ in water–alcohol mixtures: spherulitic growth, polymorph stabilization and morphology change, *Cryst. Growth Des.*, 2012, **12**, 845–853.

Professor Roberts answered: What do you feel is the mechanistic origin of this effect? Do you feel that the mixed solvent system provides a mechanism for the formation of a colloidal dispersion and that the confined solution environment provided by this may play a role in mediating the formation of both the spherulite morphologies and the lower stability aragonite phase?

Dr Dove communicated in reply: Yes, this is an interesting new paper and supports the structure-based idea that the higher coordination number about calcium in aragonite is favored by environments with lower water activity. We have conducted a few experiments that used similar solutions to also explore the effect of reduced water activity on polymorph selection. However, our solutions were within the realm of conditions that are most relevant to earth environments and we did not cross the transition to aragonite that is reported here. It would seem plausible that lower water activities are an additional path to lowering the barrier to aragonite formation.

Dr Rodríguez Blanco addressed Dr Dove: In the (geo)chemical community we usually describe the precursor phase of all crystalline $CaCO_3$ as ACC, amorphous calcium carbonate. However, we are probably too simplistic or generous when we use the name "ACC", because there could be many types of ACC with different local structures, compositions, solubility products, *etc.*, in the same way as there are different amorphous calcium phosphates (Combes and Rey, 2010). All amorphous calcium carbonates could seem to be similar when characterized with X-ray diffraction (XRD) or Fourier transform infrared spectroscopy (FTIR), but their local structures may be different.

Gebauer *et al.* (2008) suggested that at neutral- or high-pH, ACC may exhibit a local structure similar to calcite or vaterite, respectively. This local structure would be a function of the synthesis pH and would control its crystallization pathway. Following this idea, ACC precursors which crystallize to aragonite, monohydrocalcite or ikaite may also have different local structures. Furthermore, it is well known that ACC can incorporate variable amounts of Mg (and H_2O), transforming to Mg-calcite or monohydrocalcite with different Mg contents. Therefore there is no reason to discard that Mg-doped ACC is different from ACC.

Therefore, instead of "one" ACC, should we not talk about many ACCs with different local structures depending on simple parameters (composition, pH, ionic strength, supersaturation, temperature, *etc.*)?

References:

Combes and Rey, *Acta Biomater.*, 2010, **6**, 3362–3378.
Gebauer *et al. Science*, 2008, **322**, 1819–1822.

Dr Dove communicated in reply: It is possible that ACC structures are different but I would like to see a careful spectroscopic study of this question. TGA analysis of Mg-ACC shows evidence of an increasing water content in ACC with higher Mg levels (Radha *et al.*, *Geochim. Cosmochim. Acta*, 2012, **90**, 83–95). Our studies of how pH and carbonate level affects Mg uptake into ACC also suggests shifts in composition. To my knowledge, structural evidence is minimal. One paper that comes to mind is Michel *et al.*, *Chem. Mater.*, 2008, **20**, 14) and new NMR work from the Kirkpatrick research group.

Professor De Yoreo asked: Is there evidence for ACC transforming directly into aragonite, or is a separate nucleation event required?

Dr Dove responded: As you know, these experiments do not have the resolution to see nucleation events but we never observed evidence for aragonite forming from the initial ACC precipitate. Nor did we see evidence for co-existing aragonite and ACC for any of the experimental conditions used here.

Dr Wolf enquired: Concerning the question of why we typically observe calcite after ACC formation and more seldom aragonite or vaterite, could it be that ACC is capable of heterogenously inducing nucleation of calcite? We observed comparable behaviour in our experiments employing the Kitano method under levitated conditions[1] and Shen *et al.* reported a related templating action in case of calcium carbonate spheres.[2] Doubtless, the templating effect may depend on the short-range order of the respective ACC.

1. Stephan E. Wolf, Jork Leiterer, Michael Kappl, Franziska Emmerling and Wolfgang Tremel, *J. Am. Chem. Soc.*, 2008, **130**(37), 12342–7.
2. Qiang Shen *et al.*, *J. Phys. Chem. B*, **110**(7), 2994–3000.

Professor De Yoreo communicated in reply: Yes, I agree that may be the secret. The other obvious possibility is that the driving force for making calcite is much greater than for making aragonite and the kinetic limitations of attaining order, which are not there in making the initial ACC are an obstacle to both crystalline phases.

Dr Rodríguez Blanco addressed Professor Joester and Dr Dove: Some participants have asked a question about whether ACC is a precursor of aragonite or not. The answer is yes. Mixing two (pre-heated) solutions of $CaCl_2$ and Na_2CO_3 at temperatures above ~50 °C and concentrations above the saturation level for ACC will result in the instantaneous formation of ACC which will rapidly (seconds) transform to a mixture of aragonite and vaterite. The higher the temperature, the higher the aragonite/vaterite ratio. Above ~70 °C no vaterite will form. Both vaterite and aragonite will transform into calcite after several hours of reaction by a dissolution–reprecipitation process, as described by Ogino *et al.*, *Geochim. Cosmochim. Acta*, 1987, **51**, 2757–2767.

Dr Beck answered: We also found that aragonite can be produced at a high temperatures, either directly or *via* the transformation from the amorphous precursor phase.[1]

1. R. Beck and J.-P. Andreassen, *J. Cryst. Growth*, 2010, **312**, 2226–2238.

Dr Dove communicated in reply: It has been extensively shown that aragonite will precipitate at elevated temperatures and there is no doubt that high enough saturation states ACC will form. But it is important to consider the role of ACC at these elevated temperatures. Is the ACC simply a kinetic phase that forms as a result of a liquid–liquid separation process and is consumed by aragonite due to dissolution/ reprecipitation? Or is the ACC a true structural precursor to aragonite and plays an important role in the formation of aragonite at elevated temperatures?

Dr Gower addressed Dr Dove, Professor De Yoreo and Dr Wang: In your paper, you state that "Because the single Asp residue has small effects on mineralization compared to larger carboxylated macromolecules 15, 22 found in eutrophic environments, further studies of biomolecule effects are needed." I would just like to point out that there have been some studies along these lines.[1] In fact, polyaspartate has a pronounced affect on the incorporation of magnesium to yield very high levels (up to 30% under our conditions). My question is, if organics and polymer can mediate the incorporation of Mg where it is quite possible that different organics cause a different level of Mg incorporation, then won't this make it very difficult or impossible to use this as a means to reconstruct paleoenvironmental conditions?

1. X. G. Cheng, P. L. Varona, M. J. Olszta and L. B. Gower, Biomimetic synthesis of calcite films by a polymer-induced liquid-precursor (PILP) process 1. Influence and incorporation of magnesium, *J. Cryst. Growth*, 2007, **307**, 395–404.

Dr Dove communicated in reply: Yes, your question is a valid one and we have suggested that the macromolecule influences of tissues are a possible explanation for the origin of 'vital effects' that are observed for Mg signatures in biominerals (Dove and Weiner, *Rev. Min. Geochem.*, 2003, **54**, 1–31; Stephenson *et al.*, *Science*, 2008, **322**, 274–277). Briefly, the term 'vital effect' was introduced during the 1950's to describe unexplained offsets in isotopic or compositional signatures that are measured in the carbonate skeletons (of some species) from what is predicted by inorganic controls. The offsets measured in 'badly behaved' species were traditionally assumed to arise from kinetic or taxonomic differences but what does this mean? We have shown and you also point out that some organic molecules can also induce significant increases in Mg content in calcite (Stephenson *et al.*, *Science*, 2008) and ACC (Wang *et al.*, *Proc. Natl. Acad. Sci.*, 2009). An added twist is the realization that some species also involve ACC as an intermediate that could also affect the signatures contained in the final products. Currently, the most successful paleoenvironmental proxies are based upon data from the 'well-behaved' species and there are organisms that provide remarkably good calibrations. Another challenge for the biomineralization community is to understand the chemical basis for species-specific differences in signatures and how to accurately interpret chemical signals from a wider variety of organisms.

Professor Gilbert answered: Did you conduct the experiments only on silica glass, or also on other substrates? Do you expect the results to be substrate-dependent?

Dr Dove responded: Thanks for this question. In addition to the silica glass substrates, we conducted experiments using gold surfaces that were functionalized with carboxlyated (MUA) surface assembled monolayers. This is of interest because previous studies suggest that carboxyl-rich biomolecules in sediments influence the carbonate polymorph and compositions that form. We did not observe, however, any differences. The carboxylated surfaces result in the same the ACC and calcite morphologies or composition (Supplementary Fig. S3) that are observed on the silica glass.

PAPER

Aragonite crystal orientation in mollusk shell nacre may depend on temperature. The angle spread of crystalline aragonite tablets records the water temperature at which nacre was deposited by *Pinctada margaritifera*

Ian C. Olson[a] and Pupa U. P. A. Gilbert*[b]

Received 6th March 2012, Accepted 6th June 2012
DOI: 10.1039/c2fd20047c

Nacre, or mother-of-pearl, is a lamellar composite of aragonite ($CaCO_3$) tablets with a broad distribution of angular orientations. The angle spread is the full-width of this distribution. Here we analysed the angle spread as a function of position in the nacre layer of one *Pinctada margaritifera* shell, compared the results with temperature data, and found these two parameters to be highly correlated. This result suggests that one could calculate the temperature at which nacre formed by measuring only its angle spread. Validation of the correlation in modern and ancient nacre from other species is necessary, but if confirmed, nacre could provide a physical proxy for temperature in modern and ancient climates.

1 Introduction

Nacre is a composite of aragonite ($CaCO_3$) tablets and organic sheets, alternating to form a lamellar structure at the inner side of several mollusc shells,[1] including those made by gastropods (for instance, trochus, turbo, abalone, but not whelk, nor conch, nor cowrie, nor conus), one cephalopod (*Nautilus*), and a few bivalves (all *Mytilus* mussels, *Pinctada* pearl oysters, pen shells, but not clams or scallops). Previously nacre was believed to have perfectly co-oriented crystalline tablets, with their crystallographic *c*-axes parallel to the normal to the nacre layers. In recent years our group[2] and others[3] have observed that nacre tablets have a wide distribution of *c*-axis orientations, centred about the normal. The width of this distribution, termed angle spread, varies across species.[2f] In Fig. 1 we present three polarization-dependent imaging contrast maps, or PIC-maps, from different shells, and their *c'*-axis distributions.

If the tablet *c*-axes in nacre were co-oriented as previously believed, in PIC-maps such as those in Fig. 1 they would exhibit no contrast and all tablets would appear homogeneously gray.

The difference in angle spread across species stimulated our interest and a quest for the environmental parameters that might correlate with angle spread in nacre tablets. We found a strong direct correlation ($R = 0.77$) between the angle spread and the maximum temperature at which a mollusc species lives. Specifically, we found that the greater the maximum temperature the greater the angle spread measured in nacre.[2f] However, the meaning and scope of the observed correlation

[a]*Department of Physics, University of Wisconsin, Madison, WI, 53706, USA. E-mail: iolson@wisc.edu; Fax: +1 608 265 2334; Tel: +1 608 265 3767*
[b]*Departments of Physics and Chemistry, University of Wisconsin, Madison, WI, 53706, USA. E-mail: pupa@physics.wisc.edu; Fax: +1 608 265 2334; Tel: +1 608 262 5829*

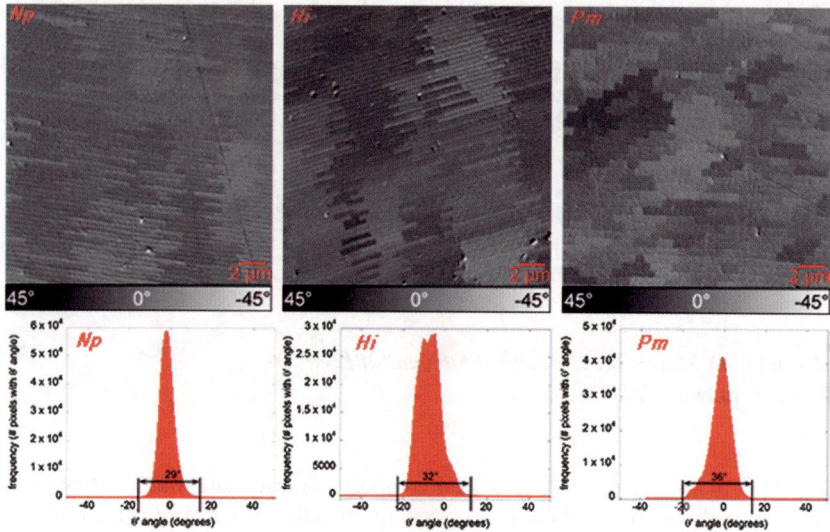

Fig. 1 Polarization-dependent imaging contrast maps (PIC-maps) from three mollusc shells: the cephalopod *Nautilus pompilius* (*Np*), the gastropod *Haliotis iris* (*Hi*), and the bivalve *Pinctada margaritifera* (*Pm*). In a PIC-map different gray levels represent different crystal *c*-axis orientations.[4] PIC-mapping has a spatial resolution of 20 nm, and ~2° resolution for discriminating different orientations of the *c'*-axis, that is, the angle θ' formed by the projection of the aragonite crystal *c*-axis onto the plane in which the polarization vector of the illuminating X-rays rotates.[2f,5] Within this resolution each nacre tablet behaves as a single crystal. Columns of tablets stacked on top of one another share the same orientation in gastropod and cephalopod columnar nacre, *e.g.* in *Np* and *Hi*. Conversely, stacks of co-oriented tablets appear to be staggered laterally in sheet nacre from bivalves, *e.g.* in *Pm*. The histogram under each PIC-map shows the frequency at which each θ' angle or gray level is measured, across all the $10^3 \times 10^3$ pixels in the PIC-map. The footprint of the distribution provides a measurement of how much the *c'*-axes can vary in their orientations, which we define here as the angle spread. The angle spread varies across species, as shown here in the histograms in *Np*, *Hi*, and *Pm* and in previous work across 8 species.[2f]

remained unclear. It was possible that the angle spread became larger as a result of adaptation to varying temperature over long time periods, as climates changed, as molluscs themselves migrated to different climates, or as continental drift or other geologic phenomena brought the molluscs into new environmental temperatures. It was also possible that the angle spread reacted rapidly to changing water temperature, during the life of a single animal. To test these hypotheses we did the experiments described here, in which one mollusc shell was analysed with secondary ion mass spectrometry (SIMS) and then the same shell locations measured with SIMS were analysed with PIC-mapping to measure the angle spread.

2 Results

In Fig. 2 we present the *Pinctada margaritifera* shell analysed with both SIMS and PIC-mapping. In Fig. 3 we present three specific areas α, β, and γ, located respectively at the beginning, the middle, and the end of the region analysed by PIC-mapping and SIMS, and also shown in Fig. 2. The PIC mapping results for angle spread as a function of position in the shell, acquired from these and many more locations are presented in Fig. 4, where they are also compared with the temperature data recorded by a National Oceanic and Atmospheric Administration (NOAA) buoy,[6] 25 m under the water surface, from a location less than 1000 km from French Polynesia, where the shell grew.

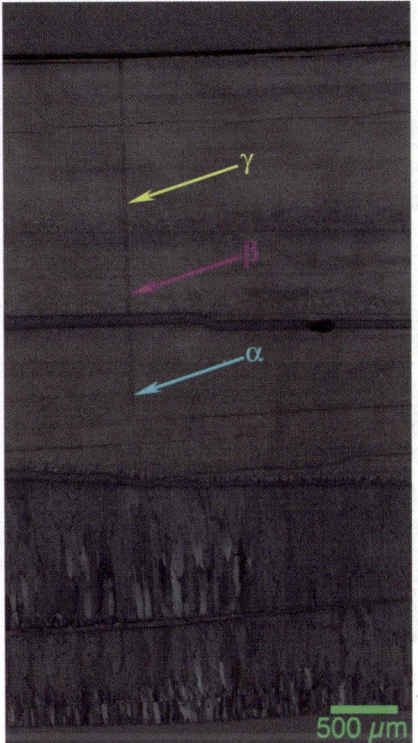

Fig. 2 Cross-polarizer, reflected light micrograph of a *Pinctada margaritifera* shell, embedded in epoxy and polished to expose a cross-section with the calcite prismatic layer at the bottom third, and the nacre layer in the top two thirds of the shell. Beginning at the inner surface of the shell (top), the nacre layer was analysed for oxygen isotope ratio with SIMS, with one 10-μm SIMS pit every 10 μ, in a zig-zag pattern to avoid overlapping, extending ~2.4 mm into the shell.[2f] The ~240 SIMS pits are visible in this image as a faint vertical gray line. A shorter distance, extending ~1.3 mm was also analysed with PIC-mapping. The arrows point towards three regions at the beginning (α) the middle (β) and the end (γ) of the region analysed by both SIMS and PIC-mapping. The nacre layer has a gap between α and β, where the shell was broken during sample preparations.

Based on the striking match of angle spread and temperature data, we may have established a correspondence between position in the shell and its time of formation. This is quantitatively described by the equation in Fig. 4 caption. We must now verify that the rest of the data deduced by the visual matching of the two curves are reasonable. We do so by first comparing in Table 1 the complete set of data measured in positions α, β, and γ, using these data to measure the nacre growth rate, then using the match in Fig. 4 to plot $\delta^{18}O$ and angle spread, or $\delta^{18}O$ and temperature data as a function of time (Fig. 5, 6), and finally plotting temperature as a function of angle-spread (Fig. 7).

Using the data in Table 1, we can see that between point α and point γ the distance is 2358 − 1043 = 1315 μm, and these points formed 3457 − 1514 = 1943 days apart from one another. Therefore the shell grew precisely 1315/1943 = 0.68 μm day^{-1}. This number is in remarkable agreement with the tablet thicknesses described by Olson *et al.*: 0.66 μm, 0.67 μm, and 0.68 μm average tablet thicknesses in different regions of *Pm*.[2f] This result strongly suggests that one tablet-layer per day is deposited by *Pm*. Furthermore, it confirms that the match of Fig. 4 is not unreasonable, as nacre layers may very well be correlated with the mollusc's circadian rhythm.

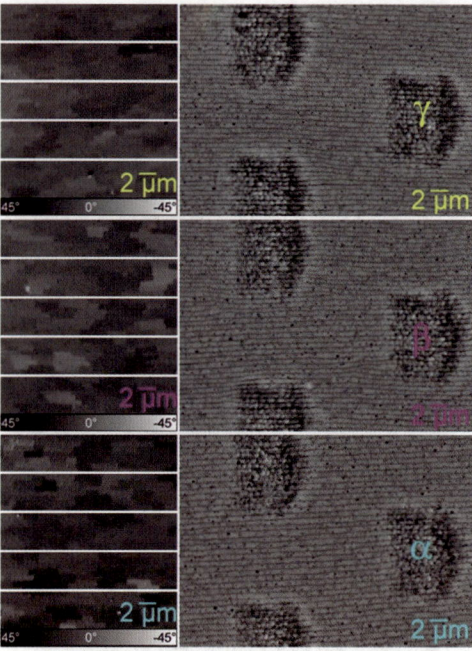

Fig. 3 On the right hand side we show back-scattered electron SEM micrographs of the pits left in nacre by SIMS analysis, each approximately 10 μm in size, from which aragonite desorbed and the corresponding stable oxygen isotope concentrations were measured. The pits labelled α, β, and γ are the same shown at lower magnification in Fig. 2. On the left hand side we present PIC-maps, acquired in 20 μm × 20 μm regions immediately adjacent to the α, β, and γ pits, each subdivided into 5 slices of 4 μm × 20 μm to provide more data points with angle spread analysis.

Assuming that the time-position relationship in Fig. 4 is correct, we can now plot the SIMS results as a function of date, and compare them with the angle spread results also plotted as a function of date, as presented in Fig. 5.

In Fig. 6 we compare $\delta^{18}O$ data with temperature. The $\delta^{18}O$ is expected to be anti-correlated with temperature in abiotic[7] and biomineral systems.[7e,8]

In order to compare $\delta^{18}O$ and temperature data in Fig. 6 we assumed that the ratio of oxygen isotope concentrations ($[^{18}O]/[^{16}O]$, expressed as $\delta^{18}O$) of the seawater in which the shell was formed did not change during the shell formation time. This assumption, which is in fact not true experimentally,[7b-e] is used here as a simplification, and because the necessary reference $\delta^{18}O$ data from the seawater were not available. If they were available, the non-constant $\delta^{18}O$ would remove fluctuations observed in Fig. 6 that are not real temperature changes. With the imperfect assumption that $\delta^{18}O$ of seawater did not change, however, the minima in $\delta^{18}O$ correspond to the maxima in temperature,[7a,7c,9] and vice versa. The data of Fig. 6 show such correspondence in 7 time points and their surroundings. This is a worse match than that in Fig. 4. The discrepancy, however, may be due to variations in seawater $\delta^{18}O$.

Since the angle spread and temperature curves as a function of time match so nicely (Fig. 4), we attempt here a data regression, that is, a plot of these two parameters versus one another. In Fig. 7 we present such data: the temperature data as a function of angle spread.

The strong correlation of water temperature and angle spread is evidenced by the linear fit in Fig. 7 and the high correlation coefficient ($R = 0.704$). In addition, the fit in Fig. 7 provides an equation according to which one can calculate the temperature

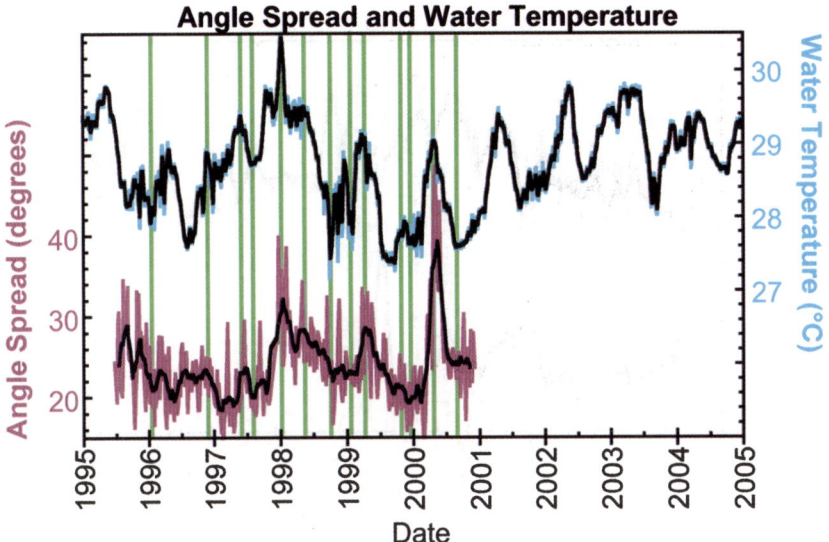

Fig. 4 Angle spread results as a function of position in the shell (magenta curve, bottom), compared with temperature data as a function of time (cyan curve, top), recorded once per day by a NOAA buoy. Each date year is placed in the position of January 1st of that year. Notice that the growth rate of the shell was unknown, and so was the mollusc's death time. The only certain data point was that the shell was acquired by one of us in 2004, and must therefore have formed before then. We smoothed both the angle-spread curve and the temperature curve over 9 points (black curves) to facilitate identification of slopes, peaks, dips, and similitude of the two curves. We then stretched the horizontal scale of the angle-spread curve, and shifted it, until the two curves matched, as judged visually. The stretching done was then quantified according to the equation: Time = -44.65 days + $1.48 \frac{\text{days}}{\mu\text{m}} \times$ position [in μm], where Time is the time of nacre formation, measured in days before January 1st, 2005. Notice that the temperature data are not perfectly reproducible in subsequent years, nor are the angle spread data, and both curves have a great deal of noise. The two curves however have a remarkable similitude. At the time points in which we observe alignment of maxima or minima in the two curves, we placed a green vertical line behind the curves. There are 13 time points in which peak-peak or dip-dip similitude is observed, and around these time points the slopes of the two curves are in striking agreement.

Table 1 Complete set of data measured directly in PIC-mapping or SIMS (angle spread, or δ^{18}O, respectively) or deduced using the match in Fig. 4 (date, temperature)

Parameter measured	Point α	Point β	Point γ
Days before Jan 1st, 2005	3457	2489	1514
Date (mm/dd/yyyy)	07/15/1995	03/09/1998	11/08/2000
Position	2358 μm	1703 μm	1043 μm
Angle spread (smoothed over 9pt)	24.0°	26.6°	23.6°
SIMS pit number	386	296	212
δ^{18}O (‰, smoothed over 5pt)	-3.2	-2.7	-2.7
T (at 25m depth)	28.86 °C	29.32 °C	27.78 °C

at which the nacre formed (T) from the measured angle spread (AS). The equation is: $T = 26.738 + 0.050\ AS$. Using this simple linear equation nacre in *Pinctada maragaritifera* can be used as a thermometer. In Fig. 7 we only plotted 15 months of data. If we plot the entire range of data available (5.3 years) the linear fit is similar (the

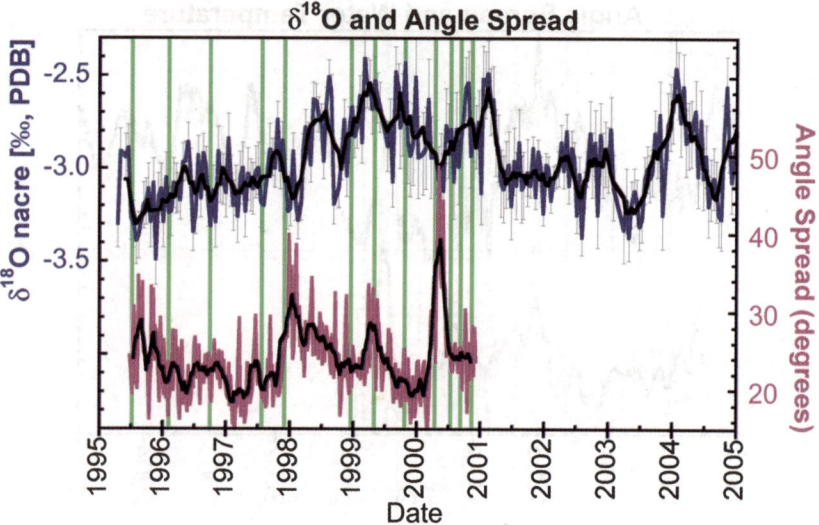

Fig. 5 Comparison of the stable oxygen isotope concentration ratio [^{18}O]/[^{16}O], expressed as δ^{18}O of nacre. To facilitate comparison, the black lines indicate smoothed curves, over 9 and 5 points for angle spread and δ^{18}O, respectively. The δ^{18}O increases when the temperature decreases, hence peaks in δ^{18}O should correspond to dips in angle spread and *vice versa*. Indeed we observe good anti-correlation at the 12 time points indicated by green vertical lines.

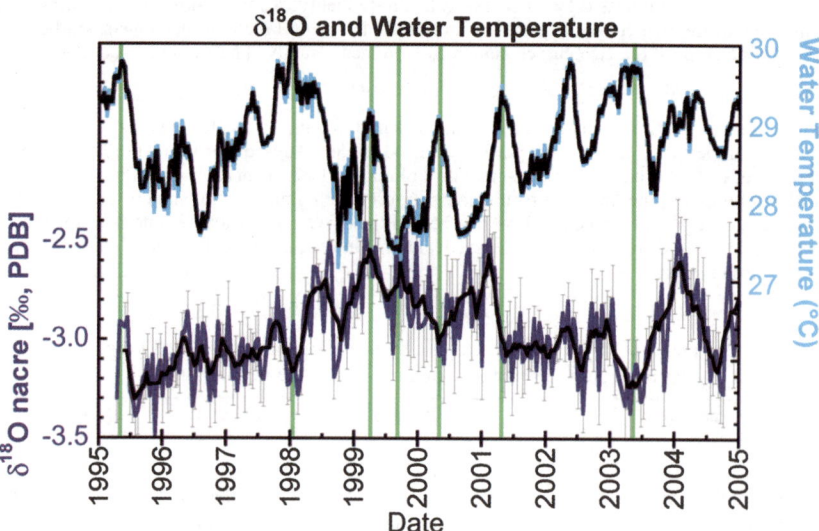

Fig. 6 Comparison of δ^{18}O of nacre and water temperature data, both as a function of date. Again we have smoothed both curves (black lines) over 5 and 9 points respectively. The δ^{18}O is expected to increase as the temperature decreases, hence the maxima and minima in these two curves should align. The two curves are far from being the mirror image of one another. However there is acceptable anti-correlation at the 7 time points indicated by green vertical lines.

intercept is 27.631 °C and the slope 0.035 °C/°), but the correlation coefficient is far worse ($R = 0.29$). We believe that this problem results from acquiring angle spread data from regions 4 μm × 20 μm (Fig. 2). Such small images contain too few tablets to provide a statistically representative histogram of angles and angle spread. The angle spread data, therefore, are under-sampled. Much larger fields of view must

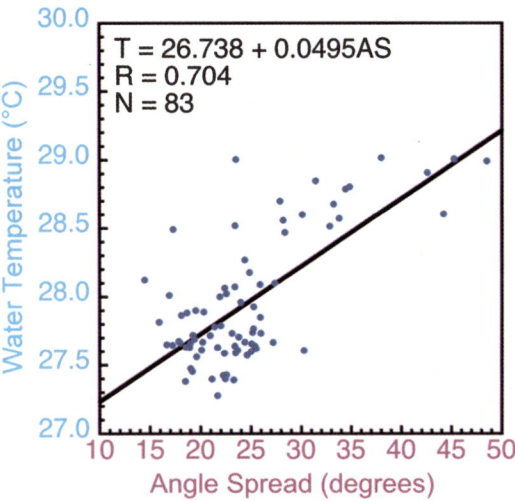

Fig. 7 The water temperature data plotted *versus* the angle spread data for dates between 8/7/1999 and 11/8/2000. Note that the 83 data points are fitted by a line with slope 0.050 °C/°, and intercept 26.738 °C. The fit gives a high correlation coefficient $R = 0.704$.

be used in future PIC-maps to resolve this problem. Acquiring such data is entirely feasible, but has not been done. Only the detailed analysis done here revealed the existence of a sampling problem, and paved the way for possible future development of nacre-thermometry. Whether the above equation applies to other modern shells, or to ancient nacre from extinct mollusc species remains to be determined.

3 Discussion

We have explored for the first time the possibility that a structural parameter in nacre, that is, the angle spread of its aragonite tablets, is correlated with the water temperature at the time in which the mollusc deposited the shell mineral. We have found a remarkably good agreement between the angle spread and the temperature curves, both plotted as a function of time as shown in Fig. 4. By comparison, the values of $\delta^{18}O$ here gave worse agreement with both angle spread and temperature data as a function of time (Fig. 5 and 6). If the match of angle spread and temperature data attempted in Fig. 4 is correct, then the angle spread is highly correlated with water temperature. This correlation strongly suggests that nacre angle spread could be used to determine the water temperature at the time a mollusc deposited its shell. The data regression of Fig. 7 further supports this suggestion. The temperature-angle spread correlation is certainly true in one *Pm* shell, as shown here, but must be confirmed on many more shells, from different species and locations, with known times of death and local temperatures, or grown in aquaria with controlled conditions. Larger fields of view (*e.g.* 80 μm × 40 μm) will improve angle-spread sampling at each location in the shell, corresponding to each formation time and temperature. In addition, mollusc shells formed in locations or aquaria in which the temperatures vary by a greater interval than that tested here (3°) will provide much stronger evidence and more stable linear fits, with slope and intercept not fluctuating depending on sampling. It is possible that, after all this future work is done, the linear equation correlating angle spread and temperature will differ from the one reported here in Fig. 7.

The question raised above on the time-scale of the angle-spread and temperature correlation is addressed by the data presented here, and these data are sufficient to rule out the possibility that adaptations in angle spread took place over a very long time period of mollusc evolution in changing climates. The increase/decrease in angle spread takes place each day during the life of the mollusc, in response to increasing/decreasing water temperature.

Why does the angle spread increase when the temperature increases? Abiotically carbonate crystals grow faster at higher temperature.[10] It is possible that this is the case also in biogenic aragonite in nacre, but this is not known. If crystals grow faster at higher temperatures, it is possible that the proteins controlling crystal growth in specific directions,[11] and therefore crystal orientations,[2f] cannot be produced fast enough by the organism, and therefore greater disorder and angle spread ensue. The opposite is also possible: an increased temperature provokes increased metabolic rate, which makes the organism produce more and better organic molecules. This would result in greater angle spread if this was advantageous for the organism. It is unknown at present whether co-orientation or mis-orientation of tablets provides an evolutionary advantage.

The angle spread was measured in this work using PIC-mapping. However, this may not be the best method to measure angle spreads, as using a microscope and then clumping together all tablet orientation angles and measuring their maximum angle spread is unnecessarily complex. More directly averaged angle-spread results could be obtained in X-ray diffraction experiments by measuring the footprint of a rocking curve,[2c,12] or the footprint of a pole figure,[13] or using the March-Dollase approach.[14] In scanning electron microscopy angle spreads can be measured by electron back-scattered diffraction,[15] or by direct imaging of partly etched tablets.[16] This has not been tested, but in principle one or more of these methods could provide nacre angle-spread proxies for temperature.

Future work will determine if nacre can be used as a proxy for temperature, as suggested by the data here. Most other temperature proxies are chemical in nature, that is, they measure ratios of concentrations of different minerals (*e.g.* aragonite/calcite ratio[17]) or different elements (*e.g.* [Mg]/[Ca][18]), or ratios of concentrations of different isotopes of the same element (*e.g.* $\delta^{18}O$[19]), or clumped isotopes of multiple elements.[20] If confirmed and validated on other shells and different growth temperatures, this could be one of only a few physical proxies, along with porosity in foraminifera, which depends on temperature.[21]

A physical proxy may have advantages and disadvantages. It is more sensitive than a chemical proxy to diagenetic structural changes. Diagenesis in nacre manisfests as dissolution[22] or dissolution and recrystallization as calcite[23] or hydroxyapatite.[24] However, when either of these phenomena occur, nacre's layering and physical structure is poorly preserved at the microscopic scale, hence these problems will be evident from images and spectra in the photoelectron emission spectromicroscopy (PEEM) experiment used for PIC-mapping. It is thereby easy to select only layered, aragonite regions of well-preserved nacre[25] for the measurement of angle spread. An advantage of a physical proxy is that, if the structure is preserved, chemical changes such as elemental losses or enrichments do not affect the angle-spread measurement, thus this method is expected to be less artifact-prone compared to chemical proxies.

We have shown here that angle-spread is a proxy for temperature in *modern* nacre from one shell, and, if validated in other modern and ancient shells, could become a valuable temperature proxy. This possibility is most intriguing when one considers that nacre is abundant in the fossil record (*e.g.* in ammonites) spanning 450 million years.[26] If the same *fossil* nacre can be analysed with the present technique and other temperature proxies, nacre could be validated not only as a thermometer, but also as a paleothermometer. Unlike oxygen isotope ratios here, a new method termed "clumped isotope thermometry" is not dependent on knowing or assuming the oxygen isotope composition of the water at the time a mineral or biomineral grew.[20] Validation of the angle-spread proxy in ancient fossil nacre, therefore, should be done with clumped isotope thermometry and angle spread measurements on the same sample. If the results are found to be in agreement, nacre-paleothermometry will be established.

4 Experimental

4.1 Samples

The *Nautilus pompilius* shell (*Np*, 183 mm maximum length) originated off the coast of Siquijor Island, Philippines, and was purchased from Conchology Inc., Philippines. The *Np* sample was collected from the outer wall of the largest chamber, not from a septum. *Haliotis iris* (*Hi*, 107 mm maximum length), the paua shell, or blackfoot abalone from New Zealand, was purchased from Australian Seashells PTY Ltd.

The *Pinctada margaritifera* shell (*Pm*, 90 mm maximum length) was purchased at the Gauguin Pearl Farm, Rangiroa, French Polynesia.

All three shells in Fig. 1 were cut with hammer and chisel, to a final size of approximately 1 cm. One side of each square sample was coarsely polished with sandpaper, and then adhered to double-stick tape for accurate vertical mounting of the cross-sections. The samples were then embedded in epoxy (EpoThin, Buehler, IL) and polished with decreasing size alumina grit down to 50 nm (MasterPrep, Buehler, IL). Sample surfaces were coated using a sputter coater (208HR, Cressington, UK) with 40 nm of platinum while the region to be analyzed by PEEM was masked off, and a final coating of 1 nm platinum was applied to the entire surface to prevent charging.[27]

The *Pm* shell sample analyzed with SIMS and PEEM was extracted with hammer and chisel and one side was coarsely polished with sandpaper for accurate vertical mounting of the cross-section. The sample was then baked for 1 h at 310 °C to remove the organic material. The baked shell sample was embedded in epoxy (EpoxyCure, Buehler, IL), with grains of UWC-3 calcite standard[8] cast in the center of the sample mount, and the sample was polished with sandpaper and alumina grit. Care was taken to reduce surface topography and polishing relief, including applying a thin layer of epoxy to the partially polished surface to fill cracks and bubbles.

4.2 XANES-PEEM analysis

X-Ray photoelectron emission spectromicroscopy was performed using the PEEM-3 microscope, on the 11.0.1 beamline at the Advanced Light Source in Berkeley, CA. The elliptically polarizing undulator (EPU) at this beamline was calibrated to provide precise linear X-ray polarization and reproducible intensities at polarization angles between 0–90° with a 5° step size.

4.3 PIC-maps

The methods for PIC-mapping are described in detail in ref. 2*f* and 5. Briefly, we define the c'-axis as the projection of the $CaCO_3$ c-axis onto the EPU polarization plane. Separate PEEM images were acquired at each EPU polarization angle from 0–90° by 5° steps, a total of 19 images for each nacre region. For all the images the sample voltage was −15 kV and the photon energy was kept constant at 290.3 eV, the carbon K-edge π* peak,[28] which is the peak most sensitive to the linear polarization angle.[5] All images were 20 μm × 20 μm in size, with 20-nm pixels. For each pixel, if the intensities in each of these 19 images are plotted against the corresponding EPU polarization angle, the resulting curve follows a cosine-squared function where the position of the curve maximum indicates the angle at which the c'-axis is aligned parallel to the EPU polarization angle.[5] Fitting these data to the curve $y = A + B \cos^2(EPU° + \theta')$, with the Levenberg–Marquardt least-squares analysis method, yields accurate identification of the position of the curve maximum and thereby determination of c'-axis orientation. Performing this fit for every 20-nm pixel in the stack of 1030 × 1054-pixel images obtained from PEEM-3, and composing the results into a single gray scale image yields a PIC-map, in which the c'-axis orientation angle is quantitatively represented by a gray level between 0–255. These gray levels correspond to the possible 180°-range of orientations between −90° and +90°, with 0° corresponding to a vertical c'-axis. Note that all c'-axes point around the normal, thus only the ±45° range is shown in Fig. 1 and 2.

The procedures used to obtain the PIC-maps were developed by one of us (I.C.O.) in WaveMetrics Igor Pro 6.2®, and are available to any interested users free of charge on our web site (see "GG Macros").[29]

4.4 Angle spread

The physical quantity we measure in PIC-maps is the projection (termed c'-axis) of the c-axis onto the EPU polarization plane.[2f,5] If the c-axes of two nacre tablets are perfectly co-oriented, their c'-axes will also be co-oriented. If the c-axes of two nacre tablets are differently oriented, then their c'-axes are almost always differently oriented. The only ambiguous case is the following: if the c-axes of two tablets point in different directions in three-dimensional space, and these directions are in a plane perpendicular to the plane in which the X-ray polarization rotates, then their c'-axes will appear the same in PIC-maps, whereas the c-axes are in reality distinct. This is the only ambiguous case because nacre tablets have their axes preferentially oriented perpendicular to the nacre planes, and depart from that direction by a maximum angle of ±50.5° (the maximum spread measured never exceeded 101°[2f]). If these angles were as large as ±90°, then two c-axes nearly perpendicular to the EPU plane could yield very large spreads in θ' angles, even though they are nearly co-oriented in three-dimensional space. Since this is never the case for a nacre cross-section sample, and the samples are always mounted with the nacre lines horizontal, the aforementioned is the only ambiguous case.

All data were collected in nacre regions far from the nacre-prismatic boundary in order to avoid the greater angle spreads that near-boundary regions would exhibit, resulting from the gradual ordering mechanism found in *Haliotis rufescens* nacre.[30] We note that the angle spread is a *maximum* spread between c'-axes orientations, and therefore any angles between the minimum and the maximum θ' angle (see histograms of Fig. 1) are included and consistently found in the data.

Measuring the "footprint" of the distribution of θ' angles in a PIC-map as shown in Fig. 1 is effective for representing the total spread of aragonite c'-axis angles across nacre tablets within a region. Footprint measurements were done from histograms in Igor Pro®, displayed as "levels" for PIC-map images.[29] To increase the number of points in which the angle spread is measured, we sliced each 20 μm × 20 μm region into five 4 μm × 20 μm regions, as shown in Fig. 2. These were 68 20-μm maps, which then became 340 PIC-maps. In retrospect, this was not a good choice, as the data turned out to be under-sampling the angle spreads, because there are too few tablets in a 4 μm × 20 μm region to give stable and meaningful values of angle spreads. In other words, the histograms of smaller regions are not as symmetric and statistically representative as those shown in Fig. 1. Evidence of appropriate sampling for angle spread will manifest itself as single-peaked, symmetric histograms, approaching a Gaussian line shape much more than those obtained from the 20-μm fields of view in Fig. 1. This is why we wrote above that 80 μm × 40 μm fields of view are required.

4.5 Oxygen isotope measurements by secondary ion mass spectrometry (SIMS)

Oxygen isotope measurements were performed by SIMS on the WiscSIMS CAMECA ims-1280 high-resolution, multi-collector ion microprobe, at the University of Wisconsin-Madison[2f,7d,31] using a $^{133}Cs^+$ beam with an intensity of 1.3 nA, focused to approximately 10 μm beam spot-size. Charging of the sample surface was compensated by Au coating, and by using an electron flood gun. The general conditions were similar to those reported in ref. 7e. The secondary $^{16}O^-$ and $^{18}O^-$ ions were collected simultaneously by Faraday cup detectors with typical $^{16}O^-$ ion intensity of 2.3×10^9 cps. The total analytical time per spot was about 4 min, including pre-sputtering (10 s), automatic centering of the secondary ion image in the field aperture (ca. 1.5 min) and analysis (ca. 2 min).

Grains of UWC-3 calcite standard ($\delta^{18}O = 12.49‰$ [VSMOW])[8] were measured in at least four spots before and after every 8–18 sample analyses, and the resulting average value of bracketing the samples was used for bias correction. The average precision (reproducibility) for a set of bracketing standards is $\pm 0.28‰$ (2 SD, spot-to-spot).

Oxygen isotope ratios of marine carbonates are traditionally expressed relative to PDB. Therefore, final data were converted from $\delta^{18}O$ on the VSMOW to the PDB scale by the equation of Coplen et al.:[7b]

$$\delta^{18}O[‰\ PDB] = 0.97002 \times \delta^{18}O[‰\ VSMOW] - 29.98.$$

Acknowledgements

We thank beamline scientists Andreas Scholl, Anthony Young and Andrew Doran for their expert technical support during the PEEM-3 experiments. We thank Reinhard Kozdon and John W. Valley for their collaboration on the SIMS experiments and for reading this manuscript. We also thanks Brian Hess and John Fournelle for SIMS sample preparations and imaging. This work was supported by NSF awards CHE-0613972 and DMR-1105167, DOE Award DE-FG02-07ER15899, and UW-Hamel Award to PUPAG. The PEEM experiments were performed at the Berkeley Advanced Light Source, supported by DOE under contract DE-AC02-05CH11231. WiscSIMS is partially supported by NSF-EAR awards 0319230, 0744079, 1053466.

5 References

1 (a) H. A. Lowenstam, *Science*, 1981, **211**, 1126–1131; (b) H. A. Lowenstam and S. Weiner, *On Biomineralization*, Oxford University Press, New York, 1989, p; (c) S. Mann, *Biomineralization: Principles and Concepts in Bioinorganic Materials Chemistry*, Oxford University Press, New York, 2001, p.
2 (a) R. A. Metzler, M. Abrecht, R. M. Olabisi, D. Ariosa, C. J. Johnson, B. H. Frazer, S. N. Coppersmith and P. U. P. A. Gilbert, *Phys. Rev. Lett.*, 2007, **98**, 268102; (b) P. U. P. A. Gilbert, R. A. Metzler, D. Zhou, A. Scholl, A. Doran, A. Young, M. Kunz, N. Tamura and S. N. Coppersmith, *J. Am. Chem. Soc.*, 2008, **130**, 17519–17527; (c) R. A. Metzler, D. Zhou, M. Abrecht, J.-W. Chiou, J. Guo, D. Ariosa, S. N. Coppersmith and P. U. P. A. Gilbert, *Phys. Rev. B: Condens. Matter Mater. Phys.*, 2008, **77**, 064110–064111/064119; (d) D. Zhou, R. A. Metzler, T. Tyliszczak, J. Guo, M. Abrecht, S. N. Coppersmith and P. U. P. A. Gilbert, *J. Phys. Chem. B*, 2008, **112**, 13128–13135; (e) R. A. Metzler, J. S. Evans, C. E. Killian, D. Zhou, T. H. Churchill, N. P. Appathurai, S. N. Coppersmith and P. U. P. A. Gilbert, *J. Am. Chem. Soc.*, 2010, **132**, 6329–6334; (f) I. C. Olson, R. Kozdon, J. W. Valley and P. U. P. A. Gilbert, *J. Am. Chem. Soc.*, 2012, **134**, 7351–7358.
3 (a) E. DiMasi and M. Sarikaya, *J. Mater. Res.*, 2004, **19**, 1471–1476; (b) K. Gries, R. Kröger, C. Kübel, M. Schowalter, M. Fritz and A. Rosenauer, *Ultramicroscopy*, 2009, **109**, 230–236; (c) A. G. Checa, J. H. E. Cartwright and M. G. Willinger, *J. Struct. Biol.*, 2011, **176**, 330–339.
4 (a) Y. R. Ma, B. Aichmayer, O. Paris, P. Fratzl, A. Meibom, R. A. Metzler, Y. Politi, L. Addadi, P. U. P. A. Gilbert and S. Weiner, *Proc. Natl. Acad. Sci. U. S. A.*, 2009, **106**, 6048–6053; (b) C. E. Killian, R. A. Metzler, Y. T. Gong, I. C. Olson, J. Aizenberg, Y. Politi, L. Addadi, S. Weiner, F. H. Wilt, A. Scholl, A. Young, S. N. Coppersmith and P. U. P. A. Gilbert, *J. Am. Chem. Soc.*, 2009, **131**, 18404–18409.
5 P. U. P. A. Gilbert, A. Young and S. N. Coppersmith, *Proc. Natl. Acad. Sci. U. S. A.*, 2011, **108**, 11350–11355.

6 NOAA in *Pacific Marine Environmental Laboratory Tropical Atmosphere Ocean Project. Data buoy at 155W 8S, Vol.* 2011.
7 (*a*) H. C. Urey, *J. Chem. Soc.*, 1947, **1**, 562–581; (*b*) T. B. Coplen, C. Kendall and J. Hopple, *Nature*, 1983, **302**, 236–238; (*c*) E. L. Grossman and T.-L. Ku, *Chem. Geol.: Isot. Geosci. Sect.*, 1986, **59**, 59–74; (*d*) N. T. Kita, T. Ushikubo, B. Fu and J. W. Valley, *Chem. Geol.*, 2009, **264**, 43–57; (*e*) R. Kozdon, D. C. Kelly, N. T. Kita, J. H. Fournelle and J. W. Valley, *Paleoceanography*, 2011, **26**, PA3206.
8 R. Kozdon, T. Ushikubo, N. T. Kita, M. Spicuzza and J. W. Valley, *Chem. Geol.*, 2009, **258**, 327–337.
9 S.-T. Kim and J. R. O'Neil, *Geochim. Cosmochim. Acta*, 1997, **61**, 3461–3475.
10 (*a*) J. W. Morse, J. J. Zullig, L. D. Bernstein, F. J. Millero, P. Milne, A. Mucci and G. R. Choppin, *Am. J. Sci.*, 1985, **285**, 147–185; (*b*) P. Zuddas and A. Mucci, *Geochim. Cosmochim. Acta*, 1994, **58**, 4353–4362; (*c*) P. Zuddas and A. Mucci, *Geochim. Cosmochim. Acta*, 1998, **62**, 757–766.
11 L. Addadi and S. Weiner, *Proc. Natl. Acad. Sci. U. S. A.*, 1985, **82**, 4110–4114.
12 (*a*) D. Chateigner, C. Hedegaard and H. R. Wenk, *J. Struct. Geol.*, 2000, **22**, 1723–1735; (*b*) J. H. E. Cartwright and A. G. Checa, *J. R. Soc. Interface*, 2007, **4**, 491–504; (*c*) D. Chateigner, S. Ouhenia, C. Krauss, C. Hedegaard, O. Gil, M. Morales, L. Lutterotti, M. Rousseau and E. Lopez, *Mater. Sci. Eng., A*, 2010, **528**, 37–51.
13 (*a*) C. M. Zaremba, A. M. Belcher, M. Fritz, Y. L. Li, S. Mann, P. K. Hansma, D. E. Morse, J. S. Speck and G. D. Stucky, *Chem. Mater.*, 1996, **8**, 679–690; (*b*) A. G. Checa and A. B. Rodríguez-Navarro, *Biomaterials*, 2005, **26**, 1071–1079.
14 E. Zolotoyabko, *J. Appl. Crystallogr.*, 2009, **42**, 513–518.
15 A. Pérez-Huerta, Y. Dauphin, J. P. Cuif and M. Cusack, *Micron*, 2011, **42**, 246–251.
16 H. Mutvei, *Biomineralisation*, 1972, **4**, 81–86.
17 (*a*) H. A. Lowenstam, *J. Geol.*, 1954, **62**, 284–322; (*b*) H. A. Lowenstam, *Proc. Natl. Acad. Sci. U. S. A.*, 1954, **40**, 39–48.
18 (*a*) Y. Rosenthal and G. P. Lohmann, *Paleoceanography*, 2002, **17**; (*b*) H. Elderfield, J. Yu, P. Anand, T. Kiefer and B. Nyland, *Earth Planet. Sci. Lett.*, 2006, **250**, 633–649.
19 (*a*) P. L. Koch, *Annu. Rev. Earth Planet. Sci.*, 1998, **26**, 573–613; (*b*) S. Schouten, E. C. Hopmans, A. Forster, Y. van Breugel, M. M. M. Kuypers and J. S. S. Damste, *Geology*, 2003, **31**, 1069–1072.
20 R. A. Eagle, T. Tutken, T. S. Martin, A. K. Tripati, H. C. Fricke, M. Connely, R. L. Cifelli and J. M. Eiler, *Science*, 2011, **333**, 443–445.
21 (*a*) A. W. H. Be, *Science*, 1968, **161**, 881–884; (*b*) W. E. Frerichs, M. E. Heiman, L. E. Borgman and A. W. H. Be, *J. Foraminiferal Res.*, 1972, **2**, 6–13.
22 Y. Dauphin, *Curr. Opin. Colloid Interface Sci.*, 2002, **7**, 133–138.
23 (*a*) H. Mutvei, *N. Jb. Geol. Paläont. Abh.*, 1967, **129**, 157–166; (*b*) C. A. McRoberts and J. G. Carter, *Journal of Paleontology*, 1994, **68**, 1405–1408.
24 (*a*) B. Runnegar, *Alcheringa: An Australasian Journal of Palaeontology*, 1985, **9**, 245–257; (*b*) C. M. Zaremba, D. E. Morse, S. Mann, P. K. Hansma and G. D. Stucky, *Chem. Mater.*, 1998, **10**, 3813–3824.
25 S. M. Antao, *RSC Adv.*, 2012, **2**, 526–530.
26 (*a*) H. Mutvei, *Lethaia*, 1983, **16**, 233–240; (*b*) H. Mutvei and E. Dunca, *Paläontologische Zeitschrift*, 2010, **84**, 457–465; (*c*) M. J. Vendrasco, A. G. Checa and A. V. Kouchinsky, *Palaeontology*, 2011, **54**, 825–850.
27 (*a*) B. Gilbert, R. Andres, P. Perfetti, G. Margaritondo, G. Rempfer and G. De Stasio, *Ultramicroscopy*, 2000, **83**, 129–139; (*b*) B. H. Frazer, B. Gilbert, B. R. Sonderegger and G. De Stasio, *Surf. Sci.*, 2003, **537**, 161–167; (*c*) G. De Stasio, B. H. Frazer, B. Gilbert, K. L. Richter and J. W. Valley, *Ultramicroscopy*, 2003, **98**, 57–62; (*d*) P. U. P. A. Gilbert, B. H. Frazer and M. Abrecht in *The organic-mineral interface in biominerals*, Vol. 59 ed.: J. F. Banfield, K. H. Nealson and J. Cervini-Silva, Mineralogical Society of America, Washington DC, 2005, p. 1570185.
28 R. A. Metzler, M. Abrecht, R. M. Olabisi, D. Ariosa, C. J. Johnson, B. H. Frazer, S. N. Coppersmith and P. Gilbert, *Phys. Rev. Lett.*, 2007, **98**, 268102.
29 GG-Macros, http://home.physics.wisc.edu/gilbert/ 2011.
30 P. U. P. A. Gilbert, R. Metzler, D. Zhou, A. Scholl, A. Doran, A. Young, M. Kunz, N. Tamura and S. Coppersmith, *J. Am. Chem. Soc.*, 2008, **130**, 17519–17527.
31 J. W. Valley and N. T. Kita in *In situ oxygen isotope geochemistry by ion microprobe*, Vol. 41, ed. M. Fayek, 2009, pp. 16–63.

PAPER

Merging models of biomineralisation with concepts of nonclassical crystallisation: is a liquid amorphous precursor involved in the formation of the prismatic layer of the Mediterranean Fan Mussel *Pinna nobilis*?

Stephan E. Wolf,[*a] Ingo Lieberwirth,[b] Filipe Natalio,[c] Jean-Francois Bardeau,[d] Nicolas Delorme,[d] Franziska Emmerling,[e] Raul Barrea,[f] Michael Kappl[b] and Frédéric Marin[a]

Received 5th March 2012, Accepted 24th April 2012
DOI: 10.1039/c2fd20045g

The calcitic prisms of *Pinna nobilis* (*Pinnidae*, Linnaeus 1758) are shown to be perfect examples of a mesocrystalline material. Based on their ultrastructure and on the occurrence of an amorphous transient precursor during the early stages of prism formation, we provide evidence for the pathway of mesocrystallisation proposed by Seto *et al.* (2012), which proceeds not by self-organized oriented attachment of crystalline nano-bricks but by aggregation of initially amorphous nanogranules which later transform by epitaxial nucleation to a three-dimensional array of well aligned nanocrystals. We further fathom the role of a liquid amorphous calcium carbonate in biomineralisation processes and provide strong evidence for the occurrence of PILP-like intermediates during prism formation. We develop a new scenario of prism formation based on the presented findings presented findings and discuss the implications of a speculative liquid amorphous calcium carbonate (LACC) intermediate *in vivo*.

Introduction

Our increasing knowledge of the structure and formation of biominerals challenged and altered our classical concepts of crystallisation considerably in the recent years. The delicate and task-optimized morphologies and intricate ultrastructures of biominerals taught us that their formation is a remarkably concerted process which cannot be explained within the framework of classical crystallization models.[1] Today, two different models of nonclassical crystallization are mainly discussed, the pathway of mesocrystallisation and the pathway of liquid amorphous intermediates which is commonly termed the polymer-induced liquid precursor process (PILP).[2-5]

[a]*Centre National de la Recherche Scientifique (CNRS) UMR 6282 Biogéosciences, Université de Bourgogne, 6 Boulevard Gabriel, 21000 Dijon, France. E-mail: stephan.wolf@u-bourgogne.fr; frederic.marin@u-bourgogne.fr; Fax: +33 3 80 39 63 87; Tel: +33 3 80 39 63 72*
[b]*Max Planck Institute for Polymer Research, Ackermannweg 10, 55128 Mainz, Germany*
[c]*Johannes Gutenberg University, Institute for Inorganic and Analytical Chemistry, Duesbergweg 10-14, 55128 Mainz, Germany*
[d]*CNRS UMR 6283 Institut des Molécules et Matériaux du Mans, Faculté des Sciences, Université du Maine, 72085 Le Mans CEDEX 9, France*
[e]*Federal Institute of Materials Research and Testing (BAM), Richard-Willstätter – Straße 11, D-12489 Berlin, Germany*
[f]*Argonne National Laboratory, Advanced Photon Source, BioCAT, Argonne, Illinois 60439, USA*

To begin with the pathway of *mesocrystallisation*, it turned recently out that mesocrystals can form *via* different mechanisms. The first one proceeds by self-organized oriented attachment of crystalline nano-bricks which may form from primary transient amorphous particles by ripening.[2,6] The ability of self-recognition of these building blocks arises from a functionalization of the particle surface with (typically) polymeric additives. The second mechanism was proposed quite recently by Seto *et al.* as a result of their ultrastructure analysis of mesocrystalline mature calcitic spines of the sea urchin "*Authoeidaris erassispina*"—apparently an erroneous spelling of *Anthocidaris crassispina* (A. Agassiz, 1863, recently renamed to *Heliocidaris crassispina*) since no *Authoeidaris erassispina* is taxonomically known. The mechanism proposed by Seto *et al.* proceeds "*via* the crystallization of a dense array of amorphous calcium carbonate (ACC) precursor particles".[7] By now, only a small number of biominerals were *de facto* proven to be mesocrystals, *i.e.* nacre of the pearl oyster *Pinctada fucata* or the spines of the sea urchins *Anthocidaris crassispina, Echinometra mathaei* and *Heterocentrotus mammillatus*.[7–9] However since the reports of Dauphin *et al.*,[10] it seems reasonable that the prismatic calcite needles from the prismatic shell layer of the noble pen shell *Pinna nobilis* (*Pinnidae*, Linnaeus 1758) are mesocrystals as well: they appear like single crystals under crossed polarizers although they are composed of nanosized crystallites but a thorough analysis is still due. Dauphin *et al.* proposed in their AFM studies on polished transverse cuts that the nanogranules, which build up the calcitic prisms of *Pinna nobilis*, are enwrapped in organic material.[10] Jacob *et al.* proofed the existence of such a coating of nanogranules by an elaborate TEM and analytical electron microscopy study on nacreous platelets of cultured pearls and revealed that this organic coating is in fact a spongy, pumiceous organic network, which permeates the biomineral and whose meshes are filled with calcium carbonate.[11]

The second nonclassical crystallisation pathway involves a transient *liquid-amorphous mineral phase* whose existence was firstly proposed by Gower *et al.* in order to explain the formation of thin calcium carbonate films in presence of tiny amounts of small anionic polymers.[3] Accordingly, Gower *et al.* dubbed this pathway the "polymer-induced liquid precursor" process (PILP) and assumed that acidic biopolymers may generate a PILP route during biomineral formation *in vivo*.[4,5,12] We revealed later that the transient liquid mineral phase does exist irrespective of the presence of polymeric additives.[13] This exceptional liquid phase is at the moment best documented for the case of liquid amorphous calcium carbonate (LACC) and other divalent metal carbonates[14,15] and its behaviour can be well explained in colloidochemical terms since LACC behaves like a classical electrostatically stabilised emulsion, be it in absence or presence of additives: Sheering events lead to coalescence and the presence of salt screens the surface potential and leads consequently to an aggregation of the droplets.[13,14,16] We could evidence that the droplets carry a negative surface charge as the addition of high concentrated (bio)polymers (here 7.5 mg mL^{-1}) either stabilises or destabilizes the emulsified state depending of the charge of the respective additive.[16] The findings were convincingly corroborated by the evidence that the preceding carbonato calcium complexes are all negatively charged.[14] On the basis of these recent results, we rationalize the PILP process on a colloidochemical basis as follows: at low concentration, the additive breaks the emulsion by depletion destabilization and the sedimenting droplets then produce the characteristic film-like products of PILP processes.

Both nonclassical mineralisation processes, the process of mesocrystallisation and the one involving a liquid mineral precursor, could be combined with the accepted but controversial biological model of the *extrapallial space*. This space represents a narrow liquid-filled compartment between the epithelial mantle cells and the shell in which the mantle cells secrete the organic and inorganic constituents of the growing shell and it thus contains the mineral precursors at a saturated level. The mineralisation of the shell does not occur in the direct vicinity of the mantle cells and thus has to proceed in a completely independent and self-organized manner.

In this contribution, we re-investigate the ultrastructure of the calcitic prisms extracted from the giant fan mussel *Pinna nobilis* employing a broad range of methods: wide-angle X-ray scattering, X-ray absorption spectroscopy, transmission electron microscopy analysis in STEM microprobe mode and atomic force microscopy. We conducted further biomimetic crystallization experiments in order to assess the impact of the two unusual acidic intraprismatic proteins caspartin (17 kDa) and calprismin (38 kDa) of *Pinna nobilis*[17] on the transient liquid amorphous calcium carbonate phase. For this, we extracted both unusual acidic proteins by preparative SDS-PAGE and performed crystallization experiments under ultrasonic levitation and analysed the precipitates by both transmission and scanning electron microscopy.

Experimental

Extraction and purification of prism and proteins

The prismatic shell layer (*pl* of left specimen in Fig. 1a) of *Pinna nobilis* (roughly 25 years old as deduced from the total height according to Moreteau *et al.*,[18] collected near the coast of Villefranche-s-Mer, Département Alpes-Maritimes, France) was scrupulously cleaned to remove epibionts and was cut into small pieces which were then immersed in dilute sodium hypochlorite solution (0.26 g of active chlorine/100 ml of water) for four days. This treatment completely destroys the periprismatic sheaths (*i.e.* the interprismatic organic matrix), removes the periostracum, and disassembles the prismatic shell layer into separated prisms (see Fig. 1b). The prisms were collected by filtration (5 μm, Millipore) and extensively rinsed. The prisms were then re-suspended in cold water and slowly dissolved with cold diluted acetic acid until the pH reached four (5% v/v, 4 °C; Schott Instruments Titronic Universal). Residual insoluble matrix was separated by centrifugation (10 min, 3900g) and the supernatant was filtered (0.45 μm, Millipore). The volume of the supernatant was reduced by ultrafiltration (Amicon YM, MWCO = 10 kDa, Millipore) before the remaining clear solution was completely desalted by extensive dialysis (Spectra-Por, MWCO = 1 kDa, SpectrumLabs). After lyophilisation (TelStar CryoDos), the intracrystalline acid-soluble matrix was fractionated by means of a preparative SDS-PAGE (12%, 250 V for 1 h, followed by 300 V overnight; Bio-Rad Prep Cell model 491). The eluted fractions containing caspartin and calprismin were immunologically detected employing a dot blot (Bio-Dot, Bio-Rad) since none of the

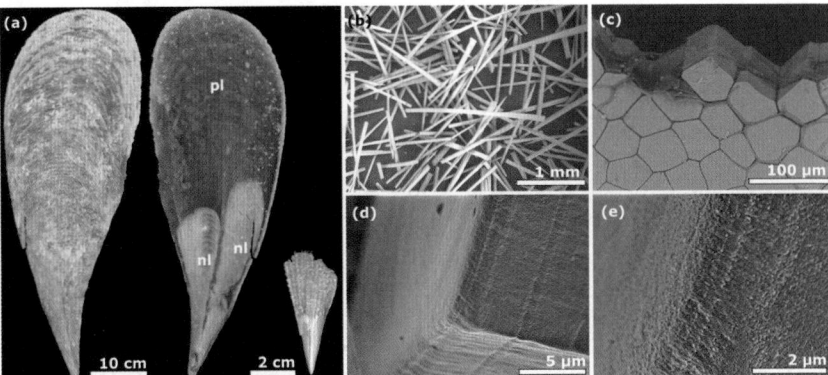

Fig. 1 (a) Two specimens of *Pinna nobilis* of different ages (left 25 years, right 6 months). The shell consists of two different layers: the prismatic layer (*pl*) and the nacreous layer (*nl*). (b) Mature calcitic prisms of the left specimen after separation. (c) Inward view on the prismatic layer of the juvenile specimen. (d, e) High resolution micrographs of separated juvenile prisms which are built up from individual nanogranules.

fractions absorbed in the UV range. After vacuum-blotting, the developing of the membrane followed a standard Western blot protocol using anti-caspartin and anti-calprismin as primary antibodies. The fractions which showed to contain either caspartin or calprismin were combined separately. Both solutions were subjected to ultrafiltration in order to reduce to volume followed by extensive dialysis against water (5 d, SpectraPor, MWCO = 1kDa, SpectrumLabs). Finally the solutions were lyophilised and their purity was assessed by a silver-stained SDS-PAGE: a single strong band of either 17 Da in case of caspartin or 38 kDa in case of calprismin was observed. The above described extraction and purification steps were conducted according to Marin et al.[17]

WAXS and XANES analysis of calcite prisms

Wide-angle X-ray scattering (WAXS) experiments were performed at the μSpot beamline at the synchrotron facility of BESSY II (Helmholtz Centre Berlin for Materials and Energy), featuring a beam diameter of 20 μm ($\lambda = 1.00257$ Å, calibrated with corundum). Further information concerning the general setup is given by Paris et al.[19] Due to the small beam diameter, mathematical 'desmearing' of the experimental scattering intensity function was not required.

An X-ray absorption near edge structure (XANES) was probed at the undulator beamline 18ID of the Advanced Photon Source at Argonne National Laboratory, for details about the beamline cf. Fischetti et al.[20] All spectra were collected in fluorescence mode at room temperature. Data reduction was accomplished with Athena v0.8.061. Calcite (p.a., Sigma Aldrich) and X-ray amorphous calcium carbonate, which was prepared following Xu et al.,[21] served as reference compounds.

Crystallization experiment and EM analysis

A suspension of $CaCO_3$ (p.a., Sigma Aldrich) in ultrapure water (Elga PureLab Option-Q7, 18.2 MΩ·cm^{-1}) was treated with carbon dioxide (Westfalen AG) overnight. The obtained saturated solution of calcium bicarbonate was filtered with a cascade of syringe filters, which consisted of a 0.1 μm Millipore Millex VV followed by 20 nm Millipore Anotop in series. Afterwards, the filtrate was treated again with carbon dioxide to dissolve nuclei with diameters below 20 nm. Shortly before levitation, an aliquot of the respective protein (either caspartin or calprismin) was added to the saturated solution of calcium bicarbonate to yield the desired protein concentration. Then, one droplet of the solution with a volume of approximately 4 μL was manually injected in the ultrasonic levitator (Tec5, Oberursel, Germany). The crystallization was followed by in situ wide angle X-ray scattering at the μSpot beamline at the synchrotron facility of BESSY II and showed for all samples the formation of calcite (space group $R\bar{3}c$). Samples were taken at different times manually by transferring the droplet to a lacey-coated TEM grid (Plano, Germany). TEM investigations were carried out at a Phillips EM 420 running at 120 kV, equipped with an ORCA-ER Camera (1024 × 1024 pixel) and run with AMT Image Capture Engine v5.42.540a. SEM investigations were performed using a FEI xT Nova 600 Nanolab.

STEM analysis of FIB-cuts of calcite prisms

For transmission electron microscopy (TEM) examination a small lamella was cut from the intermediate area between two prisms intersecting the periprismatic sheaths of a juvenile shell (Fig. 1a, right) using a FEI Nova 600 Nanolab focused ion beam. TEM examination was done using a FEI Tecnai F20 transmission electron microscope operated at an acceleration voltage of 200 kV.

Conventional brightfield imaging was performed in normal TEM mode and images were acquired on a 2k Gatan Ultrascan 1000 CCD camera. For spatial resolved diffraction analysis the microscope was operated in a parallel beam nano-area diffraction mode, using a 10 μm C2 condenser aperture in combination with a large excitation of the first condenser lens (high spot size). Under these conditions, the beam current was determined to 59 pA by measuring the counts on the CCD camera. For diffraction measurements, the beam diameter at the sample was approximately 100 nm, giving a dose rate of 400 e-/Å^{2}s. This low intensity, parallel beam probe was scanned over the sample using the deflection coils of the microscopes scanning unit in order to obtain a high angular annular dark field (HAADF) image in combination with a series of spatial resolved diffraction patterns. For FEI transmission electron microscopes this operation mode is known as "STEM microprobe" mode.

To obtain a HAADF image of the sample target area, the beam was focused to a spot diameter of approximately 10 nm. The camera length was set to 1000 mm or more in order to increase the signal intensity and the beam was scanned over the sample area using a dwell time of 1.6 μs in search mode and 10.4 μs for image acquisition. The acquired image was then used to locate the target area for subsequent diffraction analysis. The diffraction patterns were acquired using the bottom mount CCD camera operated at a binning factor of 4, the camera length of the microscope was set to 320 mm and the beam was defocused using the second condenser lens to give a beam diameter of approximately 100 nm. The dwell time for diffraction pattern acquisition was set between 1 and 4 s, respectively.

For data analysis and visualization of the diffraction pattern series a home-written script for Gatan Digital Micrograph software was used which projects the intensity of a chosen diffraction spot for the stack of diffraction patterns to the plane of the sampled area.

AFM analysis

Transverse cuts of mature prisms were obtained by directly mirror-polishing the inward face of the prismatic layer without preceding treatment. After extensive mirror-polishing (Buehler Micropolish B, 0.05 μm γ-alumina), different treatments were tested for cleaning the surface: incubation in water, in dilute sodium hypochlorite solution (0.1 g of active chlorine/100 ml of water) and in solutions of EDTA for varying time and concentrations. All AFM images were recorded under ambient conditions with either a Veeco Nanoscope IIIa (fresh fracture surface) or an Agilent 5500 AFM (mirror-polished samples). All images were realized in tapping mode using Nanosensors PointProbePlus Silicon-SPM-Sensor tips (PPP-NCHR-W, r < 10 nm) resp. Olympus silicon cantilevers (OMCL-AC240TS, r < 10 nm) which were plasma-cleaned prior to application. Image processing and roughness analysis were performed with the open-source software Gwyddion v2.25, *cf.* http://gwyddion.net/.[22]

Results and discussion

Structural analysis of mature calcite prisms from *Pinna nobilis*: determination of mesocrystallinity

Following the definition of mesocrystals given by Cölfen *et al.*,[2,6,23] a crystalline material has to feature two distinct characteristics to be *de facto* a mesocrystal: (a) The crystal has to scatter X-rays like a single crystal. (b) On the mesoscale, it must not be a compact single crystal but a colloidal crystal which is built up from individual nanocrystals which are "aligned in a common crystallographic register".[2] In this section, we will show that mature prisms of *Pinna nobilis* pass the examination of their mesocrystallinity with distinction.

Firstly, we investigated the crystallinity of a single mature prism by means of wide-angle X-ray scattering at a synchrotron beamline equipped with a microspot setup and we mapped every 100 μm the X-ray scattering profile of singles prisms (~2 mm, Fig. 1b). The experiment revealed that the prisms scatter X-rays like a calcite single crystal without a variation of their scattering pattern over their full length.

As amorphous calcium carbonate (ACC) is not detectable by X-ray diffraction, we probed for ACC in mature prisms by means of X-ray absorption spectroscopy on the Ca-K edge. We compared XANES spectra of mature prisms, crystalline calcite and X-ray amorphous calcium carbonate and found that the spectrum of the prisms resembles the one of crystalline calcite (Fig. 2). Crystalline calcite shows one prominent peak at 4047.2 eV and a broader weaker one around 4058 eV which both originate from diverse Ca–O first-shell scattering paths.[24] An additional shoulder appears at 4043 eV which arises from a 1s → 4p transition.[24–26] Further, a 1s → 3d pre-edge transition due to the trigonal symmetry of calcite is present.[25–27] Turning to the spectrum of the ACC standard, an overall decrease in features indicates an increase in structural disorder. One prominent and intense signal is present at 4048.6 eV, which is due to Ca–O scattering within the first coordination shell, and a pre-edge 1s → 3d transition,[25–27] which is more pronounced as in the case of calcite or mature prisms. In the XANES spectrum of mature prisms, all features of the calcite spectrum are quite well developed but there are subtle differences which indicate the presence of an additional phase: comparing the spectrum of the prisms to the one of calcite, the white line is slightly shifted and its maximum is displaced by 0.1 eV towards higher energies. Further, the spectrum of the prisms shows distinct differences to the spectrum of calcite around 4056 eV. XANES is a method which is very sensitive to the coordination environment of the probed element and can thus serve as a fingerprinting technique for chemical speciation and phase analysis by linear combination analysis of the contributing phases.[28–31] Since we found no other crystalline phases besides calcite by synchrotron X-ray diffraction, we employed both reference samples calcite and synthetic ACC as the basis set for the linear combination analysis. With this set, we estimated the ACC content to be about 6.9 at.%. These findings coincide notably with those recently reported by Seto *et al.* who estimated 8 at.% ACC by NMR and 4.9 at.% by calculating the area of pores in annealed spines of *Anthocidaris crassispina*.[7]

The second condition for mesocrystallinity, the colloidal character of the material, is already revealed by high resolution scanning electron microscopy of juvenile prisms (Fig. 1c–d) which were extracted from a 6 month old specimen (see Fig. 1a on the right) by treatment with sodium hypochlorite: a nanogranular substructure is apparent. In case of mature prisms we could not observe these nanogranules because the prisms have obviously undergone a maturation process. In order to assess the colloidal character of mature prisms, we thus performed atomic force microscopy on differently prepared samples. First, the samples were mirror-polished perpendicular to their long axis and then treated in various manners (water, sodium hypochlorite, EDTA) to assess the impact of the cleaning procedure. The micrograph, which is given in Fig. 3a, shows clearly that the prisms do not have a compact structure like a single crystal but they consist of elongated and oblique blob-like nanoparticles, whose diameter varies roughly from 20–50 nm. The contrast at the border in the phase images (see Fig. 3a on the left) of the particles suggests that they are enwrapped by organic material. Our findings are consistent with those reported by Dauphin *et al.*,[10] who applied a similar sample treatment. However this approach by mirror-polishing may contain a severe flaw: during polishing, the interprismatic and insoluble matrix is abraded as well and could be deposited in the grain boundaries of the crystallites and could thus feign an organic envelope. To eliminate such doubts, we conducted further experiments on freshly cleaved fragments of the prismatic shell layer. The phase image of the fresh fracture face (see Fig. 3b, left) readily shows a phase contrast at the grain boundaries. Additionally, a remarkable number of particles seem to still be covered by a sheet of organic material (*e.g.* the particles marked by an arrow in Fig. 3b, left).

Recapitulating, the mature prisms of *Pinna nobilis* are a prime example of mesocrystals as they fulfil both requirements stated above. They are well crystalline, contain only minor amounts of amorphous material and are built up by aggregated nanoparticles providing a pronounced structure on the mesoscale. We further affirm

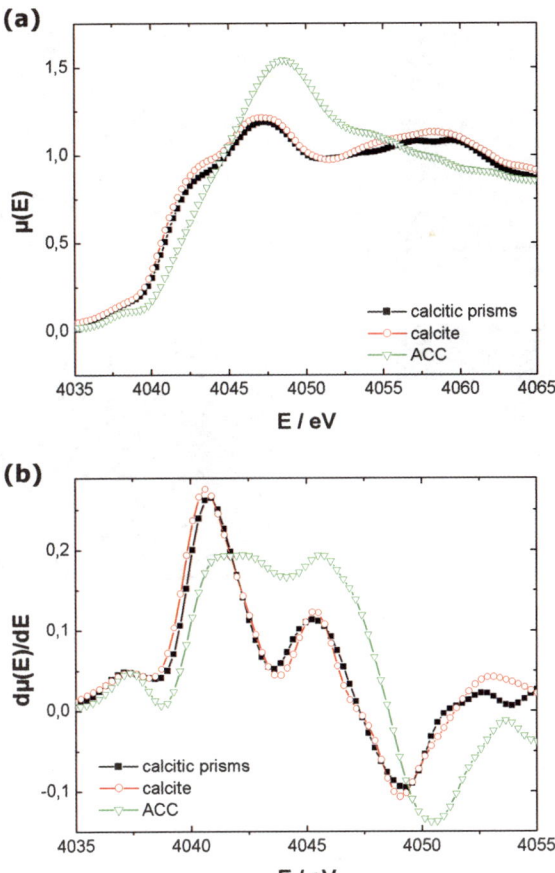

Fig. 2 (a) Comparison of the normalized XANES spectra at the Ca-K-edge of the calcitic prisms, synthetic calcite and synthetic amorphous calcium carbonate. (b) First derivative of the three XANES spectra.

and strongly corroborate the findings that the oblique nanoparticles, which build up the prisms, are surrounded by a thin organic coating.

Structural analysis of juvenile calcite prisms from *Pinna nobilis*: detection of transient amorphous phases

Nowadays, numerous studies have corroborated the assumption that amorphous intermediates play an important role in biomineralisation processes in general.[4,32] However if we turn to ACC as an original building material in molluscs, the occurrence of ACC was mainly reported for juvenile stages of the shell development, *i.e.* for larval shells[33,34] and juvenile bivalve shells.[35] Nassif *et al.* showed in case of the nacreous layer of *Haliotis laevigata* that mature platelets are continuously coated by a 5 nm thick amorphous layer of ACC, the formation of which was speculated to be due to ACC-stabilizing impurities, which were expelled during growth of the aragonite platelets.[36] In this study it remained unclear whether ACC served transiently as initial building material of the platelets. Recent contributions of Jacob *et al.* provided new evidence that ACC may in fact serve as a building material of hard tissues even in adult specimens.[37] They reported the systematic occurrence of amorphous calcium carbonate in the narrow zone between periostracum and the prismatic layer of adult specimens of *Hyriopsis cumingii* and of *Diplodon chilensis*

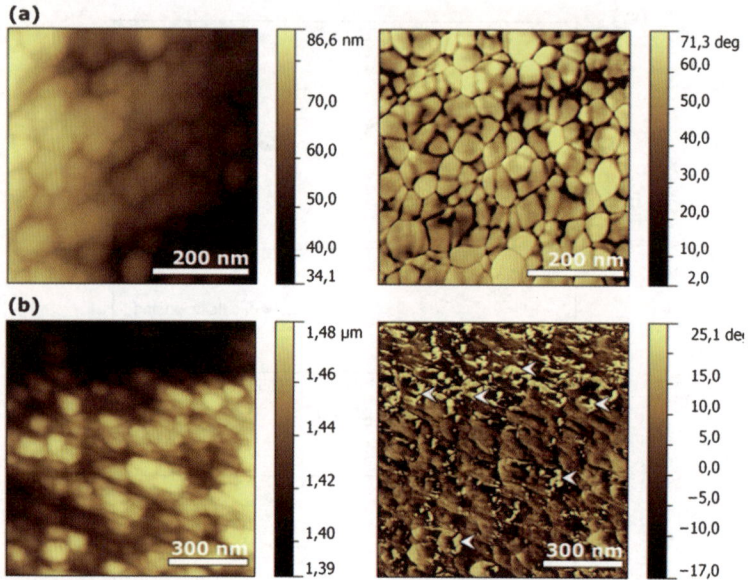

Fig. 3 Atomic force micrographs of (a) transverse mirror-polished prisms and (b) freshly cleaved prisms. In both phase images on the right, the phase contrast indicates organic material which covers the nanogranules. In the phase images of (b), some granules, which are still covered by organic material, are marked by arrows.

patagonicus, two bivalves which belong to two different families (*Unionidae* resp. *Hyriidae*) in the *Palaeoheterodonta* subclass.

Our XAS analysis on mature prisms already indicated the presence of minor amounts of ACC in mature prisms. In order to elucidate the role of ACC during early stages of prism formation and its distribution within a prism, we prepared FIB cuts of a juvenile specimen of *Pinna nobilis* collected at the age of six months and analysed them by scanning transmission electron microscopy (STEM). The FIB cut, which was prepared from the periostracal side of the shell, traversed two juvenile prisms (*p* in Fig. 4a) and the organic periprismatic sheaths in-between (*ps* in Fig. 4a). The latter are easy to distinguish from the mineralized prisms because the prisms appear brighter in HAADF mode due to the stronger scattering intensity of the calcium-containing mineral. On the upper right side, a layer of platinum (*Pt* in Fig. 4a), which was sputtered onto the periostracum (*ps* in Fig. 4a) during FIB preparation, marks the shell surface. The top left and the bottom show the two intersected juvenile prisms (*p* in Fig. 4a) whose distance gradually diminishes with increasing distance to the periostracum. The edges of their nucleating face, *i.e.* the face towards the periostracum, are rounded just as observed in SEM micrograph of separated juvenile prisms (Fig. 1d–e). The periostracum (*ps* in Fig. 4a), which is a thin, pliable and fibrous membrane which protects the shell towards the sea, runs along the very border of the platinum layer (*Pt* in Fig. 4a). It serves during prism formation as substrate for newly forming prisms. It can be removed by sodium hypochlorite and leaves then an imprint on the periostracal face of the prism, *e.g.* the cast of the membrane's folds on prism in Fig. 5a.

We analysed the crystallinity of the juvenile prisms by selected area electron diffraction (SAED). The diffractogram, which is given in Fig. 4b, represents the superposition of 1750 single parallel nanobeam probe electron diffractograms which were taken by scanning the marked area on the left. The inset in Fig. 4a represents a relative crystallinity map by displaying the integral scattering intensity of all reflections occurring in this zone and reveals that juvenile prisms are in fact not fully crystalline.

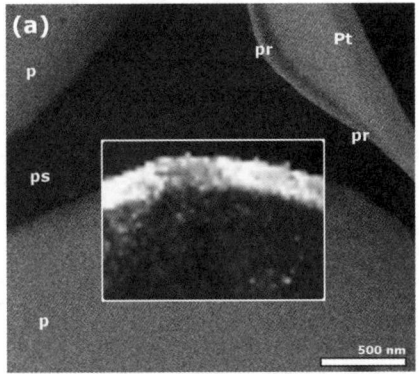
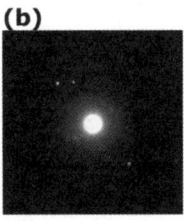

Fig. 4 High angular annular dark field STEM micrographs of FIB cut in the prismatic layer of a juvenile *Pinna nobilis* (6 months) shows the organic periprismatic sheaths (*ps*) confined by the periostracum (*pr*, along the border of *Pt*), on which a layer of platinum (*Pt*) was sputtered during FIB preparations, and two juvenile prisms (*p*). The selected area electron diffractogram (b) is the superposition of 1750 single selected area electron diffractograms acquired in the marked area in (a). The inset in (a) represents a relative crystallinity map by displaying the integral scattering intensity of all reflections which were present in the diffractogram on the right.

Except for some randomly distributed and small areas of crystallinity, the core of the juvenile prisms is still much more amorphous in contrast to the prism border. A remarkable 200 to 250 nm-thick zone of higher crystallinity is located along the border of the prisms, which shows different crystal orientations (not shown). These findings demonstrate clearly that the calcitic prisms of *Pinna nobilis* are formed *via* the aggregation of particles of an amorphous transient phase. At first sight, it is astonishing that only the outer layer of the prisms is crystalline whereas in case of the nacreous layer of *Haliotis laevigata* quite the reverse is true: the nacre platelets are surrounded by a fine layer of amorphous calcium carbonate.[36] We propose that the crystallization of the prisms is heterogeneously induced by the interface of interprismatic matrix as the SAED patterns show that the crystalline areas of the prism are not well aligned which is due to several independent nucleation events. But which mechanism leads to the later well aligned crystal orientation in mature prisms? High resolution SEM images of periostracum of a juvenile *Pinna nobilis* shows at the centre of each prism a knob in the periostracum which, as we believe, serves as a nucleation centre of the latter dominating crystal orientation. A treatment with sodium hypochlorite separates the prisms, removes the periostracum and reveals the complement, the imprint of the nucleation side of the periostracum

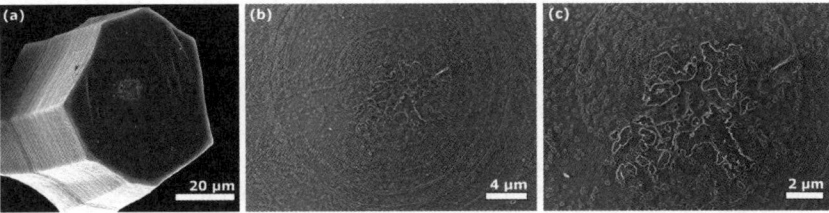

Fig. 5 HR-SEM micrographs of single juvenile prisms after treatment with sodium hypochlorite in order to remove the surrounding organic matrix, *i.e.* the interprismatic matrix and the periostracum. (a) The periostracal face of the prisms shows the imprint of the probable nucleation knob and the mold of foldings of the periostracum. (b–c) Detailed view on the periostracal face of the prism. The morphology of the mineral at the probable nucleation knob reminds of a coalesced droplet. This knob is surrounded by concentric rings which proceed on the prism's side face and reveal the layer-by-layer growth of the prism.

on the periostracal face of the prism. The probable nucleation side is surrounded by irregular concentric rings (Fig. 5b–c) which continue as stripes on the side faces of the prisms. These rings and lines reflect the layer-wise spherulitic growth of the prism: the growth rings originate from very early growth stages whereas the growth lines correspond to the later one. Strikingly, the mineral morphology at the nucleation side itself reminds of coalesced droplets on a flat support: this suggests that the first mineral layer of the prism was still in a liquid-like resp. PILP-like state when it settled on the periostracum and solidified later.

Crystallization under levitation: effect of the intraprismatic proteins on liquid amorphous calcium carbonate (LACC)

We conducted crystallization experiments to investigate the effect of both intraprismatic proteins caspartin and calprismin in early stages of calcium carbonate formation. We chose the Kitano method,[38,39] which precipitates calcium carbonate by slow evaporation of water and concomitant slow release of carbon dioxide from a saturated solution of calcium bicarbonate, because the inherent carbonate buffer leads to a constant and almost neutral pH which makes the precipitation more comparable to *in vivo* conditions. We further employed crystallisation under ultrasonic levitation because of three main advantages: (i) the sample volumes are small thus the required amounts of extracted protein can be kept at a reasonable level. (ii) The crystallization occurs under contactless conditions, only the air/water interface remains. Thus, unwanted and misleading heterogeneous effects which could disturb the crystallization process are ruled out. (iii) The crystallization under levitation permits the study of the liquid amorphous intermediate of calcium carbonate quasi *in situ* and it represents at the moment apparently the only approach to study the impact of additives on this remarkable mineral phase.

In absence of proteins, we observed the formation of an emulsified liquid amorphous calcium carbonate phase: the TEM micrograph of a saturated calcium bicarbonate solution levitated for 400 s shows individual liquid droplets which consist of highly hydrated calcium carbonate (Fig. 6a–b), as we discussed earlier.[13,16] The crystallisation of these droplets, which yielded calcite, could be triggered by an increased irradiative stress. We attribute this behaviour to a loss of hydration water by the coaction of the irradiative stress and ultrahigh vacuum.[13] The low contrast variation of the droplets gives evidence of their liquid-like character: solid spherical particles would show a distinct increase in their contrast from their particle boundary to their centre since the electron beam has traversed more material (thickness contrast). After 400 s, crystallization gradually commenced: sampling after 400 s showed already the first scattered crystallites.

Adding 100 μg mL^{-1} calprismin to the mother solution changed the behaviour of the emulsion: at earlier sampling times, we observed distinctly aggregation and strong coalescence of the droplets after 360 s of levitation (Fig. 6c). After 400 s, a mineral PILP film covered completely the grid and cracks, which occurred due to drying, made the film easy to identify (Fig. 6d). After 600 s, the film no longer covered the grid completely but the density resp. thickness of the film increased, as suggested by the contrast (Fig. 6e). It seems that due to the longer time of levitation, the droplets have already begun to solidify slowly in such a way that the droplet which sedimented on the grid could not coalesce as good as after 400 s of levitation. Calprismin at low concentrations thus seems to behave commensurably to the prediction of the PILP model: it induces the formation of a mineralized film just as predicted for acidic proteins.[3] Let us again rephrase this behaviour in colloido-chemical terms: due to the fact that LACC behaves like a classical electrostatically stabilised emulsion, the presence of calprismin at sufficiently low concentrations induces the de-emulsification by depletion destabilization of the LAAC emulsion. The sedimentation of the LACC droplets and their subsequent coalescence on the support—here the TEM grid—produces the mineral film which is characteristic of PILP processes.

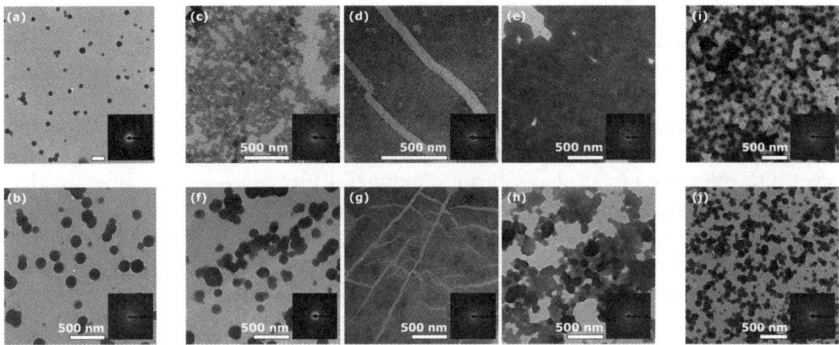

Fig. 6 TEM micrographs and the respective SAE diffractograms of a calcium bicarbonate solution after different levitation times and in presence resp. absence of protein: in absence of protein after (a) 360 s and (b) 400 s; in presence of 100 μg mL^{-1} calprismin after (c) 360 s, (d) 400 s, (e) 600 s; in presence of 100 μg mL^{-1} caspartin after (f) 360 s, (g) 400 s, (h) 600 s; in presence of (i) 50 μg mL^{-1} caspartin and 50 μg mL^{-1} calprismin and of (j) 100 μg mL^{-1} caspartin and 100 μg mL^{-1} calprismin after 600 s.

Turning to caspartin, we observed comparable behaviour but caspartin appeared to be less effective than calprismin at the chosen concentration. Aggregated and moderately coalesced droplets were found after 360 s (Fig. 6f), which vanish 40 s later. Then, distinct film formation was observed much like in the case of calprismin (Fig. 6g). After 600 s of levitation, the film formation seemed to cease as again individual droplets are distinguishable although as well aggregated and partly coalesced (Fig. 6h). We believe that this effect can be described by an up-concentration of the protein during levitation. In course of evaporation of the droplet, not only the supersaturation increases, which leads at a given concentration to phase separation or crystallization, but also does the concentration of the proteinaceous additive. Either caspartin is less effective for depletion destabilization due to its chemical composition or we hit in this special case at 600 s a critical concentration of caspartin which represents the switch from destabilization to a weak depletion stabilization of the LAAC emulsion.

We investigated finally the co-action of caspartin and calprismin on the LACC phase and tested two different concentrations, 50 and 100 μg mL^{-1}, respectively, of both protein and sampled after 600 s of levitation (Fig. 6i and 6j). In this case, the switch from depletion destabilization to depletion stabilization is much more pronounced: at the lower concentration (100 μg mL^{-1} total, Fig. 6i), strong coalescence and aggregation is observed. However at higher concentration (200 μg mL^{-1} total, Fig. 6j) individual droplets appear again, although aggregated but not coalesced. In Fig. 7, SEM images of the lower concentration are given and show a film of individual coalescing beads, which suffered from drying and developed cracks. Some of the droplets coalesced completely which strongly resemble the morphology which we found on the nucleation side of juvenile prisms (Fig. 5b and 5c).†

Conclusions

Mesocrystallisation can occur *via* two different pathways: (a) by the formation of a colloidal mosaic crystal by self-recognition and 3-D oriented attachment of primary

---

† Likewise, a striking morphological resemblance between a precipitate obtained from a liquid amorphous precursor and a biogene calcium carbonate can be found by comparing Fig. 3 in Jacob *et al.*[11] from 2008 with Fig. 4 in Wolf *et al.*[13] from the same year or Fig. 3 in Wolf *et al.*[16] from 2011.

nanoparticles or (b) by crystallisation of a dense array of amorphous particles. Our findings presented here evidence that the calcitic prisms of *Pinna nobilis* form *via* the latter mechanism, by maturation of a colloidal assembly of amorphous primary nanoparticles. Thus the well aligned crystal orientations of the individual primary particles are achieved by epitaxial crystallization and not by oriented attachment. Based on the actual available data which predicts the universal occurrence of a transient amorphous mineral phase during biomineral formation *in vivo*, it seems that this mesocrystallisation pathway by epitaxial maturation is the biologically preferred pathway. These insights in a novel mesocrystallisation pathway will open up new perspectives for our understanding of biomineralisation processes *in vivo* and—with particular regard to the remarkable properties biominerals typically exhibit—for the development and synthesis of new mesocrystalline functional materials.

We reported the effect of both intraprismatic proteins caspartin and calprismin from *Pinna nobilis* on liquid amorphous calcium carbonate (LAAC) and demonstrated their capability to break the emulsified state of LACC by depletion destabilization. Based on our findings, it seems that calprismin is a more potent de-emulsification agent than caspartin, a property which we attribute to the different constitution and size of the proteins. Calprismin is an unusually acidic 38 kDa protein which carries additionally post-translational phosphorylation and glycosylation (2 kDa of the protein's molar weight) whereas the 17 kDa-sized caspartin neither glycosylated nor phosphorylated but was actually unusually acidic due to

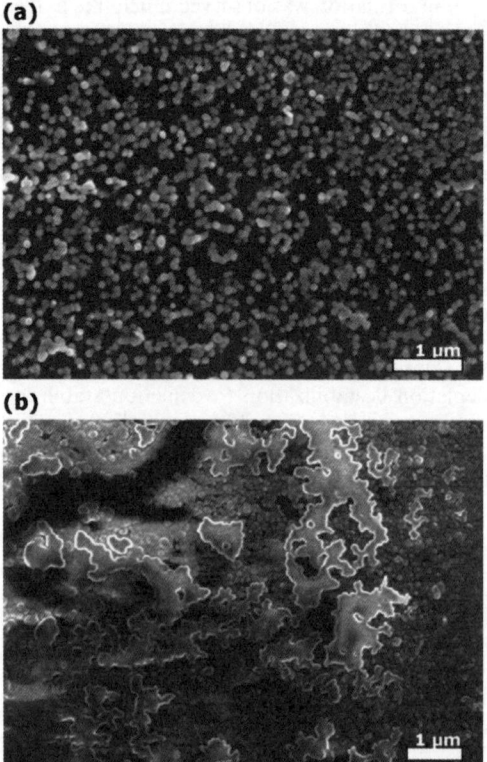

Fig. 7 SEM micrographs of a calcium bicarbonate solution which was levitated for 600 s and then transferred on a flat support. The solution contained (a) 100 μg mL^{-1} caspartin and 100 μg mL^{-1} calprismin; (b) 100 μg mL^{-1} caspartin and 50 μg mL^{-1} calprismin. The latter shows a frapant resemblance to the morphology of the mineral at the probable nucleation knob on the periostracal face of young prisms (see Fig. 5c).

its primary sequence (~69% Asx).[17] Already based on this information, it is reasonable to assume that the structure of caspartin in solution is mainly a random coil, whereas calprismin is expected to have a more defined globular shape. As the effect of depletion destabilization is caused by a reduction of conformational entropy of the additive,‡ it becomes reasonable that the de-emulsification potential of the bigger and conformationally more defined calprismin exceeds that of caspartin. Although both proteins are unusually acidic, we thus suppose them to fulfil different tasks during prism formation beside the commonly accepted task of reducing the solution's supersaturation.[4]

The SEM investigations, both on juvenile prisms (Fig. 5) and on the precipitation in presence of caspartin and calprismin (Fig. 6), lend strong support to the statement that a PILP-like mineral state occurred during the formation of the prisms. The oblate shape of the nanogranules imaged by AFM further suggests that the calcium carbonate particles behave like a viscous liquid during their attachment to the growing prism:[40] they seem to spread (oblate shape in AFM measurements, Fig. 3b) or coalesce (Fig. 5c vs. 7b), depending on their viscosity during attachment. Attached to the surface, the droplets solidify and crystallise during a later maturation process of the prism, as described above.¶ The crystallization occurs centripetally; it starts from the periprismatic sheaths, which seem to induce the initial phase transformation, and progresses towards the inward of the prism. After completion of this maturation process, one can find only a minor amount of ACC in mature prisms. We believe that this minor fraction of ACC is co-located with the protein moieties at the surface of the nanogranules. The high protein concentration at the particle's surface is capable of poisoning a phase transformation of the amorphous layer to crystalline calcite and could thus stabilize thin layers of ACC during maturation of the prisms.[36] We conjecture that the formation of the proteinaceous coating of the granules may be caused by one (or both) of the two following potential separation processes: (a) Tensioactive proteins enrich at the interface of the liquid-amorphous precursor during a still emulsified state thus before settlement. (b) The protein moiety experiences expulsion during solidification and crystallisation of the nanogranules just like the transition bars which form during crystallisation of a mineral film obtained by a PILP process.[41]

Let us now consider the advantages of such a liquid mineral phase *in vivo* and develop a possible scenario for prism formation. The formation of the liquid amorphous phase of calcium carbonate occurs by further increasing the concentration of saturated calcium bicarbonate solution. In our approach we achieve this by evaporation of a levitated droplet of calcium bicarbonate solution. The organism would definitely energetically profit from a still liquid mineral precursor as compared to a solid one: the interfacial energy of a two-phase liquid/liquid system is much lower, the concentration gradient, which is needed to induce the formation of such a phase is lower and finally it is less energy-consuming to optionally re-dissolve a liquid intermediate. Finally, a droplet of the still liquid mineral precursor—even if it would be already quite viscous due to densification—would adopt the exactly fitting shape for the hollow at which is attaches to the growing mineral. Yang *et al.* showed recently that sea urchin biominerals are space-filling, whereas synthetic mesocrystals are found to be porous without exception.[47] Consequently; they proposed a "space-filling ACC" as a structural precursor which corresponds to our proposal of a liquid

‡ The emulsified negatively charged droplets generate a layer of thickness d in which nonabsorbing polymers, *e.g.* acidic proteins, experience a reduction in their conformational entropy. If the interdroplet distance is considerably longer than d, the solution is stable. If the distance is in the range of $2d$, then the polymer will be forced to leave the interdroplet space in order to gain conformational entropy. Because of the resulting concentration gradient, the solvent diffuses out of the interparticle space; the droplets start to aggregate resp. coalesce.

¶ Note added after discussion: The AFM micrographs of Pinna nobilis prisms resemble strongly those of PILP precipitates, *cf.* ref. 46.

resp. viscous amorphous calcium carbonate. *In vivo*, the formation of such a liquid intermediate could be accomplished simply by ion pumps of vesicles in the mantle cells.[42] Such vesicles, which contain mineral precursors in a still liquid state, are already known from siliceous sponges:[43] in silicasomes, the silica esterase silicatein ensures a low-condensed and thus liquid state of the mineral precursor, before the content of the silicasome is released into the extracellular space in which the final silica spicule is formed. In our present case, a comparable scenario is conceivable: in special calcium carbonate-enriching "lithosomes"§ which we call "calcosomes", the mineral precursor is up-concentrated and acidic proteins at high concentrations serve as stabilizing agent in two ways: (a) by reducing the supersaturation by sequestering calcium cations from the solution[40] and (b) by depletion stabilization,[16] as discussed above. If the content of the calcosome is then released to the extrapallial space, the concentration of the stabilizing protein drops as it is diluted by the extrapallial fluid and the LACC emulsion experiences now depletion destabilization which eases the attachment to the growing prism. The AFM analysis showed that the nanogranules were presumably viscous during attachment; the densification of the droplets could occur either on the way through the extrapallial space or already earlier in the calcosomes. The phase transformation of the prism to a final crystalline state is triggered heterogeneously by the periprismatic sheaths and proceeds thenceforth in centripetal manner. The single nanogranules crystallise one after the other in a random path by epitaxial nucleation. The final single crystallinity of a single prism is a result of a process which can be well described in terms of the Grigorév model.[44] Recent results obtained by XANES-PEEM by Gilbert *et al.* showed almost all calcite prisms of *Pinctada fucata* to be not fully single-crystalline but to exhibit intraprismatic domains of differently oriented nanocrystals.[48] These findings indicate that a "frustration" of the prism ripening may occur and single-crystallinity may not be reached.

In summary, we re-visited the ultrastructure of the calcitic prisms of *Pinna nobilis* and demonstrated that they are a perfect example of a mesocrystalline material. We suggest that the prisms consist of protein-enwrapped nanogranules which are surrounded by a thin layer of amorphous calcium carbonate. This aggregation of initially amorphous nanogranules transforms by epitaxial nucleation to a three-dimensional array of aligned nanosized crystals, *i.e.* a well ordered mesocrystal. We thus corroborate strongly the quite recently reported pathway of mesocrystallisation by Seto *et al.*[7] Finally, we develop a novel model of prism formation which accommodates the newly provided strong evidence for the occurrence of liquid amorphous mineral precursor *in vivo*.

Acknowledgements

We warmly thank Nathalie Guichard, Maren Müller, Katrin Kirchoff, Simone Rolf and the whole BioCAT team for their excellent technical support. S.E.W. and F.M. thank further Sébastien Motreuil for providing the shells of *P. nobilis*. This work was supported by grants of the European Commission (PITN-GA-2008-215507 – BIOMINTEC). S.E.W. is grateful to the German Research Foundation (DFG) for a postdoctoral research fellowship (Wo 1712/1-1). A complementary support was provided by SRO 2012 project of the OSU THETA (Bourgogne-Franche-Comté). Use of the Advanced Photon Source, an Office of Science User Facility operated for the U.S. Department of Energy (DOE) Office of Science by Argonne National Laboratory, was supported by the U.S. DOE under Contract No. DE-AC02-06CH11357. This project was supported by grants from the National Center for

---

§ The expression lithosome shall embrace mineral-bearing vesicles which are involved in biomineral formation such as silicomes,[43] magneto-somes,[45] or the calcosomes proposed herein. Although the term was already used as a technical term in lithostratigraphy, a sub-discipline of stratigraphy, we believe that no danger of confusion is present.

Research Resources (2P41RR008630-17) and the National Institute of General Medical Sciences (9 P41 GM103622-17) from the National Institutes of Health. Last but not least, we are thankful to Dorrit Jacob for an inspiring and helpful discussion.

Notes and references

1 *Handbook of Biomineralization—Biological Aspects and Structure Formation*, ed. E. Bäuerlein, Wiley-VCH Verlag GmbH, 2007.
2 H. Cölfen and M. Antonietti, *Mesocrystals and Nonclassical Crystallization*, Wiley-VCH Verlag GmbH, 2008.
3 L. B. Gower and D. A. Tirell, *J. Cryst. Growth*, 1998, **191**, 153–160.
4 L. B. Gower, *Chem. Rev.*, 2008, **108**, 4551–627.
5 M. J. Olszta, D. J. Odom, E. P. Douglas and L. B. Gower, *Conn. Tiss. Res.*, 2003, **44**, 326–334.
6 H. Cölfen and M. Antonietti, *Angew. Chem., Int. Ed.*, 2005, **44**, 5576–91.
7 J. Seto, Y. Ma, S. A. Davis, F. Meldrum, A. Gourrier, Y.-Y. Kim, U. Schilde, M. Sztucki, M. Burghammer, S. Maltsev, C. Jäger and H. Cölfen, *Proc. Natl. Acad. Sci., U. S. A.*, 2012, Early Edit.
8 Y. Oaki and H. Imai, *Small*, 2006, **2**, 66–70.
9 Y. Oaki and H. Imai, *Angew. Chem.*, 2005, **117**, 6729–6733.
10 Y. Dauphin, *J. Biol. Chem.*, 2003, **278**, 15168–77.
11 D. Jacob, A. Soldati, R. Wirth, J. Huth, U. Wehrmeister and W. Hofmeister, *Geochim. Cosmochim. Acta*, 2008, **72**, 5401–5415.
12 M. Olszta, S. Gajjeraman, M. Kaufman and L. B. Gower, *Chem. Mater.*, 2004, **16**, 2355–2362.
13 S. E. Wolf, J. Leiterer, M. Kappl, F. Emmerling and W. Tremel, *J. Am. Chem. Soc.*, 2008, **130**, 12342–7.
14 S. E. Wolf, L. Müller, R. Barrea, C. J. Kampf, J. Leiterer, U. Panne, T. Hoffmann, F. Emmerling and W. Tremel, *Nanoscale*, 2011, **3**, 19–21.
15 S. J. Homeijer, R. A. Barrett and L. B. Gower, *Cryst. Growth Des.*, 2010, **10**, 1040–1052.
16 S. E. Wolf, J. Leiterer, V. Pipich, R. Barrea, F. Emmerling and W. Tremel, *J. Am. Chem. Soc.*, 2011, **133**, 12642–12649.
17 F. Marin, R. Amons, N. Guichard, M. Stigter, A. Hecker, G. Luquet, P. Layrolle, G. Alcaraz, C. Riondet and P. Westbroek, *J. Biol. Chem.*, 2005, **280**, 33895–908.
18 J. C. Moreteau and N. Vicente, *Malacologia*, 1982, **22**, 341–345.
19 O. Paris, C. Li, S. Siegel, G. Weseloh, F. Emmerling, H. Riesemeier, A. Erko and P. Fratzl, *J. Appl. Crystallogr.*, 2006, **40**, 466–s470.
20 R. Fischetti, S. Stepanov, G. Rosenbaum, R. Barrea, E. Black, D. Gore, R. Heurich, E. Kondrashkina, A. J. Kropf, S. Wang, K. E. Zhang, T. C. Irving and G. B. Bunker, *J. Synchrotron Radiat.*, 2004, **11**, 399–405.
21 X.-R. Xu, A.-H. Cai, R. Liu, H.-H. Pan, R.-K. Tang and K. Cho, *J. Cryst. Growth*, 2008, **310**, 3779–3787.
22 D. Nečas, *Cent. Eur. J. Phys.*, 2011, **10**, 181–188.
23 M. Niederberger and H. Cölfen, *Phys. Chem. Chem. Phys.*, 2006, **8**, 3271.
24 J. L. Fulton, S. M. Heald, Y. S. Badyal and J. M. Simonson, *J. Phys. Chem. A*, 2003, **107**, 4688–4696.
25 Y. Politi, Y. Levi-Kalisman, S. Raz, F. Wilt, L. Addadi, S. Weiner and I. Sagi, *Adv. Funct. Mater.*, 2006, **16**, 1289–1298.
26 R. S. K. Lam, J. M. Charnock, A. Lennie and F. C. Meldrum, *CrystEngComm*, 2007, **9**, 1226.
27 Y. Levi-Kalisman, S. Raz, S. Weiner, L. Addadi and I. Sagi, *Adv. Funct. Mater.*, 2002, **12**, 43.
28 Y. U. T. Gong, C. E. Killian, I. C. Olson, N. P. Appathurai, A. L. Amasino, M. C. Martin, L. J. Holt, F. H. Wilt and P. U. P. A. Gilbert, *Proc. Natl. Acad. Sci., U. S. A.*, 2012, Early Edit.
29 F. E. Sowrey, L. J. Skipper, D. M. Pickup, K. O. Drake, Z. Lin, M. E. Smith and R. J. Newport, *Phys. Chem. Chem. Phys.*, 2004, **6**, 188.
30 E. Piskorska, K. Lawniczak-Jablonska, R. Minikayev, A. Wolska, W. Paszkowicz, P. Klimczyk and E. Benko, *Spectrochim. Acta, Part B*, 2007, **62**, 461–469.
31 A. Bianconi, J. Garcia and M. Benfatto, *Topics Curr. Chem.*, 1988, 145.
32 L. Addadi, S. Raz and S. Weiner, *Adv. Mater.*, 2003, **15**, 959–970.
33 B. Hasse, H. Ehrenberg, J. C. Marxen, W. Becker and M. Epple, *Chem.–Eur. J.*, 2000, **6**, 3679–3685.

34 I. M. Weiss, N. Tuross, L. Addadi and S. Weiner, *J. Exp. Zool.*, 2002, **293**, 478–491.
35 A. Baronnet, J. P. Cuif, Y. Dauphin, B. Farre and J. Nouet, *Mineral. Mag.*, 2008, **72**, 617–626.
36 N. Nassif, N. Pinna, N. Gehrke, M. Antonietti, C. Jäger and H. Cölfen, *Proc. Natl. Acad. Sci. U. S. A.*, 2005, **102**, 12653–5.
37 D. E. Jacob, R. Wirth, A. L. Soldati, U. Wehrmeister and A. Schreiber, *J. Struct. Biol.*, 2011, **173**, 241–9.
38 Y. Kitano, *Bull. Chem. Soc. Jpn.*, 1962, **35**, 1980–1985.
39 Y. Kitano, *Bull. Chem. Soc. Jpn.*, 1962, **35**, 1973–1980.
40 L. B. Gower and D. J. Odom, *J. Cryst. Growth*, 2000, **210**, 719–734.
41 L. Dai, X. Cheng and L. B. Gower, *Chem. Mater.*, 2008, **20**, 6917–6928.
42 S. Mann, J. Webb, and R. J. P. Williams, ed., *Biomineralization—Chemical and Biochemical Perspectives*, VCH Publishers, Inc., New York, 1989.
43 W. E. G. Müller, S. E. Wolf, U. Schlossmacher, X.-H. Wang, A. Boreiko, D. Brandt, W. Tremel and H. C. Schröder, *FEBS J.*, 2008, **275**, 362–70.
44 D. P. Grigorév, *Ontogeny of Minerals*, Israel Program for Scientific Translation, Jerusalem, Israel, 1965.
45 C. Jogler and D. Schüler, in *Handbook of Biomineralization*, VCH-Verl.-Ges., Weinheim, Germany, 2007, 145–162.
46 Y.-Y. Kim, E. P. Douglas and L. B. Gower, *Langmuir*, 2007, **23**, 4862–4870.
47 L. Yang, C. E. Killian, M. Kunz, N. Tamura and P. U. P. A. Gilbert, *Nanoscale*, 2011, **3**, 603–609.
48 P. U. P. A. Gilbert, A. Young and S. N. Coppersmith, *Proc. Natl. Acad. Sci. U. S. A.*, 2011, **108**, 11350–11355.

Oligomer formation, metalation, and the existence of aggregation-prone and mobile sequences within the intracrystalline protein family, Asprich†

Moise Ndao, Christopher B. Ponce and John Spencer Evans‡*

Received 6th April 2012, Accepted 24th April 2012
DOI: 10.1039/c2fd20064c

The formation of crack propagation—resistant single crystal calcite in the prismatic layer of the mollusk shell involves the participation of a number of different proteins, some of which form intracrystalline organic inclusions. One protein family, Asprich (*Atrina rigida*), participate in the ACC formation/transformation process and become occluded within calcite. However, these two phenomena are poorly understood. Here, we experimentally establish that the Asprich "3" protein oligomerizes in solution over a wide pH range and that this oligomerization process is enhanced by the presence of Ca^{2+} ions, which form complexes with this protein as verified by ESI-MS. Bioinformatics analyses confirm that intrinsic disorder and unstable interactive domains constitute the entire length of each Asprich sequence. In addition, an amyloid- or prion-like aggregation-prone region was identified within the highly conserved 12 AA N-terminal sequence, KPVFKRSLSDPS. Together, these bioinformatics findings suggest that different sequence elements of the Asprich family contribute to the observed oligomerization process. Solution NMR studies of Asprich "3" protein assemblies document that portions of the 12 AA N-terminal domain are mobile within these assemblies. This mobility is driven by Pro imido ring *cis-trans* isomerization at P11, and this induces multiple conformational states for V3, R6, and L8 which lie upstream of this Pro residue. This Asprich "3" Pro-induced conformational exchange phenomena parallels other findings obtained for self-assembling nacre layer- and tooth enamel-specific biomineralization proteins. In conclusion, we have identified the presence of self-association, unusual sequence behavior, and ion clustering phenomena in Asprich "3" and we believe that these are key factors in the ability of Asprich proteins to control ACC formation and form organic inclusions that resist crack propagation.

Laboratory for Chemical Physics, Division of Basic Sciences and Craniofacial Biology, New York University College of Dentistry, 345 E. 24th Street, NY, NY 10010. E-mail: jse1@nyu.edu; Fax: +1-212-995-4087; Tel: +1-212-998-9605

† Electronic supplementary information (ESI) available: Table S1, ESI-MS—detected protein–metal ion adducts. See DOI: 10.1039/c2fd20064c

‡ This research was supported by the U.S. Department of Energy, Office of Basic Energy Sciences, Division of Materials Sciences and Engineering under Award DE-FG02-03ER46099, and represents contribution number 64 from the Laboratory for Chemical Physics, New York University.

1. Introduction

The mollusk shell is a true biocomposite that, in some species, is comprised of two physically distinct layers of calcium carbonate that co-exist with biomacromolecules.[1-9] The calcite-containing prismatic layer is mechanically and structurally distinct from the adjacent fracture resistant nacre (aragonite) layer.[1-10] The prismatic layer consists of long, parallel prismatic columns of calcite (single crystals, 100–200 micrometers in cross-section).[1-3] The prismatic layer has a lower fracture toughness compared to the nacreous layer, and is more brittle.[1-3] Under indentation forces, the prismatic layer experiences radial crack propagation, which serves as a protective shield that resists piercing from predators (*i.e.*, puncture-resistance).[5,6] Under compressive forces, fracture does occur alongside prismatic crystals but does not completely propagate through the prismatic layer (*i.e.*, crack propagation resistance).[5,6] Thus, the calcitic prismatic layer offers an interesting mechanism for defeating catastrophic material failure of the shell, and represents an important biological model system for materials science.

The material and biological properties of single crystal prismatic calcite have been linked to the presence of specific matrix proteins,[11-17] some of which exist as randomly distributed aggregates within calcite.[18] Several protein classes, including "acidic" proteins[11-15] have been identified as components within this mineralized framework. There is evidence that these "acidic" proteins are involved in the formation of amorphous calcium carbonate (ACC), the precursor phase of prismatic calcite.[13,14,19] An example of an "acidic" protein family is Asprich (*Atrina rigida*).[13] This library of ten proteins ("a"–"g"; "1"–"3") are intrinsically disordered[19-24] and multi-domain in nature.[13,19-21] These proteins are believed to be part of the randomly distributed organic deposit network within the calcite prisms,[18] and are therefore thought to play a significant role in the material properties of prismatic calcite.[14,19-21] In fact, *in vitro* mineralization studies of Asprich "g"[14,21] and Asprich "3"[19] confirm that both proteins stabilize ACC and control the subsequent growth of calcite. However, we do not yet understand the underlying molecular mechanisms of Asprich-mediated ACC formation nor how randomized intracrystalline protein deposits form within calcite prisms.

We report new evidence that explains how "acidic" proteins such as Asprich might regulate ACC formation and participate in the formation of intracrystalline organic deposits. Using biophysical techniques, we confirm that Asprich "3" oligomerizes in solution to form heterogeneously-sized protein assemblies, and that this oligomerization process is enhanced by Ca^{2+} ions. This Ca^{2+} enhancement results from the presence of multiple metal ion binding sites within Asprich "3", which were confirmed by ESI-MS experiments. Bioinformatics studies of the ten member Asprich family confirm two interesting sequence features that potentially drive the self-assembly process: the presence of highly unstable, interactive disordered regions and the existence of a putative amyloid- or prion-like aggregation motif located within the highly conserved 12 AA N-terminal cationic domain, KPVFKRSLSDPS, found in all Asprich sequences.[13] NMR experiments reveal that the Asprich "3" apo-assemblies feature stabilizing interchain interactions, yet exhibit Pro-induced mobile polypeptide segments similar to those found in other self-assembled biomineralization proteins.

2. Experimental section

Synthesis and sample preparation of Asprich "3"

We employed a tBoc-based stepwise solid-phase synthesis to create Asprich "3".[19] Cleaved Asprich "3" was then solubilized with mixtures of water and trifluoroacetic acid (TFA) (Sigma-Aldrich) and purified by preparative reverse phase high-performance liquid chromatography (RP-HPLC) using a Waters DeltaPak C18 RP-HPLC column, with 0.1% TFA–water mobile phase and elution with 85% acetonitrile–0.1% TFA–water linear gradient (0–85% acetonitrile, 20 min).[19] The column elution was monitored at 230 nm and individual HPLC fractions were analyzed using MALDI-TOF mass spectrometry (Applied Biosystems 4700 Proteomics Analyzer). The purity of Asprich "3" was determined to be 94%, based upon RP-HPLC re-chromatography and MALDI-TOF of the purified protein. The experimental mass of Asprich "3" was observed to be 6628.3 Da, which corresponds well with the theoretical mass of 6627.3 Da.[19]

Electrospray ionization mass spectrometry (ESI-MS)

We utilized ESI-MS[20,25–27] to detect protein–metal ion and protein–anion complexes since the ESI process does not disrupt non-covalent interactions, thus allowing these interactions to be detected in the gas phase. For cation binding experiments, 10 micromolar Asprich "3" stock solutions were created using unbuffered deionized distilled water (UDDW). The appropriate amounts of 5 M metal salt stock solutions ($CaCl_2$, $LaCl_3$, $EuCl_3$, 99.99% pure, Sigma-Aldrich Chemicals, in UDDW) were then added to each peptide sample to create final peptide : metal ion mole ratios of 1 : 10. ESI-MS experiments were conducted on an Agilent LC/MSD 1100 ESI-ion trap mass spectrometer with an electrospray ionization source. Samples were injected using a 250 μL syringe loaded into a stepper motor that delivered sample to the instrument at a rate of 15 μL min^{-1}. The experiments were done in UltraScan mode with a N_2 nebulizing gas pressure of 15 psi and a N_2 drying gas flow rate of 4 L min^{-1} at 325 °C. Negative ionization mode was used with a skimmer voltage of 40 V and a capillary exit voltage of 140 V, resulting in a capillary exit potential of 100 V.

ESI-MS generates ion species that are either positive (multiple protonation) or negative (multiple deprotonation).[25–27] In the case of an n-fold deprotonated protein species a charge-state deconvolution is performed on the profile spectra by selecting mass peaks considered representative of the envelope and applying a mathematical algorithm to determine the molecular weight of the protein and its charge-state assignment of the m/z peaks observed in the envelope, which generally takes the formula[20,25,27]

$$\frac{m}{z} = \frac{(M - nH)}{n} \tag{1}$$

where M is the observed mass of the protein (6627.3 Da) and n is the number of H protons. In the presence of metal ions this same protein becomes a metal ion adduct and the m/z peaks will shift upward by iX, where i is the number of bound metal ions in any given charge state and X is the mass of the metal ion in atomic mass units, as given by the equation,[20,25,27]

$$\frac{m}{z} = \frac{(M - nH)}{n} + iX \tag{2}$$

Dynamic light scattering (DLS)

For pH-dependent studies of apo-Asprich "3" oligomerization, polypeptide samples were mixed in various 10 mM buffers (pH 4.0–6.0, sodium acetate-acetic acid buffer; pH 7.0–8.0, Tris-HCl buffer, pH 9.0, sodium bicarbonate-sodium carbonate) with final polypeptide concentrations of 100 micromolar.[28–32] For Ca^{2+}-dependent oligomerization studies, the Asprich "3" samples were measured at pH 8.0 in 10 mM Tris-HCl buffer containing stoichiometric amounts of 99.9% $CaCl_2$ (Sigma-Aldrich) (Ca^{2+} : protein = 1 : 1, 2 : 1, 10 : 1, 20 : 1, 50 : 1). In addition Asprich "3" samples containing 12.5 mM $CaCl_2$ were also measured, and this Ca^{2+} concentration corresponds to that utilized in our previously reported Asprich "3" mineralization assay studies.[19]

The hydrodynamic radii (R_H) of all Asprich "3" samples were measured using a DynaPro MS/X dynamic light scattering instrument (Protein Solutions, Inc.) at 16 °C.[28-32] Prior to DLS measurements all buffer/protein samples were pre-filtered using 0.22 micron polyvinylidene fluoride syringe filters (Fisher Scientific) and then placed into quartz cuvettes. The samples were incubated at 16 °C for 10 min prior to measurements. Ten acquisitions were taken per trial. Analysis of the data and determination of hydrodynamic radius were performed using the regularization analysis in the Dynamics v6.0 software provided with the instrument. By measuring the fluctuations in the laser light intensity scattered by the sample, the instrument is able to detect the speed (diffusion coefficient) at which the particles are moving through the medium. This value is converted to the sphere-equivalent hydrodynamic radius (R_H) using the Stokes–Einstein relation:[28]

$$D = \frac{kT}{(6\pi\eta R_H)} \quad (3)$$

where D is the is the diffusion coefficient, k is the Boltzmann constant, T is the absolute temperature, and η is the viscosity.[28]

Bioinformatics. The ten Apsrich protein sequences, "a" through "g" (accession numbers AAU04812, AAU04809, AAU04806, AAU04810, AAU04808, AAU04807, and AAU04811, respectively) and "1" through "3" (accession numbers AAU04814, AAU04813, and AAU04815, respectively) were obtained from the National Center for Biotechnology Information (NCBI). All Asprich sequences were initially analyzed using the SignalP 4.0 algorithm[33] to identify the N-terminal signal peptide region, which was subsequently deleted from each sequence. Regions of intrinsic disorder were then identified in each Asprich sequence using the IUPRED (Intrinsically Unstructured Protein pREDiction) algorithm, which estimates the capacity of a polypeptide sequence (0 = low, 1 = high) to form stabilizing contacts.[34] The ANCHOR protein–protein binding algorithm was then applied to identify disordered regions that cannot form sufficient intrachain interactions and thus have the propensity (0 = low, 1 = high) to energetically partner with another entity, such as a protein.[35] The IUPRED and ANCHOR probabilities were then co-plotted to identify sequence regions within the Asprich family that are disordered and are potentially interactive with their environment. We also performed algorithm searches of each Asprich sequence for aggregation propensities using the BETASCAN,[36] AGGRESCAN,[37] and FOLD-AMYLOID[38] programs. BETASCAN calculates the likelihood scores for potential beta strands and strand pairing based upon correlations observed in parallel beta sheets.[36] AGGRESCAN identifies aggregation-prone segments in protein sequences using a aggregation-propensity scale.[37] FOLD-AMYLOID detects amyloidogenic sequence regions using expected probability of backbone hydrogen bond formation and expected packing density of residues.[38]

Nuclear magnetic resonance experiments

We examined the molecular features of Asprich "3" oligomers *via* solution NMR experiments performed on a 100 micromolar Asprich "3" sample [90% v/v UDDW, 10 v/v 99.9% D_2O (Cambridge Isotope Labs, Inc)] at pH 7.0, 298 K, using a 900 MHz ultrashield Bruker US2 AVANCE spectrometer (21.14 T field strength) outfitted with a 5 mm HXY triple resonance Bruker cryoprobe. We chose pH 7 conditions since pH 8.0 conditions are heavily skewed towards base-catalyzed solvent–amide proton exchange which leads to backbone proton signal broadening issues. The reader should note that our solution conditions are optimized for intrinsically disordered proteins, *i.e.*, low ionic strength conditions with no added buffer (*i.e.*, the protein acts as its own buffer).[20,32] For all experiments we utilized a 1-H 90 degree pulse of 9.35 microseconds, a relaxation delay of 1.5 s, and a spectral width of 12 ppm. Water suppression was achieved using gradient excitation sculpting methods. TOCSY experiments were performed at 50, 60, 70 and 80 ms mixing times, while NOESY experiments were performed at 50, 100, 150 and 200 ms mixing times. The data were accumulated for 2048 complex points, processed, and visualized using NMRPipe (National Institute of Health, Bethesda, MD) and Sparky (SPARKY 3, University of California, San Francisco, CA) software packages.

3. Results

Mapping the Asprich "3" oligomerization process

Previous studies revealed that intrinsically disordered nacre polypeptides oligomerize in solution as part of the mechanism of protein mediated aragonite formation.[29,30,39] However, it is not clear if calcite-associated proteins, such as Asprich "3", also utilize self-association as a means of controlling the ACC mineralization process in the prismatic layer, or, as a route towards the synthesis of intracrystalline organic inclusions. To clarify this, we used dynamic light scattering (DLS) methods to monitor the oligomerization of apo-Asprich "3". We utilized Asprich "3" concentrations (100 micromolar) that were known to induce ACC formation *in vitro*.[19] Furthermore, from previous experiments we know that the initial pH of the mineralization assay solutions containing Asprich "3" and 12.5 mM $CaCl_2$ is 4.0, and these solutions attain a pH value of 8.0–8.3 at the conclusion of the assay as the solution becomes saturated with carbonate vapor.[19,20,29,30,39] Hence, to mimic these mineralization conditions, we mapped out apo-protein oligomerization over the pH range of 4 to 9 (Fig. 1A), and subsequently repeated DLS measurements at pH 8.0 in the presence of different calcium concentrations.

We find that apo-Asprich "3" spontaneously oligomerizes over the tested pH range, with R_H = 53 to 89 nm (Fig. 1B). The polydispersity values are >30% and this indicates that heterogeneously sized protein particles exist under these

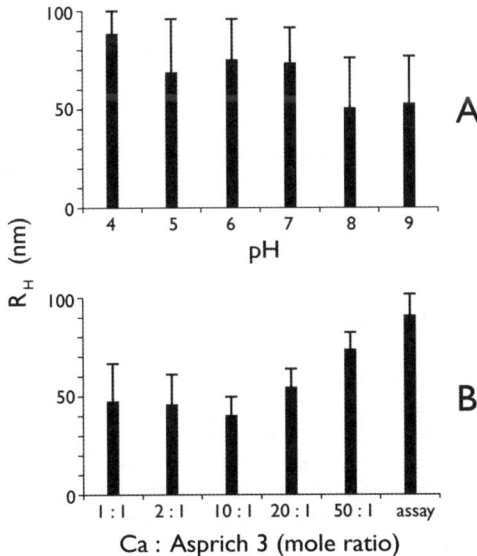

Fig. 1 Histogram comparison of DLS particle size distributions obtained for (A) apo-Asprich 3 as a function of pH and (B) Asprich 3 in the presence of Ca^{2+} at pH 8.0. Sample temperature for both experiments was 16 °C (seawater temperature). The error bars represent the skewed polydispersity of the particles. In (B), "assay" stoichiometry point refers to 12.5 mM Ca^{2+} concentration utilized in published mineralization assay studies of Asprich "3" (Ref. 19).

conditions.[28-32] As pH increases, we note a slight trend towards smaller oligomer dimensions, and this indicates that protein electrostatics may play a role in the oligomerization process. In particular, we note that from pH 4 to 9 we are traversing the pK_a range of His imidazole (pK_a = 6–7.5),[40] and we note that there is a lone His residue located at the C-terminal end of Asprich "3" (H61). Thus, we explain the pH-dependence of oligomerization as follows: as the system pH increases, the H61 residue deprotonates and this change in protein electrostatics affects the apo-Asprich "3" oligomerization process in some fashion (Fig. 1A).

Next, we examined the oligomerization process as a function of Ca^{2+} ion concentration at pH 8.0 (Fig. 1B), with the upper limit of Ca^{2+} concentration being equivalent to that utilized in published Asprich "3" mineralization assays (i.e., 12.5 mM).[19,20] Once again, we find that Asprich "3" oligmerization is spontaneous under these conditions. However, the presence of Ca^{2+} induces an 25–95% increase in R_H values relative to the apo-state at pH 8.0. We note that a similar Ca^{2+} induction trend was observed for nacre protein assemblies under identical conditions.[41] Our explanation for this Ca^{2+} induced increase in oligomer dimension is as follows. Given that the amino acid composition of Asprich "3" is 42.6% Asp and 8.2% Glu,[13] Ca^{2+} ions are binding to these residues, inducing polyelectrolyte chain collapse via charge neutralization[20,42,43] and possibly fostering Ca^{2+} bridging interactions between Asprich "3" protein molecules[20,42,43] which would lead to larger-sized oligomers.

The metalation of intrinsically disordered Asprich "3"

Recent studies indicate that protein–metal ion complexation (i.e., metalation) can induce folding, self-association, or promote protein–protein interactions in intrinsically disordered proteins.[19,20,44,45] In some cases, this metalation effect is cation-specific.[44,45] Since Ca^{2+} ions induce larger oligomer dimensions (Fig. 1B), we next investigated the ability of intrinsically disordered Asprich "3" to bind Ca^{2+}. These metalation capabilities were assessed using ESI-MS and metal ion-induced mass shift experiments.[20,25-27] For metals, we employed Ca^{2+} and the Ca^{2+} metal ion analogs La^{3+} and Eu^{3+} (Fig. 2; Table S1, Electronic Supplementary Information†).[20,25-27] As described in earlier ESI-MS Asprich fragment studies, we utilized La^{3+} and Eu^{3+} as surrogate probes to induce larger mass shifts (i.e., >100 Da) upon binding to Asprich "3" and thus assist in the interpretation of our Asprich "3" Ca^{2+} dataset.[20]

The estimated net charge of Asprich "3" at pH 7 is calculated to be -28, assuming that all carboxylate groups are fully deprotonated. Hence, we expect to see Asprich "3" adduct species with high negative charge values. Using ESI-MS, we observe that this indeed is the case, with the charged states of apo Asprich "3" range from -6 to -12 (Fig. 2; Table S1, Electronic Supplementary Information†), which indicates that there are numerous protons associated with negatively charged Asprich "3" molecules. Typically, each adduct species consists of a centroid peak with multiple peaks exist on either side of the centroid peak arising primarily from naturally occurring $^{12}C/^{13}C$ isotope distributions within the Asprich "3" protein (Fig. 2).[20,25,46]

When metal ions are introduced to Asprich "3", there are dramatic changes in the corresponding ESI-MS spectra (Fig. 2). Asprich "3" forms stable complexes with all three metal ions, and the binding of each cation induces loss of proton species and detectable changes in net adduct charge and m/z values relative to the apo sample (Fig. 2; Table S1, Electronic Supplementary Information†). Presumably, the metalation of Asprich "3" involves direct interactions of metal ions with multiple electron donors (e.g., Asp, Glu, and possibly Arg) with the subsequent displacement of protons.[20,25,46] We note that similar results were obtained with the Asprich "a"–"g" conserved N- (F1) and C- (F2) terminal sequences.[20] We were able to identify monoisotopic Asprich "3" : Eu^{3+} and Asprich "3" : La^{3+} species (1 : 1 to 1 : 3) and Asprich "3" : Ca^{2+} species (1 : 1 to 1 : 4) (Fig. 2; Table S1, Electronic Supplementary Information†). This indicates that Asprich "3" contains multiple metalation sites,

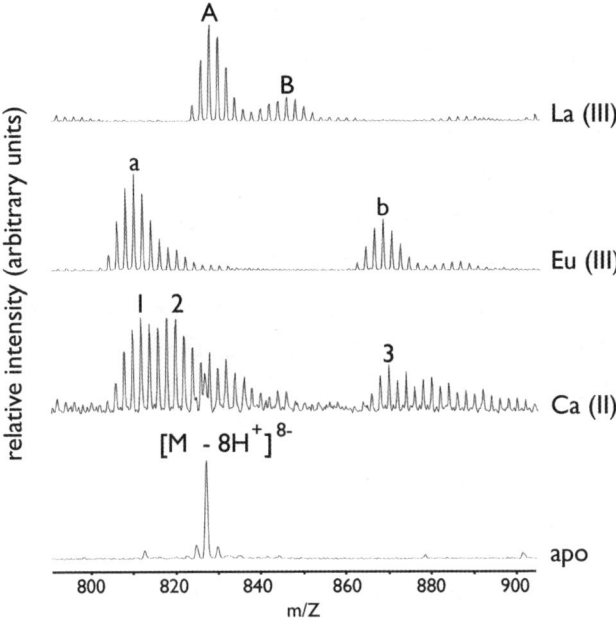

Fig. 2 Representative ESI-MS ion trap mass spectra of Asprich "3" in the apo state and in the presence of 10 : 1 (mole excess) CaCl$_2$, LaCl$_3$, and EuCl$_3$. Monoisotopic peak identifications are labeled and presented in Table S1, Electronic Supplementary Information.† Legend to monoisotopic centroid peaks: Ca^{2+}: 1 = [M + 2Ca^{2+} − 12H$^+$]$^{8-}$; 2 = [M + 4Ca^{2+} − 12H$^+$]$^{4-}$; 3 = [M + 4Ca^{2+} − 25H$^+$]$^{17-}$. Eu^{3+}: a = [M + Eu^{3+} − 13H$^+$]$^{10-}$; b = [M + Eu^{3+} − 19H$^+$]$^{16-}$. La^{3+}: A = [M + 2La^{3+} − 51H$^+$]$^{45-}$; B = [M + 2La^{3+} − 33H$^+$]$^{27-}$.

some of which may serve as sites for Ca^{2+} bridging interactions that enhance the oligomerization process (Fig. 1B), and others which may be relevant for the ACC formation/transformation processes.[19,20]

Bioinformatics and the identification of aggregation sites in Asprich sequences

The fact that apo-Asprich "3" oligomerizes in solution (Fig. 1A) implies that there are key features at the primary sequence level that facilitate protein–protein interactions between intrinsically disordered, highly charged Asprich "3" molecules. However, identifying these regions within an intrinsically disordered protein assembly represents a challenging experimental task. For this reason, we turn to bioinformatic prediction algorithms which can identify regions that are prone to aggregation within Asprich "3" and the other members of the Asprich sequence library.[13] This data can then be used to focus our future experimental efforts on specific regions of potential interest.

The first method we employed was the IUPRED intrinsic disorder[34] and the ANCHOR protein–protein binding[35] algorithms (Fig. 3). This tandem method first identifies intrinsically disordered sequence regions based upon amino acid composition (IUPRED), then, identifies regions within these disordered segments that can be modulated by their environment to fold (ANCHOR). This approach was originally utilized to identify sensitive disordered regions which would bind and fold when contacting a globular protein,[34,35] but in our instance, we are using these methods to probe the potential folding sensitivity of Asprich sequence domains without regard for the nature of the stabilizing target. As shown in Fig. 3, all ten members of the Asprich family were identified as intrinsically disordered (*i.e.*, IUPRED disorder probability > 0.5), with no evidence of globular folding. These findings are

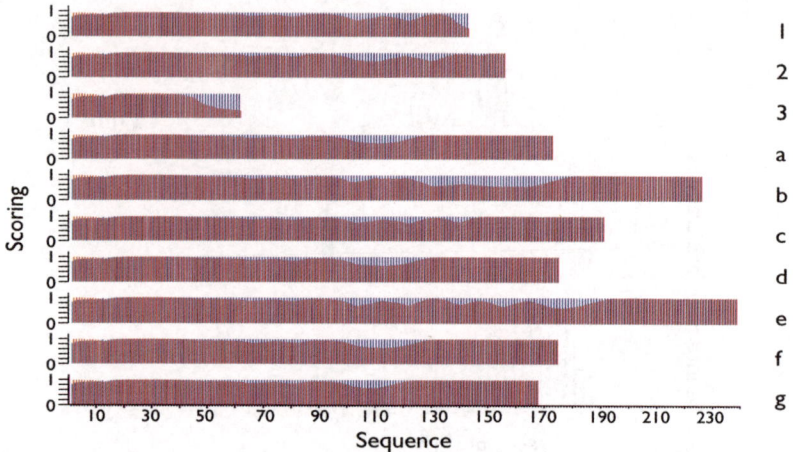

Fig. 3 Overlay histogram plot of IUPRED (red) long range disorder and ANCHOR (blue) calculated propensities for the Asprich family. Protein identifiers are shown on the right hand side of the plot.

consistent with earlier Asprich bioinformatics studies.[20] The ANCHOR method found that the folding sensitivities for all Asprich proteins are very high (0.9–1.0) and comprise the entire length of each primary sequence. This is in agreement with earlier studies that demonstrated that the structure-stabilizing solvent, 2,2,2-trifluoroethanol, induced significant folding induction within the entire apo-Asprich "3" molecule,[19] and with the results conducted on other polyelectrolyte biomineralization protein sequences.[42,43] We conclude that all Asprich protein sequences are highly disordered and conformationally sensitive to their environment and thus prone to interaction and folding induction processes. As a consequence, individual Asprich "3" molecules would be prone to aggregate with each other in order to achieve internal stability (Fig. 1A).

However, there may exist other molecular features within all Asprich sequences that can also contribute to oligomerization. In particular, it is known that short 4–6 AA sequence regions found in amyloid and prion polypeptides can induce cross beta strand pairing that leads to fibril formation.[36–38] Interestingly, we know from previous CD studies that apo- and Ca^{2+} bound Asprich "3"[19] and Asprich fragments[20] both contain minor beta strand content. In addition, NMR experiments established that extended beta strand structures existed within the highly conserved 12 AA N-terminal Asprich sequence.[20] Thus, we employed a suite of bioinformatic prediction algorithms (BETASCAN,[36] AGGRESCAN,[37] and FOLD-AMYLOID[38]) (Fig. 4) to scan all Asprich protein sequences for aggregation-prone beta strand regions. The AGGRESCAN algorithm identified an above-threshold aggregation-prone tetrad sequence (-FKRS-) within the highly conserved 12 AA N-terminal cationic domain (KPVFKRSLSDPS) that is common to all Asprich sequences (Fig. 4A, B). BETASCAN identified -KPVFKRSL- in this same N-terminal region as a putative beta strand forming sequence with a propensity for parallel strand–strand interprotein interaction (Fig. 4C).[36] The BETASCAN algorithm also identified two other parallel beta strand sequence regions within Asprich "3" but with lower probabilities: D19-E32 and D43-D46, which occur within the highly unstable fold induction region identified by ANCHOR (Fig. 3). FOLD-AMYLOID, an algorithm that scores amyloidogenic or prion-like aggregation propensities within polypeptide sequences,[38] also identified a putative aggregation dipeptide region, K5-R6, within this same conserved 12 AA N-terminal sequence (Fig. 4D). Thus, three different search methods yield the same results: the 12 AA N-terminal region of the Asprich family contains a putative amyloid- or prion-like beta strand

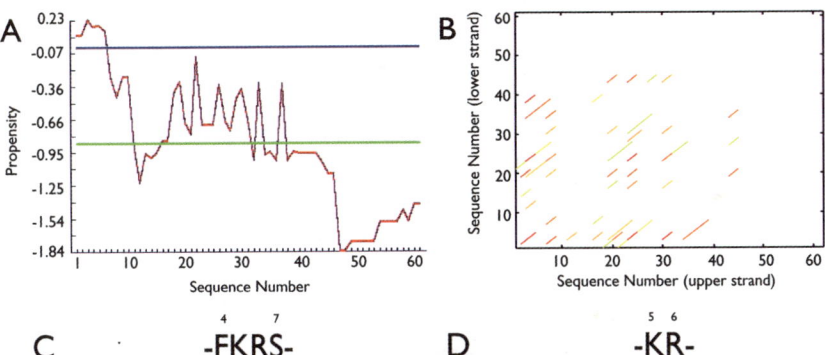

Fig. 4 (A) Asprich "3" AGGRESCAN aggregation propensity scale plot. The red plot represents the calculated aggregation propensities for each Asprich "3" residue. Propensity values that exceed the aggregation threshold (blue line) have a high propensity for aggregation. The green line represents the overall average aggregation propensity of the entire Asprich "3" sequence. (B) Asprich "3" BETASCAN local maximum likelihood beta-strand pair 2-D plot based on pairwise probabilities. The probability score is color-coded and ranges from highest probability (red) to lowest probability (yellow). (C) Asprich family AGGRESCAN-identified high propensity aggregation sequence motif. (D) Asprich family FOLD-AMYLOID identified amyloid-like sequence. The complete, leader sequence–deleted Asprich "3" sequence is presented at the top of the Figure.

aggregation site.[36–38] These bioinformatic results correlate with the CD and NMR data obtained for the 12 AA N-terminal region.[19,20] Combined with the results obtained by IUPRED and ANCHOR, we conclude that all Asprich sequences possess aggregation-prone regions that could drive some stage of the oligomerization process (Fig. 1).

Molecular features of Asprich "3" assemblies

Recent studies of monomeric tooth enamel amelogenin[32] and nacre layer AP7[47] proteins reveal that Pro imido *cis-trans* ring interconversion exists within these proteins and induces conformational exchange phenomena in nearby sequence blocks.[32,47] This Pro interconversion process may represent an important molecular feature for the function of these proteins in their respective biomineral-specific systems. Intriguingly, the highly conserved 12 AA N-terminal segment found in all Asprich sequences also contains two Pro residues (P2, P11), and it has been shown in a model Asprich fragment peptide that P11 undergoes *cis-trans* isomerization and induces conformational exchange phenomenon for upstream residues F4, S7, and S9.[20] However, this process has yet to be confirmed in a full-length Asprich protein, let alone an Asprich protein assembly.

With this in mind, we utilized ultra-high field ^{1}H NMR solution experiments (TOCSY, NOESY) to determine the molecular configuration of these assemblies under very dilute conditions (Fig. 5, 6). We tentatively assigned the proton spectra based upon the amino acid composition of Asprich "3"[13,19] and the known ^{1}H NMR chemical shifts for amino acids in Asprich fragment peptides.[20] We discovered two interesting molecular traits. The first trait is the low number of detectable TOCSY sidechain crosspeaks (30 out of the expected 61 crosspeaks, or < 50%, Fig. 5). This is far fewer than the number one would expect for the monomeric form of Asprich "3", and we believe that the absence of expected NMR signals reflect a combination of two phenomena: (1) the attenuation of scalar and NOE crosspeak intensities due to the formation of intermolecular contacts and restricted chain

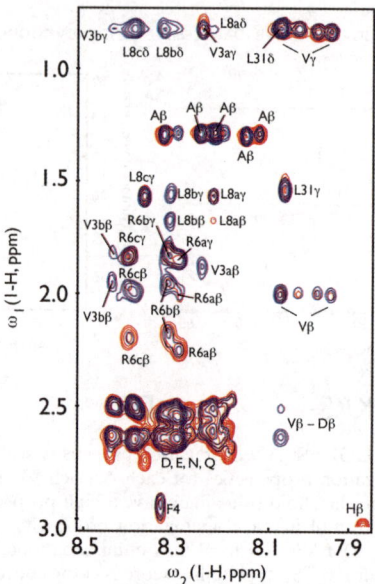

Fig. 5 900 MHz 1-H NMR TOCSY (red, scalar coupling-specific resonances)—NOESY (blue, dipolar coupling and conformational exchange resonances) overlay spectra of 100 micromolar Asprich 3. The overlay spectra portray the backbone NH⟨ –sidechain CH fingerprint region. Spin assignments are tentative and based upon Asprich "3" amino acid composition and known 1-H NMR resonances found in conserved Asprich fragment peptides (Ref. 20). Multiple crosspeaks representing multiple conformational states of V3, L8, and R6 spin systems are labeled in lowercase letters in the spectra. The collective, unassigned Asp, Glu, Asn, Gln (D,E,N,Q) spin system region is shown.

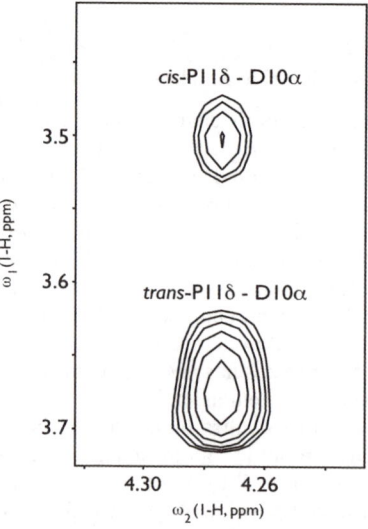

Fig. 6 900 MHz 1-H NOESY spectra of 100 micromolar Asprich 3, focusing on the P11 imido ring CHδ fingerprint region and showing the *cis-trans* NOE crosspeaks.

mobility within certain regions of assembled Asprich "3" molecules.[31,32,42,43] (2) The high redundancy of Asp and Glu residues in the Asprich "3" sequence which increases resonance frequency overlap in the NMR spectral region for Asp, Glu,

Asn, and Gln (Fig. 5).[20,42,43] The reader should note that these two problems prevented us from obtaining complete sequence-specific resonance assignments and structural details for Asprich "3" protein molecules at this time. Nonetheless, our NMR data verifies that Asprich "3" oligomers are stabilized by extensive intermolecular protein–protein contacts, in agreement with the ANCHOR predictions (Fig. 3). In addition, these assemblies still retain some degree of molecular motion, which may be important for Asprich function.

The second trait is the presence of resolvable ^{1}H NMR resonances (Fig. 5), which indicates that the Asprich "3" oligomers possess mobile backbone segments whose ^{1}H spins exhibit nearly equal degrees of freedom in all directions (*i.e.*, isotropic). We believe that these mobile segments arise as a result of two separate molecular phenomena: (1) electrostatic repulsion arising from Asp, Glu carboxylate groups, whose negative charges would induce sidechain–sidechain repulsion under low ionic strength conditions,[42,43] and (2) Pro *cis-trans* isomerization and conformational exchange processes[32,47] occurring within the 12-AA N-terminal region.[20]

Further analyses verifies that Pro *cis-trans* isomerization is one source of the segmental mobility in Asprich "3" assemblies. We successfully identified the P11 residue of the conserved 12 AA N-terminal region and NOESY analysis confirmed that P11 undergoes *cis-trans* isomerization (Fig. 6), with calculated *cis* : *trans* populations of 27% and 73% respectively. These calculated ratios are slightly different from the expected 20% : 80% ratio typically found in most folded proteins,[48–50] and we attribute this discrepancy to the unusual conformational behavior exhibited by intrinsically disordered highly charged proteins.[32,42,43,47] Thus, P11 is conformationally active not only in a peptide fragment,[20] but also within a protein assembly, and this may have functional significance.

Our NOESY experiments also detected multiple sidechain and backbone spin populations for V3 (a, b), R6 (a,b,c), and L8 (a,b), which reside upstream of P11 (Fig. 5). Interestingly, R6 and L8 are predicted to be integral components of the aggregation-prone motif within the N-terminus (Fig. 4). As reported elsewhere, Pro imido ring interconversion phenomenon can give rise to duplicated NMR crosspeaks for residues on the N-terminal side of the mobile Pro residue.[32,47] To check this, we quantitated the V3, R6, L8, and NOE crosspeak intensity ratios and found them to be in close agreement with the calculated P11 *cis* : *trans* ratio of 27% : 73% (data not shown). This confirms that these duplicated spin populations (Fig. 5) arise from conformational exchange processes induced by downstream P11 *cis-trans* interconversion (Fig. 6). Note that there may be other upstream residues within the 12 AA N-terminal domain experiencing conformational exchange as well, such as K1 and K5, but we are unable to detect these at present. From this data, we conclude the Pro induced *cis-trans* interconversion and upstream conformational exchange processes occur within the 12 AA N-terminal domain of apo-Asprich fragment sequences[20] and apo-Asprich "3" protein assemblies.

4. Discussion

Recent studies of nacre-associated protein sequences have established the presence of self-assembling protein supramolecular complexes that regulate the formation of aragonite.[29,30,39,47] Now, our present study reveals that a similar scenario exists for prismatic layer-associated proteins. Under mineralization assay conditions Asprich "3" oligomerizes to form heterogeneously-sized protein particles (Fig. 1A). This process is pH dependent and coincides with the pK_a (protonation/deprotonation) of His imidazole, which resides at position 61 at the C-terminal end. These oligomers increase in dimension (25–95%) when Ca^{2+} ions are present (Fig. 1B), and given that Asprich "3" can form protein–metal ion clusters in solution (Fig. 2), it is clear that Ca^{2+} ion binding promotes Asprich "3" supramolecular assembly formation, possibly through Ca^{2+}-mediated carboxylate sidechain–sidechain bridging interactions.[20,42,43] The strong sequence and intrinsic disorder similarities between Asprich "3" and the other nine members of the Asprich family (Fig. 4)[13] suggest that oligomerization may be a common trait within this prismatic protein family. If true, then the presence of randomly distributed intracrystalline inclusions may arise as a result of Asprich protein oligomerization during some phase of prismatic calcite formation.[18]

Our ESI-MS experiments document the ability of Asprich "3" molecules to bind Ca^{2+} (Fig. 2; Table S1, Supporting Information†). The putative sites for metal ion binding would be the numerous Asp, Glu carboxylate residues that populate the Asprich sequence family.[13,19] If individual molecules of Asprich "3" are capable of forming protein–metal ion clusters (Fig. 2), then it stands to reason that the same process occurs within Asprich oligomeric assemblies as well. This would lead to a substantial protein–Ca^{2+} clustering effect within these assemblies, which, in turn, could have a profound impact on the formation of ACC by (a) kinetically regulating the availability of free Ca^{2+} for ACC pre-nucleation cluster formation,[50,51] and/or (b) controlling the fate of ACC clusters during the post-nucleation process.[50,51] This is an intriguing concept that will be followed up in subsequent research reports using pH-controlled Ca^{2+} and carbonate ion titration measurements.

What are the driving forces that are involved in Asprich protein oligomerization (Fig. 1A)? Our research suggests that there are two major factors. First, the high degree of intrinsic disorder within the Asprich sequence family generates conformational instability and a high interaction potential for other molecular species, such as other Asprich molecules or mineral clusters,[50,51] in order to achieve internal stability (Fig. 3). These bioinformatics findings are supported by our earlier solvent-stabilization Asprich "3" experiments[20] and by our current NMR experiments, which demonstrate that >50% of the protein molecule is inhomogeneously broadened by intermolecular contacts (Fig. 5). Thus, Asprich sequences are conformationally unstable but can achieve internal stability *via* intermolecular contacts within oligomeric assemblies (Fig. 5). The second factor is a truly unique finding: the identification of small aggregation-prone beta strand motifs (Fig. 4) within the highly conserved 12 AA N-terminal region of each Asprich protein. This correlates with earlier experimental studies that established beta strand traits in this sequence segment[20] and within Asprich "3" itself.[19] It is known that small amyloid- or prion-like regions are important sites for strand–strand contact that drives polypeptide "nucleation" or conformational rearrangements in amyloid sequences.[36–38] In the case of the Asprich protein family, these short aggregation prone or amyloid-like regions may play a similar role, either in the early nucleation stages of protein oligomer formation, or, later on during the final stages of oligomer assembly. Other interactions, such as Ca^{2+}–protein metalation (Fig. 1B, 2),[19,20,27] intrinsic disorder (Fig. 3),[22–24] or other non-bonding intermolecular actions would further contribute to the stabilization of these protein oligomers in solution over time. Additional studies will be required to deduce the full range of intermolecular interactions that drive and stabilize Asprich oligomer formation.

Our NMR studies (Fig. 5, 6) reveal that the oligomeric form of Asprich "3" possesses a mobile 12 AA N-terminal region that is common to all Asprich proteins.[13,19,20] We identified that P11 imido ring *cis-trans* interconversion occurs within this 12 AA N-terminal domain (Fig. 6). This isomerization phenomenon involves slightly different *cis-trans* populations (27% *cis* : 73% *trans versus* the typical 30% : 70% found in most proteins)[48–50] which may be related to the intrinsic disorder and electrostatics of this protein. The fact that Pro *cis-trans* imido ring isomerization has now been identified in a third aggregation-prone biomineralization protein[32,47] strongly argues for the participation of Pro *cis-trans* isomerization in either the protein assembly process or in some feature of the assembly itself.[32,47] Additional experiments will be performed in order to firmly establish the role that this dynamic process plays in the biomineralization scheme.

Our NMR experiments identified multiple conformationally distinct V3, R6, and L8 spin populations within Asprich "3" oligomers that quantitatively correspond to the P11 *cis-trans* isomerization process (Fig. 5). Although R6 and L8 have been identified as labile species within Asprich "3" protein assemblies, they are also predicted to be key components in the N-terminal aggregation-prone region within Asprich proteins (Fig. 4). This apparent contradiction can be explained in the following way. The formation of amyloid fibrils is a multistep process in which a number of intermediate aggregates are formed, and the process involves the coalescence of monomers to form small oligomeric aggregates.[52–54] These small oligomers then grow further in size and complexity over time.[52–54] Considering that we are examining oligomeric Asprich "3" assemblies that have had time to mature (Fig. 5, 6), it is plausible that R6 and L8 may have played a role in the early stages of the protein assembly process, but are no longer involved in stabilizing the assembly itself, due to the presence of other stabilizing interactions (*e.g.*, intrinsic disorder, ion pairing, hydrogen bonding) that are now maintaining the integrity of the assembly. In this scenario, R6 and L8 would be available for other purposes. Alternatively, the 12 AA N-terminal amyloid-like sequence may not form a true cross-beta spine intermolecular structure,[36–38] but instead forms a structure that allows R6 and L8 to move more freely within the assembly. Obviously, we are still in the early stages of understanding Asprich "3" oligomerization and function, and thus additional studies will be required to resolve these issues.

5. Conclusions

Asprich "3" oligomerizes in solution to form heterogeneously-sized protein assemblies, and this oligomerization process may explain the presence of randomized organic intracrystalline deposits within prismatic calcite rods in *Atrina rigida*. The ability of this protein to bind Ca^{2+} not only leads to enhancement of this oligomerization process, but also provides a plausible means for Asprich assemblies to regulate the availability of Ca^{2+} ions and influence either the formation of ACC prenucleation clusters or some other aspect of ACC post-nucleation development. The oligomerization process draws upon two unique sequence features that are found in all Asprich sequences: the widespread presence of intrinsic disorder and environmentally sensitive domains which are driven to bind and subsequently fold, and, the presence of short, amyloid- or prion-like beta strand regions that are aggregation-prone. The actual Asprich "3" apo-assemblies exhibit unique, highly mobile polypeptide chain regions, and in particular we identified a mobile 12 AA N-terminal segment that is driven by P11 *cis-trans* isomerization. This interconversion process induces conformational exchange upstream at V3, R6, and L8, with R6 and L8 predicted to be key residues located within the putative cross beta strand amyloid- or prion-like aggregation region. Collectively, these results document that an ACC-associated prismatic protein, Asprich "3", is quite complex and utilizes intrinsic disorder and amyloid-like sequences to self-assemble into oligomers that could contribute to ACC formation and prismatic calcite crack-propagation resistance.

Notes and references

1 X. Li, W. C. Chang, Y. J. Chao, R. Wang and M. Chang, *Nano Lett.*, 2004, 613.
2 X. Li and P. Nardi, *Nanotechnology*, 2004, **15**, 211.
3 Q. L. Feng, H. B. Li, G. Pu, D. M. Zhang, F. Z. Cui and H. D. Li, *J. Mater. Sci.*, 2000, **35**, 3337.
4 E. Zolotoyabko and B. Pokroy, *Cryst. Eng. Commun.*, 2007, **9**, 1156.
5 B. Pokroy, A. N. Fitch and E. Zolotoyabko, *Adv. Mater.*, 2006, **18**, 2363.
6 B. Pokroy, M. Kapon, F. Marin, N. Adir and E. Zolotoyabko, *Proc. Natl. Acad. Sci. U. S. A.*, 2007, **104**, 7337.
7 Y. Dauphin, *J. Biol. Chem.*, 2003, **278**, 15168.
8 L. Addadi, J. Moradian, E. Shay, N. G. Marroudas and S. Weiner, *Proc. Natl. Acad. Sci. U. S. A.*, 1987, **84**, 2732.
9 B. Pokroy, A. N. Fitch, F. Marin, M. Kapon, N. Adir and E. Zolotoyabko, *J. Struct. Biol.*, 2006, **155**, 96.
10 H. A. Lowenstam and S. Weiner, *On biomineralization*, Oxford Press, New York, NY, 1989.

11 D. Tsukamoto, I. Sarashina and K. Endo, *Biochem. Biophys. Res. Commun.*, 2004, **320**, 1175.
12 F. Marin, R. Amons, N. Guichard, M. Stigter, A. Hecker, G. Luquet, P. Layrolle, G. Alcaraz, C. Riondet and P. Westboek, *J. Biol. Chem.*, 2005, **280**, 33895.
13 B. A. Gotliv, N. Kessler, J. L. Sumerel, D. E. Morse, N. Tuross, L. Addadi and S. Weiner, *ChemBioChem*, 2005, **6**, 304.
14 F. Nudelman, H. H. Chen, H. A. Goldberg, S. Weiner and L. Addadi, *Faraday Discuss.*, 2007, **136**, 9.
15 S. W. Lee, Y. M. Kim, H. S. Choi, J. M. Yung and C. S. Choi, *Protein J.*, 2006, **25**, 288.
16 C. Zhang, L. Cie, J. Huang, X. Liu and R. Zhang, *Biochem. Biophys. Res. Commun.*, 2006, **344**, 735.
17 M. Yano, K. Nagai, K. Morimoto and H. Miyamoto, *Comp. Biochem. Physiol., Part B: Biochem. Mol. Biol.*, 2004, **144**, 254.
18 H. Li, H. L. Xin, M. E. Kunitake, E. C. Keene, D. A. Muller and L. A. Estroff, *Adv. Funct. Mater.*, 2011, **21**, 2028.
19 M. Ndao, E. Keene, F. A. Amos, G. Rewari, C. B. Ponce, L. Estroff and J. S. Evans, *Biomacromolecules*, 2010, **11**, 2539.
20 K. Delak, S. Collino and J. S. Evans, *Biochemistry*, 2009, **48**, 3669.
21 K. Delak, J. Giocondi, C. Orme and J. S. Evans, *Cryst. Growth Des.*, 2008, **8**, 4481.
22 A. K. Dunker, C. J. Oldfield, J. Meng, P. Romero, J. Y. Yang, J. W. Chen, V. Vacic, Z. Obradovic and V. N. Uversky, *BMC Genomics*, 2008, **9**, 1.
23 P. Tompa, *Trends Biochem. Sci.*, 2002, **27**, 527.
24 C. J. Oldfield, Y. Cheng, M. S. Cortese, C. J. Brown, V. N. Uversky and A. K. Dunker, *Biochemistry*, 2005, **44**, 1989.
25 K. Strupat, *Methods Enzymol.*, 2005, **405**, 1.
26 Q. A. De Paula, J. B. Mangrum and N. P. Farrell, *J. Inorg. Biochem.*, 2009, **103**, 1347.
27 J. A. Loo, *Int. J. Mass Spectrom.*, 2001, **204**, 113.
28 W. Schärtl, *Light scattering from polymer solutions and nanoparticle dispersions*, Springer-Verlag, Heidelberg, Germany, 1st edn, 2007.
29 F. F. Amos, M. Ndao, C. B. Ponce and J. S. Evans, *Biochemistry*, 2011, **50**, 8880.
30 C. B. Ponce and J. S. Evans, *Cryst. Growth Des.*, 2011, **11**, 4690.
31 M. Ndao, K. Dutta, K. Bromley, Z. Sun, R. Lakshminarayanan, G. Rewari, J. Moradian-Oldak and J. S. Evans, *Protein Sci.*, 2011, **20**, 724.
32 K. Delak, C. Harcup, R. Lakshminarayanan, S. Zhi, Y. Fan, J. Moradian-Oldak and J. S. Evans, *Biochemistry*, 2009, **48**, 2272.
33 T. N. Petersen, S. Brunak, G. von Heijne and H. Nielsen, *Nat. Methods*, 2011, **8**, 785.
34 Z. Dosztányi, V. Csizmók, P. Tompa and I. Simon, *Bioinformatics*, 2005, **21**, 3433.
35 B. Mészáros, I. Simon and Z. Dosztányi, *PLoS Comput. Biol.*, 2009, **5**, 1.
36 A. W. Bryan, M. Menke, L. J. Cowen, S. L. Lindquist and B. Berger, *PLoS Comput. Biol.*, 2009, **5**, 1.
37 O. Conchillo-Sole, N. S. de Groot, F. X. Aviles, J. Vendrell, X. Daura and S. Ventura, *BMC Bioinformatics*, 2007, **8**, 65.
38 S. O. Garbuzynskiy, M. Y. Lobanov and O. V. Galzitskaya, *Bioinformatics*, 2010, **26**, 326.
39 F. F. Amos and J. S. Evans, *Biochemistry*, 2009, **48**, 1332.
40 G. J. Bartlett, C. T. Porter, N. Borkakoti and J. M. Thornton, *J. Mol. Biol.*, 2002, **324**, 105.
41 F. F. Amos, C. B. Ponce and J. S. Evans, *Biomacromolecules*, 2011, **12**, 1883.
42 J. S. Evans, T. Chiu and S. I. Chan, *Biopolymers*, 1994, **34**, 1359–1375.
43 J. S. Evans and S. I. Chan, *Biopolymers*, 1994, **34**, 507.
44 S. Yi, B. L. Boys, A. Brickenden, L. Konermann and W. Y. Choy, *Biochemistry*, 2007, **46**, 13120.
45 V. N. Uversky, S. E. Permyakov, V. E. Zagranichny, I. L. Rodionov, A. L. Fink, A. M. Cherskaya and E. A. Permyakov, *J. Proteome Res.*, 2002, **1**, 149.
46 Note that in previous studies of Asprich conserved N- and C-terminal sequences (F1, F2 polypeptides), these isotope distribution peaks were also observed but erroneously attributed by us to sodium–peptide adduct species (see Ref. 20).
47 S. Collino, I. W. Kim and J. S. Evans, *Biochemistry*, 2008, **47**, 3745.
48 L. K. Nicholson and K. P. Lu, *Mol. Cell*, 2007, **25**, 483.
49 A. H. Andreotti, *Biochemistry*, 2003, **42**, 9515.
50 D. Gebauer and H. Colfen, *Nano Today*, 2011, **6**, 564, 564. 1819.
51 D. Gebauer, A. Volkel and H. Colfen, *Science*, 2008, 322.
52 P. Sarkar, C. Reichman, T. Saleh, R. Birge and C. G. Kalodimos, *Mol. Cell*, 2007, **25**, 413.
53 M. Amaro, D. J. S. Birch and O. J. Rolinski, *Phys. Chem. Chem. Phys.*, 2011, **13**, 6434.
54 M. J. Thompson, S. A. Sievers, J. Karanicolas, M. I. Ivanova, D. Baker and D. Eisenberg, *Proc. Natl. Acad. Sci. U. S. A.*, 2006, **103**, 4074.

PAPER

GSP-37, a novel goldfish scale matrix protein: identification, localization and functional analysis

Kousei Miyabe,[†a] Hiroki Tokunaga,[†a] Hirotoshi Endo,[†a] Hirotaka Inoue,[a] Michio Suzuki,[ab] Naoaki Tsutsui,[a] Naoki Yokoo,[b] Toshihiro Kogure[b] and Hiromichi Nagasawa[*a]

Received 16th March 2012, Accepted 11th May 2012
DOI: 10.1039/c2fd20051a

A novel noncollagenous acidic protein was identified from the scales of goldfish (*Carassius auratus*), a freshwater teleost. Using an *in vitro* calcium phosphate crystallization assay, the EDTA-soluble fraction from these scales was screened for crystallization inhibitory activity, and a highly phosphorylated glycoprotein, named goldfish scale protein (GSP)-37, was isolated through 5 HPLC purification steps. The cDNA for GSP-37 has an open reading frame encoding a precursor protein, consisting of a signal peptide and GSP-37, with 19 and 137 amino acid residues, respectively. The C-terminal region of GSP-37 contains the RGD consensus sequence for cell adhesion. Although native GSP-37 strongly inhibited crystallization, alkaline phosphatase treatment dramatically reduced its inhibitory activity. Reverse transcription-PCR analysis revealed that *GSP-37* is expressed only in scales but not in other calcified tissues, bones or pharyngeal teeth. *In situ* hybridization demonstrated that *GSP-37*-expressing cells were localized in the central regions of regenerating scales, where organic matrices were actively synthesized and were not stained with either alkaline phosphatase or tartrate-resistant acidic phosphatase, osteoblastic and osteoclastic cell markers, respectively. Immunohistochemical analyses showed that GSP-37 is localized in the uppermost region of the bony layer of the scale, which is thought to correspond to the enamel or enameloid layer of vertebrate teeth. All these data strongly indicate that GSP-37 is deeply associated with calcification in fish scales.

1. Introduction

The teleost fish scale shares a common origin with 2 other vertebrate mineralized tissues, bones and teeth. These 3 hard tissues are mineralized with hydroxyapatite (HA) crystals, predominantly containing type-I collagen. HA, $Ca_{10}(PO_4)_6(OH)_2$, is the most stable form of calcium phosphate crystal and is specifically observed in mineralized hard tissues, whereas type-I collagen is widely distributed in most soft tissues. Therefore, calcium phosphate precipitation in those tissues is assumed to be initiated and regulated by noncollagenous proteins (NCPs). Over the past 3 decades, many studies have been conducted on NCPs such as osteocalcin (OCN), osteopontin (OPN), osteonectin (ON), bone sialoprotein (BSP), dentin matrix

---

[a]*Department of Applied Biological Chemistry, Graduate School of Agricultural and Life Sciences, University of Tokyo, Tokyo 113-8657, Japan. E-mail: anagahi@mail.ecc.u-tokyo.ac.jp*
[b]*Department of Earth and Planetary Science, Graduate School of Science, University of Tokyo, Tokyo 113-0033, Japan*

† These authors contributed equally.

protein 1 (DMP1), dentine sialo-phosphoprotein (DSPP), enamelin (ENAM), amelogenin (AMEL) and others.[1]

Among NCPs, small integrin-binding ligand, *N*-linked glycoprotein (SIBLING) family proteins are principally expressed in bones and teeth.[2] This family includes BSP, DMP1, DSPP, ENAM, OPN and matrix extracellular phosphoglycoproteins. All of these family members share common features such as post-translational modifications (*i.e.*, glycosylation and phosphorylation), predominant expression in mineralized tissues, a C-terminal RGD (Arg-Gly-Asp) motif that is responsible for integrin binding, and similar gene organization (exon-intron structure).[2]

Contrary to the many studies identifying and characterizing NCPs in bones and teeth, to date, only 2 NCPs have been identified in teleost scales. OCN was directly identified from the extracts of bluegill (*Lepomis macrochirus*)[3] and carp (*Cyprinus carpio*) scales.[4] This protein, also referred to as bone gla protein (BGP or BGLAP), exhibits high affinity to HA.[5] OCN is principally synthesized by the osteoblast, and its deficiency promotes aberrant bone formation.[6] ON, also referred to as secreted protein acidic and cysteine-rich (SPARC), is another major noncollagenous constituent of bone expressed in the scales of goldfish (*Carassius auratus*).[7] This protein has high affinity for both HA and type-I collagen.[8] Recently, the spatiotemporal expressions of *sparc* and *bgp* were examined during regeneration of goldfish scales.[9] Expressions of these 2 genes were preceded by an osteoblast differentiation factor, *runt-related gene 2* (*runx2*), suggesting that the molecular mechanism regulating scale mineralization is similar to that of mammalian bone mineralization.[9]

The teleost scale is generally composed of 3 layers: the outer limiting layer, external layer and fibrillary plate, in this order from the apical to the basal surfaces. Calcification profiles of these layers are fundamentally different from one another, and significant calcification is limited only to the 2 outer layers.[10] The external layer, the first layer to be calcified, consists of randomly oriented fine collagen fibrils mineralized with HA crystals. Calcification in this layer occurs in the presence of small matrix vesicles (MVs), in which the initial crystal deposition occurs. Such vesicle-associated mineralization is also observed in bones and dentine.[11] The outer limiting layer is also extensively mineralized with HA, but has a characteristic ultrastructure different from that of the external layer and is nearly or completely absent from collagen fibrils. In addition, mineralization takes place without mediation of MVs. These features more closely resemble the enamel or enameloid layer, rather than the bone or dentine layer. In contrast to these 2 outer layers, the fibrillary layer, underlying the external layer, is principally composed of regularly organized collagen fibrils and is partially calcified.[12–14]

Teleost scales can be readily regenerated and replaced by new ones when normal (ontogenetic) scales are spontaneously removed or artificially ablated. Therefore, detailed observations of the regeneration process have been achieved in various species (reviewed by Bereiter-Hahn and Zylberberg, 1993).[10] In the case of goldfish, multiple undifferentiated cells first form clusters in the scale pocket, and the synthesis of scale matrices begins. The secreted materials for the external layer then grow and merge into one thin plate, and the differentiated cells surround the plate. Next, calcification of the external layer takes place at multiple sites, and the basal cells begin to synthesize the fibrillary plate.[15] Although Ohira and his colleagues did not discuss the outer limiting layer, reviews on the topic conclude that this layer must be formed in the late regenerating stage.[10,16]

Based on structural observations, it is generally accepted that the teleost scale is evolutionarily closer to the tooth than to bone.[17] In some aspects, however, the scale also has several features analogous to the bone. Persson *et al.* (1995) reported that the expression of an mRNA encoding ON is suppressed by estradiol-17β, suggesting that hormonal regulation of calcium homeostasis is achieved using calcium reserved in scales like bones in tetrapoda.[18] Despite such studies, it is still difficult to discern why and how teleosts fabricate bones, teeth and scales separately. This is partially because we have not identified scale-specific organic matrices, which would allow us to speculate the evolutionary relationship among the 3 mineralized tissues.

In this study, to investigate the molecular mechanism regulating teleost fish scale formation and calcification, we attempted to identify novel matrix proteins in the scales of goldfish (*C. auratus*).

2. Experimental

2.1 Experimental animals

Goldfish, *Carassius auratus*, were purchased from a local commercial source and kept in a 12 L tank at 20 °C with gentle aeration until sampled. Mature specimens (8–12 cm in body length) of both sexes were indiscriminately used in all the experiments in the present study.

2.2 Calcium phosphate crystallization assay

Inhibitory activities of the crude extracts or the collected fractions at each purification step were measured according to the method reported previously[19] with some modifications. Briefly, crude extracts or purified samples were dissolved in 50 μL of a buffer (150 mM NaCl/50 mM HEPES/10 mM Na_2HPO_4 (pH 7.5)), and mixed with 100 μL of a buffer containing Ca^{2+} (150 mM NaCl/50 mM HEPES/10 mM $CaCl_2$ (pH 7.5)) at 37 °C. Changes in the level of turbidity of the mixed solution were monitored every minute for 5 min by the absorbance at 570 nm. In the control experiment, samples were omitted from the solution.

2.3 Preparation of EDTA-soluble matrices and purification of GSP-37

GSP-37 was extracted from the goldfish scales and purified by 5 steps of HPLC as described below. The outline of the purification scheme is shown in Fig. 1. At each step, crystallization inhibitory activity was examined using the calcium phosphate crystallization assay described above.

Goldfish scales were washed with distilled water (DW) by vortex mixing for 3 times and washed in 20 volumes of 10% NaCl with gentle stiring at 4 °C for 2 days. Then, the scales were decalcified in 0.5 M EDTA (pH 8.0) at 4 °C for 2 days. The EDTA-insoluble materials were removed by centrifugation, and the supernatant was desalted using a Sep-pak C_{18} column (Waters, Massachusetts, USA), lyophilized, and dissolved in 20 mM Tris-HCl (pH 8.0). The EDTA-soluble matrices were

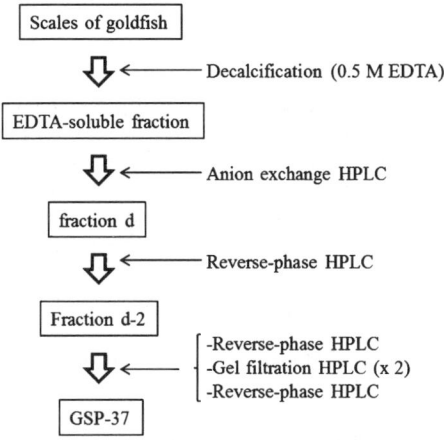

Fig. 1 Purification scheme of GSP-37.

then subjected to anion-exchange high performance liquid chromatography (HPLC) with a SuperQ-5PW column (7.5 × 7.5 mm, TOSOH, Tokyo, Japan). Elution was performed successively with 20 mM Tris-HCl (pH 8.0) for 5 min, 0–0.3 M NaCl/20 mM Tris-HCl (pH 8.0) for 45 min, then 0.3–1.0 M NaCl/20 mM Tris-HCl (pH 8.0) for 2 min, and 1.0 M NaCl/20 mM Tris-HCl (pH 8.0) for 8 min at a flow rate of 1 mL min^{-1}. The elution was monitored by the absorbance at 225 nm. Fractions were collected manually, and the fraction with the highest inhibitory activity (fraction d, retention time of 55–58 min) was subjected to reverse-phase HPLC with an ODP-50 column (4.6 × 150 mm, Shodex, Showa Denko, Tokyo, Japan). Elution was performed successively with 10 mM ammonium acetate (pH 8.8) for 5 min, 0–48% acetonitrile/10 mM ammonium acetate (pH 8.8) for 30 min, 48–80% acetonitrile/10 mM ammonium acetate (pH 8.8) for 2 min, and 80% acetonitrile/10 mM ammonium acetate (pH 8.8) for 8 min at a flow rate of 0.8 mL min^{-1}. Fractions were collected manually and the fraction with the highest inhibitory activity (fraction d2, retention time of 16–24 min) was subjected to the second round of reverse-phase HPLC with a PEGASIL-300 C4P column (4.6 × 150 mm, Senshu Kagaku, Tokyo, Japan). Elution was done successively with 0.05% trifluoroacetic acid (TFA) for 5 min, 0–48% acetonitrile/0.05% TFA for 30 min, 48–80% acetonitrile/0.05% TFA for 2 min, and 80% acetonitrile/0.05% TFA for 8 min at a flow rate of 1 mL min^{-1}. The fraction with the highest inhibitory activity (fraction d2-1, retention time of 19–23 min) was collected and subjected to gel filtration HPLC with a PROTEIN KW-803 column (8.0 × 300 mm, Shodex) in 150 mM NaCl/50 mM HEPES/10 mM Na$_2$HPO$_4$ (pH 7.5). This gel filtration step was repeated once more, and at each step, the materials from the most abundant peaks were collected. The collected fraction was put on the final round of purification using a CAPCELL-PAK C$_{18}$ (2.0 × 250 mm, SHISEIDO, Kanagawa, Japan). Elution was performed with 0.05% TFA for 5 min, 0–40% acetonitrile/0.05% TFA for 50 min, 40–80% acetonitrile/0.05% TFA for 2 min, and 80% acetonitrile/0.05% TFA for 5 min at a flow rate of 0.2 mL min^{-1}. Fractions were collected every minute from 44 to 53 min after sample injection.

2.4 Preparation of deglycosylated GSP-37 (ΔN-GSP-37) and dephosphorylated GSP-37 (ΔP-GSP-37)

To obtain ΔN-GSP-37, the intact GSP-37 was treated with PNGase F (New England Biolabs, Massachusetts, USA) according to the instructions from the manufacturer. The concentration of protein was measured by Lowry's method using bovine serum albumin (BSA) as a control. The deglycosylated protein was purified using a CAPCELL-PAK C$_{18}$ column (2.0 × 250 mm, SHISEIDO). Elution was done gradually with 0.05% TFA for 5 min, 0–40% acetonitrile/0.05% TFA for 50 min, 40–80% acetonitrile/0.05% TFA for 2 min at a flow rate of 0.2 mL min^{-1}, and fractions were collected manually. For ΔP-GSP-37, calf intestinal alkaline phosphatase (CIAP) (Promega, Wisconsin, USA) was used for dephosphorylation. The ΔP-GSP-37 was purified using a CAPCELL-PAK C$_{18}$ column (2.0 × 250 mm, SHISEIDO). Elution was done successively with 0.05% TFA for 5 min, and 0–80% acetonitrile/0.05% TFA for 50 min at a flow rate of 0.2 mL min^{-1}, and fractions were collected manually. ΔP- GSP-37 was applied to an SDS-PAGE and stained with CBB or ProQ Diamond Phosphoprotein Gel Stain (Invitrogen).

2.5 Enzymatic digestion and amino acid sequencing

For amino acid sequencing, the purified GSP-37 and ΔN-GSP-37 were digested as follows. For pepsin digestion, 25 µg of the purified protein was dissolved in 300 µL of 10 mM HCl. To this solution, 1 µg of pepsin (Nacalai Tesque, Kyoto, Japan) was added and the mixture was incubated at 37 °C for 30 min. For thermolysin digestion, 25 µg of the purified protein was dissolved in 300 µL of 20 mM

NH$_4$HCO$_3$ (pH 8.0)/2 mM CaCl$_2$, to which 1 μg of thermolysin (Sigma) was added. The mixture was incubated at 60 °C for 3 h. The reaction was stopped by addition of 200 μL of 1 M HCl. The purified GSP-37 (25 μg) was also digested with Proteinase K (Nacalai Tesque) (3 μg) in a solution containing 100 μL of 150 mM NaCl/10 mM Tris-HCl (pH 8.0)/10 mM EDTA/0.1% SDS at 16 °C for 1 h, and the digestion was terminated by heating at 95 °C at 10 min. For digestion of ΔN-GSP-37, 25 μg of the deglycosylated protein was dissolved in 300 μL of 1% TritonX-100/20 mM Tris-HCl (pH 8.0), to which 1 μg of endoproteinase Asp-N (Roche diagnostics, Mannheim, Germany) was added. The mixture was incubated at room temperature for 16 h, and 200 μL of 1 M HCl was added to stop the digestion.

Each digest was separated by reverse-phase HPLC with a CAPCELL-PAK C$_{18}$ column (2.0 mm × 250 mm, SHISEIDO). Elution was done gradually with 0.05% TFA for 5 min, 0–60% acetonitrile/0.05% TFA for 60 min, 60–80% acetonitrile/0.05% TFA for 2 min at a flow rate of 0.2 mL min^{-1}. Elution was monitored by the absorbance at 225 nm, and each peak material was collected manually. The N-terminal amino acid sequences of intact GSP-37 or enzymatic digests were analyzed on a protein sequencer, Applied Biosystems models 492 HT or 491cLC (Applied Biosystems, California, USA).

2.6 cDNA cloning of *GSP-37*

Total RNA was extracted from the scales of a mature goldfish using ISOGEN (Nippon gene, Tokyo, Japan). The first strand cDNA was generated from 1 μg of the total RNA using a SMART RACE cDNA Amplification Kit (Clontech, California, USA) according to the instructions from the manufacturer. Degenerated PCR and subsequent nested PCR were performed with primer pairs 37-F and 37-R, 37-NF and 37-NR (Table 1), respectively, using the following PCR program: initial denaturation at 94 °C for 3 min; 40 cycles of reaction at 94 °C for 30 s, 48 °C for 30 s and 72 °C for 30 s; and a final extention at 72 °C for 7 min. The amplified fragment was subcloned into a pCR 2.1 vector (Invitrogen) and sequenced using an automatic DNA squencer ABI 310 (Applied Biosystems).

The 3′ RACE was also carried out in the following 2 steps using the same cDNA sample. The first PCR and nested PCR were performed with primer pairs 37-3F and RTG, 37-3FN and RTGN (Table 1), respectively, using the following program: initial denaturation at 94 °C for 3 min; 35 cycles for the first PCR or 40 cycles for nested PCR of reaction at 94 °C for 30 s, 55 °C for 30 s and 72 °C for 2 min; and a final extention at 72 °C for 5 min. The amplified fragment was sequenced as described above. For 5′ RACE, another cDNA sample was prepared from several mature goldfish as described above. The first PCR and nested PCR were performed with primer pairs GSP-37-R1 and UPM, GSP-37-R2 and NUP (Table 1), respectively, using the following program: for the first PCR, initial denaturation at 96 °C for 3 min; 5 cycles of reaction at 96 °C for 5 s; 72 °C for 3 min; another 5 cycles of 96 °C for 5 s, 70 °C for 10 s, 72 °C for 3 min; 25 cycles of 96 °C for 5 s, 68 °C for 10 s, 72 °C for 3 min; and for the nested PCR, initial denaturation at 96 °C for 3 min; 35 cycles of 96 °C for 30 s, 57 °C for 30 s, 72 °C for 2 min; and a final extention at 72 °C for 7 min. The amplified fragment was subcloned into a pGEM-T easy vector (Promega) and sequenced as described above. Finally, to determine the entire ORF of *GSP-37*, PCR was carried out with primers GSP-37-F1 and GSP-37-R3 (Table 1) which were designed based on the 5′ and 3′ UTR. The PCR used the following cycle conditions: initial denaturation at 96 °C for 3 min; 35 cycles of reaction at 96 °C for 30 s, 56 °C for 30 s, 72 °C for 2 min. The amplified fragment was subcloned and sequenced as described above.

2.7 Preparation of recombinant GSP-37 (rGSP-37)

For preparation of rGSP-37, using the entire cDNA of *GSP-37* subcloned in pGEM-T easy vector as a template, a partial fragment encoding mature GSP-37 (A^{128} to A^{541}

Table 1 Primers used in the present study

name	nucleotide sequence
37-F	5′- YTSACNGARCARGARGTNGT -3′
37-NF	5′- ACNGARCARGARGTNGTNGA -3′
37-R	5′- TGCARNCCYTGNACRTTDAT -3′
37-NR	5′- CCCTGNACRTTDATNCCNGG -3′
37-3F	5′- ATGAATATAAGCATTGTTCCTCT -3′
37-3FN	5′- ATAAGCATTGTTCCTCTTGAACC -3′
GSP-37-F1	5′- AGGAGTTTGAACAAACAAGGGAG -3′
GSP-37-F2	5′- CCCGGATCCACACCGGTTAATCATAATGAC -3′
GSP-37-F3	5′- TTGAGCTCTTCAGAGGAGGTCAC -3′
GSP-37-R1	5′- ACAGGTTTTTCCTCAGGACTGGTGAGC -3′
GSP-37-R2	5′- CGTGTACAATTGCAGTTCTGAGACGTTT -3′
GSP-37-R3	5′- TAGTAAATACTCAGCACACTGGC -3′
GSP-37-R4	5′- GGGTCTAGATTAAATGTTGTCTCCTCTGAC -3′
GSP-37-R5	5′- CGTGTACAATTGCAGTTCTGAGA -3′
β-actin-F	5′- ACTGTACCCATCTACGAGGGTTA -3′
β-actin-R	5′- GTTGAAGGTGGTCTCATGGATAC -3′
RTG	5′- AACTGGAAGAATTCGCGGCCG -3′
RTGN	5′- TGGAAGAATTCGCGGCCGCAG -3′
UPM	5′- CTAATACGACTCACTATAGGGCAAG CAGTGGTATCAACGCAGAGT-3′
	5′- CTAATACGACTCACTATAGGGC -3′
NUP	5′- AAGCAGTGGTATCAACGCAGAGT -3′
NotI dT	5′- AACTGGAAGAATTCGCGGCCGCAGGAA(T)$_{17}$ -3′

in Fig. 6A) was amplified with primers GSP-37-F2 and GSP-37-R4 (Table 1). PCR was performed as follows: 96 °C for 3 min; 35 cycles of reaction at 96 °C for 30 s, 58 °C for 30 s and 72 °C for 2 min; and a final extention at 72 °C for 7 min. The amplified fragment was subcloned into the *Bam*HI - *Xba*I site of a pProEx HTb vector (Life Technologies, Inc., New York, USA) and bacterial expression was carried out in the *E. coli* BL 21 (DE3) cells. The recombinant GSP-37 with a histidine-tag (His-tag) and the N-terminal additional 5 amino acid residues (Gly-Ala-Met-Gly-Ser) derived from the vector was purified with a Ni-NTA Agarose column (Invitrogen, California, USA). The His-tag was then enzymatically removed with AcTEV Protease (Invitrogen) following the instructions from the manufacturer and the residual protein was purified using a CAPCELL-PAK C_{18} column (2.0 mm × 250 mm, SHISEIDO). Elution was done successively with 0.05% TFA for 5 min, 30–80% acetonitrile/0.05% TFA for 40 min at a flow rate of 0.2 mL min^{-1}, and fractions were collected manually.

2.8 Reverse transcription-PCR (RT-PCR)

The following tissues were used for expression analysis; scale, hemocyte, brain, gill, heart, liver, ovary, testis, intestine, kidney, muscle, skull, notochord, caudal fin and pharyngeal teeth. Total RNA was extracted as described above. To 1 μg of the total RNA, 5 pmol of *Not*I dT primer (Table 1) was annealed, and cDNA was synthesized. The cDNA equivalent to 10 ng of total RNA was subjected to the following PCR reaction using primer pairs, GSP-37-F1 and GSP-37-R3 for *GSP-37* and β-actin-F and β-actin-R for *β-actin* (Table 1): initial denaturation at 96 °C for 3 min; 30 or 28 cycles of reaction at 96 °C for 30 s, 56 °C for 30 s and 72 °C for 1 min for *GSP-37* or *β-actin*, respectively; and a final extension at 72 °C for 5 min.

2.9 *In situ* hybridization

For *in situ* hybridization, regenerating scales were sampled from mature goldfish 12–15 days after ablation of normal scales. Whole mount *in situ* hybridization was performed on the regenerating scale according to the method previously reported[20] with slight modifications: 2.5% glutaraldehyde in a phosphate buffered saline (PBS) and riboprobes at the concentration of 1 µg mL^{-1} were used for fixation and hybridization, respectively. For section *in situ* hybridization, the regenerating scales were fixed in 4% paraformaldehyde (PFA) in PBS and then decalcified with 0.5 M EDTA (pH 8.0) at 4 °C for 16 h. The fixed specimens were then dehydrated with graded ethanol series, penetrated with xylene, and embedded in paraffin according to the conventional protocol. The tissue sections were cut at 6 µm thickness and deparaffinized and rehydraded with xylene and graded ethanol series. Hybridization was carried out according to the previously reported method[20] except that proteinase K treatment was performed at the concentration of 20 µg mL^{-1} followed by post-fixation in 4% PFA in PBS at room temperature for 20 min. Riboprobes for hybridization were prepared as follows. Using the entire cDNA of *GSP-37* subcloned in the pGEM-T easy vector as a template, the partial fragment of *GSP-37* correspond to T^{191}-G^{490} in Fig. 6A was amplified with primers GSP-37-F3 and GSP-37-R5 (Table 1). PCR was performed as follows: 96 °C for 3 min; 30 cycles of reaction at 96 °C for 15 s, 56 °C for 30 s and 68 °C for 1 min; and a final extension at 72 °C for 7 min. The amplified fragment was subcloned into a pGEM-T easy vector, and the digoxygenin-11-dUTP labeled anti-sense and sense probes were generated with DIG-RNA labeling mix (Roche) and T7 RNA polymerase (TaKaRa, Shiga, Japan). The latter was used as a negative control. Hybridization signals were detected with NBT/BCIP solution (0.03% NBT and 0.02% BCIP in 100 mM Tris-HCl (pH 9.5)/100 mM NaCl/50 mM $MgCl_2$). For detection of alkaline phosphatase (ALP)- and tartrate resistant acidic phosphatase (TRAP)-positive cells, the adjacent sections were stained with NBT/BCIP solution and an Acid Phosphatase, Leukocyte kit (Aldrich, Missouri, USA), respectively.

2.10 Localization analysis of GSP-37

To raise a polyclonal antibody against GSP-37, the purified rGSP-37 was injected to a rabbit. Collection and preparation of an anti-serum was ordered to ProteinPurity Ltd. (Gunma, Japan). To confirm the specificity of the anti-serum, immune-dot blot analysis was carried out. Each of 250, 125 and 62.5 ng of rGSP-37, GSP-37, BSA and casein were blotted onto a Hybond-C membrane (GE Healthcare, Connecticut, USA) and baked at 80 °C for 1 h. The membrane was blocked in 5% skim milk in TTBS (137 mM NaCl/2.7 mM KCl/25 mM Tris/0.05% (v/v) Tween-20) at room temperature for 1 h and treated with the anti-GSP-37 anti-serum diluted 1 : 1000 in TTBS. The membrane was then washed 3 times with double-distilled water and reacted with Anti-Rabbit IgG (Fc) AP-Conjugate (Promega) diluted 1 : 3000 in TTBS at 4 °C for 16 h. The membrane was washed 3 times with TTBS, and treated with CDP Star (GE healthcare). Luminescent signals were detected with a FAS image analyzer (FUJI FILM, Tokyo, Japan).

For light microscopic observation, normal scales were sampled from mature goldfish. Paraffin sections were prepared as described above. For preparation of undecalcified sections, after fixation and dehydration, specimens were embedded in JB-4 resin (Ted Pella, Inc., California, USA) following the instructions from the manufacturer. The tissue sections were cut at 4 µm thickness using an automatic microtome Reichert Jung 2050 (Leica, Wetzler, Germany). Immunohistochemical staining was carried out according to the procedure described previously[21] with the following minor modifications: the anti-GSP-37 antiserum and anti-rabbit IgG were diluted at 50-fold and 200-fold, respectively. To detect calcified regions in the scale, 5% $AgNO_3$ was dropped on the section and the section was left under fluorescent light for 10 min. For scanning electron microscopic analysis, normal scales were fixed as described above, dehydrated with ethanol series followed by absolute butanol, and embedded in Epo-Cure resin (Buehler, Illinoi, USA). The sample was cut with a diamond cutter. Cross-sections (5 mm thickness) were polished with grinding powder (C-1000, −3000 and GC-6000 Corundum) and Mastermet2 colloidal silica (Buehler) for 5 min and washed with distilled water in an ultrasonicator. The sections were etched with 1 mM EDTA (pH 8.0) for 5 min prior to immunohistochemical reaction. The immunohistochemical analysis was performed according to the method of Suzuki et al. (2011) using an scanning electron microscope (S-4500, Hitachi, Tokyo, Japan) with following minor modifications: preimmune serum, anti-GSP-37 antiserum and 40 nm gold particle-conjugated anti-rabbit antibody (Funakoshi, Tokyo, Japan) were diluted at 1 : 250 with TTBS.[22]

3. Results

3.1 Purification of GSP-37

EDTA-soluble matrices were extracted from goldfish scales and stained separately with CBB and Stains-all on SDS-PAGE gels (Fig. 2A). Since the Stains-all staining detects negatively-charged substances as blue bands, these 2 staining methods showed distinguishable staining patterns. A band of approximately 37 kDa appeared as a major component in Stains-all staining, but was scarcely visible on the gel stained with CBB. As shown in Fig. 2B, calcium phosphate crystallization was inhibited in a dose-dependent manner in the EDTA-soluble fraction, and the material extracted from 20 mg of goldfish scales almost completely inhibited crystal formation. EDTA-soluble materials were then subjected to an anion-exchange column (Fig. 2C). Four fractions were collected according to the retention time, and fraction d showed the highest inhibitory activity (data not shown). This fraction was subsequently applied to reverse-phase HPLC and separated into 7 fractions (fractions d1–d7, Fig. 3A). Among these fractions, fraction d2 showed the highest inhibitory activity, and Stains-all staining showed that this fraction contained a large amount of a negatively charged material, with a molecular weight of approximately 37 kDa, which was not stained with CBB (Fig. 3B and C). This result suggests that the 37 kDa-highly acidic material must be responsible for the inhibitory activity. Since this band was susceptible to proteinase K (data not shown), we believed it to be a protein. Fraction d2 was then subjected to a second reverse-phase HPLC, and the highest inhibitory activity was observed in fraction d2-1 (Fig. 4A). This fraction was then applied to 2 steps of gel filtration HPLC (data not shown). Finally, the purified fraction was subjected to reverse-phase HPLC (Fig. 4B). At this step, fractions were collected every min from 44 to 53 min, and these fractions were subjected to SDS-PAGE. As shown in Fig. 4C, fractions 1–6 contained 2 bands, while fractions 7–9 appeared to be a single band and were found to retain crystallization inhibitory activity (see functional assay of native GSP-37 described in 3.3). Therefore, the protein from fractions 7–9 was named goldfish scale protein (GSP)-37. Fig. 1 summarizes the purification scheme for GSP-37.

3.2 Sequence analysis of GSP-37

N-terminal amino acid sequence analysis revealed 13 amino acid residues, including 3 unidentified residues (Fig. 5). Pepsin, thermolysin, proteinase K and endoproteinase Asp-N digestions afforded 15 partial fragment peptides, which were merged into 4 larger segments (Fig. 5). These partial sequences also contained a considerable number of unidentified residues, many of which were supposed to be phosphorylated Ser or Thr residues, as described below. Since these fragments could not be assembled into one, we designed degenerate primers based on the amino acid sequence of the longest fragment (partial sequence 4 in Fig. 5) and performed degenerate RT-PCR followed by 5′ and 3′ RACE. As a result, a full-length cDNA

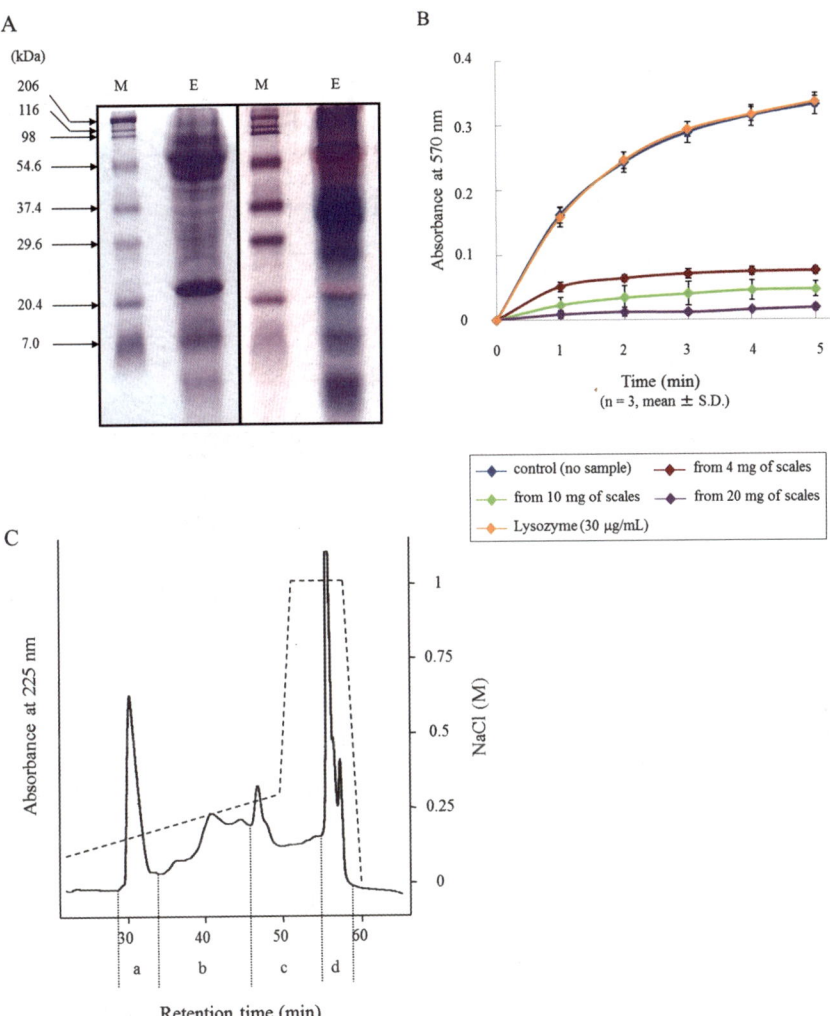

Fig. 2 Purification of GSP-37 (Part 1). A: EDTA-soluble substances extracted from goldfish scales stained with CBB (left) and Stains-all (right). Lanes M and E represent molecular weight markers and extracted substances, respectively. B: Change in the turbidity of solutions containing EDTA-soluble matrices or lysozyme as anion active control. C: Anion-exchange HPLC profile of the EDTA-soluble matrices. Fraction-d was subjected to further purification (see text for details). The concentration of NaCl is indicated by a dashed line.

encoding GSP-37 was obtained (Fig. 6A). *GSP-37* encoded a precursor protein consisting of 156 amino acid residues, and the 19 residues comprising the N-terminal sequence were predicted to represent a signal peptide by the SignalP 4.0 server. Since the N-terminal amino acid sequence of native GSP-37 started with the Val^{22} residue, the 2 residues preceding this residue, Thr^{20} and Pro^{21}, must be removed after signal cleavage. The hypothetical mature GSP-37 protein (Thr^{20}-Ile^{156}) consists of 137 amino acid residues, and the calculated molecular weight and isoelectric point were 14,700 and 4.24, respectively. The mature GSP-37 contained a highly acidic region (Ser^{31}-Glu^{46}), which was rich in phosphorylated Ser (see below) and Glu residues, and an integrin-binding motif (Arg^{152}-Gly^{153}-Asp^{154}) at the C-terminal region. As shown in Fig. 4C, the apparent molecular weight of GSP-37 was significantly higher than the calculated value, and we therefore

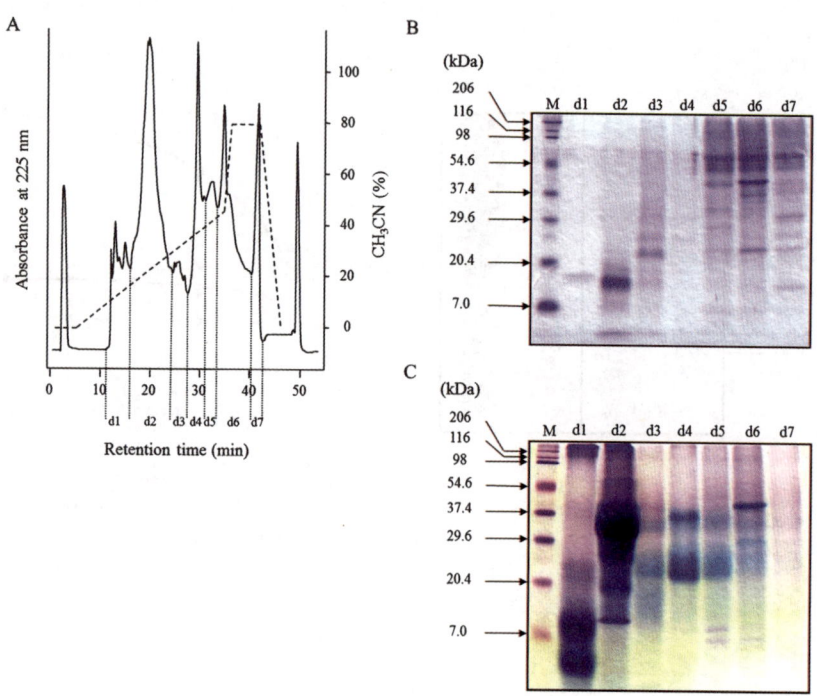

Fig. 3 Purification of GSP-37 (Part 2). A: Reverse-phase HPLC profile of fraction d purified by an anion-exchange HPLC. The concentration of acetonitrile is indicated by a dashed line. B and C: Fractions d1–d7 stained with CBB (B) and Stains-all (C). Lane M represents molecular weight markers. Fraction d-2 was subjected to further purification (see text for details).

assumed that this difference might be due to its acidic nature and post-translational modifications.

The deduced amino acid sequence of GSP-37 contained all 4 of the partial sequences (Fig. 6A), and we were able to identify most of the undetectable amino acid residues. There were some differences in the sequences; specifically, the first Lys residue in the partial sequence 2 and the 44th Gly residue in the partial sequence 4 in Fig. 5 were identified as Arg^{51} and Glu^{135}, respectively, in the deduced amino acid sequence, which might be due to sequence polymorphism. A comparison of the deduced and partial amino acid sequences of GSP-37 provides information on post-translational modifications. Two Asn residues (Asn^{102} and Asn^{107}) were postulated as Asp in the Asp-3 fragment, which was derived from ΔN-GSP-37. Since deglycosylation invokes conversion of N-glycosylated Asn to Asp, these 2 residues were proven to be glycosylated. Besides these residues, all of the other unidentified amino acid residues were serines or threonines, implicating phosphorylation or O-linked glycosylation. Using a phosphorylation prediction program, we predicted that 12 Ser and 7 Thr residues were phosphorylated. Indeed, the intact GSP-37 was strongly stained with ProQ Diamond (Fig. 7B), indicating that GSP-37 was highly phosphorylated. Although BLAST search failed to find proteins with significant similarity to GSP-37, a hypothetical protein from *Pimephales promelas* (fat head minnow, Accession No. DT101330) showed moderate similarity (31% identity in the gap-allowed alignment, Fig. 6B). It should be noted that the putative acidic region and RGD consensus sequence were almost or completely conserved in these 2 proteins.

3.3 Crystallization inhibitory activity of GSP-37, rGSP-37 and ΔP-GSP-37

In order to examine the effects of post-translational modifications on calcium phosphate crystallization inhibitory activity, recombinant GSP-37 and dephosphorylated

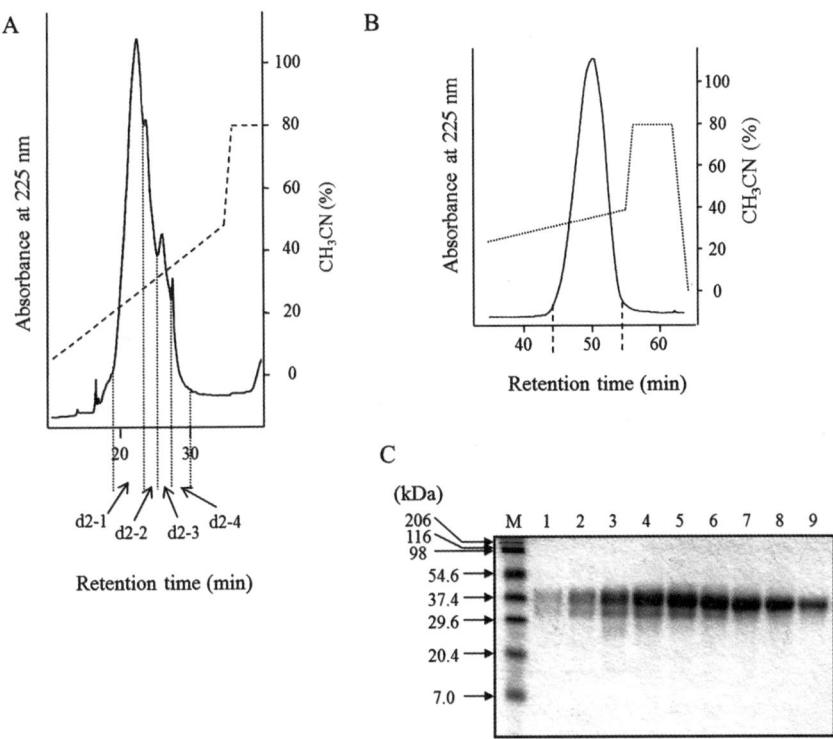

Fig. 4 Purification of GSP-37 (Part 3). A: Reverse-phase HPLC profile of fraction d2. B: Reverse-phase HPLC profile of fraction d2-1 after purification of 2 steps of gel filtration HPLC (see text for details). The concentration of acetonitrile is indicated by a dashed line. C: Fractions 1–9 purified in the final step of reverse-phase HPLC (Fig. 4B) stained with Stains-all. Lane M represents molecular weight markers.

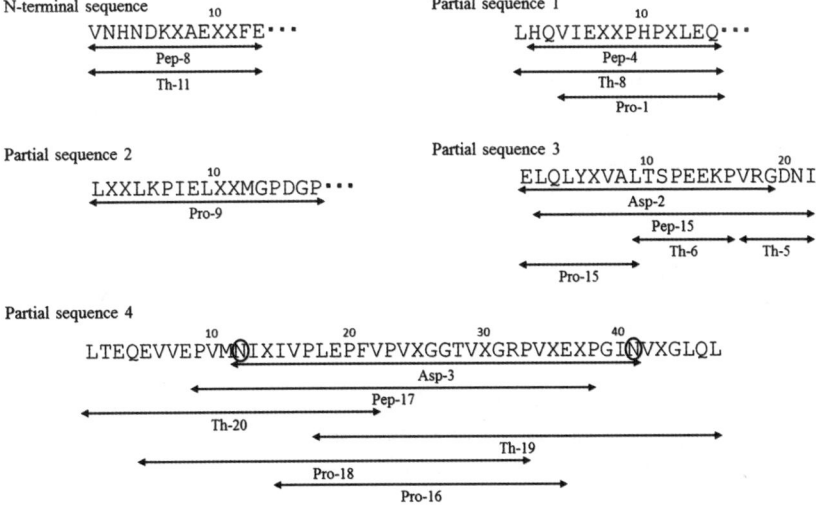

Fig. 5 N-terminal and internal partial amino acid sequences of GSP-37. Pep-x, pepsin digests; Pro-x, proteinase K digests; Th-x; thermolysin digests; Asp-x, endoproteinase Asp-N digests. Asp-N digests were derived from ΔN-GSP-37, while others were from intact GSP-37. Putative glycosylated Asn residues are circled.

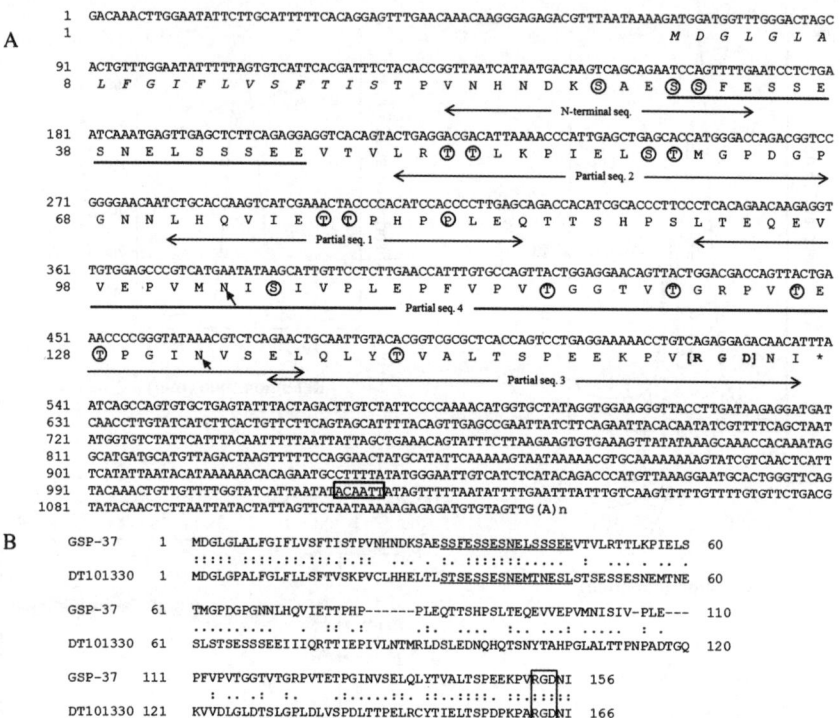

Fig. 6 Sequence analysis of GSP-37. A: Nucleotide sequence and its deduced amino acid sequence of *GSP-37*. N-terminal signal peptide is italicized. A putative highly acidic region is underlined. N-terminal and partial amino acid sequences in Fig. 5 are indicated by bi-directional arrows. The residues which could not be detected in amino acid sequence analysis are circled. The C-terminal RGD motif is bolded and bracketed. Arrows indicate *N*-glycosylated Asn residues. Polyadenylation signal is boxed. B: Amino acid sequence alignment of GSP-37 and a hypothetical protein of *Pimephales promelas* (fat head minnow) (Accession No. DT101330). ":" and "." represent identical and similar amino acid residues, respectively. A putative highly acidic region is underlined. The RGD motif is boxed.

natural GSP-37 (ΔP-GSP-37) were prepared. Fig. 7A shows a comparison of the activities of native GSP-37 and rGSP37. Native GSP-37 almost completely inhibited crystallization at a dose of 5 μg, while the activity of rGSP-37 was much lower, even when a 9-fold increase in protein was added to the solution. Then we prepared dephosphorylated GSP-37 and compared its activity to that of native GSP-37. As shown in Fig. 7B, ΔP-GSP-37 completely lost fluorescence in ProQ Diamond staining, indicating successful dephosphorylation. When ΔP-GSP-37 was subjected to the assay, inhibitory activity was dramatically reduced (Fig. 7C). These results suggest that post-translational phosphorylation is the critical modification that allows GSP-37 to interact with calcium phosphate crystals.

3.4 Expression of the *GSP-37* gene

To examine the tissue distribution of *GSP-37*, RT-PCR and *in situ* hybridization were performed. In RT-PCR analysis, *GSP-37* transcripts were detected in total RNA prepared only from the scales, but not from any other tissues, including 4 other calcified tissues (skull, notochord, caudal fin and pharyngeal teeth; Fig. 8A). We performed *in situ* hybridization using 12- to 15-day-old regenerating scales. In the whole mount hybridization, expression signals were detected only on the apical side of the scale (the surface of the body). The signals were concentrated in the

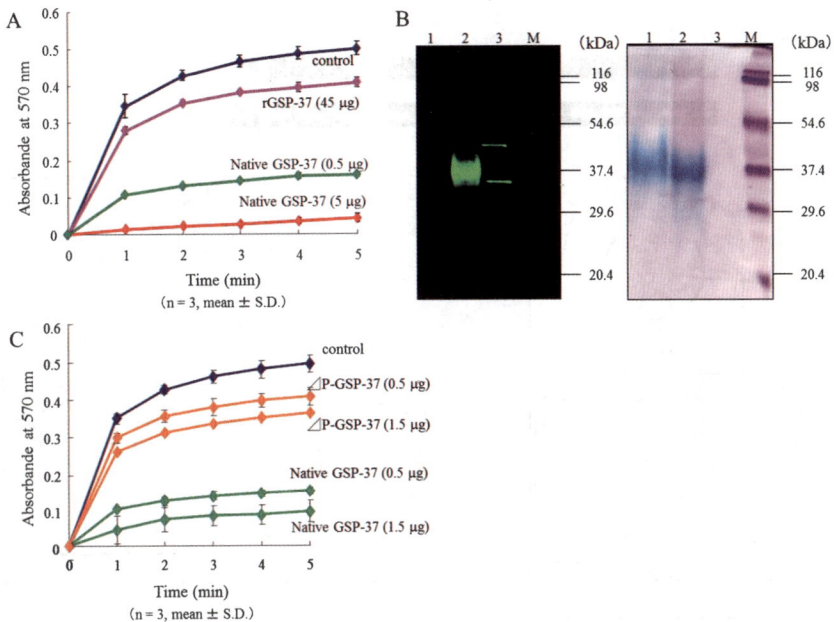

Fig. 7 Analysis of post-translational modification of GSP-37. A: Changes in the turbidity of solutions containing native GSP-37 and rGSP-37. In control experiment, proteins were omitted from the solution. B: Dephosphorylation of native GSP-37. Lane M, molecular markers; lane 1, ΔP-GSP-37; lane 2, native GSP-37; lane 3, positive controls for ProQ Diamond-staining. The upper and lower protein bands correspond to ovalbumin and β-casein, respectively. The proteins were stained with ProQ Diamond (left) or Stains-all (right) C: Changes in the turbidity of solutions containing native GSP-37 and ΔP-GSP-37. In control experiment, proteins were omitted from the solution.

central region (Fig. 8B) and were not observed in the peripheral region (data not shown). During the early stage of scale regeneration, "hollow-like" structures were frequently observed, and *GSP-37*-expressing cells were colocalized with these hollows (Fig. 8B). To examine the *GSP-37* expressing cells more thoroughly, hybridization was performed over the sagittal sections of the nascent regenerating scale. At this stage, the external layer and fibrillary plate were synthesized, but we could not capture the limiting layer. As in the case of the whole mount experiment, expression signals were detected only in the central region of the scale (Fig. 8C). The cells beside the hollow-like structure and flattened cells on the external layer were strongly stained, but the epidermal and basal cells involved in fibrillary plate synthesis were not stained (Fig. 8C). Since previous histological studies had confirmed that osteoblast- and osteoclast-like cells are present in teleost scales,[18,23,24] we attempted to examine ALP and TRAP activities of the cells in the scale. ALP activity was detected only in the epidermal cells (Fig. 8C-4), while TRAP activity was also predominantly detected in the epidermis (Fig. 8C-5). *GSP-37* expression was not detected in this region, indicating that *GSP-37*-expressing cells are different from osteoblasts and osteoclasts.

3.5 Localization of GSP-37

Prior to immunohistochemical studies, we examined the specificity of anti-GSP-37 antiserum using dot blot analysis. The antiserum reacted specifically with native GSP-37 and rGSP-37, but did not react with BSA or casein, which were used as negative controls (data not shown). Localization of GSP-37 was investigated using

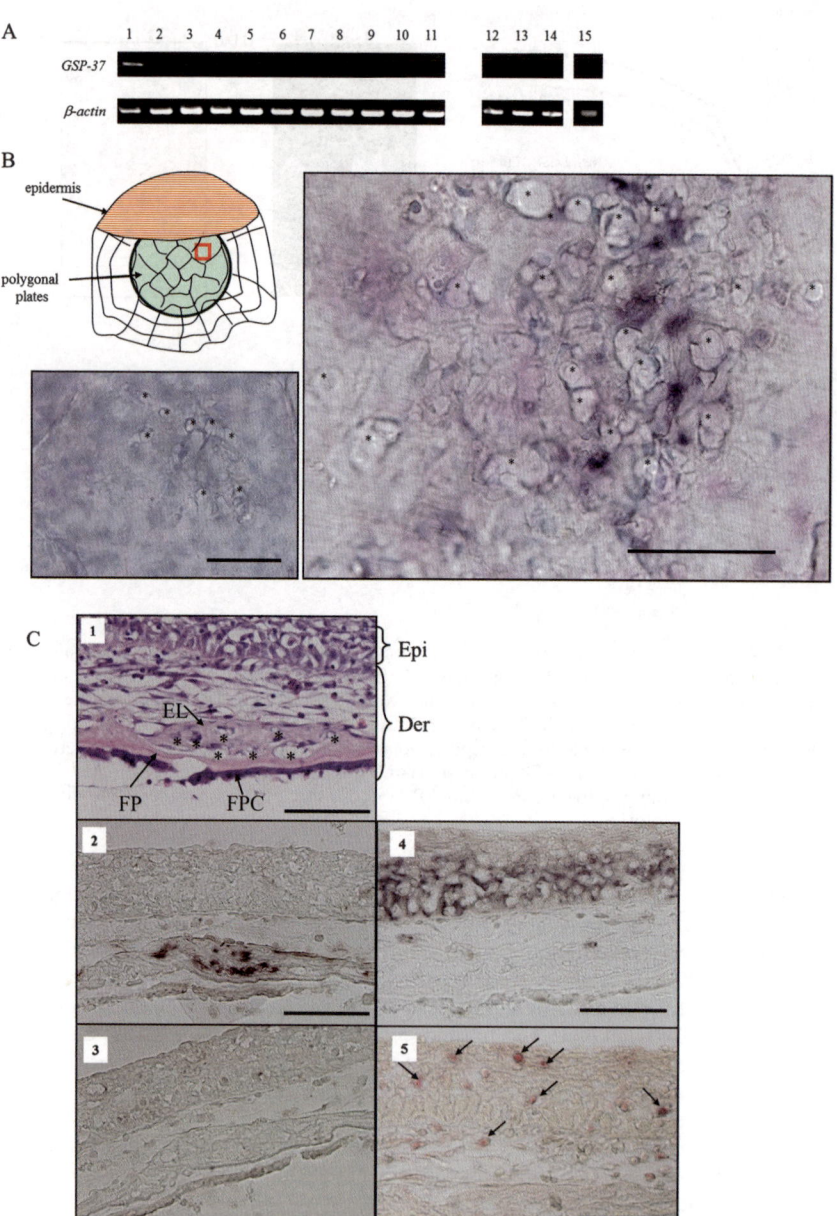

Fig. 8 Expression of *GSP-37*. A: Expression of *GSP-37* and *β-actin* transcripts: 1, scale; 2, hemocyte; 3, brain; 4, gill; 5, heart; 6, liver; 7, ovary; 8, testis; 9, intestine; 10, kidney; 11, muscle; 12, skull; 13, notochord; 14, caudal fin; 15, pharyngeal teeth. B: Whole mount *in situ* hybridization of *GSP-37*. A schematic representation of a regenerating scale (upper, left). The enclosed area corresponds to the magnified microscopic images. Expression of *GSP-37* was detected only in the central region (right), while signals were not detected in a negative control experiment using sense probe (lower left). Astarisks represent the "hollow"-like structure (see text for details). Scale bars, 50 μm. C: Section *in situ* hybridization of *GSP-37*. 1, HE-staining; 2, anti-sense probe; 3, sense-probe; 4, ALP-staining, ALP-positive cells are stained blue to purple; 5, TRAP-staining, TRAP-positive cells are indicated by arrows. EL, external layer; FP, fibrillary plate; FPC, fibrillary plate-forming cell; Epi, epidermins; Der, dermis. Scale bar, 50 μm.

light and electron microscopes. First, we performed immunohistochemical staining using decalcified tissue sections of normal scales. As shown in Fig. 9A, immunoreactive signals were predominantly detected on the surface of the scale but not in the fibrillary plate. Next, to eliminate the possibility that the decalcification process may affect the *de novo* distribution of the protein, undecalcified sections were also examined. As a result, the stained area became clearer than that in the decalcified specimen, and a thin layer of immunostaining was apparent on the scale surface (Fig. 9B). Additionally, silver nitrate staining revealed that abundant localization of GSP-37 was limited to the strongly calcified regions (Fig. 9B-3). No significant staining was observed in control experiments using a preimmune serum in both cases. In an immunoelectron microscopic analysis using back-scattered electron imaging, gold particles were observed in the outermost region of the scale with high density (Fig. 10). In goldfish scales, this outermost region of the scale is recognized as an "outer limiting layer," where calcium phosphate crystals are highly deposited and collagen fibrils are absent.[12,14] Although we could not identify the border between the outer limiting layer and the external layer, immunoreactive signals were specifically detected in this region with a width of 1–2 μm. Therefore, it is reasonable to assume that GSP-37 is predominantly localized in the outer limiting layer.

4. Discussion

In the present study, we identified a novel noncollagenous highly phosphorylated protein, named GSP-37, from the scales of the goldfish. GSP-37 was present in EDTA-soluble matrices in high abundance and was also the most major component in the acidic materials, suggesting that this protein has a central role in scale formation and calcification.

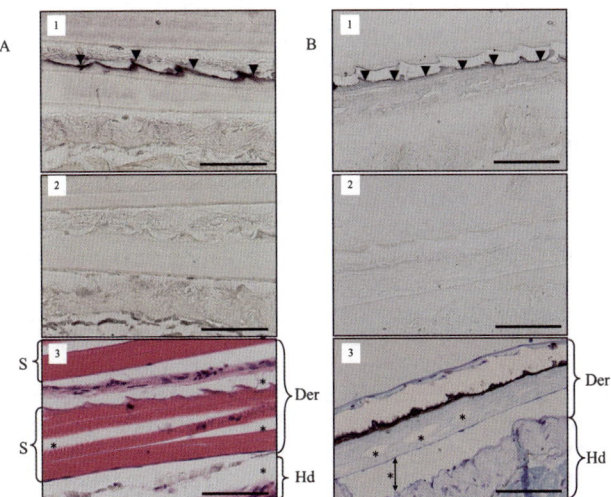

Fig. 9 Immunohistochemical staining for GSP-37. A: An immunostaining for GSP-37 in decalcified tissue sections embedded in paraffin. Immunoreactive region is indicated by arrowheads. 1, Anti-GSP-37 antiserum; 2, A negative control with a preimmune serum; 3, HE-staining. B: An immunostaining for GSP-37 in undecalcified tissue sections embedded in resin. Immunoreactive region is indicated by arrowheads. 1, Anti-GSP-37 antiserum; 2, A negative control with a preimmune serum; 3, Silver nitrate and toluidine blue-staining. The calcified region is stained dark brown. Scale bars, 50 μm. S, Scale; Der, dermis; Hd, hypodermis; asterisk, artificial loss of the specimen during sample preparation.

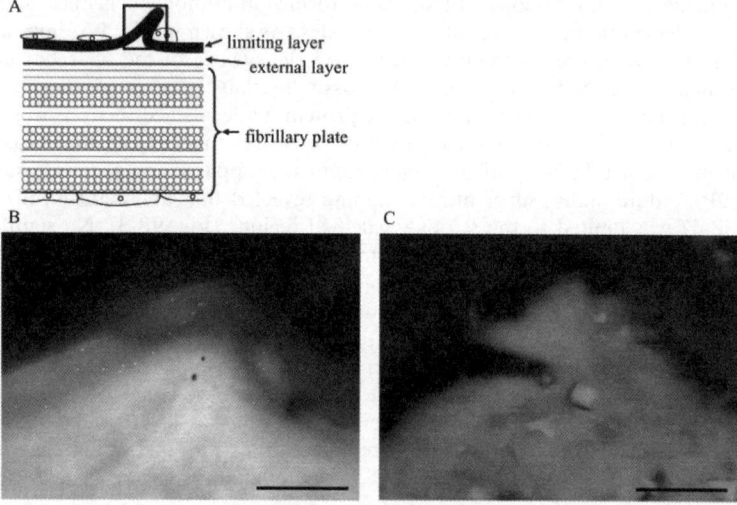

Fig. 10 An immunohistochemical imaging of GSP-37 with gold nanoparticles. A schematic representation of the cross section of fish scale (A). The enclosed area corresponds to the magnified microscopic images, B and C. Tissue sections were reacted with an anti-GSP-37 antiserum (B) or a preimmune serum (C). Small white dots represent localization of GSP-37 in scanning electron microscope. Scale bars, 3 μm.

We developed a new method for investigating calcium phosphate crystallization inhibitory activity based on the calcium carbonate crystallization assay previously reported by Inoue *et al.* (2001).[19] In this assay, proteins with affinity to calcium phosphate crystals would attach on the surface of the nascent crystals and inhibit further growth. Thus, the inhibitory activity reflected the ability of the proteins to bind to the crystals. In the present study, purified GSP-37 strongly inhibited crystallization at 2.5 μg mL^{-1}. This concentration is comparable to those reported in previous studies[19,25] examining calcium carbonate crystallization. During this experiment, we found a few fractions other than the GSP-37-containing fraction that also possessed inhibitory activity, suggesting that GSP-37 is not the only protein involved in calcification of the scale.

Enzymatic dephosphorylation dramatically reduced the inhibitory activity of GSP-37. Since the phosphorylated group can concentrate counter ions such as Ca^{2+} by virtue of their negative charge, prosphorylation plays pivotal roles in biomineralization. In crustacean exoskeleton and gastrolith formation, phosphoenolpyruvate and 3-phosphoglycerate are continuously secreted during gastrolith formation and stabilize calcium carbonate in the amorphous phase.[26] In addition to such low-molecular weight compounds, phosphorylation gives rise to significant changes in the chemical characteristics and/or function of proteins. For calcification-related proteins, phosphorylation accelerates the interactions between the proteins and inorganic crystals in many situations. Calcification associated peptide (CAP)-1, a crustacean exoskeletal peptide, represents a typical example. CAP-1 strongly inhibits calcium carbonate crystallization *in vitro via* 2 acidic regions, which are implicated in the deposition of calcium carbonate precipitates in crayfish exoskeletons. The C-terminal acidic region contains 6 consecutive Asp residues and a phosphorylated Ser residue. Bacterially expressed, unphosphorylated recombinant CAP-1, as well as mutated CAP-1, in which the phosphorylated Ser was replaced by an Asp residue, showed less than 70% of the calcium carbonate crystallization inhibitory activity of native CAP-1.[27] Another notable example is also found in calcium phosphate-deposited biominerals. Dentine phosphophoryn (DPP) is the most abundant NCP in dentine, and approximately 45% of the total amino acid residues are serines, most of which are phosphorylated.[28] This protein is recognized as an important initiator of dentine mineralization, and dephosphorylated DPP loses its Ca^{2+}-binding ability, nucleation activity, collagen-binding activity and HA-inducing activity.[29,30] On the other hand, we confirmed that removal of N-linked oligosaccharides did not affect the inhibitory activity of GSP-37. Thus, the phosphorylated residues of GSP-37 contribute to its calcium phosphate-binding properties. However, we cannot exclude the possibility that O-linked glycosylation may also be involved in the activity of the protein, since GSP-37 has several potential O-linked glycosylation sites. However, the contribution of O-glycosylation might be low, considering that dephosphorylation of GSP-37 significantly decreased its inhibitory activity.

BLAST search failed to find any proteins with significant sequence similarity to GSP-37, except for a hypothetical protein in the fat head minnow, which shows moderate sequence identity (Fig. 6B). We presume that this protein would be a homolog of GSP-37, partly because it contains an N-terminal signal sequence and an RGD motif, and partly because the putative phosphorylated Ser and Thr residues are well conserved (7 out of 15 for Ser and 5 out of 7 for Thr residues). If this is the case, GSP-37 would be conserved, at least in the order Cypriniformes.

Recent evolutionary genomics studies allow us to speculate on the evolutionary aspects of NCPs. For example, the construction and evolution of the secretory calcium-binding phosphoprotein (SCPP)-related genes have become understandable.[31,32] SCPPs are widely distributed in various vertebrates from sarcopterygians (stem-bony fish) to mammals. Kawasaki (2009, 2011) proposed that the SCPP gene family, containing all the SIBLING genes, arose by sequential gene duplication of a common ancestral gene, *SPARCL1*, encoding secreted protein acidic and cysteine-rich like 1.[31,32] SCPPs can be classified into 2 subgroups: acidic SCPPs, which generally contain more than 25% Glu, Asp and phosphorylated Ser residues, and Pro/Gln-rich SCPPs, which generally contain more than 20% Pro and Gln residues. Interestingly, in mammals, acidic SCPPs are predominantly involved in bone or dentine mineralization, while enamel mineralization is enabled by Pro/Gln-rich SCPPs. GSP-37 cannot satisfy either of these criteria, since Glu, Asp and phospho-Ser residues comprise 23% of the total amino acids (including hypothetically phosphorylated residues), while Pro/Gln residues comprise less than 15%. Thus, genomic sequence analysis, including exon-intron structure, is required to speculate the evolutionary origin of *GSP-37*.

Expression of the gene encoding GSP-37 was detected only in scales but not in any other tissues tested so far. We also confirmed that the bone matrix was not immunoreactive against GSP-37 (data not shown), indicating that GSP-37 is a scale-specific component. Many NCPs are found in multiple tissues. For example, BSP is found not only in bones, but also in teeth; OP is expressed in the bone, kidney and placenta; and *DMP1* is also detected in the brain.[33] Immunohistochemical analyses revealed that GSP-37 is localized primarily at the outermost region of the scale, the outer limiting layer. This layer is thought to correspond to the enamel or enameloid layer of the tooth, which is hypermineralized compared to the dentine and is devoid of collagen fibrils.[12] During the early stages of enamel synthesis, 3 enamel matrix proteins, amelogenin, enamelin and ameloblastin, are secreted by the ameloblast. These proteins are then degraded by enamel protein-specific proteinases during the maturation of the enamel layer and are replaced by HA crystals.[34] Since the fish genome lacks the genes encoding these 3 proteins, GSP-37 may work as a counterpart to these enamel matrix proteins. Furthermore, *in situ* hybridization has indicated that the expression of *GSP-37* was initiated prior to the synthesis of the limiting layer (Fig. 8C), and therefore, GSP-37 may also be involved in calcification of the external layer.

We still have not elucidated which kinds of cells are involved in scale formation and calcification. Previous reports have demonstrated that *bgp*, *sparc* and the differentiation factor *runx2* are expressed in the basal cells underlying the fibrillary plate.[9]

This result strongly suggests that these cells are the main participants in scale formation. However, our present ALP-staining results indicate that inorganic phosphate is supplied by the epidermal cells, but not by the basal cells. Thus, although the patterns of cell differentiation and the nature of matrix proteins during scale formation may be similar to those in bone or dentine formation, many questions on these topics remain unanswered.

In summary, we have identified a novel highly phosphorylated protein, GSP-37, from the teleost scale. This protein shares common features with the members of the SIBLING family. Based on the present results, we propose that GSP-37 is expressed in the early stages of scale formation and attaches to HA crystals through phosphorylated groups, thereby regulating the calcification process. Thus, GSP-37 may be a key molecule in the mechanism of scale calcification.

Acknowledgements

The authors are grateful to Prof. Toyoji Kaneko and Dr Soichi Watanabe from the University of Tokyo, to Prof. Yasuaki Takagi from Hokkaido University for use of microtomes, and also to Mr. Kurin Iimura from Hokkaido University for fruitful discussion. This work was supported by Grants-in-Aid for scientific research (Nos.17GS0311, 22248037 and 22228006) from the Ministry of Education, Science, Sports and Technology of Japan.

References

1 E. Bonucci, *Biological Calcification-Normal and Pathological Processes in the Early Stages*, Springer, New York, 2007.
2 L. W. Fisher and N. S. Fedarko, *Connect. Tissue Res.*, 2003, **44**, 33.
3 S. K. Nishimoto, N. Araki, F. D. Robinson and J. H. Waite, *J. Biol. Chem.*, 1992, **267**, 11600.
4 S. K. Nishimoto, J. H. Waite, M. Nishimoto and R. W. Kriwacki, *J. Biol. Chem.*, 2003, **14**, 11843.
5 P. V. Hauschka and F. H. Wians, Jr., *Anat. Rec.*, 1989, **224**, 180.
6 P. Ducy, C. Desbois, B. Boyce, G. Pinero, B. Story, C. Dunstan, E. Smith, J. Bonadio, S. Goldstein, C. Gundberg, A. Bradley and G. Karsenty, *Nature*, 1996, **382**, 448.
7 D. B. Lehane, N. McKie, R. G. Russel and I. W. Henderson, *Gen. Comp. Endocrinol.*, 1999, **114**, 80.
8 R. W. Romberg, P. G. Werness, P. Lollar, B. L. Riggs and K. G. Mann, *J. Biol. Chem.*, 1985, **260**, 2728.
9 K. Iimura, H. Tohse, K. Ura and Y. Takagi, *J. Exp. Zool.* (DOI: 10.1002/jez.b.22005), in press.
10 J. Bereiter-Hahn and L. Zylberberg, *Comp. Biochem. Physiol., Part A: Mol. Integr. Physiol.*, 1993, **105**, 625.
11 E. E. Golub, *Biochim. Biophys. Acta, Gen. Subj.*, 2009, **1790**, 1592.
12 A. A. Schönbörner, G. Boivin and C. A. Baud, *Cell Tissue Res.*, 1979, **202**, 203.
13 L. Zylberberg and G. Nicolas, *Cell Tissue Res.*, 1982, **223**, 349.
14 L. Zylberberg, J. Bereiter-Hahn and J. Y. Sire, *Cell Tissue Res.*, 1988, **253**, 597.
15 T. Ohira, M. Shimizu, K. Ura and Y. Takagi, *Fish. Sci.*, 2007, **73**, 46.
16 J. Y. Sire and M. A. Akimenko, *Int. J. Dev. Biol.*, 2004, **48**, 233.
17 J. Y. Sire, P. C. J. Donoghue and M. K. Vickaryous, *J. Anat.*, 2009, **214**, 409.
18 P. Persson, Y. Takagi and B. T. Björnsson, *Fish Physiol. Biochem.*, 1995, **14**, 329.
19 H. Inoue, N. Ozaki and H. Nagasawa, *Biosci., Biotechnol., Biochem.*, 2001, **65**, 1840.
20 L. Li, H. Katsuyama, S. N. Do, M. Saito, H. Tanii and K. Saijoh, *J. Toxicol. Sci.*, 2007, **32**, 359.
21 H. Endo, Y. Takagi, N. Ozaki, T. Kogure and T. Watanabe, *Biochem. J.*, 2004, **384**, 159.
22 M. Suzuki, A. Iwashima, N. Tsutsui, T. Ohira, T. Kogure and H. Nagasawa, *ChemBioChem*, 2011, **12**, 2478.
23 E. I. Sidorova, M. A. Lange, M. T. Dobrynina and T. V. Vasil'eva, *Ontogenez*, 2006, **37**, 469.
24 K. Azuma, M. Kobayashi, M. Nakamura, N. Suzuki, S. Yashima, S. Iwamuro, M. Ikegami, T. Yamamoto and A. Hattori, *Biochem. Biophys. Res. Commun.*, 2007, **362**, 594.
25 N. Ozaki, S. Sakuda and H. Nagasawa, *Biochem. Biophys. Res. Commun.*, 2007, **357**, 1172.

26 A. Sato, S. Nagasaka, K. Furihata, S. Nagata, I. Arai, K. Saruwatari, T. Kogure, S. Sakuda and H. Nagasawa, *Nat. Chem. Biol.*, 2011, **7**, 197.
27 H. Inoue, T. Ohira and H. Nagasawa, *Peptides*, 2007, **28**, 566.
28 W. Richardson, E. C. Munksgaard and W. T. Butler, *J. Biol. Chem.*, 1978, **253**, 8042.
29 G. He, A. Ramachandran, T. Dahl, S. George, D. Schultz, D. Cookson, A. Veis and A. George, *J. Biol. Chem.*, 2005, **280**, 33109.
30 M. Milan, R. V. Sugars, G. Embery and R. J. Waddington, *Eur. J. Oral Sci.*, 2006, **114**, 223.
31 K. Kawasaki, A. V. Buchanan and K. M. Weiss, *Annu. Rev. Genet.*, 2009, **43**, 119.
32 K. Kawasaki, *Cells Tissues Organs*, 2011, **194**, 108.
33 C. Qin, O. Baba and W. T. Butler, *Crit. Rev. Oral Biol. Med.*, 2004, **15**, 126.
34 G. Fincham, J. Moradian-Oldak and J. P. Simmer, *J. Struct. Biol.*, 1999, **126**, 270.

PAPER

CaCO₃/Chitin hybrids: recombinant acidic peptides based on a peptide extracted from the exoskeleton of a crayfish controls the structures of the hybrids†

Hiromu Kumagai,[a] Ryou Matsunaga,[b] Tatsuya Nishimura,*[a] Yuya Yamamoto,[a] Satoshi Kajiyama,[a] Yuya Oaki,[a] Kei Akaiwa,[a] Hirotaka Inoue,[c] Hiromichi Nagasawa,[c] Kohei Tsumoto[bd] and Takashi Kato*[a]

Received 27th March 2012, Accepted 6th June 2012
DOI: 10.1039/c2fd20057k

Functional peptides play an important role in the formation of biominerals. Here we focus on peptides involved in mineralization of the exoskeleton of a crayfish. New recombinant peptides are designed and synthesized on the basis of an acidic functional peptide, CAP-1, derived from the exoskeleton of a crayfish. We have examined the effect of these peptides on the hybrid formation of calcium carbonate and organic molecules in aqueous solution. In the presence of these recombinant peptides having chitin binding moieties, platelike tripodal calcite crystals are formed on the chitin matrix with unidirectional crystallographic orientation. Two kinds of interactions, (1) interactions between an acidic part of the peptides and calcium ions and (2) specific binding interactions of the peptides to chitin, are essential for the formation of the oriented crystals and the induction of specific morphologies. Moreover, the sizes of crystallites decrease with increasing the number of acidic parts of mutational peptides. The bioinspired design of peptides that control the interface of inorganic and organic domains may pave the way for an opportunity to build new hybrid structures.

Introduction

Biominerals are inorganic/organic hybrids with elaborate structures exhibiting a variety of functions.[1] Living organisms produce biominerals under mild conditions and use them as hard tissues such as teeth, bones, lens, shells, magnetite, and exoskeletons.[1,2] For example, the nacre of a seashell forms a layered hybrid structure of thin films oriented $CaCO_3$ and organic layers.[1] These brick-and-mortar structures

---

[a]*Department of Chemistry and Biotechnology, School of Engineering, University of Tokyo, Bunkyo-ku, Tokyo 113-8656, Japan. E-mail: kato@chiral.t.u-tokyo.ac.jp; tatsuya@chembio.t.u-tokyo.ac.jp*
[b]*Department of Medical Genome Science, School of Frontier Sciences, University of Tokyo, Kashiwa, Chiba 277-8562, Japan*
[c]*Department of Applied Biological Chemistry, Graduate School of Agricultural and Life Sciences, University of Tokyo, Bunkyo-ku, Tokyo 113-8657, Japan*
[d]*Medical Proteomics Laboratory, Institute of Medical Science, University of Tokyo, Minato-ku, Tokyo 108-8639, Japan*

† Electronic Supplementary Information (ESI) available: SEM images of a platelike tripodal hybrids (Fig. S1) and circular dichroism (CD) spectra of the recombinant peptides (Fig. S2). See DOI: 10.1039/c2fd20057k

achieve mechanical strength higher than those of single components.[3] Exoskeletons of crayfishes and lobsters are also good examples of targets for our material design.[4,5] Fig. 1 shows a cross-sectional scanning electron microscopy (SEM) image of the exoskeleton of a crayfish with a helical 3D structure that induces the characteristic physical and mechanical properties such as ductility, toughness, and lightweight.[4] Their hierarchical structures and their mild formation processes have been recognized as attractive strategies for materials synthesis.[6] We assume that the interface is important for this hybrid formation to achieve high mechanical strength. Calcium carbonate and chitin fibers should interact with each other at the interface. An acidic peptide, CAP-1, is believed to bind between calcium carbonate and chitin fibers for bonding them (Fig. 1, right). Biomacromolecules such as proteins and polysaccharides precisely control the crystallization of inorganic substances and make the interface of both components.[7]

New useful materials are being developed through biomineralization-inspired processes and structures. For example, we have prepared thin-film $CaCO_3$/organic polymer hybrids of through bio-inspired self-organization processes.[8] In these processes, the combination of insoluble polymer matrices and soluble acidic polymers affects the formation of thin-film hybrids (Fig. 2).

Our strategy is to control the crystallization of calcium carbonate through use of functional molecules such as designed peptides, and to build new hybrid materials with hierarchical structures. For the application of hybrid materials, it is important to control the size, orientation, and morphologies of inorganic crystals in the hybrids. We have shown that oriented thin-films of $CaCO_3$ crystals tens of microns in size are obtained in the presence of the acidic peptide CAP-1 isolated from the exoskeleton of a crayfish on chitin fibers (Fig. 1, right).[5] Analysis using transmission electron microscopy (TEM) and electron diffraction revealed that the thin films were assemblies of nanocrystallites 100 nm in size with a single crystalline orientation. The stabilization of the amorphous phases before nucleation might be a key process in the formation of the macroscopically aligned crystals.[9] CAP-1 has Rebers and Riddiford (RR) consensus sequence[10] capable of binding chitin. This peptide has an acidic property due to seven acidic amino acids at the C-terminus (Fig. 3a).[11] In particular, a phosphoserine residue at the 70th position may stabilize the amorphous state due to its stronger acidity.[12] These functional parts collaborate to collect calcium ions on the surface of the chitin matrix and to stabilize amorphous precursors on the matrix. Subsequently, the carboxylic groups in the successive aspartic acid sequence at the C-terminus serve as a template for the calcite nucleation of the amorphous precursor. In addition, we showed that a recombinant peptide rCAP-1 lacking a phosphate group also induced the oriented $CaCO_3$ crystals on a chitin matrix. However, the surface morphologies of the obtained crystals for CAP-1 were different from those for rCAP-1. In the crystallization from the amorphous phase which is stabilized and disturbed crystal nucleation by a phosphate

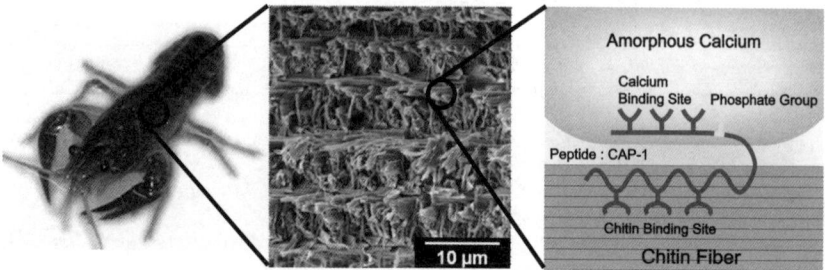

Fig. 1 Photograph of a crayfish, *Procambarus clarkii*, and a SEM image of the fractured surface of its exoskeleton. Schematic illustration of the interface structure of the exoskeleton of a crayfish.

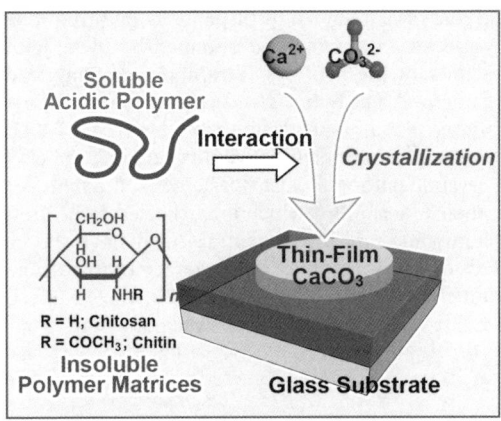

Fig. 2 Schematic illustration of the formation process for thin-film CaCO₃/organic polymer hybrids through a combination of soluble acidic polymers and insoluble polymers.

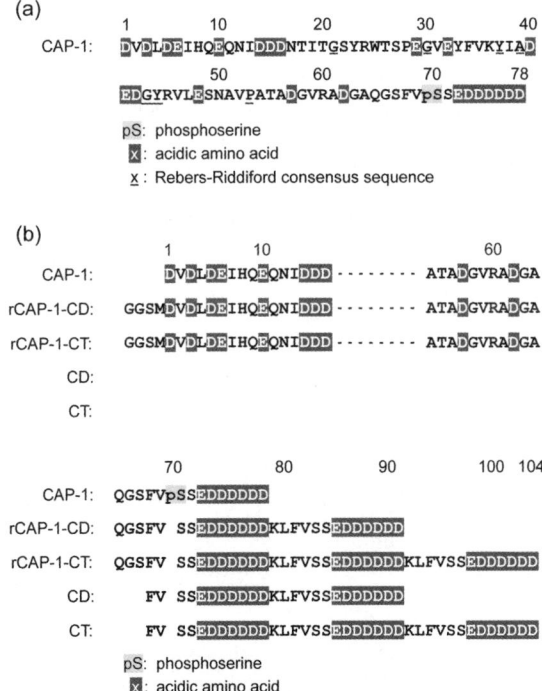

Fig. 3 (a) Sequence of the amino acids of CAP-1. (b) Comparison of sequences of CAP-1 and its related peptides. The full amino acid sequences of these peptides are presented in the supporting information.†

group, some nucleation points were generated and crystals grew rapidly after a certain induction time for crystallization, leading to the formation of the thin-film crystals having single crystallographic orientation with large domain. These results show that aligned acidic groups have great effect on the crystallization of CaCO₃.

We focus on the successive acidic moiety of the C-terminus of CAP-1 and rCAP-1. If the functions of these peptides are tuned by mutation, we expect to obtain more information on the role of acidic groups of peptides involved in biomineralization. Moreover, new hybrid structures may be obtained by these acidic peptides based on the tuning of acidity of the peptides. This approach may lead to development of new $CaCO_3$/organic polymer hybrids and related inorganic/organic hybrids.

In the present study, recombinant acidic peptides rCAP-1-CD and rCAP-1-CT (Fig. 3b) have been designed and synthesized to examine the effect of acidic parts of peptides on the crystallization of $CaCO_3$. These recombinant peptides have acidic tandem repeats at their C-terminus, which are expected to be useful to enhance the ability of the calcium binding. These peptides also have the chitin-binding RR sequence. The effect of these peptides that interact both calcium ions and chitin fibers on the calcium crystallization has been examined. The effect of simpler oligopepiedes CD and CT (Fig. 3b) has also been studied.

Experimental

Materials

Acetic acid, boric acid, calcium carbonate, calcium chloride, chitosan, dimethylacetamide, lithium chloride and 1-methyl-2-pyrrolidone were used as received from Wako (Tokyo). Chitin was also purchased from Wako. It was washed with 0.2 M HCl and 0.2 M NaOH and equilibrated with distilled water before use. Disodium dihydrogen ethylenediamine tetraacetate dihydrate was obtained from Kanto Chemicals (Tokyo) and used without further purification. Oligopeptides (CD and CT) were obtained from Bio-synthesis (Texas). The purified water used in this study was produced by an Autopure WT100 system (Yamato, Tokyo) and had a relative resistivity of about 18×10^6 Ω cm.

Cloning, expression and purification of recombinant peptides

The nucleotide sequence of rCAP-1 was codon optimized for *Escherichia Coli* (*E. Coli*) expression and chemically synthesized by GENEART (Germany). The rCAP-1 sequence was subsequently inserted into a modified pET28b vector containing a His_6-tag and a Tobacco etch virus (TEV) protease cleavage site at the N-terminal side. rCAP-1-CD and rCAP-1-CT expression vectors were produced from the rCAP-1 expression vector using a KOD-Plus-Mutagenesis Kit (TOYOBO, Osaka). The peptide expression was carried out in *E. coli* BL21 (DE3) cells transformed with each expression plasmid vector. *E. coli* cells were incubated in LB medium at 37 °C until the optical density at 600 nm reached 0.6. Peptide expression was induced with 0.5 mM isopropyl β-D-1-thiogalactopyranoside at 28 °C. Cells were harvested at 16 h after induction, and lysed by the sonication method. The cell lysate was centrifuged, and the supernatant was used for the following purification steps. Purification of these peptides was carried out using immobilized metal affinity chromatography (IMAC). Each supernatant was loaded onto an IMAC column. After washing the column, each target peptide was eluted with a buffer containing imidazole. The His_6-tag at the N-terminal of each peptide was cleaved with TEV protease. His_6-tag and TEV protease were removed by a second IMAC step. The resulting peptide-containing fractions were collected and dialyzed against water. After dialysis, the purified peptides were freeze-dried and stored until further use.

Preparation of polymer matrices

Chitin (1 wt%) was dissolved in a mixed solvent of *N*,*N*-dimethylacetamide and *N*-methyl-2-pyrrolidone (50/50: w/w) containing LiCl (3 wt%). The solution was spin coated on a glass substrate. The substrate was then washed with 2-propanol, water, and hot water (80 °C), in order. Chitosan (1 wt%) was dissolved in water with acetic acid (1 wt%). The chitosan matrix was obtained by spin coating and dried at 110 °C for 1 h.

Crystallization of $CaCO_3$ on polymer matrices

Calcium carbonate was crystallized by using its supersaturated aqueous solution. Preparation of the supersaturated solution was according to previous work.[5] Calcium carbonate (1.8 g L^{-1}) was suspended in deionized water. Carbon dioxide gas was bubbled into a stirred suspension for 3 h. The remaining solid $CaCO_3$ was then removed by filtration. The concentration of calcium ions in the solution was determined by standard titration with ethylenediaminetetraacetate. Each peptide solution of 2 μL (0.2 mg mL^{-1}) was placed on chitin and chitosan matrices. Calcium carbonate solutions of 11.3 μL (7.65 mM) were added to the peptide solutions. The final concentration of calcium ions was 6.5 mM and that of peptides was 3×10^{-3} wt%. The crystallization temperature was maintained at 25 °C in a thermostatted chamber.

Binding assay to polymer matrices

Chitin and chitosan binding assays were performed essentially according to a previous report.[13] The general procedure for the chitin-binding assay is described below. After the peptides (100 μg) were dissolved in 350 μL of 50 mM borate buffer, each solution was kept at 5 °C for 15 h with 25 mg of chitin powder. Then supernatant was collected after centrifugation. The residue was washed three times with 100 μL of 10 mM $CaCl_2$ solution. The three washings were combined and the measured absorption spectra ranged from 190 to 400 nm at 25 °C with a 1 mm quartz cell.

Circular dichroism spectral analysis

The solutions of recombinant peptides (2.0 mg mL^{-1}) were prepared with 50 mM borate buffer (pH 6.8). For circular dichroism titration, the solutions were adequately diluted with borate buffer containing $CaCl_2$.[14] The final concentration of each peptide was 0.1 mg mL^{-1}. Circular dichroism and absorption spectra were recorded from 190 to 250 nm at 25 °C with a 1 mm quartz cell.

Characterization

Polarizing optical microscopy images were taken using an Olympus BX51 polarizing optical microscope (Olympus, Tokyo). SEM images were obtained using a Hitachi S-4700 field-emission SEM (Hitachi, Tokyo) operated at 3 kV. The samples were platinum coated using a Hitachi E-1030 ion sputter. Circular dichroism spectra were obtained on a Jasco J-820 spectropolarimeter (Jasco, Tokyo). The absorption spectra were recorded on a Jasco V-670. TEM images were taken with a Jeol JEM-2010HC at 200 kV (Jeol, Tokyo). The mass spectra (MALDI-TOF mass) were recorded on a Voyager-DE STR spectrometer (AB Sciex Japan, Tokyo). Laser Raman spectra were obtained with a Jasco Model NR-1800 spectrometer with excitation by Nd:YAG laser (532 nm).

Results and discussion

Design and preparation of acidic peptides

A matrix peptide of the exoskeleton of a crayfish, named CAP-1 has a successive aspartic acid sequence ($FVSSED_6$) at the C-terminus[11a] (Fig. 3a). This C-terminus has great effect on $CaCO_3$ crystallization[15] because of its characteristic acidic nature. For example, when we used recombinant peptide that lacks these 17 amino acids at the C-terminal part, only granular nanocrystals were formed on the chitin matrix.[14] For further study on the function of the C-terminal acidic region on $CaCO_3$ crystallization, recombinant peptides, rCAP-1-CD and rCAP-1-CT were designed on the basis of CAP-1 with interest in their acidic nature, as shown in Fig. 3b.

These peptides have elongated C-terminal acidic moieties to increase interactions between the peptide and calcium ions. Two and three tandem repeats of the acidic region were introduced into the C-termini of both peptides, respectively (Fig. 3). It should be noted that each of the peptides presents a chitin binding RR sequence in its middle region. It is postulated that this motif is involved in specific interactions between chitin fibrils and peptides.[16] The RR sequence is often found in peptides of insect cuticular.[10]

We constructed the expression systems of the recombinant acidic peptides. The acidic recombinant peptides were expressed in *E. coli* cells and purified by an immobilized metal affinity chromatography column. MALDI-TOF mass analysis of these purified peptides indicated that all peptides were correctly translated and expressed, and had four extra amino acids (GGSM) at the N-terminus of each peptide due to a technical reason on His-tag purification. The effect of the artificial sequence on the function is expected to be small. The amount of obtained rCAP-1-CD and rCAP-1-CT from a 1 L culture was approximately 8 mg and 5 mg, respectively. We also obtained acidic oligopeptides without the chitin binding RR sequence, simply named as CD and CT, which have the sequence consisting of the C-terminal acidic region of each recombinant peptide.

Effect of the combination of peptides and the chitin matrix on $CaCO_3$ crystallization

In the presence of the recombinant acidic peptides rCAP-1-CD and rCAP-1-CT, the platelike tripodal calcite hybrids were formed on the chitin matrix. Both of the peptides affected the morphologies of the hybrids significantly (Fig. 4). $CaCO_3$ was crystallized from a supersaturated calcium hydrogen carbonate aqueous solution by slow evaporation of CO_2 on the chitin matrices spin-coated on glass substrates in the presence of peptides.[3c,5] In the presence of 3.0×10^{-3} wt% of the recombinant peptides, small grains of calcite (about 3 μm in size) started to grow after 10 min. The grains continued to grow (5 μm after 1 h and 10 μm after 12 h) and gradually formed platelike tripodal crystals. These crystals increased to 20 μm in size on the chitin matrix after 20 h, as shown in the polarizing optical microscopic images (Fig. 4a, b) and SEM images (Fig. 4c–f). It should be noted that two kinds of unidirectional orientations of the obtained platelike tripodal crystals are observed under polarizing optical microscopy observation (Fig. 4a, b). The platelike tripodal crystal in solid circle (type A) changed from bright to dark when the sample was rotated by 45° under crossed polarizers. In contrast, the platelike tripodal crystals in the dotted circle (type B) remained dark during sample rotation. These observations suggest that each crystal has unidirectional orientation of its *c*-axis, parallel (type A) or perpendicular (type B) to the surface of the chitin matrix. The number of crystals of type A was almost four times larger than that of type B (A : B is about 4 : 1).

The surface morphologies of the crystals have been examined using SEM as shown in Fig. 4c–f. The magnified images of the surfaces reveal that both type A and type B crystals are assemblies of small crystallites with a diameter of 100 nm. They were aligned to form unidirectional orientation for the whole of one tripodal calcite. On the basis of the magnified SEM image of the front area of a platelike tripodal calcite crystal (Fig. S1),† these crystals intimately associate with chitin fibrils and form calcium carbonate/chitin hybrids. The thickness of the obtained calcite was about 1 μm. Interestingly, hierarchical self-similar structures are observed in the magnified SEM image of the platelike tripodal crystal with perpendicular orientation (type B). The platelike tripodal calcite was composed of small nanocrystallites with a tripodal shape about 100 nm in size as shown in Fig. 4d and 3f. There are only a couple of reports about self-similar crystallization of $CaCO_3$.[17] Imai *et al.* reported that hierarchical self-similar calcite crystals with tripodal crystallites about 100 nm in size could be formed in silica gels. Cölfen *et al.* also reported self-similar calcite formation by using a block copolymer as additive.

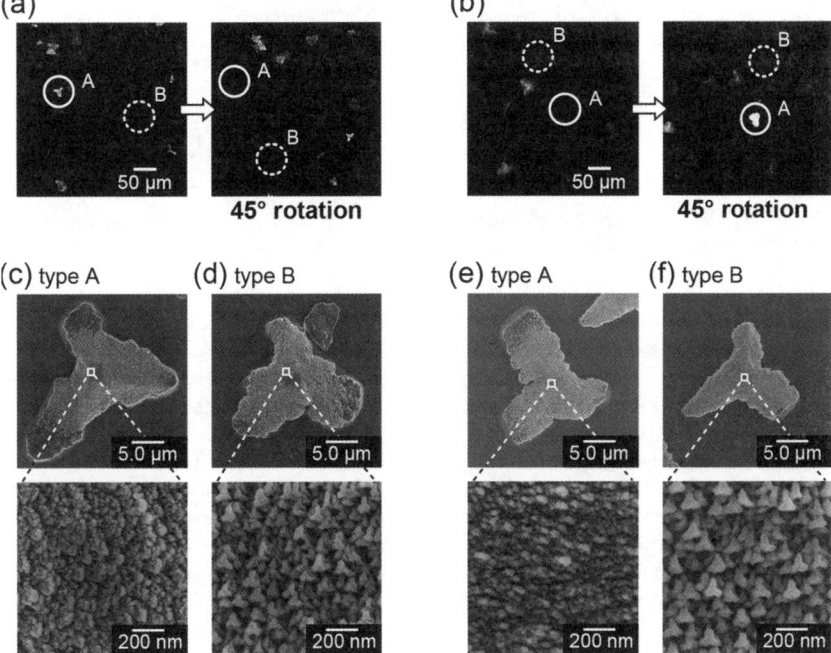

Fig. 4 Polarizing optical microscopic images of $CaCO_3$ crystals grown on the chitin matrix in the presence of (a) rCAP-1-CD and (b) rCAP-1-CT; the right-hand image is a 45° rotation of the sample in the image on the left. Solid circle: a platelike tripodal crystal which changes the birefringence under rotation (type A), dotted circle: a platelike tripodal crystal which remains dark (type B). (c, d) SEM images of the platelike tripodal calcite crystals grown on the chitin matrix in the presence of rCAP-1-CD. Both types of platelike tripodal crystals are shown (c) as type A and (d) as type B. Bottom: Magnified SEM images of the crystal surface in (c) and (d), respectively. (e, f) SEM images of the obtained calcite crystals formed in the presence of rCAP-1-CT.

TEM and electron diffraction were performed for these crystals on a tripodal calcite to examine the crystallographic orientation (Fig. 5 and 6). For this measurement, both types of platelike tripodal calcite were measured.

Fig. 5a shows a bright-field TEM image of a crystallite that changes birefringence under sample rotation (type A). The crystallite is also an assembly of small nanocrystals about 50 nm in size. The (110) plane of the calcite crystal is approximately parallel to the surface. The selected-area electron diffraction (SAED) pattern (Fig. 5b) showed that the crystals aligned with its c-axis parallel to the surface of the chitin matrix were same as the crystals obtained in the presence of peptide

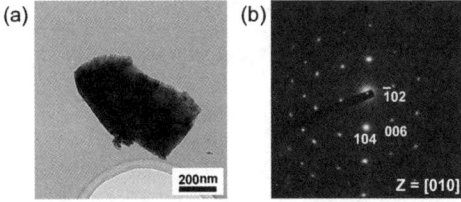

Fig. 5 (a) TEM image of the crystallite isolated from platelike tripodal calcite (type A). (b) Selected-area electron diffraction of the dotted area in (a).

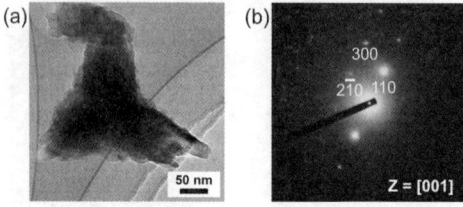

Fig. 6 (a) TEM image of the crystallite isolated from platelike tripodal calcite (type B). (b) Selected-area electron diffraction of the dotted area in (a).

CAP-1. The successive sequence of aspartic acids of rCAP-1-CT at the C-terminus might interact with the {110} face of calcite, and it grew with its {110} face parallel to the substrates. The selective adsorption of the C-terminal region of the peptide on {100} and {001} faces through stereo chemical recognition might lead to the formation of platelike tripodal crystals.

For type B, the bright-field TEM images and the corresponding SAED patterns (Fig. 6) of a crystallite show that they have single crystallographic orientation, while being assembles of small calcite plates. The *c*-axes of calcite crystalline assemblies are oriented perpendicular to the surface of chitin (type B). These observations suggest that the recombinant peptides induce oriented crystallizations of $CaCO_3$ through molecular specific interactions with particular surfaces.

$CaCO_3$ crystallization on chitosan matrices in the presence of acidic peptides

One of the characteristics of CAP-1 and its related peptides such as rCAP-1 is the specific chitin-binding ability.[5,11a,14,15] We previously showed that the synthetic oligopeptides without RR sequence induce the formation of non-oriented small grains on the chitin matrix.[4] In the present study, we have examined the $CaCO_3$ crystallization in the presence of the recombinant acidic peptides rCAP-1-CD and rCAP-1-CT on chitosan matrices. The purpose of this study is to undersand the effect of the specific binding between chitin and the RR sequence on the $CaCO_3$ crystallization by comparing them with the effect of weaker interactions between chitosan and the RR sequence.

$CaCO_3$ was crystallized on chitosan matrices from supersaturated calcium hydrogen carbonate aqueous solution by slow evaporation of CO_2. In the presence of rCAP-1-CD and rCAP-1-CT, aragonite disk-like thin films were formed on a chitosan matrix. The polymorph was confirmed using Raman spectroscopy. Fig. 7 shows images of the polarizing optical micrographs and SEM images for the thin-film crystals obtained for rCAP-1-CD (Fig. 7, left) and rCAP-1-CT (Fig. 7, right), respectively. Both thin films show the crossed extinction patterns due to the radial orientation of the *c*-axes (Fig. 7a, b). The size of these films is about 30 μm estimated from the SEM images (Fig. 7c, d, top). The magnified SEM image (Fig. 7c, d, bottom) clarified that grains 100 nm in size were self-assembled in the chitosan matrix to form a thin film hybrid. These results are similar to our previous results on the effect of the interactions between chitosan and poly(aspartic acid) on the $CaCO_3$ crystallization.[18] The effect of recombinant peptides rCAP-1-CD and rCAP-1-CT on crystallization is similar to that of simple acidic polymers such as poly(aspartic acid) in these conditions. These results suggest that the interactions between chitin and the RR sequence are specific and exert strong effect on $CaCO_3$ crystallization.

We assume that the specific interactions of peptides and chitin can form arranged and aligned acidic surfaces on the chitin matrix. This arrangement by the interactions between peptides and matrices is critical for the morphologies of the thin-film crystals because the binding manner determines the C-terminal activity to the

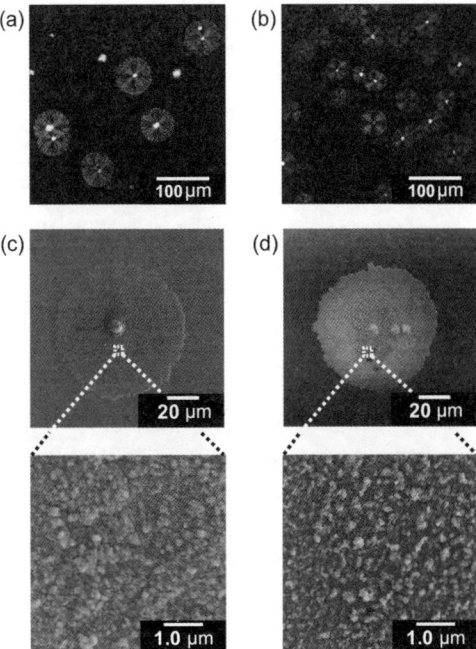

Fig. 7 Polarizing optical microscopy images of the thin-film crystals formed on chitosan matrices in the presence of (a) rCAP-1-CD and (b) rCAP-1-CT. SEM images of the thin-film crystals formed in the presence of (c) rCAP-1-CD and (d) rCAP-1-CT. Bottom: Magnified SEM images of the crystal surface in (c) and (d), respectively.

crystal surface. To confirm the binding abilities of the recombinant peptides to chitin and chitosan, the binding assays were performed using absorption spectroscopy (Fig. 8). The peptide solutions were stored in a thermostatted chamber with powdered chitin or chitosan. After centrifugation, the pellets were washed with 10 mM $CaCl_2$. The amounts of bound peptides were detected by absorption spectra.

The results of these experiments suggest that most of the peptide is bound strongly to chitin and only a small amount of peptides could be removed from the chitin using 10 mM $CaCl_2$ buffer (Fig. 8a, c). Although these peptides are also bound to chitosan, most of them elute in the buffer (Fig. 8b, d). The ability to bind to chitosan was lower than that of the chitin. These results suggest that the interactions between chitosan and peptides were too weak to form stable complexes, which disturbs the preparation of oriented crystals.

Function of the C-terminal acidic parts

Recombinant peptides rCAP-1-CD and rCAP-1-CT containing the elongated C-terminal parts have acidic regions that are longer than that of CAP-1. We expected these acidic parts to affect the morphologies of calcium carbonate crystals. We have used peptides CD and CT, which have only the acidic tandem repeat sequences to examine the effect of the C-terminus of the corresponding recombinant peptides rCAP-1-CD and rCAP-1-CT. Peptides CD and CT have no chitin binding moieties. In the presence of these peptides, the crystallization of $CaCO_3$ was performed in supersaturated calcium hydrogen carbonate aqueous solution with chitin matrices spin-coated on glass substrates. Dome structures with tripodal calcite crystallites were formed on the chitin matrix in the presence of both CD and CT (Fig. 9),

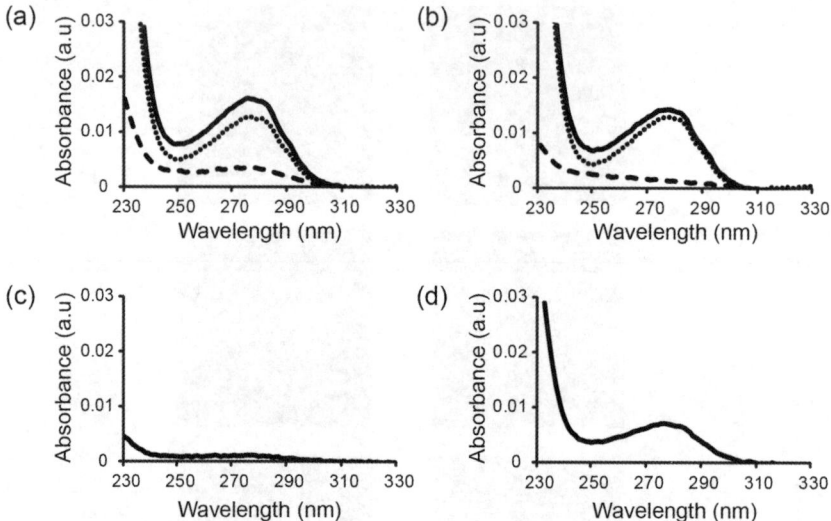

Fig. 8 Binding abilities of the recombinant peptides to (a) chitin and (b) chitosan; solid line: absorption spectra of rCAP-1-CT in borate buffer (pH = 6.8) at 25 °C (0.1 mg mL^{-1}) before mixing with chitin or chitosan; dashed line: absorption spectra of supernatants of the mixtures after storage for 15 h; dotted line: differences between the the two spectra, indicating the adsorption amount. (c, d) Absorption spectra of the rinse solutions containing calcium chloride ([Ca^{2+}] = 10 mM), which include rCAP-1-CT, extracted from (c) chitin and (d) chitosan.

while platelike tripodal crystals were obtained in the presence of recombinant peptides rCAP-1-CD and rCAP-1-CT (Fig. 4). No orientation of the crystals was observed for crystallization with CD and CT. We previously reported that an oligopeptide having only seven acidic residues induced peanut-shaped crystals with block-shaped crystallites about 600 nm in size on the chitin matrix.[7] In the presence of oligopeptides CD and CT having tandem repeating of the seven acidic residues, smaller crystallites about 200 nm in size are formed. These results suggest that the interactions between the crystals and oligopeptides partially suppress the crystal growth of CaCO$_3$.

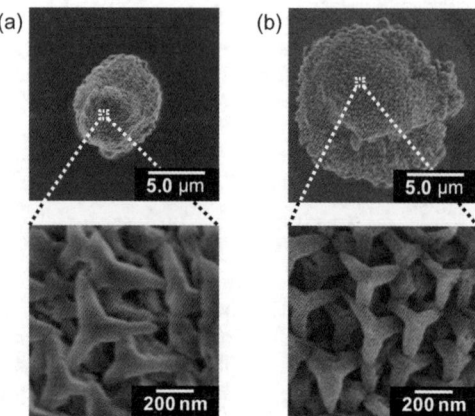

Fig. 9 SEM images of CaCO$_3$ crystals formed on the chitin matrix in the presence of (a) CD and (b) CT. Bottom: Magnified SEM images of the crystal surface.

Conclusions

We have examined the effect of the structures of the recombinant acidic peptides on the morphologies of $CaCO_3$ crystals in the crystallization processes to understand the structure–function relationships of the peptides working in biomineralization. The recombinant peptides have been designed on the basis of CAP-1 extracted from the exoskeleton of a crayfish. In the presence of these peptides, the formation of platelike tripodal calcite crystals was observed on the chitin matrix. The specific interactions of the crystal faces of $CaCO_3$ with negatively charged carboxylate at the C-terminal parts of the recombinant peptides play an important role in the formation of the tripodal morphology. Moreover, the chitin bindings of acidic peptides through RR consensus sequence control the crystallographic orientations of the crystals. The combination of these interactions may lead to forming crystals with hierarchical self-similar morphology. The results obtained in the present study are useful for the development of future organic/inorganic materials in which structure formation is controlled by organic molecules under mild conditions.

Acknowledgements

This work was partially supported by Grant-in-Aid for Scientific Research (No. 22107003) on the Innovative Areas: "Fusion Materials: Creative Development of Materials and Exploration of Their Function through Molecular Control" (Area no. 2206) (T.K.) and Global COE Program for Chemistry Innovation (T.K.) from the Ministry of Education, Culture, Sports, Science, and Technology (MEXT). We also thank the Center for Nano Lithography & Analysis of the University of Tokyo for the TEM measurements.

References

1. (a) *Handbook of Biomineralization*, ed. E. Bäuerlein, P. Behrens, and M. Epple, Wiley-VCH, Weinheim, 2007; (b) L. Addadi and S. Weiner, *Angew. Chem., Int. Ed. Engl.*, 1992, **31**, 153; (c) S. Mann, in *Inorganic Materials* 2nd Edition, ed. D. W. Bruce and D. O'Hare, John Wiley & Sons, New York, 1996, pp. 254–311; (d) Y. Levi-Kalisman, G. Falini, L. Addadi and S. Weiner, *J. Struct. Biol.*, 2001, **135**, 8.
2. (a) M. J. Olszta, X. G. Cheng, S. S. Jee, R. Kumar, Y.-Y. Kim, M. J. Kaufman, E. P. Douglas and L. B. Gower, *Mater. Sci. Eng., R*, 2007, **58**, 77; (b) J. Aizenberg, A. Tkachenko, S. Weiner, L. Addadi and G. Hendler, *Nature*, 2001, **412**, 819; (c) K. Shimizu, J. Cha, G. E. Stucky and D. E. Morse, *Proc. Natl. Acad. Sci. U. S. A.*, 1998, **95**, 6234; (d) A. Arakaki, H. Nakazawa, M. Nemoto, T. Mori and T. Matsunaga, *J. R. Soc. Interface*, 2008, **5**, 977.
3. (a) A. P. Jackson, J. F. V. Vincent and R. M. Turner, *Proc. R. Soc. London, Ser. B*, 1988, **234**, 415; (b) K. Okumura and P.-G. de Gennes, *Eur. Phys. J. E: Soft Matter Biol. Phys.*, 2001, **4**, 121; (c) T. Kato, T. Suzuki and T. Irie, *Chem. Lett.*, 2000, 186.
4. (a) D. F. Travis, *Ann. N. Y. Acad. Sci.*, 2006, **109**, 177; (b) D. Raabe, P. Romano, C. Sachs, A. Al-Sawalmih, H.-G. Brokmeier, S.-B. Yi, G. Servos and H. G. Hartwig, *J. Cryst. Growth*, 2005, **283**, 1; (c) D. Raabe, C. Sachs and P. Romano, *Acta Mater.*, 2005, **53**, 4281; (d) A. Becker, A. Ziegler and M. Epple, *Dalton Trans.*, 2005, 1814.
5. A. Sugawara, T. Nishimura, Y. Yamamoto, H. Inoue, H. Nagasawa and T. Kato, *Angew. Chem., Int. Ed.*, 2006, **45**, 2876.
6. (a) H. Cölfen and S. Mann, *Angew. Chem., Int. Ed.*, 2003, **42**, 2350; (b) F. C. Meldrum and H. Cölfen, *Chem. Rev.*, 2008, **108**, 4332; (c) N. A. J. M. Sommerdijk and G. de With, *Chem. Rev.*, 2008, **108**, 4499; (d) Y.-Y. Kim, E. P. Douglas and L. B. Gower, *Langmuir*, 2007, **23**, 4862; (e) C.-L. Chen and N. L. Rosi, *Angew. Chem., Int. Ed.*, 2010, **49**, 1924; (f) M. Umetsu, M. Mizuta, K. Tsumoto, S. Ohara, S. Takami, H. Watanabe, I. Kumagai and T. Adschiri, *Adv. Mater.*, 2005, **17**, 2571; (g) T. Kato, *Adv. Mater.*, 2000, **12**, 1543; (h) K. Naka and Y. Chujo, *Chem. Mater.*, 2001, **13**, 3245; (i) H. Imai and Y. Oaki, *MRS Bull.*, 2010, **35**, 138; (j) T. Kato, T. Sakamoto and T. Nishimura, *MRS Bull.*, 2010, **35**, 127.
7. (a) F. Marin and G. Luquet, in *Handbook of Biomineralization: Biological Aspects and Structure Formation*, ed. E. Bäuerlein, Wiley-VCH, Weinheim, 2007, vol. 1, ch. 16, pp. 273–290; (b) F. C. Meldrum, *Int. Mater. Rev.*, 2003, **48**, 187; (c) M. Suzuki,

K. Saruwatari, T. Kogure, Y. Yamamoto, T. Nishimura, T. Kato and H. Nagasawa, *Science*, 2009, **325**, 1388; (d) S. Yang, G. Chen, M. Megens, C. K. Ullal, Y.-J. Han, R. Rapaport, E. L. Thomas and J. Aizenberg, *Adv. Mater.*, 2005, **17**, 435; (e) T. Kato, A. Sugawara and N. Hosoda, *Adv. Mater.*, 2002, **14**, 869.
8 (a) T. Kato, T. Suzuki, T. Amamiya, T. Irie, M. Komiyama and H. Yui, *Supramol. Sci.*, 1998, **5**, 411; (b) N. Hosoda and T. Kato, *Chem. Mater.*, 2001, **13**, 688; (c) A. Sugawara, T. Ishii and T. Kato, *Angew. Chem., Int. Ed.*, 2003, **42**, 5299; (d) T. Sakamoto, A. Oichi, T. Nishimura, A. Sugawara and T. Kato, *Polym. J.*, 2009, **41**, 522; (e) T. Nishimura, T. Ito, Y. Yamamoto, M. Yoshio and T. Kato, *Angew. Chem., Int. Ed.*, 2008, **47**, 2800.
9 (a) S. Raz, S. Weiner and L. Addadi, *Adv. Mater.*, 2000, **12**, 38; (b) S. Tugulu, M. Harms, M. Fricke, D. Volkmer and H.-A. Klok, *Angew. Chem., Int. Ed.*, 2006, **45**, 7458; (c) J. Aizenberg, D. A. Muller, J. L. Grazul and D. R. Hamann, *Science*, 2003, **299**, 1205; (d) L. B. Gower, *Chem. Rev.*, 2008, **108**, 4551; (e) E. M. Pouget, P. H. H. Bomans, J. A. C. M. Goos, P. M. Frederik, G. de With and N. A. J. M. Sommerdijk, *Science*, 2009, **323**, 1455; (f) B. Yeom and K. Char, *Chem. Mater.*, 2010, **22**, 101; (g) Y. Yamamoto, T. Nishimura, T. Saito and T. Kato, *Polym. J.*, 2010, **42**, 583.
10 J. E. Rebers and L. M. Riddiford, *J. Mol. Biol.*, 1988, **203**, 411.
11 (a) H. Inoue, N. Ozaki and H. Nagasawa, *Biosci., Biotechnol., Biochem.*, 2001, **65**, 1840; (b) H. Inoue, T. Ohira, N. Ozaki and H. Nagasawa, *Biochem. Biophys. Res. Commun.*, 2004, **318**, 649.
12 A. Hecker, O. Testenière, F. Marin and G. Luquet, *FEBS Lett.*, 2003, **535**, 49.
13 J. Folders, J. Tommassen, L. C. van Loon and W. Bitter, *J. Bacteriol.*, 2000, **182**, 1257.
14 H. Inoue, T. Ohira and H. Nagasawa, *Peptides*, 2007, **28**, 566.
15 Y. Yamamoto, T. Nishimura, A. Sugawara, H. Inoue, H. Nagasawa and T. Kato, *Cryst. Growth Des.*, 2008, **8**, 4062.
16 (a) S. O. Andersen, P. Højrup and P. Roepstorff, *Insect Biochem. Mol. Biol.*, 1995, **25**, 153; (b) V. A. Iconomidou, G. D. Chryssikos, V. Gionis, J. H. Willis and S. J. Hamodrakas, *Insect Biochem. Mol. Biol.*, 2001, **31**, 877.
17 (a) H. Imai, T. Terada and S. Yamabi, *Chem. Commun.*, 2003, 484; (b) A.-W. Xu, M. Antonietti, S.-H. Yu and H. Cölfen, *Adv. Mater.*, 2008, **20**, 1333.
18 A. Sugawara, A. Oichi, H. Suzuki, Y. Shigesato, T. Kogure and T. Kato, *J. Polym. Sci., Part A: Polym. Chem.*, 2006, **44**, 5153.

General discussion

Professor Meldrum opened the discussion of the paper by Professor P. U. P. A. Gilbert by communicating: In your paper you state that the growth of the nacre layer is correlated to the circadian rhythm of the mollusk. Given that not all animals have a circadian rhythm, how do you know that this mollusk does?

Professor Gilbert communicated in reply: I don't know that at all, and I don't know that nacre grows one layer per day. All we say in our article[2] is that it is not unreasonable. The growth rate of nacre in *Pinctada margaritifera* shells is not known. In *Pinctada margaritifera* pearls it has been reported[1] to be greater than 3.45 micron/day, which given the layer thickness of 0.67 μm reported in ref. 2 corresponds to 5.2 layers/day. However, that was measured in a completely different environment (the open Gazi Bay, Kenya rather than the inner lagoon of Rangiroa, French Polynesia), temperature range, and most importantly in a different biomineral: a human-seeded pearl in the mollusk gonad,[3] rather than the naturally forming mollusk shell nacre.

More circumstantial evidence: other non-nacre-forming mollusk shells have been reported to follow circadian rhythms[4,5] hence it is not unreasonable that one nacre layer per day is formed in *Pinctada margaritifera*. We note that in *Haliotis rufescens* nacre a very similar growth rate was reported: 0.5 micron/day, layer thickness is 0.44, thus this is just under 1 layer/day.[6,7]

1. K. M. Mavuti, E. N. Kimani and T. Mukiama, *Afr. J. Marine Sci.*, 2005, **27**.
2. I. C. Olson, R. Kozdon, J. W. Valley and P. U. P. A. Gilbert, *J. Am. Chem. Soc.*, 2012, **134**.
3. J. P. Cuif, A. D. Ball, Y. Dauphin, B. Farre, J. Nouet, A. Perez-Huerta, M. Salome and C. T. Williams, *Microsc. Microanal.*, 2008, **14**.
4. B. R. Schone, S. D. Houk, A. D. F. Castro, J. Fiebig, W. Oschmann, I. Kroncke, W. Dreyer and F. Gosselck, *Palaios* 2005, **20**, 78.
5. G. R. Clark, *Ann. Rev. Earth Planet. Sci.*, 1974, **2**.
6. A. Lin and M. A. Meyers, *Mater. Sci. Eng. A*, 2005, **390**.
7. A. Y. M. Lin, P. Y. Chen and M. A. Meyers, *Acta Biomater.*, 2008, **4**.

Dr Nudelman asked: It is known that it is during the summer that mollusks actively grow their shells. How does this seasonal growth correlate with your model?

Professor Gilbert replied: Although some shells only grow in the summer in certain locations, this is not the case for the shell analyzed here: *Pinctada margaritifera*, grown in French Polynesia. The temperature fluctuations throughout the year are within only 3 °C, and the shells grow all year round. In other mollusk shells, *e.g. Haliotis rufescens*, thick organic layers termed "growth lines" are formed when nacre stops being deposited.[1] No such things are ever observed in *Pinctada margaritifera* shells.

1. X. W. Su, A. M. Belcher, C. M. Zaremba, D. E. Morse, G. D. Stucky and A. H. Heuer, *Chem. Mater.*, 2002, **14**.

Professor Van Blaaderen remarked: The angle spread of crystalline aragonite tablets found to correlate with the growth temperature is over a very small temperature range. Would it be possible to check this correlation under better controlled conditions, *e.g.* by having the mollusks grow in a temperature controlled tank?

Professor Gilbert answered: Indeed, that is a very easy experiment to do, although we will have to wait at least a year, and preferably many years, to have a significant sample of temperature variations while new nacre is deposited by mollusks. In nature temperature variations occur spontaneously, they are accurately recorded by buoys in many locations around the world, and at various depths in each location. Thus collecting shells from a nearby buoy is more efficient, more informative, and more representative than nacre formed in an aquarium, which could be prone to artifacts. But we will definitely do both experiments, in nature and in the aquarium.

Dr Christenson enquired: Being very skeptical, I would say that if the four data points with angle spreads above 40° were excluded from Fig. 7 of the paper it is possible that most of the correlation would be lost. It would be interesting to calculate the correlation coefficient without these points.

Professor Gilbert replied: I have re-done the linear fit after excluding the 4 points, and the correlation coefficient is lower: $R = 0.61$, but the correlation still stands ($R > 0.4$).

Professor Van Blaaderen opened the discussion of the paper by Dr Stephan E. Wolf: A colloidal crystal is defined exactly like an atomic or molecular crystal, except for the fact that the crystallizing units are colloidal particles (roughly particles with dimensions between a few nm and a few micrometers). The use of the wording 'colloidal crystal' in this paper is not compatible with this definition.

Dr Wolf replied: Our phrasing originates from the definition of a mesocrystal given by Cölfen *et al.*, *e.g.* in 2006 "Mesocrystals are colloidal crystals composed of individual nanocrystals that are aligned in a common crystallographic fashion, exhibiting scattering properties similar to a single crystal." (ref. 1) or in 2008 "The notation Mesocrystal is an abbreviation for Mesoscopically structured crystal. We define mesocrystals as colloidal crystals which are built up from individual nanocrystals, which are aligned in a common crystallographic register." (ref. 2). Accepting the common definition you provided, we agree with your comment. This shows that the definition of a mesocrystal is inadequate and needs revision.

1. M. Niederberger and H. Cölfen, *Phys. Chem. Chem. Phys.*, 2006, **28**, 3271–3287.
2. H. Cölfen and M. Antonietti, *Mesocrystals and Nonclassical Crystallization*, Wiley 2008, p. 96.

Dr Nudelman commented: In Fig. 5 of your paper you show SEM images taken from a calcitic prism from *Pinna*, where you say that the imprint seen on the surface suggests that it formed from a PILP-like phase. I am concerned that there is no real evidence for that. The fact that this imprint looks similar to what you observed from *in vitro* experiments (Fig. 7b) is no proof. I believe that looking at the periostracal side of the prism—where it nucleated from—is not very informative, since the mineral has already formed and crystallized. How can you find evidence for a PILP phase then? If you want to look for evidence of a PILP-like phase, it has to be done on the other side of the prism, which is the one where new material is being deposited. I understand, though, that it is technically very difficult to isolate and observe such a liquid precursor phase.

In my previous work (Nudelman *et al.*, *Faraday Discuss.*, 2007, **136**, 9–25), I have observed the deposition of 50 nm calcium carbonate globules on the growing surface of a prism. Of course, I cannot make conclusions whether this is a liquid precursor or not, but nevertheless how does this conform with your model of prism formation through a PILP-phase?

Dr Wolf remarked: Firstly, the SEM images we provide are no proof if taken alone. Our AFM analyses show also strong similarity with the nanogranular structures of PILP products.[1] Based on our data, we developed a model how these structures may be formed and how the space-filling characteristics of the biogene mesocrystal could be simply achived which is in principle atypical for mesocrystals formed by aggregation of already solidified particles.[2] We present here our speculative model but we know for a literal proof it is still a long way to go.

Looking at the periostracal face of the prism shows us the very first layer of deposited material, which settled on a reasonably smooth surface, the periostracum. Why should this be incomparable to recent *in vitro* experiments, in which the PILP process was traced by the morphology of the mineral, although "the mineral has already formed and crystallized"? The PILP process was initially detected just by strange crystal morphologies which were explained by the settlement of a liquid intermediate,[3] but the identification of a pure liquid condensed phase of amorphous calcium carbonate came much later.[4,5] Thus, it can be informative to look at already formed and crystallized mineral.

Your observation of 50 nm calcium carbonate globules does conform very well with our proposed model, your estimated size complies well with the size of our nanogranules characterized by AFM. The crucial idea of our model is that these nanogranules are still liquid resp. viscous when they attach to the growing surface of the prism! This idea is *inter alia* supported by the oblate shape of the nanogranules (Fig 3b). If they were already crystalline or solid during attachment, a complete space-filling could not be achieved but this is characteristic of biogene mesocrystals.[2]]

1. Y.-Y. Kim, E. P. Douglas and L. B. Gower, *Langmuir*, 2007, **23**, 4862–4870.
2. L. Yang, C. E. Killian, M. Kunz, N. Tamura and P.U.P.A. Gilbert, *Nanoscale*, 2011, **3**, 603–609.
3. L. B. Gower and D. A. Tirell, *J. Cryst. Growth*, 1998, **191**, 153–160.
4. J. Rieger *et al.*, *Faraday Discuss.*, 2007, **136**, 265.
5. S. E.Wolf *et al.*, *J. Am. Chem. Soc.*, 2008, **130**(37), 12342–7.

Dr Christenson remarked: I had trouble understanding the statement on p. 10 of your paper, at the end of the second-last paragraph. Surely, the low contrast variation of the droplets is not directly due to a liquid-like character, but rather shows that the edges are as thick as the centre. For example, a hemispherical liquid droplet on a surface would show exactly the same contrast variation as an undistorted, solid (or liquid) sphere. The low contrast variation thus indicates a thinly spread (and therefore almost certainly liquid-like) layer. I hope that I am not being too pedantic here!

Dr Wolf communicated in reply: This is exactly what we wanted to state. Thank you for your comment, it is not pedantic at all as it is crucial for understanding our contribution. We rephrased the statement in question.

Professor Meldrum asked: In your paper you very clearly show using AFM that the prisms have a granular structure. Is this unique to biogenic calcite, or can synthetic calcite (for example formed *via* a PILP phase) show this too?

Dr Wolf communicated in reply: To begin with geological calcite, which was used, *e.g.*, in growth and dissolution experiments, it does not show such a substructure;[1–3] cleavage along {104} yields surfaces which are in principle atomically flat.

Dr Beck replied: According to the authors, Fig 5 (b) and (c) show a nucleation knob that is reminiscent of a coalesced droplet.[1] In literature there is a crystal formation mechanism that can explain the formation of the observed islands on the crystal surface. The growth mechanism polynuclear growth can explain the observations. Polynuclear growth leads to 2D-nuclei that form on the crystal surface that spread out with time forming islands. This is a general growth mechanism occurring for many different substances.[2] Generally speaking, at very low supersaturations, the prevailing formation mechanism is spiral growth. Increasing the supersaturation leads to a shift in the growth mechanism to 2D-nucleation—a category to which also polynuclear growth belongs. Increasing the supersaturation further leads to dendritic growth, and at even higher supersaturation spherulitic growth can be observed.[3,4] Is it possible that in the current case the polynuclear growth mechanism is responsible for the observed features on the crystal surface?

1. S. E. Wolf, I. Lieberwirth, F. Natalio, J.-F. Bardeau, N. Delorme, F. Emmerling, R. Barrea, M. Kappl and F. Marin, *Faraday Discuss.*, **159**, 2012.
2. I. Sunagawa, *Crystals: Growth, morphology and perfection*, Cambridge University Press, Cambridge, 2005, pp 37–53.
3. J. P. Andreassen, *J. Cryst. Growth*, 2005, **274**, 256–264.
4. R. Beck and J. P. Andreassen, *Cryst. Growth Des.*, 2010, **10**, 2934–2947.

Dr Wolf communicated in reply: That is basically correct, the polynuclear crystal growth mechanism may indeed explain such rounded, island-like morphologies on a crystal's growing surface. But at the moment I am not aware of a single literature example with a comparable morphology at a lengthscale of micrometers and which was surely obtained by a polynuclear crystal growth mechanism. We believe that the polynuclear crystal growth mechanism can not explain well the nanoscaled composite structure: calcium carbonate nanogranules which are coated by an organic moiety whose overall distribution in the biomineral is quite well described by a 3D vesicular texture, *i.e.* a sponge-like and pumiceous network.[1]

1. D. Jacob, A. Soldati, R. Wirth, J. Huth, U. Wehrmeister and W. Hofmeister, *Geochim. Cosmochim. Acta*, 2008, **72**, 5401–5415.

Professor Gilbert asked: Another paper showed evidence of nano-components in the prisms of mollusk shells, and must be cited: P. U. P. A. Gilbert, A. Young and S. N. Coppersmith, Proc. Natl. Acad. Sci. U. S. A., 2011, **108**.

Dr Wolf replied: It is a pleasure for us to cite your paper. We included now as well a citation of your contribution in *Nanoscale* [L. Yang, C. E. Killian, M. Kunz, N. Tamura and P.U.P.A. Gilbert, *Nanoscale*, 2011, **3**, 603–609, in which you propose a space-filling amorphous calcium carbonate based on comparisons between synthetic and biogenic mesocrystals employing BET. We believe strongly that the liquid resp. viscous amorphous calcium carbonate precursor, which we proposed here, is exactly this space-filling amorphous calcium carbonate.

Dr Schenk asked: Your paper addresses the functions of the soluble fraction of bioproteins extracted from mollusc shell prisms. However, it has previously been demonstrated that calcite crystals from the prismatic layer of the mollusc *Atrina rigida* additionally contain a substantial amount of insoluble chitin fibres.[1] Does such an intracrystalline framework exist in *Pinna nobilis* prisms as well? If yes, can you comment on the role that the insoluble matrix plays in the process of calcium carbonate precipitation? Is the specific chemistry of the insoluble fibres important for mineral nucleation or do they merely provide a scaffold for the deposition of the precursor material such that an anisotropic orientation of the occluded organic material along the *c*-axis is achieved as seen by Li *et al.* in prisms extracted from *Atrina rigida*[2] and by Gilow *et al.* in *Pinna nobilis* prisms?[3]

1. Fabio Nudelman, Hong H. Chen, Harvey A. Goldberg, Steve Weiner and Lia Addadi, *Faraday Discuss.*, 2007, **136**, 9–25.
2. Hanying Li, Huolin L. Xin, Miki E. Kunitake, Ellen C. Keene, David A. Muller and Lara A. Estroff, *Adv. Funct. Mater.*, 2011, **21**(11), 2028–2034.
3. Christoph Gilow, Emil Zolotoyabko, Oskar Paris, Peter Fratzl, and Barbara Aichmayer, *Cryst. Growth Des.*, 2011, **11**(6), 2054–2058.

Dr Wolf responded: We do not know so far whether this chitinous network exists as well in the prisms of *Pinna nobilis*. Based on the present data, I would assume that the chitin fibres enhance distinctly the mechanical properties of the prisms but are not explicitly involved in the nucleation process of calcite. Our HAADF/SAED analysis of young prism in our paper shows that the nucleation of the crystalline phase starts on the interface of the prism and its surrounding periprismatic sheaths. If a fibrinous network embedded in the mineral steers the nucleation, we should observe not a centripetally crystal ripening but a crystal nucleation which is at least randomly distributed over the whole mineral volume.

From our actual point of view, the anisotropic distribution of the organic material is not due to a fibrinous network but due to the layerwise deposition of the nanogranules which are covered by a proteinacous matrix: they represent an organic matrix which was occluded between two consecutively deposited layers of nanogranules.

Dr Schenk enquired: In your contribution you suggest a model describing the prisms as being composed of "protein-enwrapped nanogranules which are surrounded by a thin layer of amorphous calcium carbonate". In a recent publication Gilow et al.[1] investigated the ultrastructure of calcite prisms extracted from *Pinna nobilis* shells by means of small-angle x-ray scattering (SAXS) as well as microbeam simultaneous small-and wide-angle X-ray scattering. The authors obtained SAXS profiles showing deviations from the Porod law which can be interpreted either as roughnesses at the organic–inorganic interfaces in the calcium carbonate prisms or as electron density heterogeneities within the biopolymer and/or mineral phases. In your point of view, can the amorphous material you observe even after maturation of the prisms give an explanation for the findings reported by Gilow et al.?

1. Christoph Gilow, Emil Zolotoyabko, Oskar Paris, Peter Fratzl, and Barbara Aichmayer, *Cryst. Growth Des.*, 2011, **11**(6), 2054–2058.

Dr Wolf communicated in reply: Yes, I think so. In fact, the scattering curves of Gilow et al. resemble those of Seto et al.[1] of sea urchin spines, who—in contrast to the interpretation of Gilow et al.—attributed the observed behaviour to the presence of scattering objects with two different size distributions. Especially the fact that the deviation from the Porod law vanishes after annealing the prisms to 300° C is indicative.

1. Seto, J. et al., *Proc. Natl. Acad. Sci. U. S. A.*, 2012, **109**(10), 3699–3704.

Mr Seto remarked: The idea of merging models of biomineralization is interesting and exciting, specifically the inclusion of nonclassical crystallization mechanisms in forming these elaborate structures. I do have a question regarding the composition of these so-called 'liquid-like' precursors. Are these constituents chemically similar to what Gower et al. observes in composition to calcium carbonate related PILP? If so, what degrees of hydration do these precursors have in relation to ACC1, ACC2, etc...? Do we know anything about the organic components which may/may not stabilize these 'liquid-like' phases?

Dr Wolf responded: We believe that the liquid amorphous calcium carbonate (LACC, or as termed by Laurie Gower the liquid condensed phase their paper) is rich in water[1] and much more hydrated than ACC1 or ACC2. We cannot currently provide an exact composition but we are sure that in course of time LACC loses water until it reaches a monohydrated state, thus becoming normal ACC and later eventually anhydrous ACC. Acidic (bio)polymers can extend the lifespan of such a liquid ACC phase considerably due to reduction of the supersaturation and depletion stabilization whereas basic ones break the emulsion and lead to an earlier formation of a crystalline phase.[2] The exact constitution of the intracrystalline proteins are diverging from specimen to specimen but both types of proteins, basic and unusual acidic ones (pI = 4) can be found in the organic shell matrix.[3,4] There are two types of unusual acidic proteins: (a) acidic amino acids dominate in the primary structure of the respective protein. (b) The respective protein experienced post-translational modifications which introduced additional acidic groups to the peptide core, e.g., phosphorylation or glycosylation with sialic (e.g. AP8 from *Haliotis rufescens*).[4] But how the proteins directly affect the LACC phase is still a question to answer.

1. S. E. Wolf et al., *J. Am. Chem. Soc.*, 2008, **130**(37), 12342–7.
2. S. E. Wolf et al., *J. Am. Chem. Soc.*, 2011, **133**(32), 12642–12649.
3. F. Marin and G. Luquet, Unusually Acidic Proteins in Biomineralization, in: *Handbook of Biomineralization: Biological Aspects and Structure Formation*, ed. E. Bäuerlein. Wiley-VCH, 2007.
4. F. Marin, G. Luquet, B. Marie and D. Medakovic, *Curr. Top. Dev. Biol.*, 2008, **80**, 209–76.

Professor Gale said: With regard to the discussion of the use of Pair Distribution Functions to probe the structure of ACC by Goodwin et al. (*Chem. Mater.*, 2010, **22**, 3197–3205], it should be noted that one of the two structures obtained by Reverse Monte Carlo contains unphysical overlaps between carbonate groups. Combining the use of the energy from a force field method with the Reverse Monte Carlo might help to improve the structures obtained. It is also worth noting that a recent molecular dynamics study has probed the properties of the sensible starting structure and found that there is "significant structural reorganisation" during the relaxation (J. W. Singer et al., *Chem. Mater.*, 2012, **24**, 1828–1836).

Dr Wolf enquired: Where do we draw the thin line between nonclassical and classical crystallization? Let us consider the PILP process: for me, I am sure that the PILP process can be basically explained by a liquid phase separation *via* either a spinodal or binodal decomposition—just as Faatz proposed in *Adv. Mater.* in 2004 (ref. 1), in our contribution in *J. Am. Chem. Soc.*, 2008 (ref. 2) and you in the concluding remarks of Jim. This would be well covered by the theoretical works of Cahn and Hilliard.[3] The main precondition would be a sufficient high supersaturation. Then, in the subsequent emulsified state, we can affect the liquid condensed phase by additives—just like a classical emulsion: coalescence may occur due to sheering, salts screen the interaction potential and lead to aggregation,[4] polymers may induce depletion stabilization and destabilization (DOI: 10.1039/c2fd20045g). This would be covered completely by the classical colloidochemical theories. So, if we accept a liquid phase separation *via* either a spinodal or binodal decomposition, the PILP process is covered fully by classical theories, thus the PILP process is not so nonclassical as it firstly seems. Would you agree? Which theories do we define as the fundamental, thus classical theories of crystallization and nucleation?

1. M. Faatz et al., *Adv. Mater.*, 2004, **16**(12), 996–1000.
2. S. E. Wolf et al., *J. Am. Chem. Soc.*, 2008, **130**(37), 12342–7.
3. J. W. Cahn and J. E. Hilliard, *J. Chem. Phys.*, 1958, **44**, 258–267; J. W. Cahn, *J. Chem. Phys.*, 1959, **30**, 1121–1124; J. W. Cahn and J. E. Hilliard, *J. Chem. Phys.*, 1959, **31**, 688–699.
4. S. E. Wolf et al., *Nanoscale* 2011, **3**, 19–21.
5. S. E. Wolf et al., *J. Am. Chem. Soc.*, 2011, **133**(32), 12642–12649.

Dr Gower answered: I think this is a bit of an oversimplification because the polymer not only stabilizes the phase, as an emulsifier would, but actively produces far more of the phase (see Fig. 8 of the *Faraday* paper). In addition, the stabilization of the phase by polymer directs it through the ACC pathway, rather than dissolution and recrystallization (which would lead to normal calcite rhombs for example). There seems to be dehydration of the PILP to a more solid ACC phase, so without a distinct nucleation event leading to the ACC phase, this seems non-classical. Lastly, one then gets the pseudo-solid-state transformation of the ACC phase. So you no longer have ion-by-ion addition as in classical crystal growth. So if the crystal growth in the ACC phase is not classical, is the crystal nucleation of the ACC phase classical? Overall, you have multiple steps: "nucleation" *via* condensation of a distinct liquid phase, densification of this phase to a more solid amorphous phase, and then crystal nucleation within the solid ACC phase followed by solid-state transformation. I don't know—does that fit with classical theory?

Mr Bewernitz responded: This is a wonderful point. When is "non-classical" nucleation really non-classical? Could it be due to various classical processes occurring in solution in parallel that, when forced to appear in series (single pathway), seem complex (non-classical)? If there is an initial classical liquid–liquid separation, followed by a classical solid nucleation within one of the liquid phases, as Stephan Wolf has suggested above, we would have a classical solid nucleation within a classically liquid nucleated phase which would be classical, even if there is an additional classical nucleation event simultaneously occurring in the mother solution.

However, LCP and PILP systems are an interesting frontier where non-classical nucleation might exist in a true sense. If the LCP and/or the PILP phase are dehydrating from liquid to viscous liquid to solid without discontinuity in behavior, this could be a "nucleationless" transition which, by definition I believe, may be considered to be a strong candidate for non-classical nucleation. According to some preliminary data given by Jim in his concluding remarks, the AFM indentation modulus of PILP-generated calcium carbonate material is hardening (dehydrating?) with time. If this study could be conducted closer and closer to PILP formation (without adsorption to a solid surface) and continues to show a continuous change in modulus from liquid to solid, then there is a strong case that the overall PILP process may be non-classical.

Dr Staniland enquired: How does the *in vitro* experiments inform what happens in nature?

Dr Wolf answered: This is a very far-ranging question. If we can show a certain fact *in vitro*, *e.g.* an protein–protein interaction, certain protein folding behaviour or, as in the present case, a certain result during biomimetic crystallization experiments, we can infer that it **might** happen as well *in vivo* and that the drawn conjecture appears to be plausible. Remembering Sir Karl Raimund Popper's "The Logic of Scientific Discovery", a scientific theory can not be literally proven by experiments but only disproven. Translated to our present case, our findings led us to the conjecture of the possible existence of a liquid amorphous calcium carbonate *in vivo* and we propose this idea. We do not claim to provide a decisive proof but distinct indications: *e.g.* the space-filling nanoscaled substructure which is identical to those of PILP products and the fact that similar mineral morphologies could be obtained *in vitro*.

Professor Evans opened the discussion of the paper by Professor John Spencer Evans: Do you have any additional information regarding the location of Asprich proteins and the beta chitin polysaccharide in *Atrina rigida* prismatic crystals?

Dr Nudelman communicated in reply: From what we observed, the Asprich proteins seem to be associated to calcium carbonate granules when they are deposited on the surface of the prisms at the growth front. These granules are deposited on the chitin, and hence upon etching of the mineral or extraction of the intracrystalline material the Asprich is found strongly bound to the chitin. We cannot tell whether the Asprich is located in specific regions in the prism or adsorbed to a specific crystal plane. It has been proposed by Gotliv *et al.* (*ChemBioChem*, 2005, **6**, 304–314) that this protein may be adsorbed to the (001) plane of a growing prism, but there is no evidence for that. Regarding whether the chitin has any particular structure or organization inside the prisms, it is difficult to say. Chitin can only be observed after etching the mineral, so its structure is likely to be altered after the calcium carbonate is partially dissolved and the sample dehydrated and prepared for scanning electron microscopy analysis. In the light of this, we cannot make any conclusions regarding a specific location or organization of chitin inside the prisms.

Dr Staniland said: A very nice paper. Can you give some detail on structure prediction algorithms used? I had not come across programs for estimating order and disorder.

Another question to add if I may: You said all your oligomers were produced synthetically. Can they be produced within a host organism? Is there evidence of assembly within an organism?

Professor Evans answered: The algorithms for estimating intrinsic disorder are GLOBPLOT, DISOPRED, and IUP, but there are others that have been developed. Rather than describe their details here, I would direct you to the original papers (see Reference section) where the basis for identifying disordered regions are given. As to assembly within a host organism, I cannot say that we have done this. However, there have been studies done of sea urchin spicule matrices and enamel matrix where protein assembly has been noted and these might be worth looking up if you are interested in this area.

Dr Sear remarked: Would it be useful to study modified versions of the Asprich proteins in which the domain believed to be responsible for the aggregation into oligomers was either deleted or mutated to remove binding? The idea is that if the ability to oligomerise was removed, the effect on Asprich function of oligomerisation could be tested.

Professor Evans responded: Again, we have yet to perform experiments of that type, but we do plan to do so.

Dr Sear asked: Could what you call oligomers, whose radii of gyration are shown in Fig. 1, also be possibly described as micelles? Would it be helpful to test this idea by looking for a cmc (critical micelle concentration), *i.e.*, to dilute your protein solutions and see if below a certain concentration (the cmc) the clusters dissociate?

If they do, that would both provide evidence that the the clusters are in dynamic equilibrium with monomers in solution, and give an estimate of the free energy of formation of these clusters. If there is no cmc, then this may suggest that they are produced by some irreversible aggregation mechanism.

Professor Evans answered: We haven't done those experiments but that is an excellent idea worth trying. We do not know at this time if the protein complexes are micelle-like or adopt other morphologies in solution.

Dr Zhang said: From our recent work, we have seen that the multivalent salts have a strong association effect to the acidic residues (ASP and GLU) on the protein surfaces, which have been used to tune the effective protein interactions as well as their phase behavior, such as protein aggregation, crystallization and liquid–liquid phase separation.[1-4] In the ESI-MS experiments, you have also used trivalent salts such as $LaCl_3$ and $EuCl_3$ with a fixed ratio of protein : salt. I would expect the values of the metal ion binding sites may vary by changing the ratio. Here the question is: how do the possible phases affect the resulting metallization from ESI-MS experiments?

1. F. Zhang *et al.*, Charge-controlled metastable liquid–liquid phase separation in protein solutions as a universal pathway towards crystallization, *Soft Matter*, 2012, **8**(5), 1313–1316.
2. F. Zhang *et al.*, Reentrant condensation of proteins in solution induced by multivalent counterions. *Phys. Rev. Lett.*, 2008, **101**, 148101.

3. F. Zhang *et al.*, Universality of protein re-entrant condensation in solution induced by multivalent metal ions, *Proteins Struct. Funct. Bioinform.*, 2010, **78**, 3450–3457.
4. F. Zhang *et al.*, Novel approach to controlled protein crystallization through ligandation of yttrium cations, *J. Appl. Cryst.*, 2011, **44**, 755–762.

Professor Evans replied: I don't really know. The use of La and Eu was meant to introduce metal induced mass shifts that were larger than those generated by Ca as a means of helping us detect metal-bound Asprich "3" species. We did not explore protein phases created by these ions in this paper.

Professor De Yoreo remarked: Is there evidence that, regardless of whether they have a particular structure, they undergo conformational changes to form extended structures with a longer-range order as with collagen or S-layers, either in solution or on surfaces?

Professor Evans answered: The presence of long-range ordering has not been ascertained in solution, but there is ample evidence that these assemblies are highly dynamic on a short-range scale (NMR), which would lead me to believe that there would be very little long-range ordering, too.

Dr Nudelman enquired: Do you think that the function of Asprich is dependent on the formation of aggregates, rather than on individual proteins, so that upon aggregation its effect on ACC stabilization will change?

It is interesting that there are 10 different forms of Asprich, can we assume that the aggregates are a mixture of all the 10 isoforms, or is it more likely that each one has a different function? I would not be surprised if this system is redundant and they all have the same role, do you think that it is possible?

Professor Evans responded: The 10 member family is intriguing on a number of levels. Given that all Asprich have identical N-terminal sequences and the "a" through "g" isoforms have identical C-terminal sequences, your suggestion of redundancy for survivability is an excellent one. Alternatively, the slight sequence variations may also be intended for slightly different functions, or, perhaps different Asprich proteins are localized in different regions of the prism and the sequence variations dictate the location. It is pretty early in the Asprich scenario at this point to really know what the sequence variations mean, so we will have to await future experiments to determine the facts.

Mr Giuffre opened the discussion of the paper by Professor Hiromichi Nagasawa: How do your isolation and extraction methods influence protein structure, such as functional group chemistry and folding properties? If we use these extracted materials in new environments such as *in vitro* experiments, do protein properties and resulting mineralization mechanisms change significantly as compared to biological environments?

Professor Nagasawa responded: In general, there are various types of proteins, and the protein properties depend on each protein. Proteins related to biomineralization are usually extracellular proteins and most proteins do not form a rigid tertiary structure like hormones, enzymes, and so on. So, most matrix proteins in biominerals are rather flexible in nature. However, some matrix proteins have conserved consensus sequences partially within their whole sequence, which may be related to their function and have a definite secondary or tertiary structure for interacting with other molecules. When these matrix proteins are under severe conditions such as low pH, high temperature, detergent solution, or reducing conditions, some proteins denature and cannot form a native structure after returning to the physiological conditions, while some proteins can recover. For example, CAP-1 (a crayfish cuticle peptide) has the Rebers-Riddiford consensus sequence for chitin binding. Both the purified CAP-1 and recombinant CAP-1 retain a comparable chitin-binding ability, even though natural CAP-1 was extracted with an SDS solution containing dithiothreitol (a reducing agent) in a boiling water bath.

Dr Nudelman addressed Professors Nagasawa and Evans: In my work on the prismatic layer (Nudelman et al., Faraday Discuss., 2007, **136**, 9–25) I have also obsreved that the Asprich protein was strongly bound to the intracrystalline chitin, which seems to be similar to the CAP-1 protein. Upon extraction of the intracrystalline matrix, the Asprich protein was found together with the insoluble fraction, which was mainly constituted of chitin. We could also localize the Asprich on the chitin fibers usinjg polyclonal antibodies against the protein. What is interesting is that Asprich does not possess any conserved chitin-binding domain, so we don't know what the nature of the interactions between Asprich and chitin is. Do you have any ideas?

Professor Evans communicated in reply: There is no direct evidence of a beta chitin binding site in Asprich. However, I should add that we have noted chitin binding in a nacre peptide, N16, and this peptide can discriminate between the alpha and beta forms. It may be that Asprich proteins can bind to chitin but perhaps *via* another sequence. This will eventually be tested.

Professor Nagasawa responded: Two famous consensus sequences for chitin binding have been known; one is the Rebers–Riddiford sequence often found in cuticle proteins in insects and crustaceans and the other is chitin-recognizing sequence in chitinases. Both bind to chitin by hydrophobic interaction. There may be other chitin-binding sequences, but it has not been examined which residues contribute to the binding. Like Asprich, two proteins, GAMP (gasttrolith matrix protein, from crayfish gastrolith; Tsutsui et al., Zool. Sci., 1999, **16**, 619-628) and Prismalin-14 (from the prismatic lyaer of the Japanese pearl oyster) we identified, have the ability to bind to chitin, but we do not know which residues are really important. In the latter case (Prismalin-14), we could specify the partial sequence, the Gly/Tyr-rich region, for chitin binding (Suzuki and Nagasawa, FEBS J., 2007, **274**, 5158-5166), but still could not identify which residues are most imprtant for chitin binding. In case of Asprich, it is difficult to estimate the sequence responsible for chitin binding. To know this, it is necessary to prepare some peptides containing partial sequences and test their chitin-binding ability.

Dr Sommerdijk opened the discussion of the paper by Professor Takashi Kato: Does the adhesion of the recombinant acidic peptide for chitin depend on the number of hydrophilic repeat units?

Professor Kato replied: The adhesion properties of the recombinant peptides, rCAP-1-CD and rCAP-1-CT for chitin are not so different from that of the original CAP-1. However, if we extend the acidic moiety more, the binding ability may be affected by stronger hydrophilicity of the extended chain.

Mr Ihli addressed Professor Kato and Dr Nishimura: A lot of the discussion revolved around the site of nucleation of the thin film crystals formed on the chitosan matrices. As far as I understand the matrices were formed by spin coating, thus representing a rough surface due to the arrangement of the deposited strands. It would be great if the authors could elaborate why nucleation occurs directly on top of a fiber but not at potentially favorable edge sites between two strands, providing an increased access to calcium binding sites per area.

Dr Nishimura communicated in reply: Because the recombinant peptides do not have specific binding abilities to the chitosan matrices, the thin-film crystals were formed in the chitosan matrix with spherulitic orientation as shown in Fig. 7a,b. Nucleation might occur in the center of the thin-film crystals. However, the details of the nucleation point are still unclear.

Dr Nudelman remarked: As I understand from your work, the orientation of the chitin fibers is translated to the crystal, probably through the CAP, which is adsorbed on the fibers. Since the chitin fibers are randomly oriented, is it possible that two crystals nucleate on different fibers that have different orientations and are just a few nm apart, and upon growing they fuse, generating a polycrystalline structure?

Have you tried to lift the crystals from the chitin matrix and look at their bottom surface (the one that was in contact with the chitin), and see whether the fibers left an imprint on the crystal?

Professor Kato responded: We also observed the formation of thin-film crystals that have two or more domains although they are not dominantly obtained. We assume that these crystals have two or more nucleation points for each film. We have not yet tried to lift the crystal to observe the crystal surface on the chitin matrix.

Professor De Yoreo enquired: I am confused by the tri-axial morphology. Why do you think the crystals grow preferentially along the a-axis and, given how rough the surfaces are, what can be the driving force for net growth in these directions?

Professor Kato answered: For the self-similar nanocrystallites isolated tripod crystals shown in Fig. 6a, the peptide may be adsorbed on the (110) surface through molecule/crystal recognition because the electron diffraction (Fig. 6b) of the nanocrystallites shows that the crystals grow preferentially along the a-axis or 4-4-1.

This specific absorption results in the suppression of the (110) surface and the formation of tripodal morphology.

Professor De Yoreo remarked: Because supersaturation was created by evaporation, the supersaturation at which nucleation occurs is different for different surfaces, and thus the morphology, number density and size can be completely different. Consequently, it is risky to draw conclusions about the differences in control by two different amino acid sequences. Mixing of solutions to achieve a uniform and constant supersaturation is required to make comparisons.

Professor Kato said: We also think that the environment should be the same when we compare the results. One reason we employed this method is that the experiment of the peptide is micro-gram scale and it is difficult to use the mixing solution method.

Ms Asenath-Smith said: Simply fixing reagent concentrations in a solution-based crystallisation system may not ensure identical supersaturation conditions present at nucleation in the two different (chitin, chitosan) systems. While fixing the concentration of the starting reagent solutions may ensure similar supersaturation in the bulk solutions of both systems, solutes will interact with each matrix differently due to the chemical differences between chitin and chitosan. In the presence of such fibrous, porous matrices, the *apparent* supersaturation at the point of nucleation (on the matrix) will certainly be different for each system. This is similar to crystal growth in hydrogel systems.

Dr Andreassen remarked: Proper quantification of supersaturation is a requirement in order to analyze crystal growth mechanisms and their kinetics. The majority of biomimetic mineralization generate supersaturation by gas desorption of calcium bicarbonate solutions (Kitano method) or in-diffusion of carbon dioxide and ammonia. However, in these methods the value of the supersaturation is inaccessible (in time) and unevenly distributed over the gas/liquid interface. So why are these methods applied? Is it because they better simulate the situation *in vivo*? Would it not be better to mix solutions of calcium and carbonate (at supersaturation low enough to allow for proper mixing before precipitation starts)?

Dr Sear responded: I would also agree, and I think the point you make in your first sentence is important. It will be difficult or impossible to develop predictive models without quantitative data on growth and nucleation rates, and this data must be at known values of the parameters that affect growth and nucleation rates, such as concentration and pH. If these parameters are varying in time and space, then this makes the task of understanding very difficult.

Professor Roberts answered: I agree with this point but it is not really true that (say) gaseous infusion cannot yield uniform supersaturations as on balance this should be easier than with liquid/liquid mixing where slow micro-mixing can become an issue.[1] Nonetheless, determination of the kinetic mechanism does demand careful measurement of the growth rates of the individual habit faces defined by their Miller indices (hkl) together with simultaneous measurement of the solution supersaturation. The latter can be derived *via* measurement of pH, Ca^{2+} ion concentration together with solution state modelling. When face specific measurements are difficult then crystal size can be measured and the growth rates de-convoluted drawing upon a knowledge of the crystal morphology. Crystal size measurement can be a challenge for sizes less than 10 microns or when there is substantial problems with crystal agglomeration.

1. See, *e.g.*, On the influence of mixing on crystal precipitation processes—application of the segregated feed model (SFM), R. Zauner and A. G. Jones, *Chem. Eng. Sci.*, 2002, **57**, 821 and papers cited therein.

Dr Wolf replied: Concerning your question why the community often uses the slow diffusion technique and comparable crystallization methods although supersaturation is not well controlled or defined, I think that one important motivation to choose diffusion-driven systems is simply to avoid mixing of educt solution under turbulent conditions. This was at least why we decided to use the Kitano method in our levitation experiments. For instance, Jens Rieger elaborately discussed the issue of insufficient mixing under turbulent conditions during the Faraday Discussion 136 on *Crystal Growth and Nucleation* and pointed out that such insufficient mixing can fake emulsion-like structures: the system may start reacting at the interface of the two intermixing educts solution before a state of homogeneous supersaturation is reached.[1–3]

1. J. Rieger, T. Frechen, G. Cox, W. Heckmann, C. Schmidt and J. Thieme, *Faraday Discuss.*, 2007, **136**, 265–277.
2. D. Horn and J. Rieger, *Angew. Chem., Int. Ed.*, 2001, **40**, 4340–4361.
3. H. Haberkorn, D. Franke, T. Frechen, W. Goesele, J. Rieger, *J. Colloid Interf. Sci.*, 2003, **259**, 112–126.

Professor Meldrum asked: Possible reasons for the wide use of the ammonia diffusion method is that it often gives interesting morphologies. It is also possible to work at much higher concentrations of calcium than for the other methods, giving a higher yield of calcium carbonate. This is very valuable if you need to do measurements such as TGA and BET which require relatively large amounts of sample.

Professor De Yoreo said: That makes good sense. On the other hand, if one wants to work out mechanisms of nucleation and growth, this is an awful method to rely on as the solution speciation and supersaturation are unknown and ammonia is not just a spectator ion.

Dr Rodríguez Blanco addressed Dr Andreassen: Some of you were discussing the problem of solution mixing. In our recent work (Rodríguez-Blanco et al., 2012) we have seen that the crystallization of ACC can follow very different pathways depending on solution mixing. The result of adding a $CaCl_2$ solution to a Na_2CO_3 solution is completely different than adding a Na_2CO_3 solution to a $CaCl_2$ solution. In the first case ACC transforms to calcite *via* a vaterite intermediate, while in the second case there is a direct transformation of ACC to calcite and no vaterite forms at all. This is a problem related with the pH mixing, because depending on how the solutions are mixed, ACC starts forming from an acid or basic solution and this results into the formation of ACC with either a vaterite or a calcite structure (Gebauer et al., 2008). In order to (partially) avoid this problem, one method we have used for some of our experiments is to bubble CO_2 in both solutions prior to mixing. Once mixed, the pH is so low that no $CaCO_3$ forms. Then, in order to increase the supersaturation and favor the formation of $CaCO_3$, we let the solution to degas on its own (so the pH slowly increases) or we bubble N_2 when we need to accelerate the process.

1. D. Gebauer et al., *Science*, 2008, **322**, 1819–1822.
2. J. D. Rodríguez-Blanco et al., *J. Alloys Compd.*, 2012, DOI:10.1016/j.jallcom.2011.11.057

Dr Beck communicated in reply: At high supersaturation, ACC forms directly from solution without any induction time. In this case, the time it would require to mix the solution consituents is large as compared to the induction time. This leads to spacial differences in supersaturation with respect to calcium and carbonate ions and also fluctions in concentrations of other ions. If, however, the supersaturation is tuned so that the induction time is large as compared to the mixing time, more uniform outcomes regarding both polymorphism and morphology should be expected when mixing $CaCl_2$ and Na_2CO_3 solutions.

It is true that the total amount of carbon species (CO_2(aq); HCO_3^-; CO_3^{2-}; $CaCO_3^0$; $CaHCO_3^+$) that can be dissolved is much higher when the solution is sparged with CO_2 leading to a low solution pH. Though the pH is low, supersaturation can be created by increasing the total amount of calcium species and increasing the alkalinity[2] of the solution by adding, for example, sodium carbonate or sodium hydroxide.[3] Sparging with 100% CO_2 is an appropriate method to specify the supersaturation in the solution based on speciation of constituents in the gas and liquid phase (based on thermodynamic equilibria) and based on fugacity and activity coefficients.[2] This allows, for example, for the measurements of growth rates in dependence of supersaturation and the extraction of growth rate constants from this dependence.

Degassing the solution (after it has been sparged with 100% CO_2) or sparging with nitrogen increases the supersaturation (as long as there are still carbon species present) and as a consequence induces nucleation and growth. However, is it possible to quantify the supersaturation during this degassing process and correlate it to the growth and morphology of the crystals? In my opinion it is better to achieve the supersaturation in a more controlled way so that it can be quantified. If the solution is not sparged with 100% CO_2, it can also be sparged with a quantified percentage of for example N_2 and CO_2 (for example 5% CO_2) provided that the thermodynamic equilibria are achieved fast.

If it is not desired to mix solutions of Na_2CO_3 and $CaCl_2$ directly, a controlled supersaturation can also be achieved by titration of a known concentration of, for example, Na_2CO_3 (at a predefined rate) into a solution of $CaCl_2$. The advantage of this method as compared to the Kitano method is that the supersaturation in the solution can be quantified more easily.

1. E. M. Flaten, M. Seiersten, J.-P. Andreassen, *J. Cryst. Growth*, 2010, **312**, 953–960.
2. B. Kaasa, K. Sandengen, T. Østvold, SPE International Symposium on Oilfield Scale 95075, 2005, pp. 1–13.
3. R. Beck, M. Seiersten and J.-P. Andreassen, The Constant Composition Method for Crystallization of Calcium Carbonate at Constant Supersaturation, manuscript in preparation, 2012.

Ms Asenath-Smith remarked: Crystallization systems based on double diffusion through a hydrogel column can be used to achieve controlled mixing. In addition, being that mass transport in these systems is solely by diffusion, the concentration (supersaturation) gradients that evolve with time can be modeled and effectively predicted using diffusion theory. A very recent review by J. Dorvee, A. Boskey and L. Estroff has just come out in *CrystEngComm*,[1] and is a very useful demonstration of this.

1. J. D. Dorvee, A. L. Boskey and L. A. Estroff, *CrystEngComm*, 2012, **14**, 5681–5700.

Dr Beck answered: During the so far mentioned methods, supersaturation changes during the crystallization process. One method that allows for the study of growth kinetics and particle evolution at controlled constant supersaturation is the constant composition method introduced by Tomson and Nancollas.[1]

1. M. B. Tomson and G. H. Nancollas, Mineralization kinetics: A constant composition approach. *Science*, 1978, **200**, 1059–1060.

Professor Rodger asked Professor De Yoreo: Do your conclusions about the utility and validity of classical nucleation theory for $CaCO_3$ systems change if the structure at the core of the clusters changes with cluster size for sizes spanning the critical cluster size (*i.e.* when the favourable bulk energy does not scale with volume)? Do your simulations provide any evidence that such a bulk-core is found in $CaCO_3$ solid nucleation?

Professor De Yoreo replied: Certainly any deviation from a a flat free energy landscape (*i.e.*, a size dependent dG/dn) will have a strong effect on the predicted free energy barrier to nucleation. However, the first term in the nucleation rate equation will always have a volume term, because it is simply $dg/dn \times \Delta n$, and Δn is always proportional to the dimension of the nucleus cubed while dg/dn can always be written as a constant term times a deviation from constancy. That deviation then shows up in what I wrote as the excess free energy ?gex. That can either be positive or negative. Whether or not my conclusions about the utility of the classical nucleation picture changes depends on how drastically dg/dn deviates from constancy. Having said that, the experimental facts are: 1) direct nucleation of calcite in bulk solution or on unfavorable surfaces like OH-terminated SAMs does not appear to occur, at least not before ACC nucleates, and 2) direct nucleation of calcite on COOH-terminated, SH-terminated and PO4-terminated SAMs appears to occur directly, the dependence on supersaturation follows that expected from the classical analysis and the interfacial energies extracted from the data are reasonable. That is all pretty strong evidence for the sensibility of the analysis.

PAPER

The thermodynamics of calcite nucleation at organic interfaces: Classical vs. non-classical pathways†

Q. Hu,‡[ab] M. H. Nielsen,‡[ac] C. L. Freeman,[d] L. M. Hamm,[e] J. Tao,[a] J. R. I. Lee,[f] T. Y. J. Han,[f] U. Becker,[b] J. H. Harding,[d] P. M. Dove[e] and J. J. De Yoreo*[a]

Received 24th August 2012, Accepted 29th August 2012
DOI: 10.1039/c2fd20124k

Nucleation in the natural world often occurs in the presence of organic interfaces. In mineralized tissues, a range of macromolecular matrices are found in contact with inorganic phases and are believed to direct mineral formation. In geochemical settings, mineral surfaces, which are often covered with organic or biological films, surround the volume within which nucleation occurs. In the classical picture of nucleation, the presence of such interfaces is expected to have a profound effect on nucleation rates, simply because they can reduce the interfacial free energy, which controls the height of the thermodynamic barrier to nucleation of the solid phase. However, the recent discovery of a nearly monodisperse population of calcium carbonate clusters—so called pre-nucleation clusters—and the many observations of amorphous precursor phases have called into question the applicability of classical descriptions. Here we use *in situ* observations of nucleation on organothiol self-assembled monolayers (SAMs) to explore the energetics and pathways of calcite nucleation at organic interfaces. We find that carboxyl SAM-directed nucleation is described well in purely classical terms through a reduction in the thermodynamic barrier due to decreased interfacial free energy. Moreover, the differences in nucleation kinetics on odd and even chain-length carboxyl SAMs are attributable to relative differences in these energies. These differences arise from varying degrees of SAM order related to oxygen-oxygen interactions between SAM headgroups. In addition, amorphous particles formed prior to or during crystal nucleation do not grow and are not observed to act as precursors to the crystalline phase. Instead, calcite appears to nucleate independently. These results imply that the recently proposed model of calcite formation as a non-classical process, one which proceeds *via* aggregation of stable pre-nucleation clusters that form an amorphous precursor from which the crystalline phase emerges, is not applicable

[a]*Molecular Foundry, Lawrence Berkeley National Laboratory, Berkeley, CA 94720. E-mail: jjdeyoreo@lbl.gov*
[b]*Department of Earth and Environmental Sciences, University of Michigan, Ann Arbor, MI 48109*
[c]*Department of Materials Science and Engineering, University of California, Berkeley, CA 94720*
[d]*Department of Materials Science and Engineering, University of Sheffield, Sheffield, U.K. S1 3JD*
[e]*Department of Geosciences, Virginia Tech, Blacksburg, VA 24061*
[f]*Physical and Life Sciences Directorate, Lawrence Livermore National Laboratory, Livermore, CA 94511*

† Electronic supplementary information (ESI) available. See DOI: 10.1039/c2fd20124k
‡ These authors contributed equally.

to template-directed nucleation on carboxyl SAMs and does not provide a universal description of calcite formation.

1. Introduction

Macromolecular matrices play a key role in establishing the architectural complexity and mechanical properties of biominerals by directing the organization of the mineralized component.[1-3] The ability of the matrix to perform this function is determined by both its structural relationship with the incipient nucleus[4] and the changes to the energy landscape it imposes upon the mineralizing constituents.[5] A number of studies have explored the structural aspect,[1-4,6,7] but little is known about the energetic controls. Moreover, the recent discovery that calcium carbonate[8] and phosphate[4] solutions contain clusters prior to nucleation—*i.e.*, pre-nucleation clusters—that seem to be stable relative to the free ions[8] combined with observations of non-equilibrium amorphous precursors in numerous biomineral[1,9,10] and biomimetic systems,[11-13] raises the question of whether the classical description[5] of nucleation dynamics is applicable to matrix-directed mineralization. This same question arises when considering mineral nucleation in geochemical settings where a surrounding mineral matrix, which is often coated with biofilms or other organic layers, is likely to influence nucleation kinetics. While these issues are difficult to address in the context of three-dimensional biological matrices or geological reservoirs, self-assembled monolayers (SAMs) of organothiols on noble metal surfaces, which can template mineral nucleation on distinct crystallographic planes with a high degree of specificity, offer an excellent 2D model.[12,14-17]

Here we use carboxyl- and hydroxyl-terminated SAMs to investigate the energetics and formation pathways during templated nucleation of $CaCO_3$. We first develop the basic relationships between the rate of calcite nucleation and the supersaturation for three classes of free energy landscapes, including both size independent and size dependent excess free energies, as well as one in which local or global minima create a population of pre-nucleation clusters. We then utilize an *in situ* optical microscopy method to measure nucleation rates as a function of supersaturation on SAMs, from which we derive the effective interfacial energies. We compare the resulting free energy barriers for heterogeneous nucleation on carboxyl-terminated SAMs containing carbon chains of odd and even length to that expected for homogeneous nucleation in bulk solution. Molecular dynamics simulations are used to understand the structural source of the differences in nucleation rates observed for the odd and even SAMs. Finally a combination of *in situ* optical microscopy and atomic force microscopy (AFM) observations along with Raman and transmission electron microscopy (TEM) analyses are employed to follow the pathway of calcite formation on both the carboxyl- and hydroxyl-terminated SAMs.

Theoretical analysis shows that homogeneous nucleation of calcite is highly unlikely even at concentrations approaching the solubility limit of amorphous calcium carbonate (ACC). However, introduction of a size dependent interfacial energy, the introduction of low-energy surfaces and a population of metastable clusters can all significantly reduce the barrier. Based on our measurements of nucleation rates, we find that nucleation on carboxyl-terminated SAMs is described well in purely classical terms through a reduction in the thermodynamic barrier due to decreased interfacial free energy. The differences in nucleation kinetics on carboxyl-terminated SAMs of odd and even parity—*i.e.*, an odd number (11) *vs.* an even number (16) of carbons in the alkyl chain—are attributable to relative differences in these energies that arise from varying degrees of SAM order related to oxygen-oxygen interactions between SAM headgroups. In addition, amorphous particles observed to form prior to crystal nucleation on hydroxyl SAMs and after crystal nucleation on carboxyl SAMs—even well below the accepted bulk solubility limit for amorphous calcium carbonate (ACC)—do not grow and are not observed to be precursors to the crystalline phase. Instead, calcite appears to nucleate independently. We discuss how these results can be reconciled with the recently proposed non-classical picture of calcite formation that is based on aggregation of stable or metastable pre-nucleation clusters.[8,11]

2. Theoretical analysis of nucleation rates

2.1 Homogeneous nucleation of calcite and ACC

In principle, the energetic effect of any surface on nucleation can be determined by measuring the dependence of nucleation rate on supersaturation.[5,18] In all nucleation events, two important energetic parameters influence rates. The first is the excess free energy associated with the newly formed phase. This is an ensemble property that creates a thermodynamic barrier Δg_c due to the collective behavior of the ions in the solid and liquid phases. The second is an effective kinetic barrier E_A arising from individual reactions such as desolvation of solute ions, attachment to the forming nucleus, and structural rearrangements. Both barriers appear exponentially in the expression for the rate of nucleation J through:[5]

$$J = A\, e^{-E_A/kT} e^{-\Delta g_c/kT} \tag{1}$$

where A is a pre-factor that is determined by geometric factors and material-dependent parameters (*e.g.* density) and Δg_c is a decreasing function of the chemical potential $\Delta\mu = \sigma/kT$, where σ is the supersaturation, k is Boltzmann's constant and T is the temperature. (The form of the dependence of Δg_c on σ is model dependent as discussed below.)

While the exponential dependence of J on σ through the free energy barrier is a universal hallmark of nucleation that, in essence, distinguishes it from a simple chemical reaction, the exact form of A and Δg_c are model dependent. The source of Δg_c is a positive excess free energy Δg_{ex} of the solid phase that adds to the change in free energy for a simple chemical reaction, which is simply given by $(dg/dn)\Delta n$ where Δn is the number of molecules passing from the solution to the solid phase. Note that for a supersaturated solution, $dg/dn < 0$. Thus without Δg_{ex} there would be no barrier and precipitation would happen spontaneously at infinitesimal supersaturation *without nucleation*. In classical nucleation theory (CNT), Δg_{ex} arises from the free energy of the interface between the mineral and the surrounding solvent and substrate (Fig. 1A). When the free energy landscape is flat—that is, the excess free energy is simply determined by the surface area times the interfacial free energy α, which is independent of size—then Δg_c is given by:

$$\Delta g_c = B\frac{\alpha^3}{\sigma^2} \tag{2}$$

where B is a constant that depends on the shape and density of the nucleating solid (see ESI for details†). Based on the literature value of 109 mJ m^{-2} for the interfacial free energy of calcite in solution,[19,20] the predicted classical barrier to homogeneous nucleation of a calcite rhomb is formidable (Fig. 1A and Figure 1E, blue curve), ranging from 175 kT to 93 kT for CaCl$_2$ and NaHCO$_3$ concentrations between 10 mM and 29 mM—the latter marking the literature value for the solubility limit of ACC.[20] (Note that, at 300 K, 1 kT = 2.6 kJ mol^{-1} = 0.62 kcal mol^{-1}.)

Eqn (2) also reveals the extreme level of supersaturation needed to reduce the free energy barrier for nucleation of ACC below that for calcite. Based on the scaling of interfacial free energy with solubility,[18,19] the ratio of α for ACC to that of calcite is of order 0.75. Taking into account the differences in the parameter B for calcite and ACC, we find that the free energy barrier to forming calcite will be less than the barrier to forming ACC until the solution concentration is increased to the point where the supersaturation relative to ACC exceeds ~65% of the supersaturation

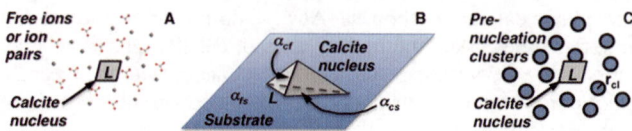

Fig. 1 (A) Schematic showing three different nucleation scenarios considered here. (A) Homogeneous nucleation of a calcite rhomb of edge length L from ions in solution. (B) Heterogeneous nucleation of a calcite rhomb of edge length L on an arbitrary face and against a substrate. The interfacial free energies for the crystal–fluid, crystal–substrate and fluid–substrate interfaces are α_{cf}, α_{cs}, and α_{fs}, respectively. (C) Nucleation of a calcite rhomb of edge length L from clusters of radius r_{cl} with excess free energy Δg_{ex}.

relative to calcite. Because the solubility of ACC is so much higher than that of calcite, for equal mixtures of $CaCl_2$ and $NaHCO_3$, this condition can only be reached if the solution Ca^{2+} concentration is far in excess of 100 mM. Even if the solubility of ACC is considerably smaller than reported in the literature, as seems likely based on recent titration studies[8] as well as from the data reported below— we find we can induce bulk ACC formation at less than half the literature value of the solubility (*i.e.*, 13 mM *vs.* 28 mM final Ca^{2+} concentration)—the required concentration for ACC to be favored thermodynamically is still extreme. For example, even if we take the equilibrium solubility to be as small as 10 mM (final Ca^{2+} concentration), a Ca^{2+} concentration of approximately 100 mM is required before there is a crossover in barriers. Even at that concentration, the classical barrier is still in excess of 53 kT. The clear conclusion of this analysis is that, for a flat energy landscape, the concentrations required to achieve homogeneous nucleation of either phase are extreme and inconsistent with experimental observations of both calcite and ACC nucleation at significantly lower concentrations. Hence, either nucleation is heterogeneous or other pathways of precipitation that avoid this barrier must be at work.

2.2 Heterogeneous nucleation

As eqn (1) shows, the thermodynamic barrier depends upon the cube of the interfacial energy. Consequently, heterogeneous nucleation on surfaces (Fig. 1B) that reduce the interfacial energy can proceed at dramatically altered rates. In this case, α becomes an effective interfacial energy α_{het} that depends on the interfacial energies of the crystal–fluid, fluid–substrate, and crystal–substrate interfaces through:

$$\alpha_{het} = \alpha_{cf} - h(\alpha_{fs} - \alpha_{cs}) \qquad (3)$$

where h is a factor that depends on the aspect ratio of the nucleus (see ESI for details†). As long as $\alpha_{cs} < \alpha_{fs}$, the value of α_{het} will be reduced from that for the homogeneous nucleus. However, even if the effective α_{fs} equals α_{cs}, that is, the interfacial energies for the crystal–substrate and fluid–substrate interfaces are equal, the barrier will already be reduced by a factor of 1.6 (Fig. 2E, red curve) simply because a surface that would have been generated during homogeneous nucleation is now a crystal–substrate interface that carries no energy penalty. A further reduction in α by only 20 to 50% due to $\alpha_{cs} < \alpha_{fs}$ would lead to a decrease in the barrier by a factor of 3 to 13 (Fig. 2E, green and orange curves). Given that nucleation rate depends exponentially on this barrier, these large reductions mean that surfaces have the potential to completely alter the dynamics and pathways of calcite formation.

2.3 Deviations from a flat energy landscape: Cluster aggregation and size dependent α

In truth, the free energy landscape is unlikely to be flat (blue curve in Fig. 2D) at small sizes. Δg_{ex} must approach zero at the size of a molecule (shown schematically in Fig. 2D by red curve) and probably exhibits local minima and maxima at very small cluster size as certain configurations expose more or less favorable coordination geometries for the surface ions (shown schematically in Fig. 2D by green curves). While these variations in Δg_{ex} are easy to account for by expressing α as a function of size, they do little to change the basic physics of the nucleation process. Nonetheless, they can potentially have significant effects on the magnitude of the barrier (Fig. 2B and 2C) if the size at which that barrier is reached—*i.e.* the critical size—becomes comparable to the dimensions at which size effects begin to emerge (red curve in Fig. 2D), or where local (or global) minima (solid (or dashed) green curves in Fig. 2D) in the free energy landscape create a population of metastable (or stable) clusters that can aggregate to form a critical nucleus (Fig. 1C). (For an analysis of how these features impact eqn (1) and (2), see the ESI.†)

Unfortunately, not much is known about the size dependence of of the interfacial free energy α. What little data do exist suggest a slight rise with decreasing size, followed by the beginning of a decrease in magnitude,[22] but those data do little to constrain the dependence in the region below 5 nm diameter, which is greater than the 1–3 nm critical size seen in Fig. 2A. Theoretical treatments suggest that even a single formula unit already possesses much of the energetic features of the bulk.[23] This suggests the fall-off in interfacial energy may not occur until diameters below 1 nm, though these simulations were performed for molecular solids and can not be directly translated to ionic crystals like calcite. However, metadynamics simulations of equilibrium calcite structure suggest the energetic features of the bulk are still manifest below 2 nm.[24] Indirect evidence for a complex dependence on size comes from both cryoTEM[11] and ultra-centrifugation data[8] that suggest there is indeed a population of sub-critical clusters (commonly referred to as pre-nucleation clusters) with a tight size distribution, which implies there is indeed a minimum in the free energy *vs.* size, as shown schematically in Fig. 2D (green curves). In fact, titration-based studies on the amount of calcium inferred to be bound in these clusters concluded that they occupy a global minimum, *i.e.*, the free energy of the pre-nucleation clusters lies *below* that of the free ions.[8]

Deviations from a flat landscape will change the dependence of Δg_c on σ (eqn (2)). In the case of a size dependent α, the change can be complex and depends on the form of the size dependence. For nucleation by aggregation of clusters that occupy a local minimum in the free energy, the dependence becomes:

$$\Delta g_c = B \frac{\alpha^3}{(\sigma \pm C)^2} \quad (4)$$

where C is a constant that depends on the shape factor, the cluster radius and the excess free energy of the cluster, and the plus or minus sign depends on whether the minimum in Δg is local or global (see ESI for details).† If it is a local minimum (Fig. 2D, solid green line), then the clusters are metastable, they carry excess free energy above the free ions, the plus sign applies and the barrier is reduced. If it is a global minimum (Fig. 2D, dashed green line), then the clusters lie below the free ions, the minus sign applies and nucleation by cluster aggregation brings with it an extra energy cost (see ESI for details).† This is the case for pre-nucleation clusters, which were found to lie about 18 kJ mol^{-1} below the free ions.[25] Thus creation of a super-critical nucleus by aggregation of pre-nucleation clusters would bring with it a larger barrier than aggregation of free ions, regardless of whether the end product is an amorphous or crystalline phase.

One interesting implication of eqn (4) is that the impact of clusters—stable or metastable—is to effectively alter the supersaturation from σ to $\sigma \pm C$, making it larger for metastable clusters and smaller for stable clusters. The result is that the existence of metastable clusters *increases* the probability of nucleating ACC rather than calcite, while the existence of stable clusters *decreases* that probability.

In the case of calcite nucleation by metastable clusters, for cluster radii below 2 nm and a reasonable range of excess free energies—such as those used in Fig. 2C, the magnitude of C in eqn (4) is less than 10% of σ over the range used in this study. The small magnitude of C relative to σ has two consequences for the current study. First, measurements of the dependence of calcite nucleation rate on supersaturation will not distinguish between ion-by-ion and cluster-by-cluster addition; either way the data will appear to follow the dependence of rate on supersaturation predicted by the classical expressions. Second, because the effect of clusters is likely to be too small to detect, if the classical dependence is not observed, then size dependence of α is the likely source of the deviation.

3. Calcite nucleation rates on carboxyl-terminated SAMs

SAMs of 16-mercaptohexadecanoic acid (MHA) and 11-mercaptoundecanoic acid (MUA), both of which are carboxyl-terminated but differ in the length and parity of the carbon chain, were prepared on Au (111) substrates using previously described methods (See ESI for details).[12] The SAMs were suspended upside down in a custom-built flow cell in the focal plane of an inverted optical microscope (see ESI, Fig. S1 for details).† A mixture of $CaCl_2$ and $NaHCO_3$ solutions with equal final concentrations of between 20 and 35 mM were flowed through the cell at constant rates under conditions that ensured nucleation was controlled by the reaction kinetics at the SAM surface rather than by diffusion or mixing (see ESI for details).†

For each concentration, the number of crystals in a fixed area was determined as a function of time (Fig. 3A–D). Plots of the number of nuclei vs. time (Fig. 3E) produced S-shaped curves exhibiting a linear rise and an approach to saturation marking the time when the density of nuclei became too great for subsequent events to be independent. The slope of the linear region gave the steady-state nucleation rate J (number of nuclei per unit area per unit time). (σ is defined as $\ln[(\{Ca^{2+}\}\{CO_3^{2-}\})/K_{sp}]$ where K_{sp} is the solubility product and $\{Ca^{2+}\}$ and $\{CO_3^{2-}\}$ are the Ca^{2+} and CO_3^{2-} activities, respectively.)

While qualitatively similar behavior was observed for both the MHA (C-16) and MUA (C-11) films, we found the nucleation rate was greater on MHA over the entire supersaturation range explored here. In addition, in accord with previous reports, nucleation occurred on distinct crystallographic planes for the two different SAMs.[15] The even parity MHA SAM induced nucleation almost exclusively on the (012) plane (Fig. 4A), while on the odd parity MUA SAM nucleation occurred primarily on the (013) face (Fig. 4B), though 30–40% of the crystals exhibited orientations between (012) and (015), and also included sporadic (104) and (001) orientations (Fig. 4B insets). In contrast, under identical conditions, nuclei on SAM-free gold films were few in number and exhibited random orientations (Fig. 4C).

Analysis of nucleation data such as those given in Fig. 3E shows that J exhibits the dependence on σ expected from CNT through eqn (1) and (2) (Fig. 3F). From the slope of $\ln(J)$ vs. σ^{-2} we obtain values for α of 72 mJ m^{-2} for MHA and 81 mJ m^{-2} for MUA, both of which are substantially smaller than the value of 109 mJ m^{-2} for calcite in bulk solution[19,20]. These differences in interfacial energy have a dramatic impact on nucleation rates. For example, in the middle of the supersaturation range explored here, the corresponding free energy barriers for nucleation on MHA, MUA and in bulk solution are found to be $19kT$, $27kT$ and $105kT$, respectively. All other factors being equal, these differences alone would correspond to relative nucleation rates $J_{MHA} : J_{MUA} : J_{sol}$ of $1 : 3.4 \times 10^{-4} : 4.5 \times 10^{-38}$, although the advantage of the MHA film over that of the MUA film is somewhat reduced because it also produces a larger value of E_A by about $7 \pm 3kT$ (as can be seen from the smaller value of the y-intercept for MHA when extrapolated to $\sigma^{-2} = 0$). These results show that calcite nucleation on these canonical SAMs proceeds as expected from CNT and that both the enhancement of nucleation on the SAMs

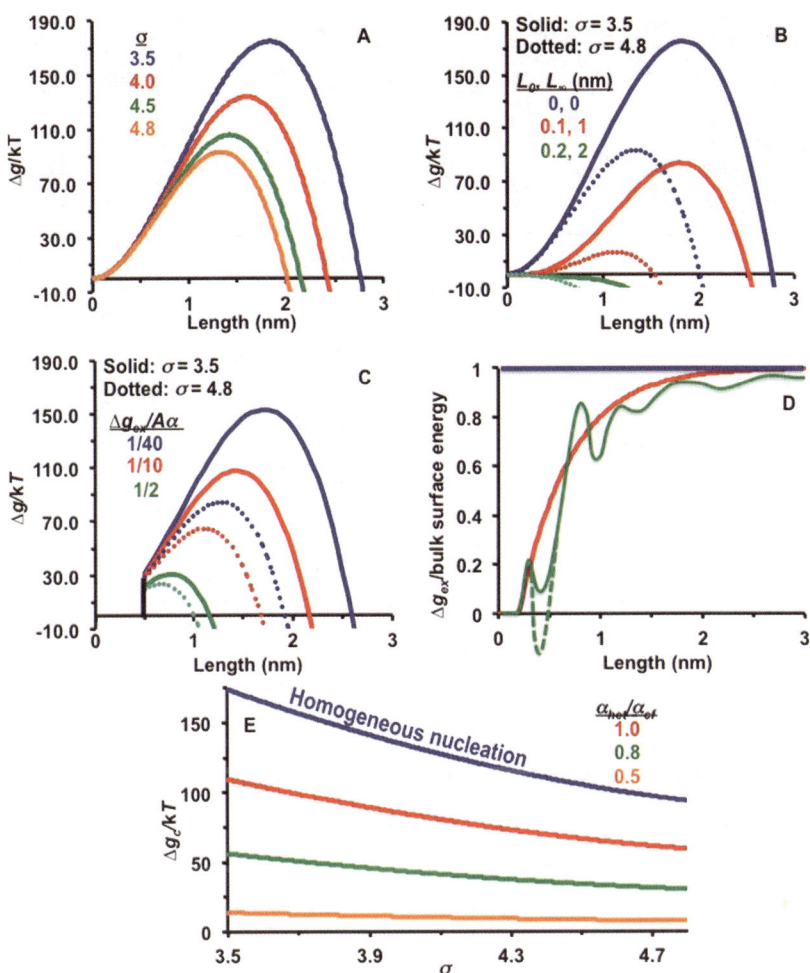

Fig. 2 (A) Dependence of free energy Δg on the length L of one side of an equilateral calcite rhombohedron for homogeneous nucleation from solutions formed from equal mixtures of CaCl$_2$ and NaHCO$_3$ for the indicated values of supersaturation. The corresponding final Ca^{2+} concentrations are: Blue – 10 mM; Red – 15 mM; Green – 23 mM; Orange – 28 mM. (B) Effect of size dependent excess free energy on Δg_c, where $\alpha = \alpha_\infty\{1 - \exp[-(L - L_0)/L_\infty]\}$ and α_∞ is the bulk interfacial energy, for two different supersaturations and the indicated values for L_0 and L_∞ where L_0 is the length at which Δg_{ex} reaches zero in Fig. 1D and L_∞ is the size at which it reaches $1 - \exp[-(1 - L_0/L_\infty)]$ of its bulk value. (C) Effect on Δg of aggregation by clusters of size $L = 0.5$ nm occupying a local minimum in excess free energy for two different supersaturations and the indicated values for the ratio of excess free energy to bulk surface energy. Note that values of Δg make no sense below $L = 0.5$ nm, because clusters of this size are assembled to make the critical nucleus. (D) Dependence of excess free energy (Δg_{ex}) on particle size. Blue line – flat energy landscape; Red line – simple size dependence; Green line – Size dependence with local minima and maxima; Green dashed line – global minimum defining stable population of clusters with narrow size distribution. (E) Dependence of free energy barrier Δg_c on σ for homogeneous nucleation of a calcite rhomb with $\alpha = 109$ mJ m^{-2} (blue line) and heterogeneous nucleation of a calcite rhomb on an (012) face with the indicated values of α_{het}/α_{cf}.

relative to bulk solution and the advantage of the SAM with even parity over that with odd parity can be explained in purely classical terms through differences in interfacial energy.

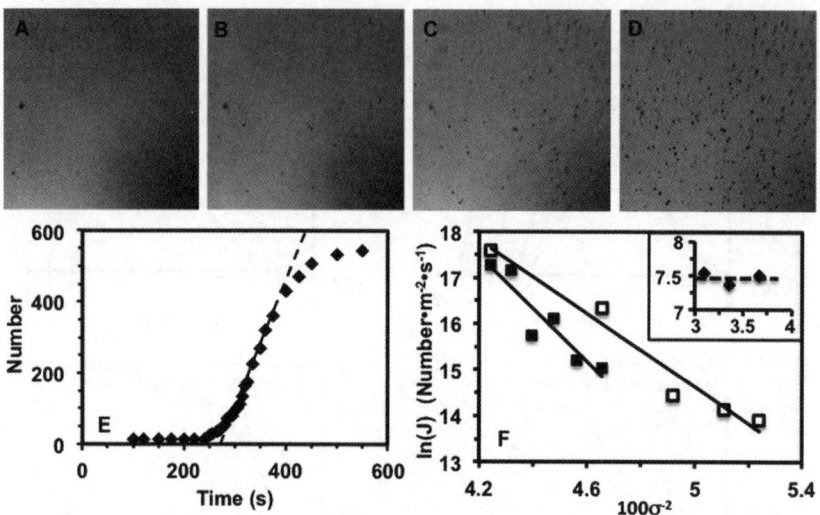

Fig. 3 Rate of calcite nucleation on MHA and MUA SAMs. (A–D) Sequential optical images collected at $t = 100$, 300, 350 and 450 s. Each image is: 0.49 × 0.49 mm. (Raw data were collected over 0.65 mm × 0.49 mm area.) (E) Typical dependence of number of nuclei on time. (F) Dependence of $\ln(J)$ on σ^{-2} showing that MHA films produce shallower slope and lower intercept than MUA films. Inset – Same as in F, but for nucleation on the OH-terminated mercapto-undecanol (MUO) SAM in solutions produced by mixing solutions of $CaCl_2$ and Na_2HCO_3 at a pH of 10.55.

4. Simulations of SAM structure and interfacial energy

To understand the source of these differences we performed molecular dynamics (MD) simulations on odd (C-15) and even (C-16) parity carboxylated SAMs using previously developed methods (See ESI for details) and examined the differences in SAM structure, the resulting calcite orientations, and the interfacial energies. Here we used 15- and 16-C chain monomers instead of the 11- and 16-C, so that the simulations are comparable with previous crystallization simulations and that we could isolate the effect of the parity only, though the results indicate that little difference would be observed for an 11-C and 16-C film. The radial distribution functions (RDFs) for the C–C atoms within the monomer chains are, as expected, identical for both SAMs (Fig. 5a). However, while the first peak (*i.e.* the nearest neighbor separation) for the headgroup-carbon-to-headgroup-carbon is also identical, there is an extra peak at ∼6.5 Å in the odd SAM that is absent in the even SAM (Fig. 5b). Also, the RDF for the headgroup-C-to-O (not shown) shows the same peaks for both the even and odd SAMs but the second peak at ∼5.8 Å is larger for the odd SAM.

These structural differences can be observed through visualization of the SAMs, which shows that the headgroups of the even SAM maintain six nearest neighbors throughout the simulations (Fig. 5C), while those of the odd SAM occasionally produce five nearest neighbors with the sixth headgroup pushed to a further separation (Fig. 5D). This effect originates from the O–O interactions between the headgroups. In the even SAM, the vector pointing from one oxygen to the other oxygen within a single headgroup (the O–O vector) exhibits an even distribution about 0°, *i.e.*, its average is parallel to the substrate surface (Fig. S2).† In the odd SAM, however, the angular distribution of the O–O vector is more complex and has an average that is non-zero, peaking at about 22°, demonstrating that there is a preference for one O to be pointing out of the SAM more than the other (Fig. S2).† Thus the oxygens of the odd SAM can find themselves pointing directly at each other and thereby generating an energetically unfavorable Coulombic repulsion. This, in turn, causes headgroups to be pushed out of the nearest neighbor shell and leads to the 5 + 1 arrangement seen in the RDF.

Simulations of calcite nucleation on these SAMs reflect these differences in SAM order. Because simulating nucleation directly from ions in solution is not feasible with current computing resources, even using techniques like metadynamics, anhydrous ACC in contact with the SAMs was used as the starting point. These simulations modeled the formation and growth of a calcite critical nucleus. Therefore the

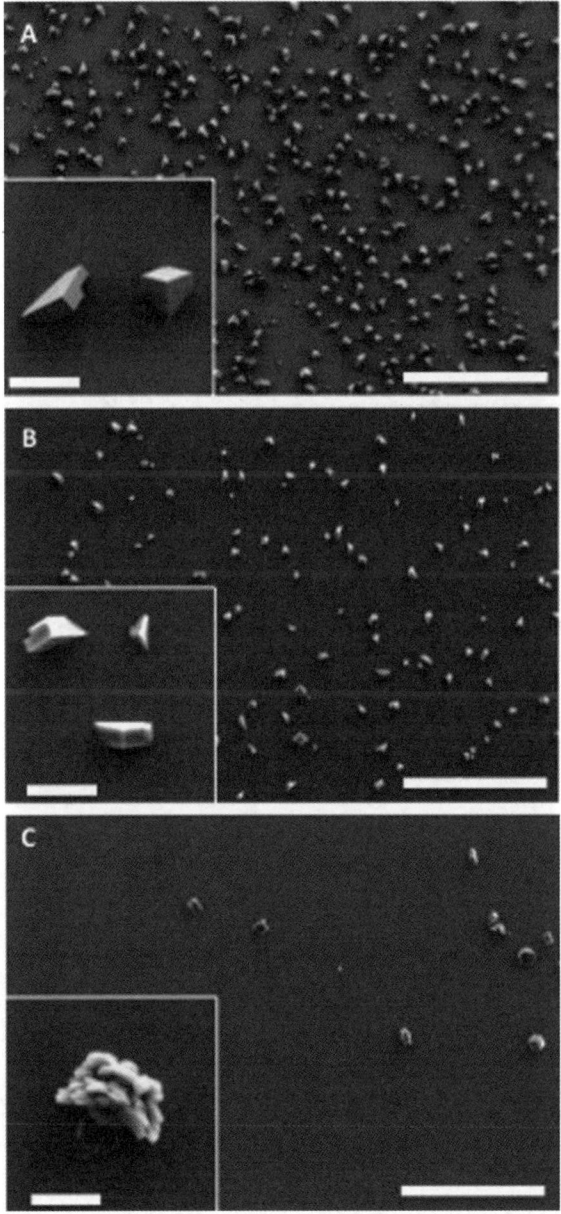

Fig. 4 SEM images showing number density and orientation of calcite crystals on (A) MHA, (B) MUA, and (C) bare gold surfaces. Scale bars are: main images – 100 μm; insets – 10 μm.

main difference between the simulations and the experiments is the arrival process of the ions, which should not influence the equilibrium structure and energetics of the interface with the SAM. The simulations identify the most energetically accessible crystal-SAM interface and are therefore complementary to the experiment.

Analyzing the results of previous crystallization simulations[26] that used an identical simulation setup to our own, we find for the even parity SAM, crystals are predicted to nucleate on the (012) plane in accordance with the experiments. However, on the odd parity SAM the plane of nucleation is defined by a mix of polar calcite faces such as (001) and (01x) and no single crystal plane of nucleation is formed, once again giving reasonable agreement with experiment. Here the inability to select a single orientation is due to the structural disorder in this film. The predicted interfacial energies are also in reasonable agreement with the experiments, giving 56 and 102 ± 6 mJ m^{-2} for the even and odd SAMs, respectively. The trends observed here are expected to be maintained over a wide range of monomer carbon chain lengths, though the exact values of the energies will differ due to the increased or decreased effect of the substrate for shorter and longer monomers, respectively.

5. Nucleation pathways

These findings support the conclusion that the classical viewpoint of nucleation control through minimization of crystal-SAM interfacial energy can describe nucleation in this system and appear to be in conflict with the proposed model for calcite

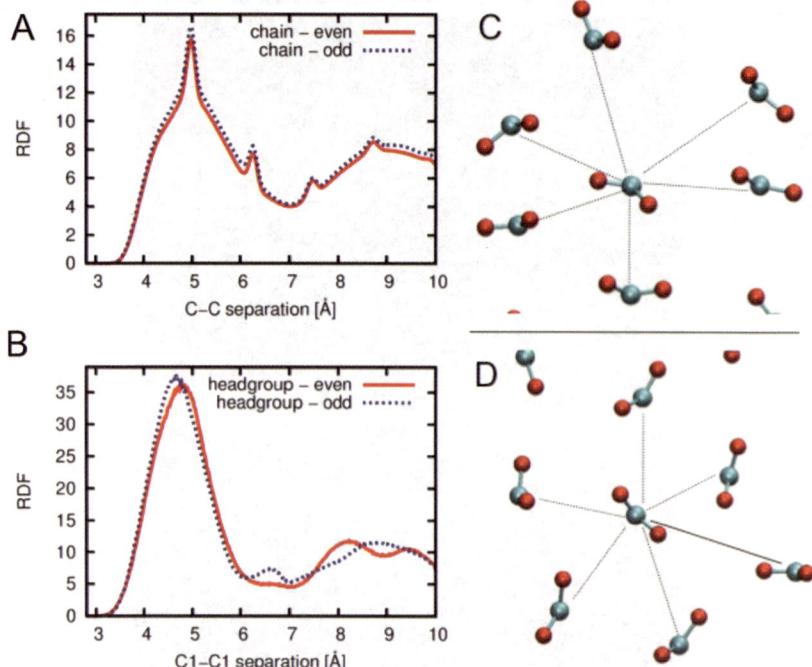

Fig. 5 (A, B) Radial distribution functions from MD simulations of odd and even parity SAMs for (A) the carbon in the chains with the other carbon in the chains and (B) the carbon in the headgroups of the monolayers with the other carbon in the headgroups for the even SAM (dashed magenta) and odd SAM (solid black). (C, D) Snapshot from simulations demonstrating the different arrangement of the headgroups by viewing the monolayer from above the surface for (A) even and (B) odd SAMs. The nearest neighbor C–C separations are indicated with a dotted line while the increased separation between two carbons in the odd SAM is highlighted as a solid line.

formation as a non-classical process.[8,11,12,25,27] In particular, they raise the question of how our results can be reconciled with those of previous studies that concluded: 1) nucleation of calcite occurs *via* an ACC precursor and 2) $CaCO_3$ solutions contain pre-nucleation clusters that aggregate to form this precursor. To address these apparent discrepancies we investigated the pathway of calcite formation on SAMs using Raman spectroscopy, *in situ* AFM imaging, optical microscopy and TEM analysis.

Raman spectroscopy provides an unambiguous means for identifying the phase of calcium carbonate precipitates. Raman spectra collected on samples that were quenched during the nucleation rate experiments by switching the incoming fluid from $CaCO_3$ solution to ethanol corresponded to that of calcite regardless of particle size investigated (Fig. 6). We are certain that the ethanol itself did not cause conversion of ACC to calcite, because we routinely preserve ACC *via* this technique using OH films in both the optical and AFM fluid cells, as well as centrifuges and button filters. However, sub-100 nm particles present on the MHA films as seen in SEM images (Fig. S3)† were below the threshold for obtaining useful Raman spectra. To assess these particles, we reproduced the conditions of the optical experiments using an AFM in place of the optical microscope (See ESI for details).† When these experiments were performed using MHA SAMs as substrates, all particles that appeared remained intact and grew in size. Even at the earliest stage of formation captured by the AFM, these particles possessed the typical rhombic shape of calcite and exhibited the orientation seen at larger size that results from nucleation on an (012) face (Fig. 7A–D).

When the same experiments were performed using an OH terminated SAM (mercapto-undecanol, MUO), at all supersaturations investigated—including concentrations that were highly undersaturated with respect to ACC based on the accepted bulk solubility—within the first minute of imaging we observed the formation of roughly spherical nanoparticles characteristic of ACC whose number and size were dependent on the solution concentration (Fig. 7E). (Raman spectra and SEM data from MUO samples (not shown) readily confirm the presence of ACC.) Unlike the result with MHA SAMs, these particles did not continue to grow in size. Instead, after a short period of time they began to dissolve (Fig. 7E–G) in response to the formation of the more stable calcite phase elsewhere in the cell. In parallel experiments using identical solution mixtures, these nanoparticles were collected on filters and examined by high resolution TEM, which revealed them to indeed be amorphous (Fig. 7H).

Despite *in situ* AFM observations on many tens of these ACC particles, in no instance did we observe their direct transformation into calcite. Rather, only dissolution was observed. However, the AFM only samples small areas (< 100 × 100 μm²) so we may have simply missed a transformation event that occurred out of the field of view. To circumvent this limitation, we utilized the fact that ACC particle size is

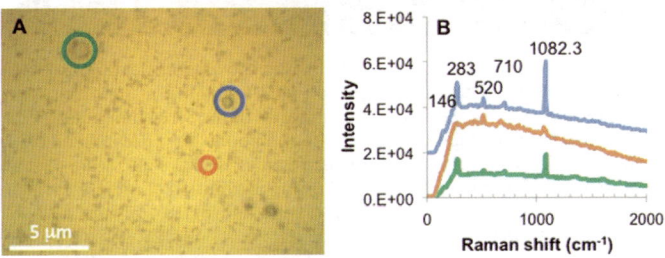

Fig. 6 Typical Raman spectra of particles formed right after the incubation time for 30 mM solutions. (A) optical microscopy image of of $CaCO_3$ particles collected on gold 75 s after mixing the solutions, (B) Raman spectra where color of curve corresponds to that of solid circle in (A).

dependent on supersaturation, which can be driven to high values by introducing carbonation *via* gaseous diffusion from an ammonium carbonate source, in order to generate a film of ACC particles with diameters of 100s of nm (see ESI, Fig. S4 for details).† As expected, when we used the OH–terminated MUO on Au we observed the formation of ACC, most of which fell on the SAM from the solution. However, this film of ACC particles rapidly dissolved back into solution as calcite rhombs nucleated and grew at their expense (Fig. 8A–C, Movie S1).†

We note that in no instance could we definitively conclude that a calcite rhomb formed through direct transformation of a pre-existing ACC particle. In a number of cases, the rhomb formed from a distinct particle that deposited on the MUO surface and immediately began to grow. The fact that ACC formation occurred before any calcite surface nucleation events took place explains why rates of calcite nucleation on these films showed no dependence on the initial solution supersaturation (Inset, Fig. 3F). Once ACC formed homogenously, the solute concentration immediately became fixed at the solubility of ACC. Thus, in essence, all calcite nucleation occurred at the same supersaturation, regardless of the initial solution conditions.

When the carbonate diffusion experiments were performed using the MHA SAMs, the progression of events was completely reversed from that seen on MUO (Fig. 8D–F, Movie S2).† The first particles to appear formed directly on the SAM surface and grew into calcite rhombs. None of these particles ever underwent dissolution. Well after they could be clearly identified as rhombs, ACC began to form in solution and deposit on the surrounding Si substrate. Some of this ACC also deposited on the SAM, but dissolved immediately due to presence of the growing calcite crystals. As the Ca^{2+} level in the surrounding solution decreased, even the ACC outside of the SAM began to dissolve due to the continued growth of the calcite rhombs. When SAMs were used that extended across the entire substrate, no ACC was observed, because the formation of calcite across the full extent of the

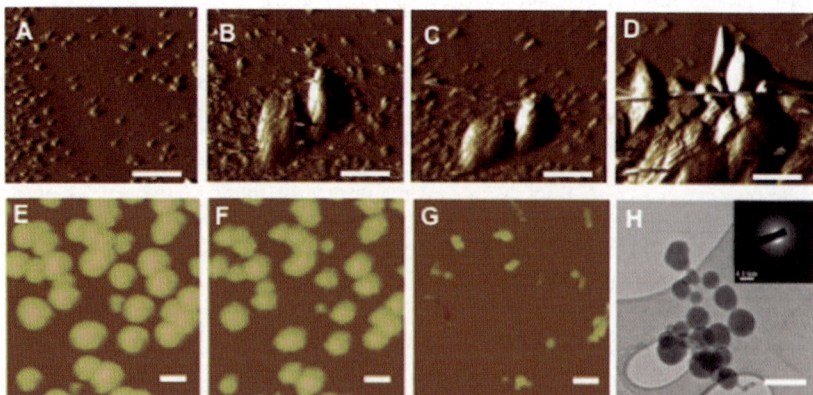

Fig. 7 (A–D) Series of sequential *in situ* AFM images showing nucleation on MHA SAM in 25 mM solution with pH 8.4. The features seen in Fig. 7A are not $CaCO_3$ precipitates; they are present even in pure water and are likely to be aggregates of the SAM monomers. First particles to appear have typical morphology seen for calcite rhombohedra nucleating on (012) face and grow in number and size with time. Time between frames is 93.2 s and scale bars are 1.0 μm. (D–F) Series of sequential *in situ* AFM images showing dissolution of initially formed nanoparticles in 13 mM $CaCO_3$ solution at room temperature and pH 8.4 on the surface of MUO. Times at which these frames were captured are (D) 68 s, (E) 128 s and (F) 190 s, where $t = 0$ corresponds to the moment when the calcium and bicarbonate solutions were mixed. The average height of nanoparticles decreases from (D) 64.2 ± 9.7 nm to (E) 53.2 ± 12 nm and to (F) 33.2 ± 14 nm. Scale bars are 200 nm. (G) TEM image and diffraction pattern (inset) of nanoparticles captured from 25 mM $CaCO_3$ using methodology described in the ESI and showing particles are amorphous. Scale bar is 100 nm.

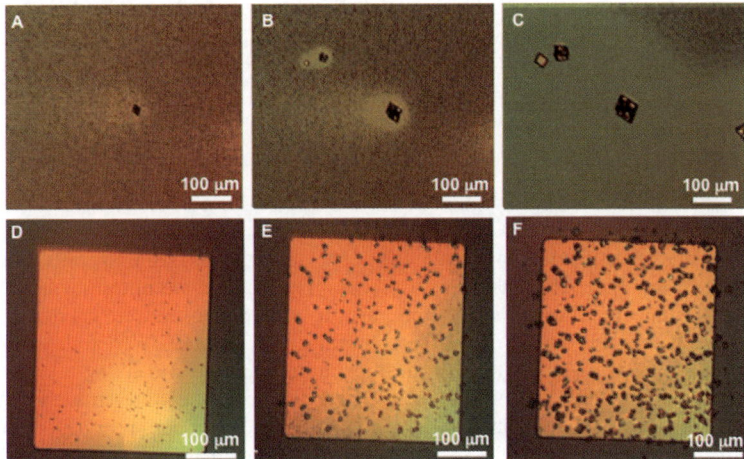

Fig. 8 (A–C) Still images from movie† showing the pathway of calcite formation during diffusion of carbonate into CaCl$_2$ solution on an MUO SAM. The appearance of ACC is followed by nucleation of calcite and concomitant dissolution of the ACC film. Frame times are 830 s, 1,892 s and 3,742 s. (D–F) Still images from movie† showing the pathway of calcite formation during diffusion of carbonate into CaCl$_2$ solution on an MHA SAM. Calcite first appears only on the MHA film. As it grows, ACC then begins to deposit from solution until the Ca^{2+} is sufficiently depleted that ACC dissolves as calcite continues to grow. Frame times are 280 s, 700 s and 1500 s.

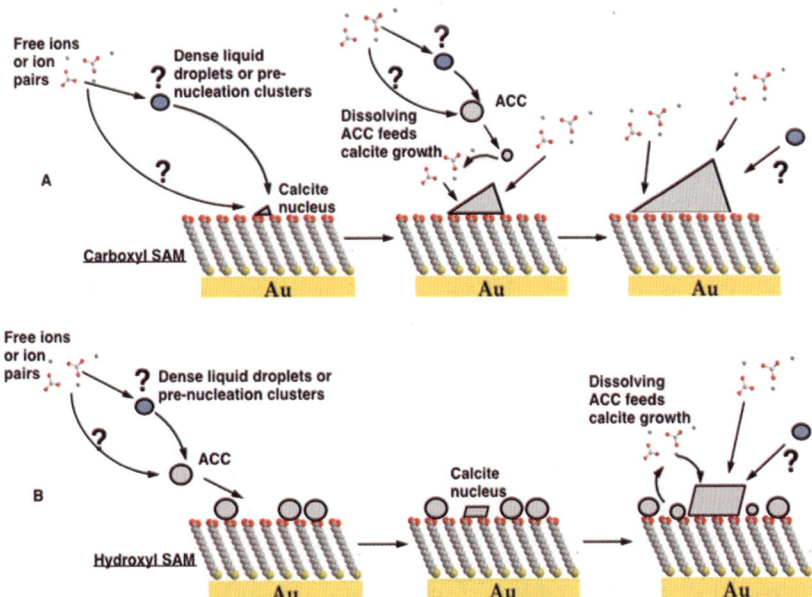

Fig. 9 Schematic showing pathways of calcite formation on (A) carboxyl- and (B) hydroxyl-terminated SAMs. With carboxyl-terminated SAMs, free ions and ion pairs, pre-nucleation clusters or dense liquid droplets aggregate directly on the SAM to form a critical nucleus of calcite, which then grows by addition of these species. ACC then subsequently forms in solution and dissolves as it falls towards the SAM, thereby feeding the growing crystal. With hydroxyl-terminated SAMs, free ions and ion pairs, pre-nucleation clusters or dense liquid droplets first aggregate to form ACC nanoparticles in the solution. The ACC particles then land on the SAM, forming an ACC film. Subsequent nucleation of calcite either directly on the SAM or in the surrounding solution then leads to dissolution of the ACC film as the calcite grows.

film prevented the supersaturation from reaching the required level for ACC formation (Movie S3).† Fig. 9 shows a schematic representation of the pathways inferred from these experiments for both the carboxyl- and hydroxyl-terminated SAMs.

6. Discussion

The results presented suggest that ACC nanoparticles did not serve as direct precursors to calcite in the experiments with carboxyl SAMs. Rather calcite formation resulted from distinct nucleation events. While this result would seem to contradict the previous claims that ACC serves as a precursor to calcite, we note that nearly all previous studies in which this two-step pathway was proposed were carried out at supersaturations well in excess of the solubility limit of ACC[8,11,12] and, with one exception,[11] the transformation of an amorphous particle into a crystal was not directly observed, rather it was inferred from the sequence of events, $i.e.$, ACC formed first and was eventually replaced by one of the crystalline phases. In the case of experiments that utilized carbonate diffusion into a $CaCl_2$ solution,[12] as well as those based on titration of a carbonate buffer,[8] the supersaturation increased continually to the point of nucleation. Thus the large barrier to homogeneous nucleation, which is still in excess of $90kT$ at the literature value of ACC solubility, may simply have prevented calcite nucleation before conditions that favored ACC formation were reached. Whether the eventual conversion to a crystalline phase occurred by direct transformation from ACC or through dissolution and re-precipitation on crystal nuclei that formed independently is simply not known.

In a series of cryoTEM studies, formation of vaterite from ACC precursor particles was deduced from analysis of ex $situ$ images that revealed vaterite nanoparticles within larger amorphous particles in contact with a Langmuir monolayer.[11,28,29] However, the interfacial energy between the Langmuir monolayer and calcium carbonate crystalline phases is unknown and may simply be too high to induce formation of any crystalline phase on the timescale of ACC to vaterite under the conditions of the experiments. One may well find that, at lower supersaturation the situation is reversed and vaterite forms directly, albeit on a longer timescale than observed for ACC at the supersatration used in the experiments. Finally, even though the calcite crystals in our study appear to nucleate before ACC forms in the case of MHA and perhaps even separately from the initial ACC nanoparticles in the case of MUO, we cannot rule out the possibility that the calcite nuclei are themselves constructed from clusters below 10 nm in size, whether crystalline or amorphous. Even the earliest nuclei we succeeded in capturing by AFM had already grown to ~20 nm along the shortest dimension. As pointed out in section 2, for a reasonable range of cluster excess free energies, their effect would be undetectable in measurements of J $vs.$ σ. Thus their dynamics would only be reflected in the pre-factors A and E_A. As long as the super-critical nucleus that emerges is that of calcite, the rate data will reflect the calcite-SAM interface through the classical expressions. Consequently, there is no obvious contradiction between these studies.

With respect to the pre-nucleation clusters,[25] there are two possible scenarios that are consistent with all of the observations. First, the clusters are like any other solution species; they behave as simple ionic complexes, constantly forming unstable sub-critical nuclei that spontaneously fluctuate in size until, by chance, they exceed the critical nucleus size either by ion addition or cluster aggregation. Second, they play no role in nucleation at the relatively low concentrations of these experiments; nucleation is ion-by-ion because the number and diffusivity of the free ions are much greater than those of the clusters or because the kinetic barriers to building an ordered nucleus from ions are much less than those to desolvating and ordering the clusters. A firm conclusion will have to wait until experimental tools enable direct characterization of pre-nucleation clusters in low concentration solutions at surfaces.

Acknowledgements

This research was supported by the U.S. Department of Energy, Office of Basic Energy Sciences, Division of Chemical, Biological and Geological Sciences through Lawrence Berkeley National Laboratory and as part of the Center for Nanoscale Control of Geologic CO_2, an Energy Frontier Research Center under contract No. DE-AC02-05CH11231. Measurements were performed at the Molecular Foundry, Lawrence Berkeley National Laboratory with support from the Office of Science, Office of Basic Energy Sciences of the U.S. Department of Energy under Contract no. DE-AC02-05CH1123. This research was made with Government support under and awarded by DoD, Air Force Office of Scientific Research, National Defense Science and Engineering Graduate (NDSEG) Fellowship, 32 CFR 168a. CLF and JHH would like to thank funding from UK EPSRC grant number EP/I001514/1; this UK programme grant funds the Materials Interface with Biology consortium. Portions of this work were performed under the auspices of the U.S. Department of Energy by Lawrence Livermore National Laboratory under Contract DE-AC52-07NA27344. The research was also supported by awards to PMD from the US Dept. of Energy (DOE BES-FG02-00ER15112) and the National Science Foundation (NSF OCE-1061763).

References

1. L. Addadi, D. Joester, F. Nudelman and S. Weiner, *Chem.–Eur. J.*, 2006, **12**, 980–987.
2. O. Braissant, G. Gailleau, C. Dupraz and E. P. Verrecchia, *J. Sediment. Res.*, 2003, **73**, 485–490.
3. S. Weiner, W. Traub and H. D. Wagner, *J. Struct. Biol.*, 1999, **126**, 241–255.
4. F. Nudelman, *et al.*, *Nat. Mater.*, 2010, **9**, 1004–1009.
5. J. J. De Yoreo and P. G. Vekilov, *Rev. Mineral. Geochem.*, 2003, **54**, 57–93.
6. I. Jager and P. Fratzl, *Biophys. J.*, 2000, **79**, 1737–1746.
7. P.-A. Fang, J. F. Conway, H. C. Margolis, J. P. Simmer and E. Beniash, *Proc. Natl. Acad. Sci. U. S. A.*, 2011, **108**, 14097–14102.
8. D. Gebauer, A. Volkel and H. Colfen, *Science*, 2008, **322**, 1819–1822.
9. Y. Politi, *et al.*, *Science*, 2004, **306**, 1161–1164.
10. J. Mahamid, A. Sharir, L. Addadi and S. Weiner, *Proc. Natl. Acad. Sci. U. S. A.*, 2008, **105**, 12748–12753.
11. E. M. Pouget, *et al.*, *Science*, 2009, **323**, 1455–1458.
12. J. R. I. Lee, *et al.*, *J. Am. Chem. Soc.*, 2007, **129**, 10370–10381.
13. A. Day, P. H. H. Bomans, F. A. Müller, J. Will, P. M. Frederik, G. de With and N. A. J. M. Sommerdijk, *Nat. Mater.*, 2010, **9**, 1010–1014.
14. J. Aizenberg, A. J. Black and G. M. Whitesides, *Nature*, 1999, **398**, 495–498.
15. A. M. Travaille, J. J. J. M. Donners, J. W. Gerritsen, N. A. J. M. Sommerdijk, R. J. M. Nolte and H. van Kempen, *Adv. Mater.*, 2002, **14**, 492–495.
16. Y.-J. Han and J. Aizenberg, *Angew. Chem., Int. Ed.*, 2003, **42**, 3668–3670.
17. C. L. Freeman, J. H. Harding and D. M. Duffy, *Langmuir*, 2008, **24**, 9607–9615.
18. A. F. Wallace, J. J. DeYoreo and P. M. Dove, *J. Am. Chem. Soc.*, 2009, **131**, 5244–5250.
19. O. Sohnel and J. W. Mullin, *J. Cryst. Growth*, 1978, **44**, 377–382.
20. O. Sohnel, *J. Cryst. Growth*, 1982, **57**, 101–108.
21. L. Brecevic and A. E. Nielsen, *J. Cryst. Growth*, 1989, **98**, 504–510.
22. H. Zhang, B. Chen and J. F. Banfield, *Phys. Chem. Chem. Phys.*, 2009, **11**, 2553–2558.
23. G. V. Gibbs, T. D. Crawford, A. F. Wallace, D. F. Cox, R. M. Parrish, E. G. Hohenstein and C. D. Sherril, *J. Phys. Chem. A*, 2011, **115**, 12933–12940.
24. J. Harding, *J. Chem. Phys.*, 2011, **134**, 044703.
25. D. Gebauer and H. Coelfen, *NanoToday*, 2011, **6**, 564–584.
26. D. Quigley, *et al.*, *J. Chem. Phys.*, 2009, **131**, 094703.
27. A. Navrotsky, *Proc. Natl. Acad. Sci. U. S. A.*, 2004, **101**, 12096–12101.
28. E. M. Pouget, P. H. H. Bomans, A. Dey, P. M. Frederik, G. de With and N. A. J. M. Sommerdijk, *J. Am. Chem. Soc.*, 2010, **132**, 111560–11565.
29. E. M. Pouget, P. H. H. Bomans, A. Dey, P. M. Frederik, G. de With and N. A. J. M. Sommerdijk, *J. Am. Chem. Soc.*, 2008, **130**, 4034–4040.

Poster titles

The influence of additives during CaCO$_3$ crystallization: from pre-nucleation to post-nucleation stages, **D Gebauer**, *University of Konstanz, Germany*

Phase behaviour of protein solutions: effects of additives, **D Wagner, F Evers, J Hansen, Z Yang, M Martinsons, I Brück and S U Egelhaaf**, *Heinrich-Heine-University Dusseldorf, Germany*

On mimiking the structural anisotropy found in the material bone, **J Seto, A Rao and H Cölfen**, *University of Konstanz, Germany*

Core–shell nanoparticles to measure carbonate dissolution *in situ*: an environmental method, **C Davis, G A Attard, P R Davies and S Barker**, *Cardiff University, UK*

A new precipitation pathway for calcium sulfate dihydrate (Gypsum) via amorphous and hemihydrate intermediates, **Y Wang, YY Kim, H K Christenson and F C Meldrum**, *University of Leeds, UK*

ZnO spherical particles grown in ethylene glycol and water solvent, **N Saito, K Matsumoto, K Watanabe, I Sakaguchi and H Haneda**, *National Institute for Materials Science, Japan*

Crystal nucleation and topography, **J M Campbell, J L Holbrough, F C Meldrum and H K Christenson**, *University of Leeds, UK*

Mesocrystal formation of aragonite?, **K K Sand, A Verch, E Makovicky, R Van de Locht, I Morrison, J D Rodriguez-Blanco, L G Benning, R Kröger and S L S Stipp**, *University of Copenhagen, Denmark*

Dipole field directed growth of bio-inspired ZnO twin crystals and hexagonal prism vaterite, **H Greer, W Zhou, M H Liu, Y H Tseng and C Y Mou**, *University of St Andrews, UK*

Calcite nanorods crystallisation and nucleation, **R Darkins, A Côté and D Duffy**, *University College London, UK*

Controls of polysaccharide chemistry on kinetics and thermodynamics of calcium carbonate nucleation, **A J Giuffre, L M Hamm, N Z Han, J J DeYoreo and P M Dove**, *Virginia Tech, USA*

Removal of direct blue dye from industrial waste water using lemon peel waste, **S Abid**, *University of the Punjab, Pakistan*

Biomimetic synthesis of facet-controllable ZnO mesocrystals, **M H Liu, Y H Tseng, H Greer, W Zhou and C Y Mou**, *National Taiwan University, Taiwan*

Modelling of the crystal morphology of the polymorphic forms of para aminobenzoic acid based on their crystallographic structure, **X Lai, K J Roberts and I Rosbottom**, *University of Leeds, UK*

In-situ XRD study of the formation and polymorphic conversion of para-aminobenzoic acid, **X Lai, K J Roberts and T D Turner**, *University of Leeds, UK*

Amyloid fibrillation kinetics: insight from atomistic nucleation theory, **R Cabriolu, D Kashchiev and S Auer**, *University of Leeds, UK*

Fish-gut amorphous calcium carbonates and ocean chemistry, **E Foran, M Fine and S Weiner**, *Interuniversity Institute for Marine Sciences, Israel*

Biomineralisation on carboxylate-terminated self-assembled monolayers: a metadynamics study, **A S Côté, R Darkins, C L Freeman, D Quigley and D M Duffy**, *University College London, UK*

The role of water in the crystallization of amorphous calcium carbonate (ACC), **J Ihli, E H Noel, Y Y Kim, A Kulak and F C Meldrum**, *University of Leeds, UK*

Amino acid incorporation into different mineral systems, **D Turner, B Demachi, K E M Pentman, Y Y Kim and F C Meldrum**, *University of Leeds, UK*

Two-step adsorption of antifreeze protein molecules on ice interface and its effect for ice crystal growth, **Y Furukawa, G Sazaki, S Zepeda, Y Uda and E Yokoyama**, *Hokkaido University, Japan*

In vitro formation, evolution and properties of biomimetic nanocrystalline apatites: significance for bone mineral, **C Combes, C Rey, C Drouet, D Grossin, S Sarda and S Cazalbou**, *ENSIACET, France*

In vitro crystallisation of hydrated calcium pyrophosphate phases of biological interest, **P Gras, C Rey, S Sarda and C Combes**, *ENSIACET, France*

Incorporation of magnetite nanoparticles in the single crystals, **A Kulak, M Semsarilar, S Armes and F Meldrum**, *University of Leeds, UK*

Precipitation of calcium carbonate in ethanolic solution mediated by a saturated fatty alcohol, **N R Bailey, M Fernandes, F A Paz and V de Zea Bermudez**, *University of Trás-os-Montes e Alto Douro, Portugal*

Morphology map of calcium carbonate formed in the presence of saturated fatty alcohols, **M Fernandes, F A Paz and V de Zea Bermudez**, *University of Trás-os-Montes e Alto Douro, Portugal*

Inorganic matrix-mediated formation of hierarchically structured iron oxide, **E Asenath-Smith and L A Estroff**, *Cornell University, USA*

In-situ experiments on the crystalisation of $CaCO_3$, **A Verch, R van de Locht, I Morrison, L Lari and R Kröger**, *University of York, UK*

Colloidal stabilisation of $CaCO_3$ prenucleation clusters with silica, **M Kellermeier, D Gebauer, W Kunz, H Cölfen**, *University of Konstanz, Germany*

Precipitation of calcium phosphate nanorods in confinement, **B Cantaert, Y Y Kim and F C Meldrum**, *University of Leeds, UK*

Development of thin-film polymer/$CaCO_3$ hybrids: effects of ion additives, **T Nishimura, F Zhu, H Tomono, H Nada and T Kato**, *University of Tokyo, Japan*

A simple model for crystallisation from vapour without supersaturation, **T Moorsom and H K Christenson**, *University of Leeds, UK*

A critical analysis of calcite mesocrystal structures, **F C Meldrum, G Hyett, Y Y Kim, J Ihli, A Schenk and N Hetherington**, *University of Leeds, UK*

Biomimetic control over size, shape and aggregation in magnetic nanoparticles, **C L Altan, J Lenders, P H H Bomans, V Dmitrovic, H Zopec, H Friedrich, A Kros, S Bucak and N A J M Sommerdijk**, *Eindhoven University of Technology, The Netherlands*

Biosynthetic routes to precision magnetic nanoparticles, **S Staniland, A Rawlings, J Bramble, M Tanaka, R Brown, N Hondow, Arakaki and T Matsunaga**, *University of Leeds, UK*

Deleterious mineral scale crystallisation processes in oil industry, **M Barber, A Neville and K J Roberts**, *University of Leeds, UK*

The significance of Cl- "spectator ion" in the formation of $CaCO_3$ precursors, **M A Bewernitz, J R Long and L B Gower**, *University of Florida, USA*

Adaptive resolution mesoscale modelling of biomineral–solvent interactions, **P J Kiley and J A Elliott**, *University of Cambridge, UK*

The role and effect of Mg on the crystallization of ACC nanoparticles, **J D Rodriguez-Blanco, S Shaw, C Wood, A P Brown and L G Benning**, *University of Leeds, UK*

Adsorbed molecular layers at Au(111) as alternative seed toward improved crystal morphology of Entacapone, **A Kwokal and K J Roberts**, *Pliva Croatia Ltd, Croatia*

The Skinner Prize for the best poster was jointly awarded to Mr Mark Bewernitz of the University of Florida, USA, for his poster on the significance of Cl- "spectator ion" in the formation of $CaCO_3$ precursors and Mr Johannes Ihli of the University of Leeds, UK, for his poster on the role of water in the crystallization of amorphous calcium carbonate (ACC).

List of participants

Dr J. Andreassen, *NTNU Norwegian University of Science and Technology, Norway*
Mrs N. Ansir, *Bangor University, United Kingdom*
Ms E. Asenath-Smith, *Cornell University, U.S.A.*
Miss M. Barber, *University of Leeds, United Kingdom*
Dr R. Beck, *NTNU Norwegian University of Science and Technology, Norway*
Dr A. Berman, *Ben-Gurion University, Israel*
Mr M. Bewernitz, *University of Florida, Dept. of Biomedical Engineering, U.S.A.*
Miss R. Cabriolu, *University of Leeds, United Kingdom*
Mr J. Campbell, *University of Leeds, United Kingdom*
Mr B. Cantaert, *University of Leeds, United Kingdom*
Dr H. Christenson, *University of Leeds, United Kingdom*
Professor H. Cölfen, *University of Konstanz, Germany*
Professor C. Combes, *CIRIMAT, France*
Dr A. Côté, *University College London, United Kingdom*
Professor R. Davey, *University of Manchester, United Kingdom*
Mr C. Davis, *Cardiff University, United Kingdom*
Professor J. De Yoreo, *Lawrence Berkeley National Lab., U.S.A.*
Professor V. De Zea Bermudez, *University of Trs-os-Montes E Alto Douro, Portugal*
Dr P. Dove, *Virginia Tech, U.S.A.*
Miss K. Dryden-Holt, *Royal Society of Chemistry, United Kingdom*
Dr D. Duffy, *UCL, United Kingdom*
Professor J. Evans, *New York University, U.S.A.*
Dr F. Evers, *Heinrich-Heine-University Düsseldorf, Germany*
Ms M. Fernandes, *UTAD, Portugal*
Mr A. Finney, *Centre for Scientific Computing, United Kingdom*
Ms E. Foran, *Interuniversity Institute of Marine Sciences in Eilat, Israel*
Professor D. Frenkel, *University of Cambridge, United Kingdom*
Professor Y. Furukawa, *Instirute of Low Temperature Science, Hokkaido University, Japan*
Professor J. Gale, *Curtin University, Australia*
Dr D. Gebauer, *University of Konstanz, Germany*
L. Gilbert, *Royal Society of Chemistry, United Kingdom*
Professor P. Gilbert, *University of Wisconsin - Madison, U.S.A.*
Mr A. Giuffre, *Virginia Tech, U.S.A.*
K. Gould, *Royal Society of Chemistry, United Kingdom*
Dr L. Gower, *University of Florida, U.S.A.*
Miss H. Greer, *University of St Andrews, United Kingdom*
Professor B. Grzybowki, *Northwestern University, U.S.A.*
Mr G. Hazell, *University of Bath, United Kingdom*
Professor G. Hutchings, *Cardiff University, United Kingdom*
Mr J. Ihli, *University of Leeds, United Kingdom*
Prof Dr D. Joester, *Northwestern University, U.S.A.*
Professor T. Kato, *The University of Tokyo, Japan*
Dr M. Kellermeier, *Physical Chemistry, University of Konstanz, Germany*
Mr S. Khan, *GlaxoSmithKline, United Kingdom*
Dr P. Kiley, *University of Cambridge, United Kingdom*
Mr J. Kirkwood, *University Of York, United Kingdom*
Mr J. Komeh, *University of Cape Coast, Ghana*
A. Kulak, *University of Leeds, United Kingdom*
Dr G. Langer, *Department of Earth Sciences, Cambridge University, United Kingdom*

Mr M. Liu, *Department of Chemistry, National Taiwan University, Taiwan*
Dr N. Loges, *BASF Construction Chemicals GmbH, Germany*
Professor A. Madhukar, *University of Southern California, U.S.A.*
Professor V. Manoharan, *Harvard University, U.S.A.*
Professor F. Meldrum, *University of Leeds, United Kingdom*
Dr J. Mithen, *University of Surrey, United Kingdom*
Mr T. Moorsom, *The University of Leeds, United Kingdom*
Professor H. Nagasawa, *University Of Tokyo, Japan*
Dr G. Nehrke, *Alfred Wegener Institute for Polar and Marine Research, Germany*
Mrs M. Nergaard, *NTNU Norwegian University of Science and Technology, Norway*
Mr M. Nielsen, *Lawrence Berkeley National Lab, U.S.A.*
Dr T. Nishimura, *The University of Tokyo, Japan*
Dr F. Nudelman, *Eindhoven University of Technology, The Netherlands*
Professor L. Penn, *University of Minnesota, U.S.A.*
Dr M. Platt, *Loughborough University, United Kingdom*
Miss R. Quine, *Royal Society of Chemistry, United Kingdom*
Mr A. Rao, *Konstanz University, Germany*
Professor K. Roberts, *University of Leeds, United Kingdom*
Professor P. Rodger, *University of Warwick, United Kingdom*
Dr J. Rodriguez Blanco, *School of Earth and Environment, University of Leeds, United Kingdom*
Mr I. Rosbottom, *University of Leeds, United Kingdom*
Dr N. Saito, *National Institute for Materials Science, Japan*
Dr K. Sand, *University of Copenhagen, Denmark*
Dr A. Schenk, *University of Leeds, United Kingdom*
Dr R. Sear, *University of Surrey, United Kingdom*
Mr J. Seto, *Universität Konstanz, Germany*
Ms T. Smekal, *Royal Society of Chemistry, United Kingdom*
Dr N. Sommerdijk, *Eindhoven University of Technology, The Netherlands*
Dr S. Staniland, *University of Leeds, United Kingdom*
Dr D. Toroz, *University of Leeds, United Kingdom*
Mr D. Turner, *University of Leeds, United Kingdom*
Mr T. Turner, *University of Leeds, United Kingdom*
Professor A. Van Blaaderen, *University of Utrecht, The Netherlands*
Professor P. Van Der Schoot, *Technical University Eindhoven, The Netherlands*
Professor P. Vekilov, *University of Houston, United Kingdom*
Dr A. Verch, *University of York, United Kingdom*
Mrs C. Virone, *DSM, The Netherlands*
Dr D. Wang, *National Institute of Standards and Technology, U.S.A.*
Miss Y. Wang, *University of Leeds, United Kingdom*
Dr S. Wolf, *CNRS/UMR 6282 Biogosciences, France*
Y. Yeoun Kim, *University of Leeds, United Kingdom*
Dr F. Zhang, *University of Tübingen, Germany*

Index of contributors*

Akaiwa, K., **483**
Anders, U., **23**
Andreassen, J-P., **247**, 139, 277, 495
Asenath-Smith, E., 277, 495
Barber, M., 139, 277
Bardeau, J-F., **433**
Barrea, R., **433**
Bewernitz, M. A., **291**, 387, 495
Beck, R., **247**, 139, 387, 495
Bomans, P. H. H., **357**
Burrows, N. D., **235**
Byelov, D. V., **181**
Campbell, J., 277
Cantaert, B., 139
Christenson, H. K., **123**, 139, 277, 387, 495
Cölfen, H., **23**, **291**, 139, 277, 387
Davey, R., 139
Davis, C., 139, 277
Delorme, N., **433**
Demichelis, R., **61**
de With, G., **357**
De Yoreo, J. J., **105**, **371**, **509**, 139, 277, 387, 495
Dimiduk, T. G., **211**
Do, T. A., **235**
Dorsaz, N., **9**
Dove, P., **371**, 387
Echigo, T., **371**
Emmerling, F., **433**
Endo, H., **463**
Evans, J. S., **449**, 495
Filion, L., **9**
Finney, A. R., **47**
Frenkel, D., **9**, 139, 387
Fung, J., **211**
Gale, J. D., **61**, 139, 495
Gebauer, D., **61**, **291**, 139, 277, 387
George, A., **357**
Gilbert, P. U. P. A., **421**, 139, 387, 495
Giuffre, A., **371**, 495
Gower, L. B., **291**, 139, 387, 495
Grotzinger, J., **371**
Grzybowski, B. A., **201**, 277
Hamm, L., **371**
Han, T. Y-J., **105**
Hu, Q., **105**
Ihli, J., 139, 387, 495
Imhof, A., **181**
Inoue, H., **463**, **483**
Jacobs, M. J., **313**

Joester, D., **345**, 387
Kajiyama, S., **483**
Kappl, M., **433**
Kato, T., **483**, 139, 277, 387, 495
Kellermeier, M., **23**, **61**, 139
Khan, S., 277, 387
Kiley, P., 277
Kim, Y-Y., **327**
Kogure, T., **463**
Kovács, T., **123**
Kowalczyk, B., **201**
Kros, A., **327**
Kuijk, A., **181**
Kumagai, H., **483**
Lagzi, I., **201**
Lee, J. R. I., **105**
Lieberwirth, I., **433**
Long, J., **291**
Lubchenko, V., **87**
Manoharan, V., **211**, 277
Marin, F., **433**
Matsunaga, R., **483**
Meldrum, F., 387, 495
Meng, G., **211**
Miyabe, K., **463**
Moise, A., **23**
Nagasawa, H., **463**, **483**, 495
Natalio, F., **433**
Ndao, M., **449**
Nergaard, M., **247**, 277
Nielsen, M. H., **105**, 139, 387
Nishimura, T., **483**, 495
Nudelman, F., **357**, 387, 495
Oaki, Y., **483**
Olson, I. C., **421**
Pan, W., **87**
Penn, R. L., **235**, 139, 277, 387
Perry, R. W., **211**
Petukhov, A. V., **181**
Ponce, C. B., **449**
Przbylski, M., **23**
Quigley, L. B., **61**
Raiteri, P., **61**
Rimsidt, J. D., **371**
Roberts, K., 139, 387, 495
Rodger, P. M., **47**, 139, 495
Rodríguez Blanco, J., 139, 387, 495
Roosen-Runge, F., **313**
Rosenberg, R., **23**
Rosbottom, I., 139

Roth, R., **313**
Sand, K., 387
Sauter, A., **313**
Sear, R., **263**, 139, 277, 387, 495
Seto, J., 387
Schenk, A. S., **327**, 387, 495
Schreiber, F., **313**
Skoda, M. W. A., **313**
Smallenburg, F., **9**
Soltis, A., **235**
Sommerdijk, N. A. J. M., **327**, **357**, 139, 277, 387
Staniland, S., 277, 495
Suzuki, M., **463**
Sztucki, M., **313**
Tester, C. C., **345**
Tokunaga, H., **463**
Toroz, D., 277
Tretiakov, V., **201**

Tsumoto, K., **483**
Tsutsui, N., **463**
Uzunova, V., **87**
van Blaaderen, A., **181**, 139, 277, 495
Vekilov, P. G., **87**, 139
Verch, A., 387
Virone, C., 139, 277
Walker, D. A., **201**
Walsh, T. R., **61**
Wang, D., **201**, **371**
Weigand, S., **345**
Wolf, S. E., **433**, 139, 387, 495
Wright, L. B., **61**
Wu, C-H., **345**
Yamamoto, Y., **483**
Yokoo, N., **463**
Yuwono, V. M., **235**
Zhang, F., **313**, 139, 277, 387
Zope, H., **327**

*The page numbers in **bold** type indicate papers submitted for discussions.